MODERN NONLINEAR OPTICS
Part 3
Second Edition

ADVANCES IN CHEMICAL PHYSICS

VOLUME 119

EDITORIAL BOARD

MODERN NONLINEAR OPTICS

Part 3

Second Edition

ADVANCES IN CHEMICAL PHYSICS
VOLUME 119

Edited by

Myron W. Evans

Series Editors

I. PRIGOGINE

Center for Studies in Statistical Mechanics and Complex Systems
The University of Texas
Austin, Texas
and
International Solvay Institutes
Université Libre de Bruxelles
Brussels, Belgium

and

STUART A. RICE

Department of Chemistry
and
The James Franck Institute
The University of Chicago
Chicago, Illinois

AN INTERSCIENCE® PUBLICATION
JOHN WILEY & SONS, INC.

For ordering and customer service, call 1-800-CALL-WILEY

Library of Congress Catalog Number: 58-9935

ISBN 0-471-38932-3

Printed in the United States of America.

10 9 8 7 6 5 4 3 2 1

CONTRIBUTORS TO VOLUME 119
Part 3

NILS ABRAMSON, Industrial Metrology and Optics, Department of Production Engineering, Royal Institute of Technology, Stockholm, Sweden

PETAR K. ANASTASOVSKI, Department of Physics, Faculty of Technology and Metallurgy, Saints Cyril and Methodius University, Skopje, Republic of Macedonia

TERENCE W. BARRETT, BSEI, Vienna, VA

FABIO CARDONE, Departimento di Fisica, Univeritá de L'Aquila, Italy

LAWRENCE B. CROWELL, The Alpha Foundation, Institute of Physics, Rio Rancho, NM

M. W. EVANS, 50 Rhyddwen Road, Craigcefnparc, Swansea, Wales, United Kingdom

HAROLD L. FOX, Editor, *Journal of New Energy*, and President, Emerging Energy Marketing Firm, Inc., Salt Lake City, UT

DAVID B. HAMILTON, U.S. Department of Energy, Washington, DC

S. JEFFERS, Department of Physics and Astronomy, York University, Toronto, Ontario, Canada

I. A. KHOVANOV, Department of Physics, Saratov State University, Saratov, Russia

D. G. LUCHINSKY, Department of Physics, Lancaster University, Lancaster LA1 4YB, United Kingdom and Russian Research Institute for Metrological Service, Moscow, Russia

R. MANNELLA, Dipartimento di Fisica, Universita di Pisa and Istituto Nazionale Fisica della Materia UdR Pisa, Pisa, Italy and Department of Physics, Lancaster University, Lancaster, United Kingdom

P. V. E. MCCLINTOCK, Department of Physics, Lancaster University, Lancaster, United Kingdom

MILAN MESZAROS, The Alpha Foundation, Institute of Physics, Budapest, Hungary

ROBERTO MIGNANI, Dipartimento di Fisica "E. Amaldi," Universitá degli Studi "Roma Tre," Roma, Italy

PAL R. MOLNAR, The Alpha Foundation, Institute of Physics, Budapest, Hungary

HÉCTOR A. MÚNERA, Department of Physics, Universidad Nacional de Colombia, Bogota D.C., Colombia

ANTONIO F. RAÑADA, Departamento de Electricidad y Electronica, Universidad Complutense, Madrid, Spain

DONALD REED, Chektowage, NY

D. F. ROSCOE, Department of Applied Mathematics, Sheffield University, Sheffield S3 7RH, United Kingdom

SISIR ROY, Physics and Applied Mathematics Unit, Indian Statistical Institute, Calcutta, India

JOSÉ L. TRUEBA, ESCET, Universidad Rey Juan Carlos, Madrid, Spain

INTRODUCTION

Few of us can any longer keep up with the flood of scientific literature, even in specialized subfields. Any attempt to do more and be broadly educated with respect to a large domain of science has the appearance of tilting at windmills. Yet the synthesis of ideas drawn from different subjects into new, powerful, general concepts is as valuable as ever, and the desire to remain educated persists in all scientists. This series, *Advances in Chemical Physics*, is devoted to helping the reader obtain general information about a wide variety of topics in chemical physics, a field that we interpret very broadly. Our intent is to have experts present comprehensive analyses of subjects of interest and to encourage the expression of individual points of view. We hope that this approach to the presentation of an overview of a subject will both stimulate new research and serve as a personalized learning text for beginners in a field.

I. PRIGOGINE
STUART A. RICE

PREFACE

This volume, produced in three parts, is the Second Edition of Volume 85 of the series, *Modern Nonlinear Optics*, edited by M. W. Evans and S. Kielich. Volume 119 is largely a dialogue between two schools of thought, one school concerned with quantum optics and Abelian electrodynamics, the other with the emerging subject of non-Abelian electrodynamics and unified field theory. In one of the review articles in the third part of this volume, the Royal Swedish Academy endorses the complete works of Jean-Pierre Vigier, works that represent a view of quantum mechanics opposite that proposed by the Copenhagen School. The formal structure of quantum mechanics is derived as a linear approximation for a generally covariant field theory of inertia by Sachs, as reviewed in his article. This also opposes the Copenhagen interpretation. Another review provides reproducible and repeatable empirical evidence to show that the Heisenberg uncertainty principle can be violated. Several of the reviews in Part 1 contain developments in conventional, or Abelian, quantum optics, with applications.

In Part 2, the articles are concerned largely with electrodynamical theories distinct from the Maxwell–Heaviside theory, the predominant paradigm at this stage in the development of science. Other review articles develop electro-dynamics from a topological basis, and other articles develop conventional or U(1) electrodynamics in the fields of antenna theory and holography. There are also articles on the possibility of extracting electromagnetic energy from Riemannian spacetime, on superluminal effects in electrodynamics, and on unified field theory based on an SU(2) sector for electrodynamics rather than a U(1) sector, which is based on the Maxwell–Heaviside theory. Several effects that cannot be explained by the Maxwell–Heaviside theory are developed using various proposals for a higher-symmetry electrodynamical theory. The volume is therefore typical of the second stage of a paradigm shift, where the prevailing paradigm has been challenged and various new theories are being proposed. In this case the prevailing paradigm is the great Maxwell–Heaviside theory and its quantization. Both schools of thought are represented approximately to the same extent in the three parts of Volume 119.

As usual in the *Advances in Chemical Physics* series, a wide spectrum of opinion is represented so that a consensus will eventually emerge. The prevailing paradigm (Maxwell–Heaviside theory) is ably developed by several groups in the field of quantum optics, antenna theory, holography, and so on, but the paradigm is also challenged in several ways: for example, using general relativity, using O(3) electrodynamics, using superluminal effects, using an extended electrodynamics based on a vacuum current, using the fact that longitudinal waves may appear in vacuo on the U(1) level, using a reproducible and repeatable device, known as the *motionless electromagnetic generator*, which extracts electromagnetic energy from Riemannian spacetime, and in several other ways. There is also a review on new energy sources. Unlike Volume 85, Volume 119 is almost exclusively dedicated to electrodynamics, and many thousands of papers are reviewed by both schools of thought. Much of the evidence for challenging the prevailing paradigm is based on empirical data, data that are reproducible and repeatable and cannot be explained by the Maxwell–Heaviside theory. Perhaps the simplest, and therefore the most powerful, challenge to the prevailing paradigm is that it cannot explain interferometric and simple optical effects. A non-Abelian theory with a Yang–Mills structure is proposed in Part 2 to explain these effects. This theory is known as O(3) *electrodynamics* and stems from proposals made in the first edition, Volume 85.

As Editor I am particularly indebted to Alain Beaulieu for meticulous logistical support and to the Fellows and Emeriti of the Alpha Foundation's Institute for Advanced Studies for extensive discussion. Dr. David Hamilton at the U.S. Department of Energy is thanked for a Website reserved for some of this material in preprint form.

Finally, I would like to dedicate the volume to my wife, Dr. Laura J. Evans.

MYRON W. EVANS

Ithaca, New York

CONTENTS

The Royal Swedish Academy of Sciences (KVA) hereby confirms its support to a joint production with John Wiley and Sons, Inc. of the collective scientific works by Professor Jean-Pierre Vigier. It is thereby understood that this endorsement only concerns and objective estimate of these works, and implying no economic obligation from the side of the Academy.

MODERN NONLINEAR OPTICS
Part 3
Second Edition

ADVANCES IN CHEMICAL PHYSICS

VOLUME 119

THE PRESENT STATUS OF THE QUANTUM THEORY OF LIGHT

M. W. EVANS AND S. JEFFERS

Department of Physics and Astronomy, York University, Toronto, Ontario, Canada

CONTENTS

Modern Nonlinear Optics, Part 3, Second Edition, Advances in Chemical Physics, Volume 119, Edited by Myron W. Evans. Series Editors I. Prigogine and Stuart A. Rice.
ISBN 0-471-38932-3   © 2001 John Wiley & Sons, Inc.

1

I. INTRODUCTION

If one takes as the birth of the quantum theory of light, the publication of Planck's famous paper solving the difficulties inherent in the blackbody spectrum [1], then we are currently marking its centenary. Many developments have occurred since 1900 or so and are briefly reviewed below. (See Selleri [27] or Milloni [6] for a more comprehensive historical review). The debates concerning wave–particle duality are historically rooted in the seventeenth century with the publication of Newton's *Optiks* [2] and the *Treatise on Light* by Christian Huygens [3]. For Huygens, light was a form of wave motion propagating through an ether that was conceived as a substance that was "as nearly approaching to perfect hardness and possessing a springiness as prompt as we choose." For Newton, however, light comprised material particles and he argues, contra Huygens, "Are not all hypotheses erroneous, in which Light is supposed to consist of Pression, or Motion propagated through a Fluid medium?" (see Newton [2], Query 28). Newton attempts to refute Huygens' approach by pointing to the difficulties in explaining double refraction if light is simply a form of wave motion and asks, "Are not the Rays of Light very small bodies emitted from shining substances? For such bodies will pass through uniform Mediums in right Lines without bending into Shadow, which is the Nature of the Rays of Light?" (Ref. 2, Query 29). The corpuscular theory received a major blow in the nineteenth century with the publication of Fresnel's essay [4] on the diffraction of light. Poisson argued on the basis of Fresnel's analysis that a perfectly round object should diffract so as to produce a bright spot on the axis behind it. This was offered as a reductio ad absurdum argument against wave theory. However, Fresnel and Arago carried out the actual experiment and found that there is indeed a diffracted bright spot. The nineteenth century also saw the advent of accurate methods for the determination of the speed of light by Fizeau and Foucault that were used to verify the prediction from Maxwell's theory relating the velocity of light to known electric and magnetic constants. Maxwell's magnificent theory of electromagnetic waves arose from the work of Oersted, Ampère, and Faraday, which proved the intimate interconnection between electric and magnetic phenomena.

This volume discusses the consequences of modifying the traditional, classical view of light as a transverse electromagnetic wave whose electric and magnetic field components exist only in a plane perpendicular to the axis of propagation, and posits the existence of a longitudinal magnetic field component. These considerations are of relatively recent vintage, however [5].

The corpuscular view was revived in a different form early in twentieth century with Planck's solution of the blackbody problem and Einstein's adoption of the photon model in 1905. Milloni [6] has emphasized the fact that Einstein's famous 1905 paper [7] "Concerning a heuristic point of view toward the emission and transformation of light" argues strongly for a model of light that *simultaneously* displays the properties of waves and particles. He quotes Einstein:

> The wave theory of light, which operates with continuous spatial functions, has worked well in the representation of purely optical phenomena and will probably never be replaced by another theory. It should be kept in mind, however, that the optical observations refer to time averages rather than instantaneous values. In spite of the complete experimental confirmation of the theory as applied to diffraction, reflection, refraction, dispersion, etc., it is still conceivable that the theory of light which operates with continuous spatial functions may lead to contradictions with experience when it is applied to the phenomena of emission and transformation of light.
>
> According to the hypothesis that I want here to propose, when a ray of light expands starting from a point, the energy does not distribute on ever increasing volumes, but remains constituted of a finite number of energy quanta localized in space and moving without subdividing themselves, and unable to be absorbed or emitted partially.

This is the famous paper where Einstein, adopting Planck's idea of light quanta, gives a complete account of the photoelectric effect. He predicts the linear relationship between radiation frequency and stopping potential: "As far as I can see, there is no contradiction between these conceptions and the properties of the photoelectric effect observed by Herr Lenard. If each energy quantum of the incident light, independently of everything else, delivers its energy to electrons, then the velocity distribution of the ejected electrons will be independent of the intensity of the incident light. On the other hand the number of electrons leaving the body will, if other conditions are kept constant, be proportional to the intensity of the incident light."

Textbooks frequently cite this work as strong empirical evidence for the existence of photons as quanta of electromagnetic energy localized in space and time. However, it has been shown that [8] a complete account of the photoelectric effect can be obtained by treating the electromagnetic field as a classical Maxwellian field and the detector is treated according to the laws of quantum mechanics.

In view of his subsequent discomfort with dualism in physics, it is ironic that Einstein [9] gave a treatment of the fluctuations in the energy of electromagnetic waves that is fundamentally dualistic insofar that, if the Rayleigh–Jeans formula is adopted, the fluctuations are characteristic of electromagnetic waves. However, if the Wien law is used, the fluctuations are characteristic of particles. Einstein made several attempts to derive the Planck radiation law without invoking quantization of the radiation but without success. There was no alternative but to accept the quantum. This raised immediately the difficult question as to how such quanta gave rise to interference phenomena. Einstein suggested that perhaps light quanta need not interfere with themselves, but might interfere with

other quanta as they propagated. This suggestion was soon ruled out by interference experiments conduced at extremely low light levels. Dirac, in his
well-known textbook [10] on quantum mechanics, stated "Each photon inter-
feres only with itself. Interference between two different photons never occurs."
The latter part of this statement is now known to be wrong [11]. The advent of
highly coherent sources has enabled two-beam interference with two separate
sources. In these experiments, the classic interference pattern is not observed
but rather intensity correlations between the two beams are measured [12].
The recording of these intensity correlations is proof that the electromagnetic
fields from the two lasers have superposed. As Paul [11] argues, any experiment
that indicates that such a superposition has occurred should be called an interference experiment.

Taylor [13] was the first to report on two-beam interference experiments undertaken at extremely low light levels such that one can assert that, on average,
there is never more than one photon in the apparatus at any given time. Such
experiments have been repeated many times. However, given that the sources
used in these experiments generated light beams that exhibited photon bunching
[14], the basic assumption that there is only ever one photon in the apparatus at
any given time is not sound. More recent experiments using sources that emit
single-photon states have been performed [15–17].

In 1917 Einstein [18] wrote a paper on the dualistic nature of light in which
he discusses emission "without excitation from external causes," in other words
stimulated emission and also spontaneous absorption and emission. He derives
Planck's formula but also discusses the recoil of molecules when they emit
photons. It is the latter discussion that Einstein regarded as the most significant
aspect of the paper: "If a radiation bundle has the effect that a molecule struck
by it absorbs or emits a quantity of energy $h\nu$ in the form of radiation (ingoing
radiation), then a momentum $h\nu/c$ is always transferred to the molecule. For an
absorption of energy, this takes place in the direction of propagation of the
radiation bundle; for an emission, in the opposite direction."

In 1923, Compton [19] gave convincing experimental evidence for this process: "The experimental support of the theory indicates very convincingly that a
radiation quantum carries with itself, directed momentum as well as energy."
Einstein's dualism raises the following difficult question: If the particle carries
all the energy and momentum then, in what sense can the wave be regarded as
real? Einstein's response was to refer to such waves as "ghost fields" (Gespen-
sterfelder). Such waves are also referred to as "empty" - a wave propagating in
space and time but (virtually) devoid of energy and momentum. If described
literally, then such waves could not induce any physical changes in matter.
Nevertheless, there have been serious proposals for experiments that might
lead to the detection of "empty" waves associated with either photons [20]
or neutrons [21]. However, by making additional assumptions about the nature

of such "empty" waves [22], experiments have been proposed that might reveal their actual existence. One such experiment [23] has not yielded any such definitive evidence. Other experiments designed to determine whether empty waves can induce coherence in a two-beam interference experiment have not revealed any evidence for their existence [24], although Croca [25] now argues that this experiment should be regarded as inconclusive as the count rates were very low.

Controversies still persist in the interpretation of the quantum theory of light and indeed more generally in quantum mechanics itself. This happens notwithstanding the widely held view that all the difficult problems concerning the correct interpretation of quantum mechanics were resolved a long time ago in the famous encounters between Einstein and Bohr. Recent books have been devoted to foundational issues [26] in quantum mechanics, and some seriously question Bohrian orthodoxy [27,28]. There is at least one experiment described in the literature [29] that purports to do what Bohr prohibits: demonstrate the simultaneous existence of wave and particle-like properties of light.

Einstein's dualistic approach to electromagnetic radiation was generalized by de Broglie [30] to electrons when he combined results from the special theory of relativity (STR) and Planck's formula for the energy of a quantum to produce his famous formula relating wavelength to particle momentum. His model of a particle was one that contained an internal periodic motion plus an external wave of different frequency that acts to guide the particle. In this model, we have a wave–particle unity—both objectively exist. To quote de Broglie [31]: "The electron ... must be associated with a wave, and this wave is no myth; its wavelength can be measured and its interferences predicted." De Broglie's approach to physics has been described by Lochak [32] as quoted in Selleri [27]:

> Louis de Broglie is an intuitive spirit, concrete and realist, in love with simple images in three-dimensional space. He does not grant ontological value to mathematical models, in particular to geometrical representations in abstract spaces; he does not consider and does not use them other than as convenient mathematical instruments, among others, and it is not in their handling that his physical intuition is directly applied; faced with these abstract representations, he always keeps in mind the idea of all phenomena actually taking place in physical space, so that these mathematical modes of reasoning have a true meaning in his eyes only insofar as he perceives at all times what physical laws they correspond to in usual space.

De Broglie's views are not widely subscribed to today since as with "empty" waves, there is no compelling experimental evidence for the existence of physical waves accompanying the particle's motion (see, however, the discussion in Selleri [27]). Models of particles based on de Broglian ideas are still advanced by Vigier, for example [33].

As is well known, de Broglie abandoned his attempts at a realistic account of quantum phenomena for many years until David Bohm's discovery of a solution of Schrödinger's equation that lends itself to an interpretation involving a physical particle traveling under the influence of a so-called quantum potential. As de Broglie stated:

> For nearly twenty-five years, I remained loyal to the Bohr-Heisenberg view, which has been adopted almost unanimously by theorists, and I have adhered to it in my teaching, my lectures and my books. In the summer of 1951, I was sent the preprint of a paper by a young American physicist David Bohm, which was subsequently published in the January 15, 1952 issue of the Physical Review. In this paper, Mr. Bohm takes up the ideas I had put forward in 1927, at least in one of the forms I had proposed, and extends them in an interesting way on some points. Later, J.P. Vigier called my attention to the resemblance between a demonstration given by Einstein regarding the motion of particles in General Relativity and a completely independent demonstration I had given in 1927 in an exercise I called the "theory of the double solution."

A comprehensive account of the views of de Broglie, Bohm, and Vigier is given in Jeffers et al. [34]. In these models, contra Bohr particles actually do have trajectories. Trajectories computed for the double-slit experiment show patterns that reproduce the interference pattern observed experimentally [35]. Furthermore, the trajectories so computed never cross the plane of symmetry so that one can assert *with certainty* through which the particles traveled. This conclusion was also reached by Prosser [36,37] in his study of the double-slit experiment from a strictly Maxwellian point of view. Poynting vectors were computed whose distribution mirrors the interference pattern, and these never cross the symmetry plane as in the case of the de Broglie–Bohm–Vigier models. Prosser actually suggested an experimental test of this feature of his calculations. The idea was to illuminate a double-slit apparatus with very short microwave pulses and examine the received radiation at a suitable point off-axis behind the double slits. Calculations showed that for achievable experimental parameters, one could detect either two pulses if the orthodox view were correct, or only one pulse if the Prosser interpretation were correct. However, further investigation [38] showed that the latter conclusion was not correct. Two pulses would be observed, and their degree of separation (i.e., distinguishability) would be inversely related to the degree of contrast in the interference fringes.

Contemporary developments include John Bell's [39] discovery of his famous inequality that is predicated on the assumptions of both locality and realism. Bell's inequality is violated by quantum mechanics, and consequently, it is frequently argued, one cannot accept quantum mechanics, realism, and locality. Experiments on correlated particles appear to demonstrate that the Bell inequalities are indeed violated. Of the three choices, the most acceptable one is to abandon locality. However, Afriat and Selleri [40] have extensively reviewed both the current theoretical and experimental situation regarding the status of Bell's inequalities. They conclude, contrary to accepted wisdom, that one can construct local and realistic accounts of quantum mechanics that violate Bell's inequalities, and furthermore, there remain several loopholes in the experiments that have not yet been closed that allow for local and realist interpretations. No actual experiment that has been performed to date has *conclusively* demonstrated that locality has to be abandoned. However, experiments that approximate to a high degree the original gedanken experiment discussed by David Bohm, and that potentially close all known loopholes, will soon be undertaken. See the review article by Fry and Walther [41]. To quote these authors: "Quantum mechanics, even 50 years after its formulation, is still full of surprises." This underscores Einstein's famous remark: "All these years of conscious brooding have brought me no nearer to the answer to the question "What are light quanta?" Nowadays, every Tom, Dick, and Harry thinks he knows it, but he is mistaken."

II. THE PROCA EQUATION

The first inference of photon mass was made by Einstein and de Broglie on the assumption that the photon is a particle, and behaves as a particle in, for example, the Compton and photoelectric effects. The wave–particle duality of de Broglie is essentially an extension of the photon, as the quantum of energy, to the photon, as a particle with quantized momentum. The Beth experiment in 1936 showed that the photon has angular momentum, whose quantum is $\hbar$. Other fundamental quanta of the photon are inferred in Ref. 42. In 1930, Proca [43] extended the Maxwell–Heaviside theory using the de Broglie guidance theorem:

$$\hbar\omega_0 = m_0 c^2 \tag{1}$$

where m_0 is the rest mass of the photon and $m_0 c^2$ is its rest energy, equated to the quantum of rest energy $\hbar\omega_0$. The original derivation of the Proca equation therefore starts from the Einstein equation of special relativity:

$$p^\mu p_\mu = m_0^2 c^2 \tag{2a}$$

The usual quantum ansatz is applied to this equation to obtain a wave equation:

$$En = i\hbar \frac{\partial}{\partial t}; \qquad p = -i\hbar\nabla \tag{2b}$$

This is an example of the de Broglie wave–particle duality. The resulting wave equation is

$$\left(\Box + \frac{m_0^2 c^4}{\hbar^2}\right)\psi = 0 \tag{3}$$

where ψ is a wave function, whose meaning was first inferred by Born in 1926. If the wave function is a scalar, Eq. (3) becomes the Klein–Gordon equation. If ψ is a 2-spinor, Eq. (3) becomes the van der Waerden equation, which can be related analytically to the Dirac equation, and if ψ is the electromagnetic 4-potential A^μ, Eq. (3) becomes the Proca equation:

$$\Box A^\mu = -\left(\frac{m_0 c^2}{\hbar}\right)^2 A^\mu \tag{4}$$

So A^μ can act as a wave function and the Proca equation can be regarded as a quantum equation if A^μ is a wave function in configuration space, and as a classical equation in momentum space.

It is customary to develop the Proca equation in terms of the vacuum charge current density

$$\Box A^\mu = -\left(\frac{m_0 c^2}{\hbar}\right)^2 A^\mu = -\kappa^2 A^\mu = \frac{1}{\varepsilon_0} J^\mu(\text{vac}) \tag{5}$$

The potential A^μ therefore has a physical meaning in the Proca equation because it is directly proportional to $J^\mu(\text{vac})$. The Proca equations in the vacuum are therefore

$$\partial_\mu F^{\mu\nu} + \left(\frac{m_0 c^2}{\hbar}\right)^2 A^\nu = 0 \tag{6}$$

$$\partial_\mu A^\mu = 0 \tag{7}$$

and, as described in the review by Evans in Part 2 of this compilation [44], these have the structure of the Panofsky, Phillips, Lehnert, Barrett, and O(3) equations, a structure that can also be inferred from the symmetry of the Poincaré group [44]. Lehnert and Roy [45] self-consistently infer the structure of the Proca equations from their own equations, which use a vacuum charge and current.

The problem with the Proca equation, as derived originally, is that it is not gauge-invariant because, under the U(1) gauge transform [46]

$$A^\mu \to A^\mu + \frac{1}{g}\partial^\mu \Lambda \tag{8}$$

the left-hand side of Eq. (4) is invariant but an arbitrary quantity $\frac{1}{g}\partial^\mu \Lambda$ is added to the right-hand side. This is paradoxical because the Proca equation is well founded in the quantum ansatz and the Einstein equation, yet violates the fundamental principle of gauge invariance. The usual resolution of this paradox is to assume that the mass of the photon is identically zero, but this assumption leads to another paradox, because a particle must have mass by definition, and the wave-particle dualism of de Broglie becomes paradoxical, and with it, the basis of quantum mechanics.

In this section, we suggest a resolution of this >70-year-old paradox using O(3) electrodynamics [44]. The new method is based on the use of covariant derivatives combined with the first Casimir invariant of the Poincaré group. The latter is usually written in operator notation [42,46] as the invariant $P_\mu P^\mu$, where P^μ is the generator of spacetime translation:

$$P^\mu = i\partial^\mu = \frac{p^\mu}{\hbar} \tag{9}$$

The ordinary derivative in gauge theory becomes the covariant derivative

$$\partial_\mu \rightarrow D_\mu = \partial_\mu - igA_\mu \tag{10}$$

for all gauge groups. The generator D_μ is a generator of the Poincaré group because it obeys the Jacobi identity

$$\sum_{\sigma,\nu,\mu} [D_\sigma, [D_\nu, D_\mu]] \equiv 0 \tag{11}$$

and the covariant derivative (10) can be regarded as a sum of spacetime translation generators.

The basic assumption is that the photon acquires mass through the invariant

$$D_\mu D^{\mu*}\psi = 0 \tag{12}$$

for any gauge group. This equation can be developed for any gauge group as

$$(\partial_\mu - igA_\mu)(\partial^\mu + igA^{\mu*})\psi = 0 \tag{13}$$

and can be expressed as

$$\begin{aligned}
&\Box\psi - igA_\mu\partial^\mu\psi + ig\partial_\mu(A^\mu\psi) + g^2A_\mu A^\mu\psi \\
&= 0 \\
&= \Box\psi - igA_\mu\partial^\mu\psi + ig\psi\partial_\mu A^\mu + igA^\mu\partial_\mu\psi + g^2A_\mu A^\mu\psi \\
&= (\Box + ig\partial_\mu A^\mu + g^2A_\mu A^\mu)\psi \\
&= 0
\end{aligned} \tag{14}$$

This equation reduces to

$$(\Box + \kappa^2)\psi = -ig\partial_\mu A^\mu\psi \qquad (15)$$

for any gauge group because

$$g = \frac{\kappa}{A^{(0)}}; \qquad A_\mu A^\mu = A^{(0)2} \qquad (16)$$

In the plane-wave approximation:

$$\partial_\mu A^\mu = 0 \qquad (17a)$$

and the Proca equation for any gauge group becomes

$$(\Box + \kappa^2)\psi = 0 \qquad (17b)$$

for any gauge group.

Therefore Eq. (18) has been shown to be an invariant of the Poincaré group, Eq. (12), and a product of two Poincaré covariant derivatives. In momentum space, this operator is equivalent to the Einstein equation under any condition. The conclusion is reached that the factor g is nonzero in the vacuum.

In gauge theory, for any gauge group, however, a rotation

$$\psi' = e^{i\Lambda}\psi \equiv S\psi \qquad (18)$$

in the internal gauge space results in the gauge transformation of A_μ as follows

$$A'_\mu = SA_\mu S^{-1} - \frac{i}{g}(\partial_\mu S)S^{-1} \qquad (19)$$

and to construct a gauge-invariant Proca equation from the operator (16), a search must be made for a potential A_μ that is invariant under gauge transformation. It is not possible to find such a potential on the U(1) level because the inhomogeneous term is always arbitrary. On the O(3) level, however, the potential can be expressed as

$$A_\mu = A_\mu^{(2)}e^{(1)} + A_\mu^{(1)}e^{(2)} + A_\mu^{(3)}e^{(3)} \qquad (20)$$

if the internal gauge space is a physical space with O(3) symmetry described in the complex circular basis ((1),(2),(3)) [3]. A rotation in this physical gauge space can be expressed in general as

$$\psi' = \exp(iM^a\Lambda^a(x^\mu))\psi \qquad (21)$$

where M^a are the rotation generators of O(3) and where $\Lambda^{(1)}, \Lambda^{(2)}$, and $\Lambda^{(3)}$ are angles.

Developing Eq. (13), we obtain

$$(\partial_\mu - igA_\mu^{(1)})(\partial^\mu + igA^{\mu(2)})\psi = 0$$
$$(\partial_\mu - igA_\mu^{(2)})(\partial^\mu + igA^{\mu(1)})\psi = 0 \qquad (22)$$
$$(\partial_\mu - igA_\mu^{(3)})(\partial^\mu + igA^{\mu(3)})\psi = 0$$

The eigenfunction ψ may be written in general as the O(3) vector

$$\psi \equiv A^\nu \qquad (23)$$

and under gauge transformation

$$A^{\nu'} = \exp(iM^a\Lambda^a(x^\mu))A^\nu \qquad (24)$$

from Eq. (21). Here, $\Lambda^{(1)}, \Lambda^{(2)}$, and $\Lambda^{(3)}$ are angles in the physical internal gauge space of O(3) symmetry.

Therefore Eqs. (22) become

$$\Box^2 A^\nu = -\kappa^2 A^\nu = \frac{1}{\varepsilon_0}J^\nu(\text{vac}) \qquad (25)$$

where

$$J^\nu = \left(\rho^{(i)}, \frac{J^{(i)}}{c}\right) \qquad i = 1, 2, 3 \qquad (26)$$

and Eqs. (25) become

$$\Box A^{\nu(1)} = -\kappa^2 A^{\nu(1)} = \frac{J^{\nu(1)}}{\varepsilon_0} \qquad (27)$$

$$\Box A^{\nu(2)} = -\kappa^2 A^{\nu(2)} = \frac{J^{\nu(2)}}{\varepsilon_0} \qquad (28)$$

$$\Box A^{\nu(3)} = 0 \qquad (29)$$

It can be seen that the photon mass is carried by $A^{\nu(1)}$ and $A^{\nu(2)}$, but not by $A^{\nu(3)}$. This result is also obtained by a different route using the Higgs mechanism in Ref. 42, and is also consistent with the fact that the mass associated with $A^{\nu(3)}$ corresponds with the superheavy boson inferred by Crowell [42], reviewed in

Ref. 42 and observed in a LEP collaboration [42]. The effect of a gauge transformation on Eqs. (27)–(29) is as follows:

$$\Box\left(A_\mu^{(1)} + \frac{1}{g}\partial_\mu\Lambda^{(1)}\right) = -\kappa^2\left(A_\mu^{(1)} + \frac{1}{g}\partial_\mu\Lambda^{(1)}\right) \tag{30}$$

$$\Box\left(A_\mu^{(2)} + \frac{1}{g}\partial_\mu\Lambda^{(2)}\right) = -\kappa^2\left(A_\mu^{(2)} + \frac{1}{g}\partial_\mu\Lambda^{(2)}\right) \tag{31}$$

$$\Box\left(A_\mu^{(3)} + \frac{1}{g}\partial_\mu\Lambda^{(3)}\right) = 0 \tag{32}$$

Equations (30) and (31) are eigenequations with the same eigenvalue, $-\kappa^2$, as Eqs. (27) and (28). On the O(3) level, the eigenfunctions $A_\mu^{(1)} + \frac{1}{g}\partial_\mu\Lambda^{(1)}$ are not arbitrary because $\Lambda^{(1)}$ and $\Lambda^{(2)}$ are angles in a physical internal gauge space. The original Eq. (12) is gauge-invariant, however, because on gauge transformation

$$g^2 A_\mu A^{\mu*} \rightarrow g^2 A'_\mu A^{\mu*'}; \qquad g' = \frac{\kappa}{A^{(0)'}} \tag{33}$$

and

$$D_\mu D^{\mu*}\psi \rightarrow D_\mu D^{\mu*}(S\psi) = \psi D_\mu D^{\mu*}S + S D_\mu D^{\mu*}\psi = 0 \tag{34}$$

because S must operate on ψ.

In order for Eq. (34) to be compatible with Eqs. (30) and (31), we obtain

$$\Box(\partial_\mu\Lambda^{(1)}) = -\kappa^2(\partial_\mu\Lambda^{(1)}) \tag{35}$$

$$\Box(\partial_\mu\Lambda^{(2)}) = -\kappa^2(\partial_\mu\Lambda^{(2)}) \tag{36}$$

which are also Proca equations. So the >70-year-old problem of the lack of gauge invariance of the Proca equation is solved by going to the O(3) level.

The field equations of electrodynamics for any gauge group are obtained from the Jacobi identity of Poincaré group generators [42,46]:

$$\sum_{\sigma,\mu,\nu} [D_\sigma, [D_\mu, D_\nu]] \equiv 0 \tag{37}$$

If the potential is classical, the Jacobi identity (37) can be written out as

$$D_\sigma G_{\mu\nu} + D_\mu G_{\nu\sigma} + D_\nu G_{\sigma\mu} - G_{\mu\nu}D_\sigma - G_{\nu\sigma}D_\mu - G_{\sigma\mu}D_\nu \equiv 0 \tag{38}$$

This equation implies the Jacobi identity:

$$[A_\sigma, G_{\mu\nu}] + [A_\mu, G_{\nu\sigma}] + [A_\nu, G_{\sigma\mu}] \equiv 0 \tag{39}$$

which in vector form can be written as

$$A_\mu \times \tilde{G}^{\mu\nu} = A^\sigma \times G^{\mu\nu} + A^\mu \times G^{\nu\sigma} + A^\nu \times G^{\sigma\mu}$$
$$\equiv 0 \tag{40}$$

As a result of this Jacobi identity, the homogeneous field equation

$$D_\mu \tilde{G}^{\mu\nu} \equiv 0 \tag{41}$$

reduces to

$$\partial_\mu \tilde{G}^{\mu\nu} \equiv 0 \tag{42}$$

for all gauge group symmetries. The implication is that instantons or pseudo-particles do not exist in Minkowski spacetime in a pure gauge theory, because magnetic monopoles and currents vanish for all internal gauge group symmetries. Therefore, the homogeneous field equation of electrodynamics, considered as a gauge theory of any internal symmetry, can be obtained from the Jacobi identity (42) of the Poincaré group of Minkowski spacetime. The homogeneous field equation is gauge-covariant for any internal symmetry. Analogously, the Proca equation is the mass Casimir invariant (12) of the Poincaré group of Minkowski spacetime.

There are several major implications of the Jacobi identity (40), so it is helpful to give some background for its derivation. On the U(1) level, consider the following field tensors in $c = 1$ units and contravariant covariant notation in Minkowski spacetime:

$$\tilde{F}^{\mu\nu} = \begin{bmatrix} 0 & -B^1 & -B^2 & -B^3 \\ B^1 & 0 & E^3 & -E^2 \\ B^2 & -E^3 & 0 & E^1 \\ B^3 & E^2 & -E^1 & 0 \end{bmatrix}; \quad \tilde{F}_{\mu\nu} = \begin{bmatrix} 0 & B_1 & B_2 & B_3 \\ -B_1 & 0 & E_3 & -E_2 \\ -B_2 & -E_3 & 0 & E_1 \\ -B_3 & E_2 & -E_1 & 0 \end{bmatrix}$$

$$F_{\rho\sigma} = \begin{bmatrix} 0 & E_1 & E_2 & E_3 \\ -E_1 & 0 & -B_3 & B_2 \\ -E_2 & B_3 & 0 & -B_1 \\ -E_3 & -B_2 & B_1 & 0 \end{bmatrix}; \quad F^{\rho\sigma} = \begin{bmatrix} 0 & -E^1 & -E^2 & -E^3 \\ E^1 & 0 & -B^3 & B^2 \\ E^2 & B^3 & 0 & -B^1 \\ E^3 & -B^2 & B^1 & 0 \end{bmatrix}$$

$$\tag{43}$$

These tensors are generated from the duality relations [47]

$$\tilde{G}^{\mu\nu} = \frac{1}{2}\varepsilon^{\mu\nu\rho\sigma}G_{\rho\sigma}; \quad G^{\mu\nu} = -\frac{1}{2}\varepsilon^{\mu\nu\rho\sigma}\tilde{G}_{\rho\sigma}$$
$$\tilde{G}_{\mu\nu} = \frac{1}{2}\varepsilon_{\mu\nu\rho\sigma}G^{\rho\sigma}; \quad G_{\mu\nu} = -\frac{1}{2}\varepsilon_{\mu\nu\rho\sigma}\tilde{G}^{\rho\sigma}$$

$$\tag{44}$$

where the totally antisymmetric unit tensor is defined as

$$\varepsilon^{0123} = 1 = -\varepsilon_{0123} \tag{45}$$

and result in the following Jacobi identity:

$$\partial_\mu \tilde{F}^{\mu\nu} = \partial^\sigma F^{\mu\nu} + \partial^\mu F^{\nu\sigma} + \partial^\nu F^{\sigma\mu} \equiv 0 \tag{46}$$

It also follows that

$$\partial_\mu F^{\mu\nu} = \partial_\sigma \tilde{F}^{\mu\nu} + \partial_\mu \tilde{F}^{\nu\sigma} + \partial_\nu \tilde{F}^{\sigma\mu} \tag{47}$$

The proof of the Jacobi identity (46) can be seen by considering a development such as

$$\partial_\mu \tilde{F}^{\mu\nu} = \frac{1}{2} \partial_\mu \left(\varepsilon^{\mu\nu\rho\sigma} F_{\rho\sigma} \right)$$

$$= \frac{1}{2} \partial_\mu \left(\varepsilon^{\mu\nu01} F_{01} + \varepsilon^{\mu\nu02} F_{02} + \varepsilon^{\mu\nu03} F_{03} + \varepsilon^{\mu\nu10} F_{10} + \varepsilon^{\mu\nu20} F_{20} + \varepsilon^{\mu\nu30} F_{30} \right.$$

$$\left. + \varepsilon^{\mu\nu12} F_{12} + \varepsilon^{\mu\nu13} F_{13} + \varepsilon^{\mu\nu21} F_{21} + \varepsilon^{\mu\nu31} F_{31} + \varepsilon^{\mu\nu23} F_{23} + \varepsilon^{\mu\nu32} F_{32} \right) \tag{48}$$

If $\nu = 0$, then

$$\partial_1 \tilde{F}^{10} + \partial_2 \tilde{F}^{20} + \partial_3 \tilde{F}^{30} = -\partial_1 F^{23} - \partial_2 F^{13} - \partial_3 F^{12} \equiv 0 \tag{49}$$

Equation (47) may be proved similarly. On the O(3) level there exist the analogous equations (40) and

$$A_\mu \times G^{\mu\nu} = A_\sigma \times \tilde{G}^{\mu\nu} + A_\mu \times \tilde{G}^{\nu\sigma} + A_\nu \times \tilde{G}^{\sigma\mu} \tag{50}$$

which is not zero in general.

It follows from the Jacobi identity (40) that there also exist other Jacobi identities such as [42]

$$A_\lambda^{(2)} \times (A_\mu^{(1)} \times A_\nu^{(2)}) + A_\mu^{(2)} \times (A_\nu^{(1)} \times A_\lambda^{(2)}) + A_\nu^{(2)} \times (A_\lambda^{(1)} \times A_\mu^{(2)}) \equiv 0 \tag{51}$$

The Jacobi identity (40) means that the homogeneous field equation of electrodynamics for any gauge group is

$$\partial_\mu \tilde{G}^{\mu\nu} \equiv 0 \tag{52}$$

If the symmetry of the gauge group is O(3) in the complex basis ((1),(2),(3)) [42,47], Eq. (52) can be developed as three equations:

$$\partial_\mu \tilde{G}^{\mu\nu(1)} \equiv 0 \tag{53}$$

$$\partial_\mu \tilde{G}^{\mu\nu(2)} \equiv 0 \tag{54}$$

$$\partial_\mu \tilde{G}^{\mu\nu(3)} \equiv 0 \tag{55}$$

Now consider a component of the Jacobi identity (39)

$$\varepsilon_{(1)(2)(3)} A_\sigma^{(2)} G_{\mu\nu}^{(3)} + \varepsilon_{(1)(2)(3)} A_\mu^{(2)} G_{\nu\sigma}^{(3)} + \varepsilon_{(1)(2)(3)} A_\nu^{(2)} G_{\sigma\mu}^{(3)} \equiv 0 \tag{56}$$

and consider next the following cyclic permutation:

$$A_0^{(2)} G_{23}^{(3)} - A_0^{(3)} G_{23}^{(2)} + A_2^{(2)} G_{30}^{(3)} - A_2^{(3)} G_{30}^{(2)} + A_3^{(2)} G_{02}^{(3)} - A_3^{(3)} G_{02}^{(2)} \equiv 0 \tag{57}$$

This gives the result

$$B_X^{(2)} + \frac{E_Y^{(2)}}{c} - A_Y^{(2)} E_Z^{(3)} \equiv 0 \tag{58}$$

Using Eq. (54), we obtain the result

$$E_Z^{(3)} \equiv 0 \tag{59}$$

thus $\boldsymbol{E}^{(3)}$ vanishes identically in O(3) electrodynamics. The third equation (55) therefore becomes the following identity:

$$\frac{\partial \boldsymbol{B}^{(3)}}{\partial t} \equiv 0 \tag{60}$$

In other words, $\boldsymbol{B}^{(3)}$ is identically independent of time, a result that follows from its definition [42,47]

$$\boldsymbol{B}^{(3)} \equiv -ig\boldsymbol{A}^{(1)} \times \boldsymbol{A}^{(2)} \tag{61}$$

The ansatz, upon which these results are based, is that the configuration of the vacuum is described by the doubly connected group O(3), which supports the Aharonov–Bohm effect in Minkowski spacetime [46]. More generally, the vacuum configuration could be described by an internal gauge space more general than O(3), such as the Lorentz, Poincaré, or Einstein groups. The O(3) group is the little group of the Poincaré group for a particle with identically nonzero mass, such as the photon. If the internal space were extended from O(3) to the Poincaré group, there would appear boost and spacetime translation operators in the gauge transform (36), as well as rotation generators. The Poincaré group is the most general group of special relativity, and the Einstein group, that of general relativity. Both groups are defined in Minkowski spacetime. In all these groups, there would be no magnetic monopole or current in Minkowski spacetime because of the Jacobi identity (37) between any group generators. The superiority of O(3) over U(1) electrodynamics has been demonstrated in several ways using empirical data [42,47–61] such as those available in the Sagnac effect, so its seems logical to extend the internal space to the Poincaré group. The widespread use of a U(1) group for electrodynamics is a historical accident. The use of an O(3) group is an improvement, so it is expected that the use of a Poincaré group would be an improvement over O(3).

Meanwhile, the Jacobi identity (40) implies, in vector notation, the identities

$$A^{(2)} \cdot B^{(3)} - B^{(2)} \cdot A^{(3)} \equiv 0$$
$$A^{(3)} \cdot B^{(1)} - B^{(3)} \cdot A^{(1)} \equiv 0 \tag{62}$$
$$A^{(1)} \cdot B^{(2)} - B^{(1)} \cdot A^{(2)} \equiv 0$$

and

$$cA_0^{(3)}B^{(2)} - cA_0^{(2)}B^{(3)} + A^{(2)} \times E^{(3)} - A^{(3)} \times E^{(2)} \equiv 0$$
$$cA_0^{(1)}B^{(3)} - cA_0^{(3)}B^{(1)} + A^{(3)} \times E^{(1)} - A^{(1)} \times E^{(3)} \equiv 0 \tag{63}$$
$$cA_0^{(2)}B^{(1)} - cA_0^{(1)}B^{(2)} + A^{(1)} \times E^{(2)} - A^{(2)} \times E^{(1)} \equiv 0$$

It has been shown elsewhere [42] that the identities (63) correspond with the B cyclic theorem [42,47–61] of O(3) electrodynamics:

$$B^{(1)} \times B^{(2)} = iB^{(0)}B^{(3)*} \tag{64}$$
$$\cdots$$

which is therefore also an identity of the Poincaré group. Within a factor, the B cyclic theorem is the rotation generator Lie algebra of the Poincaré group. In terms of the unit vectors of the basis ((1),(2),(3)), the B cyclic theorem reduces to

$$e^{(1)} \times e^{(2)} = ie^{(3)*} \tag{65}$$
$$\cdots$$

which is the frame relation itself. This relation is unaffected by a Lorentz boost and a spacetime translation. A rotation produces the same relation (65). So the B cyclic theorem is invariant under the most general type of Lorentz transformation, consisting of boosts, rotations, and spacetime translations. Similarly, the definition of $B^{(3)}$, Eq. (61), is Lorentz-invariant.

The Jacobi identities (63) reduce to the B cyclic theorem (64) because of Eqs. (53)–(55), and because $E^{(3)}$ vanishes identically [42,47–61], and the B cyclic theorem is self-consistent with Eqs. (53)–(55). The identities (62) and (63) imply that there are no instantons or pseudoparticles in O(3) electrodynamics, which is a dynamics developed in Minkowski spacetime. If the pure gauge theory corresponding to O(3) electrodynamics is supplemented with a Higgs mechanism, then O(3) electrodynamics supports the 't Hooft–Polyakov magnetic monopole [46]. Therefore Ryder [46], for example, in his standard text, considers a form of O(3) electrodynamics [46, pp. 417ff.], and the 't Hooft–Polyakov magnetic monopole is a signature of an O(3) electrodynamics with its symmetry broken spontaneously with a Higgs mechanism. In the pure gauge theory, however, the magnetic monopole is identically zero. It is clear that the theory of 't Hooft and Polyakov is O(3) electrodynamics plus a Higgs mechanism, an important result.

In order to show that the Proca equation from gauge theory is gauge-invariant, it is convenient to consider the Jacobi identity

$$D_\mu \tilde{G}^{\mu\nu} \equiv 0 \tag{66}$$

which is gauge-invariant in all gauge groups. Now use

$$D_\mu G^{\mu\nu} = D_\sigma \tilde{G}^{\lambda\kappa} + D_\kappa \tilde{G}^{\sigma\lambda} + D_\lambda \tilde{G}^{\kappa\sigma} \tag{67}$$

and let two indices be the same on the right-hand side. This procedure produces

$$D_\mu G^{\mu\nu} = D_\sigma \left(\tilde{G}^{\sigma\kappa} + \tilde{G}^{\kappa\sigma} \right) = 0 \tag{68}$$

showing that:

$$D_\mu G^{\mu\nu} = 0 \tag{69}$$

is also gauge-invariant for all gauge groups. Finally, expand Eq. (69) as

$$D_\mu G^{\mu\nu} = D_\mu (D^\mu A^\nu - D^\nu A^\mu) = 0 \tag{70}$$

to obtain

$$D_\mu D^\mu A^\nu = 0 \tag{71}$$

which is also gauge-invariant for all gauge groups.

On the U(1) level, for example, the structure of the Lehnert [45] and gauge-invariant Proca equations is obtained as follows:

$$(\Box + \kappa^2)A^\nu = 0 \tag{72}$$

$$\left(\partial_\mu + igA^*_\mu\right)G^{\mu\nu} = 0 \tag{73}$$

These are regarded as eigenequations with eigenfunctions A^ν and $G^{\mu\nu}$ in configuration space. In this method, there is no need for the Lorenz condition. The equivalent of Eq. (72) in momentum space is the Einstein equation (2), and this statement is true for all gauge group symmetries. Comparing Eqs. (6) and (7) with Eqs. (72) and (73), the following equation is obtained on the U(1) level:

$$\kappa^2 A^\nu = igA^*_\mu G^{\mu\nu} \tag{74}$$

This equation may be developed as follows:

$$\kappa^2 A^{(0)} = igA^* \cdot \frac{\boldsymbol{E}}{c} \tag{75}$$

In the plane-wave approximation

$$\kappa A^{(0)} = \frac{E^{(0)}}{c} = B^{(0)} \tag{76}$$

and it is seen that condition (74) is true on the U(1) level. Equation (73) can be written as

$$\partial_\mu G^{\mu\nu} = -igA^*_\mu G^{\mu\nu} \equiv \frac{J^\mu}{\varepsilon_0} \tag{77}$$

in the vacuum, and this is the Lehnert equation [42,45]. The latter gives longitudinal or axisymmetric solutions and can describe physical situations that the Maxwell–Heaviside theory cannot.

On the O(3) level, one can write the Proca equation in the following form (22):

$$(\Box + g^2 A^{(1)}_\mu A^{\mu(2)})A^{\nu(1)} = 0$$

$$(\Box + g^2 A^{(2)}_\mu A^{\mu(1)})A^{\nu(2)} = 0 \tag{78}$$

$$(\Box + g^2 A^{(3)}_\mu A^{\mu(3)})A^{\nu(3)} = 0$$

The third equation of (22) reduces to a d'Alembert equation

$$\Box A^{\nu(3)} = 0 \tag{79}$$

because $A_\mu^{(3)} A^{\mu(3)} = 0$ in O(3) electrodynamics. Equation (79) is consistent with the fact that $A_\mu^{(3)}$ is phaseless by definition in O(3) electrodynamics. The first two equations of the triad (78) are complex conjugate Proca equations of the form

$$
\begin{aligned}
(\Box + \kappa^2) A^\nu &= 0 \\
(\Box + \kappa^2) A^{\nu*} &= 0
\end{aligned}
\tag{80}
$$

so we obtain the U(1) Proca equation, but with the advantages of O(3) electrodynamics inbuilt.

In summary, the structure of the Proca equation on the O(3) level is as follows:

$$D_\mu G^{\mu\nu} = 0 \tag{81}$$

which is equivalent to

$$\partial_\mu G^{\mu\nu} = -g A_\mu \times G^{\mu\nu} \tag{82}$$

The latter equation can be expanded in the basis ((1),(2),(3)) as [42]

$$\nabla \cdot D^{(1)*} = ig(A^{(2)} \cdot D^{(3)} - D^{(2)} \cdot A^{(3)})$$

$$\nabla \cdot D^{(2)*} = ig(A^{(3)} \cdot D^{(1)} - D^{(3)} \cdot A^{(1)})$$

$$\nabla \cdot D^{(3)*} = ig(A^{(1)} \cdot D^{(2)} - D^{(1)} \cdot A^{(2)})$$

$$\nabla \times H^{(1)*} - \frac{\partial D^{(1)*}}{\partial t} = -ig(cA_0^{(2)} D^{(3)} - cA_0^{(3)} D^{(2)} + A^{(2)} \times H^{(3)} - A^{(3)} \times H^{(2)})$$

$$\nabla \times H^{(2)*} - \frac{\partial D^{(2)*}}{\partial t} = -ig(cA_0^{(3)} D^{(1)} - cA_0^{(1)} D^{(3)} + A^{(3)} \times H^{(1)} - A^{(1)} \times H^{(3)})$$

$$\nabla \times H^{(3)*} - \frac{\partial D^{(3)*}}{\partial t} = -ig(cA_0^{(1)} D^{(2)} - cA_0^{(2)} D^{(1)} + A^{(1)} \times H^{(2)} - A^{(2)} \times H^{(1)})$$

$$\tag{83}$$

It can be seen that, in general, there are extra Noether charges and currents that define the photon mass gauge invariantly. The magnetic field strength and electric displacement is used in Eq. (83) because, in general, there may be vacuum polarization and magnetization, defined respectively as

$$
\begin{aligned}
D &= \varepsilon_0 E + P \\
B &= \mu_0 (H + M)
\end{aligned}
\tag{84}
$$

There may be a vacuum charge on the O(3) level provided that the term

$$\nabla \cdot \boldsymbol{D}^{(3)*} = ig(\boldsymbol{A}^{(1)} \cdot \boldsymbol{D}^{(2)} - \boldsymbol{D}^{(1)} \cdot \boldsymbol{A}^{(2)}) \qquad (85)$$

is not zero. For this to be the case, the vacuum polarization must be such that the displacement $\boldsymbol{D}^{(1)}$ is not the complex conjugate of the displacement $\boldsymbol{D}^{(2)}$. It can be seen as follows that for this to be the case, polarization must develop asymmetrically as follows:

$$\begin{aligned}
\boldsymbol{D}^{(1)} &= \varepsilon_0 \boldsymbol{E}^{(1)} + a\boldsymbol{P}^{(1)} \\
\boldsymbol{D}^{(2)} &= \varepsilon_0 \boldsymbol{E}^{(2)} + b\boldsymbol{P}^{(2)}
\end{aligned} \qquad (86)$$

If there is no vacuum polarization, then the photon mass resides entirely in the vacuum current.

In the preceding analysis, commutators of covariant derivatives always act on an eigenfunction, so, for example:

$$\begin{aligned}
[D_\mu, D_\nu]\psi &= [\partial_\mu - igA_\mu, \partial_\nu - igA_\nu]\psi \\
&= (\partial_\mu\partial_\nu - \partial_\nu\partial_\mu)\psi - igA_\mu\partial_\nu\psi + ig\partial_\nu(A_\mu\psi) \\
&\quad - ig\partial_\mu(A_\nu\psi) + igA_\nu\partial_\mu\psi - g^2[A_\mu, A_\nu]\psi \\
&= -igA_\mu\partial_\nu\psi + ig\partial_\nu A_\mu\psi + igA_\mu\partial_\nu\psi - ig(\partial_\mu A_\nu)\psi \\
&\quad - igA_\nu\partial_\mu\psi + igA_\nu\partial_\mu\psi - g^2[A_\mu, A_\nu]\psi \\
&= -ig(\partial_\mu A_\nu - \partial_\nu A_\mu - ig[A_\mu, A_\nu])\psi
\end{aligned} \qquad (87)$$

giving the field tensor for all gauge groups:

$$G_{\mu\nu} = \partial_\mu A_\nu - \partial_\nu A_\mu - ig[A_\mu, A_\nu] \qquad (88)$$

In the literature, the operation $[D_\mu, D_\nu]\psi$ is often written simply as $[D_\mu, D_\nu]$ but this shorthand notation always implies that the operators act on the unwritten ψ.

On the O(3) level, the clearest insight into the meaning of the Jacobi identity (37) is obtained by writing the covariant derivative in terms of translation (P) and rotation (J) generators of the Poincaré group:

$$\begin{aligned}
D_\sigma &= \partial_\sigma - igA_\sigma = \partial_\sigma - ig(A_\sigma^X J_X + A_\sigma^Y J_Y + A_\sigma^Z J_Z) \\
\partial_\sigma &= -iP_\sigma
\end{aligned} \qquad (89)$$

where J_X, J_Y, and J_Z are the rotation generators. The translation generator is defined [42,46] as

$$P_\sigma = i\partial_\sigma \qquad (90)$$

The Jacobi identity of operators (37) therefore becomes, after index matching

$$[P_\sigma + gA_\sigma^X J_X, [P_\kappa + gA_\kappa^Y J_Y, P_\lambda + gA_\lambda^Z J_Z]]$$
$$= [P_X + gA_X^X J_X, [P_Y + gA_Y^Y J_Y, P_Z + gA_Z^Z J_Z]] \qquad (91)$$

Now consider the component

$$[P_X, [P_Y + gA_Y^Y, P_Z + gA_Z^Z J_Z]]$$
$$= [P_X, [P_Y, P_Z]] + gA_Y^Y[J_Y, P_Z] + gA_Z^Z[P_Y, J_Z] + g^2 A_Y^Y A_Z^Z[J_Y, J_Z]] \quad (92)$$

and use the Lie algebra [46]

$$\begin{array}{ll} [J_Y, P_X] = -iP_X & [P_X, P_X] = 0 \\ [P_Y, J_Z] = iP_X & [P_X, J_X] = 0 \\ [J_Y, J_Z] = iJ_X & \end{array} \qquad (93)$$

to find that it vanishes. In vector notation, this result implies Eq. (52)

$$\partial_\mu \tilde{G}^{\mu\nu} \equiv \mathbf{0} \qquad [\mathrm{O}(3) \text{ level}] \qquad (94)$$

and the result

$$[A_\sigma, G_{\kappa\lambda}] + [A_\kappa, G_{\lambda\sigma}] + [A_\lambda, G_{\sigma\kappa}] \equiv 0 \qquad (95)$$

which can be developed as

$$[\partial_\mu, \tilde{G}^{\mu\nu}]\psi \equiv 0 \qquad (96)$$

giving Eq. (94) again self-consistently. Similarly

$$[A_\mu, \tilde{G}^{\mu\nu}]\psi \equiv 0 \qquad (97)$$

giving Eq. (40). In operator form, this is

$$[gA_\sigma^X J_X, [P_\kappa + gA_\kappa^Y J_Y, P_\lambda + gA_\lambda^Z J_Z]] \equiv 0 \qquad (98)$$

and the factor $[A_\mu, \tilde{G}^{\mu\nu}]$ is a simple multiplication operation on ψ.

The overall result is that the homogeneous field equation for all group symmetries is the result of the Lie algebra of the Poincaré group, the group of special relativity. The Jacobi identity can be derived in turn from a round trip or holonomy in Minkowski spacetime, as first shown by Feynman [46] for all gauge groups. The Jacobi identity is Lorentz- and gauge-invariant.

III. CLASSICAL LEHNERT AND PROCA VACUUM CHARGE CURRENT DENSITY

In this section, gauge theory is used to show that there exist classical charge current densities in the vacuum for all gauge group symmetries, provided that the scalar field of gauge theory is identified with the electromagnetic field [O(3) level] or a component of the electromagnetic field [U(1) level]. The Lehnert vacuum charge current density exists for all gauge group symmetries without the Higgs mechanism. The latter introduces classical Proca currents and other terms that represent energy inherent in the vacuum. Some considerable mathematical detail is given as an aid to comprehension of the Lagrangian methods on which these results depend.

The starting point is the Lagrangian that leads to the vacuum d'Alembert equation for an electromagnetic field component, such as a scalar magnetic flux density component, denoted B, of the electromagnetic field. The identification of the scalar field, usually denoted ϕ [46], of gauge theory with a scalar electromagnetic field component was first made in the derivation [62,63} of the 't Hooft–Polyakov monopole. In principle, ϕ can be identified with a scalar component of the vacuum magnetic flux density (B), or electric field strength (E), or the Whittaker scalar magnetic fluxes G and F [64,65] from which all potentials and fields can be derived in the vacuum. The treatment is classical, and the field is regarded as a function of the spacetime coordinate x^μ, and not as an eigenfunction of quantum mechanics. The general mathematical method used is a functional variation on a given Lagrangian, and so it is helpful to illustrate this method in detail as an aid to understanding. The basic concept is that there exists, in the vacuum, an electromagnetic field whose scalar components are B and E, or G and F, scalar components that obey the d'Alembert, or relativistic wave, equation in the vacuum. The Lagrangian leading to this equation by functional variation is set up, and this Lagrangian is subjected to a local gauge transformation, or gauge transformation of the second kind [46]. Local gauge invariance leads directly to the inference, from the first principles of gauge field theory, of a vacuum charge current density first introduced phenomenologically by Lehnert [45]. Inclusion of spontaneous symmetry breaking with the Higgs mechanism leads to several more vacuum charge current densities on the U(1) and O(3) levels, and in general for any gauge group symmetry. Each of these charge current densities in vacuo provides energy inherent in the vacuum.

The method of functional variation in Minkowski spacetime is illustrated first through the Lagrangian (in the usual reduced units [46])

$$\mathcal{L} = -\frac{1}{4} F^{\mu\nu} F_{\mu\nu} \tag{99}$$

where $F^{\mu\nu}$ is the field tensor on the U(1) level [46–61]. The relevant Euler–Lagrange equation is

$$\partial_\nu \left(\frac{\partial \mathscr{L}}{\partial(\partial_\nu A_\mu)} \right) = \frac{\partial \mathscr{L}}{\partial A_\mu} \tag{100}$$

Consider the component

$$\partial_0 \left(\frac{\partial \mathscr{L}}{\partial(\partial_0 A_\mu)} \right) = 0 \tag{101}$$

For indices $\nu = 0$ and $\mu = 1$, summation over repeated indices gives

$$F^{\mu\nu} F_{\mu\nu} = F^{10} F_{10} + F^{01} F_{01} \tag{102}$$

Therefore

$$
\begin{aligned}
F^{10} F_{10} &= (\partial^1 A^0 - \partial^0 A^1)(\partial_1 A_0 - \partial_0 A_1) \\
&= (\partial^1 A^0)(\partial_1 A_0) - (\partial^0 A^1)(\partial_1 A_0) - (\partial^1 A^0)(\partial_0 A_1) + (\partial^0 A^1)(\partial_0 A_1) \\
&= -\partial_X A_0 \partial_X A_0 + \partial_0 A_X \partial_X A_0 + \partial_X A_0 \partial_0 A_X - \partial_0 A_X \partial_0 A_X
\end{aligned}
\tag{103}
$$

using contravariant–covariant notation. In the same notation, we have

$$\frac{\partial}{\partial(\partial_0 A_1)} = -\frac{\partial}{\partial(\partial_0 A_X)} \tag{104}$$

so

$$\frac{\partial(F^{10} F_{10})}{\partial(\partial_0 A_1)} = -\partial_X A_0 - \partial_X A_0 + \partial_0 A_X + \partial_0 A_X \tag{105}$$

Using the additional minus sign in the Lagrangian (99), we obtain

$$\frac{\partial(-F^{10} F_{10}/2)}{\partial(\partial_0 A_1)} = F^{10} \tag{106}$$

and repeating with the term

$$
\begin{aligned}
F^{01} F_{01} &= (\partial^0 A^1 - \partial^1 A^0)(\partial_0 A_1 - \partial_1 A_0) \\
&= -\partial_0 A_X \partial_0 A_X + \partial_X A_0 \partial_0 A_X + \partial_0 A_X \partial_X A_0 - \partial_X A_0 \partial_X A_0
\end{aligned}
\tag{107}
$$

gives the same as Eq. (103). So the final result of the functional variation is

$$\partial_\nu F^{\mu\nu} = 0 \tag{108}$$

which is the vacuum inhomogeneous field equation in the Maxwell–Heaviside theory. This equation is widely accepted, but it violates causality, because there is a field (effect) without a source (cause). This flaw is usually overlooked by stating that the field is in a source-free region, or that the field is infinitely distant from its source. Both explanations are unsatisfactory.

Another example of functional variation is the Lagrangian

$$\mathscr{L} = -\frac{1}{4} F_{\mu\nu} F^{\mu\nu} + \frac{1}{2} m^2 A_\mu A^\mu \tag{109}$$

which leads to the Proca equation in the received view [46]. The obvious problem with this Lagrangian is that for identically nonzero m, the product $A_\mu A^\mu$ is not gauge-invariant on the U(1) level. Setting that problem aside for the sake of argument, contravariant–covariant notation gives

$$A_\mu A^\mu = A_0^2 - A_X^2 - A_Y^2 - A_Z^2 \tag{110}$$

so that functional variation proceeds as follows:

$$\frac{\partial \mathscr{L}}{\partial A_0} = \frac{2m^2 A_0}{2}; \quad -\frac{\partial \mathscr{L}}{\partial A_X} = -\frac{2m^2 A_X}{2}; \quad -\frac{\partial \mathscr{L}}{\partial A_Y} = -\frac{2m^2 A_Y}{2}; \quad -\frac{\partial \mathscr{L}}{\partial A_Z} = -\frac{2m^2 A_Z}{2} \tag{111}$$

The overall result is

$$\frac{\partial \mathscr{L}}{\partial A_\mu} = m^2 A^\mu \tag{112}$$

giving the received Proca equation [46]:

$$\partial_\mu F^{\mu\nu} + m^2 A^\nu = 0 \tag{113}$$

The Lagrangian (109) is not gauge-invariant, so Eq. (113) is not gauge-invariant. However, the foregoing illustrates the method of functional variation that will be used throughout this section.

In order to derive field equations in the vacuum that are self-consistent, cause must precede effect and the classical current of the Proca current must be gauge-invariant. The starting point for the development is the concept of scalar field

[46], which is usually denoted ϕ. The basic idea [46] behind the existence of the scalar field ϕ is a transition from a point particle at coordinate $x(t)$ to a field

$$\phi(x^\mu) = \phi(X, Y, Z, t) \tag{114}$$

which is a function of X, Y, Z and t in Minkowski spacetime. The scalar field ϕ is a classical concept and is governed by the Euler–Lagrange equation:

$$\frac{\partial \mathcal{L}}{\partial \phi} = \partial_v \left(\frac{\partial \mathcal{L}}{\partial(\partial_v \phi)} \right) \tag{115}$$

The source of electric charge in this view is a symmetry of the action in Noether's theorem, a symmetry that means that ϕ must be complex, that is, that there must be two fields:

$$\phi = \frac{1}{\sqrt{2}} (\phi_1 + i\phi_2) \tag{116}$$

$$\phi* = \frac{1}{\sqrt{2}} (\phi_1 - i\phi_2) \tag{117}$$

These fields are regarded as independent functions in the method of functional variation. In developing their concept of a magnetic monopole, 't Hooft and Polyakov identified ϕ with a scalar component of the electromagnetic field, a component that they denoted F [46]. It is convenient for our purposes to identify ϕ with a scalar component B of the electromagnetic field in the vacuum. Therefore, there are two independent magnetic flux density components:

$$B = \frac{1}{\sqrt{2}} (B_1 + iB_2) \tag{118}$$

$$B* = \frac{1}{\sqrt{2}} (B_1 - iB_2) \tag{119}$$

The Lagrangian governing these scalar components is

$$\mathcal{L} = (\partial_\mu B)(\partial^\mu B^*) \tag{120}$$

and is invariant under global gauge transformation, also known as "gauge transformation of the first kind"

$$B \to e^{-i\Lambda} B; \qquad B^* \to e^{i\Lambda} B^* \tag{121}$$

where Λ is any real number. The Euler–Lagrange equation

$$\frac{\partial \mathcal{L}}{\partial B} = \partial_v \left(\frac{\partial \mathcal{L}}{\partial(\partial_v B)} \right) \tag{122}$$

with the Lagrangian (120) gives the d'Alembert equations:

$$\Box B = 0 \tag{123}$$

$$\Box B^* = 0 \tag{124}$$

which are the relativistic wave equations in the vacuum satisfied by B and B^*. For example, if B and B^* are components of a plane wave, they satisfy the d'Alembert equations (123) and (124).

However, in special relativity, the number Λ is a function of the spacetime coordinate x^μ. This property defines the local gauge transformation

$$B \rightarrow e^{-i\Lambda(x^\mu)}B; \qquad B^* \rightarrow e^{i\Lambda(x^\mu)}B^* \tag{125}$$

$$\mathscr{L} = (\partial_\mu B)(\partial^\mu B^*) - ig(B^* \partial^\mu B - B\partial^\mu B^*)A_\mu + g^2 A_\mu A^\mu B^* B - \frac{1}{4}F^{\mu\nu}F_{\mu\nu}$$

$$= (\partial_\mu B + igA_\mu B)(\partial^\mu B^* - igA^\mu B^*) - \frac{1}{4}F^{\mu\nu}F_{\mu\nu} \tag{126}$$

or gauge transformation of the second kind. The Lagrangian (120) is invariant under the local gauge transformation (125) if it becomes [46]: The 4-potential becomes

$$A_\mu \rightarrow A_\mu + \frac{1}{g}\partial_\mu\Lambda \tag{127}$$

where Λ is any number and the derivative ∂_μ becomes the covariant derivatives:

$$D_\mu B = (\partial_\mu + igA_\mu)B \tag{128}$$

$$D_\mu B^* = (\partial_\mu - igA_\mu)B^* \tag{129}$$

acting respectively on B and B^*. The Lagrangian (126) is gauge invariant under a U(1) gauge transformation that introduces the electromagnetic field tensor $F^{\mu\nu}$. Using the Euler–Lagrange equation (100) gives the vacuum field equation:

$$\partial_\nu F^{\mu\nu} = -ig(B^*\partial^\mu B - B\partial^\mu B^*) + 2g^2 A^\mu |B|^2$$

$$= ig(B^* D^\mu B - BD^\mu B^*) \tag{130}$$

$$\equiv -gJ^\mu(\text{vac})$$

where

$$J^\mu(\text{vac}) = i(B^* D^\mu B - BD^\mu B^*) \tag{131}$$

Therefore $J^\mu(\text{vac})$ is a covariant conserved charge current density in the vacuum. The coefficient g of the covariant derivative has the units [47–61] of $\kappa/A^{(0)}$ in the vacuum. Using

$$g = \frac{\kappa}{A^{(0)}} \tag{132}$$

has been shown recently [47–61] to explain the Sagnac effect and interferometry in general using an O(3) invariant electrodynamics. The coefficient g is the same on the U(1) and O(3) levels.

In SI units, Eq. (130) is

$$\partial_\nu F^{\mu\nu} = -igc(B^* D^\mu B - B D^\mu B^*)Ar \tag{133}$$

and shows that the electromagnetic field in the vacuum has its source in the conserved $J^\mu(\text{vac})$, which is divergentless.

In Eq. (133), Ar is the area of the electromagnetic beam, c the vacuum speed of light and μ_0 is the vacuum permeability in SI units.

The analysis can be repeated by identifying the scalar field ϕ with a scalar component A of the vacuum four potential A^μ. Thus Eqs. (118) and (119) become

$$A = \frac{1}{\sqrt{2}}(A_1 + iA_2) \tag{134}$$

$$A^* = \frac{1}{\sqrt{2}}(A_1 - iA_2) \tag{135}$$

and the Lagrangian (120) becomes

$$\mathcal{L} = (\partial_\mu A)(\partial^\mu A^*) \tag{136}$$

Local gauge transformation is defined as

$$\begin{aligned} A &\rightarrow \exp(-i\Lambda(x^\mu))A \\ A^* &\rightarrow \exp(i\Lambda(x^\mu))A^* \end{aligned} \tag{137}$$

and the gauge-invariant Lagrangian (126) becomes

$$\mathcal{L} = (\partial_\mu A + igA_\mu A)(\partial^\mu A - igA^\mu A^*) - \frac{1}{4}F^{\mu\nu}F_{\mu\nu} \tag{138}$$

Finally, the inhomogeneous field equation in the vacuum becomes

$$\partial_\nu F^{\mu\nu} = -igc(A^* D^\mu A - A D^\mu A^*) \tag{139}$$

in SI units. This form has the advantage of eliminating any geometric variable such as Ar from the vacuum charge current density. The covariant derivatives (128) and (129) become

$$D_\mu A = (\partial_\mu + igA_\mu)A \tag{140}$$

$$D_\mu A^* = (\partial_\mu - igA_\mu)A^* \tag{141}$$

indicating the presence of self-interaction in the terms $A_\mu A$ and $A_\mu A^*$. This self-interaction is observed empirically [47–61] in a number of ways, including the inverse Faraday effect and the third Stokes parameter defining the circular polarization of electromagnetic radiation.

So it is also possible to use the form (139) for the vacuum charge current density, a form that eliminates any geometric unit such as Ar that is not fully relativistic. However, A is, strictly speaking, a potential energy difference and not a field.

Using the Euler–Lagrange equation (122) with the Lagrangian (126) produces the two complex conjugate equations (reduced units):

$$\Box B = -ig(B\partial^\mu A_\mu + A_\mu\partial^\mu B) + g^2 A_\mu A^\mu B \tag{142a}$$

$$\Box B^* = ig(B^*\partial^\mu A_\mu + A_\mu\partial^\mu B^*) + g^2 A_\mu A^\mu B^* \tag{142b}$$

or their representation in terms of the scalar A:

$$\Box A = -ig(A\partial^\mu A_\mu + A_\mu\partial^\mu A) + g^2 A_\mu A^\mu A \tag{143a}$$

$$\Box A^* = ig(A^*\partial^\mu A_\mu + A_\mu\partial^\mu A^*) + g^2 A_\mu A^\mu A^* \tag{143b}$$

Equations (133) and (142) or (139) and (143) can be solved simultaneously, because they are each two equations in two unknowns (B and A^μ) or (A and A^μ).

It can be shown on this U(1) level that the introduction of a Higgs mechanism [46], namely, spontaneous symmetry breaking, produces three more vacuum charge current densities in addition to the Lehnert-type charge current density (133) or (139). One of these is a Proca vacuum charge current density that is gauge-invariant on the classical level. The Higgs mechanism is introduced by considering the usual Lagrangian [46]

$$\mathscr{L} = T - V = (\partial_\mu\phi)(\partial^\mu\phi^*) - m^2\phi^*\phi - \lambda(\phi^*\phi)^2 \tag{144}$$

and adapting it for the electromagnetic field in the vacuum by writing it as

$$\mathscr{L} = T - V = (\partial_\mu B)(\partial^\mu B^*) - m^2 B^* B - \lambda(B^* B)^2 \tag{145}$$

or

$$\mathscr{L} = T - V = (\partial_\mu A)(\partial^\mu A^*) - m^2 A^* A - \lambda(A^* A)^2 \tag{146}$$

depending on whether ϕ is chosen to be B or A. The appearance of three new currents occurs for both choices and of course B is related to A through the vector equation:

$$B = \nabla \times A \tag{147}$$

In Eq. (144), it is well known that the mass m is regarded as a parameter that can become negative and that λ premultiplies the self-interaction term. The adaptation of the Higgs mechanism for the vacuum electromagnetic field therefore automatically implies that scalar components of that field self-interact. The self-interaction of electromagnetic fields on the received U(1) level is observable in the Stokes parameters, energy and Poynting vector for example, and in nonlinear optical phenomena of various kinds [47–61].

Considering Eq. (145), we obtain

$$\frac{\partial V}{\partial B} = m^2 B^* + 2\lambda B^* (B^* B) \tag{148}$$

and if $m^2 < 0$, there is a local maximum at $B = 0$ and a minimum at

$$a^2 \equiv |B|^2 = -\frac{m^2}{2\lambda}; \qquad \text{i.e.,} \quad a = |B| \tag{149}$$

The scalar fields B and B^* therefore become

$$B(x^\mu) = a + \frac{1}{\sqrt{2}}(B_1 + iB_2) \tag{150}$$

$$B^*(x^\mu) = a + \frac{1}{\sqrt{2}}(B_1 - iB_2) \tag{151}$$

so the Lagrangian becomes

$$\mathscr{L} = \partial_\mu(a + B)\partial^\mu(a + B^*) - m^2(a + B^*)(a + B) - \lambda((a + B^*)(a + B))^2 \tag{152}$$

It is interesting to develop this expression as

$$\begin{aligned}
\mathscr{L} &= BB^*(m^2 - \lambda BB^*) + \cdots \\
&= -\lambda BB^*(2a^2 + BB^*) + \cdots
\end{aligned} \tag{153}$$

which can be expressed algebraically as

$$\mathscr{L} = -\lambda \left(a^2 + \frac{2a}{\sqrt{2}}B_1 + \frac{1}{2}(B_1^2 + B_2^2)\right)\left(3a^2 + \frac{2a}{\sqrt{2}}B_1 + \frac{1}{2}(B_1^2 + B_2^2)\right) + \cdots$$

$$= \partial_\mu B \partial^\mu B^* - 2\lambda a^2 B_1^2 - \sqrt{2}\lambda B_1(B_1^2 + B_2^2) - \frac{\lambda}{4}(B_1^2 + B_2^2)^2 - 3\lambda a^4 \qquad (154)$$

In contemporary thought, the Higgs mechanism has acted in such a way as to produce a field component B_1 with mass, specifically, a scalar field with mass that is gauge-invariant. Therefore, spontaneous symmetry breaking of the vacuum introduces fields with effective mass.

Considering a local gauge transformation of the Lagrangian (145) produces the gauge-invariant Lagrangian:

$$\mathscr{L} = (\partial_\mu + igA_\mu)(a + B)(\partial^\mu - igA^\mu)(a + B^*) - m^2(a + B)(a + B^*)$$

$$- \lambda(a + B)^2(a + B^*)^2 - \frac{1}{4}F^{\mu\nu}F_{\mu\nu} \qquad (155)$$

Using this Lagrangian in Eq. (100) produces the following result (reduced units) by functional variation:

$$\partial_\nu F^{\mu\nu} = -ig(B^* D^\mu B - B D^\mu B^*) - \frac{g^2 m^2}{\lambda}A^\mu + 2\sqrt{2}g^2 a B_1 A^\mu + \sqrt{2}ag\partial^\mu B_2$$

$$(156)$$

The term $-g^2 m^2 A^\mu/\lambda$ implies that the electromagnetic 4-potential A^μ has acquired mass. Simultaneously there appear two other terms. All four vacuum charge current densities produce vacuum energy through the equation

$$En(\text{vac}) = \int J^\mu(\text{vac})A_\mu \, dV \qquad (157)$$

Alternatively, Eq. (156) can be written from Eq. (146) in terms of the scalar A:

$$\partial_\nu F^{\mu\nu} = -ig(A^* D^\mu A - A D^\mu A^*) - g^2 m^2 \frac{A^\mu}{\lambda} + 2\sqrt{2}g^2 a A_1 A^\mu + \sqrt{2}ag\partial^\mu A_2$$

$$(158)$$

Therefore, spontaneous symmetry breaking of the vacuum on the U(1) level produces new vacuum charge current densities that act as sources for the electromagnetic field and produce energy inherent in the topology of the vacuum. The topology is described by gauge theory and group theory.

In an O(3) electromagnetic sector [47–61], the Lagrangian (120) becomes

$$\mathscr{L} = \frac{1}{2}\partial_\mu B_i \partial^\mu B^i \tag{159}$$

where there are internal indices i to indicate the existence of an internal gauge group of O(3) symmetry. In the complex basis ((1),(2),(3)), the Lagrangian can be expressed in terms of the physical magnetic field:

$$\boldsymbol{B} = B^{(2)}\boldsymbol{e}^{(1)} + B^{(1)}\boldsymbol{e}^{(2)} + B^{(3)}\boldsymbol{e}^{(3)} \tag{160}$$

In vector notation, the Lagrangian (159) can be written as

$$\mathscr{L} = \frac{1}{2}\partial_\mu \boldsymbol{B} \cdot \partial^\mu \boldsymbol{B} \tag{161}$$

and using the Euler–Lagrange equation

$$\frac{\partial \mathscr{L}}{\partial \boldsymbol{B}} = \partial_\nu \left(\frac{\partial \mathscr{L}}{\partial_\nu \boldsymbol{B}} \right) \tag{162}$$

produces the vacuum d'Alembert equation

$$\Box \boldsymbol{B} = \boldsymbol{0} \tag{163}$$

which in component form becomes

$$\Box B^{(i)} = 0; \qquad i = 1, 2, 3 \tag{164}$$

The Lagrangian (161) is invariant under a global O(3) transformation

$$\boldsymbol{B}' = e^{iJ_i\Lambda_i}\boldsymbol{B} \tag{165}$$

where J_i are rotation generators of the O(3) group, and where Λ_i are angles in the physical internal space ((1),(2),(3)).

The local O(3) transformation corresponding to Eq. (165) is

$$\boldsymbol{B}' = e^{iJ_i\Lambda_i(x^\mu)}\boldsymbol{B} \tag{166}$$

and the Lagrangian (161) is invariant under this if it becomes

$$\mathscr{L} = D_\mu \boldsymbol{B} \cdot D^\mu \boldsymbol{B} - \frac{1}{4}\boldsymbol{G}_{\mu\nu} \cdot \boldsymbol{G}^{\mu\nu} \tag{167}$$

where the field B and the electromagnetic field $G_{\mu\nu}$ are vectors of the internal gauge space and where $G_{\mu\nu}$ is a tensor of Minkowski spacetime. Field equations are obtained from the Lagrangian (167) by functional variation using Euler–Lagrange equations such as

$$\partial^{\nu}\left(\frac{\partial\mathscr{L}}{\partial(\partial^{\nu}A^{\mu})}\right) = \frac{\partial\mathscr{L}}{\partial A^{\mu}} \tag{168}$$

where A^{μ} is a vector in the internal gauge space and a 4-vector in Minkowski spacetime. The field tensor in O(3) is defined [46–61] as

$$G^{\mu\nu} = \partial^{\mu}A^{\nu} - \partial^{\nu}A^{\mu} + gA^{\mu} \times A^{\nu} \tag{169}$$

In analogy with the Lagrangian (99), the factor $-\frac{1}{4}$ is needed because of double summation over repeated indices. So functional variation of the term $-\frac{1}{4}G_{\mu\nu}\cdot G^{\mu\nu}$ gives $\partial^{\nu}G_{\mu\nu}$. However, on the O(3) level, we must consider the additional terms

$$\mathscr{L}_{1} = -\frac{1}{4}g(G^{\mu\nu}\cdot A_{\mu} \times A_{\nu} + A^{\mu} \times A^{\nu}\cdot G_{\mu\nu})$$
$$= -\frac{1}{4}g(A_{\mu}\cdot(G^{\mu\nu} \times A_{\nu}) + A^{\mu}\cdot(G_{\mu\nu} \times A^{\nu})) \tag{170}$$

which have the same premultiplier $-\frac{1}{4}$ due to double summation over repeated indices. From the terms (170)

$$\frac{\partial\mathscr{L}}{\partial A^{\mu}} = gG_{\mu\nu} \times A^{\nu} = -gA^{\nu} \times G_{\mu\nu} \tag{171}$$

So the sum of terms (which appear on the left-hand side of the field equation) from variation in the term $-\frac{1}{4}G_{\mu\nu}\cdot G^{\mu\nu}$ in the Lagrangian (167) is

$$D^{\nu}G_{\mu\nu} \equiv \partial^{\nu}G_{\mu\nu} + gA^{\nu} \times G_{\mu\nu} \tag{172}$$

which is a covariant derivative in electrodynamics invariant under a local O(3) transformation. We must also consider functional variation of the term

$$\mathscr{L}_{3} = D_{\mu}B\cdot D^{\mu}B = (\partial_{\mu} + gA_{\mu}\times)B\cdot(\partial^{\mu} + gA^{\mu}\times)B \tag{173}$$

which can be expressed as

$$\mathscr{L}_{3} = \partial_{\mu}B\cdot\partial^{\mu}B + gA_{\mu}\cdot(B \times \partial^{\mu}B) + gA^{\mu}\cdot(B \times \partial_{\mu}B)$$
$$+ g^{2}((A_{\mu}\cdot A^{\mu})(B\cdot B) - (A_{\mu}\cdot B)(B\cdot A^{\mu})) \tag{174}$$

We obtain

$$\frac{\partial \mathcal{L}_3}{\partial A^\mu} = g(\boldsymbol{B} \times \partial_\mu \boldsymbol{B}) + g^2(A_\mu(\boldsymbol{B} \cdot \boldsymbol{B}) - (A_\mu \cdot \boldsymbol{B})\boldsymbol{B})$$
$$= g(\boldsymbol{B} \times \partial_\mu \boldsymbol{B}) + g^2 \boldsymbol{B} \times (A_\mu \times \boldsymbol{B}) \tag{175}$$

So the complete field equation obtained from the Lagrangian (167) by functional variation is

$$D^\nu G_{\mu\nu} = -g(D_\mu \boldsymbol{B}) \times \boldsymbol{B} \equiv -g\boldsymbol{J}_\mu(\text{vac}) \tag{176}$$

This equation in vector notation for the internal gauge space can be developed as three equations in reduced units

$$\partial_\mu G^{\mu\nu(1)} = ig(A_\mu^{(2)} G^{\mu\nu(3)} - A_\mu^{(3)} G^{\mu\nu(2)} - B^{(2)} D^\nu B^{(3)} + B^{(3)} D^\nu B^{(2)}) \tag{177}$$

$$\partial_\mu G^{\mu\nu(2)} = ig(A_\mu^{(3)} G^{\mu\nu(1)} - A_\mu^{(1)} G^{\mu\nu(3)} - B^{(3)} D^\nu B^{(1)} + B^{(1)} D^\nu B^{(3)}) \tag{178}$$

$$\partial_\mu G^{\mu\nu(3)} = ig(A_\mu^{(1)} G^{\mu\nu(2)} - A_\mu^{(2)} G^{\mu\nu(1)} - B^{(1)} D^\nu B^{(2)} + B^{(2)} D^\nu B^{(1)}) \tag{179}$$

where a covariant derivative acting on a component such as $B^{(1)}$ is

$$D^\nu B^{(1)} = \partial^\nu B^{(1)} - ig(A^{\nu(2)} B^{(3)} - A^{\nu(3)} B^{(2)}) \tag{180}$$

Therefore there are several more vacuum current terms on the O(3) than on the U(1) level. The factor g is, however, the same on both levels. In SI units, the Eqs. (177)–(179) become

$$\partial_\mu G^{\mu\nu(1)} = ig(A_\mu^{(2)} G^{\mu\nu(3)} - A_\mu^{(3)} G^{\mu\nu(2)})$$
$$- igc(B^{(2)} D^\nu B^{(3)} - B^{(3)} D^\nu B^{(2)})Ar \tag{181}$$

$$\cdots$$

If the field ϕ is identified with the space components of $\boldsymbol{A}$ in the basis ((1),(2),(3)), the following three vacuum equations are obtained

$$\partial_\mu G^{\mu\nu(1)} = ig(A_\mu^{(2)} G^{\mu\nu(3)} - A_\mu^{(3)} G^{\mu\nu(2)})$$
$$- igc(A^{(2)} D^\nu A^{(3)} - A^{(3)} D^\nu A^{(2)}) \tag{182}$$

$$\cdots$$

in which the vacuum currents have no geometric factor.

The structure of these vacuum charge current densities can be developed as follows in terms of time-like, longitudinal and transverse components. In this development, we take the real parts of A and A_μ. The complete inhomogeneous field equation in the vacuum is

$$\partial^\nu G_{\mu\nu} + g A^\nu \times G_{\mu\nu} = -g(D_\mu A) \times A \tag{183}$$

where the right-hand side can be expanded as

$$J_\mu(\text{vac}) \equiv g \partial_\mu A \times A + g^2 A \times (A \times A_\mu) \tag{184}$$

The longitudinal current density in vacuo is investigated, first in the plane-wave first approximation, by taking the real part of the potential

$$A = \frac{A^{(0)}}{\sqrt{2}} (ii + j) e^{i(\omega t - \kappa Z)} \tag{185}$$

which is

$$\text{Re} \, A = \frac{A^{(0)}}{\sqrt{2}} (-i \sin\phi + J \cos\phi) \tag{186}$$

where

$$\phi \equiv \omega t - \kappa Z \tag{187}$$

The longitudinal current density is (in SI units)

$$J_3 = \frac{g}{\mu_0 c} \partial_3 A \times A + \frac{g^2}{\mu_0 c} A \times (A \times A_3) \tag{189}$$

and the vector magnitude is

$$A^{(0)} = |A| = (A_1^2 + A_2^2)^{1/2} \tag{189}$$

In general, the vacuum current density has a definite structure in the vacuum that is much richer than in the first plane-wave approximation: a structure that has to be computed because analytical solutions to Eq. (183) are not available.

In the plane-wave first approximation, the current density is therefore

$$J^{(3)}(\text{vac}) = \frac{2\kappa}{\mu_0 c} B^{(3)} \tag{190}$$

in SI units and is directly proportional to the vacuum $B^{(3)}$ field. The structure of Eq. (190) was first derived by considering the inverse Faraday effect as Eq. (243) of Ref. 42. Equation (190) (above) was first derived phenomenologically on the

O(3) level in Ref. 51 and first developed phenomenologically in Ref. 59. Equation (190) is its rigorous first-principles description in the vacuum. The first principles of gauge field theory therefore produce vacuum charge current densities in the vacuum for all gauge group symmetries. There are several experimental reasons [42,47–61] for preferring O(3) over U(1) for electrodynamics.

The vacuum charge density is also structured in general, but in the plane wave, first approximation is given by

$$J_0 = \frac{\kappa^2 A_0}{\mu_0 c} \tag{191}$$

because by definition, the time component of the vector A is zero. This is how it differs from the 4-vector A^μ, and why it is an independent variable in the method of functional variation used to derive Eq. (183) from an O(3) invariant Lagrangian.

The vacuum transverse current densities are also structured, and in general they are

$$J_1 = \frac{g}{\mu_0 c} \partial_1 A \times A + \frac{g^2}{\mu_0 c} A \times (A \times A_1) \tag{192}$$

$$J_2 = \frac{g}{\mu_0 c} \partial_2 A \times A + \frac{g^2}{\mu_0 c} A \times (A \times A_2) \tag{193}$$

In the plane-wave first approximation, they reduce to

$$J_1 = -g^2 A_1 A_2^2 i \tag{194}$$
$$J_2 = g^2 A_1^2 A_2 j \tag{195}$$

using the vector triple products:

$$A \times (A \times A_1) = -A_1 A_2^2 i \tag{196}$$

$$A \times (A \times A_2) = -A_1^2 A_2 j \tag{197}$$

In SI units, the transverse vacuum current densities are given in the plane-wave first approximation by

$$J_1 = -g^2 \frac{A_1 A_2^2}{\mu_0 c} i \tag{198}$$

$$J_2 = g^2 \frac{A_1^2 A_2}{\mu_0 c} j \tag{199}$$

It is emphasized, however, that there is no reason to assume plane waves. These are used as an illustration only, and in general the vacuum charge current densities of O(3) electrodynamics are richly structured, far more so than in U(1) electrodynamics, where vacuum charge current densities also exist from the first principles of gauge theory as discussed already.

The complete vacuum inhomogeneous equation is

$$\partial^{\nu}\mathbf{G}_{\mu\nu} = -g\mathbf{A}^{\nu} \times \mathbf{G}_{\mu\nu} - g(D_{\mu}\mathbf{A}) \times \mathbf{A} \qquad (200)$$

If $\mu = 2$ and $\nu = 1$, the left-hand side vanishes because G_{21} contains only B_3, which is phaseless. The right-hand side gives the equation

$$B_3 = gA_1A_2 \qquad (201)$$

which reduces in the notation that we have been using to

$$\mathbf{B}^{(3)} = -ig\mathbf{A}^{(1)} \times \mathbf{A}^{(2)} \qquad (202)$$

In the usual complex circular basis used for O(3) electrodynamics [42], this is the definition of the field $\mathbf{B}^{(3)}$.

Therefore, a check for self-consistency has been carried out for indices $\mu = 2$ and $\nu = 1$. It has been shown, therefore, that in pure gauge theory applied to electrodynamics without a Higgs mechanism, a richly structured vacuum charge current density emerges that serves as the source of energy latent in the vacuum through the following equation:

$$En = \int \mathbf{J}^{\mu} \cdot \mathbf{A}_{\mu} \, dV \qquad (203)$$

Therefore, on the O(3) level, there are several sources of energy latent in the vacuum. This conclusion is gauge-invariant because the Lagrangian is O(3) invariant. It is concluded that potentials can give rise to physical effects in the vacuum on both the U(1) and O(3) levels. These effects are reviewed experimentally by Barrett [50]. The best known is the Aharonov–Bohm effect, which Barrett has shown [50] to be supported self-consistently only by O(3) electrodynamics and not by U(1) electrodynamics. Both the O(3) and the U(1) group are non-singly connected, the O(3) group being doubly connected in topology [50]. The latter dictates the structure of the field equations in gauge theory applied to classical electrodynamics.

The wave equation in the vacuum for O(3) electrodynamics can be obtained by functional variation in the Euler–Lagrange equation

$$\frac{\partial \mathcal{L}}{\partial \mathbf{A}} = \partial_{\mu}\left(\frac{\partial \mathcal{L}}{\partial(\partial_{\mu}\mathbf{A})}\right) = \partial^{\mu}\left(\frac{\partial \mathcal{L}}{\partial(\partial^{\mu}\mathbf{A})}\right) \qquad (204)$$

with the gauge-invariant Lagrangian

$$\mathscr{L} = D_\mu A \cdot D^\mu A - \frac{1}{4} G_{\mu\nu} \cdot G^{\mu\nu} \tag{205}$$

obtained by a local gauge transformation on the Lagrangian:

$$\mathscr{L} = \frac{1}{2} \partial_\mu A \cdot \partial^\mu A \tag{206}$$

The only assumption therefore is that the Maxwell vector potential A exists in the physical internal space of O(3) symmetry. The gauge-invariant Lagrangian (205) can be developed as

$$\mathscr{L} = \partial_\mu A \cdot \partial^\mu A + g(A_\mu \times A \cdot \partial^\mu A + \partial_\mu A \cdot A^\mu \times A) + g^2 (A_\mu \times A) \cdot (A^\mu \times A) \tag{207}$$

$$\mathscr{L} = \partial_\mu A \cdot \partial^\mu A + g(A \cdot (\partial^\mu A \times A_\mu) + A \cdot (\partial_\mu A \cdot A^\mu)) + g^2 (A_\mu \times A) \cdot (A^\mu \times A) \tag{208}$$

Using the vector identity

$$A_\mu \times A \cdot A^\mu \times A = (A_\mu \cdot A^\mu)(A \cdot A) - (A_\mu \cdot A)(A \cdot A^\mu) \tag{209}$$

gives the results

$$\frac{\partial \mathscr{L}}{\partial A} = g \partial^\mu A \times A_\mu + g \partial_\mu A \times A^\mu + 2g^2 A (A_\mu \cdot A^\mu)$$
$$- g^2 (A_\mu (A \cdot A^\mu) - (A \cdot A_\mu) A^\mu) \tag{210}$$

and

$$\partial_\mu \left(\frac{\partial \mathscr{L}}{\partial (\partial_\mu A)} \right) = 2\partial_\mu \partial^\mu A + 2g \partial_\mu (A^\mu \times A) \tag{211}$$

The vacuum wave equation in O(3) electrodynamics is therefore

$$\Box A = -g \partial_\mu (A^\mu \times A) + g(\partial_\mu A) \times A^\mu + g^2 (A(A_\mu \cdot A^\mu) - A_\mu (A \cdot A^\mu)) \tag{212}$$

Using

$$A^\mu \times (A \times A_\mu) = A(A^\mu \cdot A_\mu) - A_\mu (A^\mu \cdot A) \tag{213}$$

Eq. (212) simplifies to

$$\Box A + g\partial_\mu(A^\mu \times A) = g(\partial_\mu A) \times A^\mu - g^2(A \times A_\mu) \times A^\mu \quad (214)$$

which can be written as

$$\partial_\mu((\partial^\mu + gA^\mu \times)A) = g(\partial_\mu A) \times A^\mu + g^2 A^\mu \times (A \times A_\mu) \quad (215)$$

This form further simplifies to

$$\partial_\mu(D^\mu A) = g((\partial_\mu + gA_\mu \times)A) \times A^\mu \quad (216)$$

which becomes

$$\partial_\mu(D^\mu A) = g(D^\mu A) \times A_\mu \quad (217)$$

Therefore, we finally obtain the wave equation of O(3) electrodynamics in the form

$$D_\mu(D^\mu A) = \mathbf{0} \quad (218)$$

which is a d'Alembert equation for A with O(3) covariant derivatives.

The derivation of Eq. (218) from Eq. (206) follows from local gauge invariance, and it is always possible to apply a local gauge transform to the vector A, the Maxwell vector potential. The ordinary derivative of the d'Alembert wave equation is replaced by an O(3) covariant derivative. The U(1) equivalent of Eq. (218) in quantum-mechanical (operator) form is Eq. (13), and Eq. (212) is the rigorously correct form of the phenomenological Eq. (25). It can be seen that Eq. (212) is richly structured in the vacuum and must be solved numerically. The vacuum currents present in Eq. (218) can be computed from the right-hand side of the wave equation (212), and these vacuum currents follow from local gauge invariance.

On the U(1) level, the starting Lagrangian is

$$\mathscr{L} = \partial_\mu A \partial^\mu A^* \quad (219)$$

which on local gauge transformation becomes

$$\mathscr{L} = (\partial_\mu A + ig A_\mu A)(\partial^\mu A^* - ig A^\mu A^*) - \frac{1}{4} F^{\mu\nu} F_{\mu\nu} \quad (220)$$

Using the Euler–Lagrange equations

$$\frac{\partial \mathscr{L}}{\partial A} = \partial_\mu\left(\frac{\partial \mathscr{L}}{\partial(\partial_\mu A)}\right); \qquad \frac{\partial \mathscr{L}}{\partial A^*} = \partial_\mu\left(\frac{\partial \mathscr{L}}{\partial(\partial_\mu A^*)}\right) \quad (221)$$

we obtain

$$\Box A^* = ig(\partial_\mu A^\mu)A^* + igA_\mu(\partial^\mu A^*) + g^2 A_\mu A^\mu A^* \tag{222}$$

which is Eq. (143), showing a richly structured vacuum charge current density. Equation (222) can be developed as

$$\partial_\mu(\partial^\mu A^* - igA^\mu A^*) = igA_\mu(\partial^\mu A^*) + g^2 A_\mu A^\mu A^* \tag{223}$$

that is

$$\partial_\mu(D^\mu A^*) = igA_\mu(\partial^\mu A^* - igA^\mu A^*) \tag{224}$$
$$D_\mu(D^\mu A^*) = 0 \tag{225}$$

which is a vacuum d'Alembert equation with U(1) covariant derivatives. To obtain Eq. (225) from Eq. (219), the only assumption is that the Lagrangian is invariant under the local U(1) gauge transform:

$$A \rightarrow \exp(-i\Lambda(x^\mu))A \tag{226}$$

Similarly, we obtain

$$\partial^\mu D_\mu A = -igA^\mu(\partial_\mu A + igA_\mu A) \tag{227}$$

and the d'Alembert equation

$$D^\mu(D_\mu A) = 0 \tag{228}$$

with covariant derivatives.

A possible solution of Eq. (228) is:

$$D_\mu A = 0 \tag{229}$$

specifically

$$\partial^\mu = -igA_\mu \tag{230}$$

Define

$$\kappa_\mu \equiv gA_\mu = \frac{\kappa}{A^{(0)}} A_\mu \tag{231}$$

and Eq. (230) becomes the following quantum ansatz:

$$\partial_\mu = -i\kappa_\mu = -\frac{i}{\hbar}p_\mu \tag{232}$$

On the quantum level, Eq. (229) becomes an operator equation, and, using the quantum ansatz, we obtain

$$D^{\mu*}D_\mu A = 0; \qquad \text{i.e.,} \qquad \Box A = -\kappa^\mu \kappa_\mu A \tag{233}$$

which is Eq. (12) (above). In fully covariant form, Eq. (233) becomes the gauge invariant Proca equation:

$$\Box A^\mu = -\kappa^\mu \kappa_\mu A^\mu = -\kappa^2 A^\mu = -\left(\frac{m_0 c}{\hbar}\right)^2 A^\mu \tag{234}$$

Note that the Proca equation requires

$$\kappa^\mu \kappa_\mu \neq 0 \tag{235}$$

and has been obtained without the use of the Lorenz condition.

The equivalent procedure on the O(3) level is to choose a particular solution

$$D^\mu A = (\partial^\mu + g A^\mu \times) A = 0 \tag{236}$$

which, in the general notation of gauge field theory, is

$$\partial^\mu \psi = ig A^{\mu(3)} \psi \tag{237}$$

giving again the quantum ansatz on the O(3) level. In the complex circular basis

$$\begin{aligned} A^\mu &= A^{(2)} e^{(1)} + A^{(1)} e^{(2)} + A^{(3)} e^{(3)} \\ A &= A^{(1)} + A^{(2)} + A^{(3)} \end{aligned} \tag{238}$$

and Eq. (236) becomes

$$(\partial^\mu + g a^\mu \times)(A^{(1)} + A^{(2)} + A^{(3)}) = 0 \tag{239}$$

This equation can be developed as

$$\frac{\partial}{\partial Z} A^{(1)} = -g A^{(3)} \times A^{(1)} \cdots \tag{240}$$

in other words, as

$$i\kappa A^{(2)*} = A^{(3)} \times A^{(1)} \tag{241}$$

which gives self-consistently the definition

$$B^{(2)*} = -igA^{(3)} \times A^{(1)} \qquad (242)$$

Similarly, we obtain

$$\frac{\partial}{\partial Z} A^{(2)} = gA^{(2)} \times A^{(3)} \qquad (243)$$

which gives the following definition:

$$B^{(1)*} = -igA^{(2)} \times A^{(3)} \qquad (244)$$

Using the relation $g = \kappa/A^{(0)}$ in Eqs. (242) and (244) gives two equations of the B cyclic theorem [42,47–61]:

$$\begin{aligned} B^{(3)} \times B^{(1)} &= iB^{(0)}B^{(2)*} \\ B^{(2)} \times B^{(3)} &= iB^{(0)}B^{(1)*} \end{aligned} \qquad (245)$$

It follows from the quantum ansatz (237) that

$$\begin{aligned} -\frac{\partial}{\partial X}(A^{(1)} + A^{(2)} + A^{(3)}) &= gA_X^{(3)} \times (A^{(1)} + A^{(2)} + A^{(3)}) = 0 \\ -\frac{\partial}{\partial Y}(A^{(1)} + A^{(2)} + A^{(3)}) &= gA_Y^{(3)} \times (A^{(1)} + A^{(2)} + A^{(3)}) = 0 \end{aligned} \qquad (246)$$

which is self-consistent because

$$A_X^{(3)} = A_Y^{(3)} = 0 \qquad (247)$$

Finally, the time-like component of Eq. (236) is

$$\frac{1}{c}\frac{\partial}{\partial t}(A^{(1)} + A^{(2)} + A^{(3)}) = gA_0^{(3)} \times (A^{(1)} + A^{(2)} + A^{(3)}) \qquad (248)$$

which gives again Eqs. (242) and (244).

Therefore the Proca equation can be recovered on the O(3) level from the special solution (236) as the operator equation:

$$\partial_\mu \partial^\mu \psi = -g^2 A^{\mu(3)} A_\mu^{(3)} \psi \qquad (249)$$

This result is given in Eq. (22) of the preceding section.

A Lagrangian such as Eq. (219) is made up purely of a kinetic energy term:

$$\mathscr{L} = T = \partial_\mu A \partial^\mu A^* \tag{250}$$

and a local gauge transformation on the Lagrangian produces

$$\mathscr{L} = T - V = \partial_\mu A \partial^\mu A^* + ig(A_\mu A \partial^\mu A^* - A^\mu A^* \partial_\mu A) + g^2 A_\mu A A^\mu A^* - \frac{1}{4} F^{\mu\nu} F_{\mu\nu} \tag{251}$$

where V is a potential energy term. In field theory [46], the ground state is the vacuum, and the ground state is obtained by minimizing the potential energy V with respect to a variable such as A or A^μ. The minimum of V in Eq. (251) with respect to A_μ is the vacuum charge current density, which is a ground state of the field theory and that is obviously a property of the vacuum itself. The ground state defined by the minimum

$$\frac{\partial V}{\partial(\partial_\nu A_\mu)} = F^{\mu\nu} \tag{252}$$

is the electromagnetic field, which is also a vacuum property. So the inhomogeneous field equation

$$\partial_\nu F^{\mu\nu} = \frac{J^\mu(\text{vac})}{\varepsilon_0} \tag{253}$$

is a relation between ground states of the field theory, or a relation between vacuum states. Similarly, a ground state such as

$$\frac{\partial \mathscr{L}}{\partial A} = ig A_\mu D^\mu A^* \neq 0 \tag{254}$$

is a vacuum property. It can be seen that Eq. (254) is a minimum because

$$\frac{\partial^2 \mathscr{L}}{\partial A^2} = 2g^2 |A|^2 \tag{255}$$

is always greater than zero.

The source of the potential energy V in Eq. (251) is local gauge transformation, and so the source of V is the vacuum itself, as described by special relativity and gauge theory. The kinetic energy T appearing in Eq. (250) has no role in defining the ground state of the field theory, because the ground state is defined by the minimum of V with respect to a given variable, as just argued. In these equations, the physical A and A^* are excitations above the ground state or vacuum, and the vacuum gives no contribution to the global Lagrangian (250). The potential energy V is part of the locally gauge-invariant Lagrangian that gives the field equation (253), a relation between vacuum properties. The vacuum charge current density gives energy latent in the vacuum, and rate of doing work by the vacuum. These are given respectively by

$$En = \int J^\nu(\text{vac})A_\mu\, dV \qquad (256)$$

and by

$$\frac{\partial W}{\partial t} = \int J(\text{vac}) \cdot E \, dV \qquad (257)$$

The volume V is arbitrary and, from Eq. (257) standard methods [66], give the vacuum Poynting theorem

$$\frac{\partial U}{\partial t}(\text{vac}) + \nabla \cdot S(\text{vac}) = -J(\text{vac}) \cdot E \qquad (258)$$

or law of conservation of energy and momentum for various vacuum properties. The vacuum energy flow is represented by the Poynting vector $S(\text{vac})$:

$$\nabla \cdot S(\text{vac}) = -J(\text{vac}) \cdot E \qquad (259)$$

Integrating this equation gives

$$S(\text{vac}) = -\int J(\text{vac}) \cdot E \, dr + \text{ constant of integration} \qquad (260)$$

where the constant of integration represents a physical component of energy flow whose magnitude is not limited by any concept in gauge field theory. The physical object $J(\text{vac})$ also emanates from the vacuum, and its magnitude is not limited because the magnitude of A^μ is not limited by vacuum topology. The energy flow represented by $S(\text{vac})$ is electromagnetic energy flow generated by vacuum topology, and can be converted, in principle, to other forms of energy with suitable laboratory devices.

The physical meaning of the vacuum Poynting theorem [46] in Eq. (258) is that the time rate of change of electromagnetic energy within an arbitrary volume V, combined with the energy flowing out through the boundary surfaces of the volume per unit time, is equal to the negative of the total work done by the field (a vacuum property) on the source, interpreted as vacuum charge current density. This is a statement of conservation of energy applied within the vacuum and in the absence of matter (electrons). In the received view

$$J^\mu(\text{vac}) = 0 \tag{261}$$

and there is no vacuum Poynting theorem, but as argued already, the received view violates gauge invariance, special relativity, and causality. In the correctly gauge-invariant Eq. (253), work is done by the source (a vacuum property) on the field (another vacuum property), work that can be transmitted to rate of change of mechanical energy as follows [46]:

$$\frac{dEn}{dt}(\text{mech}) = \int J(\text{vac}) \cdot E \, dV \tag{262}$$

In general relativity, gravity is curvature of spacetime, and so the ordinary potential energy mgh emanates ultimately from the vacuum topology itself. Here m is mass, g is the acceleration due to gravity, and h is a difference in height. The electromagnetic field is orders of magnitude stronger than the gravitational field. Special relativity is a special case of general relativity, and sometimes A^μ is known [46] as a connection, in analogy with the affine connection of general relativity. The gravitational field is the vacuum, and the electromagnetic field is the vacuum. Mass and gravitational field, and charge and electromagnetic field, are therefore all consequences of relativity and vacuum topology.

In this view, the structures of the vacuum and matter currents are identical:

$$J^\mu(\text{vac}) = -\frac{ig}{\mu_0 c}(A^* D^\mu A - A D^\mu A^*); \qquad g = \frac{\kappa}{A^{(0)}}$$
$$J^\mu(\text{matter}) = -\frac{ig}{\mu_0 c}(A^* D^\mu A - A D^\mu A^*); \qquad g = \frac{e}{\hbar} \tag{263}$$

and one is transformed into the other for one electron and one photon by the relation

$$\frac{\kappa}{A^{(0)}} = \frac{e}{\hbar} \tag{264}$$

Therefore, the momentum of one photon is transformed to the electron momentum

$$\hbar\kappa = eA^{(0)} \tag{265}$$

and the photon momentum and energy emanate from the vacuum itself, as just argued. In this way, the elementary charge e on the proton also becomes a topological property, arguing in analogy with the way in which mass in general relativity is a property of the vacuum. Again, in analogy with general relativity, photons are formed out of the vacuum as gravitons are formed out of the vacuum. The relation (265) is true for all internal gauge group symmetries. In the foregoing, we happen to have been arguing on the U(1) level, but the concepts are the same on the O(3) level. Therefore, charge e is the result of the field, which is a vacuum property.

The above is a pure gauge field theory. The Higgs mechanism on the U(1) level provides further sources of vacuum energy as discussed already. On the O(3) level, the Higgs mechanism can also be applied, resulting in yet more sources of energy.

Gauge theory of any symmetry must have two mathematical spaces: Minkowski spacetime and the internal gauge space. If electromagnetic theory in the vacuum is a U(1) symmetry gauge field symmetry, there is a scalar internal space of U(1) symmetry in the vacuum. This internal space is the space of the scalar A and A^* used in the foregoing arguments. In geometric form

$$A = A_1 i + A_2 j \tag{266}$$

is a vector in a two-dimensional space with orthonormal basis vectors i and j. This space is the internal gauge space of the U(1) gauge field theory applied to vacuum electromagnetism. A global gauge transform is a rotation of A through an arbitrary angle Λ. Such a process is described [46] by the O(2) group of rotations in a plane, homomorphic with U(1). The invariance of action under the same global gauge transformation results in a conserved charge Q and a divergentless current:

$$\frac{dQ}{dt} = 0; \qquad Q = \int J^0 \, dV; \qquad \partial_\mu J^\mu = 0 \tag{267}$$

These concepts stem from a variational principle applied to the action

$$S = \int \mathscr{L}(A \partial_\mu A) \, d^4 x \tag{268}$$

which is stationary [46] under the condition

$$\frac{\partial \mathscr{L}}{\partial A} = \partial_\mu \left(\frac{\partial \mathscr{L}}{\partial (\partial_\mu A)} \right) = 0 \tag{269}$$

which is the Euler–Lagrange equation for A in the internal U(1) gauge space of electromagnetic theory in the vacuum. The action is considered [46] in Noether's theorem to be unchanged by re-parameterization of x^μ and A, that is, is invariant under some group of transformations on x^μ and A. It follows [46] that there exist conserved quantities that are combinations of fields and derivatives, which are invariant under these transformations: energy, momentum, angular momentum, and charge.

For example, it can be shown that the energy momentum tensor due to A is [46]

$$\theta^\mu_\nu = \partial^\mu A \partial_\nu A - \frac{1}{2} \delta^\mu_\nu \partial_\sigma A \partial^\sigma A \qquad (270)$$

For translation of the origin of space and time [46], Noether's theorem gives

$$J^\mu_\nu = -\theta^\mu_\nu = -\partial^\mu A \partial_\nu A + \frac{1}{2} \delta^\mu_\nu \partial_\sigma A \partial^\sigma A \qquad (271)$$

The conserved quantity in this case is the energy momentum

$$\frac{d}{dt} \int \theta^0_\nu d^3x = 0 \qquad (272)$$

in the internal gauge space. The energy and momentum of the field in the internal gauge space are given by

$$En = \int \theta^0_0 d^3x; \quad p = \int \theta^0_1 d^3x \qquad (273)$$

Under the local gauge transformation (226) of the Lagrangian (219), the action is no longer invariant [46], and invariance must be restored by adding terms to the Lagrangian. One such term is

$$\mathscr{L}_1 = -gJ^\mu A_\mu \qquad (274)$$

where g is a parameter such that gA_μ has the units of ∂_μ. It is important to realize that this is true under all conditions, including the vacuum, so if electromagnetic theory in the vacuum is a U(1) gauge theory, then both g and A_μ must be introduced in the vacuum. It is clear that

$$g = \frac{\kappa}{A^{(0)}} \qquad (275)$$

satisfies the requirement that gA_μ have the same units as ∂_μ. The 4-potential A_μ is introduced from Minkowski spacetime and, under local U(1) gauge transformation

$$A_\mu \rightarrow A_\mu + \frac{1}{g} \partial_\mu \Lambda \qquad (276)$$

where Λ is arbitrary. Local gauge transformation therefore results in the total Lagrangian (251) that is needed to render the action invariant.

Therefore the Lehnert equation (253) correctly conserves action under a local U(1) gauge transformation in the vacuum. Such a transformation leads to a vacuum charge current density as the result of gauge theory itself, because U(1) gauge theory has a scalar internal space that supports A and A^*. These must be complex in order to define the globally conserved charge:

$$Q = \int J^0 dV \qquad (277)$$

from the globally invariant current:

$$J^\mu = i(A^* \partial^\mu A - A \partial^\mu A^*) \qquad (278)$$

in the internal U(1) space of the gauge theory.

The existence of a vacuum charge current density in the vacuum was first introduced phenomenologically by Lehnert [45,49], and it has been shown that the Lehnert equations can describe phenomena that the Maxwell–Heaviside equations are unable to describe. The reason for this is now clear. The vacuum Maxwell–Heaviside equations do not conserve action under a local gauge transformation in the internal scalar space of a U(1) gauge field theory. In order to conserve action, a locally gauge-invariant charge current density of the type appearing in Eq. (253) is needed in the vacuum, and it has just been argued that such a conclusion has a solid basis in gauge theory. If the charge current density were absent, there would be no scalar internal space for U(1) gauge theory applied in the vacuum to electromagnetism. It follows, as argued already, that the vector potential A_μ and the electromagnetic field tensor $F^{\mu\nu}$ are the result of local gauge transformation and originate in the vacuum topology.

There is empirical evidence that electrons and positrons annihilate to give photons, and this process is represented symbolically by

$$e^- + e^+ = 2\gamma \qquad (279)$$

This process cannot be described classically, because positrons are the result of the Dirac equation, but it illustrates the fact that a vacuum current (of photons) is made up of the interaction of two Dirac currents, one for the electron, one for the positron, and these are both matter currents. Therefore, there is a transmutation of matter current to vacuum current. On the classical level, this can be described in the scalar internal gauge space as

$$\phi \rightleftharpoons A \qquad (280)$$

where ϕ is a matter field and A is the scalar component of an electromagnetic potential. As shown in Eqs. (263), the matter and vacuum fields have the same structure. The coefficient g in the vacuum field is $\kappa/A^{(0)}$ and is $e/\hbar$ in the matter field. The process

$$\hbar\kappa \rightarrow eA^{(0)} \qquad (281)$$

is therefore a transfer of photon linear momentum to an electron, as in the Compton effect. As soon as $\hbar$ is introduced, Planck quantization is also introduced. Since e is a property of neither the electromagnetic field nor the Dirac electron, the equation

$$\hbar\kappa = eA^{(0)} \qquad (282)$$

can be regarded [47–61] as a Planck quantization of the factor g in the vacuum:

$$g = \frac{\kappa}{A^{(0)}} = \frac{e}{\hbar} \qquad (283)$$

The Lehnert equations are a great improvement over the Maxwell–Heaviside equations [45,49] but are unable to describe phenomena such as the Sagnac effect and interferometry [42], for which an O(3) internal gauge space symmetry is needed.

IV. DEVELOPMENT OF GAUGE THEORY IN THE VACUUM

Gauge theory can be developed systematically for the vacuum on the basis of material presented in Section II. Before doing so, recall that, on the U(1) level, A^μ exists in Minkowski spacetime and there is a scalar internal gauge space that can be denoted

$$\boldsymbol{A} = A_1\boldsymbol{i} + A_2\boldsymbol{j} = A_X\boldsymbol{i} + A_Y\boldsymbol{i} \qquad (284)$$

The internal gauge space has local symmetry, and is a physical space. In complex circular notation, the vector in the internal gauge space can be written as

$$\boldsymbol{A} = A^{(2)}\boldsymbol{e}^{(1)} + A^{(1)}\boldsymbol{e}^{(2)} \qquad (285)$$

indicating two states of circular polarization. Therefore, we have $A^{\mu(1)}$ and $A^{\mu(2)}$ in the vacuum. Circular polarization becomes a prerequisite for the conserved Q of Eq. (277). In the notation of Eq. (285)

$$A^{(1)} = \frac{1}{\sqrt{2}}(A_X - iA_Y); \qquad A^{(2)} = \frac{1}{\sqrt{2}}(A_X + iA_Y) \qquad (286)$$

Circular polarization appears in general if

$$A_X = A^{(0)} \exp\left(-i(\omega t - \kappa Z)\right) \tag{287}$$

$$A_Y = A^{(0)} \exp\left(-i(\omega t - \kappa Z)\right) \tag{288}$$

where we have included the electromagnetic phase on the U(1) level. The scalar internal space in the vacuum is therefore described by the following two vectors:

$$A = \frac{1}{\sqrt{2}}(A_X + iA_Y); \qquad A^* = \frac{1}{\sqrt{2}}(A_X - iA_Y) \tag{289}$$

Global gauge transformation on these vectors produces a shift in the electromagnetic phase

$$A_X \rightarrow A^{(0)} \exp\left(-i(\omega t - \kappa Z + \Lambda)\right) \tag{290}$$

$$A_Y \rightarrow A^{(0)} \exp\left(-i(\omega t - \kappa Z + \Lambda)\right) \tag{291}$$

where Λ is an arbitrary number. So under global gauge transformation, the electromagnetic phase in the vacuum is defined only up to an arbitrary Λ. Under local gauge transformation

$$A_X \rightarrow A^{(0)} \exp\left(-i(\omega t - \kappa Z + \Lambda(x^\mu))\right) \tag{292}$$

$$A_Y \rightarrow A^{(0)} \exp\left(-i(\omega t - \kappa Z + \Lambda(x^\mu))\right) \tag{293}$$

and the U(1) electromagnetic phase is defined up to an arbitrary number Λ, which is a function of the spacetime coordinate x^μ. In consequence, it has been shown elsewhere [42,47–61] that U(1) gauge theory applied to electromagnetism does not describe interferometry or physical optics in general.

There is an interrelation between the A and A^μ vectors of the scalar internal gauge space and components of $A^{\mu(1)}$ and $A^{\mu(2)}$ in the vacuum

$$\boldsymbol{A}^{(1)} = iA_X \boldsymbol{e}^{(1)} \tag{294}$$

$$\boldsymbol{A}^{(2)} = -iA_Y \boldsymbol{e}^{(2)} \tag{295}$$

so that $\boldsymbol{A}^{(1)} = \boldsymbol{A}^{(2)*}$ is a vacuum plane wave. It can be seen that, on the U(1) level, local and global gauge transformation introduce arbitrariness into the electromagnetic phase factor:

$$\gamma = \exp\left(-i(\omega t - \kappa Z)\right) \tag{296}$$

Dirac attempted to remedy this flaw on the U(1) level by defining the electromagnetic phase factor by [42]

$$\gamma = \exp\left(ig \oint A_\mu(x^\mu)\,dx^\mu\right) \tag{297}$$

On the O(3) level, vacuum gauge theory is defined by a Clifford algebra

$$A_\mu = A_\mu^{(2)}e^{(1)} + A_\mu^{(1)}e^{(2)} + A_\mu^{(3)}e^{(3)} \tag{298}$$

$$A = A^{(2)}e^{(1)} + A^{(1)}e^{(2)} + A^{(3)}e^{(3)} \tag{299}$$

where A_μ is a vector in the internal gauge space of O(3) symmetry and a 4-vector in Minkowski spacetime. In the internal gauge space, the Maxwell vector potential is defined as

$$A = A_X i + A_Y j + A_Z k = A^{(2)}e^{(1)} + A^{(1)}e^{(2)} + A^{(3)}e^{(3)} \tag{300}$$

indicating by ansatz the existence of a nonzero $A^{(3)}$ in the vacuum. The latter describes the Sagnac effect with precision as demonstrated elsewhere [42] using a non-Abelian Stokes theorem. On the O(3) level, the electromagnetic phase factor is a Wu–Yang phase factor denoted

$$\gamma = P\exp\left(ig \oint A_\mu(x^\mu)\,dx^\mu\right) \tag{301}$$

where parallel transport is implied [42] with O(3) covariant derivatives. In the vacuum, the factor g is given by Eq. (275) for all gauge group symmetries. There is again a relation between the internal vector A and components in the vacuum of the four vector A^μ. For example

$$A^{(1)} = iA_X e^{(1)}; \qquad A^{(2)} = -iA_Y e^{(2)}; \qquad A^{(3)} = A_Z k \tag{302}$$

So it becomes clear that the description of the vacuum in gauge theory can be developed systematically by recognizing that, in general, A is an n-dimensional vector. On the U(1) level, it is one-dimensional; on the O(3) level, it is three-dimensional; and so on. The internal gauge space in this development is a physical space that can be subjected to a local gauge transform to produce physical vacuum charge current densities.

So in the general case where A is an n-dimensional vector [46], a local gauge transform on this vector is represented in the vacuum by

$$\begin{aligned} A(x^\mu) \rightarrow A'(x^\mu) &= \exp(iM^a\Lambda^a(x^\mu))A(x^\mu) \\ &\equiv S(x^\mu)A(x^\mu) \end{aligned} \tag{303}$$

where M^a are the generators of the group that describes the symmetry of the internal gauge space, and where the index a is summed from 1 to 3 when the internal gauge group is O(3). It follows that

$$\partial_\mu A' = S(\partial_\mu A) + (\partial_\mu S)A \tag{304}$$

so $\partial_\mu A$ does not transform covariantly. This is the basis of the gauge principle and the principle of parallel transport in the vacuum for any gauge group symmetry. Parallel transport in the vacuum produces the vector δA, where

$$\delta A = ig\,M^a A_\mu^a dx^\mu A \tag{305}$$

So the product $gM^a A_\mu^a$ is the result of special relativity in the vacuum, and g is adjusted for correct units. Ryder [46] simply describes A_μ^a as "an additional field or potential;" Feynman describes it as "the universal influence." Therefore, as argued in the foregoing section, both the potential and the electromagnetic field in the vacuum originate in local gauge transformation, which, in turn, originates in special relativity itself.

The covariant derivative in the vacuum for any internal gauge group symmetry is therefore defined by

$$D_\mu A = (\partial_\mu - igM^a A_\mu^a)A \tag{306}$$

and is valid for an n-dimensional A and for any internal gauge group whose generators are represented by matrices M^a [46]. The U(1) covariant derivative in the vacuum is given by $M = -1$, resulting in

$$D_\mu A = (\partial_\mu + igA_\mu)A \tag{307}$$

On the O(3) level, the covariant derivative in the vacuum is given by

$$D_\mu A = \partial_\mu A + gA_\mu \times A \tag{308}$$

Considering a rotation $A = SA$ in the vacuum, the covariant derivative transforms as

$$D_\mu A \rightarrow D'_\mu A' = SD_\mu A \tag{309}$$

that is

$$(\partial_\mu - igA'_\mu)A' = S(\partial_\mu - igA_\mu)A \tag{310}$$

which [42] leads to the law governing A_μ under gauge transformation in any gauge group:

$$A'_\mu = SA_\mu S^{-1} - \frac{i}{g}(\partial_\mu S)S^{-1} \tag{311}$$

It is also possible to consider the holonomy of the generic A in the vacuum. This is a round trip or closed loop in Minkowski spacetime. The general vector A is transported from point A, where it is denoted $A_{A,0}$ around a closed loop with covariant derivatives back to the point $A_{A,0}$ in the vacuum. The result [46] is the field tensor for any gauge group

$$G_{\mu\nu} = \frac{i}{g}[D_\mu, D_\nu] = \partial_\mu A_\nu - \partial_\nu A_\mu - ig[A_\mu, A_\nu] \tag{312}$$

and the field tensor is the result of rotating the vector A in the internal space of the gauge theory in the vacuum. It is seen that the field tensor is a commutator of covariant derivatives, and therefore originates in local gauge transformation. On the U(1) level, the field tensor in the vacuum is

$$F_{\mu\nu} = \partial_\mu A_\nu - \partial_\nu A_\mu \tag{313}$$

and on the O(3) level is

$$\boldsymbol{G}_{\mu\nu} = \partial_\mu \boldsymbol{A}_\nu - \partial_\nu \boldsymbol{A}_\mu + g\boldsymbol{A}_\mu \times \boldsymbol{A}_\nu \tag{314}$$

The field tensor transforms covariantly [46] because

$$\begin{aligned} A_{A,0} &\rightarrow A'_{A,0} = SA_{A,0} \\ A_{A,1} &\rightarrow A'_{A,1} = SA_{A,1} \end{aligned} \tag{315}$$

in the vacuum.

Similarly, transport of the generic A around a three-dimensional closed loop [46] produces the Jacobi identity

$$\sum_{\text{cyclic}} [D_\sigma, [D_\mu, D_\nu]] = 0 \tag{316}$$

for any gauge group symmetry in the vacuum. On the U(1) level, it is the homogeneous field equation

$$\partial_\mu \tilde{F}^{\mu\nu} \equiv 0 \tag{317}$$

and on the O(3) level, the homogeneous field equation:

$$D_\mu \tilde{G}^{\mu\nu} \equiv 0 \tag{318}$$

The complete set of vacuum field and wave equations on the U(1) level is therefore

$$\partial_\mu \tilde{F}^{\mu\nu} \equiv 0 \tag{319}$$

$$\partial_\mu F^{\mu\nu} = \frac{J^\mu(\text{vac})}{\varepsilon_0} \tag{320}$$

$$D^\mu D_\mu A = 0 \tag{321}$$

and the complete set on the O(3) level is

$$\partial_\mu \tilde{G}^{\mu\nu} \equiv 0 \tag{322}$$

$$D^\nu G_{\mu\nu} = -gc(D_\mu A) \times A \tag{323}$$

$$D_\mu(D^\mu A) = 0 \tag{324}$$

All these results are derived essentially by considering a rotation of the general vector A in the internal space of the gauge theory in the vacuum.

In order to demonstrate that spontaneous symmetry breaking can affect the energy inherent in the vacuum, consider the globally invariant Higgs Lagrangian:

$$\mathcal{L} = \partial_\mu(a + A)\partial^\mu(a + A^*) - m^2(a + A^*)(a + A) - \lambda((a + A^*)(a + A))^2 \tag{325}$$

It has been demonstrated already that local gauge transformation on this Lagrangian leads to Eq. (153), which contains new charge current density terms due to the Higgs mechanism. For our present purposes, however, it is clearer to use the locally invariant Lagrangian obtained from Eq. (325), specifically

$$\mathcal{L} = (\partial_\mu + igA_\mu)(a + A)(\partial^\mu - igA^\mu)(a + A^*)$$
$$- m^2(a + A)(a + A^*) - \lambda(a + A)^2(a + A^*)^2 - \frac{1}{4}F^{\mu\nu}F_{\mu\nu} \tag{326}$$

with the Euler–Lagrange equations:

$$\frac{\partial \mathcal{L}}{\partial A} = \partial_\mu\left(\frac{\partial \mathcal{L}}{\partial(\partial_\mu A)}\right); \qquad \frac{\partial \mathcal{L}}{\partial A^*} = \partial_\mu\left(\frac{\partial \mathcal{L}}{\partial(\partial_\mu A^*)}\right) \tag{327}$$

Such a procedure produces the equations:

$$D_\mu D^\mu A^* = -m^2 A^* - 2\lambda A^* (AA^*)$$
$$D_\mu D^\mu A = -m^2 A - 2\lambda A (A^* A) \tag{328}$$

where we have used $a = a^*$. So the effect of the Higgs mechanism is to generate the inhomogeneous wave equations (328) from the homogeneous wave equations (225) and (228) by spontaneous symmetry breaking [46] of the vacuum. The charge current densities on the right-hand side of Eq. (328) can be used to generate the equivalent matter charge current densities as discussed later in this section.

Without the Higgs mechanism, the Lagrangian (325) is

$$\mathcal{L} = \partial_\mu A \partial^\mu A^* - m^2 A^* A - \lambda A^* A A^* A \tag{329}$$

and using Eqs. (327) produces the wave equations:

$$\Box A^* = -(m^2 + 2\lambda A^* A)A^*$$
$$\Box A = -(m^2 + 2\lambda AA^*)A \tag{330}$$

At the Higgs minimum

$$a^2 = |A|^2 = -\frac{m^2}{2\lambda} \tag{331}$$

Eqs. (330) become

$$\Box A^* = 0$$
$$\Box A = 0 \tag{332}$$

At the local Higgs maximum [46] for $m^2 < 0$, that is, at $m = 0$, Eqs. (330) become

$$\Box A^* = -2\lambda(A^* A)A^*$$
$$\Box A = -2\lambda(AA^*)A \tag{333}$$

and Eqs. (328) become

$$D_\mu D^\mu A^* = -2\lambda A^*(AA^*)$$
$$D_\mu D^\mu A = -2\lambda A(A^* A) \tag{334}$$

So both the globally and locally invariant equations of motion of the internal gauge space [the Euler–Lagrange equations (327)] are different at the Higgs maximum and minimum. The minimum and local maximum are different ground states of the field, and are different vacuum states. The difference between the Higgs maximum and minimum represents potential energy difference within the vacuum itself. The Higgs mechanism is well known to lead to electroweak theory and to the existence of the Higgs boson, so it is well established that in the vacuum, there is a usable difference of potential energy, the different minima of which lead to different ground states of the field theory and to different vacua. In nineteenth-century classical electromagnetism, on which a text such as that by Jackson [66] is based, such concepts do not exist. There is no vacuum charge current density, and there are no potential energy maxima or minima in the vacuum itself.

It is well known that there is an interesting analogy between spontaneous symmetry breaking of the vacuum and the Landau–Ginzburg free energy in superconductors. The latter is obtained from the locally invariant Lagrangian (325) in the static limit [46]

$$\partial_0 A = 0 \tag{335}$$

where the mass term is defined as $m^2 = a(T - T_c)$ near the critical temperature T_c. At $T > T_c$, $m^2 > 0$ and the minimum free energy is at $|A| = 0$. When $T < T_c$, $m^2 < 0$ and the minimum free energy is at $|A|^2 = -(m^2/2\lambda) > 0$. This is an analogy with the case of spontaneous symmetry breaking in the vacuum, where there is a difference of free energy (or latent free energy) on the classical level that can be used for practical devices.

The effect of the Higgs mechanism can be seen most clearly by minimizing the Lagrangian (251) with respect to A:

$$\frac{\partial \mathscr{L}}{\partial A} = ig A_\mu D^\mu A^* = 0 \tag{336}$$

This minimum value defines the ground state and the true vacuum through the equation

$$D^\mu A^* = 0$$
$$D^\mu A = 0 \tag{337}$$

This means, however, that the vacuum charge current density disappears:

$$J^\mu(\text{vac}) = -\frac{ig}{\mu_0 c}(A^* D^\mu A - A D^\mu A^*) = 0 \tag{338}$$

It thus becomes clear that the vacuum charge current density introduced by Lehnert is an excitation above the true vacuum in classical electrodynamics. The true vacuum is defined by Eq. (337). It follows that in the true classical vacuum, the electromagnetic field also disappears.

Using the Higgs Lagrangian (326) however, the true vacuum is defined by

$$\frac{\partial \mathscr{L}}{\partial t} = igA_\mu D^\mu A^* - m^2 A^* - 2\lambda A^*(AA^*) = 0 \qquad (339)$$

and the true vacuum itself carries a charge current density. The charge current density in the true vacuum is described by Eq. (339), which is consistent with the fact that the Lehnert charge current density implies photon mass, as does the Higgs mechanism.

The transfer of the energy associated with this true vacuum charge current density to a matter current is achieved by adjusting the value of the coupling constant g such that the vacuum value $g = \kappa/A^{(0)}$ becomes $e/\hbar$ in matter. The resulting equation is

$$g = \frac{\kappa}{A^{(0)}} = \frac{e}{\hbar} \qquad (340)$$

specifically

$$\hbar\kappa = eA^{(0)} \qquad (341)$$

which classically gives the minimal prescription:

$$\boldsymbol{p} = e\boldsymbol{A} \qquad (342)$$

The momentum $\boldsymbol{p}$ is derived from a limit of general relativity, and so is derived from the structure of spacetime. Therefore $e\boldsymbol{A}$ is also derived from the structure of spacetime, or from the vacuum itself. The meaning of e is reinterpreted as the minimum value of

$$e = \hbar \frac{\kappa}{A^{(0)}} \qquad (343)$$

and this minimum value is the charge on the proton.

At the Higgs minimum, the Lagrangian in the internal space of the U(1) gauge theory is

$$\mathscr{L} = \partial_\mu a \partial^\mu a^* - m^2 a^* a - \lambda(a^* a)^2 \qquad (344)$$

which, on local gauge transformation, becomes

$$\mathscr{L} = (\partial_\mu + igA_\mu)a(\partial^\mu - igA^\mu)a^* - m^2aa^* - \lambda(aa^*)^2 - \frac{1}{4}F^{\mu\nu}F_{\mu\nu} \quad (345)$$

The equations of motion of the field at the Higgs minimum (the minimum potential energy of the vacuum) are the Euler–Lagrange equations

$$\frac{\partial\mathscr{L}}{\partial a} = \partial_\mu\left(\frac{\partial\mathscr{L}}{\partial(\partial_\mu a)}\right); \qquad \frac{\partial\mathscr{L}}{\partial a^*} = \partial_\mu\left(\frac{\partial\mathscr{L}}{\partial(\partial_\mu a^*)}\right) \quad (346)$$

$$\frac{\partial\mathscr{L}}{\partial A_\mu} = \partial_\mu\left(\frac{\partial\mathscr{L}}{\partial(\partial_\nu A_\mu)}\right) \quad (347)$$

and using the globally invariant Lagrangian (344) in Eqs. (346) gives the result

$$\Box a^* = -(m^2 + 2\lambda a^*a)a^* = 0$$
$$\Box a = -(m^2 + 2\lambda aa^*)a = 0 \quad (348)$$

and, using the locally invariant Lagrangian (345) in Eqs. (346) gives the result

$$D_\mu(D^\mu a^*) = -(m^2 + 2\lambda a^*a)a^* = 0$$
$$D_\mu(D^\mu a) = -(m^2 + 2\lambda aa^*)a = 0 \quad (349)$$

Equation (348) is the globally invariant wave equation defining a, and Eq. (349) is its locally invariant equivalent. Using the locally invariant Lagrangian (345) in Eq. (347) gives the inhomogeneous field equation (SI units)

$$\partial_\nu F^{\mu\nu} = -igc(a^*D^\mu a - aD^\mu a^*) \quad (350)$$

where the charge current density on the right-hand side is obtained from the pure vacuum by local gauge transformation and local gauge invariance. Both the left- and right-hand sides of Eq. (350) are defined by the minimum of potential energy, and by the minimum value that A can attain. This minimum value is a, and is the vacuum expectation value of A [46], associated with a nonzero potential energy that gives rise to A^μ and $F^{\mu\nu}$ by local gauge invariance. Therefore the source of an electromagnetic field propagating in the vacuum is the Higgs minimum value of A, which is denoted a. If we do not use a Higgs mechanism, then the vacuum expectation value of A in the internal gauge space of the U(1) gauge theory is zero, and the globally invariant Lagrangian disappears.

Therefore, in the presence of a Higgs mechanism

$$|\langle 0|A|0\rangle|^2 = a^2 \tag{351}$$

and in its absence:

$$|\langle 0|A|0\rangle|^2 = 0 \tag{352}$$

The Lagrangian (345) can be written as [see Eq. (158)]

$$\mathscr{L} = g^2 a^2 A_\mu A^\mu + \cdots \tag{353}$$

and if the photon mass is identified as

$$m_p^2 \equiv 2g^2 a^2 \tag{354}$$

the Lagrangian (353) gives a Proca equation that is locally gauge invariant on the U(1) level. Therefore, application of the Higgs mechanism in this way has produced one massive photon from one massless photon. The scalar field a remains unaffected, so degrees of freedom are conserved. Therefore, this theory identifies photon mass as the result of local gauge invariance applied at the Higgs minimum, that is, the minimum value that the potential energy of the globally invariant Lagrangian can take in the vacuum.

This minimum value provides the true vacuum energy

$$En(\text{vac}) = \int J^\mu(\text{vac}) A_\mu \, dV \tag{355}$$

and a rate of doing work:

$$\frac{dW}{dt}(\text{vac}) = \int \boldsymbol{J}(\text{vac}) \cdot \boldsymbol{E} \, dV \tag{356}$$

The Poynting theorem for the true vacuum can be developed as in Eqs. (258)–(262). The true vacuum energy (355) comes from the vacuum current in Eq. (350), which is transformed into a matter current by a minimal prescription as discussed already. This matter current in principle provides an electromotive force in a circuit. It is to be noted that the local Higgs maximum occurs at $A = 0$ [46], so the local Higgs minimum occurs below the zero value of A.

The overall conclusion is that there is no objection in principle to extracting electromotive force from the true vacuum, defined by the minimum value, a, which can be attained by A in the internal scalar space of the gauge theory, which is the theory underlying electromagnetic theory.

On the O(3) level, the globally invariant Lagrangian corresponding to Eq. (344) is

$$\mathcal{L} = T - V = \frac{1}{2}\partial_\mu a \cdot \partial^\mu a - \frac{m^2}{2}a \cdot a - \lambda(a \cdot a)^2 \tag{357}$$

with potential energy:

$$V = \frac{m^2}{2}a \cdot a + \lambda(a \cdot a)^2 \tag{358}$$

Here, a is a vector in the internal space of O(3) symmetry. The equation of motion is

$$\frac{\partial \mathcal{L}}{\partial a} = \partial_\mu \left(\frac{\partial \mathcal{L}}{\partial_\mu a}\right) = \partial^\mu \left(\frac{\partial \mathcal{L}}{\partial^\mu a}\right) \tag{359}$$

and produces, from the Lagrangian (357), the result

$$\Box a = m^2 a + \lambda a(a \cdot a) = 0 \tag{360}$$

which is a globally invariant wave equation of d'Alembert type for the three components of a. Local gauge transformation of the Lagrangian (357) produces [cf. Eq. (205)] the following equation:

$$\mathcal{L} = \frac{1}{2}D_\mu a \cdot D^\mu a - \frac{m^2}{2}a \cdot a - \lambda(a \cdot a)^2 - \frac{1}{4}G_{\mu\nu} \cdot G^{\mu\nu} \tag{361}$$

Use of Eq. (359) produces the wave equation

$$D^\mu(D_\mu a) = m^2 a + \lambda a(a \cdot a) = 0 \tag{362}$$

The Euler–Lagrange equation

$$\frac{\partial \mathcal{L}}{\partial A^\mu} = \partial^\nu \left(\frac{\partial \mathcal{L}}{\partial(\partial^\nu A^\mu)}\right) \tag{363}$$

produces the field equation

$$D^\nu G_{\mu\nu} = -g(D_\mu a) \times a \tag{364}$$

where the current on the right-hand side is a current generated by the minimum value of A in the internal O(3) symmetry gauge space. This minimum value is the vacuum and is denoted by the vector a

The Lagrangian (361) can be written as

$$\mathcal{L} = g^2 A_\mu \times \boldsymbol{a} \cdot \boldsymbol{A}^\mu \times \boldsymbol{a} + \cdots \tag{365}$$

and produces three photons with mass from the vector identity

$$(\boldsymbol{A}_\mu \times \boldsymbol{a}) \cdot (\boldsymbol{A}^\mu \times \boldsymbol{a}) = (\boldsymbol{A}_\mu \cdot \boldsymbol{A}^\mu)(\boldsymbol{a} \cdot \boldsymbol{a}) - (\boldsymbol{A}_\mu \cdot \boldsymbol{a})(\boldsymbol{a} \cdot \boldsymbol{A}^\mu) \tag{366}$$

and the term

$$\mathcal{L} = g^2 (\boldsymbol{a} \cdot \boldsymbol{a})(\boldsymbol{A}_\mu \cdot \boldsymbol{A}^\mu) + \cdots \tag{367}$$

One of these is the superheavy Crowell boson [42], associated with index (3) in the ((1),(2),(3)) basis, and the other two are massive photons associated with indices (1) and (2). The superheavy Crowell boson comes from electroweak theory with an SU(2) electromagnetic sector and may have been observed in a LEP collaboration at CERN [44,56].

On the O(3) level, the vacuum current (SI units)

$$\boldsymbol{J}^\mu(\text{vac}) = \frac{g}{\mu_0 c} (D^\mu \boldsymbol{a}) \times \boldsymbol{a} \tag{368}$$

gives the vacuum energy

$$En = \int \boldsymbol{J}^\mu(\text{vac}) \cdot \boldsymbol{A}_\mu \, dV \tag{369}$$

which can be transformed into a matter current by the minimal prescription (342). This matter current is effectively an electromotive force in a circuit. Gauge theory of any internal gauge symmetry applied to electromagnetism comes to the same result, that energy is available from the vacuum, defined as the Higgs minimum. This appears to be a substantial advance in understanding.

In order to check these results for self-consistency, the locally invariant Higgs Lagrangian, when written out in full, is

$$\begin{aligned}
\mathcal{L} = {} & \partial_\mu(a_0 + A)\partial^\mu(a_0 + A)^* - ig((a_0 + A)^*\partial^\mu(a_0 + A) - (a_0 + A)\partial^\mu(a_0 + A)^*)A_\mu \\
& + g^2 A_\mu A^\mu (a_0 + A)^2 - m^2(a_0 + A)(a_0 + A)^* \\
& - \lambda((a_0 + A)(a_0 + A)^*)^2 - \frac{1}{4} F^{\mu\nu} F_{\mu\nu}
\end{aligned} \tag{370}$$

where a_0 is the minimum value and where the complex scalar field is

$$A = \frac{1}{\sqrt{2}}(A_1 + iA_2) \tag{371}$$

in the internal space. In this Lagrangian, a_0 is a constant so the Lagrangian (370) can be written as

$$\mathcal{L} = -\frac{1}{4}F_{\mu\nu}F^{\mu\nu} + g^2 a_0^2 A_\mu A^\mu + \frac{1}{2}(\partial_\mu A_1)^2 + \frac{1}{2}(\partial_\mu A_2)^2$$
$$- 2\lambda a_0^2 A_1^2 + \sqrt{2} g a_0 A^\mu \partial_\mu A_2 + \cdots \tag{372}$$

At its minimum value, this Lagrangian is

$$\mathcal{L} = -\frac{1}{4}F_{\mu\nu}F^{\mu\nu} + g^2 a_0^2 A_\mu A^\mu \tag{373}$$

which gives the following locally gauge-invariant Proca equation:

$$\partial_\mu F^{\mu\nu} + m_p^2 A^\nu = 0 \tag{374}$$

The photon mass is identified as argued already by

$$m_p^2 = 2g^2 |a_0^2| \tag{375}$$

and if we further identify

$$g^2 \equiv \frac{\kappa}{2|a_0^2|} \tag{376}$$

we obtain the de Broglie guidance theorem in SI units:

$$\hbar\omega = m_p c^2 \tag{377}$$

So, as argued already, the photon mass is picked up from the vacuum, that is, from the minimum value of the locally invariant Higgs Lagrangian (370). This conclusion means that the Lehnert charge current density that leads to the Proca equation [45,49] is also a property of the vacuum, as argued above. In order to show this result, the constant a_0 is expressed as the product of two complex fields a and a^*. To illustrate this by analogy, one can show that the dot product of two conjugate plane waves gives a constant

$$A^{(0)2} = \frac{A^{(0)}}{\sqrt{2}}(\mathbf{i} - i\mathbf{j}) \cdot e^{i\phi} \frac{A^{(0)}}{\sqrt{2}}(\mathbf{i} + i\mathbf{j})e^{-i\phi} = \mathbf{A}^{(1)} \cdot \mathbf{A}^{(2)} \tag{378}$$

but the individual plane waves are functions of coordinates and time. Analogously, therefore, a and a^* are functions of x^μ. The vacuum Lagrangian can therefore be written as

$$\mathscr{L} = \partial_\mu a \partial^\mu a^* - ig(a^*\partial^\mu a - a\partial^\mu a^*)A_\mu$$
$$+ g^2 A_\mu A^\mu a^2 - m^2 a^* a - \lambda(a^*a)^2 - \frac{1}{4}F^{\mu\nu}F_{\mu\nu} \tag{379}$$

From Eq. (373), it is known that this Lagrangian is

$$\mathscr{L} = g^2 A_\mu A^\mu a^2 - \frac{1}{4}F^{\mu\nu}F_{\mu\nu} \tag{380}$$

There is therefore a balance between globally invariant Lagrangians:

$$\mathscr{L} = \partial_\mu a \partial^\mu a^* - m^2 a^* a - \lambda(a^*a)^2$$
$$= ig(a^*\partial^\mu a - a\partial^\mu a^*)A_\mu = gJ^\mu A_\mu \tag{381}$$

The globally invariant vacuum energy is therefore:

$$En = \int J^\mu A_\mu dV = \frac{1}{g}\int \partial_\mu a \partial^\mu a^* - m^2 a^* a - \lambda(a^*a)^2 dV \tag{382}$$

and is defined in the internal space of the gauge theory being considered [in this case of U(1) symmetry]. It can be seen that the vacuum energy is essentially a volume integration over the original globally invariant Lagrangian

$$\mathscr{L} = \partial_\mu a \partial^\mu a^* - m^2 a^* a - \lambda(a^*a)^2 \tag{383}$$

used in the Higgs mechanism. We have defined the mass of the photon by Eq. (375), and so the locally gauge-invariant Proca wave equation is

$$\Box A_\mu = -2g^2 a_0^2 A_\mu \tag{384}$$

Energy is usually written as the volume integral over the Hamiltonian, and not the Lagrangian, and Eq. (382) may be transformed into a volume integral over a Hamiltonian if we define the effective potential energy

$$V = -m^2 a^* a - \lambda(a^*a)^2 \tag{385}$$

which is negative.

The locally gauge-invariant Lehnert field equation corresponding to Eq. (374) was derived as Eq. (350). The photon picks up mass from the vacuum itself, and having derived a locally gauge-invariant Proca equation, canonical quantization can be applied to produce a photon with mass with three space dimensions.

V. SCHRÖDINGER EQUATION WITH A HIGGS MECHANISM: EFFECT ON THE WAVE FUNCTIONS

In order to measure the effect of vacuum energy in atoms and molecules, in the simplest case of the hydrogen atom, it is necessary to develop the nonrelativistic Schrödinger equation with an inbuilt Higgs mechanism. The method used in this section is to start with the Lagrangian for the Higgs mechanism in matter fields, derive a Klein–Gordon equation, and from that, an Einstein equation, then to take the nonrelativistic limit of the Einstein equation, and finally quantize that to give the Schrödinger equation with a Higgs mechanism. It turns out that the Higgs minimum is at an energy $\frac{1}{2}mc^2$ below the vacuum minimum with no Higgs mechanism, meaning that this amount of energy is available in the vacuum. Some examples of the effect of this negative potential energy on analytical solutions of the Schrödinger equation are given in this section.

The starting Lagrangian on the U(1) level for a free particle, such as an electron, is the standard Lagrangian for the Higgs mechanism:

$$\mathscr{L} = \partial_\mu \phi \partial^\mu \phi^* - m^2 \phi^* \phi - \lambda (\phi^* \phi)^2 \tag{386}$$

Using Eqs. (115) and (221), this Lagrangian gives the Klein–Gordon equations

$$(\Box + (m^2 + 2\lambda \phi^* \phi))\phi^* = 0 \tag{387}$$

$$(\Box + (m^2 + 2\lambda \phi \phi^*))\phi = 0 \tag{388}$$

in which ϕ and ϕ^* are considered to be complex-valued one-particle wave functions. It can be seen that the effect of the Higgs mechanism is to increase the mass term m^2 to $m^2 + 2\lambda \phi^* \phi$.

This additional effective mass is introduced from spontaneous symmetry breaking of the vacuum. The two Klein–Gordon equations therefore take on the form

$$(\Box + m^2)\phi^* = -2\lambda(\phi \phi^*)\phi^* \tag{389}$$

$$(\Box + m^2)\phi = -2\lambda(\phi \phi^*)\phi \tag{390}$$

The classical equivalent of these equations is the Einstein equation for one particle

$$En^2 = p^2 c^2 + m_0^2 c^4 + 2\lambda(\phi \phi^*)c^4 \tag{391}$$

The Higgs mechanism has produced an additional rest energy:

$$En_0(\text{Higgs}) = 2\lambda(\phi \phi^*)c^4 \tag{392}$$

In Eq. (391), En is the total energy, and the equation can be written as follows:

$$p^2c^2 = En^2 - En_0^2$$
$$= m_0^2c^4\left(1 - \frac{u^2}{c^2}\right)^{-1} - m_0^2c^4 - 2\lambda\langle\phi^2\rangle c^4 \qquad (393)$$

To reach the nonrelativistic limit of this equation, the right-hand side is expanded as

$$p^2c^2 = m^2c^4\frac{u^2}{c^2} - 2\lambda\langle\phi^2\rangle c^4 \qquad (u \ll c) \qquad (394)$$

which, for $u \ll c$, results in the nonrelativistic equation

$$p^2c^2 = m^2c^4\frac{u^2}{c^2} - 2\lambda\langle\phi^2\rangle c^4 = En^2 - En_0^2 \qquad (395)$$

which has the same form as the original, fully relativistic, equation (393). The nonrelativistic equation (395) can be written as

$$m^2u^2 = p^2 + 2\lambda\langle\phi^2\rangle c^2 \quad (u \ll c) \qquad (396)$$

that is

$$\frac{1}{2}mu^2 = \frac{p^2}{2m} + \frac{\lambda}{m}\langle\phi^2\rangle c^2 \qquad (397)$$

The left-hand side is the nonrelativistic kinetic energy of one particle. It can be seen that the Higgs mechanism changes the classical nonrelativistic expression

$$En = \frac{1}{2}mu^2 = \frac{p^2}{2m} = T \qquad (398)$$

to Eq. (397). The Schrödinger equation without the Higgs mechanism is obtained by applying the quantum ansatz

$$En \rightarrow i\hbar\frac{\partial}{\partial t}; \qquad \boldsymbol{p} \rightarrow -i\hbar\nabla \qquad (399)$$

to Eq. (398), giving

$$-i\hbar\frac{\partial\phi}{\partial t} = \frac{\hbar^2}{2m}\nabla^2\phi \qquad (400)$$

The Schrödinger equation in the presence of the Higgs mechanism is therefore

$$-i\hbar\frac{\partial\phi}{\partial t} = \frac{\hbar^2}{2m}\nabla^2\phi + \frac{\lambda}{m}\langle\phi^2\rangle c^2 \tag{401}$$

where $\langle\phi^2\rangle$ is the expectation value of the wave function. At the Higgs minimum, this expectation value is [46]

$$\langle\phi^2\rangle = -\frac{m^2}{2\lambda} \tag{402}$$

and so the Schrödinger equation at the Higgs minimum is

$$-i\hbar\frac{\partial\phi}{\partial t} = \frac{\hbar^2}{2m}\phi - \frac{1}{2}mc^2\phi \tag{403}$$

which can be written in the familiar form

$$En\,\phi = H\phi = (T + V)\phi$$
$$T = -\frac{\hbar}{2m}\nabla^2 \tag{404}$$

where

$$V = -\frac{1}{2}mc^2 = \min\left(\frac{\lambda}{m}\langle\phi^2\rangle c^2\right) \tag{405}$$

is a negative potential energy produced by spontaneous symmetry breaking of the vacuum. The Schrödinger equation (404) shows that the Higgs minimum (the symmetry broken vacuum) is at an energy:

$$V(\text{Higgs}) = \frac{1}{2}mc^2 \tag{406}$$

below the vacuum for the ordinary Schrödinger equation (400). The vacuum expectation value for the ordinary Schrödinger equation is

$$\langle\phi^2\rangle = 0 \tag{407}$$

We have therefore derived a nonrelativistic Schrödinger equation for a free particle with an additional negative potential energy term $V = -\frac{1}{2}mc^2$. In order to apply this method to the hydrogen atom, the relevant Schrödinger equation is

$$\left(-\frac{\hbar^2}{2\mu}\nabla^2 - V_{\text{Coulomb}} + V\right)\phi = En\,\phi \tag{408}$$

$$V_{\text{Coulomb}} = \frac{e^2}{4\pi\varepsilon_0}\frac{1}{r} \tag{408a}$$

where V_{Coulomb} is the classical Coulomb interaction between one electron and one proton and μ is the reduced mass:

$$\mu = \frac{m_e m_p}{m_e + m_p} \tag{409}$$

The Higgs mechanism is the basis of electroweak theory and other elementary particle and gauge field theories, so it can be stated with confidence that to a good approximation the energy $\frac{1}{2}mc^2$ is released from the vacuum when a shift occurs between the Higgs minimum and the ground state of the hydrogen atom. The challenge is how to find a mechanism for releasing this energy. Mills [67] has found a working device based on the postulated collapse of the H atom below its ground state. The Schrödinger equation with a Higgs mechanism shows that there is an extra negative potential energy term that may account for the energy observed by Mills [67]. This possibility will be explored later by solving Eq. (408) analytically to find the effect of V on the states of the H atom. First, however, we illustrate the effect of V on analytical solutions of the Schrödinger equation, starting with the free-particle solution.

The wave function for Eq. (404) is well known [68] to be of the form

$$\phi = A'e^{i\kappa'Z} + B'e^{-i\kappa'Z}; \qquad \kappa' = \left(\frac{2m(E-V)}{\hbar^2}\right)^{1/2} \tag{410}$$

where the particle momentum is given by $\hbar\kappa'$. The scheme in the following equation group explains the role of the two parts of the wave function:

$$\begin{aligned}
&\rightarrow p = \hbar\kappa'; \qquad \psi = A'e^{i\kappa'Z}\\
&\leftarrow p = \hbar\kappa'; \qquad \psi = B'e^{-i\kappa'Z}
\end{aligned} \tag{411}$$

In the Schrödinger equation (404), the maximum value of the vacuum potential energy is the Newton vacuum

$$V = 0 \tag{412}$$

and its minimum value is the Higgs vacuum, or minimum of the symmetry-broken vacuum:

$$V = -\frac{1}{2}mc^2 \tag{413}$$

In Newtonian mechanics, the particle cannot be found below $V = 0$, therefore Newtonian mechanics always corresponds to $V = 0$ [e.g., Eq. (398)], and this represents, classically, an insurmountable barrier to a particle such as an electron attempting to enter the Higgs region below $V = 0$. In quantum mechanics, however, an electron may enter the Higgs region by quantum tunneling, which occurs when $E < V = 0$. The wave function for this process is well known to be [68]

$$\phi = Ae^{-\kappa Z} \tag{414}$$

and has a nonzero amplitude. An electron of energy 1.6×10^{-19} J incident on a barrier of height 3.2×10^{-19} J has a wave function that decays with distance as $e^{-5.12 \ (Z/nm)}$, and decays to $1/e$ of its initial value after 0.2 nm, about the diameter of an atom [68]. Therefore, quantum tunneling is important on atomic scales. So quantum-mechanically, an electron can enter the Higgs region and gain negative energy. This means that it radiates positive energy [46]. The maximum amount of energy that can be radiated is determined by the minimum value of the Higgs region, which defines the ground state, namely, the Higgs vacuum. This is a result of Eq. (404) for a free electron. To see that negative-energy states En are possible, write Eq. (395) as

$$En^2 = p^2 c^2 + En_0^2 \tag{415}$$

and its solutions are

$$En = \pm(p^2 c^2 + En_0^2)^{1/2} \tag{416}$$

The states of the hydrogen atom must be found from Eq. (408). When $V = 0$, the ground state of the H atom is well known [68] to be determined by the expectation value

$$En = -\frac{\mu e^4}{32\pi^2 \varepsilon_0^2 \hbar^2} \frac{1}{n}; \qquad n = 1 \tag{417}$$

from the Schrödinger equation:

$$-\frac{\hbar^2}{2\mu} \nabla^2 \phi - \frac{e^2}{4\pi\varepsilon_0 r} \phi = En\, \phi \tag{418}$$

When V is not zero, Eq. (418) becomes

$$-\frac{\hbar^2}{2\mu} \nabla^2 \phi - \frac{e^2}{4\pi\varepsilon_0 r} \phi = (En - V)\phi \tag{419}$$

and the electronic orbital energy becomes

$$En = -\frac{\mu e^4}{32\pi^2 \varepsilon_0^2 \hbar^2} \frac{1}{n} + V \qquad (420)$$

Here, n is the principal quantum number. So, for $V = 0$ the electronic orbital energy in the H atom becomes less negative as n increases. However, if we add $V < 0$ from the Higgs region to the ground state of H determined by $n = 1$, the electronic orbital energy falls below its ground state. This emits energy in the same way as an electron falling from a higher to a lower electronic atomic orbital emits energy. The energy emitted by driving the H orbital below its ground state has been observed experimentally by Mills et al. [67], repeatedly and reproducibly. The Higgs mechanism on the U(1) level accounts for this energy emission.

VI.   VECTOR INTERNAL BASIS FOR
SINGLE-PARTICLE QUANTIZATION

Conventional single particle quantization is based on the quantum ansatz (399) applied to the Einstein equation (415) to produce the Klein–Gordon equation

$$(\Box + m^2)\phi = 0 \qquad (421)$$

$$(\Box + m^2)\phi^* = 0 \qquad (422)$$

where ϕ is regarded as a single-particle wave function. In the nonrelativistic limit, the Schrödinger equation is obtained as demonstrated in Section IV. Formally, the Klein–Gordon equations (421) can be obtained from the U(1) Lagrangian [46]

$$\mathscr{L} = (\partial_\mu \phi)(\partial^\mu \phi^*) - m^2 \phi \phi^* \qquad (423)$$

which is globally invariant. Usually, the Lagrangian (423) is applied to complex fields, but formally, these can also be wave functions. On the U(1) level, they take the form

$$\phi = \phi^{(1)} = \frac{1}{\sqrt{2}}(\phi_X - i\phi_Y) \qquad (424)$$

$$\phi^* = \phi^{(2)} = \frac{1}{\sqrt{2}}(\phi_X + i\phi_Y) \qquad (425)$$

On the O(3) level, there are three wave functions:

$$\phi^{(1)} = \frac{1}{\sqrt{2}}(\phi_X - i\phi_Y)$$

$$\phi^{(2)} = \frac{1}{\sqrt{2}}(\phi_X + i\phi_Y) \tag{426}$$

$$\phi^{(3)} = \phi_Z$$

and it is possible to collect these components in vector form through the relation

$$\phi = \phi^{(2)}e^{(1)} + \phi^{(1)}e^{(2)} + \phi^{(3)}e^{(3)} = \phi_X i + \phi_Y j + \phi_Z k \tag{427}$$

where ϕ_X, ϕ_Y, ϕ_Z are real-valued. The unit vectors of the circular basis are defined as

$$e^{(1)} = \frac{1}{\sqrt{2}}(i - ij)$$

$$e^{(2)} = \frac{1}{\sqrt{2}}(i + ij) \tag{428}$$

$$e^{(3)} = k$$

On the O(3) level, therefore, the probability density of the Schrödinger equation is

$$\rho = \phi^{(1)}\phi^{(2)} = \phi^{(2)}\phi^{(1)} = \phi^{(3)}\phi^{(3)*} \tag{429}$$

and there are three Schrödinger equations:

$$\frac{\hbar^2}{2m}\nabla^2\phi^{(1)} = -i\hbar\frac{\partial\phi^{(1)}}{\partial t}$$

$$\frac{\hbar^2}{2m}\nabla^2\phi^{(2)} = -i\hbar\frac{\partial\phi^{(2)}}{\partial t} \tag{429a}$$

$$\frac{\hbar^2}{2m}\nabla^2\phi^{(3)} = -i\hbar\frac{\partial\phi^{(3)}}{\partial t}$$

which identify $\phi^{(1)}, \phi^{(2)}, \phi^{(3)}$ as angular momentum wave functions. Atkins [48] has shown that angular momentum commutator relations can be used to derive the laws of nonrelativistic quantum mechanics. So the internal O(3) space, in this instance, corresponds to ordinary three-dimensional space. In a U(1) internal space, the third component $\phi^{(3)}$ of angular momentum is missing and the wave functions are $\phi^{(1)}$ and $\phi^{(2)}$. In Newtonian and nonrelativistic quantum mechanics, the internal space is therefore O(3). The probability currents of the Schrödinger equation are

$$j = i\frac{\hbar}{2m}(\phi^{(2)}\nabla\phi^{(1)} - \phi^{(1)}\nabla\phi^{(2)}) \tag{430}$$

and

$$j = i\frac{\hbar}{2m}(\phi^{(3)}\nabla\phi^{(3)*} - \phi^{(3)*}\nabla\phi^{(3)}) = 0 \tag{431}$$

in the complex circular basis. In a more general spherical harmonic [68] basis for three-dimensional space, the angular momentum wave functions are eigenfunctions such that

$$|n, m\rangle = Y(\theta, \phi) \tag{432}$$

where

$$Y_{lm_l}(\theta, \phi); \qquad l = 1; \quad m = 0, \pm 1 \tag{433}$$

are the spherical harmonics. Therefore, it is also possible to describe the internal O(3) basis of electrodynamics in terms of spherical harmonics.

The probability densities of the Klein–Gordon equation [46] in an O(3) internal basis contains terms such as

$$\rho = i\frac{\hbar}{2mc^2}\left(\phi^{(2)}\frac{\partial\phi^{(1)}}{\partial t} - \phi^{(1)}\frac{\partial\phi^{(2)}}{\partial t}\right) \tag{434}$$

This term is usually written as

$$\rho = i\frac{\hbar}{2mc^2}\left(\phi^*\frac{\partial\phi}{\partial t} - \phi\frac{\partial\phi^*}{\partial t}\right) \tag{435}$$

and in general can become negative. So the Klein–Gordon equation is abandoned in general as an equation for single-particle quantum mechanics. However, for the photon with mass, the probability density from the Klein–Gordon equation is positive definite, because it is possible to use the de Broglie wave functions:

$$\begin{aligned}\phi^* = \phi^{(2)} &= \exp\left(i(\omega t - \kappa Z)\right)\\ \phi = \phi^{(1)} &= \exp\left(-i(\omega t - \kappa Z)\right)\end{aligned} \tag{436}$$

to give

$$\rho = \frac{\hbar\omega}{mc^2} \tag{437}$$

When mass m is the rest mass, the de Broglie theorem states that

$$m_0 c^2 = \hbar\omega_0 \tag{438}$$

and $\rho = 1$. For the free photon with mass, the Klein–Gordon equation gives a positive definite probability density because the derivative $\partial\phi^{(1)}/\partial t$ is not independent of $\phi^{(2)}$. The equation shows that the free photon with mass can also take on negative energies. Therefore, the vector ϕ in this case can be interpreted as a single-particle wave function. The probability 4-vector for the photon with mass is given by [46]

$$j^\mu = i\frac{\hbar}{2mc}(\phi^* \partial^\mu \phi - (\partial^\mu \phi^*)\phi) \tag{439}$$

which for the de Broglie wave function gives

$$j_z = -\frac{\hbar\kappa}{mc} \tag{440}$$

The 4-current j^μ is conserved:

$$\partial_\mu j^\mu = i\frac{\hbar}{2mc}(\phi^* \Box \phi - \phi \Box \phi^*) \tag{441}$$

If we define

$$\begin{aligned} \boldsymbol{A} &= A^{(2)}\boldsymbol{e}^{(1)} + A^{(1)}\boldsymbol{e}^{(2)} + A^{(3)}\boldsymbol{e}^{(3)} \\ &= A_X\boldsymbol{i} + A_Y\boldsymbol{j} + A_Z\boldsymbol{k} \end{aligned} \tag{442}$$

there emerge four Klein–Gordon equations that all give a positive probability density:

$$(\Box + m^2)A^{(i)} = 0; \qquad i = 0, 1, 2, 3 \tag{443}$$

for an O(3) invariant theory. In a U(1) invariant theory, there are only two equations:

$$(\Box + m^2)A^{(i)} = 0; \qquad i = 1, 2 \tag{444}$$

The four Klein–Gordon equations are for the photon regarded as a scalar particle without spin. If the scalar components $A^{(0)}$, $A^{(1)}$, $A^{(2)}$, $A^{(3)}$ are regarded as fields and quantized, a many-particle interpretation of the photon emerges, and they are recognized as bosons, which have integral spin. Therefore, in an internal space that is globally invariant under a gauge transform, the four equations (443) give, after field quantization (second quantization), a globally gauge invariant Proca equation

$$(\Box + m^2)A^\mu = 0 \tag{445}$$

where the 4-vector is defined as

$$A^\mu = (A^{(0)}, A^{(1)}, A^{(2)}, A^{(3)}) \tag{446}$$

To an excellent approximation, the four Klein–Gordon equations (443) are d'Alembert equations, which are locally gauge-invariant.

However, there remains the problem of how to obtain a locally gauge-invariant Proca equation. To address this problem rigorously, it is necessary to use a non-Abelian Higgs mechanism applied within gauge theory.

The starting point of our derivation is the globally invariant O(3) Lagrangian of the Higgs mechanism

$$\mathscr{L} = \partial_\mu A \cdot \partial^\mu A^* - m^2 A \cdot A^* - \lambda(A \cdot A^*)^2 \tag{447}$$

where A and A^* are regarded as independent complex vectors in the O(3) internal space of the gauge theory. Application of the Euler–Lagrange equations (204) give the following results:

$$\begin{aligned}
\frac{\partial \mathscr{L}}{\partial A} &= -m^2 A^* - 2\lambda A^*(A \cdot A^*) \\
\frac{\partial \mathscr{L}}{\partial A^*} &= -m^2 A - 2\lambda A(A \cdot A^*)
\end{aligned} \tag{448}$$

Therefore, at the Higgs minimum

$$A \cdot A^* = -\frac{m^2}{2\lambda} \equiv a_0^2 \tag{449}$$

The wave equation obtained from Eqs. (204) and (448) with the Lagrangian (447) is

$$\partial^\mu \partial_\mu A = -m^2 A - 2\lambda A(A \cdot A^*) \tag{450}$$

and, at the Higgs minimum, reduces to

$$\partial^\mu \partial_\mu A = 0 \tag{451}$$

If we define:

$$A = A^{(2)} e^{(1)} + A^{(1)} e^{(2)} + A^{(3)} e^{(3)} \tag{452}$$

then four globally invariant d'Alembert equations are obtained:

$$\begin{aligned} \Box A^{(1)} &= 0 \\ \Box A^{(2)} &= 0 \\ \Box A^{(3)} &= 0 \\ \Box A^{(0)} &= 0 \end{aligned} \tag{453}$$

The locally invariant Lagrangian obtained from the Lagrangian (447) is

$$\mathscr{L} = D_\mu A \cdot D^\mu A^* - \frac{1}{4} G_{\mu\nu} \cdot G^{\mu\nu} - m^2 A \cdot A^* - \lambda (A \cdot A^*)^2 \tag{454}$$

where it is understood that

$$\begin{aligned} A &\rightarrow a_0 + A \\ A^* &\rightarrow a_0^* + A^* \end{aligned} \tag{455}$$

The following Euler–Lagrange equation is used next with the Lagrangian (454):

$$\frac{\partial \mathscr{L}}{\partial A_\mu} = \partial_\nu \left(\frac{\partial \mathscr{L}}{\partial (\partial_\nu A_\mu)} \right) \tag{456}$$

The Lagrangian (454) contains terms such as

$$\begin{aligned} D_\mu A \cdot D^\mu A^* &= (\partial_\mu + A_\mu \times) A \cdot (\partial^\mu - A^\mu \times) A^* \\ &= \partial_\mu A \cdot \partial^\mu A^* + g A_\mu \times A \cdot \partial^\mu A^* - g \partial_\mu A \cdot A^\mu \times A^* \\ &\quad - g^2 (A_\mu \times A) \cdot (A^\mu \times A^*) \end{aligned} \tag{457}$$

and a field equation emerges from the analysis by using

$$\begin{aligned} \frac{\partial \mathscr{L}}{\partial A_\mu} &= g A \times \partial^\mu A^* - g^2 (A^\mu (A \cdot A^*) - A^* (A \cdot A^\mu)) \\ &= -g \partial^\mu A^* \times A + g^2 (A^\mu \times A^*) \times A \\ &= -g D^\mu A^* \times A \end{aligned} \tag{458}$$

giving

$$D_\nu G^{\mu\nu} = -gD^\mu A^* \times A \qquad (459)$$

At the Higgs minimum, this field equation reduces to the locally gauge-invariant Proca equation

$$D_\nu G^{\mu\nu} = -g^2 a_0 \times (A^\mu \times a_0^*) \qquad (460)$$

and the Lagrangian reduces to

$$\mathscr{L} = -\frac{1}{4} G_{\mu\nu} \cdot G^{\mu\nu} - g^2 (A_\mu \times a_0) \cdot (A^\mu \times a_0^*) \qquad (461)$$

Therefore, it can be seen that the mass of the photon in this analysis is derived from the Higgs vacuum, which is the minimum of the potential energy term in the Lagrangian (454). The field equation (460) is O(3) invariant and, therefore, the existence of photon mass is made compatible with the existence of the $B^{(3)}$ field, as inferred originally by Evans and Vigier [42]. The Higgs mechanism is the basis of much of modern elementary particle theory; thus this derivation is based on rigorous gauge theory that is locally O(3) invariant.

VII. THE LEHNERT CHARGE CURRENT DENSITIES IN O(3) ELECTRODYNAMICS

We have established that, in O(3) electrodynamics, the vacuum charge current densities first proposed by Lehnert [42,45,49] take the form

$$J_\mu(\text{vac}) = g\partial_\mu A \times A + g^2 A \times (A \times A_\mu) \qquad (462)$$

In this section, we illustrate the self-consistent calculation of these charge current densities in the plane-wave approximation, using plane waves in the X, Y, and Z directions. In general, the solution of the field equation (459) must be found numerically, and it is emphasized that the plane-wave approximation is a first approximation only. In the internal space, there is the real vector:

$$A = A_X i + A_Y j + A_Z k \qquad (463)$$

and by definition

$$A_\mu = A_{\mu,X} i + A_{\mu,Y} j + A_{\mu,Z} k \qquad (464)$$

First, we consider a plane-wave propagating in the Z direction, so that

$$A = -\frac{A^{(0)}}{\sqrt{2}} \sin\phi\, i + \frac{A^{(0)}}{\sqrt{2}} \cos\phi\, j + A_Z k \qquad (465)$$

and adapt the following notation:

$$A_3 = A_{3.Z} k = -A_Z k \qquad (466)$$

Elementary vector algebra then gives

$$g^2 A \times (A \times A_3) = \kappa^2 (-A_X i - A_Y j + A_Z k) \qquad (467)$$

and

$$g \partial_Z A \times A = \kappa^2 A^{(0)} \left(k + \frac{1}{\sqrt{2}} \sin\phi\, i - \frac{1}{\sqrt{2}} \cos\phi\, j \right) \qquad (468)$$

The i and j terms must cancel, so we obtain the following, self-consistently:

$$A_X = -\frac{A^{(0)}}{\sqrt{2}} \sin\phi; \qquad A_Y = \frac{A^{(0)}}{\sqrt{2}} \cos\phi \qquad (469)$$

The Lehnert vacuum current density for a plane wave in the Z direction is therefore

$$J_Z = 2 \frac{\kappa^2 A^{(0)}}{\mu_0} k \qquad (470)$$

If this is used in the third equation of Eq. (83), the B cyclic theorem [47–61] is recovered self-consistently as follows. Without considering vacuum polarization and magnetization, the third equation of Eqs. (83) reduces to

$$\nabla \times B^{(3)} = 0 \qquad (471)$$

because B is phaseless and E is zero by definition. This must mean that there is a balance of terms on the right-hand side, giving

$$\kappa B^{(3)*} = -ig A^{(1)} \times B^{(2)}$$
$$\kappa B^{(3)*} = -ig A^{(2)} \times B^{(1)} \qquad (472)$$

so that

$$\kappa A^{(0)} B^{(3)*} = -i\kappa A^{(1)} \times B^{(2)} \qquad (473)$$

giving the B cyclic theorem self-consistently:

$$\boldsymbol{B}^{(1)} \times \boldsymbol{B}^{(2)} = iB^{(0)} \boldsymbol{B}^{(3)*} \tag{474}$$

The Lehnert charge density for a plane wave propagating in the Z direction is obtained similarly as

$$\rho = \frac{2\kappa^2 A^{(0)}}{\mu_0 c} \tag{475}$$

If a plane wave is now considered propagating in the X direction, the vector in the internal space is defined as

$$\boldsymbol{A} = -\frac{A^{(0)}}{\sqrt{2}} \sin\phi \, \boldsymbol{j} + \frac{A^{(0)}}{\sqrt{2}} \cos\phi \, \boldsymbol{k} + A_X \boldsymbol{i} \tag{476}$$

and it can be shown that the Lehnert vacuum current in the X direction is given self-consistently from Eq. (462) by

$$\boldsymbol{J}_X = 2\frac{\kappa^2 A^{(0)}}{\mu_0} \boldsymbol{i} \tag{477}$$

Finally, if we consider a plane wave propagating in the Y direction, the vector in the internal space is given by

$$\boldsymbol{A} = -\frac{A^{(0)}}{\sqrt{2}} \sin\phi \, \boldsymbol{k} + \frac{A^{(0)}}{\sqrt{2}} \cos\phi \, \boldsymbol{i} + A_Y \boldsymbol{j} \tag{478}$$

and the vacuum current density is given by

$$\boldsymbol{J}_Y = -2\frac{\kappa^2 A^{(0)}}{\mu_0} \boldsymbol{j} \tag{479}$$

Therefore, in order to obtain self-consistent results from Eq. (462), it is necessary to consider plane waves in all three directions. This is as far as an analytical approximation will go. In order to obtain solutions from the field equation (459), computational methods are required.

In summary, the Lehnert current densities in the Z, X, and Y directions, respectively, are

$$\boldsymbol{J}_Z = 2\frac{\kappa^2 A^{(0)}}{\mu_0} \boldsymbol{k}; \qquad \boldsymbol{J}_X = 2\frac{\kappa^2 A^{(0)}}{\mu_0} \boldsymbol{i}; \qquad \boldsymbol{J}_Y = -\frac{\kappa^2 A^{(0)}}{\mu_0} \boldsymbol{j} \tag{480}$$

and are accompanied by a vacuum charge density:

$$\rho = 2\frac{\kappa^2 A^{(0)}}{\mu_0} \tag{481}$$

These results are obtained self-consistently from using plane waves in the X, Y, and X directions.

VIII. EMPIRICAL TESTING OF O(3) ELECTRODYNAMICS: INTERFEROMETRY AND THE AHARONOV–BOHM EFFECT

In order to form a self-consistent description [44] of interferometry and the Aharonov–Bohm effect, the non-Abelian Stokes theorem is required. It is necessary, therefore, to provide a brief description of the non-Abelian Stokes theorem because it generalizes the ordinary Stokes theorem, and is based on the following relation between covariant derivatives for any internal gauge group symmetry:

$$\oint D_\mu \, dx^\mu = -\frac{1}{2}\int [D_\mu, D_\nu] \, d\sigma^{\mu\nu} \tag{482}$$

This expression can be expanded in general notation [46] as

$$\oint (\partial_\mu - igA_\mu) \, dx^\mu = -\frac{1}{2}\int [\partial_\mu - igA_\mu, \partial_\nu - igA_\nu] \, d\sigma^{\mu\nu} \tag{483}$$

where g is a coupling constant, and A_μ is the potential for any gauge group symmetry [44]. The coupling constant in the vacuum is

$$g = \frac{\kappa}{A^{(0)}} \tag{484}$$

as used throughout this review and the review by Evans in Part 1 of this three-volume compilation [44]. The terms

$$\oint \partial_\mu \, dx^\mu = [\partial_\mu, \partial_\nu] = 0 \tag{485}$$

are zero because by symmetry

$$\partial_\nu \partial_\mu = \partial_\mu \partial_\nu \tag{486}$$

M. W. EVANS AND S. JEFFERS

so

$$\oint \partial_\mu \, dx^\mu = -\frac{1}{2} \int [\partial_\mu, \partial_\nu] \, d\sigma^{\mu\nu} = 0 \qquad (487)$$

It can also be shown, as in the earlier part of this review, that

$$[A_\mu, \partial_\nu] = -\partial_\nu A_\mu; \qquad [\partial_\mu, A_\nu] = \partial_\mu A_\nu \qquad (488)$$

Therefore a convenient and general form of the non-Abelian Stokes theorem is

$$\oint A_\mu \, dx^\mu = -\frac{1}{2} \int G_{\mu\nu} \, d\sigma^{\mu\nu} \qquad (489)$$

where the field tensor for any gauge group is

$$G_{\mu\nu} = \partial_\mu A_\nu - \partial_\nu A_\mu - ig[A_\mu, A_\nu] \qquad (490)$$

Equation (489) reduces to the ordinary Stokes theorem when U(1) covariant derivatives are used. First, define the units of the vector potential as

$$A_\mu \equiv (\phi, c\mathbf{A}) \qquad (491)$$

and the units of the U(1) field tensor as

$$F_{\mu\nu} \equiv \begin{bmatrix} 0 & \dfrac{E_1}{c} & \dfrac{E_2}{c} & \dfrac{E_3}{c} \\[2mm] -\dfrac{E_1}{c} & 0 & B_3 & -B_2 \\[2mm] -\dfrac{E_2}{c} & -B_3 & 0 & B_1 \\[2mm] -\dfrac{E_3}{c} & B_2 & -B_1 & 0 \end{bmatrix} \qquad (492)$$

Summing over repeated indices gives the time-like part of the U(1) Stokes theorem:

$$\oint \phi \, dt = \frac{1}{2c^2} \left(\int E_X \, d\sigma^{01} + \int E_Y \, d\sigma^{02} \right) \qquad (493)$$

where the SI units on either side are those of electric field strength multiplied by area. Summing over the space indices gives

$$\oint A_i \, dx^i = -\frac{1}{2} \int F_{ij} \, d\sigma^{ij} \qquad (494)$$

which can be rewritten in Cartesian coordinates as

$$\oint A_X \, dX = \int B_X \, d\sigma^{YZ}$$
$$\oint A_Y \, dY = \int B_Y \, d\sigma^{ZX} \tag{495}$$
$$\oint A_Z \, dZ = \int B_Z \, d\sigma^{XY}$$

or as the vector relation

$$\oint \boldsymbol{A} \cdot d\boldsymbol{r} = \int \boldsymbol{B} \cdot d\boldsymbol{Ar} \tag{496}$$

which is the ordinary Stokes theorem in Maxwell–Heaviside electrodynamics. In the vacuum, $\boldsymbol{A}$ is a plane wave and is perpendicular to the propagation axis, so

$$\oint A_Z \, dZ = 0; \qquad \nabla \times A_Z \boldsymbol{k} = \boldsymbol{0} \tag{497}$$

which is self-consistent with $A_Z = 0$ for Maxwell–Heaviside electrodynamics.

If electrodynamics is a gauge theory with internal O(3) gauge group symmetry, however, there are internal indices and the vector potential becomes

$$\boldsymbol{A}_\mu = A_\mu^{(2)} \boldsymbol{e}^{(1)} + A_\mu^{(1)} \boldsymbol{e}^{(2)} + A_\mu^{(3)} \boldsymbol{e}^{(3)} \tag{498}$$

The field tensor is similarly

$$\boldsymbol{G}_{\mu\nu} = G_{\mu\nu}^{(2)} \boldsymbol{e}^{(1)} + G_{\mu\nu}^{(1)} \boldsymbol{e}^{(2)} + G_{\mu\nu}^{(3)} \boldsymbol{e}^{(3)} \tag{499}$$

where

$$\boldsymbol{e}^{(1)} \times \boldsymbol{e}^{(2)} = i\boldsymbol{e}^{(3)*} \tag{500}$$

$$\cdots$$

In O(3) electrodynamics therefore, Eq. (482) gives a term such as

$$\oint A_3^{(3)} \, dx^3 = -i\frac{g}{2} \left(\int [A_1^{(1)}, A_2^{(2)}] \, d\sigma^{12} + \int [A_2^{(1)}, A_1^{(2)}] \, d\sigma^{21} \right) \tag{501}$$

which reduces to

$$\oint A_Z^{(3)} \, dZ = -ig \int [A_X^{(1)}, A_Y^{(2)}] \, dAr = \int B_Z^{(3)} \, dAr \tag{502}$$

Both $A^{(3)}$ and $B^{(3)}$ are longitudinally directed and are nonzero in the vacuum. Both $A^{(3)}$ and $B^{(3)}$ are phaseless, but propagate with the radiation [47–62] and with their (1) and (2) counterparts. The radiated vector potential $A^{(3)}$ does not give rise to a photon on the low-energy scale, because it has no phase with which to construct annihilation and creation operators. On the high-energy scale, there is a superheavy photon [44] present from electroweak theory with an SU(2)× SU(2) symmetry. The existence of such a superheavy photon has been inferred empirically [44]. However, the radiated vector potential $A^{(3)}$ is not zero in O(3) electrodynamics from first principles, which, as shown in this section, are supported empirically with precision.

On the O(3) level, there are time-like relations such as

$$\oint A_0 \, dx^0 = -\frac{1}{2} \int \partial_0 A_v - \partial_v A_0 - ig[A_0, A_v] \, d\sigma^{(0v)} \tag{503}$$

which define the scalar potential on the O(3) level. The constant $A^{(3)}$ can be expanded in a Fourier series:

$$A_Z = A_Z \left(\frac{\pi^2}{3} - 4\left(\cos\phi - \frac{1}{4}\cos 2\phi + \frac{1}{9}\cos 3\phi + \cdots \right) \right) \tag{504}$$

where α is chosen so that

$$\phi = \omega t - \kappa Z + \alpha \tag{505}$$

is always one radian. So both the scalar and vector potentials in O(3) have internal structure.

The non-Abelian Stokes theorem gives the homogeneous field equation of O(3) electrodynamics, a Jacobi identity in the following integral form:

$$\oint D_\mu \, dx^\mu + \frac{1}{2} \int [D_\mu, D_v] \, d\sigma^{\mu v} = 0 \tag{506}$$

To prove this, we again use

$$\oint D_\mu \, dx^\mu = -\frac{1}{2} \int [D_\mu, D_v] \, d\sigma^{\mu v} \tag{507}$$

to obtain the identity

$$\frac{1}{2} \int \left([D_\mu, D_v] - [D_\mu, D_v] \right) d\sigma^{\mu v} \equiv 0 \tag{508}$$

whose integrand is the identity

$$[D_\mu, D_\nu] - [D_\mu, D_\nu] \equiv 0 \tag{509}$$

From this, we obtain the Jacobi identity

$$\sum_{\sigma,\mu,\nu} [D_\sigma, [D_\mu, D_\nu]] \equiv 0 \tag{510}$$

straightforwardly for all group symmetries, including, of course, O(3). The homogeneous field equation in O(3) can be written in differential form as

$$D_\mu \tilde{G}^{\mu\nu} \equiv \mathbf{0}$$
$$\equiv D^\lambda G^{\mu\nu} + D^\mu G^{\nu\lambda} + D^\nu G^{\lambda\mu} \tag{511}$$

and the equivalent in U(1) electrodynamics in the differential form is

$$\partial_\mu \tilde{F}^{\mu\nu} \equiv 0$$
$$\equiv \partial^\lambda F^{\mu\nu} + \partial^\mu F^{\nu\lambda} + \partial^\nu F^{\lambda\mu} \tag{512}$$

As discussed in the earlier part of this review, Eq. (511) is an identity between generators of the Poincaré group, which differs from the Lorentz group because the former contains the generator of spacetime translations

$$p = i\partial_\mu \tag{513}$$

a group generator that also obeys the Jacobi identity. So we can write

$$\sum_{\sigma,\mu,\nu} [P_\sigma, [D_\mu, D_\nu]] \equiv 0 \tag{514}$$

which is:

$$D_\mu \tilde{G}^{\mu\nu} \equiv \mathbf{0} \tag{515}$$

and it follows that Eq. (515) can be written as

$$\partial_\mu \tilde{G}^{\mu\nu} \equiv \mathbf{0}$$
$$A_\mu \times \tilde{G}^{\mu\nu} \equiv \mathbf{0} \tag{516}$$

The homogeneous field equation (515) of O(3) electrodynamics therefore reduces to

$$\nabla \times \boldsymbol{E}^{(1)} + \frac{\partial \boldsymbol{B}^{(1)}}{\partial t} = \boldsymbol{0}$$

$$\nabla \times \boldsymbol{E}^{(2)} + \frac{\partial \boldsymbol{B}^{(2)}}{\partial t} = \boldsymbol{0} \qquad (517)$$

$$\frac{\partial \boldsymbol{B}^{(3)}}{\partial t} = \boldsymbol{0}$$

Equation (515) can be expanded into the O(3) Gauss and Faraday laws

$$\nabla \cdot \boldsymbol{B}^{(1)*} = ig(\boldsymbol{A}^{(2)} \cdot \boldsymbol{B}^{(3)} - \boldsymbol{B}^{(2)} \cdot \boldsymbol{A}^{(3)}) = 0 \qquad (518)$$

$$\cdots$$

$$\nabla \times \boldsymbol{E}^{(1)*} + \frac{\partial \boldsymbol{B}^{(1)*}}{\partial t} = -ig(cA_0^{(3)}\boldsymbol{B}^{(2)} - cA_0^{(2)}\boldsymbol{B}^{(3)} + \boldsymbol{A}^{(2)} \times \boldsymbol{E}^{(3)} - \boldsymbol{A}^{(3)} \times \boldsymbol{E}^{(2)})$$

$$\cdots \qquad (519)$$

which are homomorphic with the SU(2) invariant Gauss and Faraday laws given by Barrett [50]:

$$\nabla \cdot \boldsymbol{B} = -iq(\boldsymbol{A} \cdot \boldsymbol{B} - \boldsymbol{B} \cdot \boldsymbol{A}) \qquad (520)$$

$$\nabla \times \boldsymbol{B} + \frac{\partial \boldsymbol{B}}{\partial t} = -iq([A_0, \boldsymbol{B}] + \boldsymbol{A} \times \boldsymbol{E} - \boldsymbol{E} \times \boldsymbol{A}) \qquad (521)$$

The vacuum O(3) and SU(2) field equations, on the other hand, are more complicated in structure and highly nonlinear. The O(3) inhomogeneous field equation is given in Eq. (323) and must be solved numerically under all conditions.

These field equations are therefore the result of a non-Abelian Stokes theorem that can also be used to compute the electromagnetic phase in O(3) electrodynamics. It turns out that all interferometric and physical optical effects are described self-consistently on the O(3) level, but not on the U(1) level, a result of major importance. This result means that the O(3) (or SO(3) = SU(2)/Z2) field equations must be accepted as the fundamental equations of electrodynamics.

If we define

$$A^{(0)} \equiv |A_Z^{(3)}|; \qquad g = \frac{\kappa}{A^{(0)}} \qquad (522)$$

then an equation is obtained for optics and interferometry:

$$\oint dZ = \kappa \int dAr \qquad (523)$$

which relates the line integral on the left-hand side to the area integral. Multiplying both sides of Eq. (523) by κ gives a relation between the dynamical phase and topological phase on the right-hand side [44]:

$$\kappa \oint dZ = \kappa^2 \int dAr \qquad (524)$$

Application of an O(3) gauge transform to Eq. (502) results in

$$A_Z^{(3)} \rightarrow A_Z^{(3)} + \frac{1}{g} \partial_Z \Lambda^{(3)}$$
$$B_Z^{(3)} \rightarrow SB_Z^{(3)} S^{-1} \qquad (525)$$

So after gauge transformation

$$\oint \left(A_Z^{(3)} + \frac{1}{g} \partial_Z \Lambda^{(3)} \right) dZ = \int SB_Z^{(3)} S^{-1} dAr \qquad (526)$$

and if $A_Z^{(3)}$ is initially zero (vacuum without the Higgs mechanism), the gauge transform produces the nonzero result:

$$\oint \partial_Z \Lambda^{(3)} dZ = \Delta\Lambda^{(3)} = g \int SB_Z^{(3)} S^{-1} dAr \qquad (527)$$

which is the Aharonov–Bohm effect, developed in more detail later.

The time-like part of the gauge transform gives the frequency shift [44]:

$$\omega \rightarrow \omega + \frac{\partial \Lambda^{(3)}}{\partial t} \equiv \omega + \Omega \qquad (528)$$

The left-hand side of Eq. (523) denotes a round trip or closed loop in Minkowski spacetime [46]. On the U(1) level, this is zero in the vacuum because the line integral

$$\oint dZ = \kappa \int dAr \qquad (529)$$

reduces in U(1) to a line integral of the ordinary Stokes theorem and is zero. In O(3) electrodynamics, Eq. (529) is a line integral over a closed path with O(3) covariant derivatives and is nonzero.

In the Sagnac effect, for example, the closed loop and area can be illustrated as follows:

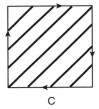   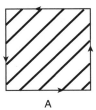

C   A

There is no Sagnac effect in U(1) electrodynamics, as just argued, a result that is obviously contrary to observation [44]. In O(3) electrodynamics, the Sagnac effect with platform at rest is given by the phase factor [44]

$$\exp\left(i\oint_{A-C} \boldsymbol{\kappa}^{(3)} \cdot d\boldsymbol{r}\right) = \exp\left(i\kappa^2 A r\right) \tag{530}$$

because on the O(3) level, there is a component $\boldsymbol{\kappa}^{(3)}$ that is directed in the path $\boldsymbol{r}$. The phase factor (530) gives the interferogram

$$\gamma = \cos\left(2\frac{\omega^2}{c^2} A r \pm 2\pi n\right) \tag{531}$$

as observed. The Sagnac effect with platform in motion is a rotation in the internal gauge space given by Eq. (528), which, when substituted into Eq. (530), gives the observed Sagnac effect to high accuracy:

$$\Delta\gamma = \cos\left(4\frac{\omega\Omega A r}{c^2} \pm 2\pi n\right) \tag{532}$$

The Sagnac effect is therefore due to a gauge transformation and a closed loop in Minkowski spacetime with O(3) covariant derivatives.

If we attempt the same exercise in U(1) electrodynamics, the closed loop gives the Maxwell–Heaviside equations in the vacuum, which are invariant under T and that therefore cannot describe the Sagnac effect [44] because one loop of the Sagnac interferometer is obtained from the other loop by T symmetry. The U(1) phase factor is $\omega t - \kappa Z + \alpha$, where α is arbitrary [44], and this phase factor is also T-invariant. The Maxwell–Heaviside equations in the vacuum are also invariant under rotation, and are metric-invariant, so cannot describe the Sagnac effect with platform in motion.

Physical optics, and interferometry in general, are described by the phase equation of O(3) electrodynamics, Eq. (524). The round trip or closed loop in Minkowski spacetime is illustrated as follows:

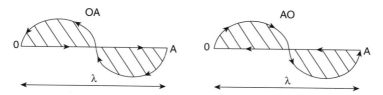

over one wavelength λ of radiation. If $k = \kappa/A^{(0)}$, the area is shown straightforwardly to be

$$Ar = \frac{\lambda^2}{\pi} \tag{533}$$

and if g is proportional to $\kappa/A^{(0)}$, the area is proportional to λ^2/π. Only the Z axis contributes to the left hand side of Eq. (524), which correctly describes all physical optical and interferometric effects. The closed loop is zero in U(1) electrodynamics because the line integral in Eq. (524) is zero from the ordinary Stokes theorem. Therefore Maxwell–Heaviside electrodynamics cannot describe optics and interferometry. The root cause of this failure is that the phase is random on the U(1) level.

The description of Young interferometry for electromagnetism is obtained immediately through the fact that the change in phase difference over trajectories 1 and 2 illustrated below

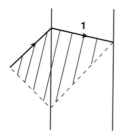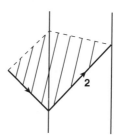

is given by

$$\Delta\delta = \frac{\kappa}{A^{(0)}} \oint_{2-1} \boldsymbol{A}^{(3)} \cdot d\boldsymbol{r} = \kappa\Delta r \tag{534}$$

where $A^{(0)} = |A^{(3)}|$, and where $A^{(3)}$ is directed along the path r in the vacuum. Equation (536) gives the correct result for Young interferometry for vacuum electromagnetism:

$$\Delta\delta = \kappa\Delta r = \frac{2\pi}{\lambda}\Delta r \qquad (535)$$

The change in phase difference of the Young experiment is related through the non-Abelian Stokes theorem to the topological

$$\Delta\delta = g \int B^{(3)} dAr \qquad (536)$$

which is an integral over the $B^{(3)}$ field of O(3) electrodynamics. The Young interferometer can therefore be regarded as a round trip in Minkowski spacetime with O(3) covariant derivatives, as can any type of interferometry or physical optical effect. If an attempt is made to describe the Young interferometer as a round trip with U(1) covariant derivatives, the change in phase difference (534) vanishes because the vector potential in U(1) electrodynamics is a transverse plane wave and is always perpendicular to the path. So on the U(1) level

$$\Delta\delta = 0 \qquad (537)$$

and there is no Young interferometry, contrary to observation. The same result occurs in Michelson interferometry and therefore in ordinary reflection [44].

The O(3) description of the Aharonov–Bohm effect relies on developing the static magnetic field of a solenoid placed between the two apertures of the Young experiment as follows

$$B^{(3)*} = -igA^{(1)} \times A^{(2)} \qquad (538)$$

where

$$A^{(1)} = A^{(2)*} = \frac{A^{(0)}}{\sqrt{2}}(ii + j)e^{i\omega t} \qquad (539)$$

are nonpropagating and transverse. On the O(3) level, the following gauge transformations occur:

$$\begin{aligned} A_\mu^{(1)} &\to A_\mu^{(1)} + \frac{1}{g}\partial_\mu\Lambda^{(1)} \\ A_\mu^{(2)} &\to A_\mu^{(2)} + \frac{1}{g}\partial_\mu\Lambda^{(2)} \end{aligned} \qquad (540)$$

This means that on O(3) gauge transformation

$$A^{(1)} \to A^{(1)} + A^{(1)'}$$
$$A^{(2)} \to A^{(2)} + A^{(2)'} \tag{541}$$

In regions outside the solenoid, the static magnetic field is represented by

$$SB^{(3)}S^{-1} = -ig A^{(1)'} \times A^{(2)'} \tag{542}$$

and is not zero. The Aharonov–Bohm effect is therefore described by

$$\Delta\delta = \frac{e}{\hbar} \int SB^{(3)}S^{-1} \cdot dS \tag{543}$$

as observed [46]. On the U(1) level, the static magnetic field is represented by

$$B = \nabla \times A \tag{544}$$

but in regions outside the solenoid

$$B = \nabla \times \left(\frac{1}{g}\nabla\Lambda\right) = 0 \tag{545}$$

and the magnetic field is zero. So there is no Aharonov–Bohm effect on the U(1) level because $B^{(3)}$ is zero in the integral (543). This has also been pointed out by Barrett [50] with an O(3) invariant electrodynamics.

Therefore, in this section, several effects have been demonstrated to be describable accurately by O(3) electrodynamic and to have no explanation at all in Maxwell–Heaviside electrodynamics. It is safe to infer, therefore, that O(3) electrodynamics must replace U(1) electrodynamics if progress is to be made.

IX. THE DEBATE PAPERS

There has been an unusual amount of debate concerning the development of O(3) electrodynamics, over a period of 7 years. When the $B^{(3)}$ field was first proposed [48], it was not realized that it was part of an O(3) electrodynamics homomorphic with Barrett's SU(2) invariant electrodynamics [50] and therefore had a solid basis in gauge theory. The first debate published [70,79] was between Barron and Evans. The former proposed that $B^{(3)}$ violates C and CPT symmetry. This incorrect assertion was adequately answered by Evans at the time, but it is now clear that if $B^{(3)}$ violated C and CPT, so would classical gauge theory, a reduction to absurdity. For example, Barrett's SU(2) invariant theory [50] would violate C and CPT. The CPT theorem applies only on the quantum level, something that Barron did not seem to realize.

In chronological order, the next critical papers to appear were by Lakhtakia [71] and Grimes [72]. Both papers are obscure, and were adequately answered by Evans [73]. Neither critical paper realized that the $B^{(3)}$ field is part of a classical gauge theory homomorphic with the SU(2) invariant theory by Barrett, published earlier in a volume edited by Lakhtakia [50] himself. This fact reflects the depth of Lakhtakia's confusion. Critical papers were published next by Buckingham and Parlett [74] and by Buckingham [75], essentially duplicating Barron's argument. If these papers were correct, then classical gauge theory would violate CPT and T, a reduction to absurdity. This has been pointed out by Evans [42] and by Evans and Crowell [76]. The next critical paper to appear was by Lakhtakia [77], answered by Evans [78]. Lakhtakia had already published Barrett's SU(2) invariant theory [50] 2 years earlier, so his critical paper is invalidated by the fact that the SU(2) and O(3) invariant theories discussed, for example, in the preceding section, are homomorphic. Then appeared a paper by Rikken [79] answered by Evans [80]. The former claimed erroneously that $B^{(3)}$ is a nonradiated static magnetic field and set about finding it experimentally on this basis. His estimate was orders of magnitude too big, as pointed out by Evans [42] and in the third volume of Ref. 42. The correct use of $B^{(3)}$ gives the empirically observed inverse Faraday effect [42].

These papers were followed by a letter by van Enk [81], answered by Evans [82]. Although not denying the possibility of a $B^{(3)}$, van Enk made the error of arguing on a U(1) level, because, again, he did not realize that $B^{(3)}$ is part of an O(3) invariant electrodynamics and does not exist on the U(1) level. All critical papers cited to this point argued on the U(1) level and are automatically incorrect for this reason. This error was next repeated by Comay [83], who was answered by Evans and Jeffers [84]. Comay attempted to apply the ordinary Abelian Stokes theorem to $B^{(3)}$ and is automatically incorrect because the non-Abelian Stokes theorem should have been applied. The Lorentz covariance of the B cyclic theorem was next challenged by Comay [85], and answered by Evans [86]. The B cyclic theorem is the basic definition of $B^{(3)}$ in an O(3) invariant gauge theory, which is therefore automatically Lorentz covariant, as are all gauge theories for all gauge group symmetries. Comay [87] then challenged the ability of $B^{(3)}$ theory to describe dipole radiation and was answered by Evans [42,88]. It is clear that an O(3) or SU(2) invariant electrodynamics can produce multipole radiation of many types. These comments by Comay are therefore trivially incorrect, not least because they argue again on the U(1) level.

Two papers by Raja et al. [89,90] erroneously claimed once more that $B^{(3)}$ is a static magnetic field and should have produced Faraday induction vacuo. These papers were answered by Evans [91,92]. In the O(3) invariant electrodynamics defining $\boldsymbol{B}^{(3)}$, the latter is a radiated, phaseless, field, and does not produce Faraday induction.

Independent confirmation of the invariance of the B cyclic theorem was next produced by Dvoeglazov [93], but he did not argue on the O(3) level as required. His argument is therefore only partially valid, but produces the correct result.

Comay [94] then repeated the earlier arguments [69,74] on C and CPT violation and was answered by Evans and Crowell [76], who showed that all gauge theories trivially conserve CPT and C on the quantum level. Comay again made the error of arguing on the U(1) and classical levels, whereas $\boldsymbol{B}^{(3)}$ exists only on the O(3) level and the CPT theorem exists only on the quantum level. The argument by Comay using the Stokes theorem [83] was next duplicated by Hunter [95], who again argued erroneously on the U(1) level. The reply to Hunter [96] pointed this out. Next in chronological order, Hunter again duplicated Comay's argument [97] and was again replied to by Evans [98], on the correct O(3) level. Additionally, Comay and Dvoeglazov [99,100] have argued erroneously on the U(1) level concerning the Lorentz covariance of the B cyclic theorem, something that follows trivially from the O(3) gauge invariance of the gauge theory that defines $\boldsymbol{B}^{(3)}$.

The preceding section, and a review in Part 1 of this compilation, supply copious empirical evidences of the fact that the $\boldsymbol{B}^{(3)}$ field is part of the topological phase that describes interferometry through a non-Abelian Stokes theorem. Therefore, the early critical papers are erroneous because they argue on a U(1) level.

X. THE PHASE FACTOR FOR O(3) ELECTRODYNAMICS

The phase factor in classical electrodynamics is the starting point for quantization in terms of creation and annihilation operators, and so it is important to establish its properties on the classical O(3) level. In this context, Barrett [50] has provided a useful review of the development of the phase factor, and Simon [101] has shown that the phase factor is in general due to parallel transport in the presence of a gauge field. On the O(3) level, therefore, the phase factor must be due to parallel transport around a closed loop in Minkowski spacetime (a holonomy) with O(3) covariant derivatives and is governed by the non-Abelian Stokes theorem, Eq. (482). This inference means that all phases in O(3) electrodynamics have their origin in topology on the classical level. This inference is another step in the evolution of understanding of topological phase effects. As pointed out by Barrett [50], the origin of such effects was the development of the Dirac phase factor by Wu and Yang [102], who argued that the wave function of a system will be multiplied by a path-dependent phase factor after its transport around a closed curve in the presence of a potential in ordinary space. This process is now understood to be the origin of the non-Abelian Stokes theorem (482) and to explain the Aharonov–Bohm effect. The phases proposed by Berry [103], Aharonov and Anandan [104], and Pancharatnam [105] are due to a closed loop in parameter or momentum space. These effects occur both on the classical and quantum levels [50].

Originally, Berry [103] proposed a geometric phase for a nondegenerate quantum state that varied adiabatically over a closed loop in parameter space. This occurred in addition to the dynamical phase. It was shown later [50] that the effect is present without the need for an adiabatic approximation, and is also present for degenerate states. Aharonov and Anandan [104] showed that the effect is present for any cyclic evolution of a quantum system, and Bhandari and Samuel [106] showed that the effect is closely related to the geometrical phase discovered by Pancharatnam [105]. The topological phase, therefore, has its origin in topology, either on the classical or quantum level, and is equivalent to a gauge potential in the parameter space of the system on the classical or quantum level.

There are at least three variations of topological phases [50]:

1. A phase arising from cycling in the direction of a beam of light
2. The Pancharatnam phase from cycling of polarization states while keeping the direction of the beam of light constant, a phase change due to polarization change
3. The phase change due to a cycle of changes in squeezed states of light

If the topological phase is denoted Φ, then it obeys the conservation law

$$\Phi(C) = -g \oint A \cdot dr \qquad (546)$$

and occurs on the classical level from polarization changes due to changes in the topological path of a light beam. The angle of rotation of linearly polarized light is a direct measure of the topological phase at the classical level. An example of this is the Sagnac effect, which can be explained using O(3) as discussed already. The Sagnac effect can be considered as one loop in the Tomita–Chiao effect [107], which is the rotation of the plane of polarization of a light beam when propagating through an optical fiber.

The next level in the evolution of understanding of the electromagnetic phase is to consider that all optical phases are derived from the non-Abelian Stokes theorem (482), so all optical phases originate in the phase factor

$$\gamma = \exp\left(ig \oint D_\mu \, dx^\mu\right) = \exp\left(-\frac{1}{2} ig \int [D_\mu, D_\nu] \, d\sigma^{\mu\nu}\right) \qquad (547)$$

which originates directly in the non-Abelian Stokes theorem (482). Therefore, on the O(3) level, all optical phases are topological in origin. We have briefly discussed how the phase factor reduces to a line integral over the dynamical phase and this property of Eq. (547) is also reviewed in Part 1 by Evans [44]. It has been argued that the most general equation (547) reduces to

$$\gamma = \exp\left(ig \oint A_\mu \, dx^\mu \right) = \exp\left(-i\frac{g}{2} \int G_{\mu\nu} \, d\sigma^{\mu\nu} \right) \qquad (548)$$

for a round trip in Minkowski spacetime for all internal gauge group symmetries. The notation used in Eq. (548) is the condensed notation used by Ryder [46], in which the field tensor is in general defined by

$$G_{\mu\nu} = \partial_\mu A_\nu - \partial_\nu A_\mu - ig[A_\mu, A_\nu] \qquad (549)$$

In free space, as argued already, the factor g is $\kappa/A^{(0)}$.

If we attempt to apply Eq. (548) on the U(1) level, relations such as

$$\gamma = \exp\left(ig \oint \mathbf{A} \cdot d\mathbf{r} \right) = \exp\left(ig \int \mathbf{B} \cdot d\mathbf{Ar} \right) \qquad (550)$$

are obtained. In free space, on the U(1) level, $\mathbf{A}$ is, however, a plane wave, and is therefore always perpendicular to the path $\mathbf{r}$ of the radiation. Therefore, on the U(1) level in free space

$$\oint \mathbf{A} \cdot d\mathbf{r} = \int \mathbf{B} \cdot d\mathbf{Ar} = 0 \qquad (551)$$

On the O(3) level in free space, however, relations such as

$$\gamma = \exp\left(ig \oint \mathbf{A}^{(3)} \cdot d\mathbf{r} \right) = \exp\left(ig \int \mathbf{B}^{(3)} \cdot d\mathbf{Ar} \right) \qquad (552)$$

are obtained, where $\mathbf{A}^{(3)}$ is parallel to the path of the radiation. Using $g = \kappa/A^{(0)}$ in free space, Eq. (552) reduces to

$$\gamma = \exp\left(i \oint \kappa^{(3)} \cdot d\mathbf{r} \right) = \exp\left(ig \int \mathbf{B}^{(3)} \cdot d\mathbf{Ar} \right) \qquad (553)$$

and the left-hand side can be recognized as a line integral over what is usually termed the *dynamical phase*. By definition, the line integral changes sign on traversing a closed loop from O to A to A to O, and this fundamental mathematical property is responsible for all optics and interferometry as argued in this review and in Ref. 44. This inference is an evolution in understanding of the phase in optics and electrodynamics.

The $B^{(3)}$ field appearing on the right-hand side of the non-Abelian Stokes theorem (553) changes sign [47–62] between left- and right-handed circularly polarized states, and a linearly polarized state is a superposition of two circularly polarized states. This inference gives rise to Pancharatnam's phase, which is due to polarization changes and also to the phase caused by the cycling of the tip of the vector in a circularly polarized electromagnetic field. Therefore, we reach the important conclusion that the $B^{(3)}$ field is an observable of the phase in all optics and electrodynamics. It has been argued briefly in this review and in Part 1 of this series [44] that the $B^{(3)}$ field provides an explanation of the Sagnac effect.

The U(1) phase factor in the received view, on the other hand, is well known to be

$$\gamma = \exp\left(i(\omega t - \mathbf{\kappa} \cdot \mathbf{r} + \alpha)\right) \tag{554}$$

where α is an arbitrary number. So the phase factor (γ) is defined only up to an arbitrary α, an unphysical result. If $\alpha = 0$ for the sake of argument, the phase factor (γ) is invariant under motion reversal symmetry (T) and parity inversion symmetry (P) [44]. Since one loop of the Sagnac effect is generated from the other by T, it follows that the received phase factor (γ) is invariant in the Sagnac effect with platform at rest and there is no phase shift, contrary to observation [44]. The phase factor (553), on the other hand, changes sign under T and produces the observed Sagnac effect. The phase factor (554) is invariant under P and cannot explain Michelson interferometry or normal reflection [44]. The phase factor (553) changes sign under P and explains Michelson interferometry as observed [44]. We have argued earlier in this review that the phase factor (553) also explains Young interferometry straightforwardly.

Therefore, the distinction between the topological and dynamical phase has vanished, and the realization has been reached that the phase in optics and electrodynamics is a line integral, related to an area integral over $B^{(3)}$ by a non-Abelian Stokes theorem, Eq. (553), applied with O(3) symmetry-covariant derivatives. It is essential to understand that a non-Abelian Stokes theorem must be applied, as in Eq. (553), and not the ordinary Stokes theorem. We have also argued, earlier, how the non-Abelian Stokes explains the Aharonov-Bohm effect without difficulty.

We also infer that, in the vacuum, there exists the topological charge

$$g_m = \frac{1}{V} \oint A_\mu \, dx^\mu \tag{555}$$

where V is a volume, and for one photon, the quantum of electromagnetic energy, the phase becomes

$$\phi = g \oint A^{(3)} \cdot dr = g \int B^{(3)} \cdot dAr = \pm 1 \tag{556}$$

where $g = \kappa/A^{(0)}$. The flux due to one photon is classically

$$\int B^{(3)} \cdot dAr = \frac{A^{(0)}}{\kappa} = \frac{\hbar}{e} \tag{557}$$

and so we have the quantum classical equivalence

$$eA^{(0)} = \hbar\kappa \tag{558}$$

which is a Planck quantization. In quantum theory, the magnetic flux of one photon is $\pm\hbar/e$, depending on the sense of circular polarization.

It can be shown that the Sagnac effect with platform at rest is the rotation of the plane of linearly polarized light as a result of radiation propagating around a circle in free space. Such an effect cannot exist in the received view where the phase factor in such a round trip is always the same and given by Eq. (554). However, it can be shown as follows that there develops a rotation in the plane of polarization when the phase is defined by Eq. (553). It is now known that the phase must always be defined by Eq. (553). Therefore, proceeding on this inference, we construct plane polarized light as the sum of left and right circularly polarized components:

$$\text{Re}(i - ij)e^{i\phi} = \cos\phi \, i + \sin\phi \, j \tag{559}$$

$$\text{Re}(i - ij)e^{-i\phi} = \cos\phi \, i - \sin\phi \, j \tag{560}$$

where the phase factor $e^{i\phi}$ is given by Eq. (553). Plane-polarized light at the beginning of the 180° round trip of the Sagnac effect is therefore

$$(i - ij)(e^{i\phi} + e^{-i\phi}) = 2i\cos\phi \tag{561}$$

The round trip of the Sagnac effect in a given—say, clockwise—direction produces the effect

$$(i - ij)e^{i(\phi+\phi_S)} + (i - ij)e^{-i(\phi-\phi_S)} \tag{562}$$

where

$$\phi_S = g \oint A^{(3)} \cdot dr = g \int B^{(3)} \cdot dAr \tag{563}$$

is generated by the round trip over 2π radians. The extra phase factor for the left circularly polarized component is ϕ_s, and the extra phase factor for the right circularly polarized component is $-\phi_s$ because $B^{(3)}$ changes sign between senses of circular polarization. The effect of the round trip in the Sagnac effect on the plane of linearly polarized light is therefore

$$(\cos(\phi + \phi_S) + \cos(\phi - \phi_S))i + (\sin(\phi + \phi_S) + \sin(\phi - \phi_S))j \qquad (564)$$

Using the angle formulas

$$\cos(A \pm B) = \cos A \cos B \mp \sin A \sin B$$
$$\sin(A \pm B) = \sin A \cos B \pm \cos A \sin B \qquad (565)$$

the effect can be expressed as

$$2\cos\phi(i\cos\phi_S - j\sin\phi_S) \qquad (566)$$

The original plane-polarized light at the beginning of the round trip is described by

$$2\cos\phi i \qquad (567)$$

so the overall effect is to rotate the plane of polarized light. Therefore, a linearly polarized laser beam sent around an optical fiber in a circle arrives back at the origin with its plane rotated as in Eq. (566). This is a description of the Sagnac effect with the platform at rest. Spinning the platform produces an extra phase shift that is described [44] by a gauge transformation of $A^{(3)}$ [a rotation in the physical O(3) internal space]. This extra phase shift produces an extra rotation in the plane of polarization of linearly polarized light.

Therefore, it becomes clear that the Sagnac effect is one loop of the Tomita–Chiao effect [107], which is the rotation of the plane of a linearly polarized light beam sent through a helical optical fiber. In both the Sagnac and Tomita–Chiao effects, the angle of rotation (or phase shift) is a direct measure of the phase factor (γ), whose origin is in topology. A circle can always be drawn out into a helix of given pitch (p), length (s), and radius (r). This can be seen by straightening out the helix into a line, and bending the line into a circle. So the Tomita–Chiao effect must reduce to the Sagnac effect for this reason. The former effect can be expressed in general as

$$\phi = 2\pi\left(1 - \frac{p}{S}\right)g\int B^{(3)} \cdot dAr \qquad (568)$$

because for one photon

$$g \int \mathbf{B}^{(3)} \cdot d\mathbf{Ar} = \pm 1 \tag{569}$$

Therefore, the Tomita–Chiao effect reduces to the Sagnac effect under the condition

$$2\pi \left(1 - \frac{p}{S}\right) = 1 \tag{570}$$

that is

$$\frac{p}{S} = 1 - \frac{1}{2\pi} \tag{571}$$

or when the pitch : length ratio of the helix is this number, which is self-consistently less than one (the length s is always greater than the pitch p).

The received view, in which the phase factor of optics and electrodynamics is given by Eq. (554), can describe neither the Sagnac nor the Tomita–Chiao effects, which, as we have argued, are the same effects, differing only by geometry. Both are non-Abelian, and both depend on a round trip in Minkowski spacetime using O(3) covariant derivatives.

Having argued thus far, it becomes clear that the phase factor (553) can be generalized and put on a rigorous footing in topology [50]. It is precisely obtained from a set of angles associated with a group element, and only one such angle can correspond to a holonomy transformation of a vector bundle around a closed curve on a sphere. For example, in a SU(2) invariant electrodynamics, there is a single angle from the holonomy of the Riemannian connection on a sphere. Thus, we infer that gauge structure appears at a very fundamental level in all optical effects that depend on the electrodynamical phase. We can also infer new effects, for example, if the helix of the Tomita–Chiao experiment is spun, an effect equivalent to the Sagnac effect should be observable. The general conclusion is that all electrodynamical phases are non-Abelian, and quantization proceeds naturally on this basis. For example, Berry's phase was first inferred in quantum mechanics. We can conclude that all phases are topological.

The properties of the phase factor (548) on O(3) gauge transformation have been shown [47] to explain the Sagnac effect with platform in motion. In condensed notation, gauge transformation produces the results

$$\begin{aligned} A'_\mu &= SA_\mu S^{-1} - \frac{i}{g}(\partial_\mu S)S^{-1} \\ G'_{\mu\nu} &= SG_{\mu\nu}S^{-1} \end{aligned} \tag{572}$$

where S is defined by

$$S = \exp(iM^a\Lambda^a(x^\mu)) \qquad (573)$$

In the O(3) gauge group, M^a are rotation generators, and Λ^a are angles in three-dimensional space, which coincides with the internal gauge space. Rotation about the Z axis leaves the $B^{(3)}$ field unaffected. In matrix notation, this can be demonstrated by

$$\begin{bmatrix} 0 & -B_Z & 0 \\ B_Z & 0 & 0 \\ 0 & 0 & 0 \end{bmatrix} = \begin{bmatrix} \cos\alpha & \sin\alpha & 0 \\ -\sin\alpha & \cos\alpha & 0 \\ 0 & 0 & 1 \end{bmatrix} \begin{bmatrix} 0 & -B_Z & 0 \\ B_Z & 0 & 0 \\ 0 & 0 & 0 \end{bmatrix} \begin{bmatrix} \cos\alpha & -\sin\alpha & 0 \\ \sin\alpha & \cos\alpha & 0 \\ 0 & 0 & 1 \end{bmatrix}$$

$$(574)$$

The gauge transformation of A_Z has been shown [44] to be given by

$$A_Z \rightarrow A_Z + \frac{1}{g}\partial_Z\alpha \qquad (575)$$

Therefore, the phase factor on O(3) gauge transformation becomes

$$\exp\left(ig\oint(A^{(3)} + \nabla\alpha)\cdot dr\right) = \exp\left(ig\int B^{(3)}\cdot dAr\right) \qquad (576)$$

and using the property

$$\oint\nabla\alpha\cdot dr = 0, \qquad \text{i.e.,} \quad \nabla\times(\nabla\alpha) = 0 \qquad (577)$$

it is seen that the phase factor is invariant under an O(3) gauge transformation. The phase factor, however, contains only the space part of the complete expression (548). Gauge transformation of the time part gives the result [44]

$$\omega \rightarrow \omega \pm \Omega; \qquad \Omega = \frac{\partial\alpha}{\partial t} \qquad (578)$$

which explains the Sagnac effect with platform in motion.

On the U(1) level, the ordinary Stokes theorem applies, and this can be written as

$$\oint A\cdot dr = \int \nabla\times A\cdot dAr \qquad (579)$$

which is gauge-invariant because of the property

$$\oint \nabla \chi \cdot d\boldsymbol{r} = 0 \tag{580}$$

which is equivalent to the fundamental vector property:

$$\nabla \times (\nabla \chi) = \boldsymbol{0} \tag{581}$$

However, as argued, A is always perpendicular to the path $\boldsymbol{r}$ on the U(1) level, and so the phase factor (548) cannot be applied on this level.

Barrett [50] has interestingly reviewed and compared the properties of the Abelian and non-Abelian Stokes theorems, a review and comparison that makes it clear that the Abelian and non-Abelian Stokes theorems must not be confused [83,95]. The Abelian, or original, Stokes theorem states that if $A(x)$ is a vector field, S is an open, orientable surface, C is the closed curve bounding S, dl is a line element of C, n is the normal to S, and C is traversed in a right-handed (positive direction) relative to n, then the line integral of A is equal to the surface integral over S of $\nabla \times A \cdot \boldsymbol{n}$:

$$\oint A \cdot dl = \int_S (\nabla \times A) \cdot \boldsymbol{n} \, da \tag{582}$$

and, as pointed out by Barrett [50], the original Stokes theorem just described takes no account of boundary conditions.

In the non-Abelian Stokes theorem (482), on the other hand, the boundary conditions are defined because the phase factor is path-dependent, that is, depends on the covariant derivative [50]. On the U(1) level [50], the original Stokes theorem is a mathematical relation between a vector field and its curl. In O(3) or SU(2) invariant electromagnetism, the non-Abelian Stokes theorem gives the phase change due to a rotation in the internal space. This phase change appears as the integrals

$$\oint A^{(3)} \cdot d\boldsymbol{r} = \int B^{(3)} \cdot d\boldsymbol{A}\boldsymbol{r} \tag{583}$$

which do not exist in Maxwell–Heaviside electromagnetism. There is a profound ontological difference therefore between the original Stokes theorem, in which $B^{(3)}$ is zero, and the non-Abelian Stokes theorem, in which $B^{(3)}$ is nonzero and of key importance. Therefore progress from a U(1) to an O(3) or SU(2) invariant electromagnetism is a striking evolution in understanding, as argued throughout Ref. 44 and references cited therein and in several reviews of this volume.

Equation (482) is a simple form of the non-Abelian Stokes theorem, a form that is derived by a round trip in Minkowski spacetime [46]. It has been adapted directly for the O(3) invariant phase factor as in Eq. (547), which gives a simple and accurate description of the Sagnac effect [44]. A U(1) invariant electrodynamics has failed to describe the Sagnac effect for nearly 90 years, and kinematic explanations are also unsatisfactory [50]. In an O(3) or SU(2) invariant electrodynamics, the Sagnac effect is simply a round trip in Minkowski spacetime and an effect of special relativity and gauge theory, the most successful theory of the late twentieth century. There are open questions in special relativity [108], but no theory has yet evolved to replace it.

By using the O(3) invariant phase factor (547), we have also removed the distinction between the topological phase and the dynamical phase, reaching, as argued earlier, a new level of understanding in all optical effects that depend on electromagnetic phase.

For example, the description of the Aharonov–Bohm effect and other types of interferometry become closely similar. The Young interferometer, for example, is described by

$$\frac{\kappa}{A^{(0)}} \oint_{2-1} A^{(3)} \cdot dr = \frac{\kappa}{A^{(0)}} \int B^{(3)} \cdot dS \tag{584}$$

and the Aharonov–Bohm effect can be described by

$$\frac{e}{\hbar} \oint_{2-1} A^{(3)} \cdot dr = \frac{e}{\hbar} \int B^{(3)} \cdot dS = \frac{e}{\hbar} \Phi^{(3)} \tag{585}$$

In both cases, the magnetic flux

$$\Phi^{(3)} = \int B^{(3)} \cdot dS \tag{586}$$

is generated by the round trip in Minkowski space with O(3) covariant derivatives (holonomy) on the left-hand side of Eqs. (584) and (585). So the original magnetic field inside the solenoid does not contribute to the Aharonov–Bohm effect, as pointed out by Barrett [50], and the U(1) invariant description [46] of the effect is erroneous. The effect is due to the magnetic field $B^{(3)}$ of O(3) electrodynamics. The Sagnac, Michelson, and Mach–Zehnder effects, and all interferometric effects are similarly described by Eq. (584), and all interferometry and optics originate in topology. The only difference between these effects and the Aharonov–Bohm effect is that in the latter, interaction with electrons takes place, so the factor $\kappa/A^{(0)}$ is replaced by $e/\hbar$ in a minimal prescription.

The interpretation of Eq. (584) is that the potential $A^{(3)}$ is defined along the integration path of the line integral. The field $B^{(3)}$ is defined as being perpendicular to the plane or surface enclosed by the line integral. Neither $A^{(3)}$ nor $B^{(3)}$ exists in a U(1) invariant electrodynamics. Effects attributed to the topological phase, such as those of Pancharatnam and Tomita and Chiao, reviewed already, do not exist in a U(1) invariant electrodynamics, but are described by Eq. (584) in an O(3) invariant theory. Equation (3) is for circularly polarized radiation propagating in a plane, and so allowance may have to be made for the geometry of a particular experiment. We have illustrated this with the Tomita–Chiao effect. The key to this evolution in understanding is that there exists in an O(3) invariant electrodynamics, an internal gauge space with index (3). The existence of this index gives rise to the non-Abelian Stokes theorem (584). The internal space on a ((1),(2),(3)) level is considered to be the physical space of three dimensions and not an isospace. Therefore, a rotation in the internal space ((1),(2),(3)) is a physical rotation in three-dimensional space. The spinning platform of the Sagnac effect is an example of one such rotation, about the axis perpendicular to the platform, and results in Eq. (578), which, as shown elsewhere [44], gives the observed Sagnac effect, again through Eq. (584). Such concepts are available in neither a U(1) invariant electrodynamics nor gauge theory, which considers the internal space as an isospace.

Therefore, it has been shown convincingly that electrodynamics is an O(3) invariant theory, and so the O(3) gauge invariance must also be found in experiments with matter waves, such as matter waves from electrons, in which there is no electromagnetic potential. One such experiment is the Sagnac effect with electrons, which was reviewed in Ref. 44, and another is Young interferometry with electron waves. For both experiments, Eq. (584) becomes

$$\oint \boldsymbol{\kappa}^{(3)} \cdot d\boldsymbol{r} = \kappa^2 Ar \tag{587}$$

and for matter waves

$$\omega^2 = c^2\kappa^2 + \frac{m_0^2 c^4}{\hbar^2} \tag{588}$$

where m_0 is the mass of the particle. The Sagnac effect in electrons [44] is therefore the same as the Sagnac effect in photons, and is given [44] by

$$\Delta\phi = \frac{Ar}{c^2}\left(((\omega+\Omega)^2 - (\omega-\Omega)^2) - \frac{m_0^2 c^4}{\hbar^2} + \frac{m_0^2 c^4}{\hbar^2}\right)$$
$$= \frac{4\omega\Omega Ar}{c^2} \tag{589}$$

from the gauge transform (578). This is the observed result [44]. The Young effect for electrons is similarly

$$\Delta\phi = \oint_{2-1} \boldsymbol{\kappa}^{(3)} \cdot d\boldsymbol{r} \tag{590}$$

and also more generally for particles such as atoms and molecules, the famous two-slit experiment.

On this empirical evidence, it is possible to reach a far-reaching conclusion that all wave functions in quantum mechanics are of the form (590). For example, the electron wave function from the Dirac equation is

$$\text{Positive energy: } \psi^{(\alpha)}(r) = u^{(\alpha)}(p) \exp\left(-i \oint \boldsymbol{p} \cdot d\boldsymbol{r}\right) \tag{591}$$

$$\text{Negative energy: } \psi^{(\alpha)}(r) = v^{(\alpha)}(p) \exp\left(i \oint \boldsymbol{p} \cdot d\boldsymbol{r}\right) \tag{592}$$

instead of the conventional [46]

$$\psi^{(\alpha)}(r) = u^{(\alpha)}(p) \exp\left(-i\boldsymbol{p} \cdot \boldsymbol{r}\right) \tag{593}$$

$$\psi^{(\alpha)}(r) = v^{(\alpha)}(p) \exp\left(i\boldsymbol{p} \cdot \boldsymbol{r}\right) \tag{594}$$

The path and area in Eq. (584) and in wave functions such as those of the photon and electron are given by the following sketch:

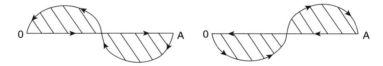

The shaded area in this sketch is not arbitrary, as it is determined by the right-hand side of Eq. (587). The line integrals OA and AO change sign, and this accounts for reflection of matter waves and for the Sagnac and Young effects in matter waves, such as electron waves. Therefore, the electron is an O(3) invariant entity, as shown by the Sagnac effect for electron waves [44]. It follows that the Dirac equation should be developed as an O(3) invariant equation.

The Fermat principle can now be reworked into an O(3) invariant form and the principles of quantum mechanics on a nonrelativistic level developed from it. In so doing, we modify the discussion by Atkins [68] for an O(3) invariant treatment. Fermat's principle of least time is the basic law governing light propagation in geometric optics. In the received view, light travels in a straight line in geometric optics, but the physical nature of light is a wave motion. These two fundamental aspects are unified in the sketch following Eq. (594), constructed in an O(3) invariant theory, in which the phase now correctly describes both the wave nature of light and the fact that it travels in a straight line in the vacuum or a uniform medium. The U(1) invariant phase shows only the latter property of light, and consequently is a number invariant under motion reversal symmetry

(*T*) and parity inversion symmetry (*P*). Similarly, particles travel in a straight line by Newton's first law, but de Broglie demonstrated that particles have a wave nature–wave particle duality. Therefore, the phase in classical electrodynamics becomes the wave function in quantum mechanics, and the general and important conclusion is reached that both the electromagnetic phase and the wave function of particles are O(3) invariant. We have already argued that this new general principle is supported by the Sagnac and Young effects in matter waves. In retrospect, it is not surprising that the wave function should reflect wave–particle duality, for both the photon and matter waves.

A simple example of the Fermat principle may be used to show the weakness inherent in a U(1) invariant phase. Fermat's principle states that the path taken by a light ray through a medium is such that its time of passage is a minimum. Following Atkins [68], consider the relation between angles of incidence and reflection. The least-time path is the one corresponding to the angle of incidence being equal to the angle of reflection, giving Snell's law. However, reflection is a parity inversion, under which the U(1) invariant phase

$$P(\omega t - \boldsymbol{\kappa} \cdot \boldsymbol{r}) = \omega t - \boldsymbol{\kappa} \cdot \boldsymbol{r} \tag{595}$$

does not change [44]. This is seen at its clearest in normal reflection. Therefore, the U(1) invariant phase cannot describe normal reflection and Snell's law, and violates Fermat's principle. The O(3) invariant phase

$$\phi = \oint \omega \, dt - \oint \boldsymbol{\kappa} \cdot d\boldsymbol{r} \tag{596}$$

on the other hand, changes sign on reflection, because of the property of the path integral

$$P\left(\oint \boldsymbol{\kappa} \cdot d\boldsymbol{r}\right) = -\oint \boldsymbol{\kappa} \cdot d\boldsymbol{r} \tag{597}$$

and so is in accordance with the Fermat principle. This conclusion is a major evolution in understanding because Fermat's principle is at the root of quantum mechanics, in particular, the time-dependent Schrödinger equation.

Following Atkins [68], the propagation of particles follows a path dictated by Newton's laws, equivalent to Hamilton's principle, that particles select paths between two points such that the action associated with the path is a minimum. Therefore, Fermat's principle for light propagation is Hamilton's principle for particles. The formal definition of action is an integral identical in structure with the phase length in physical optics. Therefore, particles are associated with wave motion, the wave–particle dualism. Hamilton's principle of least action leads directly to quantum mechanics. The final touch to this development was made by de Broglie. Therefore, a particle is also described by an amplitude $\psi(r)$, and amplitudes at different points are related by an expression of the following form [68]:

$$\psi(P_2) = e^{i(r_2 - r_1) \cdot \kappa} \psi(P_1) \tag{598}$$

If this is to be O(3) invariant, the phase in quantum mechanics must take the form (597). In the classical limit, the particle propagates along a path that makes the action S a minimum. Therefore, the O(3) invariant phase is proportional to S through the Planck constant. It is concluded that the O(3) invariant phase in quantum mechanics is given by

$$\phi = \oint \kappa \cdot dr \tag{599}$$

The amplitude describing a particle in O(3) invariant quantum mechanics is

$$\psi = \psi_0 \exp^{-i\phi} = \psi_0 \exp^{-i(S/\hbar)} \tag{600}$$

where S is the action associated with the path from P_1 (a point at x_1, t_1) to P_2 (a point at x_2, t_2). An equation of motion can be developed from this form by differentiating with respect to time t_2:

$$\frac{\partial}{\partial t} \psi(x, t) = -\frac{i}{\hbar} En \, \psi(x, t) \tag{601}$$

The rate of change of the action is equal to $-En$, where En is the total energy $T + V$:

$$\frac{\partial S}{\partial t} = -En \tag{602}$$

Therefore, the equation of motion is

$$\frac{\partial}{\partial t} \psi(x, t) = \frac{i}{\hbar} \frac{\partial S}{\partial t} \psi(x, t) \tag{603}$$

and if En is interpreted as the Hamiltonian operator H, the O(3) invariant time-dependent Schrödinger equation is obtained:

$$H\psi = i\hbar \frac{\partial \psi}{\partial t} \tag{604}$$

So, if the O(3) invariant wave function is defined as

$$\psi = \psi_0 \exp\left(-i \oint \omega \, dt - \oint \boldsymbol{\kappa} \cdot d\boldsymbol{r}\right) \qquad (605)$$

where

$$S = -\hbar\left(\oint \omega \, dt - \oint \boldsymbol{\kappa} \cdot d\boldsymbol{r}\right) \qquad (606)$$

the energy is given by

$$En = \hbar\omega = -\frac{\partial S}{\partial t} \qquad (607)$$

which is the energy for one photon. Equation (605) is the O(3) invariant de Broglie wave function.

XI.   O(3) INVARIANCE: A LINK BETWEEN ELECTROMAGNETISM AND GENERAL RELATIVITY

In order to develop a Riemannian theory of classical electromagnetism, it is necessary [109] to consider a curve corresponding to a plane wave:

$$f(Z) = (\boldsymbol{i} - i\boldsymbol{j})e^{i\phi} \qquad (608)$$

In terms of the retarded time $[t] = t - Z/c$, the U(1) phase ϕ is $\omega[t]$, and the retarded distance is $Z - Z_0 = c[t]$. The electromagnetic wave propagates along the Z axis, and the trajectory of the real part is

$$f_R(Z) = \text{Re}(f(Z)) = (\cos\phi, \sin\phi, \phi) \qquad (609)$$

which is a circular helix. The curve (609) is a function of Z with Z_0 regarded as a constant in partial differentiation of $f(Z)$ with respect to Z. More generally, a Z-dependent phase angle must be incorporated in ϕ, which becomes [42]:

$$f_R(Z) = (\cos(\kappa(Z - Z_0) + \Phi), \sin(\kappa(Z - Z_0) + \Phi), \kappa(Z - Z_0) + \Phi) \qquad (610)$$

Frenet's tangent vector (T) is obtained by differentiation:

$$\frac{\partial f_R(Z)}{\partial Z} = \kappa T = (-\kappa \sin\phi, \kappa \cos\phi, \kappa) \qquad (611)$$

In elementary differential geometry, therefore, the electromagnetic helix produces a nonzero T, and tangent vectors are characteristic of curved spacetime in general relativity. The scalar curvature in elementary differential geometry is

$$R = \left| \frac{\partial^2 f_R(Z)}{\partial Z^2} \right| = \left| \kappa^2 (\cos \phi, -\sin \phi, 0) \right| = \kappa^2 \tag{612}$$

and this is also the scalar curvature of the electromagnetic wave in general relativity, specifically, the scalar curvature of Riemann's tensor, obtained by suitable antisymmetric index contraction. The electromagnetic field therefore becomes a property of spacetime, or the vacuum.

The metric coefficient in the theory of gravitation [110] is locally diagonal, but in order to develop a metric for vacuum electromagnetism, the antisymmetry of the field must be considered. The electromagnetic field tensor on the U(1) level is an angular momentum tensor in four dimensions, made up of rotation and boost generators of the Poincaré group. An ordinary axial vector in three-dimensional space can always be expressed as the sum of cross-products of unit vectors

$$\boldsymbol{I} = \boldsymbol{i} \times \boldsymbol{j} + \boldsymbol{j} \times \boldsymbol{k} + \boldsymbol{k} \times \boldsymbol{i} \tag{613}$$

a sum that can be expressed as the metric

$$g = g_{\mu\nu}^{(A)} i^\mu j^\nu \tag{614}$$

where the $g_{\mu\nu}^{(A)}$ coefficient in three dimensions is the fully antisymmetric 3×3 matrix. This becomes the right-hand side in four dimensions. In the language of differential geometry, the field tensor becomes the Faraday 2-form [110]

$$F = \frac{1}{2} F_{\alpha\beta} \, dx^\alpha \wedge dx^\beta \tag{615}$$

where the wedge product $dx^\alpha \wedge dx^\beta$ between differential forms is an exterior product. Equation (615) translates in tensor notation into

$$F = F_{\alpha\beta} \, dx^\alpha \otimes dx^\beta \tag{616}$$

We have argued here and elsewhere [44] that the plane-wave representation of classical electromagnetism is far from complete. In tensor language, this incompleteness means that the antisymmetric electromagnetic field tensor on the O(3) level must be proportional to an antisymmetric frame tensor of spacetime, $R_{\mu\nu}^{(A)}$, derived from the Riemannian tensor by contraction on two indices:

$$R_{\mu\nu}^{(A)} = R_{\lambda\mu\nu}^\lambda \tag{617}$$

Therefore $R^{(A)}_{\mu\nu}$ is an antisymmetric Ricci tensor obtained from the index contraction from the Riemann curvature tensor. Further contraction of $R^{(A)}_{\mu\nu}$ leads to the scalar curvature R, which, for electromagnetism, is κ^2. The contraction must be

$$R = \frac{1}{12} g^{(A)}_{\mu\nu} R^{\mu\nu(A)} \tag{618}$$

The principle of equivalence between electromagnetism and the antisymmetric Ricci tensor is

$$R^{(A)}_{\mu\nu} = g G_{\mu\nu} = \frac{\kappa}{A^{(0)}} G_{\mu\nu} \tag{619}$$

whose scalar form is

$$R = g G^{(0)} \tag{620}$$

where $G^{(0)}$ is a scalar field amplitude and where $R = \kappa^2$ is the scalar curvature of vacuum electromagnetism, whose metric coefficient is antisymmetric. In this view, vacuum electromagnetism is the antisymmetric Ricci 2-form [110], and gravitation is the symmetric Ricci 2-form.

Geodesic equations can be developed for the vacuum plane wave from the starting point [110]

$$D\kappa^\mu = \frac{d\kappa^\mu}{d\lambda} + \Gamma^\mu_{\nu\sigma} \kappa^\nu \kappa^\sigma = 0 \tag{621}$$

where $\kappa^\mu = dx^\mu/d\lambda$ is the wave 4-vector and $\Gamma^\mu_{\nu\sigma}$ is the affine connection. The symbol D in Eq. (621) is therefore a covariant derivative. In the received view, on the U(1) level, Eq. (621) becomes

$$d\kappa^\mu = 0 \tag{622}$$

in which the wave-vector does not vary along its path. Equation (621), on the other hand, has a parameter that varies along the ray, and the world line is a helix. This is a conclusion reminiscent of the fact that the O(3) electromagnetic phase is described by a line integral, as developed in the previous section.

A relation is first established between κ^μ and the A^μ 4-vector:

$$\kappa^\mu = \frac{\kappa}{A^{(0)}} A^\mu \tag{523}$$

Using this equation in Eq. (621) gives

$$\frac{dA^\mu}{d\lambda} + \frac{\kappa}{A^{(0)}} \Gamma^\mu_{\nu\sigma} A^\nu A^\sigma = \frac{dA^\mu}{d\lambda} + \frac{\kappa}{A^{(0)}} A^2 \Gamma^\mu = 0 \tag{624}$$

where A is a scalar. The contracted affine connection Γ^μ is proportional to A^μ in general gauge theory, and we adopt this rule to give

$$\Gamma^\mu = \frac{\kappa}{A^{(0)}} A^\mu \tag{625}$$

which is an equivalence principle between field and frame (or vacuum) properties. Such an equivalence does not appear on the U(1) level if the ordinary derivative replaces the covariant derivative.

Equation (625) can be written as

$$\frac{dA^\mu}{d\lambda} + \kappa^2 A^\mu = 0 \tag{626}$$

where the dimensionality of λ is κ^2, the inverse of the Thomson area of a photon [42], and if $\lambda = Z^2/2$, Eq. (626) become

$$\frac{d^2 A^\mu}{dZ^2} + R A^\mu = 0 \tag{627}$$

This has the form of a geodesic equation [111], and is obeyed by a plane wave. Similarly, we obtain:

$$\frac{1}{c^2} \frac{d^2 A^\mu}{dt^2} + R A^\mu = 0 \tag{628}$$

an equation that is also obeyed by a plane wave. Now, subtract Eq. (627) from Eq. (628) to give the d'Alembert wave equation:

$$\Box A^\mu = (R - R) A^\mu = 0 \tag{629}$$

which is the Proca equation

$$\Box A^\mu = -\frac{m_0^2 c^4}{\hbar^2} A^\mu = 0 \tag{630}$$

whose right-hand side happens to be zero because we have used a plane wave to derive it. The Proca equation (629) is an equation of a spacetime or vacuum whose curvature is $R = \kappa^2$, and not zero.

Equations (627) and (628) are special cases of the usual definition of the Riemann tensor in curvilinear geometry

$$A_{\mu;\nu;\kappa} - A_{\mu;\kappa;\nu} \equiv R^\lambda_{\mu\nu\kappa} A_\lambda \tag{631}$$

where A_λ is a general 4-vector field [111]. Equation (631) can be written as

$$(D_\nu D_\kappa - D_\kappa D_\nu)A_\mu + R^\lambda_{\mu\nu\kappa}A_\lambda = 0 \tag{632}$$

and this is a geodesic equation. Multiply Eq. (632) by the antisymmetric metric coefficient $g^{(A)}_{\mu\nu}$ to obtain

$$g^{\nu\kappa}_{(A)}(D_\nu D_\kappa - D_\kappa D_\nu) + g^{\nu\kappa}_{(A)}R^\lambda_{\mu\nu\kappa}A_\lambda = 0 \tag{633}$$

and identify

$$R \equiv g^{\nu\kappa}_{(A)}R^\mu_{\mu\nu\kappa}; \qquad \frac{d^2}{dZ^2} = g^{\nu\kappa}_{(A)}(D_\nu D_\kappa - D_\kappa D_\nu) \tag{634}$$

This procedure reduces Eq. (631) to Eqs. (627) and (628), which are special cases obtained by tensor contraction.

Electromagnetism can therefore be defined geometrically in curvilinear co-ordinates, and has vacuum properties such as scalar curvature, metric coefficient, affine connection, and Ricci tensor that manifest themselves fully on the O(3) level:

$$G_{\mu\nu} = \frac{A^{(0)}}{\kappa}R^{(A)}_{\mu\nu} \tag{635}$$

This equation can be written in precise analogy with the Einstein equation

$$T^{(A)}_{\mu\nu} = \hbar\omega\left(\frac{R^{(A)}_{\mu\nu}}{R}\right) \tag{636}$$

where $T^{(A)}_{\mu\nu}$ is an antisymmetric electromagnetic energy-momentum tensor and $R = \kappa^2$ is the scalar curvature in O(3) electromagnetism. Equation (636) is therefore a rotational Einstein equation. The scalar curvature in electromagnetism is defined through the antisymmetric metric coefficient ($g^{(A)}_{\mu\nu}$):

$$R = \kappa^2 = g^{\mu\nu}_{(A)}R^{(A)}_{\mu\nu} \tag{637}$$

The analogous definition of scalar curvature in gravitation is given through the metric $g_{\mu\nu}$:

$$R(\text{grav}) = g^{\mu\nu}R^{(S)}_{\mu\nu} \tag{638}$$

and the symmetric part of the Ricci tensor $R_{\mu\nu}^{(s)}$, that is, through the equation

$$R_{\lambda\nu}^{(S)} = R_{\lambda\kappa\nu}^{\kappa} \tag{639}$$

If O(3) electromagnetism [denoted e.m. in Eq. (640)] and gravitation are both to be seen as phenomena of curved spacetime, then both fields are derived ultimately from the same Riemann curvature tensor as follows:

$$T_{\mu\nu}^{(A)}(\text{e.m.}) = \hbar\omega\frac{R_{\mu\nu}^{(A)}}{R} \tag{640}$$

$$T_{\mu\nu}^{(S)}(\text{grav.}) = \frac{c^4}{8\pi k}\left(R_{\mu\nu}^{(S)} - \frac{1}{2}g_{\mu\nu}R\right) \tag{641}$$

$$R_{\mu\nu} = R_{\mu\nu}^{(S)} + R_{\mu\nu}^{(A)} \tag{642}$$

The unification of O(3) electromagnetism and gravitation using these concepts is summarized in Table I.

TABLE I
Some Concepts in the Unified Theory of Fields

Concept of Quantity	Gravitation	Electromagnetism
Riemann tensor	$R_{\lambda\mu\nu}^{\kappa}$	$R_{\lambda\mu\nu}^{\kappa}$
Ricci tensor	$R_{\mu\nu}^{(S)} = R_{\lambda\alpha\nu}^{\alpha}$	$R_{\mu\nu}^{(A)} = R_{\alpha\mu\nu}^{\alpha}$
Metric coefficient	$g_{\mu\nu}$ (diagonal)	$g_{\mu\nu}^{(A)}$ (off-diagonal)
Scalar curvature	$R = g^{\mu\nu}R_{\mu\nu}^{(S)}$	$R = g^{\mu\nu(A)}R_{\mu\nu}^{(A)} = \kappa^2$
Einstein tensor	$R_{\mu\nu}^{(S)} - \frac{1}{2}g_{\mu\nu}R \equiv G_{\mu\nu}^{(E)}$	$R_{\mu\nu}^{(A)}$
Field equation	$G_{\mu\nu}^{(E)} = \frac{8\pi k}{c^4}T_{\mu\nu}^{(S)}$	$R_{\mu\nu}^{(A)} = \frac{\kappa^2}{\hbar\omega}T_{\mu\nu}^{(A)}$
Connection	$\Gamma_{\mu\nu}^{\lambda}$	$\Gamma_{\mu\nu}^{\lambda} = \frac{\kappa}{A^{(0)}}M_\mu M_\nu A^\lambda$
Local group	Poincaré	Poincaré
Group generator	Bianchi identity	Feynman Jacobi
Identity	$D_\rho R_{\lambda\mu\nu}^{\kappa} + D_\mu R_{\lambda\nu\rho}^{\kappa} + D_\nu R_{\lambda\rho\mu}^{\kappa} = 0$	identity($\kappa = \lambda$)
Energy-momentum tensor	$T_{\mu\nu}^{(S)}$ (translational)	$T_{\mu\nu}^{(A)} \equiv \omega J_{\mu\nu}$
		$= \frac{\hbar\omega}{R}R_{\mu\nu}^{(A)}$ (rotational)
Equivalence principle	Gravitation is a noninertial frame	Electromagnetism is a noninertial frame
Universal constant	k (Einstein's constant)	$\frac{\kappa}{A^{(0)}}$

The electromagnetic field equations on the O(3) level can be obtained from this purely geometrical theory by using Eq. (631) in the Bianchi identity

$$D_\kappa G_{\mu\nu} + D_\nu G_{\kappa\mu} + D_\mu G_{\nu\kappa} = R^\lambda_{\mu\nu\kappa}A_\lambda + R^\lambda_{\nu\kappa\mu}A_\lambda + R^\lambda_{\kappa\mu\nu}A_\lambda$$
$$= D_\rho R^\kappa_{\lambda\mu\nu} + D_\mu R^\kappa_{\lambda\nu\rho} + D_\nu R^\kappa_{\lambda\rho\mu} = 0 \quad (643)$$

with appropriate index contraction. The end result is the Feynman Jacobi identity discussed in earlier sections of this review

$$D_\mu \tilde{G}^{\mu\nu} \equiv 0 \qquad (644)$$

an identity that can be written as

$$D_\mu \tilde{\boldsymbol{G}}^{\mu\nu} \equiv \boldsymbol{0} \qquad (645)$$

The O(3) field equations can be obtained from the fundamental definition of the Riemann curvature tensor, Eq. (631), by defining the O(3) field tensor using covariant derivatives of the Poincaré group.

Equation (643) is also a Bianchi identity in the theory of gravitation because $G_{\mu\nu}$ is derived from the antisymmetric part of the Riemann tensor, whose symmetric part can be contracted to the Einstein tensor.

Similarly, Eq. (643) can be developed into an inhomogeneous equation of the unified field. First, raise indices in the Riemann tensor and field tensor:

$$G^{\nu\kappa} = g^{\nu\rho}g^{\kappa\sigma}G_{\rho\sigma}; \qquad R^{\lambda\nu\kappa}_\mu = g^{\nu\rho}g^{\kappa\sigma}R^\lambda_{\mu\rho\sigma} \qquad (646)$$

From the equivalence of $G_{\mu\nu}$ and $R^{(A)}_{\mu\nu}$ in Eq. (635), individual terms in the identity (643) can be equated:

$$D_\kappa G^{\mu\nu} = R^{\lambda\mu\nu}_\kappa A_\lambda \qquad (647a)$$

$$D_\nu G^{\kappa\mu} = R^{\lambda\kappa\mu}_\nu A_\lambda \qquad (647b)$$

$$D_\mu G^{\nu\kappa} = R^{\lambda\nu\kappa}_\mu A_\lambda \qquad (647c)$$

Consider the antisymmetric part of the Riemann tensor in Eqs. (647) by suitable contraction. In Eq. (647c), for example, the contraction is $\lambda = \mu$. The result reduces to the O(3) inhomogeneous field equation of electromagnetism in the form

$$D_\mu G^{\nu\mu} = R^{\sigma\nu\mu}_\sigma A_\mu \equiv \frac{J^\nu(\text{vac})}{\varepsilon_0} \qquad (648)$$

where the term

$$J^\nu(\text{vac}) = \varepsilon_0 R_\sigma^{\sigma\nu\mu} A_\mu \tag{649}$$

is the O(3) charge current density, which can be seen to exist in the vacuum as argued earlier.

There are well known similarities between the Riemann curvature tensor of general relativity and the field tensor in non-Abelian electrodynamics. The Riemann tensor is

$$R^\kappa_{\lambda\mu\nu} = \partial_\nu \Gamma^\kappa_{\lambda\mu} - \partial_\mu \Gamma^\kappa_{\lambda\nu} + \Gamma^\rho_{\lambda\mu} \Gamma^\kappa_{\rho\nu} - \Gamma^\rho_{\lambda\nu} \Gamma^\kappa_{\rho\mu} \tag{650}$$

and is made up of a Ricci tensor and a Weyl conformal tensor. The following contraction of indices

$$R^\kappa_{\kappa\mu\nu} = \partial_\nu \Gamma^\kappa_{\kappa\mu} - \partial_\mu \Gamma^\kappa_{\kappa\nu} + \Gamma^\rho_{\kappa\mu} \Gamma^\kappa_{\rho\nu} - \Gamma^\rho_{\kappa\nu} \Gamma^\kappa_{\rho\mu} \tag{651}$$

leads to an expression similar to the field tensor as argued. The holonomy [46] in general relativity is

$$\triangle V^\mu = \frac{1}{2} R^\mu_{\rho\sigma\lambda} V^\rho \triangle S^{\sigma\lambda} \tag{652}$$

which can be compared with the holonomy in gauge theory

$$\triangle \psi_A = -ig\triangle S^{\mu\nu} G_\mu \psi_A \tag{653}$$

In both cases, the $\triangle S^{\mu\nu}$ factor is a hypersurface. This suggests that the Ricci tensor is in general complex, and given by

$$R_{\mu\nu} = R^{(S)}_{\mu\nu} + i R^{(A)}_{\mu\nu} \tag{654}$$

where the real part is symmetric and the imaginary part is antisymmetric. Barrett [50] has pointed out that O(3) gauge theory is non-Minkowskian in general, and requires an extrapolation of twistor algebra to non-Minkowski spacetime, requiring the presence of a Weyl tensor, complex spacetime, and curved twistor space. In O(3) electrodynamics, therefore, Minkowski spacetime applies only locally, and Minkowski vector spaces are tangent spaces of spacetime events. The Weyl anti-self-dual spacetime is independent of the self-dual spacetime. There is conformally curved, complex spacetime, as reflected in the complex Ricci tensor discussed already. The Weyl tensor is not zero. A complex spacetime [50] is defined by a four-dimensional complex manifold, M, with a holomorphic metric g_{ab}. A differential function defined on an open set of complex numbers is holomorphic [50] if it satisfies the Cauchy–Riemann equations. With respect to a holomorphic coordinate basis $x^\mu = (x^0, x^1, x^2, x^3)$, the metric is a 4×4 matrix of holomorphic functions of x^μ, and its determinant is nowhere vanishing. The Ricci tensor becomes complex-valued as argued already. Self-consistently, it can be checked that the determinant of the metric

$$g_{ab} = \begin{bmatrix} 0 & -1 & -1 & -1 \\ 1 & 0 & -1 & 1 \\ 1 & 1 & 0 & -1 \\ 1 & -1 & 1 & 0 \end{bmatrix} \tag{655}$$

is nonzero, (i.e., -1). So the use of an antisymmetric Ricci tensor is justified from first principles.

XII. BASIC ALGEBRA OF O(3) ELECTRODYNAMICS AND TESTS OF SELF-CONSISTENCY

In this section, some elementary details of the complex circular basis algebra generated by ((1),(2),(3)) are given. The basis vectors are

$$\begin{aligned} e^{(1)} &= \frac{1}{\sqrt{2}}(i - ij); & i &= \frac{1}{\sqrt{2}}(e^{(1)} + e^{(2)}) \\ e^{(2)} &= \frac{1}{\sqrt{2}}(i + ij); & j &= \frac{1}{\sqrt{2}}(e^{(1)} - e^{(2)}) \\ e^{(3)} &= k \end{aligned} \tag{656}$$

Within a phase factor and amplitude, $e^{(1)} = e^{(2)*}$ is the vectorial part of the complex description of right and left circularly polarized radiation. The basis unit vectors $e^{(1)}$, $e^{(2)}$, and $e^{(3)}$ form the O(3) cyclic permutation relations:

$$\begin{aligned} e^{(1)} \times e^{(2)} &= ie^{(3)*} \\ e^{(2)} \times e^{(3)} &= ie^{(1)*} \\ e^{(3)} \times e^{(1)} &= ie^{(2)*} \end{aligned} \tag{657}$$

A closely similar complex circular basis has been described by Silver [112] for three-dimensional space. This space forms the internal gauge space in O(3) electrodynamics, as argued already. In the complex circular basis, the unit vector dot product is

$$\begin{aligned} e^{(1)} \cdot e^{(2)} &= e^{(2)} \cdot e^{(1)} = e^{(3)} \cdot e^{(3)} = 1 \\ e^{(1)} \cdot e^{(1)} &= e^{(2)} \cdot e^{(2)} = 0 \end{aligned} \tag{658}$$

as compared with the same concept in the Cartesian basis

$$i \cdot i = j \cdot j = k \cdot k = 1$$
$$i \cdot j = i \cdot k = j \cdot k = 0$$

(659)

Vectors are defined as

$$A \equiv A^{(1)} + A^{(2)} + A^{(3)}$$
$$= A^{(2)}e^{(1)} + A^{(1)}e^{(2)} + A^{(3)}e^{(3)}$$

(660)

where

$$A^{(1)} = \frac{1}{\sqrt{2}}(A_X - iA_Y) = A^{(2)*}$$
$$A^{(3)} = A_Z$$

(661)

The dot product of two vectors is therefore

$$A \cdot B = A^{(1)}B^{(2)}e^{(1)} \cdot e^{(2)} + A^{(2)}B^{(1)}e^{(2)} \cdot e^{(1)} + A^{(3)}B^{(3)}e^{(3)} \cdot e^{(3)}$$
$$= A^{(1)}B^{(2)} + A^{(2)}B^{(1)} + A^{(3)}B^{(3)}$$

(662)

The del operator in the circular basis is defined by

$$\nabla_X = \frac{\partial}{\partial X} = \frac{1}{\sqrt{2}}(\nabla^{(1)} + \nabla^{(2)}); \qquad \nabla^{(1)} = \frac{1}{\sqrt{2}}(\nabla_X - i\nabla_Y)$$

$$\nabla_Y = \frac{\partial}{\partial Y} = \frac{i}{\sqrt{2}}(\nabla^{(1)} - \nabla^{(2)}); \qquad \nabla^{(2)} = \frac{1}{\sqrt{2}}(\nabla_X + i\nabla_Y)$$

(663)

$$\nabla_Z = \frac{\partial}{\partial Z} = \nabla^{(3)}; \qquad\qquad \nabla^{(3)} = \nabla_Z$$

and the divergence of a vector is therefore

$$\nabla \cdot A = \nabla^{(1)}A^{(2)} + \nabla^{(2)}A^{(1)} + \nabla^{(3)}A^{(3)}$$

(664)

and the gradient of a scalar is

$$\nabla \phi = \nabla^{(1)}\phi e^{(2)} + \nabla^{(2)}\phi e^{(1)} + \nabla^{(3)}\phi e^{(3)}$$

(665)

The curl operator in the complex circular basis is

$$\nabla \times A = -i \begin{vmatrix} e^{(1)} & e^{(2)} & e^{(3)} \\ \nabla^{(1)} & \nabla^{(2)} & \nabla^{(3)} \\ A^{(1)} & A^{(2)} & A^{(3)} \end{vmatrix} = \begin{vmatrix} i & j & k \\ \nabla_X & \nabla_Y & \nabla_Z \\ A_X & A_Y & A_Z \end{vmatrix}$$

(666)

and the vector cross-product is

$$A \times B = -i \begin{vmatrix} e^{(1)} & e^{(2)} & e^{(3)} \\ A^{(1)} & A^{(2)} & A^{(3)} \\ B^{(1)} & B^{(2)} & B^{(3)} \end{vmatrix} = \begin{vmatrix} i & j & k \\ A_X & A_Y & A_Z \\ B_X & B_Y & B_Z \end{vmatrix} \tag{667}$$

It is helpful to exemplify the basis by calculating the vector cross-product in detail and comparing it with the Cartesian counterpart. This procedure shows that the $((1),(2),(3))$ and Cartesian representations are equivalent when correctly worked out.

The $e^{(3)}$ component can be developed as

$$- i e^{(3)} (A^{(1)} B^{(2)} - A^{(2)} B^{(1)})$$
$$= -i \left(\frac{1}{\sqrt{2}} (A_X + iA_Y) \frac{1}{\sqrt{2}} (B_X - iB_Y) - \frac{1}{\sqrt{2}} (A_X - iA_Y) \frac{1}{\sqrt{2}} (B_X + iB_Y) \right)$$
$$= A_X B_Y - A_Y B_X \tag{668}$$

and is equivalent to the Cartesian component obtained from the well-known expression

$$\begin{vmatrix} i & j & k \\ A_X & A_Y & A_Z \\ B_X & B_Y & B_Z \end{vmatrix} = (A_X B_Y - A_Y B_X) k + \cdots \tag{669}$$

The other two components are evaluated by developing the sum

$$A \times B = -i[e^{(1)} (A^{(2)} B^{(3)} - A^{(3)} B^{(2)}) - e^{(2)} (A^{(1)} B^{(3)} - A^{(3)} B^{(1)})] + \cdots$$
$$= -i \left[\frac{1}{\sqrt{2}} (i - ij) \left(\frac{1}{\sqrt{2}} (A_X + iA_Y) B_Z - \frac{A_Z}{\sqrt{2}} (B_X + iB_Y) \right) \right.$$
$$\left. - \frac{1}{\sqrt{2}} (i + ij) \left(\frac{1}{\sqrt{2}} (A_X - iA_Y) B_Z - \frac{A_Z}{\sqrt{2}} (B_X - iB_Y) \right) \right] + \cdots$$
$$= i(A_Y B_Z - A_Z B_Y) - j(A_X B_Z - A_Z B_X) + \cdots \tag{670}$$

and again we obtain a result equivalent to the Cartesian sum.

A conjugate product such as $A^{(1)} \times A^{(2)}$ is evaluated as

$$-i \begin{vmatrix} e^{(1)} & e^{(2)} & e^{(3)} \\ A^{(2)} & 0 & 0 \\ 0 & A^{(1)} & 0 \end{vmatrix} = -iA^{(0)2} k \tag{671}$$

and is the same as the Cartesian equivalent:

$$\begin{vmatrix} i & j & k \\ A_X^{(2)} & A_Y^{(2)} & 0 \\ A_X^{(1)} & A_Y^{(1)} & 0 \end{vmatrix} = -iA^{(0)2}k \tag{672}$$

In the logic of the complex circular basis, unity is expressed as the product of two complex conjugates, referred to hereinafter as *complex unity*

$$1^2 = 1^{(1)}1^{(2)} \tag{673}$$

where

$$1^{(1)} = \frac{1}{\sqrt{2}}(1 - i); \qquad 1^{(2)} = \frac{1}{\sqrt{2}}(1 + i) \tag{674}$$

Therefore, developments such as the following are possible:

$$e^{(1)} \cdot e^{(2)} = 1^{(2)}e^{(1)} \cdot 1^{(1)}e^{(2)} = 1^{(1)}1^{(2)} = 1^2 = 1$$
$$A^{(1)} \cdot A^{(2)} = A^{(2)}e^{(1)} \cdot A^{(1)}e^{(2)} = A^{(1)}A^{(2)} = A^{(0)2} \tag{675}$$

Since the product $1^{(1)}1^{(2)}$ is always unity, it makes no difference to the dot product of unit vectors or of conjugate vectors such as $A^{(1)}$ and $A^{(2)}$, but the dot product of a vector $A^{(1)}$ and a unit vector $e^{(2)}$ is

$$A^{(1)} \cdot e^{(2)} = A^{(2)}1^{(1)}e^{(1)} \cdot e^{(2)} = \frac{1}{2}(A_X - iA_Y)(1 + i)$$

$$= \frac{1}{2}(A_X - iA_Y + iA_X + A_Y) \tag{676}$$

Similarly [42], the dot product of a complex circular Pauli matrix $\sigma^{(1)}$ and a unit vector $e^{(2)}$ is

$$\sigma^{(1)} \cdot e^{(2)} = \frac{1}{2}(\sigma_X - i\sigma_Y + i\sigma_X + \sigma_Y) \tag{677}$$

leading to

$$(\sigma^{(1)} \cdot e^{(2)})(\sigma^{(2)} \cdot e^{(1)}) = e^{(1)} \cdot e^{(2)} + i\sigma^{(3)} \cdot e^{(1)} \times e^{(2)} \tag{678}$$

and the prediction of radiatively induced fermion resonance.

As we have argued, the basis $((1),(2),(3))$ defines an internal space in electrodynamics, and was first applied as such by Barrett [50] in an SU(2) invariant gauge theory. As a consequence of this hypothesis, we can write

$$A^\mu = A^{\mu(2)}e^{(1)} + A^{\mu(1)}e^{(2)} + A^{\mu(3)}e^{(3)} \tag{679}$$

so A^μ is developed as a vector in the internal space. The object $A^{\mu(1)}, A^{\mu(2)}$, and $A^{\mu(3)}$ are scalar coefficients in the internal space. The boldface character A^μ is simultaneously a vector in the basis $((1),(2),(3))$ and a 4-vector in spacetime. If we consider to start with the received view of ordinary plane waves, the boldface character in this case is a vector of three-dimensional space in the basis $((1),(2),(3))$ and so is also a vector in the internal space of O(3) electrodynamics. As we have argued, the phase factor $e^{i\phi}$ on the O(3) level is made up of a line integral, related to an area integral by a non-Abelian Stokes theorem. In order to expand the horizon of the gauge structure of electrodynamics to the O(3) level, an additional spacetime index must appear in the definition of the plane wave, and the (1) and (2) indices must become indices of the internal space. This is achieved by recognizing that

$$A^{1(1)} = A_X^{(1)} = i\frac{A^{(0)}}{\sqrt{2}}e^{-i\phi} = A^{1(2)*}$$
$$A^{2(1)} = A_Y^{(1)} = \frac{A^{(0)}}{\sqrt{2}}e^{-i\phi} = A^{2(2)*} \tag{680}$$
$$A^{0(1)} = A^{3(1)} = A^{0(2)} = A^{3(2)} = 0$$

These equations define two of the scalar coefficients of the complete 4-vector A^μ

$$A^{\mu(1)} = (0, A^{(1)})$$
$$A^{\mu(2)} = (0, A^{(2)}) \tag{681}$$

a deduction that follows from the fact that $A^{(1)} = A^{(2)*}$ are transverse and so can have X and Y components only. The scalar coefficients $A^{\mu(1)}$ and $A^{\mu(2)}$ are light-like invariants

$$A^{\mu(1)}A_\mu^{(1)} = A^{\mu(2)}A_\mu^{(2)} = 0 \tag{682}$$

of polar 4-vectors in spacetime. The third index (3) of the non-Abelian theory must therefore be in the direction of propagation of radiation and must also be a light-like invariant

$$A^{\mu(3)}A_\mu^{(3)} = 0 \tag{683}$$

in the vacuum.

One possible solution of Eq. (683) is

$$A^{\mu(3)} = (cA^{(0)}, \boldsymbol{A}^{(3)}) \tag{684}$$

where

$$cA^{(0)} = |\boldsymbol{A}^{(3)}| \tag{685}$$

Such a solution is proportional directly to the wave 4-vector

$$\kappa^{\mu(3)} \equiv (c\kappa, \boldsymbol{\kappa}e^{(3)}) = gA^{\mu(3)} \tag{686}$$

and to the photon energy momentum:

$$p^{\mu(3)} = \hbar g A^{\mu(3)} = \hbar \kappa^{\mu(3)} \tag{687}$$

in the vacuum. Therefore, the complete vector in the internal $((1),(2),(3))$ space is the light-like polar vector

$$A^{\mu} = (0, \boldsymbol{A}^{(2)})e^{(1)} + (0, \boldsymbol{A}^{(1)})e^{(2)} + (cA^{(0)}, \boldsymbol{A}^{(3)})e^{(3)} \tag{688}$$

and has time-like, longitudinal, and transverse components, which are all physical components in the vacuum. On the U(1) level, the time-like and longitudinal components are combined in an admixture [46].

Similarly, the field tensor on the O(3) level is a vector in the internal space:

$$\boldsymbol{G}^{\mu\nu} = G^{\mu\nu(2)}e^{(1)} + G^{\mu\nu(1)}e^{(2)} + G^{\mu\nu(3)}e^{(3)} \tag{689}$$

and the coefficients $G^{\mu\nu(i)}$ are scalars in the internal space. They are also antisymmetric tensors in spacetime. General gauge field theory for O(3) symmetry then gives

$$\begin{aligned}
\boldsymbol{G}^{\mu\nu(1)*} &= \partial^{\mu}A^{\nu(1)*} - \partial^{\nu}A^{\mu(1)*} - igA^{\mu(2)} \times A^{\nu(3)} \\
\boldsymbol{G}^{\mu\nu(2)*} &= \partial^{\mu}A^{\nu(2)*} - \partial^{\nu}A^{\mu(2)*} - igA^{\mu(3)} \times A^{\nu(1)} \\
\boldsymbol{G}^{\mu\nu(3)*} &= \partial^{\mu}A^{\nu(3)*} - \partial^{\nu}A^{\mu(3)*} - igA^{\mu(1)} \times A^{\nu(2)}
\end{aligned} \tag{690}$$

which is a relation between vectors in the internal space $((1),(2),(3))$. The cross-product notation is also a vector notation; for example, $A^{\mu(2)} \times A^{\nu(3)}$ is a cross-product of a vector $A^{\mu(2)}$ with the vector $A^{\nu(3)}$ in the internal space. In forming the cross-product, the Greek indices are not transmuted and the complex basis is used, so that the terms quadratic in $\boldsymbol{A}$ become natural descriptions of the empirically observable conjugate product. As we have argued, the scalar coefficient $g = \kappa/A^{(0)}$ is a scalar in both the internal gauge space and spacetime. In field–matter interaction, g changes magnitude [44]. The field tensor on the O(3) level is therefore a vector in the internal space and is nonlinear in the potential. It contains the longitudinal field $\boldsymbol{B}^{(3)}$ in the vacuum. The field tensor on the U(1) level does not define $\boldsymbol{B}^{(3)}$, which exists only on the O(3) level.

Equation (690) is a concise description that contains a considerable amount of information about the O(3) theory of electromagnetism in the vacuum: information that is available without assuming any form of field equation. It is important to give details of the correct algebraic form of reduction of Eq. (690). Consider, for example, the equation

$$G^{\mu\nu(1)*} = \partial^{\mu}A^{\nu(1)*} - \partial^{\nu}A^{\mu(1)*} - igA^{\mu(2)} \times A^{\nu(3)} \tag{691}$$

which consists of components such as

$$G^{12(1)*} = \partial^{1}A^{2(1)*} - \partial^{2}A^{1(1)*} - ig\varepsilon_{(1)(2)(3)}A^{1(2)}A^{2(3)} \tag{692}$$

where $\varepsilon_{(1)(2)(3)}$ is the Levi–Civita symbol defined by

$$\varepsilon_{(1)(2)(3)} \equiv 1 = -\varepsilon_{(1)(2)(3)} = \cdots \tag{693}$$

Now take the vector potential as defined already with

$$\partial^{\mu} = \left(\frac{1}{c}\frac{\partial}{\partial t}, -\nabla\right) \tag{694}$$

then we obtain

$$G^{12(1)*} = \partial^{1}A^{2(1)*} - \partial^{2}A^{1(1)*} - ig(A^{1(2)}A^{2(3)} - A^{1(3)}A^{2(2)})$$
$$= 0 \tag{695}$$

This is a self-consistent result because there is no Z component of $G^{\mu\nu(1)*}$, which is defined as transverse. Both the linear and nonlinear components are zero.

Consider next the element:

$$G^{13(1)*} = \partial^{1}A^{3(1)*} - \partial^{3}A^{1(1)*} - ig\varepsilon_{(1)(2)(3)}A^{1(2)}A^{3(3)}$$
$$= \partial^{1}A^{3(2)} - \partial^{3}A^{1(2)} - ig(A^{1(2)}A^{3(3)} - A^{1(3)}A^{3(2)})$$
$$= -(\partial^{3} + igA^{3(3)})A^{1(2)} = -(\partial^{3} + i\kappa)A^{1(2)} \tag{696}$$

where we have used

$$g = \frac{\kappa}{A^{(0)}}; \qquad A^{3(3)} = A_Z^{(3)} = A^{(0)} \tag{697}$$

There are two contributions to the field element $G^{13(2)}$, a magnetic component:

$$-\partial^3 A^{1(2)} \tag{697a}$$

and

$$-ig A^{3(3)} A^{1(2)} \tag{697b}$$

In vector notation, Eq. (696) is a component of

$$\begin{aligned}
2\boldsymbol{B}^{(1)} &\equiv \nabla \times \boldsymbol{A}^{(1)} - ig\boldsymbol{A}^{(3)} \times \boldsymbol{A}^{(1)} \\
&= (\nabla - ig\boldsymbol{A}^{(3)}) \times \boldsymbol{A}^{(1)} \\
&= \nabla \times \boldsymbol{A}^{(1)} - \frac{i}{B^{(0)}} \boldsymbol{B}^{(3)} \times \boldsymbol{B}^{(1)}
\end{aligned} \tag{698}$$

Furthermore:

$$\partial^3 A^{1(2)} = i\kappa A^{1(2)} \tag{699}$$

and so it follows that

$$\boldsymbol{B}^{(1)} = \nabla \times \boldsymbol{A}^{(1)} = -\frac{i}{B^{(0)}} \boldsymbol{B}^{(3)} \times \boldsymbol{B}^{(1)} \tag{700}$$

Similarly:

$$\boldsymbol{B}^{(2)} = \nabla \times \boldsymbol{A}^{(2)} = -\frac{i}{B^{(0)}} \boldsymbol{B}^{(2)} \times \boldsymbol{B}^{(3)} \tag{701}$$

Therefore, the definition of the field tensor in O(3) electrodynamics gives the first two components of the B cyclic theorem [47–62]

$$\begin{aligned}
\boldsymbol{B}^{(3)} \times \boldsymbol{B}^{(1)} &= iB^{(0)} \boldsymbol{B}^{(2)*} \\
\boldsymbol{B}^{(2)} \times \boldsymbol{B}^{(3)} &= iB^{(0)} \boldsymbol{B}^{(1)*}
\end{aligned} \tag{702}$$

together with the definition of $\boldsymbol{B}^{(1)}$ and $\boldsymbol{B}^{(2)}$ in terms of the curl of vector potentials:

$$\begin{aligned}
\boldsymbol{B}^{(1)} &= \nabla \times \boldsymbol{A}^{(1)} \\
\boldsymbol{B}^{(2)} &= \nabla \times \boldsymbol{A}^{(2)}
\end{aligned} \tag{703}$$

It is convenient to write this result as

$$H(\text{vac}) = \frac{1}{\mu_0} B - M(\text{vac}) \tag{704}$$

where $H(\text{vac})$ is the vacuum magnetic field strength and μ_0 is the vacuum permeability. The object $M(\text{vac})$ does not exist on the U(1) level and can be termed *vacuum magnetization*:

$$M^{(1)}(\text{vac}) = -\frac{1}{i\mu_0 B^{(0)}} B^{(3)} \times B^{(1)} \tag{705}$$

The objects $M^{(1)}(\text{vac})$ and $M^{(2)}(\text{vac})$ depend on the phaseless vacuum magnetic field $B^{(3)}$ and so do not exist as concepts in U(1) electrodynamics. The $B^{(3)}$ field itself is defined through

$$G^{\mu\nu(3)*} = \partial^\mu A^{\nu(3)*} - \partial^\nu A^{\mu(3)*} - ig A^{\mu(1)} \times A^{\nu(2)} \tag{706}$$

with (3) aligned in the Z axis. So, by definition, the only nonzero components are

$$G^{12(3)*} = -G^{21(3)*} = B_Z^{(3)} \tag{707}$$

It follows that

$$B_Z^{(3)} = -ig(A^{1(1)}A^{2(2)} - A^{1(2)}A^{2(1)}) \tag{708}$$

or

$$B^{(3)} = B^{(3)*} = -ig A^{(1)} \times A^{(2)} = -\frac{i}{B^{(0)}} B^{(1)} \times B^{(2)} \tag{709}$$

giving the third component of the B cyclic theorem $B^{(1)} \times B^{(2)} = iB^{(0)}B^{(3)*}$, and the vacuum magnetization:

$$M^{(3)*} = -\frac{1}{i\mu_0 B^{(0)}} B^{(1)} \times B^{(2)} \tag{710}$$

On the U(1) level, $A^{(1)} \times A^{(2)}$ is considered to be an operator [44] of nonlinear optics with no third axis, but on the O(3) level it defines $B^{(3)}$ as argued.

Therefore, on the O(3) level, the magnetic part of the complete free field is defined as a sum of a curl of a vector potential and a vacuum magnetization inherent in the structure of the B cyclic theorem. On the U(1) level, there is no $B^{(3)}$ field by hypothesis.

The following field coefficients can be calculated:

$$
\begin{aligned}
G^{01(2)} &= (\partial^0 + igA^{0(3)})A^{1(2)} = -G^{10(2)} \\
G^{02(2)} &= (\partial^0 + igA^{0(3)})A^{2(2)} = -G^{20(2)} \\
G^{03(2)} &= 0 \\
G^{13(2)} &= -(\partial^3 + igA^{3(3)})A^{1(2)} = -G^{31(2)} \\
G^{23(2)} &= -(\partial^3 + igA^{3(3)})A^{2(2)} = -G^{32(2)} \\
G^{12(2)} &= 0
\end{aligned}
\tag{711}
$$

so that

$$
\begin{aligned}
G^{01(1)} &= G^{01(2)*} = (\partial^0 + igA^{0(3)})A^{1(1)} \\
G^{12(3)*} &= -G^{21(3)*} = -ig(A^{1(1)}A^{2(2)} - A^{1(2)}A^{2(1)})
\end{aligned}
\tag{712}
$$

The three field tensors are therefore the transverse

$$
G^{\mu v(1)} = G^{\mu v(2)*} = \begin{bmatrix}
0 & -E^{1(1)} & -E^{2(1)} & 0 \\
E^{1(1)} & 0 & 0 & cB^{2(1)} \\
E^{2(1)} & 0 & 0 & -cB^{1(1)} \\
0 & -cB^{2(1)} & cB^{1(1)} & 0
\end{bmatrix}
\tag{713}
$$

and the longitudinal:

$$
G^{\mu v(3)*} = G^{\mu v(3)} = \begin{bmatrix}
0 & 0 & 0 & 0 \\
0 & 0 & -cB^{3(3)} & 0 \\
0 & cB^{3(3)} & 0 & 0 \\
0 & 0 & 0 & 0
\end{bmatrix}
\tag{714}
$$

On the O(3) level, there also exists a vacuum polarization, because the complete electric field strength is given in the vacuum by

$$
\begin{aligned}
2E^{(2)} &= -\frac{\partial A^{(2)}}{\partial t} - igcA^{(0)}A^{(2)} \\
&= -\left(\frac{\partial}{\partial t} + igcA^{(0)}\right)A^{(2)} = 2E^{(1)*}
\end{aligned}
\tag{715}
$$

Using $g = \kappa/A^{(0)}$, then

$$
E^{(2)} = -\frac{\partial A^{(2)}}{\partial t} = -ic\kappa A^{(2)} = -i\omega A^{(2)}
\tag{716}
$$

and it is convenient to express this result as

$$\boldsymbol{D}^{(2)}(\text{vac}) = \varepsilon_0 \boldsymbol{E}^{(2)} + \boldsymbol{P}^{(2)}(\text{vac}) \tag{717}$$

where $\boldsymbol{D}^{(2)}(\text{vac})$ is the electric displacement in vacuo, and where the vacuum polarization is

$$\boldsymbol{P}^{(2)}(\text{vac}) = -i\varepsilon_0 \omega \boldsymbol{A}^{(2)} \tag{718}$$

The vacuum polarization is well known to have an analog in quantum electrodynamics [46], the photon self-energy. The latter has no classical analog on the U(1) level, but one exists on the O(3) level, thus saving the correspondence principle. The classical vacuum polarization on the O(3) level is transverse and vanishes when $\omega = 0$. It is pure transverse because, as follows, the hypothetical $\boldsymbol{E}^{(3)}$ field is zero on the O(3) level

$$\begin{aligned}
G^{03(3)*} &= \partial^0 A^{3(3)*} - \partial^3 A^{0(3)*} - ig(A^{0(1)}A^{3(2)} - A^{3(2)}A^{0(1)}) \\
&= 0
\end{aligned} \tag{719}$$

giving

$$G^{03(1)} = G^{03(2)} = G^{03(3)} = 0 \tag{720}$$

in the vacuum. In the presence of field–matter interaction, this result is no longer true because of the Coulomb field, indicating polarization of matter.

In the presence of field–matter interaction [44]

$$\boldsymbol{H}^{\mu\nu(i)*} = \varepsilon_0 \boldsymbol{F}^{\mu\nu(i)*} - \boldsymbol{M}^{\mu\nu(i)*} \tag{721}$$

where $i = 1, 2, 3$. Here

$$\begin{aligned}
\boldsymbol{F}^{\mu\nu(i)*} &\equiv \partial^\mu \boldsymbol{A}^{\nu(i)} - \partial^\nu \boldsymbol{A}^{\mu(i)} \\
\boldsymbol{M}^{\mu\nu(1)} &\equiv i\varepsilon_0 g' \boldsymbol{A}^{\mu(2)} \times \boldsymbol{A}^{\nu(3)}
\end{aligned} \tag{722}$$

in cyclic permutation, with $g' \ll g$ empirically [44].

There are therefore obvious points of similarity between the O(3) theory of electrodynamics and the Yang–Mills theory [44]. Both are based, as we have argued, on an O(3) or SU(2) invariant Lagrangian. However, in O(3) electrodynamics, the particle concomitant with the field has the topological charge $\kappa/A^{(0)}$. In O(3) electrodynamics, the internal space and spacetime are not independent spaces but form an extended Lie algebra [42]. In elementary particle theory, the internal space is usually an abstract isospin space [46]. The overall structures of O(3) electrodynamics and of Yang–Mills theory are the same.

XIII.   QUANTIZATION FROM THE B CYCLIC THEOREM

The B cyclic theorem is a Lorentz invariant construct in the vacuum and is a relation between angular momentum generators [42]. As such, it can be used as the starting point for a new type of quantization of electromagnetic radiation, based on quantization of angular momentum operators. This method shares none of the drawbacks of canonical quantization [46], and gives photon creation and annihilation operators self-consistently. It is seen from the B cyclic theorem:

$$\boldsymbol{B}^{(1)} \times \boldsymbol{B}^{(2)} = iB^{(0)}\boldsymbol{B}^{(3)*}$$

$$\boldsymbol{B}^{(2)} \times \boldsymbol{B}^{(3)} = iB^{(0)}\boldsymbol{B}^{(1)*} \tag{723}$$

$$\boldsymbol{B}^{(3)} \times \boldsymbol{B}^{(1)} = iB^{(0)}\boldsymbol{B}^{(2)*}$$

that if any one of the magnetic fields $\boldsymbol{B}^{(1)}$, $\boldsymbol{B}^{(2)}$, or $\boldsymbol{B}^{(3)}$ is zero, this implies that the other two will also be zero. The B cyclic theorem can be put in commutator form by using the result that an axial vector is equivalent to a rank 2 antisymmetric polar tensor

$$B_k = \frac{1}{2}\varepsilon_{ijk}B_{ij} \tag{724}$$

where ε_{ijk} is the Levi–Civita symbol. The rank 2 tensor representation of the axial vector B_k is mathematically equivalent but has the advantage of being accessible to commutator (matrix) algebra, allowing $\boldsymbol{B}^{(1)}$, $\boldsymbol{B}^{(2)}$, and $\boldsymbol{B}^{(3)}$ to be expressed as infinitesimal rotation generators and as quantum-mechanical angular momentum operators. These methods show that the photon has an elementary longitudinal flux quantum, the photomagneton operator $B^{(3)}$, which is directly proportional to its intrinsic spin angular momentum [42].

The unit vector components of the classical magnetic fields $\boldsymbol{B}^{(1)}$, $\boldsymbol{B}^{(2)}$, and $\boldsymbol{B}^{(3)}$ in vacuo are all axial vectors by definition, and it follows that their unit vector components must also be axial in nature. In matrix form, they are, in the Cartesian basis

$$\boldsymbol{i} = \begin{bmatrix} 0 & 0 & 0 \\ 0 & 0 & 1 \\ 0 & -1 & 0 \end{bmatrix}; \quad \boldsymbol{j} = \begin{bmatrix} 0 & 0 & -1 \\ 0 & 0 & 0 \\ 1 & 0 & 0 \end{bmatrix}; \quad \boldsymbol{k} = \begin{bmatrix} 0 & 1 & 0 \\ -1 & 0 & 0 \\ 0 & 0 & 0 \end{bmatrix} \tag{725}$$

and in the circular basis

$$e^{(1)} = \frac{1}{\sqrt{2}} \begin{bmatrix} 0 & 0 & i \\ 0 & 0 & 1 \\ -i & -1 & 0 \end{bmatrix}; \quad e^{(2)} = \frac{1}{\sqrt{2}} \begin{bmatrix} 0 & 0 & -i \\ 0 & 0 & 1 \\ i & -1 & 0 \end{bmatrix}; \quad e^{(3)} = \begin{bmatrix} 0 & 1 & 0 \\ -1 & 0 & 0 \\ 0 & 0 & 0 \end{bmatrix}$$

$$(726)$$

The latter form a commutator Lie algebra, which is mathematically equivalent to the vectorial Lie algebra:

$$[e^{(1)}, e^{(2)}] = -ie^{(3)*} \tag{727}$$

$$\cdots$$

Equations (723) and (727) therefore represent a closed, cyclically symmetric, algebra in which all three space-like components are meaningful. The cyclical commutator basis can be used to build a matrix representation of the three space-like magnetic components of the electromagnetic wave in the vacuum

$$B^{(1)} = iB^{(0)}e^{(1)}e^{i\phi}$$
$$B^{(2)} = -iB^{(0)}e^{(2)}e^{-i\phi} \tag{728}$$
$$B^{(3)} = B^{(0)}e^{(3)}$$

from which emerges the commutative Lie algebra equivalent to the vectorial Lie algebra

$$[B^{(1)}, B^{(2)}] = -iB^{(0)}B^{(3)*} \tag{729}$$

$$\cdots$$

This algebra can be expressed in terms of the infinitesimal rotation generators of the O(3) group [42] in three dimensional space:

$$J^{(1)} = -ie^{(1)} = \frac{1}{\sqrt{2}} \begin{bmatrix} 0 & 0 & 1 \\ 0 & 0 & -i \\ 1 & i & 0 \end{bmatrix}; \quad J^{(2)} = ie^{(2)} = \frac{1}{\sqrt{2}} \begin{bmatrix} 0 & 0 & 1 \\ 0 & 0 & i \\ -1 & -i & 0 \end{bmatrix}$$

$$J^{(3)} = -ie^{(3)} \begin{bmatrix} 0 & -1 & 0 \\ i & 0 & 0 \\ 0 & 0 & 0 \end{bmatrix} \tag{730}$$

The magnetic field matrices and rotation generators are linked by

$$B^{(1)} = -B^{(0)} J^{(1)} e^{i\phi}$$

$$B^{(2)} = -B^{(0)} J^{(2)} e^{-i\phi} \qquad (731)$$

$$B^{(3)} = iB^{(0)} J^{(3)}$$

so the commutative algebra of the magnetic fields (729) is part of the Lie algebra of spacetime. The real and physical $\boldsymbol{B}^{(3)}$ component is directly proportional to the rotation generator $J^{(3)}$, which is a fundamental property of spacetime, in which the matrices (730) become

$$J^{(1)} = \frac{1}{\sqrt{2}} \begin{bmatrix} 0 & 0 & 1 & 0 \\ 0 & 0 & -i & 0 \\ -1 & i & 0 & 0 \\ 0 & 0 & 0 & 0 \end{bmatrix}; \qquad J^{(2)} = \frac{1}{\sqrt{2}} \begin{bmatrix} 0 & 0 & 1 & 0 \\ 0 & 0 & i & 0 \\ -1 & -i & 0 & 0 \\ 0 & 0 & 0 & 0 \end{bmatrix}$$

$$J^{(3)} = \begin{bmatrix} 0 & -i & 0 & 0 \\ i & 0 & 0 & 0 \\ 0 & 0 & 0 & 0 \\ 0 & 0 & 0 & 0 \end{bmatrix} \qquad (732)$$

It follows that magnetic fields in the vacuum on the O(3) level are directly proportional to rotation generators of the Poincaré group [42], and electric fields are directly proportional to boost generators.

The rotation generators form a commutator algebra of the following type in the circular basis:

$$[J^{(1)}, J^{(2)}] = -J^{(3)*} \qquad (733)$$

which becomes

$$[J_X, J_Y] = iJ_Z \qquad (734)$$

in the Cartesian basis, and which is, within a factor $\hbar$, identical with the commutator algebra of angular momentum operators in quantum mechanics. This inference provides a simple route to the quantization of the magnetic fields, giving the result

$$B^{(1)} = -B^{(0)} \frac{J^{(1)}}{\hbar} e^{i\phi}; \qquad B^{(2)} = -B^{(0)} \frac{J^{(2)}}{\hbar} e^{-i\phi}; \qquad B^{(3)} = iB^{(0)} \frac{J^{(3)}}{\hbar} \qquad (735)$$

where $B^{(i)}$ are now operators of quantum mechanics. Such a quantization scheme can exist only on the O(3) level. In particular, the longitudinal $B^{(3)}$ is the photomagneton operator, which is a stationary state in quantum mechanics.

These results can be generalized to electric fields using boost operators, $K^{(i)}$, which in the Poincaré group are also 4×4 matrices:

$$E^{(1)} = E^{(0)} K^{(1)} e^{i\phi}$$
$$E^{(2)} = E^{(0)} K^{(2)} e^{-i\phi} \tag{736}$$
$$iE^{(3)} = iE^{(0)} K^{(3)}$$

Therefore, electric fields are boost generators, whereas magnetic fields are rotation generators. It follows that the Lie algebra of electric and magnetic fields in spacetime is isomorphic with that of the infinitesimal generators of the Poincaré group [42]. The latter type of Lie algebra can be summarized as follows:

$$[J^{(1)}, J^{(2)}] = -J^{(3)*} \cdots$$
$$[K^{(1)}, K^{(2)}] = -ie^{(3)*} \cdots$$
$$[K^{(1)}, e^{(2)}] = -iK^{(3)*} \cdots \tag{737}$$
$$[K^{(1)}, J^{(1)}] = 0 \cdots$$

This isomorphism is conclusive evidence for the existence of the longitudinal $B^{(3)}$ in the vacuum.

There is also a relation between polar unit vectors, boost generators, and electric fields. An electric field is a polar vector, and unlike the magnetic field, cannot be put into matrix form as in Eq. (724). The cross-product of two polar unit vectors is however an axial vector k, which, in the circular basis, is $e^{(3)}$. In spacetime, the axial vector k becomes a 4×4 matrix related directly to the infinitesimal rotation generator $J^{(3)}$ of the Poincaré group. A rotation generator is therefore the result of a classical commutation of two matrices that play the role of polar vectors. These matrices are boost generators. In spacetime, it is therefore

$$[K_X, K_Y] = -iJ_Z \tag{738}$$

and cyclic permutations. In the circular basis, this algebra becomes

$$[K^{(1)}, K^{(2)}] = -ie^{(3)*} \tag{739}$$

Therefore, although polar vectors cannot be put into matrix form in three-dimensional space, they correspond to 4×4 matrices in spacetime. In three-dimensional space, the electric component of the electromagnetic field are oscillatory fields that can be written directly in terms of the unit vectors of the circular basis:

$$E^{(1)} = \frac{E^{(0)}}{\sqrt{2}}(i - ij)e^{i\phi}; \qquad E^{(2)} = \frac{E^{(0)}}{\sqrt{2}}(i + ij)e^{-i\phi} \qquad (740)$$

In spacetime, the equivalents are

$$E^{(1)} = E^{(0)}K^{(1)}e^{i\phi}; \qquad E^{(2)} = E^{(0)}K^{(2)}e^{-i\phi} \qquad (741)$$

The phase ϕ is a line integral on the O(3) level. The boost generators appearing in Eq. (741) are written in a circular basis

$$K^{(1)} = \frac{1}{\sqrt{2}}\begin{bmatrix} 0 & 0 & 0 & 1 \\ 0 & 0 & 0 & -i \\ 0 & 0 & 0 & 0 \\ -1 & i & 0 & 0 \end{bmatrix}; \qquad K^{(2)} = \frac{1}{\sqrt{2}}\begin{bmatrix} 0 & 0 & 0 & 1 \\ 0 & 0 & 0 & i \\ 0 & 0 & 0 & 0 \\ -1 & -i & 0 & 0 \end{bmatrix} \qquad (742)$$

and correspond to the complex, polar, unit vectors $e^{(1)}$ and $e^{(2)}$ in Euclidean space.

It is not possible to form a real electric field from the cross-product of $E^{(1)}$ and $E^{(2)}$, and this is self-consistent with the fact that on the O(3) level there is no real $E^{(3)}$ [42].

The complete Lie algebra of the infinitesimal boost and rotation generators of the Poincaré group can be written as we have seen either in a circular basis or in a Cartesian basis. In matrix form, the generators are

$$K_X = \begin{bmatrix} 0 & 0 & 0 & 1 \\ 0 & 0 & 0 & 0 \\ 0 & 0 & 0 & 0 \\ -1 & 0 & 0 & 0 \end{bmatrix}; \quad K_Y = \begin{bmatrix} 0 & 0 & 0 & 0 \\ 0 & 0 & 0 & 1 \\ 0 & 0 & 0 & 0 \\ 0 & -1 & 0 & 0 \end{bmatrix}; \quad K_Z = \begin{bmatrix} 0 & 0 & 0 & 0 \\ 0 & 0 & 0 & 0 \\ 0 & 0 & 0 & 1 \\ 0 & 0 & -1 & 0 \end{bmatrix}$$

$$J_X = \begin{bmatrix} 0 & 0 & 0 & 0 \\ 0 & 0 & -i & 0 \\ 0 & i & 0 & 0 \\ 0 & 0 & 0 & 0 \end{bmatrix}; \quad J_Y = \begin{bmatrix} 0 & 0 & i & 0 \\ 0 & 0 & 0 & 0 \\ -i & 0 & 0 & 0 \\ 0 & 0 & 0 & 0 \end{bmatrix}; \quad J_Z = \begin{bmatrix} 0 & -i & 0 & 0 \\ i & 0 & 0 & 0 \\ 0 & 0 & 0 & 0 \\ 0 & 0 & 0 & 0 \end{bmatrix}$$

$$(743)$$

The relation between fields and generators in spacetime can be summarized as

$$
\begin{aligned}
B^{(1)} &= -B^{(0)} J^{(1)} e^{i\phi} = iB^{(0)} e^{(1)} e^{i\phi} \\
B^{(2)} &= -B^{(0)} J^{(2)} e^{-i\phi} = -iB^{(0)} e^{(2)} e^{-i\phi} \\
B^{(3)} &= iB^{(0)} J^{(3)} = B^{(0)} e^{(3)} \\
E^{(1)} &= E^{(0)} K^{(1)} e^{i\phi} \\
E^{(2)} &= E^{(0)} K^{(2)} e^{-i\phi} \\
iE^{(3)} &= iE^{(0)} K^{(3)}
\end{aligned}
\tag{744}
$$

leading to the Lie algebra:

$$
\begin{aligned}
[B^{(1)}, B^{(2)}] &= iB^{(0)} B^{(3)*} \cdots \\
[E^{(1)}, E^{(2)}] &= iE^{(0)2} e^{(3)*} \cdots \\
[E^{(1)}, B^{(2)}] &= iB^{(0)} (iE^{(3)}) \cdots \\
[E^{(1)}, B^{(1)}] &= 0 \cdots
\end{aligned}
\tag{745}
$$

where we have used the notation

$$
\begin{aligned}
&ie^{(1)} = J^{(1)}; &&-ie^{(2)} = J^{(2)}; &&ie^{(3)} = J^{(3)} \\
&ie^{(2)} = J^{(2)}; &&-ie^{(1)} = J^{(1)}; &&ie^{(3)} = -J^{(3)}
\end{aligned}
\tag{746}
$$

This type of Lie algebra occurs on the O(3) level, but not on the U(1) level. Since $iE^{(3)}$ is purely imaginary, it has no physical meaning.

Therefore, the Lie algebra of the magnetic and electric components of the plane waves and spin fields in free space is isomorphic with that of the infinitesimal boost and rotation generators of the Poincaré group in spacetime. Experimental evidence (presented in Ref. 3 and in this review) suggests that $\boldsymbol{B}^{(3)}$ is real and physical and the theory of electromagnetism in the vacuum is relativistically rigorous if and only if the longitudinal fields $\boldsymbol{B}^{(3)}$ (physical) and $i\boldsymbol{E}^{(3)}$ (unphysical) are accounted for through the appropriate algebra. If $\boldsymbol{B}^{(3)}$ and $i\boldsymbol{E}^{(3)}$ are set to zero, as in the received view [U(1) level], then the isomorphism is lost, and electromagnetism becomes incompatible with relativity. If $\boldsymbol{B}^{(3)}$ were zero, the rotation generator $J^{(3)}$ would be zero, which is incorrect. Similarly, if $i\boldsymbol{E}^{(3)}$ were zero, the boost generator $K^{(3)}$ would be incorrectly zero.

In units of $\hbar$, the eigenvalues of the massless photon are -1 and $+1$, and those of the photon with mass are -1, 0, and $+1$. In three-dimensional space, the latter are obtained from relations such as:

$$J^{(3)}e^{(1)} = +1e^{(1)}$$
$$J^{(3)}e^{(2)} = -1e^{(2)} \tag{747}$$
$$J^{(3)}e^{(3)} = 0e^{(3)}$$

where $J^{(3)}$ is the rotation generator:

$$J^{(3)} = ie^{(3)}\times = \begin{bmatrix} 0 & -i & 0 \\ i & 0 & 0 \\ 0 & 0 & 0 \end{bmatrix} \tag{478}$$

There is no paradox [112] in the use of $e^{(3)}$ as an operator as well as a unit vector. In the same sense [112], there is no paradox in the use of the scalar spherical harmonics as operators. The rotation operators in space are first-rank T operators, which are irreducible tensor operators, and under rotations, transform into linear combinations of each other. The T operators are directly proportional to the scalar spherical harmonic operators. The rotation operators, J, of the full rotation group are related to the T operators as follows

$$T^1_{-1} = iJ^{(1)}; \qquad T^1_1 = iJ^{(2)}; \qquad T^1_0 = iJ^{(3)} \tag{749}$$

and to the scalar spherical harmonic operators by

$$Y^1_{-1} = \frac{i}{r}\left(\frac{3}{4\pi}\right)^{1/2}J^{(1)}; \qquad Y^1_1 = \frac{i}{r}\left(\frac{3}{4\pi}\right)^{1/2}J^{(2)}; \qquad Y^1_0 = \frac{i}{r}\left(\frac{3}{4\pi}\right)^{1/2}J^{(3)} \tag{750}$$

This implies that the fields $B^{(1)}$, $B^{(2)}$, and $B^{(3)}$ are also operators of the full rotation group, and are therefore irreducible representations of the full rotation group. Specifically

$$B^{(1)} = B^{(0)}r\left(\frac{4\pi}{3}\right)^{1/2}Y^1_{-1}e^{i\phi}$$
$$B^{(2)} = B^{(0)}r\left(\frac{4\pi}{3}\right)^{1/2}Y^1_1e^{-i\phi} \tag{751}$$
$$B^{(3)} = B^{(0)}r\left(\frac{2\pi}{3}\right)^{1/2}Y^1_0$$

which shows that $B^{(3)} = ? \, 0$ violates the fundamentals of group theory. Thus $B^{(1)}$, $B^{(2)}$, and $B^{(3)}$ are all nonzero components of the same rank 1 scalar spherical harmonic Y_M^1; $M = -1, 0, 1$. Furthermore, since the operators $J^{(1)}$, $J^{(2)}$, and $J^{(3)}$ are components in a circular basis of the spin, or intrinsic, angular momentum of the vector field representing the electromagnetic field, the fields $B^{(1)}$, $B^{(2)}$, and $B^{(3)}$ are themselves components of spin angular momentum. It is also clear that $J^{(1)}$ is a lowering (annihilation) operator

$$J^{(1)}e^{(2)} = +1e^{(3)}; \qquad J^{(1)}e^{(3)} = -1e^{(1)}; \qquad J^{(1)}e^{(1)} = 0e^{(2)} \qquad (752)$$

and that $J^{(2)}$ is a raising (creation) operator:

$$J^{(2)}e^{(2)} = 0e^{(1)}; \qquad J^{(2)}e^{(3)} = -1e^{(2)}; \qquad J^{(2)}e^{(1)} = +1e^{(3)} \qquad (753)$$

The total angular momentum J^2 is also an eigenoperator, for example:

$$J^2 e^{(3)} = l(l+1)e^{(3)}; \qquad l = 1 \qquad (754)$$

The operator $J^{(3)}$ is therefore also an intrinsic spin, and can be identified in this novel quantization method based on the B cyclic theorem with the intrinsic spin of a photon with mass, with eigenvalues -1, 0, and $+1$.

For a classical vector field, its intrinsic (spin) angular momentum is identifiable with its transformation properties [112] under rotations, and within a factor $\hbar$, the rotation operators are spin angular momentum operators of the spin one boson. Recognition of a nonzero $B^{(3)}$ is therefore compatible with the eigenvalues of both the massive and massless bosons. The vector spherical harmonics [112] are specific vector fields that are eigenvalues of j^2 and of j_Z where j is the operator for vector fields of infinitesimal rotations about axis (3). They have definite total angular momentum and occur in sets of dimension $(2j+1)$ that span in standard form the D representations of the full rotation group, and are therefore irreducible tensors of rank j. Defining the total angular momentum as the sum of the "orbital" angular momentum I and intrinsic (spin) angular momentum J, we have

$$j = I + J \qquad (755)$$

and the vector spherical harmonics are compound irreducible tensor operators [112]:

$$Y_{Ml1}^L \equiv \left[Y^1 \otimes l\right]_M^L \qquad (756)$$

They are formed from the scalar spherical harmonics Y_M^l, which form a complete set for scalar functions, and the $e^{(i)}$ operators, which form a complete set for any vector in three-dimensional space. Therefore, the vector spherical harmonics form a complete set for the expansion of any arbitrary classical vector field:

$$A = A_X i + A_Y j + A_Z k \tag{757}$$

in a Cartesian basis. For this vector, the I_Z operates on the A_X, A_Y, A_Z and J_Z operates on i, j and k. Thus, I_Z operates on the spatial part of the field and J_Z, on the vector part.

Therefore the operator for infinitesimal rotations about the Z axis contains two "angular momentum" operators, I and J, analogous with orbital and spin angular momentum in the quantum theory of atoms and molecules. The infinitesimal rotation is therefore formally a coupling of a set of spatial fields transforming according to $D^{(1)}$ with a set of three vector fields $[e^{(1)}, e^{(2)}, e^{(3)}]$, transforming according to $D^{(1)}$. Equation (756) is an expression of this coupling, or combining, of entities in two different spaces to give a total angular momentum. It follows, from these considerations, that the vector spherical harmonics are defined by

$$Y_{Ml1}^L = \sum_{mn} \langle l1mn|l1LM \rangle Y_m^l e_m \tag{758}$$

where $\langle l1mn|l1LM \rangle$ are Clebsch–Gordan, or coupling, coefficients [112]. For photons regarded as bosons of unit spin, it is possible to multiply Eq. (758) by $\langle 110M|11LM \rangle$ and to sum over L [112]. Using the orthogonality condition

$$\sum_j \langle j_1 m'_1 j_2 m - m'_2 | j_1 j_2 j_m \rangle \langle j_1 j_2 j_m | j_1 m_1 j_2 m - m_2 \rangle = \delta_{m_1 m'_1} \tag{759}$$

it is found that

$$Y_0^1(\theta, \phi) e_M = \sum_{L=|l-1|}^{l+1} \langle l10M|l1LM \rangle Y_{Ml1}^L \tag{760}$$

which is an expression for the unit vectors e_M in terms of sums over vector spherical harmonics, that is, of irreducible compound tensors, representations of the full rotation group.

On the U(1) level, the transverse components of e_M are physical but the longitudinal component corresponding to $M = 0$ is unphysical. This asserts two states of transverse polarization in the vacuum: left and right circular. However, this assertion amounts to $e_0 \equiv e^{(3)} = ?0$, meaning the incorrect disappearance of some vector spherical harmonics that are nonzero from fundamental group theory because some irreducible representations are incorrectly set to zero.

This point can be emphasized by expanding $B^{(3)}$ in terms of Wigner 3-j symbols [112], which yields results such as

$$B^{(3)} = B^{(0)}e^{(3)} = 2B^{(0)}\frac{Y^1_{001}}{Y^1_0} = \frac{B^{(0)}}{\sqrt{3}}\frac{\sqrt{2}Y^2_{001} - Y^0_{011}}{Y^1_0} \tag{761}$$

showing that $B^{(3)}$ is nonzero and proportional to the nonzero vector spherical harmonic Y^1_{001} on a fundamental level. Therefore, the fundamentals of group theory are obeyed on the O(3) level, but not on the U(1) level.

All three of $e^{(1)}, e^{(2)}, e^{(3)}$ can be expressed in terms of vector spherical harmonics. Thus, in addition to the nonlinear B cyclic theorem, the following linear relations occur

$$B^{(3)} = B^{(0)}e^{(3)} = \frac{\sqrt{2}}{2}aB^{(0)}(e^{(1)} + e^{(2)}) + B^{(0)}b$$

$$= -\frac{\sqrt{2}}{2}cB^{(0)}(e^{(1)} - e^{(2)}) + B^{(0)}d \tag{762}$$

where the coefficients are defined by the following combination of scalar and vector spherical harmonics:

$$a = \frac{2}{\sqrt{2}}\left(\frac{Y^1_0}{Y^1_1 - Y^1_{-1}}\right); \qquad c = -\frac{2}{\sqrt{2}}\left(\frac{Y^1_0}{Y^1_1 + Y^1_{-1}}\right)$$

$$b = \sqrt{2}\left(\frac{Y^1_{111} + Y^1_{-111}}{Y^1_1 - Y^1_{-1}}\right); \qquad d = \sqrt{2}\left(\frac{Y^1_{111} - Y^1_{-111}}{Y^1_1 + Y^1_{-1}}\right) \tag{763}$$

On the O(3) level, therefore, $B^{(3)}$ is nonzero because $B^{(1)}$ and $B^{(2)}$ are nonzero.

On the U(1) level, the plane wave is subjected to a multipole expansion in terms of the vector spherical harmonics, in which only two physically significant values of M in Eq. (761) are assumed to exist, corresponding to $M = +1$ and -1, which translates into our notation as follows:

$$e_1 = -e^{(2)}; \qquad e_{-1} = e^{(1)}; \qquad e_0 = e^{(3)} \tag{764}$$

On the O(3) level, the case $M = 0$ is also considered to be physically meaningful. In consequence, there is an additional, purely real, 2^L- pole component of the electromagnetic plane wave in vacuo corresponding to $B^{(3)}$. The vector spherical harmonics Y^L_{mL1} with $1 = L$ are no longer transverse fields, and the vector $e^{(3)}$, which is longitudinal, can also be expressed in terms of the $L = 1, M = 0$ vector spherical harmonics as in Eq. (761). The longitudinal $B^{(3)}$, according to Eq. (761), can be expanded for all integer 1 of that equation in terms of vector spherical harmonics. Each value of 1 for $M = 0$ in Y_{0l1}^L defines a different nonzero component of $\boldsymbol{B}^{(3)}$. Therefore the $L = 1$ components in the expansion of $\boldsymbol{B}^{(3)}$ are dipolar fields.

As an example of these methods, consider the B cyclic theorem for multipole radiation, which can be developed for the multipole expansion of plane-wave radiation to show that the $\boldsymbol{B}^{(3)}$ field is irrotational, divergentless, and fundamental for each multipole component. The magnetic components of the plane wave are defined, using Silver's notation [112] as

$$
\begin{aligned}
\boldsymbol{B}_1 &= B^{(0)} e^{i\phi} \boldsymbol{e}_1 \\
\boldsymbol{B}_{-1} &= B^{(0)} e^{-i\phi} \boldsymbol{e}_{-1} \\
\boldsymbol{B}_0 &= B^{(0)} \boldsymbol{e}_0
\end{aligned}
\tag{765}
$$

where the basis vectors in Silver's spherical representation are related by

$$
\boldsymbol{e}_{-1} \times \boldsymbol{e}_1 = -i\boldsymbol{e}_0
\tag{766}
$$

in cyclic permutation. The phase factor on the O(3) is a line integral, as argued in this review and elsewhere [44]. The B cyclic theorem in this notation is therefore

$$
\boldsymbol{B}_{-1} \times \boldsymbol{B}_1 = -iB^{(0)} \boldsymbol{B}_0
\tag{767}
$$

In order to develop Eq. (767) for multipole radiation, we use the following expansions [112]:

$$
\begin{aligned}
e^{ikZ} &= \sum_l i^l (2l + 1) j_l(kZ) P_l(\cos\theta) \\
\boldsymbol{e}_M &= \frac{1}{Y_0^1} \sum_{L=|l-1|}^{l+1} \langle l10M | l1LM \rangle Y_{Ml1}^L
\end{aligned}
\tag{768}
$$

where l is the lth multipole moment, j_l the lth modified Bessel function, and P_l is the lth Legendre polynomial. The basis vector $\boldsymbol{e}_m$ ($M = -1, 0, +1$) is expanded in terms of the Clebsch–Gordan coefficients $\langle l10M | l1LM \rangle$ and the vector spherical harmonics Y_{Ml1}^L, and normalized with the scalar spherical harmonic Y_0^1.

In deriving Eq. (767), we have used on the left-hand side the conjugate product of phase factors:

$$
e^{i\phi} e^{-i\phi} = 1
\tag{769}
$$

Using Eqs. (768a) and (769), it is seen that the product is unity if we sum over all multipole components with $1 \rightarrow \infty$ in Eq. (768). In all other cases, the B cyclic theorem is

$$
\boldsymbol{B}_{-1} \times \boldsymbol{B}_1 = -ixB^{(0)} \boldsymbol{B}_0
\tag{770}
$$

where x is different from unity. It is given as follows for the first few multipoles:

$$\begin{aligned}
x &= 9j_1^2 P_1^2, && \text{for} \quad l = 1 \\
&= 25j_2^2 P_2^2 && \text{for} \quad l = 2 \\
&= 49j_3^2 P_3^2 && \text{for} \quad l = 3
\end{aligned} \qquad (771)$$

In this notation

$$\begin{aligned}
P_l(\cos\theta) &= (2\pi(2l+1))^{1/2} Y_0^l(\theta) \\
j_l(kr) &= \left(-\frac{r}{k}\right)^l \left(\frac{1}{r}\frac{d}{dr}\right) j_0(kr)
\end{aligned} \qquad (772)$$

It is important to note that $\boldsymbol{B}_0$ in Eq. (770) is the same as $\boldsymbol{B}_0$ in Eq. (767): phaseless, irrotational, and divergentless. The factor x arises purely from the truncation of the infinite series (768a) in individual multipole components. As discussed by Silver [112], the $\boldsymbol{e}_m$ vectors are polarization vectors for the electromagnetic wave, but are also spin angular momentum eigenfunctions. Tautologically, therefore, Eq. (767), the B cyclic theorem, is a spin angular momentum equation for the photon, with $M = -1, 0, 1$. The photon wave function, therefore, has components $e^{ikZ}\boldsymbol{e}_1$, $e^{-ikZ}\boldsymbol{e}_{-1}$, and $\boldsymbol{e}_0$. The observables in this theory are therefore energy and $\boldsymbol{B}_0$. The complete vector fields $\boldsymbol{B}_1, \boldsymbol{B}_{-1}, \boldsymbol{B}_0$ are described in terms of the vector spherical harmonics, and the B cyclic theorem indicates the existence of an intrinsic magnetic field $\boldsymbol{B}_0$, which is described by the transformation of the frame under rotation. As is well known in classical angular momentum theory, only the $\boldsymbol{B}_0$ component remains sharply defined under rotation. The components $\boldsymbol{B}_1$ and $\boldsymbol{B}_{-1}$ are defined only within an arbitrary phase factor. Within $\hbar$, this is the quantum theory of angular momentum [112].

Since $\boldsymbol{B}^{(3)}$ is time-independent, it obeys

$$\boldsymbol{B}^{(3)} = -\nabla\Phi_B \qquad (773)$$

where Φ_B is determined by the Laplace equation:

$$\nabla^2\Phi_B = 0 \qquad (774)$$

Analogously, a Coulomb field can be expressed as the gradient of a scalar potential that obeys the Laplace equation in a source-free region such as the vacuum in conventional electrostatics. To find the general form of $\boldsymbol{B}^{(3)}$ in a multipole expansion, we therefore solve the Laplace equation for Φ_B, and evaluate the gradient of this solution

$$\Phi_B = \frac{U(r)}{r}\rho(\theta)Q(\Phi) \qquad (775)$$

in spherical polar coordinates. The general solution (775) can be written as

$$\Phi_B = (Ar^l + Br^{-2})Y_{lm}(\theta, \phi) \qquad (776)$$

where $Y_{lm}(\theta, \phi)$ are the spherical harmonics and A and B are constants. Here, m and l are integers, with l running from $-m$ to m. The solution of Laplace's equation is therefore obtained as a product of radial and angular functions. The latter are orthonormal functions, the spherical or tesseral harmonics, which form a complete set on the surface of the unit sphere for the two indices l and m. The integer l defines the order of the multipole component; $l = 1$ is a dipole, $l = 2$ is a quadrupole, $l = 3$ is an octopole, and $l = 4$ is a hexadecapole.

The most general form of $\boldsymbol{B}^{(3)}$ from the Laplace equation is therefore

$$\boldsymbol{B}^{(3)} = -\nabla(Ar^l + Br^{-2})Y_{lm}(\theta, \phi) \qquad (777)$$

This is the phaseless magnetic field of multipole radiation on the O(3) level. The solution (777) reduces to the simple

$$\boldsymbol{B}^{(3)} = B^{(0)}\boldsymbol{e}^{(3)} = B^{(0)}\boldsymbol{k} \qquad (778)$$

when $l = 1$, $m = 0$, $r = Z$, $\theta = 0$, $A = -B^{(0)}$, $B = 0$, and $\nabla = (\partial/\partial Z)\boldsymbol{k}$. More generally, there exist other irrotational forms of $\boldsymbol{B}^{(3)}$:

1. $\boldsymbol{B}^{(3)}$ for dipole radiation: $l = 1$, $m = -1, 0, 1$
2. $\boldsymbol{B}^{(3)}$ for quadrupole radiation: $l = 2$, $m = -2, \ldots, 2$ (779)
3. $\boldsymbol{B}^{(3)}$ for octopole radiation: $l = 3$, $m = -3, \ldots, 3$

The $\boldsymbol{B}^{(3)}$ fields for n-pole fields are irrotational for all n on the O(3) level.

As argued, infinitesimal field generators appear as a by-product of this novel quantization scheme, so that $\boldsymbol{B}^{(3)}$ is rigorously nonzero from the symmetry of the Poincaré group and the B cyclic theorem is an invariant of the classical field. The basics of infinitesimal field generators on the classical level are to be found in the theory of relativistic spin angular momentum [42,46] and relies on the Pauli–Lubanski pseudo-4-vector:

$$W^\lambda = -\frac{1}{2}\varepsilon^{\lambda\mu\nu\rho}p_\mu J_{\nu\rho} \qquad (780)$$

where $\varepsilon^{\lambda\mu\nu\rho}$ (with $\varepsilon^{0123} = 1$) is the antisymmetric unit 4-vector. The antisymmetric matrix of generators $J_{\nu\rho}$ is given by

$$J_{\nu\rho} = \begin{bmatrix} 0 & K_1 & K_2 & K_3 \\ -K_1 & 0 & -J_3 & J_2 \\ -K_2 & J_3 & 0 & -J_1 \\ -K_3 & -J_2 & J_1 & 0 \end{bmatrix} \qquad (781)$$

where every element is an element of spin angular momentum in four dimensions. The energy momentum polar 4-vector is defined by

$$p^\mu = (p^0, \boldsymbol{p}) = \left(\frac{En}{c}, \boldsymbol{p}\right) \tag{782}$$

The infinitesimal generators can be represented as matrices or as combinations of differential operators [46]. The Pauli–Lubanski operator then becomes a product of the $J_{\nu\rho}$ and p_μ operators. Barut [113] shows that the Lie algebra of the W^μ operators is

$$[W^\mu, W^\nu] = -i\varepsilon^{\mu\nu\sigma\rho} p_\sigma W_\rho \tag{783}$$

which is a four-dimensional commutator relation. The theory is relativistically covariant and, of course, compatible with special relativity. Equation (783) gives the Lie algebra [42] of intrinsic spin angular momentum because rotation generators are angular momentum operators within a factor $\hbar$, and this allows relativistic quantization to be considered. Similarly, translation generators are energy momentum operators within a factor $\hbar$. This development leads to Wigner's famous result that every particle is characterized by two Casimir invariants of the Poincaré group, the mass and spin invariants [46].

Our basic ansatz is to assume that this theory applies to the vacuum electromagnetic field, considered as a physical entity of spacetime in the theory of special relativity. The intrinsic spin of the classical electromagnetic field is the magnetic flux density $\boldsymbol{B}^{(3)}$. Infinitesimal generators of rotation correspond with those of intrinsic magnetic flux density in the vacuum. Boost generators correspond with intrinsic electric field strength. Translation generators correspond with the intrinsic, fully covariant, field potential. Thus, the symbols are transmuted as follows:

$$J \to B; \qquad K \to E; \qquad P \to A \tag{784}$$

In Cartesian notation, the Pauli–Lubanski vector of particle theory becomes a 4-vector of the classical electromagnetic field

$$W^\lambda = -\frac{1}{2}\varepsilon^{\lambda\mu\nu\rho} A_\mu F_{\nu\rho} \tag{785}$$

and the Lie algebra (783), a Lie algebra of the field.

If it is assumed that the electromagnetic field propagates at c in the vacuum, then we must consider the Lie algebra (783) in a light-like condition. The latter is satisfied by a choice of

$$A^\mu = (A^0, A_Z)$$
$$A^0 = A_Z \tag{786}$$

The basic ansatz is that there is a field vector analogous to the Pauli–Lubanski vector of particle physics, a field vector defined by

$$W^\lambda = \tilde{F}^{\lambda\mu} A_\mu \tag{787}$$

where $\tilde{F}^{\lambda\mu}$ is the dual of the antisymmetric field tensor. This vector has the following components:

$$
\begin{aligned}
W^0 &= -B^1 A_1 - B^2 A_2 - B^3 A_3 \\
W^1 &= B^1 A_0 + E^3 A_2 - E^2 A_3 \\
W^2 &= B^2 A_0 - E^3 A_1 + E^1 A_3 \\
W^3 &= B^3 A_1 + E^2 A_1 - E^1 A_2
\end{aligned}
\tag{788}
$$

If it assumed that for the transverse components

$$\mathbf{B} = \nabla \times \mathbf{A} \tag{789}$$

that $\mathbf{A}$ and $\mathbf{B}$ are plane waves

$$
\begin{aligned}
\mathbf{A} &= \frac{A^{(0)}}{\sqrt{2}}(i\mathbf{i} + \mathbf{j})e^{i\phi} \\
\mathbf{B} &= \frac{B^{(0)}}{\sqrt{2}}(i\mathbf{i} + \mathbf{j})e^{i\phi}
\end{aligned}
\tag{790}
$$

and that the longitudinal $\mathbf{E}^{(3)}$ is zero, then Eq. (788) reduces to

$$
\begin{aligned}
W_0 &= A_Z B_Z \\
W_X &= A_0 B_X + A_Z E_Y \\
W_Y &= A_0 B_Y - A_Z E_X \\
W_Z &= A_0 B_Z
\end{aligned}
\tag{791}
$$

These assumptions mean that

$$A^\mu = (A^0, 0, 0, A^3); \qquad A^0 = A^3 \tag{792}$$

can be used as an ansatz. Conversely, the use of this definition means that the transverse components are plane waves, and for the transverse components, $\mathbf{B} = \nabla = \mathbf{A}$.

　　In the Coulomb gauge, the vector W^μ vanishes, meaning that there is no correspondence between the particle and field theory for the Coulomb gauge, or the received view of transversality in the vacuum. The final result is therefore

$$W^\mu = A^0(B_Z, 0, 0, B_Z) \tag{793}$$

which is compatible with the Lie algebra of a light-like particle. This corresponds in the particle interpretation to the light-like translation generator:

$$p^\mu = (p^0, p_Z); \quad p^0 = p_Z \tag{794}$$

The Pauli–Lubanski pseudovector of the field in this condition is

$$\begin{aligned}
W^\mu &= (A_Z B_Z, \; A_Z E_Y + A_0 B_X, \; -A_Z E_X + A_0 B_Y, \; A_0 B_Z) \\
&= A_0(B_Z, \; E_Y + B_X, \; -E_X + B_Y, \; B_Z)
\end{aligned} \tag{795}$$

and the Lie algebra (783) becomes, in $c = 1$ units:

$$\begin{aligned}
[B_X + E_Y, \; B_Y - E_X] &= i(B_Z - B_Z) \\
[B_Y - E_X, \; B_Z] &= i(B_X + E_Y) \\
[B_Z, \; B_X + E_Y] &= i(B_Y - E_X)
\end{aligned} \tag{796}$$

which has E(2) symmetry. In the particle interpretation, Eqs. (795) and (796) correspond to

$$W^\mu = (p_Z J_Z, \; p_Z K_Y + p_0 J_X, \; -p_Z K_X + p_0 J_Y, \; p_0 J_Z) \tag{797}$$

and

$$\begin{aligned}
[J_X + K_Y, \; J_Y - K_X] &= i(J_Z - J_Z) \\
[J_Y - K_X, \; J_Z] &= i(J_X + K_Y) \\
[J_Z, \; J_X - K_Y] &= i(J_Y - K_X)
\end{aligned} \tag{798}$$

In the rest frame of a photon with mass, the field and particle Pauli–Lubanski vectors are respectively

$$W^\mu = (0, \; A_0 B_X, \; A_0 B_Y, \; A_0 B_Z) \tag{799}$$

and

$$W^\mu = (0, \; p_0 J_X, \; p_0 J_Y, \; p_0 J_Z) \tag{800}$$

The rest frame Lie algebra for field and particle is respectively (normalized $B^{(0)} = 1$ units):

$$[B_X, B_Y] = i B_Z \cdots \tag{801}$$

and

$$[J_X, J_Y] = iJ_Z \cdots \tag{802}$$

The $E^{(2)}$ field algebra is compatible with the vacuum Maxwell equations written for eigenvalues of our novel infinitesimal field operators. This can be demonstrated as follows:

$$B_Y = E_X; \qquad B_X = -E_Y \tag{803}$$

It is assumed that the eigenfunction (χ) operated on by these infinitesimal field generators is such that the same relation (803) holds between eigenvalues of the field. In order for this to be true, the eigenfunction must be the de Broglie wave function, specifically, the phase of the classical electromagnetic field. On the O(3) level, this is a line integral, as we have seen.

The relation (803) interpreted as one between eigenvalues is compatible with the plane-wave solutions

$$
\begin{aligned}
\boldsymbol{E}^{(1)} &= \boldsymbol{E}^{(2)*} = \frac{E^{(0)}}{\sqrt{2}}(\boldsymbol{i} - i\boldsymbol{j})e^{i\phi} \\
\boldsymbol{B}^{(1)} &= \boldsymbol{B}^{(2)*} = \frac{B^{(0)}}{\sqrt{2}}(i\boldsymbol{i} + \boldsymbol{j})e^{i\phi}
\end{aligned}
\tag{804}
$$

which are special cases of the O(3) invariant electrodynamics defined by

$$\boldsymbol{e}^{(1)} = \boldsymbol{e}^{(2)*} = \frac{1}{\sqrt{2}}(\boldsymbol{i} - i\boldsymbol{j}); \qquad \boldsymbol{e}^{(3)} = \boldsymbol{k} \tag{805}$$

It follows that the same analysis can be applied to the particle interpretation, giving

$$\partial_\mu J^{\mu\nu} = \partial_\mu \tilde{J}^{\mu\nu} = 0 \tag{806}$$

in the vacuum. This is a possible conservation equation (relation between spins) that is compatible with the $E^{(2)}$ symmetry of the little group of the Poincaré group. This is the little group for a massless particle. On the U(1) level, therefore, it is concluded that the vacuum Maxwell equations for the field correspond with Eq. (806) for the particle, an equation that asserts that the spin angular momentum matrix is divergentless. In vector notation, we obtain from Eqs. (803)–(806) the familiar U(1) equations

$$
\begin{aligned}
\nabla \cdot \boldsymbol{B} &= 0; & \nabla \cdot \boldsymbol{E} &= 0 \\
\nabla \times \boldsymbol{E} + \frac{\partial \boldsymbol{B}}{\partial t} &= \boldsymbol{0}; & \nabla \times \boldsymbol{B} - \frac{1}{c^2}\frac{\partial \boldsymbol{E}}{\partial t} &= \boldsymbol{0}
\end{aligned}
\tag{807}
$$

and the less familiar relation between eigenvalues of spin angular momentum in four dimensions:

$$\nabla \cdot \boldsymbol{J} = 0; \qquad \qquad \nabla \cdot \boldsymbol{K} = 0$$

$$\nabla \times \boldsymbol{J} + \frac{\partial \boldsymbol{K}}{\partial t} = 0; \qquad \nabla \times \boldsymbol{K} - \frac{\partial \boldsymbol{J}}{\partial t} = 0 \tag{808}$$

On the O(3) level, particular solutions of the $E^{(2)}$ Lie algebra (796) give a total of six commutator relations. Three of these form the B cyclic theorem ($B^{(0)} = 1$ units):

$$[B_X, B_Y] = iB_Z$$

$$[B_Y, B_Z] = iB_X \tag{809}$$

$$[B_Z, B_X] = iB_Y$$

and the other three are

$$[E_X, E_Y] = -iB_Z$$

$$[B_Z, E_X] = iE_Y \tag{810}$$

$$[E_Y, B_Z] = iE_X$$

In the particle interpretation, these are part of the Lie algebra of rotation and boost generators of the Poincaré group:

$$[J_X, J_Y] = iJ_Z \quad [K_X, K_Y] = -iJ_Z$$

$$[J_Y, J_Z] = iJ_X \quad [J_Z, K_X] = iK_Y \tag{811}$$

$$[J_Z, J_X] = iJ_Y \quad [K_Y, J_Z] = iK_X$$

From these relations, we can obtain

$$\boldsymbol{B}^{(1)} \times \boldsymbol{B}^{(2)} = iB^{(0)}\boldsymbol{B}^{(3)*}$$

$$\boldsymbol{B}^{(2)} \times \boldsymbol{B}^{(3)} = iB^{(0)}\boldsymbol{B}^{(1)*} \tag{812}$$

$$\boldsymbol{B}^{(3)} \times \boldsymbol{B}^{(1)} = iB^{(0)}\boldsymbol{B}^{(2)*}$$

$$\boldsymbol{E}^{(1)} \times \boldsymbol{E}^{(2)} = ic^2 B^{(0)}\boldsymbol{B}^{(3)*}$$

$$\boldsymbol{B}^{(3)} \times \boldsymbol{E}^{(1)} = icB^{(0)}\boldsymbol{E}^{(2)*} \tag{813}$$

$$\boldsymbol{B}^{(3)} \times \boldsymbol{E}^{(2)} = -icB^{(0)}\boldsymbol{E}^{(1)*}$$

where $\boldsymbol{B}^{(3)} = \boldsymbol{B}^{(0)}\boldsymbol{e}^{(3)}$. Similarly, in the particle interpretation, and switching from rotation generators to spin angular momentum, we obtain:

$$\boldsymbol{J}^{(1)} \times \boldsymbol{J}^{(2)} = i\hbar\boldsymbol{J}^{(3)*}$$
$$\boldsymbol{J}^{(2)} \times \boldsymbol{J}^{(3)} = i\hbar\boldsymbol{J}^{(1)*} \qquad (814)$$
$$\boldsymbol{J}^{(3)} \times \boldsymbol{J}^{(1)} = i\hbar\boldsymbol{J}^{(2)*}$$

where $\hbar$ is the quantum of spin angular momentum.

In the rest frame of a photon or particle with mass, we obtain, for field and particle, respectively, Eqs. (812) and (813); that is, there are no boost generators.

From this analysis, it is concluded that the $\boldsymbol{B}^{(3)}$ component is identically nonzero, otherwise all the field components vanish in the B cyclic theorem (812) and Lie algebra (809). If we assume Eq. (803) and at the same time assume that $\boldsymbol{B}^{(3)}$ is zero, then the Pauli–Lubanski pseudo-4-vector vanishes for all A_0. Similarly, in the particle interpretation, if we assume the equivalent of Eq. (803) and assume that $\boldsymbol{J}^{(3)}$ is zero, the Pauli–Lubanski vector W^μ vanishes. This is contrary to the definition of the helicity of the photon. Therefore, for finite field helicity, we need a finite $\boldsymbol{B}^{(3)}$.

The precise correspondence between field and photon interpretation developed here indicates that E(2) symmetry does not imply that $\boldsymbol{B}^{(3)}$ is zero, any more than it implies that $\boldsymbol{J}^{(3)} = \boldsymbol{0}$. The assertion $\boldsymbol{B}^{(3)} = \boldsymbol{0}$ is counterindicated by a range of data reviewed here and in Ref. 44, and the B cyclic theorem is Lorentz-covariant, as it is part of a Lorentz-covariant Lie algebra. If we assume the particular solutions (809) and (810) and use in it the particular solution (803), we obtain the cyclics (809) from the three cyclics Eq. (810); thus we obtain

$$[B_Y, -B_X] = iB_Z$$
$$[B_Z, B_Y] = -iB_X \qquad (815)$$
$$[B_Z, -B_X] = -iB_Y$$

This is also the relation obtained in the hypothetical rest frame. Therefore, the B cyclic theorem is Lorentz-invariant in the sense that it is the same in the rest frame and in the light-like condition. This result can be checked by applying the Lorentz transformation rules for magnetic fields term by term [44]. The equivalent of the B cyclic theorem in the particle interpretation is a Lorentz-invariant construct for spin angular momentum:

$$\boldsymbol{J}^{(1)} \times \boldsymbol{J}^{(2)} = i\hbar\boldsymbol{J}^{(3)^*} \qquad (816)$$

It is concluded that the $B^{(3)}$ component in the field interpretation is nonzero in the light-like condition and in the rest frame. The B cyclic theorem is a Lorentz-invariant, and the product $B^{(1)} \times B^{(2)}$ is an experimental observable [44]. In this representation, $B^{(3)}$ is a phaseless and fundamental field spin, an intrinsic property of the field in the same way that $J^{(3)}$ is an intrinsic property of the photon. It is incorrect to infer from the Lie algebra (796) that $B^{(3)}$ must be zero for plane waves. For the latter, we have the particular choice (803) and the algebra (796) reduces to

$$i(B_Z - B_Z) = 0 \qquad (817)$$

which does not indicate that B_Z is zero any more than the equivalent particle interpretation indicates that J_Z is zero.

In order to translate a Cartesian commutator relation such as

$$[B_X, B_Y] = iB^{(0)}B_Z \qquad (818)$$

to a $((1),(2),(3))$ basis vector equation such as

$$B^{(1)} \times B^{(2)} = iB^{(0)}B^{(3)*} \qquad (819)$$

consider firstly the usual vector relation in the Cartesian frame:

$$i \times j = k \qquad (820)$$

The unit vector i, for example, is defined by

$$i = u_X i \qquad (821)$$

where u_x is a rotation generator, in general a matrix component [46]. Therefore

$$u_X = i(J_X)_{YZ} \qquad (822)$$

The cross-product $\times j$ therefore becomes a commutator of matrices

$$[J_X, J_Y] = iJ_Z \qquad (823)$$

that is

$$\frac{1}{i}\begin{bmatrix} 0 & 0 & 0 \\ 0 & 0 & 1 \\ 0 & -1 & 0 \end{bmatrix} \frac{1}{i}\begin{bmatrix} 0 & 0 & -1 \\ 0 & 0 & 0 \\ 1 & 0 & 0 \end{bmatrix} - \frac{1}{i}\begin{bmatrix} 0 & 0 & -1 \\ 0 & 0 & 0 \\ 1 & 0 & 0 \end{bmatrix} \frac{1}{i}\begin{bmatrix} 0 & 0 & 0 \\ 0 & 0 & 1 \\ 0 & -1 & 0 \end{bmatrix}$$
$$= \begin{bmatrix} 0 & 1 & 0 \\ -1 & 0 & 0 \\ 0 & 0 & 0 \end{bmatrix} \qquad (824)$$

This can be extended straightforwardly to angular momentum operators and infinitesimal magnetic field generators. Therefore, a commutator such as Eq. (818) is equivalent to a vector cross-product. If we write $B^{(0)}$ as the scalar magnitude of magnetic flux density, the commutator (818) becomes the vector cross-product

$$(B^{(0)}i) \times (B^{(0)})j = B^{(0)}(B^{(0)}k) \tag{825}$$

which can be written conveniently as

$$(B_X B_Y)^{1/2}i \times (B_X B_Y)^{1/2}j = iB^{(0)}B_Z k \tag{826}$$

However, the Cartesian basis can be extended to the circular basis using relations between unit vectors developed in this review chapter. So Eq. (826) can be written in the circular basis as

$$(B_X B_Y)^{1/2}e^{(1)} \times (B_X B_Y)^{1/2}e^{(2)} = -B^{(0)}B_Z e^{(3)*} \tag{827}$$

which is equivalent to

$$B^{(1)} \times B^{(2)} = iB^{(0)}B^{(3)*} \tag{828}$$

where we define

$$B^{(1)} = B^{(2)} = (B_X B_Y)^{1/2}e^{(1)}; \qquad B^{(3)} = B_Z e^{(3)} \tag{829}$$

To complete the derivation, we multiply both sides of Eq. (828) by the phase factor $e^{i\phi}e^{-i\phi}$ to obtain the B cyclic theorem. The latter is therefore equivalent to a commutator relation of the Poincaré group between infinitesimal magnetic field generators. Similarly

$$[E_X, E_Y] = ic^2 B^{(0)}B_Z \tag{830}$$

is equivalent to

$$E^{(1)} \times E^{(2)} = ic^2 B^{(0)}B^{(3)} \tag{831}$$

XIV.   O(3) AND SU(3) INVARIANCE FROM THE RECEIVED FARADAY AND AMPÈRE–MAXWELL LAWS

The received Faraday and Ampère–Maxwell laws [111] in the vacuum asserts that there are fields without sources, so the laws become respectively

$$\nabla \times E + \frac{\partial B}{\partial t} = 0 \tag{832}$$

$$\nabla \times B - \frac{1}{c^2}\frac{\partial E}{\partial t} = 0 \tag{833}$$

These laws are useful but represent cause without effect, that is, fields propagating without sources, and the Maxwell displacement current is an empirical construct, one that happens to be very useful. These two laws can be classified as U(1) invariant because they are derived from a locally invariant U(1) Lagrangian as discussed already. Majorana [114] put these two laws into the form of a Dirac–Weyl equation (Dirac equation without mass)

$$W\psi_1 - ip_2\psi_3 - ip_3\psi_2 = 0$$
$$W\psi_2 - ip_3\psi_1 - ip_1\psi_3 = 0 \tag{834}$$
$$W\psi_3 - ip_1\psi_2 - ip_2\psi_1 = 0$$

in which a combination of fields (SI units) acts as a wave function

$$\psi_i = \frac{1}{c}E_i - iB_i; \qquad i = 1, 2, 3 \tag{835}$$

and in which the quantum ansatz

$$p = -i\hbar\nabla; \qquad i\hbar\frac{\partial}{\partial t} \rightarrow En \equiv W \tag{836}$$

has been used. It is shown in this section that the Majorana equations are O(3) invariant, so the received view is self-contradictory. There is something hidden inside the structure of the Faraday and Ampère–Maxwell laws that removes their U(1) invariance [44]. It can be checked straightforwardly that Eqs. (835) and (834) lead back to Eqs. (833) and (832). In condensed notation, the Majorana equations (834) have the form of the Dirac–Weyl equation:

$$(W + \alpha \cdot p)\Psi = 0 \tag{837}$$

The structure of the Dirac–Weyl equation itself is [46]

$$(\gamma^0 p_0 + \gamma^i p_i)\psi = 0 \tag{838}$$

In Eq. (837), however, the α matrix is an O(3) rotation generator matrix with components

$$\alpha_1 = \begin{bmatrix} 0 & 0 & 0 \\ 0 & 0 & -i \\ 0 & i & 0 \end{bmatrix}; \qquad \alpha_2 = \begin{bmatrix} 0 & 0 & i \\ 0 & 0 & 0 \\ -i & 0 & 0 \end{bmatrix}; \qquad \alpha_3 = \begin{bmatrix} 0 & -i & 0 \\ i & 0 & 0 \\ 0 & 0 & 0 \end{bmatrix} \tag{839}$$

obeying the O(3) invariant commutator equation

$$[\alpha_i, \alpha_k] = -i\varepsilon_{ikl}\alpha_l, \qquad (i, k, l = 1, 2, 3) \tag{840}$$

which is within a factor $\hbar$, the O(3) invariant commutator equation for angular momentum [42,44,46]. Therefore, the Majorana form of Eqs. (832) and (833), namely, Eq. (837), is O(3) invariant, not U(1) invariant. The determinant condition

$$\begin{vmatrix} W & -ip_3 & ip_2 \\ ip_3 & W & -ip_1 \\ -ip_2 & ip_1 & W \end{vmatrix} = 0 \tag{841}$$

gives the relation between energy and momentum for a massless photon, but at the same time, the Majorana equation (837) can be written as a Schrödinger equation

$$H\Psi = W\Psi \tag{842}$$

$$H \equiv -\boldsymbol{\alpha} \cdot \boldsymbol{p} \tag{843}$$

which is usually a nonrelativistic equation for a particle with mass. This is another self-inconsistency of the received Faraday and Ampère–Maxwell laws: the latter ought to be a law for a particle with mass and ought to account for the Lehnert current, as argued already. Operators such as

$$\sum = -i\boldsymbol{\alpha} \times \boldsymbol{\alpha} \tag{844}$$

are intended for the intrinsic spin of the photon, which however, must have eigenvalues $-1, 0, +1$ in order to be consistent with the O(3) angular momentum commutator equation (840). The received view [42,44,46] produces eigenvalues -1 and $+1$ only, which is another self-inconsistency.

Equation (837) can be put into the form of an O(3) covariant derivative acting on the wave function Ψ

$$(\partial_0 - igA_0)\Psi \equiv D_0\Psi = 0 \tag{845}$$

where

$$gA_0 = \boldsymbol{\alpha} \cdot \frac{\boldsymbol{p}}{\hbar} = \boldsymbol{\alpha} \cdot \boldsymbol{\kappa}$$

$$g = \frac{\kappa}{A^{(0)}} \tag{846}$$

So the simplest form of the Majorana equation is

$$D_0\Psi = 0 \tag{847}$$

and is the time-like part of an O(3) covariant derivative acting on the wave function Ψ. The form of Eq. (847) is not, however, fully covariant. The fully covariant form of the vacuum O(3) field equations, as argued already, is collectively Eqs. (318) and (323), which have a Yang–Mills structure. Therefore, the Majorana equation is part of an approximation to the O(3) invariant field equations (318) and (323). As argued already, these latter equations give photon mass through the Higgs mechanism. It does not seem possible to introduce photon mass into the Majorana equation (837), revealing that it is an approximation. This implies that the received Faraday and Ampère–Maxwell laws in the vacuum are also incomplete [42,44] and that U(1) invariant electrodynamics is incomplete. The latter is seen dramatically in interferometry, as argued in this review and elsewhere [44]. For example, a U(1) invariant electrodynamics cannot describe Sagnac interferometry, with platform either at rest or in motion; and cannot describe Michelson interferometry. An O(3) invariant electrodynamics describes both effects self consistently. Oppenheimer [115] derived the same equation as Majorana independently a few years later.

The Majorana equation (837) can also be put in the form

$$\bar{\Psi}(W + \boldsymbol{\alpha} \cdot \boldsymbol{p}) = 0 \tag{848}$$

which is analogous with the corresponding equation for Dirac–Weyl adjoint wave function. The notation of Eq. (848) means that

$$\boldsymbol{p} \equiv i\frac{\overleftarrow{\partial}}{\partial x}; \qquad \bar{\Psi} = (\Psi^*)^T \tag{849}$$

The symmetric energy-momentum tensor $(T_{\mu\nu})$ of electromagnetism in the vacuum can be defined from the Majorana equation using the matrices

$$2\alpha_{00} = 1; \qquad 2\alpha_{01} = \alpha_1; \qquad 2\alpha_{02} = \alpha_2; \qquad 2\alpha_{03} = \alpha_3$$

$$2\alpha_{11} = \begin{bmatrix} -1 & 0 & 0 \\ 0 & 1 & 0 \\ 0 & 0 & 1 \end{bmatrix}; \qquad 2\alpha_{12} = \begin{bmatrix} 0 & -1 & 0 \\ -1 & 0 & 0 \\ 0 & 0 & 0 \end{bmatrix}$$

$$2\alpha_{13} = \begin{bmatrix} 0 & 0 & -1 \\ 0 & 0 & 0 \\ -1 & 0 & 0 \end{bmatrix}; \qquad \alpha_{22} = \begin{bmatrix} 1 & 0 & 0 \\ 0 & -1 & 0 \\ 0 & 0 & 1 \end{bmatrix} \tag{850}$$

$$2\alpha_{23} = \begin{bmatrix} 0 & 0 & 0 \\ 0 & 0 & -1 \\ 0 & -1 & 0 \end{bmatrix}; \qquad 2\alpha_{33} = \begin{bmatrix} 1 & 0 & 0 \\ 0 & 1 & 0 \\ 0 & 0 & -1 \end{bmatrix}$$

where

$$\alpha_{\mu\nu} = \alpha_{\nu\mu} \qquad (\mu, \nu = 0, 1, 2, 3) \tag{851}$$

to give the result:

$$T_{\mu\nu} = \bar{\Psi} \alpha_{\mu\nu} \Psi \tag{852}$$

Only eight of the nine matrices (850) are independent, and they form a basis for the SU(3) group, which is used for strong-field theory [46]. Therefore, the energy-momentum tensor is SU(3) invariant.

Therefore, if we start from a traditionally U(1) invariant pair of equations (832) and (833), we find that they can be put into an O(3) invariant form, and that the concomitant energy-momentum tensor is SU(3) invariant. It is therefore interesting to speculate that an SU(3) invariant electrodynamics can be constructed self-consistently, and is more general than the O(3) invariant form developed here and elsewhere [44]. To view electrodynamics in the vacuum as a U(1) invariant theory is highly restrictive, self-inconsistent [44], and in contradiction with ordinary data such as those from ordinary interferometry and ordinary physical optical effects such as normal reflection [44]. Analyses by Majorana, and later Oppenheimer, show that invariance symmetries can be transmuted among each other for the same set of equations, and so it seems that there is no limit to the internal structural symmetry of electrodynamics on both classical and quantum levels. It is necessary to check each set of equations empirically as the theory is developed. The O(3) invariant electrodynamics [44], for example, has been checked extensively with interferometry and other forms of data [47–62] by several leading specialists. Broad agreement has been reached as to the fact that a paradigm shift has occurred, and that the Maxwell–Heaviside electrodynamics have been replaced by one where there can be invariance under symmetry groups different from U(1). This paradigm shift has extensive consequences throughout physics and the ontology of physics, in chemistry, and in cosmology. The next section, for example, shows that the dark matter in the universe can be thought of as being made up of photons with mass slowed to their rest frame by the Higgs mechanism. The Dirac equation itself is SU(2) invariant [46], and therefore a model of the electron must be either SU(2) or O(3) invariant. Vigier has recently developed an O(3) invariant model of the electron [116] based on the development of an O(3) invariant electrodynamics [42,45,47–62]. The Dirac equation is the relativistically correct form of the Schrödinger equation, and an example of an O(3) invariant Schrödinger equation appears in Eq. (842). We argued earlier that the phase of the Schrödinger equation must be O(3) invariant in general. Taking this line of argument to its logical conclusion, then, Newtonian dynamics are also O(3) invariant. The latter is clear from the fact that Newtonian dynamics takes place in the space of three dimensions described by the rotation group O(3). Another insight is obtained from the fact that the angular momentum commutator relations of quantum mechanics [68] are O(3) invariant.

The O(3) invariance of the Majorana equation (837) can be demonstrated clearly by the use of plane waves

$$
A = \frac{A^{(0)}}{\sqrt{2}}(ii + j)e^{i\phi}
$$

$$
B = \frac{B^{(0)}}{\sqrt{2}}(ii + j)e^{i\phi} \tag{853}
$$

$$
E = \frac{E^{(0)}}{\sqrt{2}}(i - ij)e^{i\phi}
$$

whereon

$$
-iB + \frac{1}{c}E = -2\kappa iA \tag{854}
$$

Therefore, Eqs. (834) reduce to

$$
W\frac{B_1}{\kappa} = -ip_2A_3 + ip_3A_2
$$

$$
W\frac{B_2}{\kappa} = -ip_3A_1 + ip_1A_3 \tag{855}
$$

$$
W\frac{B_3}{\kappa} = -ip_1A_2 + ip_2A_1
$$

Using the four equations

$$
W = p^0 = \hbar gA^0
$$

$$
p_i = \hbar gA_i; \qquad i = 1, 2, 3 \tag{856}
$$

we recover the O(3) invariant definition of the $B^{(3)}$ field and two other similar equations that are equations of the O(3) invariant field tensor as argued already:

$$
B_Z = -i\frac{\kappa}{A^{(0)}}(A_X A_Y^* - A_Y A_X^*) \tag{857}
$$

$$
\cdots
$$

These equations reduce in turn [42,44,47–62] to the B cyclic theorem:

$$
\boldsymbol{B}^{(1)} \times \boldsymbol{B}^{(2)} = iB^{(0)}\boldsymbol{B}^{(3)*} \tag{858}
$$

$$
\cdots
$$

showing that the Majorana equations are the B cyclic theorem. The latter is therefore O(3) gauge-invariant and Lorentz-covariant because the Majorana equations are equivalent to equations with these properties.

XV. SELF-CONSISTENCY OF THE O(3) ANSATZ

A three-way cross-check of the self-consistency of the O(3) ansatz can be carried out starting from Eq. (459), in which A is complex because the electromagnetic field in O(3) electrodynamics carries a topological charge $\kappa/A^{(0)}$. The vector field A in the internal space of O(3) symmetry must depend on x^μ by special relativity and can be written as

$$A = A^{(1)} + A^{(2)} + A^{(3)} \tag{859}$$

where

$$A^{(1)} = A^* = \frac{A^{(0)}}{\sqrt{2}} (i\mathbf{i} + \mathbf{j})e^{-i\phi} = A^{(2)*}$$
$$A^{(3)} = A^{(0)}\mathbf{k} \tag{860}$$

It is now possible to check whether Eq. (459), with its extra vacuum current, is compatible with Eq. (106) of Ref. 44, which is

$$\nabla \times B^{(3)} = \frac{1}{c^2} \frac{\partial E^{(3)}}{\partial t} - ig(A^{(1)} \times B^{(2)} - A^{(2)} \times B^{(1)}) - \frac{g}{\mu_0} D^\mu A^* \times A \tag{861}$$

It follows, from the structure adopted for A in Eq. (860), that

$$D^3 A^* \times A = \partial^3 A^{(2)} \times A^{(1)} + \partial^3 A^{(1)} \times A^{(2)}$$
$$= i\kappa A^{(1)} \times A^{(2)} - i\kappa A^{(2)} \times A^{(1)}$$
$$= igB^{(1)} \times A^{(2)} - igB^{(2)} \times A^{(1)} \tag{862}$$

and so we obtain

$$\nabla \times B^{(3)} = 0 \tag{863}$$

which is self-consistent with the fact that $B^{(3)}$ is irrotational and that $E^{(3)}$ is zero. Another consequence of Eq. (459) is that it gives a vacuum polarization

$$\nabla \cdot P^{(3)} = \rho(\text{vac}) = -\frac{g}{\mu_0} D^0 A^* \times A \tag{864}$$

where $\rho(\text{vac})$ is the vacuum charge density. The vacuum polarization $P^{(3)}$ does not appear from the field tensor [42], but appears from the vacuum charge current density term on the right-hand side of Eq. (459). This vacuum charge current density term must always be present from fundamental gauge principles on the O(3) level. So we have identified the concept of a vacuum charge density as the divergence of a vacuum polarization.

The concepts of O(3) electrodynamics developed in this review and in Ref. 44 scratch the surface of what is possible. The field equations must be solved numerically to obtain all the possible solutions, and checked against empirical data at each stage. Numerical solution of this nature has not yet been attempted. The concept of radiatively induced fermion resonance [44], which might lead to nuclear magnetic resonance and electron spin resonance without the need for permanent magnets, is one obviously useful spinoff of O(3) electrodynamics that has not been explored. These are two of several major advances that could be made within the near future. On the high-energy scale, the concept of higher-symmetry electrodynamics has led to the Crowell boson, which has been detected empirically, and, as reviewed by Crowell in this edition, leads to a novel grand unified theory. The development of O(3) electrodynamics also gives better insight into the energy inherent in the vacuum, and shows beyond reasonable doubt that all optical phenomena are O(3) invariant, a major advance in the 400-year subject of physical optics. During the course of this development, it has been shown that there are several internal inconsistencies [44] in the U(1) invariant electrodynamics, and several instances, in particular interferometry, where the U(1) theory fails. Two typical examples are the Sagnac and Michelson effects. The O(3) invariant electrodynamics succeeds in describing both effects with precision from first principles because of the use of a non-Abelian Stokes theorem for the electromagnetic phase, a theorem that shows that all interferometry is topological in nature and depends on the Evans–Vigier field $B^{(3)}$. The O(3) invariant electrodynamics carries a topological charge $\kappa/A^{(0)}$ in the vacuum, a charge that also acts as the coupling constant of the O(3) covariant derivative. The concept of vacuum charge current density has been established self-consistently on the O(3) invariant level from the first principles of gauge field theory. These are some of several major advances.

Therefore, the empirical and theoretical evidence for the superiority of an O(3) invariant over a U(1) invariant electrodynamics is overwhelming. It is clear that the process of development can be continued, for example, in quantum electrodynamics, electroweak theory, and grand unified theory, and the ontology of these developments can also be studied in parallel.

XVI. THE AHARONOV–BOHM EFFECT AS THE BASIS OF ELECTROMAGNETIC ENERGY INHERENT IN THE VACUUM

The Aharonov–Bohm effect shows that the vacuum is configured or structured, and that the configuration can be described by gauge theory [46]. The result of this experiment is that, in the structured vacuum, the vector potential A can be nonzero while the electric field strength E and magnetic flux density B can be zero. This empirical result is developed in this section by defining an inner space for the gauge theory, and by summarizing some of the results proposed earlier in this review in light of the Aharonov–Bohm effect. Therefore the non-simply connected U(1) vacuum is described by a scalar internal gauge space, and the non-simply connected O(3) vacuum, by a vector internal gauge space. The core of the idea being presented in this section is that the Aharonov–Bohm effect is a local gauge transformation of the true vacuum, where

$$A_\mu = 0 \tag{865}$$

This type of gauge transformation produces a vector potential from the true vacuum. Components of this vector potential are used for the internal gauge space whose Lagrangian is globally invariant. A local gauge transformation of this Lagrangian produces the topological charge

$$g = \frac{\kappa}{A^{(0)}} \tag{866}$$

the electromagnetic field, which carries energy, and the vacuum charge current density first proposed empirically by Lehnert [49] and developed by Lehnert and Roy [45]. These authors have also demonstrated that the existence of a vacuum charge current density implies the existence of photon mass. Empirical evidence for the existence of the vacuum charge current density is available from total internal reflection [45,49]. The source of the energy inherent in vacuo is therefore spacetime curvature introduced through the use of a covariant derivative:

$$D_\mu \equiv \partial_\mu - igA_\mu \tag{867}$$

The product gA_μ in the covariant derivative is, within a factor $\hbar$, an energy momentum. Therefore, photon mass is produced by spacetime curvature because, in a covariant derivative, the axes vary from point to point and there is spacetime curvature. Similarly, mass is produced by spacetime curvature in general relativity. Therefore, spacetime curvature in the configured vacuum implied by the Aharonov–Bohm effect is the source of electromagnetic energy momentum in the vacuum. There is no theoretical upper bound to the magnitude of this electromagnetic energy momentum, which can be picked up by devices, as reviewed in this series by Bearden and Fox (Part 2, Chapters 11 and 12; this part, Chapter 11). Therefore, devices can be manufactured, in principle, to take an unlimited amount of electromagnetic energy from the configured vacuum as defined by the Aharonov–Bohm effect, without violation of Noether's theorem.

The gauge theory developed earlier in this review is summarized for the U(1) and O(3) non-simply connected vacua using the appropriate internal gauge spaces. The earlier calculations are summarized in this section. It has been demonstrated in this series, that there are several advantages of O(3) gauge theory applied to electrodynamics over U(1) gauge theory applied to electrodynamics, but the latter can be used to illustrate the method and to produce the vacuum Poynting theorem that is an expression of Noether's theorem for the structured vacuum. The theory being used is standard gauge theory, so the Noether theorem is conserved; that is, the laws of energy/momentum and charge current conservation are obeyed. The magnitude of the energy momentum is not bounded above by gauge theory, so the Poynting theorem (law of conservation of electromagnetic energy) in the configured vacuum indicates this fact through the presence of a constant of integration whose magnitude is not bounded above. This suggests that the magnitude of the electromagnetic energy in the structured classical vacuum is, in effect, limitless.

The non-simply connected U(1) vacuum is considered first to illustrate the method as simply as possible. This is defined as earlier in this review by the globally invariant Lagrangian density

$$\mathcal{L} = \partial_\mu A \partial^\mu A^* \tag{868}$$

where A and A^* are considered to be independent complex scalar components of the vector potential obtained by gauge transformation of the true vacuum, where $A_\mu = 0$ [46]. The potentials A and A^* are complex because they are associated with a topological charge g, which appears in the covariant derivative when the Lagrangian (868) is subjected to a local gauge transformation. The topological charge g should not be confused with the point charge e on the proton. In the classical structured vacuum, g exists but e does not exist. The two scalar fields are therefore defined as complex conjugates:

$$A = \frac{1}{\sqrt{2}} (A_1 + iA_2) \tag{869}$$

$$A^* = \frac{1}{\sqrt{2}} (A_1 - iA_2) \tag{870}$$

The two independent Euler–Lagrange equations

$$\frac{\partial \mathcal{L}}{\partial A} = \partial_v \left(\frac{\partial \mathcal{L}}{\partial (\partial_v A)} \right); \qquad \frac{\partial \mathcal{L}}{\partial A^*} = \partial_v \left(\frac{\partial \mathcal{L}}{\partial (\partial_v A^*)} \right) \tag{871}$$

produce the independent d'Alembert equations of the structured vacuum:

$$\Box A = 0; \qquad \Box A^* = 0 \tag{872}$$

The Lagrangian (868) is invariant under a global gauge transformation:

$$A \to e^{-i\Lambda}A; \qquad A^* \to e^{i\Lambda}A^* \qquad (873)$$

where Λ is a number. Under a local gauge transformation, however

$$A \to e^{-i\Lambda(x^\mu)}A; \qquad A^* \to e^{i\Lambda(x^\mu)}A^* \qquad (874)$$

where Λ becomes a function of the spacetime coordinate x^μ by special relativity. Under a local gauge transformation [46] of the structured U(1) vacuum defined by the Lagrangian (868), the latter is changed to

$$\mathscr{L} = D_\mu A D^\mu A^* - \frac{1}{4}F^{\mu\nu}F_{\mu\nu} \qquad (875)$$

as argued earlier in this review. Here, $F_{\mu\nu}$ is the U(1) invariant electromagnetic field tensor

$$F_{\mu\nu} = \partial_\mu A_\nu - \partial_\nu A_\mu \qquad (876)$$

where the covariant derivatives are defined by

$$D_\mu A = (\partial_\mu + igA_\mu)A \qquad (877)$$
$$D^\mu A^* = (\partial^\mu - igA^\mu)A^* \qquad (878)$$

Here, A_μ is the vector 4-potential introduced in the vacuum as part of the covariant derivative, and therefore introduced by spacetime curvature. The electromagnetic field and the topological charge g are the results of the invariance of the Lagrangian (868) under local U(1) gauge transformation, in other words, the results of spacetime curvature.

By using the Euler–Lagrange equation

$$\partial_\nu \left(\frac{\partial \mathscr{L}}{\partial(\partial_\nu A_\mu)} \right) = \frac{\partial \mathscr{L}}{\partial A_\mu} \qquad (879)$$

with the Lagrangian (875), we obtain the field equation of the U(1) structured vacuum

$$\partial_\nu F^{\mu\nu} = -igc(A^* D^\mu A - AD^\mu A^*) \qquad (880)$$

a field equation that identifies the vacuum charge current density

$$J^\mu(\text{vac}) \equiv -igc\varepsilon_0(A^* D^\mu A - AD^\mu A^*) \qquad (881)$$

first introduced and developed by Lehnert et al. [45,49]. Equation (880) is an inhomogeneous field equation of the configured U(1) vacuum, and gives rise to the inherent energy of the configured vacuum

$$En = \int J^\mu(\text{vac}) A_\mu \, dV \tag{882}$$

and rate of doing work by the configured vacuum

$$\frac{dW}{dt} = \int \boldsymbol{J}(\text{vac}) \cdot \boldsymbol{E} \, dV \tag{883}$$

where $\boldsymbol{E}$ is the electric field strength of the field tensor $F_{\mu\nu}$. The volume V is arbitrary, and standard methods of U(1) invariant electrodynamics give the Poynting theorem of the U(1) configured vacuum:

$$\frac{dU(\text{vac})}{dt} + \nabla \cdot \boldsymbol{S}(\text{vac}) = -\boldsymbol{J}(\text{vac}) \cdot \boldsymbol{E} \tag{884}$$

Here, $\boldsymbol{S}(\text{vac})$ is the Poynting vector of the U(1) configured vacuum, representing electromagnetic energy flow, and is defined by

$$\nabla \cdot \boldsymbol{S}(\text{vac}) = -\boldsymbol{J}(\text{vac}) \cdot \boldsymbol{E} \tag{885}$$

Integrating this equation gives

$$\boldsymbol{S}(\text{vac}) = -\int \boldsymbol{J}(\text{vac}) \cdot \boldsymbol{E} \, dr + \text{constant} \tag{886}$$

where the constant of integration is not bounded above. The electromagnetic energy flow inherent in the U(1) configured vacuum is not bounded above, meaning that there is an unlimited amount of electromagnetic energy flow available in theory, for use in devices. Some of these devices are reviewed in this edition by Bearden and Fox [chapters given above, in text following Eq. (867)]. Sometimes, the constant of integration is referred to as the "Heaviside component of the vacuum electromagnetic energy flow," and the detailed nature of this component is not restricted in any way by gauge theory. The Poynting theorem (884) is, of course, the result of gauge theory.

   In the non-simply connected O(3) vacuum, the internal gauge space is a vector space rather than the scalar space of the U(1) vacuum. Therefore, we can summarize and collect earlier results of this review using the concept of an O(3) symmetry internal gauge space, a space in which there exist complex vectors A and A^*. The globally invariant Lagrangian density for this internal space is

$$\mathscr{L} = \partial_\mu A \cdot \partial^\mu A^* \tag{887}$$

and the two independent Euler–Lagrange equations are

$$\frac{\partial \mathscr{L}}{\partial A} = \partial_\nu \left(\frac{\partial \mathscr{L}}{\partial_\nu A} \right); \qquad \frac{\partial \mathscr{L}}{\partial A^*} = \partial_\nu \left(\frac{\partial \mathscr{L}}{\partial_\nu A^*} \right) \tag{888}$$

giving the d'Alembert equations

$$\square A = 0; \qquad \square A^* = 0 \tag{889}$$

Under the local O(3) invariant gauge transformation

$$A \rightarrow e^{iJ_i \Lambda_i} A; \qquad A^* \rightarrow e^{-iJ_i \Lambda_i} A^* \tag{890}$$

the Lagrangian (887) becomes, as we have argued earlier

$$\mathscr{L} = D_\mu A \cdot D^\mu A^* - \frac{1}{4} G_{\mu\nu} \cdot G^{\mu\nu} \tag{891}$$

and using the Euler–Lagrange equation

$$\frac{\partial \mathscr{L}}{\partial A_\mu} = \partial_\nu \left(\frac{\partial \mathscr{L}}{\partial (\partial_\nu A_\mu)} \right) \tag{892}$$

the inhomogeneous O(3) invariant field equation is obtained

$$D_\nu G^{\mu\nu} = -g D^\mu A^* \times A \tag{893}$$

as shown in detail earlier. The term on the right-hand side is the O(3) invariant vacuum charge current density that is the non-Abelian equivalent of the right-hand side of Eq. (880). In general, Eq. (893) must be solved numerically, but the presence of a vacuum charge current density gives rise to the energy of the O(3) configured vacuum

$$En(\text{vac}) = \int j^\mu(\text{vac}) \cdot A_\mu \, dV \tag{894}$$

whose source is curvature of spacetime introduced by the O(3) covariant derivative containing the rotation generators J_i of the O(3) group. The curvature of spacetime is also the source of photon mass, in analogy with general relativity, where curvature of spacetime occurs in the presence of mass or a gravitating object.

Therefore, in summary, the empirical basis of the development in this section is that the Aharonov–Bohm effect shows that, in regions where E and B are both zero, A can be nonzero. Therefore, the Aharonov–Bohm effect can be regarded as a local gauge transformation of the true vacuum, defined by $A_\mu = 0$, and the Aharonov–Bohm effect shows that a nonzero A_μ can be generated by a local gauge transformation from regions in which A_μ is zero. Therefore, in a structured vacuum, it is possible to construct a gauge theory whose internal space is defined by components of A_μ in the absence of an electromagnetic field. The latter is generated by a local gauge transformation of components of an A_μ which was generated originally by a local gauge transformation of the true vacuum where $A_\mu = 0$. This concept is true for all gauge group symmetries. It is well known that contemporary gauge theories lead to richly structured vacua whose properties are determined by topology [46]. The Yang–Mills vacuum, for example, is infinitely degenerate [46]. Therefore local gauge transformation can produce electromagnetic energy, a vacuum charge current density, a vacuum Poynting theorem, and photon mass, all interrelated concepts. We reach the sensible conclusion, that in the presence of a gravitating object (a photon with mass), spacetime is curved. The curvature is described through the covariant derivative for all gauge group symmetries. The energy inherent in the vacuum is contained in the electromagnetic field, and the coefficient g is a topological charge inherent in the vacuum. For all gauge group symmetries, the product gA_μ is energy momentum within a factor $\hbar$, indicating clearly that the covariant derivative applied in the vacuum contains energy momentum produced on the classical level by spacetime curvature. This energy momentum, as in general relativity, is not bounded above, so the electromagnetic energy inherent in the classical structured vacuum is not bounded above. There appear to be several devices available that extract this vacuum energy, and these are reviewed in this compilation by Bearden and Fox. In theory, the amount of energy appears to be unlimited.

The Aharonov–Bohm effect depends on the group space of the internal space used in the gauge theory. If this internal space is U(1), the group space is a circle, which is denoted in topology [46] by S^1. This group space is not simply connected because a path that goes twice around a circle cannot be continuously deformed, while staying on the circle, to one that goes around only once [46]. A curve going around the solenoid n times cannot be shrunk to one around m times, where $m \neq n$. The configuration space of the vacuum is therefore not simply connected, and this allows a gauge transform of the pure vacuum, to create what is known as a "pure gauge vacuum" [46]. In U(1) gauge theory, the mathematical reason for the Aharonov–Bohm effect is that the configuration space of the null field (pure gauge vacuum) is a ring, denoted by $S^1 \times R$ in topology [46]. The vector potential in the pure gauge vacuum is derived from a gauge function that maps the gauge space in to the configuration space. These mappings are not all deformable to a constant gauge function χ, which would give a zero $\nabla\chi$ in the pure gauge vacuum and a null Aharonov–Bohm effect. This, then, is the conventional U(1) invariant explanation of the Aharonov–Bohm effect.

The O(3) invariant explanation, as we have seen, uses an internal gauge space that is the physical space O(3). This space is doubly connected [46]. The group space of O(3) is obtained by identifying opposite points on the 3-space S^3, which is the topological description of the unit sphere in four-dimensional Euclidean space, denoted E^4. Opposite points on the 3-space S^3 correspond to the same O(3) transformation. It is possible to show that this space is doubly connected by considering closed curves S^1 in the group space of O(3). One can consider paths [46] that may be shrunk to (are homotopic to) a point and to a straight line. These are the two types of closed path S^1 in the group space of O(3), with the implication that there is one nontrivial vortex in an O(3) gauge theory.

The simplest example of the O(3) invariant Aharonov–Bohm effect is the equation of interferometry

$$\oint_J A^{(3)} \cdot dr = \int B^{(3)} \cdot dS \qquad (895)$$

used in the region outside the solenoid where the vector potential sketched below is nonzero:

$$\bigcirc \!\!\!\!\!\cdot \, \rightarrow A^{(3)} \qquad (896)$$

The line integral is defined over the circular path, exactly as in the O(3) invariant explanation of the Sagnac effect discussed earlier in this review and in Vol. 114, part 2. The key difference between the O(3) and U(1) invariant explanations of the Aharonov–Bohm effect is that, in the former, there is a magnetic field $B^{(3)}$ present at the point of contact with the electrons. Agreement with the empirical data is obtained because

$$\left| B^{(3)} \right| = |B| \qquad (897)$$

that is the total magnetic flux inside the area S must be generated by the static magnetic field B of the solenoid. The fact that we are using an O(3) gauge theory means that the configuration space of the vacuum is doubly connected. As discussed in the technical appendix, the vector potential $A^{(3)}$ in Eq. (895) can be regarded as having been generated by an O(3) gauge transformation that leaves $B^{(3)}$ invariant. Equation (895) is the consequence of a round trip in spacetime using parallel transport with O(3) covariant derivatives. Therefore, the simplest O(3) invariant explanation of the Aharonov–Bohm effect simply means that it is an interferometric effect, very similar in nature to the O(3) invariant explanation of the Sagnac effect or Michelson interferometry.

The simplest example of the generation of energy from a pure gauge vacuum is to consider the case of an electromagnetic potential plane wave defined by

$$A = A_1 i + A_2 j = \frac{A^{(0)}}{\sqrt{2}} (ii + j)e^{-i(\omega t - \kappa Z)} \tag{898}$$

The pure gauge vacuum is then defined by

$$A \neq 0; \qquad E = 0; \qquad B = 0 \tag{899}$$

and a Lagrangian density can be constructed which is proportional to

$$\mathcal{L} = \partial_\mu A \cdot \partial^\mu A \tag{900}$$

A global gauge transformation of A in the pure gauge vacuum is equivalent to a rotation of A through an angle Λ [46], producing a conserved quantity Q as the result of the invariance of the action under the global gauge transformation. It can be shown as follows that Q is proportional to conserved electromagnetic kinetic energy

$$En = \frac{1}{\mu_0} \int B^{(0)2} \, dV \tag{901}$$

generated by the global gauge transformation of the pure gauge vacuum, which, in turn, is generated from the pure vacuum by a local gauge transformation.

For plane waves

$$A_1 = i \frac{A^{(0)}}{\sqrt{2}} e^{-i(\omega t - \kappa Z)}; \qquad A_2 = \frac{A^{(0)}}{\sqrt{2}} e^{-i(\omega t - \kappa Z)} \tag{902}$$

In a U(1) invariant theory, the pure gauge vacuum is defined by a scalar internal gauge space in which there exist the independent complex scalar fields:

$$A = \frac{1}{\sqrt{2}} (A_1 + iA_2); \qquad A^* = \frac{1}{\sqrt{2}} (A_1 - iA_2) \tag{903}$$

These are complex scalar fields because there is an invariant topological charge present, defined by

$$g = \frac{\kappa}{A^{(0)}} \tag{904}$$

The Lagrangian density produced by these scalar fields is, as we have seen

$$\mathscr{L} = \partial_\mu A \partial^\mu A^* \tag{905}$$

and the global gauge transformation is defined by

$$A \rightarrow e^{-i\Lambda} A; \qquad A^* \rightarrow e^{i\Lambda} A^* \tag{906}$$

This type of transformation is not dependent on spacetime and is purely internal [46] in Noether's theorem. Under a global gauge transformation, Noether's theorem gives the conserved current

$$J^\mu = igc(A^* \partial^\mu A - A \partial^\mu A^*) \tag{907}$$

with a vanishing 4-divergence and a conserved topological charge:

$$Q = \int J^0 \, dV \tag{908}$$

From Eq. (907), the conserved topological charge Q is

$$Q = \frac{2c}{A^{(0)}} \int \kappa^2 A^{(0)2} \, dV = \frac{2c}{A^{(0)}} \int B^{(0)2} \, dV \tag{909}$$

which can be written as

$$Q = \frac{2c\mu_0}{A^{(0)}} En \tag{910}$$

where

$$En = \frac{1}{\mu_0} \int B^{(0)2} \, dV \tag{911}$$

is a conserved kinetic electromagnetic energy. For a monochromatic plane wave in the vacuum, the quantity g is also conserved because κ and $A^{(0)}$ do not change. Therefore it has been demonstrated that, in a pure gauge vacuum defined by the plane wave A, conserved electromagnetic energy density is generated by a global gauge transformation, which is a rotation of A through the angle Λ.

This is the simplest example of the generation of kinetic electromagnetic energy by a gauge transformation of a pure gauge vacuum defined initially by a nonzero A and zero E and B. The more complete description of energy generated from the pure gauge vacuum is given by a local gauge transformation, as argued already in this review on the U(1) and O(3) levels. It is to be noted that the conserved quantity Q has the following properties:

1. It is time independent.
2. It does not depend on the charge on the proton.
3. It is a classical quantity.
4. It is not integer-valued and when A is real it vanishes.

It can be shown as follows that the transition from a pure vacuum to a pure gauge vacuum is described by the spacetime translation generator of the Poincaré group. The pure vacuum on the U(1) invariant level is described by the field equations:

$$\partial_\mu \tilde{F}^{\mu\nu} \equiv 0 \tag{912}$$

$$\partial_\mu F^{\mu\nu} = 0 \tag{913}$$

with

$$\tilde{F}^{\mu\nu} = 0; \qquad F^{\mu\nu} \equiv 0 \tag{914}$$

So the kinetic electromagnetic energy term in the Lagrangian

$$\mathscr{L} = -\frac{1}{4} F_{\mu\nu} F^{\mu\nu} \tag{915}$$

is zero. In the pure gauge vacuum, the ordinary derivative is replaced by the covariant derivative, so the field equations (912) and (913) become

$$\partial_\mu \tilde{F}^{\mu\nu} = -iA_\mu \tilde{F}^{\mu\nu} \tag{916}$$

$$\partial_\mu F^{\mu\nu} = -iA_\mu F^{\mu\nu} \tag{917}$$

where A_μ is defined by

$$A_\mu = -\frac{i}{g}(\partial_\mu S)S^{-1} \tag{918}$$

but where the fields $\tilde{F}^{\mu\nu}$ and $F^{\mu\nu}$ are still zero. Therefore

$$F^{\mu\nu} = \partial^\mu A^\nu - \partial^\nu A^\mu = 0 \tag{919}$$

and the contribution of the field to the energy in a pure gauge vacuum is zero. However, there occurs an energy change from a pure vacuum to a pure gauge vacuum, an energy change proportional to gA_μ. The origin of this energy change is topological; that is, the energy change can be traced to the replacement of the ordinary derivative ∂_μ by the covariant derivative D_μ.

Essentially, this replacement means that the spacetime changes from one that is conformally flat to one that is conformally curved; in other words, the axes vary from point to point whenever a covariant derivative is used for any gauge group symmetry. It is this variation of the axes that introduces energy into a pure gauge vacuum. The covariant derivative in the latter is

$$D_\mu = \partial_\mu - igA_\mu \tag{920}$$

which can be written using the rule $i\partial_\mu = \kappa_\mu$ as

$$\kappa_\mu \rightarrow \kappa_\mu + \kappa'_\mu \tag{921}$$

This expression is equivalent [42] to

$$P_\mu \rightarrow P_\mu + P'_\mu \tag{922}$$

where P_μ is the spacetime translation generator of the Poincaré group. Within a factor $\hbar$, the spacetime translation generator is the energy-momentum 4-vector. It becomes clear that the use of a covariant derivative introduces energy momentum into the vacuum, in this case a pure gauge vacuum. Lagrangians, consisting of components of A_μ in the pure gauge vacuum when subjected to a local gauge transformation, give the electromagnetic field and its source, the vacuum charge/current density, first introduced empirically by Lehnert [49].

In the final part of this section, the method of local gauge transformation is outlined in detail to show how the electromagnetic field and conserved vacuum charge current density emerge from the local gauge transformation of the pure gauge vacuum. The illustration is given for convenience in a U(1) invariant theory, and leans heavily on the excellent account given by Ryder [46, pp. 94ff.]. We therefore consider a local gauge transformation of a pure gauge vacuum with scalar components A and A^*:

$$\begin{aligned} A &\rightarrow \exp(-i\Lambda(x^\mu))A \\ A^* &\rightarrow \exp(-i\Lambda(x^\mu))A^* \end{aligned} \tag{923}$$

For $\Lambda \ll 1$

$$\delta A = -i\Lambda A \tag{924}$$

and

$$\partial_\mu A \rightarrow \partial_\mu A - i(\partial_\mu \Lambda)A - i\Lambda(\partial_\mu A) \tag{925}$$

Therefore

$$\delta(\partial_\mu A) = -i\Lambda(\partial_\mu A) - i(\partial_\mu \Lambda)A \tag{926}$$

and

$$\begin{aligned}
\delta A^* &= i\Lambda A^* \\
\delta(\partial_\mu A^*) &= i\Lambda(\partial_\mu A^*) + i(\partial_\mu \Lambda)A^*
\end{aligned} \tag{927}$$

The effect of the local gauge transform is to introduce an extra term $\partial_\mu \Lambda$ in the transformation of the derivatives of fields. Therefore, $\partial_\mu A$ does not transform covariantly, that is, does not transform in the same way as A itself. These extra terms destroy the invariance of the action under the local gauge transformation, because the change in the Lagrangian is

$$\delta\mathcal{L} = \frac{\partial\mathcal{L}}{\partial A}\delta A + \frac{\partial\mathcal{L}}{\partial(\partial_\mu A)}\delta(\partial_\mu A) + (A \rightarrow A^*) \tag{928}$$

where $(A \rightarrow A^*)$ denotes the two additional terms in A^*. Substituting the Euler–Lagrange equation (888) into the first term, and using Eqs. (924)–(926), gives

$$\begin{aligned}
\partial\mathcal{L} &= \partial_\mu \frac{\partial\mathcal{L}}{\partial(\partial_\mu A)}(-i\Lambda A) + \frac{\partial\mathcal{L}}{\partial(\partial_\mu A)}(-i\Lambda\partial_\mu A - iA\partial_\mu \Lambda) \\
&= -i\Lambda\partial_\mu \frac{\partial\mathcal{L}A}{\partial(\partial_\mu A)} - i\frac{\partial\mathcal{L}}{\partial(\partial_\mu A)}(\partial_\mu \Lambda)A + (A \rightarrow A^*)
\end{aligned} \tag{929}$$

The first term is a total divergence, so the corresponding change in the action is zero. Using

$$\mathcal{L} = (\partial_\mu A)(\partial^\mu A^*) - m^2 A^* A \tag{930}$$

for the Lagrangian then gives

$$\delta\mathcal{L} = i\partial_\mu \Lambda(A^*\partial^\mu A - A\partial^\mu A^*) = J^\mu \partial_\mu \Lambda \tag{931}$$

where the (SI) current is given by Eq. (907), in reduced units

$$J^\mu = i(A^*\partial^\mu A - A\partial^\mu A^*) \tag{932}$$

The action is therefore not invariant under local gauge transformation. To restore invariance the four potential, A_μ must be introduced into the pure gauge vacuum to give the Lagrangian

$$
\begin{aligned}
\mathscr{L}_1 &= -gJ^\mu A_\mu \\
&= -ig(A^*\partial^\mu A - A\partial^\mu A^*)A_\mu
\end{aligned}
\tag{933}
$$

where g is the topological charge in the vacuum, defined in such a way that gA_μ has the same SI units as κ_μ. On the U(1) level, local gauge transformation means that

$$
A_\mu \rightarrow A_\mu + \frac{1}{g}\partial_\mu \Lambda
\tag{934}
$$

so that

$$
\begin{aligned}
\delta\mathscr{L}_1 &= -g(\delta J^\mu)A_\mu - gJ^\mu(\delta A_\mu) \\
&= -g(\delta J^\mu)A_\mu - J^\mu\partial_\mu\Lambda
\end{aligned}
\tag{935}
$$

The action is still not invariant under a local gauge transformation, however, because of the presence of the term $-g(\delta J^\mu)A_\mu$ on the right-hand side of Eq. (935), a term in which

$$
\begin{aligned}
\delta J^\mu &= i\delta(A^*\partial^\mu A - A\partial^\mu A^*) \\
&= 2A^*A\partial^\mu\Lambda
\end{aligned}
\tag{936}
$$

so that

$$
\delta\mathscr{L} + \delta\mathscr{L}_1 = -2gA_\mu(\partial^\mu\Lambda)A^*A
\tag{937}
$$

Therefore, another term must be added to the Lagrangian $\mathscr{L}$:

$$
\mathscr{L}_2 = e^2A_\mu A^\mu A^*A
\tag{938}
$$

Using Eq. (934), we find that

$$
\delta\mathscr{L}_2 = 2g^2A_\mu\delta A^\mu A^*A = 2gA_\mu(\partial^\mu\Lambda)A^*A
\tag{939}
$$

so that

$$
\delta\mathscr{L} + \delta\mathscr{L}_1 + \delta\mathscr{L}_2 = 0
\tag{940}
$$

The total Lagrangian $\mathscr{L} + \mathscr{L}_1 + \mathscr{L}_2$ is now invariant under the local gauge transformation because of the introduction of the 4-potential A_μ, which couples to the current J_μ of the complex A of the pure gauge vacuum. The field A_μ also contributes to the Lagrangian, and since $\mathscr{L} + \mathscr{L}_1 + \mathscr{L}_2$ is invariant, an extra term $\mathscr{L}_3$ appears, which must also be gauge-invariant. This can be so only if the electromagnetic field is introduced

$$F_{\mu\nu} = \partial_\mu A_\nu - \partial_\nu A_\mu \tag{941}$$

so that

$$\mathscr{L}_3 = -\frac{1}{4} F^{\mu\nu} F_{\mu\nu} \tag{942}$$

The total invariant Lagrangian is therefore

$$\begin{aligned}
\mathscr{L}_{\text{tot}} &= \mathscr{L} + \mathscr{L}_1 + \mathscr{L}_2 + \mathscr{L}_3 \\
&= (\partial_\mu A + i g A_\mu A)(\partial^\mu A^* - i g A^\mu A^*) - m^2 A^* A - \frac{1}{4} F^{\mu\nu} F_{\mu\nu}
\end{aligned} \tag{943}$$

The Lehnert field equation is obtained from this Lagrangian using the Euler–Lagrange equation

$$\frac{\partial \mathscr{L}}{\partial A_\mu} - \partial_\nu \left(\frac{\partial \mathscr{L}}{\partial (\partial_\nu A_\mu)} \right) = 0 \tag{944}$$

giving in SI units

$$\partial_\nu F^{\mu\nu} = -i g c (A^* D^\mu A - A D^\mu A^*) \tag{945}$$

It is noted that the Lehnert charge current density

$$J^\mu = -i \varepsilon_0 g c (A^* D^\mu A - A D^\mu A^*) \tag{946}$$

is gauge-covariant and also conserved, and thus cannot be gauged to zero by any method of gauge transformation. It is the direct result of a local gauge transformation on a pure gauge vacuum and acts as the source of the vacuum electromagnetic field $F^{\mu\nu}$, as discussed already. The covariant current (946) is conserved because

$$\partial_\mu J^\mu = 0 \tag{947}$$

XVII. INTRODUCTION TO THE WORK
OF PROFESSOR J. P. VIGIER

We append what we believe to be a comprehensive listing of the publications of Professor Jean-Pierre Vigier. They represent a wide range of topics from the interpretation of quantum mechanics, particle physics, cosmology, and relativistic physics. What is remarkable about this list is not just the breadth of topics, but the philosophical consistency that underlies the physics. Firmly rejecting the orthodox interpretation of quantum mechanics, particles of all types are, at all times, regarded as objectively existing entities with their own internal structure. Particles are guided by pilot waves, so the dualism of orthodoxy is firmly rejected in favor of realist ontology.

What follows is a brief account of Professor Vigier's life and career as related to one of us (S. Jeffers) in a series of conversations held in Paris during the summer of 1999. A more complete version of these conversations will appear in a book being compiled by Apeiron Press and the Royal Swedish Academy to mark the 80th birthday of Professor Vigier. A comprehensive biography of this remarkable man, whose life has witnessed major revolutions both in physics and in politics (his twin passions), remains to be written.

"Great physicists fight great battles"—so wrote Professor Vigier in an essay he wrote in a tribute to his old friend and mentor Louis de Broglie. However, this phrase could be applied equally well to Vigier himself. He has waged battle on two fronts—within physics and within politics. Now almost 80, he still continues to battle.

He was born on January 16, 1920 to Henri and Françoise (née Dupuy) Vigier. He was one of three brothers, Phillipe (deceased) and François, currently Professor of Architecture at Harvard University. His father was Professor of English at the École Normale Supérieure—hence Vigier's mastery of that language. He attended an international school in Geneva at the time of the Spanish civil war. This event aroused his intense interest in politics, as most of his school friends were both Spanish and Republicans. Vigier was intensely interested in both physics and mathematics, and was sent by his parents to Paris in 1938 to study both subjects. For Vigier, mathematics is more like an abstract game, his primary interest being in physics as it rests on two legs, the empirical and the theoretical.

All the young soldiers were sent to Les Chantiers de la Jeunesse, and it was there that he joined the Communist Party. The young radicals were involved in acts of sabotage near the Spanish border, such as oiling the highways to impede the progress of the fascists. At that time, the French Communist Party was deeply split concerning the level of support to be given to the Résistance. A few leaders went immediately to the Résistance, while others, like Thorez, wavered. In the period before the Nazi attack on the Soviet Union, the party equivocated with respect to the Résistance. At that time, Vigier was in a part of France controlled by the famous communist leader, Tillion, who had participated in the revolt of the sailors in the Black Sea in 1918. Tillion immediately organized groups of resistance fighters called the *Organisation Spéciale*. Vigier was involved in bombing campaigns against both the Nazis and Vichy collaborators in the Free Zone.

In Geneva, Vigier was involved in communicating between the French military communist staff and Russia, until he was arrested at the French border in the spring of 1942 and taken to Vichy. There, the French police interrogated him as he was carrying coded documents. Two police officers brought him by train from Vichy to Lyon to be delivered into the hands of the notorious Klaus Barbie. Fortunately, the train was bombed by the English, and Vigier managed to jump through a window, escaped to the mountains, and resumed his activities with the Résistance until the end of the war. He became an officer in the FTP movement (Francs-Tireurs et Partisans, meaning sharpshooters and supporters). When De Gaulle returned to France, part of the Résistance forces were converted to regular army units. The cold war started almost immediately after the defeat of the Germans. Vigier was still a member of the French General Staff while completing the requirements for a Ph.D. in mathematics in Geneva. Then the communists were kicked out of the General Staff and Vigier went to work for Joliot-Curie. He, in turn, lost his job for refusing to build an atomic bomb for the French government. Vigier became unemployed for a while and then learned, through an accidental meeting with Joliot-Curie, that Louis de Broglie was looking for an assistant. When he met De Broglie, the only questions asked were "Do you have a Ph.D. in mathematics?" and "Do you want to do physics?" He was hired immediately in 1948, and with no questions asked about his political views. Although Secretary of the French Academy of Science, de Broglie was marginalized within physics circles given his well-known opposition to the Copenhagen interpretation of quantum mechanics. Notwithstanding his Nobel prize, de Broglie had difficulty in finding an assistant. Vigier entered the CNRS (Centre national de la recherche scientifique) and worked with De Broglie until his retirement. Vigier's political involvement at that time included responsibility for the French communist student movement.

In 1952, a visiting American physicist named Yevick, gave a seminar at the CNRS on the recent ideas of David Bohm. Vigier reports that upon hearing of this work, De Broglie became radiant and commented that these ideas were first considered by himself a long time ago. Bohm had gone beyond De Broglie's original ideas however. De Broglie charged Vigier with reading all of Bohm's works in order to prepare a seminar. De Broglie went back to his old ideas, and both he and Vigier started working on the causal interpretation of quantum mechanics. At the 1927 Solvay Congress, de Broglie had been shouted down, but now, following the work of Bohm, there was renewed interest in his idea that wave and particle could coexist, eliminating the need for dualism. Vigier recalls that at that time, the catholic archbishop of Paris who exclaimed that everyone knew that Bohr was right, upbraided de Broglie, and how de Broglie could possibly believe otherwise. Although a devout Christian, he was inclined to materialist philosophy in matters of physics.

Vigier comments on his days with de Broglie that he was a very timid man who would meticulously prepare his lectures in written form—in fact, his books are largely compendia of his lectures. He also recalls one particular incident that illustrates de Broglie's commitment to physics. Vigier was in the habit of meeting weekly with de Broglie to take direction as to what papers he should be reading, and what calculations he should be focusing on. On one of these occasions, he was waiting in an anteroom for his appointment with de Broglie. Also waiting was none other than the French Prime Minister Edgar Faure who had come on a courtesy visit in order to discuss his possible membership in the French Academy. When the door finally opened, de Broglie called excitedly for Mr. Vigier to enter as he had some important calculations for him to do, and as for the prime minister, well he could come back next week! For De Broglie, physics took precedence over politicians, no matter how exalted.

De Broglie sent Vigier to Brazil to spend a year working on the renewed causal interpretation of quantum mechanics with David Bohm. Thereafter, Yukawa got in touch with de Broglie, with the result that Vigier went to Japan to work with him for a year. Vigier comments that about the only point of disagreement between him and de Broglie was over nonlocality. De Broglie never accepted the reality of nonlocal interactions, whereas Vigier himself accepts the results of experiments such as Aspect's that clearly imply that such interactions exist.

His response to the question "Why do we do science?" is that, in part, it is to satisfy curiosity about the workings of nature, but it is also to contribute to the liberation from the necessity of industrial labor. With characteristic optimism, he regards the new revolution of digital technology as enhancing the prospects for a society based on the principles enunciated by Marx, a society whose members are freed from the necessity of arduous labor—this, as a result of the application of technological advances made possible by science.

TECHNICAL APPENDIX A: CRITICISMS OF THE U(1) INVARIANT THEORY OF THE AHARONOV–BOHM EFFECT AND ADVANTAGES OF AN O(3) INVARIANT THEORY

In this appendix, the U(1) invariant theory of the Aharonov–Bohm effect [46] is shown to be self-inconsistent. The theory is usually described in terms of a holonomy consisting of parallel transport around a closed loop assuming values in the Abelian Lie group U(1) [50] conventionally ascribed to electromagnetism. In this appendix, the U(1) invariant theory of the Aharonov–Bohm effect is criticized in several ways with reference to the well–known test of the effect verified empirically by Chambers [46] and a holonomy consisting of parallel transport with O(3) covariant derivatives is applied to the Aharonov–Bohm effect, eliminating the self-inconsistencies of the U(1) invariant theory. Close similarities between the O(3) invariant theories of the Aharonov–Bohm and Sagnac effects are revealed.

It is well known that the change in phase difference of two electron beams in the Aharonov–Bohm effect is described in the conventional U(1) invariant theory by

$$\Delta\delta = \frac{e}{\hbar}\int \nabla \times A \cdot dS = \frac{e}{\hbar}\int B \cdot dS \qquad (A.1)$$

where the magnetic flux density B of the solenoid is related to the vector potential A by

$$B = \nabla \times A \qquad (A.2)$$

Outside the solenoid, however

$$B = \nabla \times A = 0 \qquad (A.3)$$

which means that the change in phase difference in Eq. (A.1) is zero, and that there is no Aharonov–Bohm effect, contrary to the observation. In the U(1) theory, an attempt is made to remedy this self-inconsistency by using the fact that A is not zero outside the solenoid, and so can be represented by a function of the type

$$A = \nabla\chi \qquad (A.4)$$

The Aharonov–Bohm effect is then described by [46]

$$\Delta\delta = \frac{e}{\hbar}\oint \nabla\chi \cdot dr = \frac{e}{\hbar}[\chi]_0^{2\pi} = \frac{e}{\hbar}\int B \cdot dS \qquad (A.5)$$

using the assertion that χ is not single-valued. The analytical form of χ is

$$\chi = \frac{BR^2}{2}\phi \qquad (A.6)$$

where B is the magnitude of the flux density B inside the solenoid, R is the radius of the solenoid, and ϕ is an angle, the ϕ component in cylindrical polar coordinates.

However, the interpretation in (A.5) is self-inconsistent in several ways:

1. Outside the solenoid, $B = 0$, so $\chi = 0$ from Eq. (A.5), and there is no Aharonov–Bohm effect, contradicting Eq. (A.5).
2. For any function χ, a basic theorem of vector analysis states that

$$\nabla \times (\nabla\chi) \equiv 0 \tag{A.7}$$

This theorem is also valid for a periodic function, so outside the solenoid

$$B = \nabla \times A = 0 \tag{A.8}$$

for χ, and from Eq. (A.1), the Aharonov–Bohm effect again disappears. For example, if we take the angle

$$\chi = \sin^{-1}\frac{x}{a}, \qquad (|x| < a) \tag{A.9}$$

then:

$$\nabla\chi = (a^2 - x^2)^{-1/2}\boldsymbol{i} \tag{A.10}$$

and

$$\nabla \times (\nabla\chi) \equiv 0 \tag{A.11}$$

or if we take the periodic function

$$\chi = \cos x; \qquad \nabla\chi = -\sin x \, \boldsymbol{i} \tag{A.12}$$

then

$$\nabla \times (\nabla\chi) = 0 \tag{A.13}$$

Another criticism of Eq. (A.5) is that the empirical result is obtained only if $\chi \to \chi + 2\pi$, whereas in general, $\chi \to \chi + 2\pi n$ for a periodic function. So the value of n has to be artificially restricted to $n = 1$ to obtain the correct analytical and empirical result.

The basic problem in a U(1) invariant description of the Aharonov–Bohm effect is that the field B is zero outside the solenoid, so outside the solenoid, $\nabla \times A$ is zero, whereas A is not zero [46]. At the same time, the U(1) Stokes theorem states that

$$\int \nabla \times A \cdot dS = \oint A \cdot dr \tag{A.14}$$

so that the holonomy $\oint A \cdot dr$ is zero and the effect again disappears for A outside the solenoid because the left-hand side in Eq. (A.14) is zero.

In the O(3) invariant theory of the Aharonov–Bohm effect, the holonomy consists of parallel transport using O(3) covariant derivatives and the internal gauge space is a physical space of three dimensions represented in the basis ((1),(2),(3)). Therefore, a rotation in the internal gauge space is a physical rotation, and causes a gauge transformation. The core of the O(3) invariant explanation of the Aharonov–Bohm effect is that the Jacobi identity of covariant derivatives [46]

$$\sum_{\sigma,\mu,\nu} [D_\sigma, [D_\mu, D_\nu]] \equiv 0 \qquad (A.15)$$

is identical for all gauge group symmetries with the non-Abelian Stokes theorem:

$$\oint D_\mu \, dx^\mu + \frac{1}{2} \int [D_\mu, D_\nu] \, d\sigma^{\mu\nu} \equiv 0 \qquad (A.16)$$

for any covariant derivative in any gauge group symmetry. In the O(3) invariant theory, the following three identities therefore exist

$$\oint A^{(i)} \cdot dr = \int B^{(i)} \cdot dS; \qquad i = 1, 2, 3 \qquad (A.17)$$

that is, one for each of the three internal indices (1), (2), and (3). The quantities in Eq. (A.17) are linked by the following vacuum definition:

$$B^{(3)*} \equiv -ig A^{(1)} \times A^{(2)} \qquad (A.18)$$

The vector potential $A^{(3)}$ and the longitudinal flux density $B^{(3)}$ are both phaseless, so Eq. (A.17) with $i = 3$ is the invariant equation needed for a description of the Aharonov–Bohm effect

$$\oint A^{(3)} \cdot dr = \int B^{(3)} \cdot dS \qquad (A.19)$$

The Aharonov–Bohm effect is therefore caused by a gauge transformation in a vacuum whose configuration space is O(3). The effect is a gauge transformation of Eq. (A.19) into the region outside the solenoid because the left- and right-hand sides of Eq. (A.19) exist only inside the solenoid. In general field theory, gauge transformations of the potential and of the field are defined through the rotation operator

$$S = \exp(iM^a \Lambda^a(x^\mu)) \qquad (A.20)$$

where M^a are the group rotation generators and Λ^a are angles that depend on the 4-vector χ^μ. Under a general gauge field transformation

$$A'_\mu = SA_\mu S^{-1} - \frac{i}{g}(\partial_\mu S)S^{-1} \tag{A.21}$$

$$G'_{\mu\nu} = SG_{\mu\nu}S^{-1} \tag{A.22}$$

In the O(3) invariant expression (A.19), the vector potential transforms according to

$$A^{(3)} \to A^{(3)} + \frac{1}{g}\frac{\partial\alpha}{\partial Z}e^{(3)} \tag{A.23}$$

and the magnetic field transforms as

$$B^{(3)} \to B^{(3)} \tag{A.24}$$

At the point of contact with the electrons, therefore, in the region outside the solenoid, the Aharonov–Bohm effect is caused by

$$\frac{1}{g}\oint\frac{\partial\alpha}{\partial Z}e^{(3)} \cdot dr = \int B^{(3)} \cdot dS \tag{A.25}$$

in other words, there is a magnetic field present at the point of contact with the electrons and the left-hand side of Eq. (A.25) is physically significant. The reason for this is that the O(3) symmetry internal space of the theory is the physical space of three dimensions: the vacuum with configuration space O(3), a nonsimply connected configuration space. Therefore, none of the self-inconsistencies present in the U(1) invariant theory are present in the O(3) invariant theory of the Aharonov–Bohm effect. Agreement with the empirical data is obtained through the O(3) invariant equation:

$$\Delta\delta = \frac{e}{\hbar}\int B^{(3)} \cdot dS \tag{A.26}$$

and this analysis clearly demonstrates the simplicity with which the novel O(3) electrodynamics removes the self-inconsistencies of the U(1) description.

TECHNICAL APPENDIX B: O(3) ELECTRODYNAMICS FROM THE IRREDUCIBLE REPRESENTATIONS OF THE EINSTEIN GROUP

In Part 1 of this three-volume set, Sachs [117] has demonstrated that electromagnetic energy is available from curved spacetime by using the irreducible representations of the Einstein group. The metric is expressed using a quaternion-valued 4-vector, q^μ, with 16 components. If we define the scalar components of q^μ as

$$q^\mu = (q^0, q^1, q^2, q^3) \tag{B.1}$$

the quaternion-valued 4-vector is defined as

$$\sigma^\mu q_\mu = (q^0 \sigma^0, q_1 \sigma^1, q_2 \sigma^2, q_3 \sigma^3) \tag{B.2}$$

In the flat spacetime limit, the q^μ is replaced by the 4-vector made up of Pauli matrices:

$$\sigma^\mu = (\sigma^0, \sigma^1, \sigma^2, \sigma^3) \tag{B.3}$$

The field tensor given by Sachs in his Eq. (4.19) contains, in general, longitudinal as well as transverse components under all conditions, including the vacuum defined as Riemannian spacetime. Sachs' Eq. (4.16) shows that the electromagnetic canonical energy-momentum tensor (T^μ) is spacetime curvature in precisely the same way that gravitational canonical energy momentum is spacetime curvature. Therefore, code must be developed to solve Sachs' Eqs. (4.16) and (4.18) in order to understand electromagnetic phenomena in general relativity for any given situation. Sachs' Eq. (4.16) shows that electromagnetic energy is available in the vacuum, defined as Riemannian curved spacetime, and can be used to power devices.

The electromagnetic field propagating through the curved spacetime vacuum always has a source, part of whose structure is the quaternion-valued T^μ. This source is the most general form of the Lehnert vacuum 4-current [45,49]. General relativity [117] also shows that there is no electromagnetic field if there is no curvature, so a field cannot propagate through the flat spacetime vacuum of Maxwell–Heaviside theory. The latter's notion of transverse plane waves propagating in the vacuum without a source is therefore inconsistent with both general relativity and causality, because there cannot be cause without effect (i.e., field without source).

In general, all the off-diagonal elements of the quaternion-valued commutator term [the fifth term in Sachs' Eq. (4.19)] exist, and in this appendix, it is shown, by a choice of metric, that one of these components is the $B^{(3)}$ field discussed in the text. The $B^{(3)}$ field is the fundamental signature of O(3) electrodynamics discussed in Vol. 114, part 2. In this appendix, we also give the most general form of the vector potential in curved spacetime, a form that also has longitudinal and transverse components under all conditions, including the vacuum. In the Maxwell–Heaviside theory, on the other hand, the vector potential in the vacuum is generally considered to have transverse components only in the radiation zone, a result that is inconsistent with general relativity, O(3) electrodynamics, and Lehnert's extended electrodynamics.

In Vol. 114, part 1, Sachs has shown that the most general form of the electromagnetic field tensor is

$$F_{\rho\gamma} = Q\left(\frac{1}{4}\left(\kappa_{\rho\lambda}q^{\lambda}q_{\gamma}^{*} + q_{\gamma}q^{\lambda*}\kappa_{\rho\lambda} + q^{\lambda}\kappa_{\rho\lambda}^{+}q_{\gamma}^{*} + q_{\gamma}\kappa_{\rho\lambda}^{+}q^{\lambda*}\right)\right.$$
$$\left. + \frac{1}{8}\left(q_{\rho}q_{\gamma}^{*} - q_{\gamma}q_{\rho}^{*}\right)R\right) \tag{B.4}$$

where $\kappa_{\rho\lambda}$ is the curvature tensor defined in terms of the spin–affine connection [117]

$$\kappa_{\rho\lambda} \equiv \partial_{\rho}\Omega_{\lambda} - \partial_{\lambda}\Omega_{\rho} - \Omega_{\lambda}\Omega_{\rho} + \Omega_{\rho}\Omega_{\lambda} \tag{B.5}$$

where $Q \equiv \Phi^{(0)}$ has the SI units of magnetic flux (Weber), and where R is the scalar curvature in inverse square meters. The asterisk in Eq. (B.4) denotes quaternion conjugate, which entails [117] reversing the sign of the time component of the quaternion-valued q^{μ}. Thus, if

$$q^{\mu} = (q^0, q^1, q^2, q^3) \tag{B.6}$$

then

$$q^{\mu*} = (-q^0, q^1, q^2, q^3) \tag{B.7}$$

The metric in the irreducible representation of the Einstein group is proportional to [117]

$$q^{\mu}q^{\nu*} + q^{\nu}q^{\mu*} \neq 0 \tag{B.8}$$

and replaces the familiar metric $g^{\mu\nu}$ generated by the reducible representations of the Einstein group and used to describe gravitation. Therefore, the replacement of reducible by irreducible representations unifies the gravitational and electromagnetic fields inside the structure of one Lie group: the Einstein group. This important result shows that electromagnetic energy is available from curved spacetime in the same way that gravitational energy is available from curved spacetime, a well-accepted concept.

The demonstration by Sachs [117] that electromagnetic energy is available from the vacuum (Riemannian curved spacetime) generates the most precise classical electromagnetic theory available. Its notable successes [42] include the ability to reproduce the Lamb shift in hydrogen without renormalization; the ability to produce the Planck distribution of blackbody radiation classically: the correct prediction of the lifetime of the muon state and electron–muon mass splitting. The Sachs theory also shows the existence of physical longitudinal and time-like components of the vector potential in the vacuum, predicts a small but nonzero neutrino and photon mass, and establishes grounds for charge quantization. These precise predictions firmly establish the possibility of obtaining electromagnetic energy from the vacuum, and firmly establish the existence of the $\boldsymbol{B}^{(3)}$ field as one of the possible longitudinal components of the tensor (B.4) in the vacuum (Riemannian curved spacetime). It follows that O(3) electrodynamics is also a theory of curved spacetime, and that the extended electrodynamics of Lehnert is a transitional theory in flat spacetime, but one that has several notable advantages over the Maxwell–Heaviside theory, as reviewed by Lehnert in Part 2 of Vol. 114. The Lehnert theory also gives the $\boldsymbol{B}^{(3)}$ field in the vacuum.

Equation (B.4) shows that the electromagnetic field in general relativity is non-Abelian, and acts as its own source. The gravitational field also acts as its own source, in that the gravitational field is a source of energy that, in turn, is gravitation. In gravitational theory, the Einstein curvature tensor is equated with the canonical energy-momentum tensor. In electromagnetic theory, the same applies, as in Sachs' Eq. (4.16). Gravitation is therefore an obvious manifestation of energy from the vacuum; electromagnetic energy from the vacuum is also available in nature, a result that has been confirmed experimentally to the precision of the Lamb shift. Therefore, there is an urgent need to develop code to solve the Sachs field equations for any given experimental setup. This code will show precisely the amount of electromagnetic energy that is available in the vacuum (Riemannian curved spacetime).

The quaternion-valued metric q^μ can be written as

$$q^\mu = \begin{bmatrix} q_0 + q_Z & q_X - iq_Y \\ q_X + iq_Y & q_0 - q_Z \end{bmatrix} \tag{B.9}$$

Therefore

$$q_X = \begin{bmatrix} 0 & q_X \\ q_X & 0 \end{bmatrix}; \qquad q_Y = \begin{bmatrix} 0 & -iq_Y \\ iq_Y & 0 \end{bmatrix} \tag{B.10}$$

and

$$q_X q_Y - q_Y q_X = i(q_X q_Y + q_Y q_X)\sigma_Z \tag{B.11}$$

Similarly

$$q_X q_Y + q_Y q_X = i(q_X q_Y - q_Y q_X)\sigma_Z \tag{B.12}$$

In order for both $q_X q_Y + q_Y q_X$ and $q_X q_Y - q_Y q_X$ to have real-valued parts, the individual scalar components q_X and q_Y must be complex-valued in general.

We recover the structure of O(3) electrodynamics in quaternion-valued form by a choice of metric

$$q_X = \frac{A_X^{(1)}}{A^{(0)}} = -ie^{i\phi}; \qquad q_Y = \frac{A_Y^{(2)}}{A^{(0)}} = e^{-i\phi} \tag{B.13}$$

where ϕ is an electromagnetic phase factor and where $\boldsymbol{A}^{(1)} = \boldsymbol{A}^{(2)*}$ is part of the vector potential of O(3) electrodynamics as described in the text, and whose phase factor is a Wu–Yang phase factor as developed in Vol. 114, part 2. The choice of metric in Eq. (B.13) leads to

$$q_X q_Y - q_Y q_X = 2\sigma_Z \tag{B.14}$$

giving the phaseless and longitudinally directed $\boldsymbol{B}^{(3)}$ field of O(3) electrodynamics

$$B^{(3)} = \pm \frac{1}{4} \Phi^{(0)} R \tag{B.15}$$

where $\Phi^{(0)}$ is a magnetic flux in webers. The two signs in Eq. (B.15) represent left and right circular polarization. Within a factor of $\frac{1}{4}$, the result (B.15) is the same as that obtained [42] using a unification scheme based on an antisymmetric Ricci tensor.

It can therefore be inferred that O(3) electrodynamics is a theory of Riemannian curved spacetime, as is the homomorphic SU(2) theory of Barrett [50]. Both O(3) and SU(2) electrodynamics are substructures of general relativity as represented by the irreducible representations of the Einstein group, a continuous Lie group [117]. The $\boldsymbol{B}^{(3)}$ field in vector notation is defined in curved spacetime by

$$\boldsymbol{B}^{(3)*} = -ig\boldsymbol{A}^{(1)} \times \boldsymbol{A}^{(2)} \tag{B.16}$$

while in the flat spacetime of Maxwell–Heaviside theory it vanishes:

$$\boldsymbol{B}^{(3)*} = -ig\boldsymbol{A} \times \boldsymbol{A} = 0 \tag{B.17}$$

From general relativity, it may therefore be inferred that the $\boldsymbol{B}^{(3)}$ field must exist, and that it is a physically meaningful magnetic flux density in the vacuum. The phaseless $\boldsymbol{B}^{(3)}$ component is one of an infinite set of longitudinal, and in general oscillatory, components of the field tensor (B.4). This result has been tested experimentally to the precision of the Lamb shift.

In general, all the off-diagonal elements of the commutator term in Eq. (B.4) exist and are nonzero. For example

$$q_0 q_z^* - q_z q_0^* = 2q_0 q_z \sigma_z \tag{B.18}$$

which is a real and physical, longitudinally directed, electric field component in the vacuum. Such a component is in general phase-dependent. If the metric is chosen so that

$$q_0 = q_z = \frac{A^{(3)}}{A^{(0)}} = 1 \tag{B.19}$$

we recover the longitudinal and phaseless electric field component:

$$E^{(3)} = \pm \frac{1}{4} c \Phi R \tag{B.20}$$

There is in-built parity violation in the Sachs theory [76], so the distinction between axial and polar vector is lost. This is the reason why the Sachs theory allows a phaseless $E^{(3)}$ to exist while O(3) electrodynamics does not. There is no parity violation in O(3) electrodynamics. The question arises as to what is the interpretation of the phaseless $E^{(3)}$ in general relativity. The empirical evidence for a radiated $B^{(3)}$ field is reviewed in Vol. 114, Part 2 and in the text of this review chapter. An example is the inverse Faraday effect, which is magnetization produced by circularly polarized radiation. However, there is no electric equivalent of the inverse Faraday effect; that is, there is no polarization produced by a circularly polarized electromagnetic field. The phaseless $E^{(3)}$ present in the vacuum in general relativity may, however, be interpretable as the Coulomb field between two charges in the radiation zone. The Coulomb field is missing in Maxwell–Heaviside theory, where the electric field is pure transverse, and as pointed out by Dirac [42], this result cannot be a proper description of the fact that there a longitudinal and phase-free Coulomb field between transmitter and receiver must always be present.

The most general form of the vector potential can be obtained by writing the first four terms of Eq. (B.4) as

$$F_{\rho\gamma,1} \equiv \partial_\rho A_\gamma^* - \partial_\gamma A_\rho^* \tag{B.21}$$

The vector potential is therefore obtained as

$$A_\gamma^* = \frac{Q}{4} \int (\kappa_{\rho\lambda} q^\lambda + q^\lambda \kappa_{\rho\lambda}^+) q_\gamma^* \, dx^\rho \tag{B.22}$$

and can be written as

$$A_\gamma^* = q_\gamma^* \left(\frac{Q}{4} \int (\kappa_{\rho\lambda} q^\lambda + q^\lambda \kappa_{\rho\lambda}^+) \, dx^\rho \right) \tag{B.23}$$

In order to prove that

$$\int q_\gamma^* dx^\rho = q_\gamma^* \int dx^\rho \tag{B.24}$$

we can take examples, giving results such as

$$\int q_1^* dx^2 = - \int q_X \, dY = -q_X \int dY \tag{B.25}$$

because q_X has no functional dependence on Y. The overall structure of the field tensor is therefore the quaternion-valued

$$F_{\rho\gamma} = C(\partial_\rho q_\gamma^* - \partial_\gamma q_\rho^*) + D(q_\rho q_\gamma^* - q_\gamma q_\rho^*) \tag{B.26}$$

where C and D are coefficients:

$$C \equiv \frac{Q}{4} \int (\kappa_{\rho\lambda} q^\lambda + q^\lambda \kappa_{\rho\lambda}^+) \, dx^\rho$$
$$D \equiv \frac{QR}{8} \tag{B.27}$$

Equation (B.26) has the structure of a quaternion-valued non-Abelian gauge field theory. If we denote

$$\frac{D}{C^2} = -ig \tag{B.28}$$

Eq. (B.26) becomes

$$F_{\rho\gamma} = \partial_\rho A_\gamma^* - \partial_\gamma A_\rho^* - ig(A_\rho A_\gamma^* - A_\gamma A_\rho^*) \tag{B.29}$$

which is a general gauge field theory where A_γ^* is quaternion-valued. The rules of gauge field theory developed in the text and in part 2 of Vol. 114 can be applied to Eq. (B.29); for example, Eq. (B.29) is derived from a holonomy in curved spacetime.

References

1. M. Planck, *Verh. Phys. Ges.* **2**, 202,237 (1900).
2. I. Newton, *Optiks*, Dover, New York, 1952.
3. C. Huygens, *Treatise on Light*, Dover, New York, 1962.

4. A. J. Fresnel, Mémoires Acad. Sci. **5**, 1821–1822 (1819).

5. M. W. Evans, "The present status of the quantum theory of light," in S. Jeffers, S. Roy, J. P. Vigier, and G. Hunter (Eds.), *Proc. Symp.* (Aug. 25–29, 1997) *in Honour of Jean-Pierre Vigier,* Kluwer, Dordrecht, 1998.

6. P. W. Milloni, *The Wave Particle Dualism—a tribute to Louis De Broglie on his 90th Birthday,* Reidel, Dordrecht, 1984.

7. A. Einstein, *Ann. Phys.* **17**, 132 (1905).

8. L. Mandel, *Prog. Opt.* **13**, 27 (1976).

9. A. Einstein, *Phys. Z.* **10**, 185 (1909).

10. P. A. M. Dirac, *The Principles of Quantum Mechanics* Clarendon, Oxford, 1930 (4th ed. by Oxford Univ. Press, London, 1958), p. 9.

11. H. Paul, *Rev. Mod. Phys.* **58**(1) (1986).

12. R. L. Pfleegor and L. Mandel, *Phys. Rev.* **159**, 1084 (1967).

13. G. I. Taylor, *Proc. Cambridge Phil. Soc.* **15**, 114 (1909).

14. R. Hanbury-Brown and R. Q. Twiss, *Nature* **177**, 27 (1956).

15. P. Grangier, G. Roger, and A. Aspect, *Europhys. Lett.* **1**, 173 (1986).

16. A. Aspect and P. Grangier, *Hyp. Int.* **37**, 3 (1987).

17. A. Aspect, in A. I. Miller (Ed.), *Sixty Two Years of Uncertainty,* Plenum, New York, 1990.

18. A. Einstein, *Phys. Z.* **18**, 121 (1917).

19. A. H. Compton, *Phys. Rev.* **21**, 483 (1923).

20. G. Tarozzi, *Lett. Nuovo Cimento* **42**, 438 (1985).

21. J. R. Croca, *Found. Phys.* **17**, 971 (1987).

22. J. R. Croca, A. Garrucio, and F. Selleri, *Found. Phys. Lett.* **1**, 101 (1988).

23. S. Jeffers and J. Sloan, *Found. Phys. Lett.* **7**, 333 (1994).

24. X. Y. Zou, T. Grayson, L. J. Wang, and L. Mandel, *Phys. Rev. Lett.* **68**, 3667 (1992).

25. J. R. Croca, "Beyond non-causal quantum mechanics," Part 2, Chap. 8, this compilation.

26. G. Garola and A. Arcangelo Rossi, *The Foundations of Quantum Mechanics-Historical Analysis and Open Questions,* Kluwer, Dordrecht, 1995.

27. F. Selleri, in A. van der Merwe (Ed.), *Quantum Paradoxes and Physical Reality,* Kluwer, Dordrecht, 1990.

28. D. Home, *Conceptual Foundations of Quantum Mechanics—an Overview from Modern Perspectives,* Plenum, New York, 1997.

29. Y. Mizobuchi and Y. Ohtake, *Phys. Lett. A* **168**, 1 (1992).

30. L. de Broglie, *Ann. Phys.* (Paris) **3**, 22 (1925).

31. L. de Broglie, Nobel lecture, 1929.

32. G. Lochak, preface to L. de Broglie, *Les incertitudes d'Heisenberg et l'interprétation probabilitiste de la mécanique ondulatoire,* Gauthier-Villars, Paris, 1982.

33. J. P. Vigier, in G. Hunter, S. Jeffers, and J. P. Vigier (Eds.), *Causality and Locality in Modern Physics,* Kluwer, Dordrecht, 1998.

34. S. Jeffers, B. Lehnert, N. Abramson, and L. Chebotarev, *Jean-Pierre Vigier and the Stochastic Interpretation of Quantum Mechanics,* Apeiron, Montreal, 2000.

35. C. Philippidis, C. Dewdney, and B. J. Hiley, *Nuovo Cimento* **52B**, 15 (1979).

36. R. D. Prosser, *Int. J. Theor. Phys.* **15**, 181 (1976).

37. R. D. Prosser, S. Jeffers, and J. Desroches, "Maxwellian analysis of reflection of refraction" (1997), in S. Jeffers et al. (Eds.), *The Present Status of the Quantum Theory of Light*, Kluwer Dordrecht, 2000.

38. S. Jeffers, R. D. Prosser, W. C. Berseth, G. Hunter, and J. Sloan in L. Carin and L. B. Felsen (Eds.), *Ultra-Wideband Short Pulse Electromagnetics*, Vol. 2, Plenum, New York, 1995.

39. J. S. Bell, *Physics* **1**, 195 (1964) (reprinted in *Speakable and Unspeakable in Quantum Mechanics*, Cambridge Univ. Press, Cambridge, UK, 1987, p. 14).

40. A. Afriat and F. Selleri, *The Einstein, Podolsky and Rosen Paradox in Atomic, Nuclear and Particle Physics*, Plenum, New York, 1998.

41. S. F. Fry and T. Walther, *Adv. Atom. Mol. Phys.* **42**, 1–27 (2000).

42. M. W. Evans, J. P. Vigier, S. Roy, and S. Jeffers, *The Enigmatic Photon*, in 5 volumes, Kluwer, Dordrecht, 1994–1999.

43. A. Proca, *Compt. Rend. Acad. Sci. Paris* **190**, 1377; **191**, 26 (1930).

44. M. W. Evans, "O(3) Electrodynamics," Part 2, this volume (i.e., this 3-vol. compilation).

45. B. Lehnert and S. Roy, *Extended Electrodynamics*, World Scientific, Singapore, 1998.

46. L. H. Ryder, *Quantum Field Theory*, 2nd ed., Cambridge Univ. Press, Cambridge, UK, 1987.

47. M. W. Evans, J. P. Vigier, and S. Roy, *The Enigmatic Photon*, Vol. 4, Kluwer, Dordrecht, 1997.

48. M. W. Evans, *Physica B* **182**, 227,237 (1992).

49. B. Lehnert, *Optik* **99**, 113 (1995); *Phys. Scripta* **59**, 204 (1996), and a review in Ref. 47.

50. T. W. Barrett, in A. Lakhtakia (Ed.), *Essays on the Formal Aspects of Electromagnetic Theory*, World Scientific, Singapore, 1993; T. W. Barrett, in T. W. Barrett and D. M. Grimes (Eds.), *Advanced Electromagnetism*, World Scientific, Singapore, 1995; T. W. Barrett, in C. R. Keys and M. W. Evans (Eds.), *The New Electrodynamics*, Apeiron, 2000.

51. M. W. Evans, *The Enigmatic Photon*, Vol. 5, Kluwer, Dordrecht, 1999.

52. M. W. Evans et al., AIAS group papers "Higher symmetry electrodynamics," Special Issue 2000 to *J. New Energy* **4**(3), Trenergy, Salt Lake City, UT, 1999.

53. M. W. Evans et al., AIAS group papers, U.S. Dept. Energy, web site http://www.ott.doe.gov/electromagnetic/.

54. M. W. Evans et al., AIAS group papers, *Phys. Scripta* **61**, 79,287,513 (2000); two papers in press (2000).

55. M. W. Evans et al., AIAS group paper, *Optik* **111**, 53 (2000); two papers in press (2000).

56. M. W. Evans, L. B. Crowell et al., AIAS group papers, *Found. Phys. Lett.* **11**, 595 (1998); **12**, 187,251,373,475,579 (1999); **13**, 179,193, in press (2000); *Found. Phys.* (in press).

57. M. W. Evans and L. B. Crowell, *Apeiron* **5**, 165 (1998).

58. M. W. Evans et al., AIAS group papers, *Frontier Perspect.* **7**, 7 (1998); **8**, 15 (1999).

59. M. W. Evans and L. B. Crowell, *Classical and Quantum Electrodynamics and the $B^{(3)}$ Field*, World Scientific, Singapore, 2000.

60. H. F. Harmuth, in T. W. Barrett and D. M. Grimes (Eds.), *Advanced Electrodynamics*, World Scientific, Singapore, 1995, pp. 506ff.; H. F. Harmuth, *Information Theory Applied to Spacetime Physics*, World Scientific, Singapore, 1993; H. F. Harmuth and M. G. M. Husain, *Propagation of Electromagnetic Signals*, World Scientific, Singapore, 1994.

61. M. W. Evans, *Found. Phys.* **24**, 1519,1671 (1994).

62. G. 't Hooft, *Nuclear Phys.* **B79**, 276 (1974).

63. A. M. Polyakov, *JETP Lett.* **20**, 194 (1974); *Sov. Phys. JETP* **41**, 988 (1976).

64. E. T. Whittaker, *Math. Ann.* **57**, 333 (1903) (On the partial differential equations of mathematical physics).

65. E. T. Whittaker, *Proc. Lond. Math. Soc.* **1**, 367 (1904) (On an expression of the electromagnetic field due to electrons by means of two scalar potentials).

66. J. D. Jackson, *Classical Electrodynamics*, Wiley, New York, 1962.

67. R. L. Mills, W. R. Good, J. Phillips, and A. I. Popov, U.S. Patent 6,024,935 (2000) (on lower-energy hydrogen methods and structures).

68. P. W. Atkins, *Molecular Quantum Mechanics*, 2nd ed., Oxford Univ. Press, Oxford, 1983.

69. L. D. Barron, *Physica B* **190**, 307 (1993).

70. M. W. Evans, *Physica B* **190**, 310 (1993).

71. A. Lakhtakia, *Physica B* **191**, 362 (1993).

72. D. M. Grimes, *Physica B* **191**, 367 (1993).

73. M. W. Evans, *Found. Phys. Lett.* **8**, 563 (1995).

74. A. D. Buckingham and L. Parlett, *Science* **264**, 1748 (1994).

75. A. D. Buckingham, *Science* **266**, 665 (1994).

76. M. W. Evans and L. B. Crowell, *Found. Phys. Lett.* **11**, 595 (1998).

77. A. Lakhtakia, *Found. Phys. Lett.* **8**, 183 (1995).

78. M. W. Evans, *Found. Phys. Lett.* **8**, 187 (1995).

79. G. L. J. A. Rikken, *Opt. Lett.* **20**, 846 (1995).

80. M. W. Evans, *Found. Phys. Lett.* **9**, 61 (1996).

81. S. J. van Enk, *Found. Phys. Lett.* **9**, 183 (1996).

82. M. W. Evans, *Found. Phys. Lett.* **9**, 191 (1996).

83. E. Comay, *Chem. Phys. Lett.* **261**, 601 (1996).

84. M. W. Evans and S. Jeffers, *Found. Phys. Lett.* **9**, 587 (1996).

85. E. Comay, *Found. Phys. Lett.* **10**, 245 (1997).

86. M. W. Evans, *Found. Phys. Lett.* **10**, 255 (1997).

87. E. Comay, *Physica B* **222**, 150 (1996).

88. M. W. Evans, *Found. Phys. Lett.* **10**, 403 (1997).

89. M. Y. A. Raja, W. N. Sisk, M. Youssaf, and D. Allen, *Appl. Phys. Lett.* **67**, 2123 (1995).

90. M. Y. A. Raja, W. N. Sisk, and D. Allen, *Appl. Phys. B* **64**, 79 (1997).

91. M. W. Evans, *Apeiron* **4**, 80 (1997).

92. M. W. Evans, *Found. Phys. Lett.* **10**, 487 (1997).

93. V. V. Dvoeglazov, *Found. Phys. Lett.* **10**, 383 (1997).

94. E. Comay, *Physica A* **242**, 522 (1997).

95. G. Hunter, *Chem. Phys.* **242**, 331 (1999).

96. M. W. Evans et al., AIAS group paper, *Phys. Scripta* (in press).

97. G. Hunter, *Apeiron* **7**, 17 (2000).

98. M. W. Evans, *Apeiron* **7**, 29 (2000).

99. V. V. Dvoeglazov, *Apeiron* **6**, 227 (1999).

100. E. Comay, *Apeiron* **6**, 233 (1999).

101. B. Simon, *Phys. Rev. Lett.* **51**, 2170 (1983).

102. T. T. Wu and C. N. Yang, *Phys. Rev. D* **12**, 3857 (1975).

103. M. V. Berry, *Proc. Roy. Soc.* **392A**, 45 (1984).

104. Y. Aharonov and J. Anandan, *Phys. Rev. Lett.* **58**, 1593 (1987).

105. S. Pancharatnam, *Proc. Indian Acad. Sci.* **44A**, 247 (1956).

106. B. Bhandari and J. Samuel, *Phys. Rev. Lett.* **60**, 1211 (1988).

107. A. Tomita and R. Y. Chiao, *Phys. Rev. Lett.* **57**, 937,940 (1986).

108. F. Selleri (Ed.), *Open Questions in Special Relativity*, Apeiron, Montreal, 1998.

109. C. Cohen-Tannoudji, J. Dupont-Roc, and G. Grynberg, *Photons and Atoms*, Wiley, New York, 1989.

110. C. W. Misner, K. S. Thorne, and J. A. Wheeler, *Gravitation*, Freeman, San Francisco, 1973.

111. L. D. Landau and E. M. Lifshitz, *The Classical Theory of Fields*, Pergamon, Oxford, 1975.

112. B. L. Silver, *Irreducible Tensor Theory*, Academic, New York, 1976.

113. A. O. Barut, *Electrodynamics and Classical Theory of Fields and Particles*, Macmillan, New York, 1964.

114. E. Majorana, manuscripts at the Domus Galileaeana, Pisa, Italy, notebooks **2**, 101; **3**, 11,160; **15**; **16**; **17**, 83,159.

115. J. R. Oppenheimer, *Phys. Rev.* **38**, 725 (1931).

116. J. P. Vigier, manuscript in preparation.

117. M. Sachs, chapter 11, Part 1, this volume.

PUBLICATIONS OF PROFESSOR JEAN-PIERRE VIGIER

1. J. P. Vigier, in S. Jeffers (Ed.), *Jean-Pierre Vigier and the Stochastic Interpretation of Quantum Mechanics*, ISBN 0-96-836895-6, Apeiron (Aug. 31, 2000).

2. J. P. Vigier, "Photon mass and Heaviside force," *Phys. Lett. A* **270**(5), 221–231 (2000).

3. J. P. Vigier, New quantum mechanical tight bound states and 'cold fusion' experiments, *Phys. Lett. A* **265**(3), 163–167 (2000).

4. J. P. Vigier, "New 'hidden' parameters describing internal motions within extended particle elements associated with a Feynman–Gellmann type causal electron model," this volume, 2000.

5. J. P. Vigier et al., "Classical electrodynamics without the Lorentz condition: Extracting energy from the vacuum," *Physica Scripta* **61**(5), 513–517 (2000).

6. J. P. Vigier et al., "Equations of the Yang–Mills theory of classical electrodynamics," *Optik* (Stuttgart) **111**(2), 53–56 (2000).

7. J. P. Vigier (co-author), The "AIAS papers" [contains approx. 60 papers on O(3) electrodynamics], Special Issue 2000 to *J. New Energy* **4**(3), Trenergy, Salt Lake City, UT, 1999.

8. B. Drag, Z. Mari, and J. P. Vigier, "Interpretation of 'cold fusion' in terms of new Bohr orbits resulting from spin-spin and spin-orbit couplings in external magnetic fields: Theoretical and experimental evidence," *Int. Conf. Cold Fusion* (ICCF-7), Vancouver, BC, Canada, (April 1998), Part 3 of proceedings [*New Energy News* **6**(3), 147 (1998), published by Fusion Information Center Inc., distributed by ENECO, 391-B, Chipeta Way, Salt Lake City, UT 84108].

9. J. P. Vigier, "The energy spectrum of the hydrogen atom with magnetic spin-orbit and spin-spin interactions," *Phys. Lett. A* **237**(6), 349–353 (1998).

10. G. Hunter, S. Jeffers, and J. P. Vigier (Eds.), *Causality and Locality in Modern Physics and Astronomy: Open Questions and Possible Solutions*; *Proc. Symp.* (Aug. 25–29, 1997) *in Honour of Jean-Pierre Vigier*; ISBN 0-79-235227-0, Kluwer, Dordrecht, 1998.

11. M. W. Evans, J. P. Vigier, S. Roy, and S. Jeffers, *The Enigmatic Photon*, Vols. 1–4, Kluwer Dordrecht, 1994–1997.

12. J. P. Vigier, "Possible consequences of an extended charged particle model in electromagnetic theory," *Phys. Lett. A* **235**(5), 419–431 (1997).

13. S. Jeffers et al. (Eds.), *The Present Status of the Quantum Theory of Light, Proc. Symp. Honour of Jean-Pierre Vigier*, Fundamental Theories of Physics, Vol. 80, ISBN 0-79-234337-9, Kluwer, Dordrecht, 1997.

14. J. P. Vigier, "New non-zero photon mass interpretation of the Sagnac effect as direct experimental justification of the Langevin paradox," *Phys. Lett. A* **234**(2), 75–85 (1997).

15. J. P. Vigier, "Relativistic interpretation (with non-zero photon mass) of the small ether drift velocity detected by Michelson, Morley and Miller," *Apeiron* **4**(2–3) (Special Issue: *The $B^{(3)}$ Field: Beyond Maxwell*) (April–July 1997).

16. J. P. Vigier, "On cathodically polarized Pd/D systems," *Phys. Lett. A* **221**(12), 138–140 (1996).

17. C. Fenech, and J. P. Vigier, "Variation of local heat energy and local temperatures under Lorentz transformations," *Phys. Lett. A* **215**(5–6) (247–253) (1996).

18. J. P. Vigier, "Derivation of inertial forces from the Einstein–de Broglie–Bohm causal stochastic interpretation of quantum mechanics," *Found. Phys.* **25**(10), 1461–1494 (1995).

19. J. P. Vigier, "Fundamental problems of quantum physics," *Apeiron* **2**(4) (1995).

20. J. P. Vigier, "Possible test of the reality of superluminal phase waves and particle phase space motions in the Einstein–de Broglie–Bohm causal stochastic interpretation of quantum mechanics," *Found. Phys.* **24**(1), 61–83 (1994).

21. C. Fenech, and J. P. Vigier, "Thermodynamical properties/description of the de Broglie–Bohm pilot waves," *Phys. Lett. A* **182**(1), 37–43 (1993).

22. R. Antanasijevic, L. Lakicevic, Z. Maric, D. Zevic, A. Zaric, and J. P. Vigier, "Preliminary observations on possible implications of new Bohr orbits (resulting from electromagnetic spin-spin and spin-orbit coupling) in 'cold' quantum mechanical fusion processes appearing in strong 'plasma focus' and 'capillary fusion' experiments, *Phys. Lett. A* **180**(1–2), 25–32 (1993).

23. M. C. Combourieu and J. P. Vigier, "Absolute space-time and realism in Lorentz invariant interpretation of quantum mechanics," *Phys. Lett. A* **175**(5), 269–272 (1993).

24. P. R. Holland and J. P. Vigier, "David Joseph Bohm: 1917–1992," *Found. Phys.* **23**(1) (1993).

25. J. P. Vigier, "From Descartes and Newton to Einstein and de Broglie," *Found. Phys.* **23**(1) (1993).

26. J. P. Vigier, "Present experimental status of the Einstein-de Broglie theory of light," in H. Ezawa and Y. Murayama (Eds.), *Quantum Control and Measurement: Proc. ISQM Satellite Workshop* (ARL, Hitachi, Hatoyama, Saitama, Aug. 28–29, 1992), ISBN 0-44-489561-2, North-Holland, Amsterdam, 1993.

27. M. Bozic, Z. Maric, and J. P. Vigier, "De Broglian probabilities in the double slit experiment," *Found. Phys.* **22**(11), 1325–1344 (1992).

28. P. Garbaczewski and J. P. Vigier, "Quantum dynamics from the Brownian recoil principle," *Phys. Rev. A* (Special Issue: *Statistical Physics, Plasmas, Fluids, and Related Interdisciplinary Topics*) **46**(8), 4634–4638 (1992).

29. P. Garbaczewski and J. P. Vigier, "Brownian motion and its descendants according to Schrödinger," *Phys. Lett. A.* **167**, 445–451 (1992).

30. N. Cufaro-Petroni and J. P. Vigier, "Single-particle trajectories and interferences in quantum mechanics, *Found. Phys.* **22**(1), 1–40 (1992).

31. J. P. Vigier, "Do quantum particles travel in real space-time? Experimental evidence and theoretical implications," in H. Atmanspacher and H. Scheinraber (Eds.), *Information Dynamics*, Plenum, New York, 1991.

32. M. Rambaut and J. P. Vigier, *Method and Device for Producing Fusion Energy from a Fusible Material*, Fr. Patent WO 91/16713 (PCT/FR91/00305).

33. M. Rambaut and J. P. Vigier, *Process and Device for Producing Fusion Energy from a Fusible Material*, Fr. Patent WO 91/15016 (PCT/FR91/00225).

34. H. Rauch and J. P. Vigier, reply to comment on "Proposed neutron interferometry test of Einstein's 'Einweg' assumption in the Bohr–Einstein controversy," *Phys. Lett. A* **157**(45), 311–313 (1991).

35. P. R. Holland and J. P. Vigier, Comment on "Experimental test of the de Broglie guided wave theory for photons" (with reply), *Phys. Rev. Lett.* **67**(3), 402–403 (1991).

36. J. V. Narlikar, J. C. Pecker, and J. P. Vigier, "Does a possible laboratory observation of a frequency anisotropy of light result from a non-zero photon mass?" *Phys. Lett. A* **154**(5–6), 203–209 (1991).

37. J. V. Narlikar, J. C. Pecker, and J. P. Vigier, "Some consequences of a spatially varying cosmological constant in a spherically symmetric distribution of matter," *J. Astrophys. Astron.* **12**(1), 7–16 (1991).

38. J. P. Vigier, "Explicit mathematical construction of relativistic non-linear de Broglie waves described by three-dimensional (wave and electromagnetic) solitons 'piloted' (controlled) by corresponding solutions of associated linear Klein–Gordon and Schrödinger equations," *Found. Phys.* **21**(2) (1991).

39. R. D. Pearson, *Intelligence Behind the Universe* (includes a recommendation for part of the gravitational section by Professor J. P. Vigier, Assessor for *Physics Letters A* and one from astrophysicist S. Nicholls, ISBN 0-94-782321-2, Bathford Publishers, Curbar Edge, 2 Rowlands Close, Bathford, Bath, Somerset, BA1 7TZ, Dec. 1990.

40. H. Rauch and J. P. Vigier, "Proposed neutron interferometry test of Einstein's 'Einweg' assumption in the Bohr–Einstein controversy," *Phys. Lett. A* **151**(6–7), 269–275 (1990).

41. M. Rambaut and J. P. Vigier, "Ampère forces considered as collective non-relativistic limit of the sum of all Lorentz interactions acting on individual current elements: Possible consequences for electromagnetic discharge stability and Tokamak behavior," *Phys. Lett. A* **148**(5), 229–238 (1990).

42. J. C. Pecker, J. P. Vigier, and T. Jaakkola, "Spatial fluctuation of the hubble 'Constant," *Apeiron* (6) (1990).

43. J. P. Vigier, "Evidence for non-zero mass photons associated with a vacuum-induced dissipative red-shift mechanism," *IEEE Trans. Plasma Sci.* **18**(1), 64–72 (1990).

44. J. P. Vigier, "Real physical paths in quantum mechanics: Equivalence of the Einstein–de Broglie and Feynman points of view on quantum particle behaviour," in S.-I. Kobayashi et al. (Eds.), *Proc. 3rd Int. Symp. Foundations of Quantum Mechanics in the Light of New Technology* (ISQM Tokyo '89), Central Research Laboratoty, Hitachi Ltd, Tokyo, Japan, Aug. 28–31, 1989), ISBN 4-89027-003-5, Physical Society of Japan, Tokyo, 1990.

45. J. P. Vigier, "Equivalence between the Einstein–de Broglie and Feynman interpretations of quantum mechanics, in J. Mizerki, A. Posiewnik, J. Pykacz, and M. Zorowski (Eds.), *Problems in Quantum Physics II; Gdansk '89 Recent and Future Experiments and Interpretations* (Sept. 18–23, 1989), ISBN 9-81-020177-X, World Scientific, Singapore, 1990. pp. 168–202.

46. J. P. Vigier, "Comments on the 'uncontrollable' character of non-locality, in E. I. Bitsakis and C. A. Nicolaides (Eds.), *Concept of Probability*, Kluwer, Dordrecht, 1989.

47. M. Rambaut and J. P. Vigier, "The simultaneous existence of EM Grassmann–Lorentz forces (acting on charged particles) and ampère forces (acting on charged conducting elements) does not contradict relativity theory, *Phys. Lett. A* **142**(8–9), 447–452 (1989).

48. P. N. Kaloyerou and J. P. Vigier, "Evolution time Klein–Gordon equation and derivation of its non-linear counterpart," *J. Phys. A* (Special Issue: *Mathematical and General*) **22**(6), 663–673, (1989).

49. J. P. Vigier, "Particular solutions of a non-linear Schrödinger equation carrying particle-like singularities represent possible models of de Broglie's double solution theory," *Phys. Lett. A* **135**(2), 99–105 (1989).

50. Z. Maric, K. Popper, J. P. Vigier, and J. Hilgevoord, "Comments on 'violation of Heisenberg's uncertainty relation. . . ,'" *Found. Phys. Lett.* **2**, 403 (1989).

51. J. P. Vigier, "Quantum particle motions in real physical space-time E_4," in L. Kostro, A. Posiewnik, J. Pykacz, and M. Zukowski (Eds.), *Problems in Quantum Physics: Gdansk '87. Recent and Future Experiments and Interpretation*, (Sept. 21–25, 1987), ISBN 9-97-150449-9, World Scientific, Singapore, 1988, pp. 317–349.

52. A. Kyprianidis and J. P. Vigier, "Action-at-a-distance: The mystery of Einstein–Podolsky-Rosen correlations," in F. Selleri (Ed.), *Quantum Mechanics versus Local Realism: The Einstein–Podolsky–Rosen Paradox*, ISBN 0-30-642739-7, Plenum, New York, 1988, p. 273.

53. J. P. Vigier, "EPR version of Wheeler's delayed choice experiment, in *Microphysical Reality and Quantum Formalism*, ISBN 9-02-772686-8, Kluwer, Dordrecht, 1988.

54. Z. Maric, K. Popper, and J. P. Vigier, "Violation of Heisenberg's uncertainty relations on individual particles within subset of gamma photons in $e^+e^- = 2\gamma$ pair creation," *Found. Phys. Lett.* **1**(4) (1988).

55. N. Cufaro-Petroni, P. Guéret, and J. P. Vigier, "Second-order wave equation for spin $^1/_2$ Fields: 8-Spinors and canonical formulation," *Found. Phys.* **18**(11) (1988).

56. C. Dewdney, P. R. Holland, A. Kyprianidis, and J. P. Vigier, "Spin and non-locality in quantum mechanics," *Nature* **336**(6199), 536–544 (1988).

57. P. R. Holland and J. P. Vigier, "The quantum potential and signaling in the Einstein–Podolsky–Rosen experiment," *Found. Phys.* **18**(7), 741–750 (1988).

58. J. P. Vigier, "New theoretical implications of Neutron interferometric double resonance experiments, *Int. Workshop on Matter Wave Interferometry in the Light of Schrödinger's Wave Mechanics* (Vienna, Austria, Sept. 14–16, 1987) *Physica B, C* **151**(1–2), 386–392 (1988), ISSN 0378-4363 (Conf. sponsor: Hitachi; Erwin Schrödinger Gesellschaft; Siemens; et al.).

59. P. N. Kaloyerou and J. P. Vigier, "Derivation of a non-linear Schrödinger equation describing possible vacuum dissipative effects," *Phys. Lett. A* **130**(4–5), 260–266 (1988).

60. C. Dewdney, P. R. Holland, A. Kyprianidis, Z. Maric, and J. P. Vigier, "Stochastic physical origin of the quantum operator algebra and phase space interpretation of the Hilbert space formalism: the relativistic spin-zero case," *Phys. Lett. A* **113A**(7), 359–364 (1988).

61. J. C. Pecker and J. P. Vigier, "A possible tired-light mechanism," *Apeiron* (2) (1988).

62. J. P. Vigier, "Einstein's materialism and modern tests of quantum mechanics," *Annal. Physik* **45**(1), 61–80 (1988).

63. J. P. Vigier, "Theoretical implications of time dependent double resonance neutron interferometry," in W. M. Honig, D. W. Kraft, and E. Panarella (Eds.), *Quantum Uncertainties, Recent and Future Experiments and Interpretations. Proc. NATO Advanced Research Workshop on Quantum Violations: Recent and Future Experiments and Interpretations*, (Bridgeport, CT, June 23–27, 1986) ISBN 0-30-642670-6, Plenum, New York, 1987, pp. 1–18.

64. J. P. Vigier, C. Dewdney, P. R. Holland, and A. Kyprianidis, in B. J. Hiley and F. D. Peat (Eds.), *Quantum Implications*, Routledge & Kegan Paul, London, 1987, pp. 169–204.

65. N. Cufaro-Petroni, C. Dewdney, P. R. Holland, A. Kyprianidis, and J. P. Vigier, "Einstein–Podolsky–Rosen constraints on quantum action at a distance: The Sutherland paradox," *Found. Phys.* **17**(8), 759–773 (1987).

66. P. R. Holland, A. Kyprianidis, and J. P. Vigier, "Trajectories and causal phase space approach to relativistic quantum mechanics," *Found. Phys.* **17**(5), 531–547 (1987).

67. A. Kyprianidis and J. P. Vigier, "Quantum properties of chaotic light in first order interference experiments," *Europhys. Lett.* **3**(7), 771–775 (1987).

68. A. Kyprianidis, S. Roy, and J. P. Vigier, "Distinguishability or indistinguishability in classical and quantum statistics," *Phys. Lett. A* **119**(7), 333–336 (1987).

69. P. R. Holland, A. Kyprianidis, and J. P. Vigier, "A non-negative distribution function in relativistic quantum mechanics," *Physica A* **139A**(2–3), 619–628 (1986).

70. A. Kyprianidis and J. P. Vigier, "Theoretical implications of neutron interferometry derived from the causal interpretation of the Pauli equation," *Hadronic J. Suppl.* **2**(3), 534–556 (1986).

71. P. R. Holland, A. Kyprianidis, and J. P. Vigier, "Causal phase space approach to fermion theories understood through Clifford algebras," *Lett. Math. Phys.* **12**(2), 101–110 (1986).

72. P. R. Holland, A. Kyprianidis, Z. Maric, and J. P. Vigier, "Relativistic generalization of the Wigner function and its interpretation in the causal stochastic formulation of quantum mechanics," *Phys. Rev. A* (Special Issue: *General Physics*) **33**(6), 4380–4383 (1986).

73. C. Dewdney, P. R. Holland, A. Kyprianidis, and J. P. Vigier, "Relativistic Wigner function as the expectation value of the PT operator," *Phys. Lett. A* **114A**(8–9), 440–444 (1986).

74. N. C. Petroni, P. Guéret, J. P. Vigier, and A. Kyprianidis, "Second order wave equation for spin$\frac{1}{2}$ fields; II. The Hilbert space of the states," *Phys. Rev. D* (Special Issue: *Particles and Fields*) **33**(6), 1674–1680 (1986).

75. C. Dewdney, P. R. Holland, A. Kyprianidis, and J. P. Vigier, "Cosmology and the causal interpretation of quantum mechanics," *Phys. Lett. A* **114A**(7), 365–370 (1986).

76. K. P. Sinha, E. C. G. Sudarshan, and J. P. Vigier, "Superfluid vacuum carrying real Einstein–de Broglie waves," *Phys. Lett. A* **114A**(6), 298–300 (1986).

77. W. Muckenheim, G. Ludwig, C. Dewdney, P. R. Holland, A. Kyprianidis, J. P. Vigier, N. Cufaro-Petroni, M. S. Bartlett, and E. T. Jaynes, "A review of extended probabilities," *Phys. Rep.* **133**(6), 337–401 (1986).

78. J. P. Vigier, "L'onde et la particule," *Science et Vie* (Feb. 1986).

79. C. Dewdney, P. R. Holland, A. Kyprianidis, J. P. Vigier, and Z. Maric, "Stochastic physical origin of the quantum operator algebra and phase space interpretation of the Hilbert space formalism: The relativistic spin zero case," *Phys. Lett. A* **113**(7), 359–364 (1986).

80. J. P. Vigier, Editors: E. I. Bitsakis and N. Tambakis, "Non-locality and space-time in quantum *N*-body systems, in *Determinism in Physics*, Gutenberg, 1985.

81. C. Dewdney, M. A. Dubois, P. R. Holland, A. Kyprianidis, L. Laurent, M. Pain, and J. P. Vigier, "Testing for non-locally correlated particle motions in the hydrogen atom," *Phys. Lett. A* **113A**(3), 135–138 (1985).

82. C. Dewdney, A. Garuccio, Ph. Guéret, A. Kyprianidis, and J. P. Vigier, "Time dependent neutron interferometry: Evidence against wave packet collapse?," *Found. Phys.* **15**(10), 1031–1042 (1985).

83. J. P. Vigier, "Causal stochastic interpretation of quantum statistics," *Pramana* **25**(4), 397–418 (1985).

84. J. P. Vigier and S. Roy, "Rauch's experiment and the causal stochastic interpretation of quantum statistics," *India J.: Hadronic J. Suppl.* **1**(3), 475–501 (1985).

85. N. Cufaro-Petroni, C. Dewdney, P. Holland, A. Kyprianidis, and J. P. Vigier, "Realistic physical origin of the quantum observable operator algebra in the frame of the casual stochastic interpretation of quantum mechanics: The relativistic spin-zero case," *Phys. Rev. D* (Special Issue: *Particles and Fields*) **32**(6), 1375–1383 (1985).

86. P. R. Holland and J. P. Vigier, "Positive probabilities and the principle of equivalence for spin-zero particles in the causal stochastic interpretation of quantum mechanics," *Nuovo Cimento B* **88B**(1) (Ser. 2), 20–28 (1985).

87. N. C. Petroni, P. Guéret, J. P. Vigier, and A. Kyprianidis, "Second order wave equation for spin $1/2$ fields," *Phys. Rev. D* (Special Issue: *Particles and Fields*) **31**(12), 3157–3161 (1985).

88. C. Dewdney, P. R. Holland, A. Kyprianidis, and J. P. Vigier, "Causal action at a distance in a relativistic system of two bound charged spinless particles: Hydrogen-like models," *Phys. Rev. D* (Special Issue: *Particles and Fields*) **31**(10), 2533—2538 (1985).

89. N. C. Petroni, P. Guéret, A. Kyprianidis, and J. P. Vigier, "An alternative derivation of the spin dependent quantum potential," *Lett. Nuovo Cimento* **42**(7) (Ser. 2), 362–364 (1985).

90. N. C. Petroni, C. Dewdney, P. Holland, A. Kyprianidis, and J. P. Vigier, "Causal space-time paths of individual distinguishable particle motions in N-body quantum systems: Elimination of negative probabilities," *Lett. Nuovo Cimento* **42**(6) (Ser. 2), 285–294 (1985).

91. P. Guéret, P. R. Holland, A. Kyprianidis, and J. P. Vigier, "Positive energy positive probability density association in second order fermion theories," *Phys. Lett. A* **107A**(8), 379–382 (1985).

92. P. R. Holland, A. Kyprianidis, and J. P. Vigier, "On the association of positive probability densities with positive energies in the causal theory of spin 1 particles," *Phys. Lett. A* **107A**(8), 376–378 (1985).

93. E. Giraud, and J. P. Vigier, "Dispersion of radial velocities in the Local Supercluster," *C. R. Acad. Sci.* (Ser. II) (Special Issue: *Mécanique, Chimie, Sciences de l'Univers, Sciences de la Terre*) **300**(1), 9–12 (1985).

94. S. Depaquit, J. C. Pecker, and J. P. Vigier, "The redshift distribution law of quasars revisited," *Astron. Nachr.* **306**(1), 7–15 (1985).

95. S. Depaquit, J. C. Pecker, and J. P. Vigier, *Astron. Nach.* **306**(1), 1 (1985).

96. J. P. Vigier, "Non-local quantum potential interpretation of relativistic actions at a distance in many-body problems," in G. Tarozzi and A. Van der Merwe (Eds.), *Open Questions in Quantum Physics. Invited Papers on the Foundations of Microphysics*, in (Bari, Italy, May 1983), ISBN 9-02-771853-9, Reidel, Dordrecht, 1985, pp. 297–332.

97. A. O. Barut, A. van der Merwe, and J. P. Vigier (Eds.), *Quantum Space and Time—the Quest Continues: Studies and Essays in Honour of Louis de Broglie, Paul Dirac, and Eugene Wigner*, Cambridge Monographs Physics. ASIN 0521-319110, Cambridge Univ. Press, Cambridge, UK, 1984 [originally published in *Found. Phys.* (1982–1983)].

98. N. Cufaro-Petroni, C. Dewdney, P. Holland, A. Kypriandis, and J. P. Vigier, "Elimination of negative probabilities within the causal stochastic interpretation of quantum mechanics," *Phys. Lett. A* **106A**(8), 368–370 (1984).

99. C. Dewdney, A. Kyprianidis, and J. P. Vigier, "Causal non-local interpretation of the double slit experiment and quantum statistics," *Epistemol. Lett.* **36**, 71 (1984).

100. A. Garuccio, A. Kyprianidis, and J. P. Vigier, "Relativistic predictive quantum potential: The N-body case, *Nuovo Cimento B* **83B**(2) (Ser. 2), 135–144 (1984).

101. C. Dewdney, A. Kyprianidis, J. P. Vigier, and M. A. Dubois, "Causal stochastic prediction of the non-linear photoelectric effects in coherent intersecting laser beams," *Lett. Nuovo Cimento* **41**(6) (Ser. 2), 177–185 (1984).

102. C. Dewdney, A. Garuccio, A. Kyprianidis, and J. P. Vigier, "The anomalous photoelectric effect: Quantum potential theory versus effective photon hypothesis," *Phys. Lett. A* **105A**(12), 15–18 (1984).

103. C. Dewdny, A. Kyprianidis, and J. P. Vigier, "Illustration of the causal model of quantum statistics," *J. Phys. A* (Special Issue: *Mathematical and General*) **17**(14), L741–L744 (1984).

104. C. Dewdney, A. Caruccio, A. Kyprianidis, and J. P. Vigier, "Energy conservation and complementarity in neutron single crystal interferometry, *Phys. Lett. A* **104A**(6–7), 325–328 (1984).

105. C. Dewdney, A. Kyprianidis, J. P. Vigier, A. Garuccio, and P. Guéret, "Time dependent neutron interferometry: Evidence in favor of de Broglie waves," *Lett. Nuovo Cimento* **40**(16) (Ser. 2), 481–487 (1984).

106. N. C. Petroni, P. Guéret, and J. P. Vigier, "Form of a spin dependent quantum potential," *Phys. Rev. D* (Special Issue: *Particles and Fields*) **30**(2), 495–497 (1984).

107. N. Cufaro-Petroni, N. Guéret, and J. P. Vigier, "A causal stochastic theory of spin ½ fields," *Nuovo Cimento* **B81**, 243–259 (1984).

108. C. Dewdney, P. Guéret, A. Kyprianidis, and J. P. Vigier, "Testing wave particle dualism with time dependent neutron interferometry," *Phys. Lett. A* **102A**(7), 291–294 (1984).

109. A. Garuccio, A. Kypriandis, D. Sardelis, and J. P. Vigier, "Possible experimental test of the wave packet collapse," *Lett. Nuovo Cimento* **39**(11) (Ser. 2) 225–233 (1984).

110. N. Cufaro-Petroni, A. Kyprianidis, Z. Maric, D. Sardelis, and J. P. Vigier, "Causal stochastic interpretation of Fermi–Dirac statistics in terms of distinguishable non-locally correlated particles," *Phys. Lett. A* **101A**(1), 4–6 (1984).

111. N. Cufaro-Petroni and J. P. Vigier, "Random motions at the velocity of light and relativistic quantum mechanics," *J. Phys. A* (Special Issue: *Mathematical and General*) **17**(3), 599–608 (1984).

112. A. Kyprianidis, D. Sardelis, and J. P. Vigier, "Causal non-local character of quantum statistics," *Phys. Lett. A* **100A**(5), 228 (1984).

113. N. Cufaro-Petroni and J. P. Vigier "Stochastic interpretation of relativistic quantum equations," in A. van der Merwe (Ed.), *Old and New Questions in Physics, Cosmology, Philosophy, and Theoretical Biology: Essays in Honor of Wolfgang Yourgrau*, ISBN 0-20-640962-3, Plenum, New York, 1983, pp. 325–344.

114. P. Guéret and J. P. Vigier, "Relativistic wave equations with quantum potential nonlinearity," *Lett. Nuovo Cimento* **38**, 125–128 (1983).

115. J. P. Vigier and R. Dutheil, "Sur une interprétation par la théorie de la relativité générale du modèle d'électron rigide de Dirac" (An interpretation of the Dirac rigid electron model by the theory of general relativity), *Bull. Soc. Roy. Sci. de Liège* (Belgium) **52**(5), 331–335 (1983).

116. Ph. Guéret and J. P. Vigier, "Relativistic wave equations with quantum potential non-linearity," *Lett. Nuovo Cimento*, **38**(4) (1983).

117. M. A. Dubois, E. Giraud, and J. P. Vigier, "A surprising feature of a set of apparent QSO-galaxy associations," *C. R. Séances Acad. Sci.* (*Ser. II*) **297**(3), 259–260 (1983).

118. N.Cufaro-Petroni and J. P. Vigier, "Random motions at the velocity of light and relativistic quantum mechanics," *J. Phys. A* (Special Issue: *Mathematical and General*) **17**(3), 599–608 (1984).

119. Ph. Guéret and J. P. Vigier, "Relativistic wave equations with quantum potential non-linearity," *Lett. Nuovo Cimento* **38**(4) (Ser. 2), 125–128 (1983).

120. E. Giraud and J. P. Vigier, "Sur l'anomalie de décalage spectral des paires mixtes de galaxies" (On the anomalous redshift of mixed pairs of galaxies), *C. R. Acad. Sci. Paris* **296**, 193 (1983).

121. J. Andrade e Silva, F. Selleri, and J. P. Vigier, "Some possible experiments on quantum waves," (*Lett. Nuovo Cimento* **36**(15) (Ser. 2), 503–508 (1983).

122. N. Cufaro-Petroni and J. P. Vigier, "Dirac's aether in relativistic quantum mechanics," *Found. Phys.* **13**(2), 253–286 (1983).

123. N. Cufaro-Petroni and J. P. Vigier, "Causal action-at-a distance interpretation of the Aspect–Rapisarda experiments," *Phys. Lett. A* **93A**(8), 383–387 (1983).

124. H. Arp, E. Giraud, J. W. Sulentic, and J. P. Vigier, "Pairs of spiral galaxies with magnitude differences greater than one," *Astron. Astrophys.* **121**(1) (Pt. 1), 268 (1983).

125. J. P. Vigier and N. Cufaro-Petroni, *Quantum Space and Time*, Cambridge Univ. Press, Cambridge, UK, 1982, p. 505.

126. Ph. Guéret and J. P. Vigier, "De Broglie's wave particle duality in the stochastic interpretation of quantum mechanics: A testable physical assumption," *Found. Phys.* **12**(11) (1982).

127. J. P. Vigier, "Louis de Broglie: Physicist and thinker," *Found. Phys.* **12**(10), 923–930 (1982).

128. P. Guéret and J. P. Vigier, "Non-linear Klein–Gordon equation carrying a non-dispersive soliton-like singularity," *Lett. Nuovo Cimento* **35**(8) (Ser. 2), 256–259 (1982).

129. P. Guéret and J. P. Vigier, "Soliton model of Einstein's 'Nadelstrahlung' in real physical Maxwell waves," *Lett. Nuovo Cimento* **35**(8) (Ser. 2), 26–34 (1982).

130. J. P. Vigier, *Astron. Nach.* **303**, 55 (1982).

131. A. Garrucio, V. Rapisarda, and J. P. Vigier, "New experimental set-up for the detection of de Broglie waves," *Phys. Lett.* **90A**(1), 17 (1982).

132. E. Giraud, M. Moles, and J. P. Vigier, "Une corrélation entre le décalage spectral et le type morphologique pour les galaxies binaires" (A correlation between the spectral shift and morphological type for binary galaxies), *C. R. Acad. Sci. Paris* **294** (1982).

133. N. C. Petroni and J. P. Vigier, "Stochastic model for the motion of correlated photon pairs," *Phys. Lett. A* **88A**(6), 272–274 (1982).

134. F. Halbwachs, F. Piperno, and J. P. Vigier, "Relativistic Hamiltonian description of the classical photon behaviour: A basis to interpret aspect's experiments," *Lett. Nuovo Cimento* **33**(11) (1982).

135. A. Garuccio, V. A. Rapisarda, and J. P. Vigier, "Superluminal velocity and causality in EPR correlations," *Lett. Nuovo Cimento* **32**, 451–456 (1981).

136. A. Garuccio, K. R. Popper, and J. P. Vigier, "Possible direct physical detection of de Broglie waves," *Phys. Lett. A* **86A**(8), 297–400 (1981).

137. C. Fenech and J. P. Vigier, "Analyse du potentiel quantique de De Broglie dans le cadre de l'intérpretation stochastique de la théorie des quanta" (Analysis of the de Broglie quantum potential in the frame of the stochastic interpretation of quantum theory), *C. R. Acad. Sci. Paris* **293**, 249 (1981).

138. A. Garuccio and J. P. Vigier, "An experiment to interpret E.P.R. action-at-a-distance: The possible detection of real de Broglie waves," *Epistemol. Lett.* (1981) (includes reply by O. Costa de Beauregard).

139. N. C. Petroni, P. DrozVincent, and J. P. Vigier, "Action-at-a-distance and causality in the stochastic interpretation of quantum mechanics," *Lett. Nuovo Cimento* **31**(12) (Ser. 2), 415–420 (1981).

140. N. C. Petroni, Z. Maric, Dj. Zivanovic, and J. P. Vigier, "Stable states of a relativistic bilocal stochastic oscillator: A new quark-lepton model," *J. Phys. A* (Special Issue: *Mathematical and General*) **14**(2), 501–508 (1981).

141. A. Garuccio and J. P. Vigier, "Description of spin in the causal stochastic interpretation of Proca–Maxwell waves: Theory of Einstein's 'ghost waves," *Lett. Nuovo Cimento* **30**(2) (Ser. 2), 57–63 (1981).

142. N. Cufaro-Petroni and J. P. Vigier, "Stochastic derivation of the Dirac equation in terms of a fluid of spinning tops endowed with random fluctuations at the velocity of light," *Phys. Lett. A* **81A**(1), 12–14 (1981).

143. Y. Fitt, A. Faire, and J. P. Vigier, *The World Economic Crisis: U.S. Imperialism at Bay*, transl. by M. Pallis, Zed Press, London, 1980.

144. J. P. Vigier, "Après le colloque de Cordoue, un accusé nommé Einstein," *Raison Présente* **56** (*La parapsychologie, oui ou non?*) (4ème trimestre), 77 (1980).

145. N. Cufaro-Petroni, Z. Maric, Dj. Zivanovic, and J. P. Vigier, "Baryon octet magnetic moments in an integer charged quark oscillator model," *Lett. Nuovo Cimento* **29**(17) (Ser. 2) 565–571 (1980).

146. J. P. Vigier, "De Broglie waves on Dirac aether: A testable experimental assumption," *Lett. Nuovo Cimento* **29**(14) (Ser. 2), 467–475 (1980).

147. A. Garuccio, K. Popper, and J. P. Vigier, *Phys. Lett.* **86A**, 397 (1980).

148. A. Garrucio and J. P. Vigier, "Possible experimental test of the causal stochastic interpretation of quantum mechanics: Physical reality of de Broglie waves," *Found. Phys.* **10**(9–10), 797–801 (1980).

149. F. Selleri and J. P. Vigier, "Unacceptability of the Pauli–Jordan propagator in physical applications of quantum mechanics," *Lett. Nuovo Cimento* **29**(1) (Ser. 2), 7–9 (1980).

150. J. and M. Andrade e Silva, *C. R. Acad. Sci. Paris* **290**, 501 (1980).

151. N. Cufaro-Petroni, A. Garuccio, F. Selleri, and J. P. Vigier, "On a contradiction between the classical (idealized) quantum theory of measurement and the conservation of the square of the total angular momentum in Einstein–Podolsky–Rosen paradox," *C. R. Acad. Sci., Sér. B* (Sciences Physiques), **290**(6), 111–114 (1980).

152. A. Garuccio, G. D. Maccarrone, E. Recami, and J. P. Vigier, "On the physical non-existence of signals going backwards in time, and quantum mechanics," *Lett. Nuovo Cimento* **27**(2) (Ser. 2), 60–64 (1980).

153. A. Garuccio, G. D. Maccarrone, E. Recami, and J. P. Vigier, *On the Physical Non-existence of Signals Going Backwards in Time, and Quantum Mechanics*, Report PP/635, Instituto Nazionale Fisica Nucleare, Catania, Italy, 1979.

154. N. Curfaro-Petroni and J. P. Vigier, "Markov process at the velocity of light: The Klein–Gordon statistic," *Int. J. Theor. Phys.* **18**(11), 807–818 (1979).

155. N. Cufaro-Petroni and J. P. Vigier, "Stochastic derivation of Proca's equation in terms of a fluid of Weyssenhoff tops endowed with random fluctuations at the velocity of light," *Phys. Lett. A* **73A**(4), 289–299 (1979).

156. J. P. Vigier, C. Phillipidis et al., "Quantum interference and the quantum potential," *Nuovo Cimento* **52B**, 25 (1979).

157. N. Cufaro-Petroni and J. P. Vigier, "Causal superluminal interpretation of the Einstein–Podolsky–Rosen paradox," *Lett. Nuovo Cimento* **26**(5) (Ser. 2), 149–154 (1979).

158. N. C. Petroni and J. P. Vigier, "On two conflicting physical interpretations of the breaking of restricted relativistic Einsteinian causality by quantum mechanics," *Lett. Nuovo Cimento* **25**(5) (Ser. 2), 151–156 (1979).

159. T. Jaakkola, M. Moles, and J. P. Vigier, "Empirical status in cosmology and the problem of the nature of redshifts," *Astron. Nach.* **300**(5), 229–238 (1979).

160. J. P. Vigier, "Model of quantum statistics in terms of a fluid with irregular stochastic fluctuations propagating at the velocity of light: A derivation of Nelson's equations," *Lett. Nuovo Cimento* **24**(8) (Ser. 2), 265–272 (1979).

161. J. P. Vigier, "Superluminal propagation of the quantum potential in the causal interpretation of quantum mechanics," *Lett. Nuovo Cimento* **24**(8) (Ser. 2), 258–264 (1979).

162. P. Guéret, P. Merat, M. Moles, and J. P. Vigier, "Stable states of a relativistic harmonic oscillator imbedded in a random stochastic thermostat," *Lett. Math. Phys.* **3**(1), 47–56 (1979).

163. C. Fenech, M. Moles, and J. P. Vigier, "Internal rotations of spinning particles," *Lett. Nuovo Cimento* **24**(2) (Ser. 2), 56–62 (1979).

164. N. Chomsky and J. P. Vigier, *Verso la terza guerra mondiale?* Collection Biblioteca di Nuova Cultura, Milan, Italy, G. Mazzotta, 1978.

165. T. Jaakkola, M. Moles, and J. P. Vigier, "Interpretation of the apparent North-South asymmetry and fluctuations of galactic rotation," *Astrophys. Space Sci.* **58**(1), 99–102 (1978).

166. J. P. Vigier, S. Depaquit, and G. Le Denmat, "Liste de supernovae de type I dont la détermination du maximum de luminosité permet l'établissement d'établissement d'échantillons homogènes," *C. R. Acad. Sci.* **285B**, 161 (1977).

167. L. Nottale and J. P. Vigier, "Continuous increase of Hubble modulus behind clusters of galaxies," *Nature*, **268**(5621), 608–610 (1977).

168. M. Moles and J. P. Vigier, "Remarks on the impact of photon scalar boson scattering on Planck's radiation law and Hubble effect (and reply)," *Astron. Nach.* **298**(6), 289–291 (1977).

169. D. Gutkowski, M. Moles, and J. P. Vigier, "Hidden parameter theory of the extended Dirac electron; I. Classical theory," *Nuovo Cimento B* **B39**, 193–225 (1977).

170. G. B. Cvijanovich and J. P. Vigier, "New extended model of the Dirac electron," *Found. Phys.* **7**(1–2), 77–96 (1977).

171. Z. Maric, M. Moles, and J. P. Vigier, "Red shifting of light passing through clusters of galaxies: A new photon property?" *Lett. Nuovo Cimento* **18**(9) (Ser. 2), 269–276 (1977).

172. M. Flato, Z. Maric, A. Milojevic, D. Sternheimer, and J. P. Vigier (Eds.), *Quantum Mechanics, Determinism, Causality, and Particles*, ISBN 9-02-770623-9, North-Holland, Dordrecht, 1976.

173. Y. Fitt, A. Farhi, and J. P. Vigier, *La crise de l'impérialisme et la troisième guerre mondiale*, F. Maspero, Paris, 1976.

174. Z. Maric, M. Moles, and J. P. Vigier, "Possible measurable consequences of the existence of a new anomalous red shift cause on the shape of symmetrical spectral lines," *Astron. Astrophys.* **53**(2) (Pt. 1), 191–196 (1976).

175. T. Jaakkola, H. Karoji, G. Le Denmat, M. Moles, L. Nottale, J. P. Vigier, and J. C. Pecker, "Additional evidence and possible interpretation of angular redshift anisotropy," *MNRAS* (*Monthly Notices of Royal Astronomical Society*) **177**, 191–213 (1976).

176. H. Karoji, L. Nottale, and J. P. Vigier, "A peculiar distribution of radial velocities of faint radiogalaxies with $13.0 < m_{corr} < 15.5$," *Astrophys. Space Sci.* **44**, 229–234 (1976).

177. L. Nottale, J. C. Pecker, J. P. Vigier, and W. Yourgrau, "La constante de Hubble mise en question," *La Recherche* **68**, 5 (1976).

178. J. C. Pecker and J. P. Vigier, "A set of working hypotheses towards a unified view of the universe," *Astrofizika* **12**(2), 315–330 (1976).

179. J. P. Vigier, "Charmed quark discovery in anti-neutrino nucleon scattering?" *Lett. Nuovo Cimento* **15**(2) (Ser. 2), 41–48 (1976).

180. J. P. Vigier, "Possible implications of de Broglie's wave mechanical theory of photon behavior," *Math. Phys. Appl. Math.* **1**, 237–249 (1976).

181. M. Flato, Z. Maric, A. Milojevic, D. Sternheimer, and J. P. Vigier, *Quantum Mechanics, Determinism, Causality, and Particles. An International Collection of Contributions in Honor of Louis De Broglie on the Occasion of the Jubilee of His Celebrated Theses*, Kluwer, Dordrecht, 1975.

182. J. Borsenberger, J. C. Pecker, and J. P. Vigier, "Calcul et tables d'une intégrale utile dans l'interprétation de phénomènes voisins du bord solaire (ou stellaire)" [Computation and tables of a useful integral for the interpretation of phenomena near the solar limb (or stellar disk)], *C. R. Acad. Roy. Liège* (Belgium) **44**(11–12), 706–716 (1975).

183. H. Karoji, L. Nottale, and J. P. Vigier, "Déplacements anormaux vers le rouge liés à la traversée des amas de galaxies par la lumière" (Observation of excess red shifts when light travels through clusters of galaxies), *C. R. Acad. Sci., Sér. B* (Sciences Physiques) **281**(1), 409–412 (1975).

184. J. C. Pecker and J. P. Vigier, "Sur les dangers d'une approximation classique dans l'analyse des décalages spectraux vers le rouge" (The dangers of a classical approximation in the analysis of spectral redshifts), *C. R. Acad. Sci., Sér. B* (Sciences Physiques) **281**, 369–372 (1975).

185. M. Flato, C. Piron, J. Grea, D. Sternheimer, and J. P. Vigier, "Are Bell's inequalities concerning hidden variables really conclusive?" *Helv. Phys. Acta* **48**(2), 219–225 (1975).

186. T. Jaakkola, H. Karoji, M. Moles, and J. P. Vigier, "Anisotropic red shift distribution for compact galaxies with absorption spectra," *Nature* **256**(5512), 24–25 (1975).

187. T. Jaakkola, M. Moles, J. P. Vigier, J. C. Pecker, and W. Yourgrau, "Cosmological implications of anomalous red shifts: A possible working hypothesis," *Found. Phys.* **5**(2), 257–269 (1975).

188. J. P. Vigier, G. Le Denmat, M. Moles, and J. L. Nieto, "Possible local variable of the Hubble constant in VanDenBergh's calibration of Sc-type galaxies," *Nature*, **257**, 773 (1975).

189. G. Le Denmat and J. P. Vigier, "Les supernovae de type I et l'anisotropie de la 'constante' de Hubble (Type I supernova and angular anisotropy of the Hubble "Constant"), *C. R. Acad. Sci., Sér. B* (Sciences Physiques) **280**(14), 459–461 (1975).

190. S. Depaquit, J. P. Vigier, and J. C. Pecker, "Comparaison de deux observations de déplacement anormaux vers le rouge observés au voisinage du disque solaire," *C. R. Acad. Sci. Paris, Sér. B* (Sciences Physiques) **280**, 113–114 (1975).

191. J. P. Vigier, *On the Geometrical Quantization of the Electric Charge in Five Dimensions and Its Numerical Determination as a Consequence of Asymptotic SO(5,2) Group Invariance*, Colloques Internationaux du CNRS (Centre National de la Recherche Scientifique), No. 237, 1974.

192. P. Péan, *Pétrole, la troisième guerre mondiale*, Preface by J. P. Vigier, "Questions d'actualité," Calmann-Lévy, Paris, 1974.

193. S. Depaquit, J. P. Vigier, and J. C. Pecker, "Comparaison de deux observations de déplacements anormaux vers le rouge observés au voisinage du disque solaire" (Comparison of two observations of abnormal red shifts observed in the solar disc region), *C. R. Acad. Sci. Paris* **279** (1974).

194. J. P. Vigier, "Three recent experiments to test experimentally realizable predictions of the hidden variables theory," *C. R. Acad. Sci., Sér. B* (Sciences Physiques) **279**(1), 1–4 (1974).

195. J. P. Vigier, "Sur trois vérifications expérimentales récentes de conséquences mesurables possibles de la théorie des paramètres cachés" (Three recent experimental verifications of possible measurable consequences of the hidden variables theory), *C. R. Acad. Sci. Paris* **279** (1974).

196. P. Merat, J. C. Pecker, J. P. Vigier, and W. Yourgrau, "Observed deflection of light by the sun as a function of solar distance," *Astron. Astrophys.* **32**(4) (Pt. 1), 471–475 (1974).

197. L. V. Kuhi, J. C. Pecker, and J. P. Vigier, "Anomalous red shifts in binary stars," *Astron. Astrophys.* **32**(1) (Pt. 2), 111–114 (1974).

198. M. Moles, and J. P. Vigier, "Possible interpretation of solar neutrino and Mont Blanc muon experiments in terms of neutrino-boson collisions, *Lett. Nuovo Cimento* **9**(16) (Ser. 2), 673–676 (1974).

199. P. Merat, J. C. Pecker, and J. P. Vigier, "Possible interpretation of an anomalous redshift observed on the 2292 MHz line emitted by Pioneer-6 in the close vicinity of the solar limb," *Astron. Astrophys.* **30**(1), (Pt. 1), 167–174 (1974).

200. P. Guéret, J. P. Vigier, and W. Tait, "A symmetry scheme for hadrons, leptons and Intermediate vector bosons," *Nuovo Cimento A* **17A**(4), 663–680 (1973).

201. J. P. Vigier, J. C. Pecker, and W. Tait, "Photon mass, quasar red shifts and other abnormal red shifts," *Nature* **241**, 338 (1973).

202. J. P. Vigier, "Calcul théorique de la valeur de $\alpha = e^2/\hbar c$ à partir du groupe d'invariance asymptotique de particules de Dirac en mouvement dans un champ extérieur constant" (Theoretical determination of $\alpha = e^2/\hbar c$ deduced from asymptotic group invariance properties of high-energy charged dirac particles in a constant external field), *C. R. Acad. Sci. Paris* **277**(9), 397–400 (1973).

203. J. P. Vigier, "Theoretical determination of $\alpha = e^2/\hbar c$ deduced from asymptotic group invariance properties of high energy charged dirac particles in a constant external vector potential, *Lett. Nuovo Cimento* **7**(12) (Ser. 2) 501–506 (1973).

204. J. P. Vigier, and M. Moles, "Conséquences physiques possibles de l'existence d'une masse non nulle du photon sur les interactions de la lumière avec la matière et la théorie du corps noir," *C. R. Acad. Sci. Paris* **276B**, 697 (1973).

205. J. P. Vigier, and G. Marcilhacy, "Are vector potentials measurable quantities in electromagnetic theory?," *Lett. Nuovo Cimento* **4**(13) (Ser. 2), 616–168 (1972).

206. J. C. Pecker, A. P. Roberts, and J. P. Vigier, "Non-velocity red shifts and photon-photon interactions," *Nature* **237**(5352), 227–279, (see also editorial on p. 193) (1972).

207. J. P. Vigier, J. C. Pecker, and A. P. Roberts, Sur une interprétation possible du déplacement vers le rouge des raies spectrales dans le spectre des objets astronomiques. Suggestions en vue d'expériences directes" (On a possible interpretation of the red shift of the spectral lines in the spectrum of astronomical objects. Suggestions toward direct experiments), *C. R. Acad. Sci. Paris.* **274B**, 1159 (1972).

208. L. de Broglie and J. P. Vigier, "Photon mass and new experimental results on longitudinal displacements of laser beams near total reflection, *Phy. Rev. Lett.* **28**(15), 1001–1004 (1972).

209. J. C. Pecker, A. P. Roberts, and J. P. Vigier, "Sur une interprétation possible du déplacement vers le rouge des raies spectrales dans le spectre des objets astronomiques" (On a possible interpretation of the red shift of the spectral lines in the spectrum of astronomical objects), *C. R. Acad. Sci. Paris* **274B**(11), 765–768 (1972).

210. S. Depaquit, Ph. Guéret, and J. P. Vigier, "Classical spin variables and classical counterpart of the Dirac–Feynman–Gellmann equation," *Int. J. Theor. Phys.* **4**(1), 1932 (1972).

211. L. de Broglie, and J. P. Vigier, "Masse du photon, Effet Imbert et effet Goos-Hänchen en lumière incidente polarisée" (Photon mass, imbert effect and Gooshanchen effect in polarized light), *C. R. Acad. Sci., Sér B* (Sciences Physiques), **273B**(25), 1069–1073 (1971).

212. J. P. Vigier, "Masse du photon et expériences de Kunz sur l'effet photoélectrique du plomb supraconducteur" (Photon mass and Kunz' experiments on the photoelectric effect of super-conducting lead), *C. R. Acad. Sci. Paris* **273**, 993 (1971).

213. J. P. Vigier, and M. Duchesne, "Sur la réciprocité de la réponse de la caméra électronique aux très faibles flux de lumière cohérente" [On the reciprocity of the response of the electron camera to weak coherent light flux (SbCs$_3$ layer properties)], *C. R. Acad. Sci. Paris* **273B**, 911 (1971).

214. P. Guéret and J. P. Vigier, "Remarks on a possible dynamical and geometrical unification of external and internal groups of motions of elementary particles and their application to SO(6,1) global dynamical symmetry," *Nuovo Cimento A* **67**(1), 238 (1970).

215. J. P. Vigier, and Ph. Guéret, "Remarks on a possible dynamical and geometrical unification of external groups of motions of elementary particles and their application to SO(6,1) global dynamical symmetry," *Nuovo Cimento* **67A**, 23 (1970).

216. Ph. Guéret, and J. P. Vigier, "Unification des quarks et des leptons dans la représentation de la base de SO(6,1)" [Sakata–Yukawa model unifying quarks and leptons in the total dynamic group SO(6,1)], *C. R. Acad. Sci. Paris* **270**(10), 653–656 (1970).

217. J. P. Vigier, P. Bozec, M. Cagnet, M. Duchesne, and J. M. Leconte, "Nouvelles expériences d'interférence en lumière faible," *C. R. Acad. Sci. Paris* **270**, 324 (1970).

218. J. P. Vigier, *Rivoluzione scientifica e imperialismo*, transl. by M. L. Rotondi De Luigi, Libreria Feltrinelli, Milano, 1969.

219. J. P. Vigier, and Ph. Guéret, "Formule de masse associée aux multiplets de SU(3) et équations d'ondes des baryons, obtenues à partir du groupe dynamique global d'unification $G = $ SO(6,1)," *C. R. Acad. Sci. Paris* **268**, 1153 (1969).

220. S. Depaguit, and J. P. Vigier, "Phenomenological spectroscopy of baryons and bosons considered as discrete quantized states of an internal structure of elementary particles," *C. R. Acad. Sci., Sér B* (Sciences Physiques) **268**(9), 657–659 (1969).

221. P. Guéret, and J. P. Vigier, "Formule de masse associée aux multiplets de SU(3) et équations d'ondes des baryons, obtenues à partir du groupe dynamique global d'unification $G = $ SO(6,1) × U(1)," *C. R. Acad. Sci. Paris* **268** (1969).

222. J. P. Vigier, "Unification on external and internal motions within SO(6,1) a possible mass splitting on SU(3) baryon multiplets without symmetry breaking," *Lett. Nuovo Cimento Sér 1* **1**, 445 (1969).

223. J. P. Vigier, and S. Depaquit, "Spectroscopie phénoménologique des baryons et des bosons considérés comme états quantifiés discrets d'une structure interne des particules élémentaires," *C. R. Acad. Sci. Paris* **268**, 657 (1969).

224. Ph. Guéret, and J. P. Vigier, "Sur une extension possible de la notion de spin de Pauli aux groupes SO(p,1) et In (SO (p,1))" [On a possible extension to the motion of Pauli spin to the groups SO(p,1) and In (SO(p,1))], *C. R. Acad. Sci., Sér B* (Sciences Physiques) **267**(19), 997–999 (1968).

225. J. P. Vigier, "Interprétation géométrique et physique de la formule du guidage en relativité générale," *C. R. Acad. Sci. Paris* **266**, 598 (1968).

226. J. P. Vigier, "Intervention à propos de l'exposé introductif de W. Heisenberg 'Déterminisme et Indéterminisme,'" in UNESCO (Ed.), *Science et Synthèse*, Collection Idées, Gallimard, Paris, 1967, pp. 270–272.

227. J. P. Vigier, "Intervention à propos de l'exposé introductif de R. Poirier 'Vers une cosmologie,'" in UNESCO (Ed.), *Science et Synthèse*, Collection Idées, Gallimard, Paris, 1967, pp. 248–252.

228. J. P. Vigier, "Intervention à propos de l'exposé introductif de L. de Broglie 'De la pluralité à l'unité,'" in UNESCO (Ed.), *Science et Synthèse*, Collection Idées, Gallimard, Paris, 1967, pp. 176–180.

229. J. P. Vigier, "Hidden parameters associated with possible internal motions of elementary particles," in M. Bunge (Ed.), *Studies in Foundations, Methodology, and Philosophy of Sciences* Vol. 2, Springer-Verlag, Berlin, 1967.

230. P. Guéret, and J. P. Vigier, "Diagonalisation de l'opérateur de masse au carré dans les modéles étendus de particules élémentaires," *C. R. Acad. Sci. Paris* **265** (1967).

231. P. Guéret, and J. P. Vigier, "Fonctions d'ondes définies sur une variété de Riemann V_5 admettant localement le groupe conforme SU(2,2) comme groupe d'isométrie," *C. R. Acad. Sci. Paris* **264** (1967).

232. J. P. Vigier, M. Flato, D. and J. Sternheimer, and G. Wathaghin, On the masses on non-strange pseudo-scalar mesons and the generalized Klein–Gordon equation (1966).

233. J. P. Vigier, "Possibilité d'unification des comportements internes et externes des particules élémentaires au moyen de groupes de mouvements isométriques sur des variétés de Riemann," *Communication au Colloque International du C.N.R.S. sur l'extension du groupe de Poincaré aux symétries internes des particules élémentaires*, CNRS, Paris, 1966.

234. F. Bon, and M. A. Burnier, *Les nouveaux intellectuels*, preface by J. P. de Vigier, Cujas, Paris, 1966.

235. P. Guéret, and J. P. Vigier, "Une représentation paramétrique de l'algèbre de Lie du groupe SU(2,2) considéré comme groupe d'isométrie sur une variete de Riemann à cinq dimensions," *C. R. Acad. Sci. Paris* **263** (1966).

236. J. P. Vigier, "Unification possible des mouvements internes et externes des particules élémentaires décrits comme groupes de mouvements isométriques de déplacements sur des variétés de Riemann, *C. R. Acad. Sci. Paris* **262**, 1239–1241 (1966).

237. H. F. Gautrin, R. Prasad, and J. P. Vigier, "Sur une généralisation conforme possible de l'équation de Dirac," *C. R. Acad. Sci. Paris* **262** (1966).

238. J. P. Vigier, "Possible external and internal motions on elementary particles on Riemannian manifolds," *Phys. Rev. Lett.* **17**(1) (1966).

239. J. P. Vigier, *L'extension du groupe de Poincaré aux symétries internes des particules élémentaires,* Colloques internationaux du CNRS (Centre national de la recherche scientifique), No. 159, 1966.

240. J. P. Vigier, "Structure des micro-objets dans l'interprétation causale de la théorie des quanta," preface by L. de Broglie, Collection *Les grands problèmes des sciences*, Gauthier-Villars, Paris, 1965.

241. J. P. Vigier, M. Flato, and G. Rideau, "Definition on PCT operators for the relativistic-rotator model and the Bronzon-low symmetry," *Nucl. Phys.* **61**, 250–256 (1965).

242. J. P. Vigier, D. Bohm, M. Flato, and D. Sternheimer, "Conformal group symmetry on elementary particles," *Nuovo Cimento* **38**, 1941 (1965).

243. D. Bohm, M. Flato, F. Halbwachs, P. Hillion, and J. P. Vigier, "On the 'space-time character' of internal symmetries of elementary particles," *Nuovo Cimento, Sér X* **36**, (1965).

244. J. P. Vigier and V. Wataghin, "High energy electron-positron scattering," *Nuovo Cimento, Sér X* **36**, 672 (1965).

245. J. P. Vigier, M. Flato, and D. Sternheimer, "Le groupe conforme comme possibilité de symmétrie unifiée en physique des interactions fortes," *C. R. Acad. Sci. Paris* **260**, 3869–3872 (1965).

246. J. P. Vigier, "Signification physique des potentiels vecteurs et abandon de la notion d'invariance de jauge en théorie des mésons vectoriels intermédiares," *C. R. Acad. Sci. Paris* **259** (1964).

247. P. Hillion, and J. P. Vigier, "Un test possible de la symmétrie d'hypercharge," *C. R. Acad. Sci. Paris* **258** (1964).

248. L. de Broglie, *Introduction to the Vigier Theory of Elementary Particles*, transl. by A. J. Knodel, Elsevier, Amsterdam, NY, 1963 (Vigier wrote one chapter).

249. J. P. Vigier, H. Yukawa, and E. Katayama, "Theory of weak interactions based on a rotator model," *Prog. Theor. Phys.* **29**, 470 (1963).

250. J. P. Vigier, H. Yukawa, and E. Katayama, "An approach to the unified theory of elementary particles," *Prog. Theor. Phys.* **29**, 468 (1963).

251. J. P. Vigier, L. de Broglie, D. Bohm, P. Hillion, F. Halbwachs, and T. Takabayasi, "Space-time model of relativistic extended particles in Minkowski space II," *Phys. Rev.* **129**, 451 (1963).

252. J. P. Vigier, L. de Broglie, D. Bohm, P. Hillion, F. Halbwachs, and T. Takabayasi, "Rotator model of elementary particles considered as relativistic extended structures in Minkowski space," *Phys. Rev.* **129**, 438 (1963).

253. J. P. Vigier, and L. de Broglie, "Application de la théorie de la fusion au nouveau modèle étendu de particules élémentaires," *C. R. Acad. Sci. Paris* **256**, 3390 (1963).

254. J. P. Vigier and L. de Broglie, "Table des particules élémentaires associées au nouveau modèle étendu des particules élémentaires," *C. R. Acad. Sci. Paris* **256**, 3351 (1963).

255. G. Barbieri, S. Depaquit, and J. P. Vigier, "Spectroscopy of baryons and resonances considered as quantized levels of a relativistic rotator," *Nuovo Cimento, Sér. X* **30** (1963).

256. L. de Broglie, F. Halbwachs, P. Hillion, T. Takabayasi, and J. P. Vigier, "Space-time model of relativistic extended particles in Minkowski space II, free particle and interaction theory," *Phys. Rev.* **129** (1963).

257. P. Hillion, and J. P. Vigier, "Théorie des résonances isobariques considérées comme états excités internes du modèle du rotateur relativiste des particules élémentaires," *C. R. Acad. Sci. Paris* **255** (1962).

258. J. P. Vigier, "Mass of the Yang–Mills vector field," *Nuovo Cimento, Sér X* **23** (1962).

259. P. Hillion, and J. P. Vigier, "Sur les équations d'ondes associées à la structure des fermions," *Annal. Institut Henri Poincaré* 229–254 (1962).

260. J. P. Vigier, "Les ondes associées à une structure interne des particules, théorie relativiste," *Annal. Institut Henri Poincaré* **111**, 149 (1962).

261. P. Hillion, and J. P. Vigier, "Une seconde extension du formalisme de Utiyama," *Cahier Phys.* 137 (Jan. 1962).

262. J. P. Vigier, "Sur le champ de Yang et Mills," *C. R. Acad. Sci. Paris* **252**, 1113 (1961).

263. J. P. Vigier and Y. P. Terletski, "On the physical meaning of negative probabilities, *J. F. T. P.* **13**, 356 (1961).

264. J. P. Vigier and P. Hillion, "Sur l'interprétation causale de l'équation non-linéaire de Heisenberg," *Cahiers Phys.* **130–131**, 315 (1961).

265. J. P. Vigier, "Sur un groupe de transformations isomorphe en tant que groupe du groupe de Lorentz," *Cahier Phys.* **127** (1961).

266. J. P. Vigier, *Représentation hydrodynamique des fonctions d'ondes (Travaux résumés dans la thèse de F. Halbwachs), théorie relativiste des fluides à spin*, Gauthier-Villars, Paris, 1960.

267. P. Hillion, T. Takabayasi, and J. P. Vigier, "Relativistic hydrodynamics of rotating fluid masses moving with the velocity of light," *Acta Physica Polonica* **XIX** (1960).

268. P. Hillion, and J. P. Vigier, "New isotopic spin space and classification of elementary particles," *Nuovo Cimento, Sér X* **18** (1960).

269. P. Hillion and J. P. Vigier, "Forme possible des fonctions d'ondes relativistes associées au mouvement et à la struture des particules élémentaires, au niveau nucléaire," *C. R. Acad. Sci. Paris* **250** (1960).

270. P. Hillion and J. P. Vigier, "Application des groupes d'invariance relativistes aux modèles de particules étendues en relativité restreinte," *C. R. Acad. Sci. Paris* **250** (1960).

271. F. Halbwachs, P. Hillion, and J. P. Vigier, "Formalisme Hamiltonien associé au rotateur de Nakano," *C. R. Acad. Sci. Paris* **250** (1960).

272. P. Hillion, and J. P. Vigier, "Les fonctions d'onde non-relativistes associées au corpuscle étendu," *C. R. Acad. Sci. Paris* **250** (1960).

273. J. P. Vigier, D. Bohm, P. Hillion, and T. Takabayashi, "Relativistic rotators and bilocal theory," *Prog. Theor. Phys.* **23**, 496 (1960).

274. J. P. Vigier, D. Bohm, and P. Hillion, "Internal quantum states of hyperspherical (Nakano) relativistic rotators," *Prog. Theor. Phys.* **16**, 361 (1960).

275. J. P. Vigier, and P. Hillion, "Elementary particle waves and irreducible representations of the Lorentz group," *Nucl. Phys.* **16**, 361 (1960).

276. F. Halbwachs, P. Hillion, and J. P. Vigier, "Internal motions of relativistic fluid masses," *Nuovo Cimento, Sér X* **15** (1960).

277. P. Hillion, B. Stepanov, and J. P. Vigier, "Interprétation de la valeur moyenne des opérateurs internes dans la théorie des masses fluides relativistes," *Cahier Phys.* 283–289 (1959).

278. P. Hillion and J. P. Vigier, "Étude mathématique des fonctions propres des moments cinétiques internes des masses fluides relativistes," *Cahier Phys.* 257–282 (1959).

279. J. P. Vigier, F. Halbwachs, and P. Hillion, "Quadratic Lagrangians in relativistic hydrodynamics," *Nuovo Cimento* **11**, 882 (1959).

280. J. P. Vigier, L. de Broglie, and P. Hillion, "Propriétés classiques et représentation bilocale du rotateur de Nakano," *C. R. Acad. Sci. Paris* **249**, 2255 (1959).

281. J. P. Vigier and F. Halbwachs, "Formalisme Lagrangien pour une particule relativiste isolée étendue," *C. R. Acad. Sci. Paris* **248**, 490 (1959).

282. J. P. Vigier, L. Pauling, S. Sakata, S. Tomonaga, and H. Yukawa, "Quelques précisions sur la nature et les propriétés des retombées radioactives résultant des explosions atomiques depuis 1945," *C. R. Acad. Sci. Paris* **245** (might be **248**), 982 (1959).

283. J. P. Vigier, P. Hillion, and T. Takabayasi, "Relativistic hydrodynamics of rotating fluid masses moving with the velocity of light," *Acta Physica Polonica* (1959).

284. J. P. Vigier and P. Hillion, "Fonctions propres des opérateurs quantiques de rotation associés aux angles d'Euler dans l'espace-temps," *Annal. Institut Henri Poincaré* **XVI**(III), 161 (1959).

285. J. P. Vigier, F. Halbwachs, and P. Hillion, "Lagrangian formalism in relativistic hydrodynamics of rotating fluid masses," *Nuovo Cimento* **90**(58), 818 (1958).

286. J. P. Vigier, and T. Takabayasi, "Description of Pauli matter as a continuous assembly of small rotating bodies," *Prog. Theor. Phys.* **18**(6), 573 (1958).

287. J. P. Vigier, "Introduction des paramètres relativistes d'Einstein–Klein dans l'hydrodynamique relativiste du fluide à spin de Weyssenhoff," *C. R. Acad. Sci.* **245**, 1891 (1957).

288. J. P. Vigier, "Unal, introduction des paramètres relativistes d'Einstein, Kramers et de Cayley–Klein dans la théorie relativiste des fluides dotés de moment cinétique interne (spin)," *C. R. Acad. Sci.* **245**, 1787 (1957).

289. J. P. Vigier, "Structure des micro-objets dans l'interprétation causale de la théorie des quanta," preface by L. de Broglie, *Les grands problèmes des sciences*, Vol. 5, Gauthier-Villars, Paris, 1956.

290. J. P. Vigier, D. Bohm, and G. Lochack, *Interprétation de l'équation de Dirac comme approximation linéaire de l'équation d'une onde se propageant dans un fluide tourbillonnaire en agitation chaotique du type éther de Dirac*, Séminaire de L. De Broglie, Institut Henri Poincaré, Paris, 1956.

291. J. P. Vigier, and M. Fuchs, "Tendance vers un état d' équilibre stable de phénomènes soumis à évolution Markovienne," *C. R. Acad. Sci.* **9**, 1120 (note) (1956).

292. J. P. Vigier, *Structure des micro-objets dans l'interprétation causale de la théorie des quanta,* thesis 3600, series. A No. 2727 [also published under Collection (*Les grands problèmes des sciences*); see above].

293. J. P. Vigier, F. Halbwachs, and G. Lochack, "Décomposition en fonction de variables dynamiques du tenseur d'énergie impulsion des fluides relativistes dotés de moment cinétique interne," *C. R. Acad. Sci.* **241**, 692 (note) (1955).

294. J. P. Vigier and D. Bohm, "Model of the causal interpretation of quantum theory in terms of a fluid with irregular fluctuations," *Phys. Rev.* **2**(96), 208–216 (1955).

295. L. de Broglie, *La physique quantique restera-t-elle indéterministe?*, Exposé du problème, followed by copies of some documents and a contribution of J. P. Vigier, Gauthier-Villars, Paris, 1953.

296. J. P. Vigier, "Forces s'exerçant sur les lignes de courant des particules de spin 0, 1/2 et 1 en théorie de l'onde pilote," *C. R. Acad Sci* **235**, 1107 (note) (1952).

297. J. P. Vigier, "Sur la relation entre l'onde à singularité et l'onde statistique en théorie unitaire relativiste," *C. R. Acad. Sci.* **235**, 869 (note) (1952).

298. J. P. Vigier, "Introduction géométrique de l'onde pilote en théorie unitaire affine," *C. R. Acad. Sci.* **233**, 1010 (note) (1951).

299. J. P. Vigier, "Remarque sur la théorie de l'onde pilote," *C. R. Acad. Sci.* **233**, 641 (note) (1951).

300. J. P. Vigier, *Rapport sur la théorie générale de la diffusion des neutrons thermiques*, internal report, CEA (Centre de l'énergie atomique), 1949.

301. J. P. Vigier, *Étude sur les suites infinies d'opérateurs Hermitiens*, thesis 1089, presented at Geneva to obtain the grade of Docteur és-Sciences Mathématiques, 1946 (also published under *Orthogonal Series*, Courbevoie, 1946).

TOPOLOGICAL ELECTROMAGNETISM WITH HIDDEN NONLINEARITY

ANTONIO F. RAÑADA

Departamento de Electricidad y Electrónica, Universidad Complutense, Madrid, Spain

JOSÉ L. TRUEBA

ESCET, Universidad Rey Juan Carlos, Madrid, Spain

CONTENTS

Modern Nonlinear Optics, Part 3, Second Edition, Advances in Chemical Physics, Volume 119, Edited by Myron W. Evans. Series Editors I. Prigogine and Stuart A. Rice.
ISBN 0-471-38932-3 © 2001 John Wiley & Sons, Inc.

I. INTRODUCTION

A. Force Lines, Vortex Atoms, Topology, and Physics

The lines of force, both electric and magnetic, were very real to Faraday when he proposed the idea of field. In his view, forged after many long hours of laboratory work, they had to be tangible and concrete, since the experiments indicated clearly that something very special occurred along them, a sort of perturbation of space of a nature still to be understood at the time [1].

Faraday's original view was maintained during most of the nineteenth century, as it is clear from the many attempts to explain the lines of force in terms of the streamlines and vorticity lines of the ether. During a long period, the electromagnetic phenomena were supposed to be a manifestation of the motion of this subtle substance, which would be understood eventually thanks to the mechanics of fluids. According to this opinion, the lines of force were associated with ether particles and had, therefore, the reality of a material substance, even if it were of a very special kind. Nobody less than Maxwell, who had argued several times in terms of this interpretation of the electric and magnetic lines [2–4], admitted as a sound and promising idea Kelvin's suggestion in 1868 that atoms were knots or links of the vortex lines of the ether, a picture presented expressively in a paper called "On vortex atoms" [5–8]. He liked the idea, as it expressed for instance in his presentation of the term "atomism" in the *Encyclopaedia Britannica* in 1875 [9,10].

Kelvin used to say: "I can never satisfy myself until I can make a mechanical model of a thing." Because of this urge, deeply engraved in his scientific style, he was reluctant to fully accept Maxwell's new electromagnetic theory. Looking for a different approach, he had applied to his topological idea the then new Helmholtz's theorems on fluid dynamics. He did find extremely unsatisfactory the then widely held view of infinitely hard point atoms or, in his own words, "the monstrous assumption of infinitely strong and infinitely rigid pieces of matter" [11]. Kelvin was much impressed by the conservation of the strength of the vorticity tubes in an inviscid fluid according to Helmholtz theorems, thinking that this was an inalterable quality on which to base an atomic theory of matter without infinitely rigid entities. We know now that this is also a trait of topological models, in which some invariant numbers characterize configurations that are rigid and can deform, distort or warp. As he put it, "Helmholtz has proved an absolutely unalterable quality in any motion of a perfect liquid . . . any portion [of it] has one recommendation of Lucretius' atoms—infinitely perennial specific quality."

Inspired by Helmholtz theorems, Kelvin understood in a striking combination of geometric insight and physical intuition that such knots and links would be extremely stable, just as matter is. Furthermore, he thought that the remarkable variety of the properties of the chemical elements could be a consequence of the many different ways in which such curves can be linked or knotted. Should he be alive today, he could have added to stability and variety two other important properties of matter, not known in his time [12]. One is transmutability, the ability of atoms to change into another kind in a nuclear reaction, which could be related to the breaking and reconnection of lines, as happens, for instance, to the magnetic lines in plasmas after disruptions in a tokamak. The other is the discrete character of the spectrum, which is also a property of the nontrivial topological configurations of a vector field, as was shown by Moffatt [13].

The reception to Kelvin's idea was good; Maxwell was impressed by its mechanical simplicity and because its success in explaining phenomena would not depend on ad hoc hypothesis. However, neither topology nor atomic phenomenology was sufficiently developed to follow this deep insight. It was soon forgotten to remain unknown for a long time.

It is ironic that, in spite of his favorable attitude to Kelvin's model, Maxwell himself contributed to the fading of the force lines with his monumental *Treatise on Electromagnetism*, after which, because of the successful developments of algebra and differential geometry, the line of force was relegated behind the concepts of electromagnetic tensor $F_{\mu\nu}$ and electromagnetic vectors E_i, B_j, A_μ. It is usually now a secondary concept, always derived from $F_{\mu\nu}$ as the integral lines of **B** and **E**. As it is used mainly in elementary presentations, it is often assumed that it is not adequate for a deep analysis of the electromagnetic interactions.

Topology appeared again in fundamental physics with Dirac's appealing and intriguing proposal of the monopole in 1931 [14] and its quantization of the electric charge because of a mechanism requiring that the fundamental electric and magnetic charges, e and g, verify the so-called Dirac relation $eg = 2\pi$ (in MKS rationalized units with vacuum permittivity equal to 1); although that new particle was never observed, the idea is certainly fertile and was later developed in other contexts [15,16]. Since 1959, when Aharonov and Bohm [17] discovered the effect that bears their name, it is known that the description of some electromagnetic phenomena does require topological considerations. Another important milestone is the sine–Gordon equation, which offers the simplest model with a conservation law of topological origin, based on the degree of a map $S^1 \mapsto S^1$ of the circle on itself. Its extension to three dimensions by Skyrme [18–20] lead to a model with topological solitons and a conserved constant proportional to the degree of a map $S^3 \mapsto S^3$ between three-dimensional spheres. Skyrme had studied with attention Kelvin's ideas on vortex atoms. As he explained he had three reasons for proposing his own model: unification, renormalization, and what he called the "fermion problem." He hoped that his skyrmion, as his basic solution came to be known, would be a fundamental boson from which all the particles would be built; because any topological theory is nonlinear, the possibility of removing the infinities seemed a realistic aim; he did not like fermions as fundamental entities so that explaining them out of bosons seemed very attractive to him.

The classification of knots and links had been attempted by Tait in 1911, when trying to develop Kelvin's model of the vortex atom. He posed the problem and formulated some conjectures, treating with success the simplest cases. However, in spite of its interest, this new branch of mathematics fell into oblivion for many years—in spite of the discovery in 1928 of the Alexander polynomials, which are invariants associated with knots and links—until the 1980s, when Jones found another set of polynomials that opened the door to the proof of some of Tait's conjectures. Simultaneously, the idea that topology will play a major role in quantum physics was progressively imposing itself. As Atiyah [12] puts it, this is not surprising, since "both topology and quantum physics go from the continuous to the discrete." Developments from pure mathematics turned out to be related to Yang–Mills field theory, such as the proposal by Witten of a topological quantum field theory that may open the way to a deeper understanding of quantum physics [21,22] or the study of configurations of vector fields [23].

B. The Aim of This Work

The aim of this report is to explain and develop a topological model of electromagnetism that was presented by one of us (AFR) in 1989 [24–26]. The main characteristics of this model are

1. It is based on the idea of "electromagnetic knot," introduced in 1990 [27–29] and developed later [30–32]. An electromagnetic knot is defined as a standard electromagnetic field with the property that any pair of its magnetic lines, or any pair of its electric lines, is a link with linking number ℓ (which is a measure of the extent to which the force lines curl themselves around one another, i.e., of the helicity of the field). These lines coincide with the level curves of a pair of complex scalar fields ϕ, θ. The physical space and the complex plane are compactified to S^3 and S^2, so that the scalars can be

interpreted as maps $S^3 \mapsto S^2$, which are known to be classified in homotopy classes characterized by the integer value of the Hopf index n, which is related to the linking number ℓ. Moreover, the Faraday 2-form and its dual are equal to the pullbacks of the area 2-form in S^2 by the two maps $\mathscr{F} = -\phi^*\sigma$, $*\mathscr{F} = \theta^*\sigma$. The topology of the force lines thus induces a topological structure in the set of the fields of the model.

2. The topological model is locally equivalent to Maxwell's standard theory in the sense that the set of electromagnetic knots coincides locally with the set of the standard radiation fields. In other words, standard radiation fields can be understood as patched-together electromagnetic knots. This can still be expressed as the statement that, in any bounded domain of spacetime, any standard radiation fields can be approximated arbitrarily enough by electromagnetic knots (except for a zero measues set). However, it is not globally equivalent to Maxwell's theory because of the special way in which the electromagnetic knots behave around the point at infinity.

3. The standard Maxwell equations are the *exact linearization by change of variables (not by truncation)* of a set of nonlinear equations referring to the scalars ϕ, θ. The fact that this change is not completely invertible produces a hidden nonlinearity, thanks to which the linearity of the Maxwell equations is compatible with the existence of topological constants of motion that are nonlinear in A^μ and $F_{\mu\nu}$.

4. In the case of empty space, one of these topological constants of the motion is the electromagnetic helicity of a knot, defined as the semisum of the magnetic and electric helicities, which turns out to be equal to the Hopf index n of the maps ϕ and θ: $\mathscr{H} = \frac{1}{2}\int(\mathbf{A}\cdot\mathbf{B} + \mathbf{C}\cdot\mathbf{E})\,d^3r = n$, where $\mathbf{B} = \nabla\times\mathbf{A}$, $\mathbf{E} = \nabla\times\mathbf{C}$. This implies an interesting interpretation of the Hopf index n, since that helicity is equal to the classical expression of the difference between the numbers of right-handed and left-handed photons contained in the field $N_R - N_L$ (defined by substituting Fourier transform functions for creation and annihilation operators in the quantum expression). In other words, $n = N_R - N_L$. This establishes a relation between the wave and the particle understanding of the idea of helicity, that is, between the curling of the force lines to one another and the difference between right- and left-handed photons contained in the field.

5. Another topological constant of motion is the electric charge (and eventually the magnetic charge as well), which is topologically quantized in such a way that any charge is always equal to an integer number times q_0: (a) the fundamental charge is $q_0 = 1$, in natural units (in the rationalized MKS system $q_0 = \sqrt{\hbar c}$; in SI units $q_0 = \sqrt{\hbar c \epsilon_0}$); (b) the number of fundamental charges inside a volume is the degree of a map between two spheres S^2. Note that $q_0 = 3.3\,e = 5.29\times10^{-19}$ C. We will argue in Section VII that this is a "good" value.

The model has, moreover, the appealing property of being completely symmetric between electricity and magnetism, to the point of having room for magnetic charges, also quantized and having the same fundamental value q_0. This might seem a negative feature at first sight, since the Dirac monopole g has a value quite different from that of the electric charge: $(2\alpha)^{-1} = 68.5$ times bigger. However, it is known that the sea of virtual electron–positron pairs of the vacuum is dielectric but must be paramagnetic, so that the observed electric charge must be smaller than the bare one, while the observed magnetic charge should be larger. As the model is classical, an intriguing idea arises: q_0 could be interpreted as the common value of the electric and the magnetic bare charges, assuming that the effect of the virtual pairs is to decrease the observed electric charge from $q_0 = 5.29 \times 10^{-19}$ C to $e = 1.6 \times 10^{-19}$ C and to increase the magnetic charge with room in the model from q_0 to $g = 1.1 \times 10^{-17}$ C $= 68.5\,e$.

The model was proposed in Ref. 24 and developed in Refs. 25,27–32, and 34.

C. Faraday's Conception of Force Lines Suggests a Topological Structure for Electromagnetism

As we said above, Faraday thought of the force lines as something real, concrete, and tangible. Let us now be faithful to his original view, representing the dynamics of the electromagnetic field by the evolution of its magnetic and electric lines or, in other words, attempting a line dynamics (For the time being, we consider only the case of empty space; point charges will be introduced later.) In order to compare with the standard formulation of electromagnetism, we need to know how to derive the electromagnetic tensor from these lines. As a simple tentative idea, let us represent the magnetic lines by the equation $\phi(t, x, y, z) = \phi_0$ where ϕ is a complex function of space and time and ϕ_0 is a constant labeling each line. This means that the magnetic lines are the level curves of $\phi(t, x, y, z)$. As the magnetic field is tangent to them, it can always be written as (bars over complex numbers indicate in this work complex conjugation)

$$\mathbf{B} = g\boldsymbol{\nabla}\bar{\phi} \times \boldsymbol{\nabla}\phi$$

where g is some function that, because $\boldsymbol{\nabla} \cdot \mathbf{B} = 0$, must depend on $(t,\mathbf{r})$ only through ϕ and $\bar{\phi}$, that is

$$\mathbf{B} = g(\phi, \bar{\phi})\boldsymbol{\nabla}\bar{\phi} \times \boldsymbol{\nabla}\phi \tag{1}$$

which can also be written as

$$B_k = -\frac{1}{2}\epsilon_{ijk} F_{ij} \tag{2}$$

where

$$F_{ij} = -g(\phi, \bar{\phi})(\partial_i \bar{\phi} \partial_j \phi - \partial_j \bar{\phi} \partial_i \phi) \tag{3}$$

Covariance implies, then, that the Faraday electromagnetic tensor must have the form

$$F_{\mu\nu} = -g(\phi, \bar{\phi})(\partial_\mu \bar{\phi} \partial_\nu \phi - \partial_\nu \bar{\phi} \partial_\mu \phi) \tag{4}$$

and the electric field is

$$\mathbf{E} = -g(\phi, \bar{\phi})(\partial_0 \bar{\phi} \boldsymbol{\nabla} \phi - \partial_0 \phi \boldsymbol{\nabla} \bar{\phi}) \tag{5}$$

Therefore, when trying a line dynamics, an antisymmetric rank 2 tensor appears. As is seen, $\mathbf{E} \cdot \mathbf{B} = 0$ or, equivalently, $\det(F_{\mu\nu}) = 0$, as the electric and the vector fields are orthogonal, which means that, with this method, the Faraday 2-form is degenerate and the field is of radiation type (also called singular, degenerate, or null).

We will admit that the total energy is finite, which implies, of course, that $\mathbf{B}$ and $\mathbf{E}$ go to zero at infinity. The simplest way for this condition to be achieved is requiring that the limit of ϕ when $r \to \infty$ does not depend on the direction or, stated otherwise, that ϕ takes only one value at infinity. There are certainly other ways; we could, for instance, ask that ϕ is real or that its real part is a function of its imaginary part at $r = \infty$. In this work the first and simplest possibility is explored and so, after assuming that the magnetic lines are the level curves of the scalar ϕ, it will be admitted also that ϕ is one-valued at infinity.

The title of this section alludes to an important consequence of this argument. It is clear that the fact that ϕ is one-valued at infinity implies that R^3 is compactified to S^3 and that $\phi(\mathbf{r}, t)$ can be interpreted at any time as a map $S^3 \mapsto S^2$, after identifying, via stereographic projection, $R^3 \cup \{\infty\}$ with S^3 and the complete complex plane C with S^2. Maps of this kind have nontrivial topological properties, so that *the attempt to describe electromagnetism by the evolution of the magnetic lines, represented as the level curves of a complex function, leads in a compelling and almost unavoidable way to the appearance of a topological structure*—and a very rich one, as we will see.

It turns out that a tensor as that of (4) is similar to an important geometric object related to the map ϕ. Let us consider the area 2-form σ in the sphere S^2, normalized to unit total area. Its pullback to $S^3 \times R$ (identified with the spacetime) is

$$\phi^* \sigma = \frac{1}{2\pi i} \frac{d\phi \wedge d\bar{\phi}}{(1 + \bar{\phi}\phi)^2} \tag{6}$$

[This means that we take the complex number $\phi(\mathbf{r}, t)$ as a coordinate in S^2.] Note that ϕ^* in (6) indicates pullback by the corresponding map ϕ and should not be mistaken for the complex conjugate of ϕ, which we denote as $\bar{\phi}$. As we see, there is a 2-form closely associated with the scalar, the level curves of which coincide with the magnetic lines. Since both $\phi^*\sigma$ and the Faraday 2-form $\mathscr{F} = \frac{1}{2}F_{\mu\nu}dx^\mu \wedge dx^\nu$ are closed, it seems natural to identify the two, up to a normalization constant factor that, for later convenience, we write as $-\sqrt{a}$. More precisely, we assume that

$$\mathscr{F} = -\sqrt{a}\, \phi^*\sigma \tag{7}$$

and, consequently

$$F_{\mu\nu} = \frac{\sqrt{a}}{2\pi i} \frac{\partial_\mu\bar{\phi}\partial_\nu\phi - \partial_\nu\bar{\phi}\partial_\mu\phi}{(1 + \bar{\phi}\phi)^2} \tag{8}$$

Note that the normalizing constant a, with the dimensions of action times velocity, must be introduced necessarily in order for $F_{\mu\nu}$ to have the right dimensions; it can always be written as the product of a pure number times the Planck constant times the light velocity (in natural units, a is a pure number; the electric and magnetic fields are then inverse square lengths). As is seen, $F_{\mu\nu}$ is of the form (4) with $g = -\sqrt{a}/(2\pi i(1 + \bar{\phi}\phi)^2)$. It should be stressed now that the assumption that ϕ has only one value at infinity leads compellingly from (1) to (8). Because $\mathscr{F}$ is closed in S^3, the second cohomology group of which is trivial, there exists a 1-form $\mathscr{A} = A_\mu dx^\mu$, such that $\mathscr{F} = d\mathscr{A}$, where the 4-vector A_μ is clearly the electromagnetic field, $F_{\mu\nu} = \partial_\mu A_\nu - \partial_\nu A_\mu$ and $\mathbf{B} = \nabla \times \mathbf{A}$ [24,25,36,37].

As long as no charges are present, we can play the same game with the electric field $\mathbf{E}$ and a scalar field θ, the level curves of which coincide with the electric lines. In that case, if the pullback of the area 2-form in S^2 by θ is

$$\theta^*\sigma = \frac{1}{2\pi i} \frac{d\theta \wedge d\bar{\theta}}{(1 + \bar{\theta}\theta)^2} \tag{9}$$

and the dual to the Faraday form is taken to be

$$*\mathscr{F} = \sqrt{a}\, \theta^*\sigma \tag{10}$$

the dual to the Faraday tensor is then

$$*F_{\mu\nu} = \frac{\sqrt{a}}{2\pi i} \frac{\partial_\mu\theta\partial_\nu\bar{\theta} - \partial_\nu\theta\partial_\mu\bar{\theta}}{(1 + \bar{\theta}\theta)^2} \tag{11}$$

so that the following duality condition must be fulfilled

$$*F^{\mu\nu} = \frac{1}{2} \epsilon^{\mu\nu\alpha\beta} F_{\alpha\beta} \tag{12}$$

which expresses the duality of $\mathscr{F}$ and $*\mathscr{F}$. The conditions for the existence of the pair ϕ, θ will be discussed later; for the moment let us say that they pose no difficulty. Equation (12) can be written also somewhat more formally as

$$*\phi^*\sigma = -\theta^*\sigma \tag{13}$$

where $*$ is the Hodge or duality operator.

It is convenient now to introduce two definitions:

1. We will say that the map $\chi: S^3 \times R \mapsto S^2$, given by a scalar field $\chi(\mathbf{r}, t)$, generates an electromagnetic field if the corresponding pullback of the area form in S^2 $\chi^*\sigma$, or its dual form $*\chi^*\sigma$, verifies the Maxwell equations in empty space.

2. A pair of maps $\phi, \theta: S^3 \times R \mapsto S^2$, given by two scalar fields $\phi(\mathbf{r}, t), \theta(\mathbf{r}, t)$, will be said to be *a pair of dual maps*, if the pullback of σ, the area form in S^2, by the first map is equal to the dual of the pullback by the second one. In other words, if

$$\phi^*\sigma = *\theta^*\sigma, \qquad \theta^*\sigma = -*\phi^*\sigma \tag{14}$$

Note that, as the square of the Hodge operator is -1, these two equations imply each other.

An equivalent definition is that ϕ and θ are dual if they define, by pullback of the area 2-form in S^2, two tensors $F_{\mu\nu}, {}^*F_{\mu\nu}$, given by Eqs. (8) and (11), which are dual in the sense of Eq. (12).

A surprising and important property appears now, expressed by the following proposition.

Proposition 1.   *If $\phi(\mathbf{r}, t), \theta(\mathbf{r}, t)$ are two scalar fields one-valued at infinity, and they form a pair of dual maps $\phi, \theta: S^3 \mapsto S^2$, the forms, $\mathscr{F} = -\sqrt{a}\,\phi^*\sigma$ and ${}^*\mathscr{F} = \sqrt{a}\,\theta^*\sigma$, verify necessarily the Maxwell equations in empty space.*

Proof. The proof is simple. The 2-forms $\mathscr{F} = -\sqrt{a}\,\phi^*\sigma$ and $*\mathscr{F} = \sqrt{a}\,\theta^*\sigma$ are exact (because the second group of cohomology of S^3 is trivial), so that they verify

$$d\mathscr{F} = 0, \qquad d*\mathscr{F} = 0 \tag{15}$$

which are the Maxwell equations in empty space. Note that the substitution of the duality condition (13) changes each one of these equations into the other, as it gives

$$d * (\phi^*\sigma) = 0, \qquad d * (\theta^*\sigma) = 0 \tag{16}$$

which are in fact $d * \mathcal{F} = 0$, $d\mathcal{F} = 0$. [In terms of the two tensors: it is easy to see from (8) and (11) that both obey automatically the first pair $\epsilon^{\alpha\beta\gamma\delta}\partial_\beta F_{\gamma\delta} = 0$, $\epsilon^{\alpha\beta\gamma\delta}\partial_\beta {}^*F_{\gamma\delta} = 0$. Substitution of (12) gives $\partial_\alpha F^{\alpha\beta} = 0$, $\partial_\alpha^* F^{\alpha\beta} = 0$, the second pair for both.]

To summarize this subsection, the description of the dynamics of the force lines as the level curves of two maps $S^3 \mapsto S^2$, given by two complex functions ϕ and θ, leads in a compelling way to a topological structure, in such a way that the mere existence of a pair of such functions guarantee that the corresponding pullbacks of the area 2-form in S^2 automatically obey the Maxwell's equations in empty space.

In other words, for any pair of complex scalar fields, dual to one another in the sense explained above, there is an electromagnetic field in empty space. This association is studied in detail in the following text.

D. The Hopf Index

As we have shown, if a scalar field $\phi(\mathbf{r})$ is one-valued at infinity (i.e. if its limit when $r \to \infty$ does not depend on the direction), it can be interpreted as a map $\phi: S^3 \mapsto S^2$. To do that, one must identify, via stereographic projection, the 3-space plus the point at infinity $R^3 \cup \{\infty\}$ with S^3, and the complete complex plane $C \cup \{\infty\}$ with the sphere S^2.

To realize these identifications, we can proceed as follows. A point P in S^2 can be represented in two convenient ways: (1) with the Cartesian coordinates n_1, n_2, n_3, such that $n^2 = \sum n_k^2 = 1$; and (2) with the spherical angles ϑ, φ, related to n_k by $n_1 + in_2 = \sin\vartheta \exp(i\varphi)$, $n_3 = \cos\vartheta$. Its stereographic projection is the complex number $\phi = \cot(\vartheta/2)\exp(i\varphi)$, which will be taken in this work as the coordinate of P in S^2 (unless otherwise stated). On the other hand, the Cartesian $u_1, \ldots, u_4$ with the condition $\sum u_k^2 = 1$ can be taken as coordinates of a point Q in S^3. Their relation with the Cartesian x_1, x_2, x_3 of its stereographic projection on the 3-space $u_4 = 0$ are $x_k = u_k/(1 - u_4)$, and the inverse equations $u_k = 2x_k/(1 + r^2)$, $u_4 = (r^2 - 1)/(r^2 + 1)$, with $r^2 = \sum x_k^2$.

In this way, a complex function $\phi(\mathbf{r})$ can be interpreted as a map $S^3 \mapsto S^2$. This is very important, since maps of this kind can be classified in homotopy classes labeled by a topological integer number called the *Hopf index*, so that the same topological property applies to any scalar field (provided that it is one-valued at infinity).

Let a map $f: S^3 \mapsto S^2$, which we suppose to be smooth, be realized by a complex function $f(\mathbf{r})$, and let us consider the pullback of the area 2-form of S^2

normalized to the unity σ, which is equal to (for convenience we introduce a minus sign)

$$F = -f^*\sigma = \frac{1}{2\pi i} \frac{d\bar{f} \wedge df}{(1+\bar{f}f)^2} \tag{17}$$

Since F is closed in S^3 whose second group of cohomology is trivial, it must also be exact or, in other words, there exists a 1-form A, well defined in S^3 and such that $F = dA$. As was shown in 1947 by Whitehead [43], the integral of the form F through Σ_a which gives the Hopf index can be written as

$$n = \int_{S^3} A \wedge F \tag{18}$$

[Note that this expression is unchanged by the minus sign introduced in Eq. (17).]

It must be stressed that the Hopf index is closely related to the linking number of any pair of lines ℓ (defined as the number of intersections of one line with a surface bounded by the other), but is a different concept. If the multiplicity of the map is m [i.e., if the level curves defined by the equation $\phi(\mathbf{r}) = \phi_0$ have m disjoint connected components], the Hopf index is $n = \ell m^2$. If a line is defined as a level curve, and there is multiplicity, it is the union of m closed loops. In that case, we could generalize the idea of linking number of the lines to design all the linkings of two sets of m loops each one, what is precisely ℓm^2. However, mathematicians seldom use that generalization [36–42].

It is convenient to consider the 2-form F in more detail. From (17) we can write

$$F = \frac{1}{2} f_{ij} dx_i \wedge dx_j = \frac{1}{4\pi i} \frac{\partial_i \bar{f} \partial_j f - \partial_j \bar{f} \partial_i f}{(1+\bar{f}f)^2} dx_i \wedge dx_j \tag{19}$$

Like any antisymmetric tensor in three dimensions, f_{ij} can be expressed in terms of a vector $\mathbf{b}(\mathbf{r})$ as

$$f_{ij} = -\epsilon_{ijk} b_k, \qquad b_k = -\frac{1}{2} \epsilon_{ijk} f_{ij} \tag{20}$$

It can be seen from (19) that $\nabla \cdot \mathbf{b} = 0$. Consequently, if $A = -a^i dx_i$, it turns out that $\mathbf{b} = \nabla \times \mathbf{a}$. It is clear that the vector $\mathbf{b}$ is always tangent to the level curves of f, which are its integral lines. It plays an important role in the description of the maps from S^3 (or R^3) to S^2 (or C). Here, it is called *the Whitehead vector* of the map f, and is noted $\mathbf{b} = \mathbf{W}_f$. The expression (18) of the Hopf index can then be written as

$$n = \int_{R^3} \mathbf{a} \cdot \mathbf{b} \, d^3 r \tag{21}$$

The quantity in the right-hand side of (21) is called *the helicity of the vector* **b** and is used in several contexts, mainly in fluid and in plasma physics. The term was coined by Moffatt in 1969 in a paper on tangled vorticity lines [44] with the velocity of a fluid **v** and its vorticity $\omega = \nabla \times \mathbf{v}$ as the vectors **a** and **b**. The magnetic helicity $h = \int \mathbf{A} \cdot \mathbf{B} d^3 r$ is useful to study the magnetic configurations in astrophysical plasmas and in tokamaks.

E.   Magnetic and Electric Helicities

In Section I.B, two maps ϕ, θ were introduced, such that its level curves are the magnetic and electric lines of an electromagnetic field. Comparing the expressions (6)–(7) and (9)–(10) with (17), we observe a close formal relation between the theory of the Hopf index and the Maxwell theory in empty space. It follows that, for the electromagnetic fields generated by pairs of dual maps $\phi, \theta: S^3 \mapsto S^2$, there are two constants of the motion of topological origin. They are *the magnetic helicity*

$$h_m = \int_{R^3} \mathbf{A} \cdot \mathbf{B} \, d^3 r = n_m a \tag{22}$$

where n_m is the Hopf index of the map ϕ, related (as explained before) to the linking number of any pair of its level curves that coincide with the magnetic lines, and *the electric helicity*

$$h_e = \int_{R^3} \mathbf{C} \cdot \mathbf{E} \, d^3 r = n_e a \tag{23}$$

where **C** is a vector potential for **E** , that is, $\nabla \times \mathbf{C} = \mathbf{E}$, and n_e is the Hopf index of the map θ, related (as explained before) to the linking number of any pair of electric lines which are the level curves of θ.

It will be shown in Section II.C that the two Hopf indices are equal, $n_m = n_e = n$, in the case of electromagnetic knots in empty space.

F.   Definition of an Electromagnetic Knot

The defining physical feature of an electromagnetic knot is that any pair of magnetic lines (or of electric lines) is a pair of linked loops (except perhaps for some exceptional lines or exceptional times) (see Fig.1). From the mathematical point of view, we define *an electromagnetic knot* to be an electromagnetic field generated by a pair of dual maps $\phi, \theta: S^3 \mapsto S^2$ verifying (14) [i.e., an electromagnetic field that can be expressed in terms of a pair of dual maps by means of equations (8) and (11)].

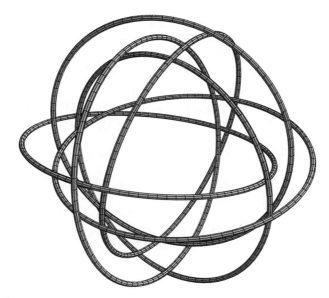

Figure 1. Schematic aspect of several force lines (either magnetic or electric) of an electromagnetic knot. Any two of the six lines shown are linked once.

This definition implies that the corresponding Faraday 2-form $\mathscr{F}$ and its dual $*\mathscr{F}$ can be written as $\mathscr{F} = -\phi^*\sigma$ and $*\mathscr{F} = \theta^*\sigma$, that is, as minus the pullback and the pullback of the area form in S^2 σ by the two maps. This will be relevant for the quantization of the charge.

A very important property is that the magnetic and electric lines of an electromagnetic knot are the level curves of the scalar fields $\phi(\mathbf{r}, t)$ and $\theta(\mathbf{r}, t)$, respectively. Another is that the magnetic and the electric helicities are topological constants of the motion, equal to the common Hopf index of the corresponding pair of dual maps ϕ, θ times a constant with dimensions of action times velocity.

In an electromagnetic knot, each line is labeled by a complex number. If there are m lines with the same label, we will say that m is the multiplicity. If all the pairs of line have the same linking number ℓ, it turns out that the Hopf index is given as $n = \ell m^2$.

An electromagnetic knot is a radiation field (i.e., $\mathbf{E} \cdot \mathbf{B} = 0$); the magnetic and electric lines are orthogonal at any point. This means that $\mathbf{E}$, $\mathbf{B}$, and the Poynting vector $\mathbf{S} = \mathbf{E} \times \mathbf{B}$ are three orthogonal vectors everywhere. The corresponding three families of curves (electric, magnetic, and energy flux lines) form three orthogonal fibrations of S^3, since each family fills all the space, in the sense that a line of each kind passes through every point (although there may be some exceptional lines, or one or several of the vectors may vanish at certain points or times). This property is allowed by the fact that S^3 is parallelizable. Indeed, an electromagnetic knot has a rich structure.

II. MEANINGS OF ELECTRIC AND MAGNETIC HELICITIES

The helicity of a divergenceless vector field was already used by Woltjer in 1958 [45] in an astrophysical context. Moreau [46] showed soon after that it is a conserved quantity in certain flows of fluid dynamics. Moffatt, in a seminal paper [44], coined the term *helicity* and clarified its topological meaning. For a pedagogical review, see Ref. 47.

Let $\mathbf{X}(\mathbf{r})$, $\mathbf{r} \in D$, be a real vector field defined in a parallelizable three-dimensional (3D) manifold D. If $\mathbf{X}$ is divergenceless, that is, if $\nabla \cdot \mathbf{X} = 0$, another vector field exists in D, at least locally: the vector potential $\mathbf{Y}(\mathbf{r})$, such that $\mathbf{X} = \nabla \times \mathbf{Y}$. The *helicity* of the divergenceless vector field $\mathbf{X}(\mathbf{r})$ in D is the integral

$$h(\mathbf{X}, D) = \int_D \mathbf{X} \cdot \mathbf{Y} \, d^3 r \tag{24}$$

We will write $h(\mathbf{X})$ or simply h if there is no risk of confusion. The helicity (24) is especially useful in two physical contexts: (1) in fluid dynamics, where $\mathbf{Y}$ is the flow velocity $\mathbf{v}(\mathbf{r}, t)$, $\mathbf{X}$ is the vorticity $\mathbf{w} = \nabla \times \mathbf{v}$, and $h(\mathbf{w}, D)$ is called *vortex helicity*; and (2) in plasma physics, or in general in electromagnetism, under the form of *magnetic helicity*, defined as

$$h_m = \int_D \mathbf{A} \cdot \mathbf{B} \, d^3 r \tag{25}$$

in terms of the magnetic field and its vector potential. Using the field equations for $\mathbf{B}$ and $\mathbf{A}$, we find that the time evolution of the magnetic helicity (25) is given by

$$\frac{\partial}{\partial t} \int_D \mathbf{A} \cdot \mathbf{B} \, d^3 r = -2 \int_D \mathbf{E} \cdot \mathbf{B} \, d^3 r - \int_{\partial D} (A^0 \mathbf{B} + \mathbf{E} \times \mathbf{A}) \cdot \mathbf{n} \, dS \tag{26}$$

where $\mathbf{n}$ is an unit vector orthogonal to the surface ∂D, the border of the manifold D, and dS is the area element in that surface. For the magnetic helicity to be time invariant, the integrals at the right hand side of (26) must vanish, this implying two conditions. The surface integral in (26) depends on the boundary conditions;

we will take fields that vanish in ∂D, so that the only requirement for the time invariance of the magnetic helicity is

$$\int_D \mathbf{E} \cdot \mathbf{B} \, d^3 r = 0 \tag{27}$$

The equation $\mathbf{E} \cdot \mathbf{B} = 0$ is Lorentz-invariant. The fields that satisfy it are called radiation fields.

The electric and magnetic fields are invariant under gauge transformations $A_\mu(x) \mapsto A_\mu(x) + \partial_\mu \Lambda(x)$. The effect of these transformations on the helicity has been treated by Marsh [49]. The variation of the magnetic helicity under a gauge transformation $\delta \mathbf{A} = -\boldsymbol{\nabla}\Lambda$ is

$$\delta h_m = -\int_D \boldsymbol{\nabla}\Lambda \cdot \mathbf{B} \, d^3 r = -\int_D \boldsymbol{\nabla} \cdot (\Lambda \mathbf{B}) \, d^3 r \tag{28}$$

If D is simply connected, its first cohomology group is trivial and, consequently, Λ is globally defined in D (it is a one-valued function). The Stokes theorem implies then that $\delta h_m = 0$; in other words, the magnetic helicity is gauge-invariant in simply connected domains, with the standard boundary conditions. If, on the other hand, D is not simply connected, a gauge transformation implies a nontrivial change of the helicity. Here, we will consider only simply connected spatial domains so that the magnetic helicity will always be gauge-invariant.

A.   Helicity and Topology of the Force Lines

We will now apply the ideas of magnetic and filamental tubes. Consider a small closed circular curve C. We define its *magnetic tube* to be the set of all magnetic lines that intersect C. It is obvious that the internal flux of any magnetic tube does not depend on the section that we use to compute it. The *strength* of the magnetic tube is defined as the flux across any section. A *filamental tube* is a magnetic tube with infinitesimal section, but with non vanishing finite strength.

In the first paper on the topological meaning of the helicity, Moffatt [44] considered closed non-self-knotted filamental tubes. Suppose that D, the region on which a divergenceless vector field $\mathbf{B}$ is defined at a certain time is a simply connected domain ($\mathbf{B}$ is here a magnetic field, but the results can be applied to any divergenceless vector field). Now consider the special situation in which $\mathbf{B}$ is zero except in two filamental tubes whose axes are two oriented, closed, and non-self-knotted magnetic lines C y C', that can, however, be linked to one another. The filamental tubes have zero section but nonvanishing strengths $\delta\Phi$, $\delta\Phi'$, respectively. Moreover, the magnetic lines run parallel to C and C', respectively, along each filamental tube. Moffatt showed that, in these conditions, the magnetic helicity can be written as

$$h_m = 2\delta\Phi\delta\Phi'\left(\frac{1}{4\pi}\oint_C\oint_{C'}(d\mathbf{r}\times d\mathbf{r}')\cdot\frac{\mathbf{r}-\mathbf{r}'}{\|\mathbf{r}-\mathbf{r}'\|^3}\right) \qquad (29)$$

where the integral between brackets in (29) is known as the *Gauss integral* and coincides with the linking number $L(C,C')$ of the closed curves C and C'. Consequently, the helicity of two filamental tubes is

$$h_m = 2L(C,C')\delta\Phi\delta\Phi' \qquad (30)$$

If there are several closed filamental tubes, the total helicity is

$$h_m = \sum_i\sum_{j\neq i}L(C_i,C_j)\delta\Phi_i\delta\Phi_j \qquad (31)$$

where $L(C_i,C_j)$ is the linking number of the tubes C_i,C_j. Equation (30) can be generalized to finite section magnetic tubes under the condition that, into each one, the magnetic lines are not linked, and the tubes are not self-knotted. We may then assume that every magnetic tube is composed of a large number of filamental tubes of infinitesimal sections. Then each pair of filamental tubes (one for each magnetic tube) contributes to the total helicity with a quantity given by (30). With addition of the contributions of all pairs, this results in

$$h_m = 2L(C_1,C_2)\Phi_1\Phi_2 \qquad (32)$$

where Φ_1 and Φ_2 are, respectively, the total strengths of the magnetic tubes and $L(C_1,C_2)$ is the linking number of two filamental tubes representing each magnetic tube.

The relationship between the linking number and the helicity of a self-knotted filamental tube has been studied by Berger and Field [50] and Moffatt and Ricca [51]. We are looking for an expression for the helicity in a filamental tube T of strength $\delta\Phi$, around a closed magnetic line C that can be self-knotted. In this case, the magnetic helicity takes the value

$$h_m = (\delta\Phi)^2(\mathscr{W}+\mathscr{T}+\mathscr{T}_0) \qquad (33)$$

where $\mathscr{W}(C)$ is the *spatial writhing number*, which is a real number defined by the limiting form of the Gauss integral (29) when $C\to C'$, $\mathscr{T}(C)$ is the total torsion of the curve, and $\mathscr{T}_0(C,\mathbf{B})$ is the intrinsic twisting number, defined as the (integer) number of times that any magnetic line in the surface of the tube T surrounds the axis C. None of the numbers $\mathscr{W}$, $\mathscr{T}$, and $\mathscr{T}_0$ is a topological invariant, and only the last one is an integer number, though all of them have a well defined geometrical meaning. But their sum, that is the important thing in (33), is, according to the *Calugareanu theorem* [52], equal to a topological invariant, the linking number of the axis C and any magnetic line C_m that surrrounds the surface of the tube T:

$$L(C, C_m) = \mathcal{W}(C) + \mathcal{T}(C) + \mathcal{T}_0(C, \mathbf{B}) \tag{34}$$

Using this theorem, we conclude that the magnetic helicity of a filamental tube T whose axe is a closed, self-knotted curve C, is given by

$$h_m = (\delta\Phi)^2 L(C, T) \tag{35}$$

where $L(C, T)$ is the linking number of the axe C and any magnetic line that rounds the surface of the tube T.

The next stage in the study of the topological meaning of the helicity is to consider a continuous distribution. In Ref. 47, the case was considered of a simply-connected domain D, and a divergenceless vector field $\mathbf{B}$ satisfying the following conditions: (1) $\mathbf{B}$ is regular and nonvanishing in D, and (2) the magnetic lines are closed curves. Under these conditions it is easy to see that the linking number of any two magnetic lines does not depend on the lines that we choose to compute it, because two closed curves cannot tie or untie under smooth deformation. Consequently, the linking number of any two magnetic lines is a property of the vector field, and can be denoted $L(\mathbf{B})$. The linking number is also invariant under smooth deformation of the field and, in particular, is time-invariant. This means that the set of divergenceless vector fields defined on a domain D satisfying conditions (1) and (2) can be classified in homotopy classes, labeled by its linking number. We split D in an infinite number of filamental tubes with strengths $\delta\Phi_i$, in such a way that they completely fill D. Now using (30) and (35), and assuming that the linking number is an invariant, the magnetic helicity in D results in

$$h_m = L(\mathbf{B}) \left(\sum_i \sum_{j \neq i} 2\delta\Phi_i \delta\Phi_j + \sum_i (\delta\Phi_i)^2 \right)$$

$$= L(\mathbf{B}) \left(\sum_i \delta\Phi_i \right)^2 = L(\mathbf{B})\Phi^2 \tag{36}$$

where Φ is the total strength of the magnetic field (the sum of strengths of all the tubes that fill D).

We have seen that the helicity of simple field configurations, in which the magnetic lines are linked, depends on the linking number of these lines. However, some configurations are possible in which the field lines are open, infinitely long, and possibly space-filling. The classical linking number above defined has no sense in this case, so it is difficult to study the meaning of the helicity related to the topology of these lines. Nonclosed lines can be treated, however, following the approach proposed by Arnold [53], who used the language of dynamical systems to define the linking invariant of the problem. Arnold defined a *asymptotic linking number* from the classical linking number of the field lines that become closed by using a family of short paths (he showed that the result is independent of the family of short paths). In this way, Arnold proved that the contribution to the total helicity of two filamental tubes around two magnetic lines C and C' is

$$h_m = 2\lambda(C, C')\delta\Phi\delta\Phi' \tag{37}$$

which is formally equal to the Moffatt equation (30), if the Gauss linking number L is replaced by the mean value λ of the asymptotic linking number. If the lines were closed, then λ would coincide with the classical linking number, and Eq. (37) would be general.

In conclusion, the helicity of a divergenceless vector field is a measure of the linking number of the field lines. Even in the case that these lines are not closed, the notion of linkage has sense, because a mean value of an asymptotic linking number can be defined, and this value coincides with the helicity.

The electromagnetic helicity has also been studied by Evans [54–57], especially its consequences for his new non-Abelian SO(3) gauge version of QED (quantum electrodynamics).

B. The Case of Maxwell's Theory in Vacuum: Electromagnetic Helicity

In standard classical electrodynamics, the Maxwell equation $d\mathcal{F} = 0$ becomes a Bianchi identity by using the electromagnetic potential $\mathcal{A}$, defined as $\mathcal{F} = d\mathcal{A}$. The dynamical equation for this field in empty space is $d*\mathcal{F} = 0$.

But the Minkowski spacetime R^4 has trivial cohomology. This means that the Maxwell equation implies that $*\mathcal{F}$ is a closed 2-form, so it is also an exact form and we can write $*\mathcal{F} = d\mathcal{C}$, where $\mathcal{C}$ is another potential 1-form in the Minkowski space. Now the dynamical equation becomes another Bianchi identity. This simple idea is a consequence of the electromagnetic duality, which is an exact symmetry in vacuum. In tensor components, with $\mathcal{A} = A_\mu dx^\mu$ and $\mathcal{C} = C_\mu dx^\mu$, we have $F_{\mu\nu} = \partial_\mu A_\nu - \partial_\nu A_\mu$ and $^*F_{\mu\nu} = \partial_\mu C_\nu - \partial_\nu C_\mu$ or, in vector components

$$\mathbf{B} = \mathbf{\nabla} \times \mathbf{A} = \frac{\partial \mathbf{C}}{\partial t} + \mathbf{\nabla} C^0$$

$$\mathbf{E} = \mathbf{\nabla} \times \mathbf{C} = -\frac{\partial \mathbf{A}}{\partial t} - \mathbf{\nabla} A^0 \tag{38}$$

Note that the equations (38) are clearly invariants under the gauge transformations $A_\mu \mapsto A_\mu + \partial_\mu \Lambda$, $C_\mu \mapsto C_\mu + \partial_\mu \Gamma$.

The electric and magnetic fields are dual to one another and have the same properties in Maxwell theory in empty space. Given the divergenceless vector field $\mathbf{B}$, we have defined the magnetic helicity as

$$h_m = \int_{R^3} \mathbf{A} \cdot \mathbf{B} \, d^3 r \qquad (39)$$

where $\mathbf{B} = \nabla \times \mathbf{A}$. Now, in vacuum, given the divergenceless vector field $\mathbf{E}$, we can also define an *electric helicity* through

$$h_e = \int_{R^3} \mathbf{C} \cdot \mathbf{E} \, d^3 r \qquad (40)$$

where $\mathbf{E} = \nabla \times \mathbf{C}$.

Equations (39) and (40) imply that the two helicities are finite if the magnetic and electric fields decrease faster than r^{-2} when $r \to \infty$. This implies that the vector potentials A^μ and C^μ must decrease faster than r^{-1} when $r \to \infty$. We will assume that our fields always satisfy these conditions at infinity.

We must consider two currents of helicity. In Ref. 47 the following magnetic helicity current was considered:

$$\mathcal{H}^\mu_m = A_\nu \, {}^* F^{\nu\mu} \qquad (41)$$

The corresponding electric helicity current is

$$\mathcal{H}^\mu_e = C_\nu F^{\mu\nu} \qquad (42)$$

As it is easy to show that

$$\partial_\mu \mathcal{H}^\mu_m = -\partial_\mu \mathcal{H}^\mu_e = -2\mathbf{E} \cdot \mathbf{B} \qquad (43)$$

the magnetic and the electric helicities

$$h_m = \int_{R^3} \mathcal{H}^0_e \, d^3 r, \qquad h_e = \int_{R^3} \mathcal{H}^0_m \, d^3 r \qquad (44)$$

are time invariants for radiation fields $\mathbf{E} \cdot \mathbf{B} = 0$ and, more generally, if the spatial integral of $\mathbf{E} \cdot \mathbf{B}$ vanishes.

Given any Maxwell field in vacuum, we define the current density of *electromagnetic helicity* $\mathcal{H}^\mu$ as one-half times the sum of the current densities of electric and magnetic helicities (41) and (42):

$$\mathcal{H}^\mu = \frac{1}{2} \left(F^{\mu\nu} C_\nu + {}^* F_{\mu\nu} A_\nu \right) \qquad (45)$$

By construction, and taking into account Eqs. (43), the density of electromagnetic helicity is a conserved current for any Maxwell field in vacuum (with the above indicated behavior at infinity):

$$\partial_\mu \mathscr{H}^\mu = 0 \tag{46}$$

This implies that the quantity

$$\mathscr{H} = \int_{R^3} \mathscr{H}^0 d^3 r = \frac{1}{2} \int_{R^3} (\mathbf{A} \cdot \mathbf{B} + \mathbf{C} \cdot \mathbf{E}) d^3 r \tag{47}$$

is a constant of the motion, called *electromagnetic* or *total helicity*. Because of the previous considerations, the electromagnetic helicity is also gauge-invariant. From now on, we will call (47) the *electromagnetic helicity*, or just the helicity, and (45) will be the *density of electromagnetic helicity*.

C. The Particle Meaning of Helicity

The helicity is gauge-invariant. In the Coulomb gauge, it is obvious that $\mathbf{A}$ and $\mathbf{C}$ satisfy the d'Alembert equation, whose solutions can be written in terms of Fourier transforms

$$\mathbf{A}(\mathbf{r}, t) = \frac{1}{(2\pi)^{3/2}} \int \frac{d^3 k}{\sqrt{2\omega}} \left(\mathbf{a}(\mathbf{k}) e^{-ik \cdot x} + \bar{\mathbf{a}}(\mathbf{k}) e^{ik \cdot x} \right) \tag{48}$$

where $k^\mu = (\omega, \mathbf{k})$ is null ($k^\mu k_\mu = \omega^2 - \mathbf{k}^2 = 0$) and $k \cdot x = k^\mu x_\mu = \omega t - \mathbf{k} \cdot \mathbf{r}$. The factor $1/\sqrt{2\omega}$ is a normalization factor that allows the measure to be Lorentz-invariant. $\bar{z}$ is the complex conjugate of z.

The divergenceless condition of the field $\mathbf{A}$ in the Coulomb gauge means that the complex vector $\mathbf{a}(\mathbf{k})$ is transverse, so that $\mathbf{k} \cdot \mathbf{a}(\mathbf{k}) = 0$. Then, for every value of $\mathbf{k}$, we can choose an orthonormal trihedron with by the real vectors $\mathbf{k}/\omega$, $\mathbf{e}_1(\mathbf{k})$ and $\mathbf{e}_2(\mathbf{k})$, and we can represent the field $\mathbf{a}$ as

$$\mathbf{a}(\mathbf{k}) = a_1(\mathbf{k}) \mathbf{e}_1(\mathbf{k}) + a_2(\mathbf{k}) \mathbf{e}_2(\mathbf{k}) \tag{49}$$

where

$$\mathbf{e}_1 \times \mathbf{e}_2 = \frac{\mathbf{k}}{\omega}, \qquad \frac{\mathbf{k}}{\omega} \times \mathbf{e}_1 = \mathbf{e}_2, \qquad \frac{\mathbf{k}}{\omega} \times \mathbf{e}_2 = -\mathbf{e}_1 \tag{50}$$

For convenience, the definition of the trihedron is completed with

$$\mathbf{e}_1(-\mathbf{k}) = -\mathbf{e}_1(\mathbf{k}), \qquad \mathbf{e}_2(-\mathbf{k}) = \mathbf{e}_2(\mathbf{k}) \tag{51}$$

We have written the field $\mathbf{A}$ as a superposition of plane waves. However, it will be more convenient to write it as a superposition of circularly polarized waves, in the same way as in quantum electrodynamics [58]. With this aim, we define the components right (R) and left (L) as

$$\mathbf{e}_R(\mathbf{k}) = \frac{\mathbf{e}_1(\mathbf{k}) + i\mathbf{e}_2(\mathbf{k})}{\sqrt{2}}, \qquad \mathbf{e}_L(\mathbf{k}) = \frac{\mathbf{e}_1(\mathbf{k}) - i\mathbf{e}_2(\mathbf{k})}{\sqrt{2}}$$

$$a_R(\mathbf{k}) = \frac{a_1(\mathbf{k}) - ia_2(\mathbf{k})}{\sqrt{2}}, \qquad a_L(\mathbf{k}) = \frac{a_1(\mathbf{k}) + ia_2(\mathbf{k})}{\sqrt{2}} \tag{52}$$

With these definitions

$$\begin{aligned}
\mathbf{k} \cdot \mathbf{e}_R = \mathbf{k} \cdot \mathbf{e}_L = 0, & \qquad \mathbf{e}_R \cdot \mathbf{e}_R = \mathbf{e}_L \cdot \mathbf{e}_L = 0 \\
\mathbf{e}_R \cdot \mathbf{e}_L = 1, & \qquad \mathbf{e}_R \times \mathbf{e}_R = \mathbf{e}_L \times \mathbf{e}_L = 0 \\
\frac{\mathbf{k}}{\omega} \times \mathbf{e}_R = -i\mathbf{e}_R, & \qquad \frac{\mathbf{k}}{\omega} \times \mathbf{e}_L = i\mathbf{e}_L, \qquad \mathbf{e}_R \times \mathbf{e}_L = -i\frac{\mathbf{k}}{\omega}
\end{aligned} \tag{53}$$

and, moreover

$$\begin{aligned}
\mathbf{e}_R(-\mathbf{k}) &= -\mathbf{e}_L(\mathbf{k}) \\
\mathbf{e}_L(-\mathbf{k}) &= -\mathbf{e}_R(\mathbf{k})
\end{aligned} \tag{54}$$

Leaving out the argument $\mathbf{k}$ in the quantities $\mathbf{e}(\mathbf{k})$ and $a(\mathbf{k})$, we arrive at

$$\mathbf{A}(\mathbf{r}, t) = \frac{1}{(2\pi)^{3/2}} \int \frac{d^3 k}{\sqrt{2\omega}} \left((\mathbf{e}_R a_R + \mathbf{e}_L a_L) e^{-ik \cdot x} + (\mathbf{e}_L \bar{a}_R + \mathbf{e}_R \bar{a}_L) e^{ik \cdot x} \right) \tag{55}$$

The Fourier components a_R and a_L are, in (55), functions of the vector $\mathbf{k}$. In QED, $a_R(\mathbf{k})$ is interpreted as a destruction operator of photonic states with energy ω, linear momentum $\mathbf{k}$ and spin $\mathbf{k}/\omega$, while the function $\bar{a}_R$ becomes the creation operator a_R^+ of such states. Analogously, $a_L(\mathbf{k})$ is a destruction operator of photonic states with energy ω, linear momentum $\mathbf{k}$ and spin $-\mathbf{k}/\omega$, and a_L^+ is the correspondent creation operator [58].

We can play the same game with $\mathbf{C}$. In this case, we can represent $\mathbf{C}$ in the form (55), but changing the functions a_R and a_L by new functions c_R and c_L. Now we must satisfy the equations (38). Doing this, we find

$$\frac{-\mathbf{k}}{\omega} \times (\mathbf{e}_R a_R + \mathbf{e}_L a_L) = \mathbf{e}_R c_R + \mathbf{e}_L c_L$$

$$\mathbf{e}_R a_R + \mathbf{e}_L a_L = \frac{\mathbf{k}}{\omega} \times (\mathbf{e}_R c_R + \mathbf{e}_L c_L) \tag{56}$$

and, using (53), they reduce to

$$c_R(\mathbf{k}) = ia_R(\mathbf{k}), \qquad c_L(\mathbf{k}) = -ia_L(\mathbf{k}) \tag{57}$$

The vector potential $\mathbf{C}$ results in

$$\mathbf{C}(\mathbf{r}, t) = \frac{i}{(2\pi)^{3/2}} \int \frac{d^3k}{\sqrt{2\omega}} \left((\mathbf{e}_R a_R - \mathbf{e}_L a_L) e^{-ik \cdot x} - (\mathbf{e}_L \bar{a}_R - \mathbf{e}_R \bar{a}_L) e^{ik \cdot x} \right) \tag{58}$$

In the Coulomb gauge, the helicity can be written as

$$\mathscr{H} = \frac{1}{2} \int_{R^3} \left(\mathbf{A} \cdot \frac{\partial \mathbf{C}}{\partial t} - \mathbf{C} \cdot \frac{\partial \mathbf{A}}{\partial t} \right) d^3 r \tag{59}$$

Introducing the expressions (55) and (58) in (59), we obtain

$$\mathscr{H} = \int (\bar{a}_R(\mathbf{k}) a_R(\mathbf{k}) - \bar{a}_L(\mathbf{k}) a_L(\mathbf{k})) \, d^3 k \tag{60}$$

This is what we were looking for. In quantum electrodynamics, the right-hand side of (60) is interpreted as the helicity operator, that is, the difference between the numbers of right-handed and left-handed photons. We can write the usual expressions

$$N_R = \int \bar{a}_R(\mathbf{k}) a_R(\mathbf{k}) \, d^3 k$$

$$N_L = \int \bar{a}_L(\mathbf{k}) a_L(\mathbf{k}) \, d^3 k \tag{61}$$

and write Eq. (60) as

$$\mathscr{H} = (N_R - N_L) \tag{62}$$

The consequence is that *the helicity that we are studying is the classical limit of the difference between right-handed and left-handed photons* [26,30,31]. Note that, in physical units (with $\hbar \neq 1$ and $c \neq 1$), Eq. (62) would be

$$\mathscr{H} = \hbar c(N_R - N_L) \tag{63}$$

This equation shows a close relation between the wave and particle aspects of the helicity. On the left side, the wave helicity is the semisum of the electric and magnetic helicities that characterizes the topology of the force lines as a function of the linking number of the pairs of electric lines and of the magnetic lines. On the right side, the particle helicity is the classical limit of the difference between right- and left-handed photons. For all this, if we have an electromagnetic field with a trivial configuration of force lines (vanishing linking number), then we will know that the classical expression for the number of right-handed photons is equal to the classical expression for the number of left-handed photons. But if we observe a non-vanishing linking number in the magnetic and the electric lines, these two numbers will be different. Here it is the wave–particle duality of the helicity.

We previously defined the radiation fields (called *singular* by mathematicians) as those electromagnetic fields that satisfy $\mathbf{E} \cdot \mathbf{B} = 0$. Now consider the case of singular fields in vacuum, with the abovementioned contour conditions, which we can summarize by stating that the helicity must be finite. In this case, the Fourier components a_R and a_L should be less singular than $\omega^{-3/2}$ when $\omega \to 0$, and they should decrease more rapidly than ω^{-2} when $\omega \to \infty$. This behavior allows us to prove the following property [31]: *The electric and magnetic helicities of any radiation field in vacuum are equal.*

For the proof, we use the representations of $\mathbf{A}$ and $\mathbf{C}$ given by (55) and (58). It is easy to see that

$$h_m - h_e = \int d^3k[(a_L(\mathbf{k})a_L(-\mathbf{k}) - a_R(\mathbf{k})a_R(-\mathbf{k}))e^{-i\omega\tau} + \text{c.c.}] \qquad (64)$$

where c.c. means that the complex conjugate should be added, and $\tau = 2t$. We compute the integral in the angular variables of the spherical coordinates in the space of vectors $\mathbf{k}$. The result is called $F(\omega)$, specifically

$$F(\omega) = \omega^2 \int d\Omega[a_L(\mathbf{k})a_L(-\mathbf{k}) - a_R(\mathbf{k})a_R(-\mathbf{k})] \qquad (65)$$

where Ω is the solid angle and $F(\omega) = F(-\omega)$. The difference of helicities is now

$$h_m - h_e = \int_0^\infty d\omega \left[F(\omega)e^{-i\omega\tau} + F(-\omega)e^{i\omega\tau} \right] \qquad (66)$$

Because of the previously stated behavior of a_R and a_L, and looking at the definition (65), it is clear that $F(\omega)$ is a square integrable function, and that

$$h_m - h_e = f(\tau) + \bar{f}(\tau) \qquad (67)$$

where $f(\tau)$ is the Fourier transform of $F(\omega)$. It is also known that

$$\frac{d}{dt}(h_m - h_e) = -4 \int_{R^3} \mathbf{E} \cdot \mathbf{B} d^3 r \qquad (68)$$

so, in the case of singular field, $\mathbf{E} \cdot \mathbf{B} = 0$ and $h_m - h_e$ is a constant. Equation (68) implies then that the real part of $f(\tau)$ does not depend on τ. Now, recalling that it is a square integrable function, it can only be zero. Consequently, $h_m - h_e = 0$ for singular fields in vacuum.

The conclusion of this subsection is that Eq. (62) for singular fields takes the form

$$\mathcal{H} = h_m = h_e = (N_R - N_L) \tag{69}$$

This will play an important role in the relationships presented in the following sections.

III. ELECTROMAGNETIC KNOTS

In a way that is completely independent of their use as a basis for the topological model presented above, electromagnetic knots are standard solutions of Maxwell equations in vacuum that have special topological properties; their helicity is topologically quantized, with $h_m = h_e = an$, where n is an integer, the Hopf invariant of the two applications from which electric and magnetic fields are constructed. This allows us to classify the set of electromagnetic knots in homotopy classes C_n, labeled by the value n of the Hopf index, which as explained in Section I.D, is related to the linking number ℓ of the lines.

Here we summarize a program to find explicitly the Cauchy data of electromagnetic knots [25,27,30–32]. Let $\phi_0, \theta_0 : S^3 \rightarrow S^2$ be two applications satisfying the following two conditions:

1. The level curves of ϕ_0 must be orthogonal, in each point, to the level curves of θ_0, since we know that electromagnetic knots are singular fields ($\mathbf{E} \cdot \mathbf{B} = 0$). This condition can be written as

$$(\nabla \bar{\phi}_0 \times \nabla \phi_0) \cdot (\nabla \bar{\theta}_0 \times \nabla \theta_0) = 0 \tag{70}$$

2. Moreover, in order to maintain the orthogonality (70) through every time, it is necessary that the Hopf index of ϕ_0 and of θ_0 be equal:

$$H(\phi_0) = H(\theta_0) \tag{71}$$

Given ϕ_0 and θ_0 with these two conditions, we can build the magnetic and electric fields in $t = 0$ as

$$\mathbf{B}(\mathbf{r}, 0) = \frac{\sqrt{a}}{2\pi i} \frac{\nabla \phi_0 \times \nabla \bar{\phi}_0}{(1 + \bar{\phi}_0 \phi_0)^2}$$

$$\mathbf{E}(\mathbf{r}, 0) = \frac{\sqrt{a}}{2\pi i} \frac{\nabla \bar{\theta}_0 \times \nabla \theta_0}{(1 + \bar{\theta}_0 \theta_0)^2} \tag{72}$$

Next, two complex functions $f(\mathbf{r})$ and $g(\mathbf{r})$ should be given, such that

$$\mathbf{\nabla}\theta_0 \times \mathbf{\nabla}\bar{\theta}_0 = f\mathbf{\nabla}\bar{\phi}_0 - \bar{f}\mathbf{\nabla}\phi_0$$
$$\mathbf{\nabla}\phi_0 \times \mathbf{\nabla}\bar{\phi}_0 = \bar{g}\mathbf{\nabla}\theta_0 - g\mathbf{\nabla}\bar{\theta}_0$$

(73)

from which we define

$$\phi_1 = \left(\frac{1 + \bar{\phi}_0\phi_0}{1 + \bar{\theta}_0\theta_0}\right)^2 f$$

$$\theta_1 = \left(\frac{1 + \bar{\theta}_0\theta_0}{1 + \bar{\phi}_0\phi_0}\right)^2 g$$

(74)

Then, the Cauchy data of the fields $\phi(\mathbf{r}, t)$ and $\theta(\mathbf{r}, t)$ are given by

$$\phi_0(\mathbf{r}) = \phi(\mathbf{r}, 0), \qquad \phi_1(\mathbf{r}) = \left.\frac{\partial\phi(\mathbf{r}, t)}{\partial t}\right|_{t=0}$$

$$\theta_0(\mathbf{r}) = \theta(\mathbf{r}, 0), \qquad \theta_1(\mathbf{r}) = \left.\frac{\partial\theta(\mathbf{r}, t)}{\partial t}\right|_{t=0}$$

(75)

and the Cauchy data of the magnetic and electric fields are

$$\mathbf{B}(\mathbf{r}, 0) = \frac{\sqrt{a}}{2\pi i}\frac{\mathbf{\nabla}\phi_0 \times \mathbf{\nabla}\bar{\phi}_0}{(1 + \bar{\phi}_0\phi_0)^2} = \frac{\sqrt{a}}{2\pi i}\frac{\bar{\theta}_1\mathbf{\nabla}\theta_0 - \theta_1\mathbf{\nabla}\bar{\theta}_0}{(1 + \bar{\theta}_0\theta_0)^2}$$

$$\mathbf{E}(\mathbf{r}, 0) = \frac{\sqrt{a}}{2\pi i}\frac{\mathbf{\nabla}\bar{\theta}_0 \times \mathbf{\nabla}\theta_0}{(1 + \bar{\theta}_0\theta_0)^2} = \frac{\sqrt{a}}{2\pi i}\frac{\bar{\phi}_1\mathbf{\nabla}\phi_0 - \phi_1\mathbf{\nabla}\bar{\phi}_0}{(1 + \bar{\phi}_0\phi_0)^2}$$

(76)

It is easy to see that Eqs. (76) are precisely the duality condition $\theta^*\sigma = -*\phi^*\sigma$, in $t = 0$, of an electromagnetic knot defined as $F = -\sqrt{a}\phi^*\sigma$, $*F = \sqrt{a}\theta^*\sigma$. As the duality condition is conserved in time, (76) thus defines an electromagnetic knot of homotopy class n (the Hopf index of both ϕ_0 and θ_0).

A.   Hopf Fibration

The group-theoretic method to find nontrivial maps $S^3 \to S^2$ is based on the isomorphism between S^3 and the group manifold $SU(2)$. Every point $g \in SU(2)$ can be written as

$$g = \exp(i\alpha^j\sigma_j)$$

(77)

where α^j are three real parameters, $j = 1, 2, 3$, and σ_j are the Pauli matrices. Every point V in Lie algebra $su(2)$ can be written as

$$V = i\alpha^j\sigma_j$$

(78)

where the manifold S^2 is simply

$$S^2 = \{V \in su(2), \det V = 1\} \tag{79}$$

In this way, every map $S^3 \to S^2$ can be viewed also as a map from the group $SU(2)$ to the subset of the Lie algebra $su(2)$ with determinant equal to 1.

Any point in $SU(2)$ can be written, in Skyrme's parametrization (analogous to the stereographic projection $S^3 = SU(2) \cup \{\infty\}$), as

$$g = \exp\left(i\frac{\alpha}{r}x^j\sigma_j\right), \qquad \alpha = 2\arctan r \tag{80}$$

Consider the set of transformations $SU(2) \to S^2$ given by

$$n^i\sigma_i = g(k^j\sigma_j)g^{-1} \tag{81}$$

where $\mathbf{n}$, $n^i n_i = 1$ are the coordinates in S^2, $\mathbf{k}$, $k^j k_j = 1$ are constant parameters (that we can choose in order to obtain different maps $S^3 \to S^2$), and $g \in SU(2)$ is given by (80).

It is easy to see that, under the applications (81), the inverse image of every point in S^2 is a closed line $SU(2)$, the fiber. The set of these fibers is called the fibration of $SU(2)$ by the map (81), and it is given by

$$g(\tau) = g\exp(i\tau k^j\sigma_j) \tag{82}$$

where τ is the evolution parameter of the fiber. For every point g there is only one fiber. We can also define the scalar product of the velocity vectors of two of these fibrations as

$$(\dot{g}_1, \dot{g}_2) = \frac{1}{2}\mathrm{Tr}(\dot{g}_1^+\dot{g}_2) = \mathbf{k}_1 \cdot \mathbf{k}_2 \tag{83}$$

Consequentely, if we choose, for example, $\mathbf{k}_1 = (0,0,-1)$, $\mathbf{k}_2 = (-1,0,0)$, and $\mathbf{k}_3 = (0,-1,0)$, we obtain, not only two, but three fibrations; its fibers are mutually orthogonal in each point. These fibrations can be written in R^3 coordinates (thanks to the Skyrme's parametrization) as

- The Hopf fibration, $g(\tau) = g\exp(-i\tau\sigma_3)$, with

$$x(\tau) = \frac{2x\cos\tau + 2y\sin\tau}{(r^2+1) - (r^2-1)\cos\tau + 2z\sin\tau}$$

$$y(\tau) = \frac{2y\cos\tau - 2x\sin\tau}{(r^2+1) - (r^2-1)\cos\tau + 2z\sin\tau} \tag{84}$$

$$z(\tau) = \frac{2z\cos\tau + (r^2-1)\sin\tau}{(r^2+1) - (r^2-1)\cos\tau + 2z\sin\tau}$$

It is easy to see that these fibers have linking number equal to one. For instance, the fibers passing through the points $(x, y, z) = (1, 0, 0)$ and $(x, y, z) = (0, 0, 1)$ are, respectively, the circle $x^2 + y^2 = 1$ and the z axis, that are obviously linked.

- A fibration orthogonal to (84), $g(\tau) = g \exp(-i\tau\sigma_1)$, with

$$x(\tau) = \frac{2x\cos\tau + (r^2 - 1)\sin\tau}{(r^2 + 1) - (r^2 - 1)\cos\tau + 2x\sin\tau}$$

$$y(\tau) = \frac{2y\cos\tau + 2z\sin\tau}{(r^2 + 1) - (r^2 - 1)\cos\tau + 2x\sin\tau} \tag{85}$$

$$z(\tau) = \frac{2z\cos\tau - 2y\sin\tau}{(r^2 + 1) - (r^2 - 1)\cos\tau + 2x\sin\tau}$$

This fibration can be obtained from the Hopf one by the change $(x, y, z) \mapsto (y, z, x)$ in the expression of the map, so its fibers are linked too, and the linking number is also one.

- A fibration orthogonal to both (84) and (85), $g(\tau) = g \exp(-i\tau\sigma_2)$, with

$$x(\tau) = \frac{2x\cos\tau - 2z\sin\tau}{(r^2 + 1) - (r^2 - 1)\cos\tau + 2y\sin\tau}$$

$$y(\tau) = \frac{2y\cos\tau + (r^2 - 1)\sin\tau}{(r^2 + 1) - (r^2 - 1)\cos\tau + 2y\sin\tau} \tag{86}$$

$$z(\tau) = \frac{2z\cos\tau + 2x\sin\tau}{(r^2 + 1) - (r^2 - 1)\cos\tau + 2y\sin\tau}$$

Once more, this corresponds to the change $(x, y, z) \mapsto (z, x, y)$ in the Hopf map, so the linking number is one.

Summarizing this subsection, the group-theoretic techniques allow us to obtain three maps $S^3 \to S^2$ whose velocity vectors are mutually orthogonal, and with the same linking number. Next, we have to build the Cauchy data of the electromagnetic knots based on these maps.

B. Cauchy Data for Electromagnetic Knots

It is convenient to work with nondimensional coordinates in the mathematical spacetime $S^3 \times R$, and in S^2. In order to do that, we define the non dimensional coordinates (X, Y, Z, T), related to the physical ones (x, y, z, t) by

$$(X, Y, Z, T) = \lambda(x, y, z, t) \tag{87}$$

and $\lambda^2 r^2 = \lambda^2(x^2 + y^2 + z^2) = X^2 + Y^2 + Z^2 = R^2$, where λ is a constant with inverse length dimensions. Now, we can perform the corresponding stereographic projections in the maps (81), with $\mathbf{k} = (0, 0, -1)$, $\mathbf{k} = (-1, 0, 0)$ and $\mathbf{k} = (0, -1, 0)$, to obtain the following maps $R^3 \cup \{\infty\} \to C \cup \{\infty\}$:

- The Hopf map

$$\phi_0 = \frac{2(X + iY)}{2Z + i(R^2 - 1)} \tag{88}$$

- The map corresponding to the change $(X, Y, Z) \mapsto (Y, Z, X)$ in (88)

$$\theta_0 = \frac{2(Y + iZ)}{2X + i(R^2 - 1)} \tag{89}$$

- The map corresponding to the change $(X, Y, Z) \mapsto (Z, X, Y)$ in (88)

$$\varphi_0 = \frac{2(Z + iX)}{2Y + i(R^2 - 1)} \tag{90}$$

Because of their construction, it is obvious that the three maps (88)–(90) have the same Hopf index. Following the methods explained in Sections I.D–I.F, it is easily shown that the three maps have Hopf index $n = 1$ and that the three fibrations are mutually orthogonal at each point. Consequently, any two of these three maps is a pair of dual maps, from which we can build an electromagnetic knot. The fibers of the third fibration are everywhere tangent to the Poynting vector of that knot. There is then a nice mathematical structure, with three fibrations that can be termed *magnetic, electric* and of the *energy flux*. This happens also in the general case.

If we choose the maps ϕ_0 and θ_0 to generate a knot, the Cauchy data for the magnetic and electric fields are

$$\mathbf{B}(\mathbf{r}, 0) = \frac{4\sqrt{a}\lambda^2}{\pi(1 + R^2)^3} (2(Y - XZ), -2(X + YZ), -1 - Z^2 + X^2 + Y^2)$$
$$\mathbf{E}(\mathbf{r}, 0) = \frac{4\sqrt{a}\lambda^2}{\pi(1 + R^2)^3} (1 + X^2 - Y^2 - Z^2, 2(-Z + XY), 2(Y + XZ)) \tag{91}$$

From (91), two vector potentials, $\mathbf{A}$ and $\mathbf{C}$, can be computed, such that $\mathbf{B} = \nabla \times \mathbf{A}$, $\mathbf{E} = \nabla \times \mathbf{C}$, with the results

$$\mathbf{A}(\mathbf{r}, 0) = \frac{2\sqrt{a}\lambda}{\pi(1 + R^2)^2} (Y, -X, -1)$$
$$\mathbf{C}(\mathbf{r}, 0) = \frac{2\sqrt{a}\lambda}{\pi(1 + R^2)^2} (1, -Z, Y) \tag{92}$$

The magnetic and electric helicities of this knot turn out to be

$$h_m = \int_{R^3} \mathbf{A} \cdot \mathbf{B} \, d^3r = h_e = \int_{R^3} \mathbf{C} \cdot \mathbf{E} \, d^3r = a \qquad (93)$$

Consequently, we have obtained the Cauchy data of an electromagnetic knot, a representative of the homotopy class C_1, for which, according to (63)

$$N_R - N_L = \frac{a}{\hbar c} \qquad (94)$$

C.   Time-Dependent Expressions

To find the electromagnetic knot, defined at every time, from the Cauchy data (91), we use the Fourier analysis. The magnetic and electric fields can be written as

$$\mathbf{B}(\mathbf{r}, t) = \frac{1}{(2\pi)^{3/2}} \int d^3k \, (\mathbf{R}_1(\mathbf{k}) \cos k \cdot x - \mathbf{R}_2(\mathbf{k}) \sin k \cdot x)$$

$$\mathbf{E}(\mathbf{r}, t) = \frac{1}{(2\pi)^{3/2}} \int d^3k (\mathbf{R}_1(\mathbf{k}) \sin k \cdot x + \mathbf{R}_2(\mathbf{k}) \cos k \cdot x) \qquad (95)$$

where $k \cdot x = \omega t - \mathbf{k} \cdot \mathbf{r}$, $\omega^2 = \mathbf{k}^2$, and the real vectors $\mathbf{R}_1$, $\mathbf{R}_2$ satisfy, in order to mantain Maxwell's equations, the relations

$$\mathbf{k} \cdot \mathbf{R}_1 = \mathbf{k} \cdot \mathbf{R}_2 = \mathbf{R}_1 \cdot \mathbf{R}_2 = 0$$

$$\mathbf{k} \times \mathbf{R}_2 = \omega \mathbf{R}_1, \mathbf{k} \times \mathbf{R}_1 = -\omega \mathbf{R}_2 \qquad (96)$$

The important point in the use of Fourier analysis is that the vectors $\mathbf{R}_1, \mathbf{R}_2$ can be computed from the Cauchy data of the electromagnetic field:

$$\mathbf{R}_1(\mathbf{k}) + i\mathbf{R}_2(\mathbf{k}) = \frac{1}{(2\pi)^{3/2}} \int d^3r (\mathbf{B}(\mathbf{r}, 0) + i\mathbf{E}(\mathbf{r}, 0)) e^{i\mathbf{k} \cdot \mathbf{r}} \qquad (97)$$

For the electromagnetic knot with Cauchy data given by (91), we find

$$\mathbf{R}_1 = \frac{\sqrt{a}}{\sqrt{2\pi}\lambda^2} \frac{e^{-\omega/\lambda}}{\omega} (k_1 k_3, \omega k_3 + k_2 k_3, -\omega k_2 - k_1^2 - k_2^2)$$

$$\mathbf{R}_2 = \frac{\sqrt{a}}{\sqrt{2\pi}\lambda^2} \frac{e^{-\omega/\lambda}}{\omega} (\omega k_2 + k_2^2 + k_3^2, -\omega k_1 - k_1 k_2, -k_1 k_3) \qquad (98)$$

Introducing these vectors in (95), the expressions, for all the times, of one electromagnetic knot representative of the homotopy class C_1 are

$$\mathbf{B}^{(1)}(\mathbf{r},t) = \frac{\sqrt{a}\lambda^2}{\pi(A^2 + T^2)^3}(Q\mathbf{H}_1 + P\mathbf{H}_2)$$

$$\mathbf{E}^{(1)}(\mathbf{r},t) = \frac{n\sqrt{a}\lambda^2}{\pi(A^2 + T^2)^3}(Q\mathbf{H}_2 - P\mathbf{H}_1)$$

(99)

where the superscript (n) indicates the homotopy class C_n of the knot, the quantities A, P, Q are defined by

$$A = \frac{R^2 - T^2 + 1}{2}, \qquad P = T(T^2 - 3A^2), \qquad Q = A(A^2 - 3T^2) \qquad (100)$$

and the vectors $\mathbf{H}_1$ and $\mathbf{H}_2$ are

$$\mathbf{H}_1 = \left(Y + T - XZ, -X - (Y+T)Z, \frac{-1 - Z^2 + X^2 + (Y+T)^2}{2}\right)$$

$$\mathbf{H}_2 = \left(\frac{1 + X^2 + Z^2 - (Y+T)^2}{2}, -Z + X(Y+T), Y + T + XZ\right)$$

(101)

This solution verifies $\mathbf{E}^{(1)} \cdot \mathbf{B}^{(1)} = 0$ and $(E^{(1)})^2 - (B^{(1)})^2 = 0$. To study its time evolution, an interesting tool is the energy density:

$$P^0(\mathbf{r},t) = \frac{(E^{(1)})^2 + (B^{(1)})^2}{2} = \frac{a\lambda^4}{4\pi^2}\frac{(1 + X^2 + (Y+T)^2 + Z^2)^2}{(A^2 + T^2)^3} \qquad (102)$$

It can be seen in this expression how the knot spreads, its energy density going to zero.

The final step to characterize this knot is to find the time evolution of the basic complex scalar fields ϕ and θ. This is not easy since these fields satisfy highly non-linear equations, the duality equations,

$$F_{\mu\nu} = \frac{\sqrt{a}}{2\pi i}\frac{\partial_\mu \bar{\phi}\partial_\nu \phi - \partial_\mu \phi \partial_\nu \bar{\phi}}{(1 + \bar{\phi}\phi)^2}$$

$$^*F_{\mu\nu} = \frac{\sqrt{a}}{2\pi i}\frac{\partial_\mu \theta \partial_\nu \bar{\theta} - \partial_\mu \bar{\theta}\partial_\nu \theta}{(1 + \bar{\theta}\theta)^2}$$

(103)

with the corresponding Cauchy data ϕ_0, θ_0. However, the basic fields have a very important property that allows us to solve (103)—their level curves evolve in time in such way that their linking number is a constant of the motion (because the magnetic and electric helicities are constants of the motion for the electromagnetic field). This stability condition is a kind of hint on the form of the basic fields. The result is that the scalar fields ϕ and θ that give way, through (103), to the electromagnetic knot (99), are

$$\phi(\mathbf{r}, t) = \frac{(AX - TZ) + i(AY + T(A - 1))}{(AZ + TX) + i(A(A - 1) - TY)}$$

$$\theta(\mathbf{r}, t) = \frac{(AY + T(A - 1)) + i(AZ + TX)}{(AX - TZ) + i(A(A - 1) - TY)}$$

(104)

where A is given by (100).

D.   A Family of Electromagnetic Knots with Hopf Indices $\pm n^2$

The electromagnetic knot given in the previous subsections, a representative of the homotopy class C_1, can be easily generalized to classes C_{n^2}. To do that, we will need a property of the Hopf index.

Consider a smooth map $f: S^3 \rightarrow S^2$. We have called the fiber of a point $p \in S^2$ to the inverse image $f^{-1}(p)$, which is generally a closed curve in S^3. Now we define the multiplicity of the fiber $f^{-1}(p)$ to the number of connected components of $f^{-1}(p)$. Consider the map $f^n: S^3 \rightarrow S^2$, where n is an integer, for f^n to be a good smooth map. The linking number of the closed curves that form the fibers of f^n is equal to the linking number of the closed curves that form the fibers of f (they are the same curves). However, the multiplicity of the fibers of f^n is equal to n times the multiplicity of the fibers of f. Consequently, the Hopf index has the following property:

$$H(f^n) = n^2 H(f)$$

(105)

Instead of the nth power, we will use a different function with the same property. If the map f is written as $f = Pe^{iq}$, we define the map $f^{(n)}$ as

$$f^{(n)} = Pe^{inq}$$

(106)

where n is an integer. The Faraday's tensor of $f^{(n)}$ is $F_{\mu\nu}(f^{(n)}) = nF_{\mu\nu}(f)$, so that

$$H(f^{(n)}) = n^2 H(f)$$

(107)

Now, instead of (104), we can use the basic scalar fields $\phi^{(n)}$ and $\theta^{(n)}$, defined by (106), and given by

$$\phi^{(n)}(\mathbf{r}, t) = \left(\frac{(AX - TZ) + i(AY + T(A - 1))}{(AZ + TX) + i(A(A - 1) - TY)} \right)^{(n)}$$

$$\theta^{(n)}(\mathbf{r}, t) = \left(\frac{(AY + T(A - 1)) + i(AZ + TX)}{(AX - TZ) + i(A(A - 1) - TY)} \right)^{(n)}$$

(108)

where A is given by (100). These two maps have Hopf index equal to n^2, n integer, and their corresponding velocity curves are mutually orthogonal (they are the same as the velocity curves of ϕ and θ, respectively). So we conclude that the complex scalar fields $\phi^{(n)}$ and $\theta^{(n)}$ give place to electromagnetic knots representatives of the homotopy classes C_{n^2}. The magnetic and the electric fields are simply

$$\mathbf{B}^{(n^2)}(\mathbf{r}, t) = n\mathbf{B}^{(1)}(\mathbf{r}, t)$$
$$\mathbf{E}^{(n^2)}(\mathbf{r}, t) = n\mathbf{E}^{(1)}(\mathbf{r}, t) \qquad (109)$$

where $\mathbf{B}^{(1)}$ and $\mathbf{E}^{(1)}$ are, respectively, the magnetic and the electric fields of the representative of the homotopy class C_1, given by (99). The electromagnetic knots (109) satisfy

$$h_m = h_e = an^2 \qquad (110)$$

so the topological charge has the value

$$N_R - N_L = \frac{a}{\hbar c} n^2 \qquad (111)$$

These particular knots have the following curious values for the energy, linear momentum, and angular momentum:

$$p^0 = \int \left(\frac{(E^{(n^2)})^2 + (B^{(n^2)})^2}{2} \right) d^3 r = 2n^2 a\lambda$$
$$\mathbf{p} = \int \mathbf{E}^{(n^2)} \times \mathbf{B}^{(n^2)} : d^3 r = n^2 a\lambda \mathbf{e}_y \qquad (112)$$
$$\mathbf{J} = \int \mathbf{r} \times (\mathbf{E}^{(n^2)} \times \mathbf{B}^{(n^2)}) d^3 r = n^2 a \mathbf{e}_y$$

These knots thus move in the y direction, and the angular momentum is along the motion direction.

It is easy to show that we can also construct electromagnetic knots with Hopf index $-n^2$ by means of the dual fields

$$\phi^{(-,n)}(\mathbf{r}, t) = \phi^{(n)}(-\mathbf{r}, -t)$$
$$\theta^{(-,n)}(\mathbf{r}, t) = \theta^{(n)}(-\mathbf{r}, -t) \qquad (113)$$

The magnetic and electric fields of the electromagnetic knot are then

$$\mathbf{B}^{(-n^2)}(\mathbf{r}, t) = \mathbf{B}^{(n^2)}(-\mathbf{r}, -t)$$
$$\mathbf{E}^{(-n^2)}(\mathbf{r}, t) = \mathbf{E}^{(n^2)}(-\mathbf{r}, -t) \qquad (114)$$

and the magnetic and electric helicities are given by

$$h_m = h_e = -an^2 \tag{115}$$

and the topological charge

$$N_R - N_L = -\frac{a}{\hbar c} n^2 \tag{116}$$

The energy, linear momentum, and angular momentum of the particular knots (114), representatives of the homotopy classes C_{-n^2}, are as follows:

$$p^0 = \int \left(\frac{(E^{(-n^2)})^2 + (B^{(-n^2)})^2}{2} \right) d^3r = 2n^2 a\lambda$$

$$\mathbf{p} = \int \mathbf{E}^{(-n^2)} \times \mathbf{B}^{(-n^2)} : d^3r = n^2 a\lambda \mathbf{e}_y \tag{117}$$

$$\mathbf{J} = \int \mathbf{r} \times (\mathbf{E}^{(-n^2)} \times \mathbf{B}^{(-n^2)}) d^3r = -n^2 a\mathbf{e}_y$$

IV.   A TOPOLOGICAL MODEL OF ELECTROMAGNETISM

The discussion in Section I, especially Section I.C, suggest the possibility of a theory of the electromagnetic field that uses as coordinates the pair of complex scalar fields ϕ, θ, whose level curves coincide with the magnetic and electric force lines. For this purpose, let us recall the definition of electromagnetic knot given in Section I.F as a solution of the Maxwell equations in empty space, such that any pair of magnetic lines (or any pair of electric lines) is a link. As was shown in Section II.C, the Hopf indices are necessarily equal, as a consequence of the Maxwell equations. It must also be stressed that the electromagnetic knots are radiation fields, in the sense that their magnetic and electric fields are orthogonal (i.e., verify the condition $\mathbf{E} \cdot \mathbf{B} = 0$).

A.   A First Model

As a first step in constructing a topological model of the electromagnetic field, let us consider the set of electromagnetic knots defined by pairs of dual scalars (ϕ, θ). If we try a theory based on these two scalars, the most natural election for the action integral is

$$\mathscr{S} = -\frac{1}{4} \int (\mathscr{F}(\phi) \wedge *\mathscr{F}(\phi) + *\mathscr{F}(\theta) \wedge \mathscr{F}(\theta)) \tag{118}$$

where $\mathscr{F}(\phi) = -\sqrt{a}\,\phi^*\sigma$, $*\mathscr{F}(\phi)$ is its dual, $*\mathscr{F}(\theta) = \sqrt{a}\,\theta^*\sigma$, and $\mathscr{F}(\theta)$ is the dual of $*\mathscr{F}(\theta)$ [as in Eqs. (9) and (12)], since this is equal to the standard action $-\int \frac{1}{2}\mathscr{F} \wedge *\mathscr{F} = -\int \frac{1}{4}F_{\mu\nu}F^{\mu\nu}\,d^4x$. As the scalars ϕ, θ must be a dual pair, they must be submitted to the duality condition or constraint $-*(\phi^*\sigma) = \theta^*\sigma$, which is written in terms of the electromagnetic tensors

$$G^{\mu\nu} = {}^*F^{\mu\nu}(\theta) - \frac{1}{2}\epsilon^{\mu\nu\alpha\beta}F_{\alpha\beta}(\phi) = 0 \tag{119}$$

[As discussed in Section I.C we will say that two scalars are dual or that they form a dual pair if they verify the duality constraint (15) or, equivalently, (119) for any given time.] According to the method of the Lagrange multipliers, let us vary, as independent fields, the two scalars ϕ and θ in the modified Lagrangian density

$$\mathscr{L}' = \mathscr{L} + \mu^{\alpha\beta}G_{\alpha\beta} \tag{120}$$

with $\mathscr{L} = -(F_{\mu\nu}(\phi)F^{\mu\nu}(\phi) - {}^*F_{\mu\nu}(\theta)\,{}^*F^{\mu\nu}(\theta))/8$, where the components of the tensor $\mu^{\alpha\beta}$ are the multipliers. A simple calculation shows that the duality constraint (119) does not contribute to the field equations, which means that, if it is satisfied by the Cauchy data, it is kept naturally in the time evolution, an interesting and consequential property. The Euler–Lagrange equations turn out to be

$$\begin{aligned} \partial_\alpha F^{\alpha\beta}(\phi)\partial_\beta\phi = 0, \qquad & \partial_\alpha F^{\alpha\beta}(\phi)\partial_\beta\phi^* = 0 \\ \partial_\alpha\,{}^*F^{\alpha\beta}(\theta)\partial_\beta\theta = 0, \qquad & \partial_\alpha\,{}^*F^{\alpha\beta}(\theta)\partial_\beta\theta^* = 0 \end{aligned} \tag{121}$$

It follows immediately that both $F_{\alpha\beta}(\phi)$ and ${}^*F_{\alpha\beta}(\theta)$ obey the Maxwell equations in empty space. In fact, the first pair for both tensors

$$\epsilon^{\alpha\beta\gamma\delta}\partial_\beta F_{\gamma\delta}(\phi) = 0, \qquad \epsilon^{\alpha\beta\gamma\delta}\partial_\beta\,{}^*F_{\gamma\delta}(\theta) = 0 \tag{122}$$

holds automatically for any arbitrary pair of dual scalars because of the definitions in Eqs. (10) and (13). On the other hand, it follows from (122) and the duality condition (119) that

$$\partial_\beta F^{\alpha\beta}(\phi) = 0, \qquad \partial_\beta\,{}^*F^{\alpha\beta}(\theta) = 0 \tag{123}$$

which is the second pair for both tensors. As $F_{\mu\nu}(\phi)$ and ${}^*F_{\mu\nu}(\theta)$ are the electromagnetic tensor and its dual, respectively, the Eqs. (122) and (123) are indeed the Maxwell equations in empty space; we thus have a model of topological electromagnetic fields.

It must be stressed that we have in fact proved two different properties, which can be stated as follows.

Property 1. In a theory based on the pair of fields (ϕ, θ) with action integral equal to (118), submitted to the duality constraint (119), both tensors $F_{\alpha\beta}$ and $^*F_{\alpha\beta}$ obey the Maxwell equations in empty space. As the duality constraint is naturally conserved in time, the same result is obtained if it is imposed just at $t = 0$.

Property 2. If two scalar fields ϕ, θ form an arbitrary pair of dual fields, in the sense of Eq. (15) [or, equivalently, if they verify (119)], the tensors $F_{\alpha\beta}$ and $^*F_{\alpha\beta}$ satisfy the Maxwell equations in empty space at any time.

Note that the property 2 is surprising and beautiful; for the Maxwell equations to hold, it is not necessary to consider any variational principle whatsoever. Given a scalar field that can be interpreted as a map $\phi: S^3 \mapsto S^2$, *the mere existence of a dual map* θ *guarantees that the two pull–backs of the area 2-form in* S^2 *obey Maxwell's equations in empty space.* This fact must be stressed—the duality condition on the two scalars implies the Maxwell equations by itself.

A better understanding of this curious property can be obtained by using, instead of (118), the following action integral

$$\mathscr{S} = -\frac{1}{2} \int \mathscr{F}(\phi) \wedge *\mathscr{F}(\theta) \tag{124}$$

which is also equal to the standard action for the electromagnetic field. The integrand in (124) has an interesting interpretation. If we now define the product map $\chi = \phi \times \theta : S^3 \mapsto S^2 \times S^2$, it turns out that it is equal to the pullback of the volume form in $S^2 \times S^2$ by the map χ, that is, to $\mathscr{V} = \chi^*(\sigma \wedge \sigma) = \phi^*\sigma \wedge \theta^*\sigma$, so that

$$\mathscr{S} = \frac{a}{2c} \int \mathscr{V} \tag{125}$$

It turns out that $\mathscr{V}$ is an exact form. As explained in Section I, there exist two 1-forms in S^3 $\mathscr{A}$ and $\mathscr{C}$, such that $\mathscr{F} = d\mathscr{A}$ and $*\mathscr{F} = d\mathscr{C}$ (because the second group of cohomology of S^3 is trivial). It is then clear that $\mathscr{V} = -(4a)^{-1}d(\mathscr{A} \wedge *\mathscr{F} + \mathscr{F} \wedge \mathscr{C})$. As a consequence, the Euler–Lagrange equations of (124) are trivial (just $0 = 0$) and the action (125) takes a stationary value for a pair of maps (or of scalar fields), even if they are not dual.

This means that, if the two scalars are dual (i.e., if they define the same electromagnetic field), the corresponding pullbacks obey the Maxwell equations and are solutions of a variational problem with the standard action expressed in terms of these scalars.

Let us now identify the Cauchy data. As the Maxwell equations are of second order in the scalars, the initial data should be the two functions ($\phi(\mathbf{r}, 0), \theta(\mathbf{r}, 0)$, plus their time derivatives $\partial_0\phi(\mathbf{r}, 0), \partial_0\theta(\mathbf{r}, 0)$). However, it is easy to show that the latter can be expressed in terms of the former, as a consequence of the duality constraint (119).

As the knots are radiation fields, the level curves of the two scalars of a dual pair ϕ, θ must be mutually orthogonal (i.e., form two fibrations of the 3-space, orthogonal to one another). This means the they must obey the differential condition

$$(\nabla\bar{\phi} \times \nabla\phi) \cdot (\nabla\bar{\theta} \times \nabla\theta) = 0 \qquad (126)$$

which is a real partial differential equation for two complex functions and has therefore an infinity of solutions. This condition (126) is conserved naturally under the time evolution. Let us write the electromagnetic tensor corresponding to the scalar χ as $f_{\mu\nu}(\chi)$ [this means that $F_{\mu\nu} = f_{\mu\nu}(\phi)$, while $^*F_{\mu\nu} = f_{\mu\nu}(\theta)$]; we will use the following notation:

$$\mathcal{B}_i(\chi) = -\frac{1}{2}\epsilon_{ijk}f_{jk}(\chi), \qquad \mathcal{E}_i = f_{0i}(\chi) \qquad (127)$$

The duality condition then takes the form

$$\mathbf{E} = \mathcal{E}(\phi) = -\mathcal{B}(\theta), \qquad \mathbf{B} = \mathcal{B}(\phi) = \mathcal{E}(\theta) \qquad (128)$$

According to (128), the electric field $-\mathcal{B}(\theta)$ is a linear combination of $\nabla\bar{\phi}$ and $\nabla\phi$, which can be written as $\mathcal{B}(\theta) = b\nabla\bar{\phi} + \bar{b}\nabla\phi$, so that the function $b(\mathbf{r}, t)$ can be expressed in terms of ϕ, θ and their space derivatives.

Substitution in (128) shows that $\partial_0\phi = 2\pi i b(1 + \phi^*\phi)^2$, an analogous expression holding for $\partial_0\theta$. Consequently, the time derivatives of the scalars can be expressed in terms of the scalars and their space derivatives. In other words, the Cauchy data are just the pair of complex functions $\phi(\mathbf{r}, 0), \theta(\mathbf{r}, 0)$ that verify the condition (126). The system therefore has two degrees of freedom with a differential constraint that is conserved naturally under the time evolution.

B. A Topological Quantization Condition

As was shown in Section I.E, the Hopf indices n_m and n_e of the maps $S^3 \mapsto S^2$ defined by the two scalars are related to the magnetic and electric helicities as

$$h_m = \int_{R^3} \mathbf{C} \cdot \mathbf{E} \, d^3r = n_m a, \qquad h_e = \int_{R^3} \mathbf{A} \cdot \mathbf{B} \, d^3r = n_e a \qquad (129)$$

where it must be recallded that the constant a has dimensions of action times velocity. This certainly has the aspect of a quantization condition. In fact, it looks similar to the conditions used in the old quantum theory

$$I_k = \oint p_k \, dq_k = n_k h \tag{130}$$

As was shown in Section II, the two Hopf indices (which are the magnetic and electric linking numbers) are equal in empty space (i.e., without charges), $n_e = n_m = n$. Let us emphasize this fact; the electromagnetic knots are classified in homotopy classes labeled by the linking number of any pair or magnetic (or electric) lines. They verify thus a quantum condition of topological origin. We will see later that this has a very intuitive and suggestive physical interpretation.

C.   The Topological Model

Electromagnetic radiation fields—also called *degenerate* or *singular* by mathematicians—are defined by the condition $\det(F_{\mu\nu}) = 0$ or, equivalently, by $\mathbf{E} \cdot \mathbf{B} = 0$, that is, by the orthogonality of the electric and magnetic vectors. As was stated above, the electromagnetic knots are of this type. This means that the model just described contains only radiation fields.

Radiation fields are especially interesting since they are usually represent photon states. Moreover, it is known that, because of the Darboux theorem [59–61], the Faraday form of any electromagnetic field $\mathscr{F}$ and its dual $*\mathscr{F}$ can be written, locally, as

$$\mathscr{F} = \sqrt{a}\,(dq_1 \wedge dp_1 + dq_2 \wedge dp_2), \qquad *\mathscr{F} = \sqrt{a}\,(dv_1 \wedge du_1 + dv_2 \wedge du_2) \tag{131}$$

where q_k, p_k, v_k, u_k are functions of spacetime that can also be chosen as coordinates of the field [62] and a is a constant with dimensions of action times velocity, introduced here in order for these functions to be dimensionless. Each of the two terms in these sums is a radiation field (i. e. verifies $\mathbf{E} \cdot \mathbf{B} = 0$). This means that any standard electromagnetic field in empty space can be expressed as the sum of two fields of radiation type, although we must note that this representation is not unique, since we can make canonical transformations to new variables $(q_k, p_k) \rightarrow (Q_k, P_k)$ [or $(v_k, u_k) \rightarrow (V_k, U_k)$] without changing the form of (131) (by "standard electromagnetic field" we mean any solution of Maxwell equations).

In physical terms, this can be understood in the following way. Take an electromagnetic field with Poynting vector $\mathbf{S} = \mathbf{E} \times \mathbf{B}$. By a suitable Lorentz transformation [with direction unit vector $\mathbf{n}$ and velocity parameter η given by $\mathbf{n} \tanh 2\eta = 2\mathbf{S}/(E^2 + B^2)$], we can change to a frame in which $\mathbf{S} = 0$ at any prescribed point P, which means that $\mathbf{E}$ and $\mathbf{B}$ are parallel there [65]. Taking their common direction as the Oz axis, the Faraday form can be written in the form (131)

$$\mathscr{F} = dt \wedge d(Ez) + d(Bx) \wedge dy \tag{132}$$

(because $\mathscr{F}$ is closed). In general, the Faraday form (or its dual) cannot be expressed in a form simpler than (131), because it is of rank 4 and also of class 4 (this means that four 1-forms and four functions, respectively, are needed to express it). However, in the important case of the radiation fields a simpler representation is possible, since the Faraday form, which is only of rank 2 and class 2, is degenerate and can be written in terms of only two functions $q(\mathbf{r}, t), p(\mathbf{r}, t)$ and two 1-forms dq, dp (a similar property holds for its dual) as follows:

$$\mathscr{F} = \sqrt{a}\, dq \wedge dp, \qquad *\mathscr{F} = \sqrt{a}\, dv \wedge du \tag{133}$$

The electric and magnetic fields then have the form

$$\begin{aligned}
\mathbf{B} &= \sqrt{a}\, \nabla p \times \nabla q = \sqrt{a}(\partial_0 u \nabla v - \partial_0 v \nabla u) \\
\mathbf{E} &= \sqrt{a}\, \nabla u \times \nabla v = \sqrt{a}(\partial_0 q \nabla p - \partial_0 p \nabla q)
\end{aligned} \tag{134}$$

Note that, in this case, the magnetic lines are contained in magnetic surfaces. There are in fact two families of them, given by the equations $p = p_0$ and $q = q_0$, where each line forms the intersection of two surfaces, one of each family (there are also two families of electric surfaces $u = u_0$ and $v = v_0$). The functions (p, q) and (u, v) are the Clebsch variables of $\mathbf{B}$ and $\mathbf{E}$, respectively [63,64]. They can be used as canonical variables [62]. As explained above, they are not uniquely defined, but may be changed by canonical transformations.

We must emphasize that, given a constant a, any electromagnetic field may be written in the form (131) and that, with this definition, the Clebsch variables are dimensionless quantities.

It is easy to express the Clebsch variables of a knot in terms of the scalars. If it derives from the scalars $\phi = S \exp(2i\pi\gamma)$ and $\theta = R \exp(2i\pi\rho)$ through Eqs. (10)–(13), it turns out that

$$p = \frac{1}{1+S^2}, \qquad q = \gamma, \qquad v = \frac{1}{1+R^2}, \qquad u = \rho \tag{135}$$

as can be seen by simple substitution in (134), and on comparison with (134) later. It must be emphasized that, as is seen, this election of Clebsch variables verifies the following two properties: (1) $0 \le p, v \le 1$ and (2) q, u are phase functions.

We can now construct a topological model of electromagnetism in empty space, which can be formalized by means of a variational principle as follows. Let us take two pairs of dual scalars ϕ_k, θ_k, where $k = 1, 2$ as fundamental fields and define an electromagnetic field by the equations

$$\mathscr{F} = -\sqrt{a}(\phi_1^* \sigma + \phi_2^* \sigma), \qquad *\mathscr{F} = \sqrt{a}(\theta_1^* \sigma + \theta_2^* \sigma) \qquad (136)$$

where the asterisk superscripts indicate pullback of the area 2-form σ in S^2 to the Minkowski spacetime $M = R^3 \times R$ (identified here with $S^3 \times R$) by the corresponding map. Note that (136) has the same form as (131), since the two terms on the right-hand sides of each of these equations can be written as exterior products of differentials of function, because they are electromagnetic knots.

It seems logical to take as action integral

$$\mathscr{S} = -\frac{1}{4} \sum_k \int \left(\mathscr{F}(\phi_k) \wedge *\mathscr{F}(\phi_k) + *\mathscr{F}(\theta_k) \wedge \mathscr{F}(\theta_k) \right) \qquad (137)$$

which coincides with the usual form $-\frac{1}{2}(\mathscr{F} \wedge *\mathscr{F})$. The duality conditions

$$-*(\phi_k^* \sigma) = \theta_k^* \sigma, \qquad k = 1, 2 \qquad (138)$$

must be imposed by means of the Lagrange multipliers method. It is very easy to show that the corresponding Euler–Lagrange equations are

$$d\mathscr{F} = 0, \qquad d*\mathscr{F} = 0 \qquad (139)$$

since again the duality conditions do not contribute to these equations. This means that they are naturally conserved under time evolution. In this way, we can extend the topological model to a theory of electromagnetism in empty space, which includes nonradiation fields, and uses electromagnetic knots instead of radiation fields. We will see below-that it is locally equivalent to Maxwell's standard theory, as will be shown in next section.

In the same way as before, we could use as action integral

$$\mathscr{S} = -\frac{1}{2} \int \left(\mathscr{F}(\phi_1) \wedge *\mathscr{F}(\theta_1) + \mathscr{F}(\phi_2) \wedge *\mathscr{F}(\theta_2) \right) \qquad (140)$$

as in the previous case of only one pair of scalars.

Note that (139) are highly nonlinear in the scalars but become exactly the linear Maxwell equations in the fields $F_{\mu\nu}$ and $^*F_{\mu\nu}$. In this sense, the Maxwell equations are the exact linearization (by change of variables, not by truncation!) of a nonlinear theory with topological properties, in which the force lines coincide with the level curves of two scalar fields. The model thus gives a line dynamics.

We end this section with a comment referring to the Cauchy data for the scalars. In standard Maxwell theory, the Cauchy data are the eight functions $A_\mu, \partial_0 A_\mu$, and there is gauge invariance. In this topological model, they are the four complex functions $\phi_k(\mathbf{r}, 0)$, $\theta_k(\mathbf{r}, 0)$, that is, eight real functions, constrained by the two conditions $(\boldsymbol{\nabla}\bar{\phi}_k \times \boldsymbol{\nabla}\phi_k) \cdot (\boldsymbol{\nabla}\bar{\theta}_k \times \boldsymbol{\nabla}\theta_k) = 0$, $k = 1, 2$, to ensure that the level curves of ϕ_k will be orthogonal to those of θ_k. It is not necessary to prescribe the time derivatives $\partial_0\phi_k$, $\partial_0\theta_k$ since they are determined by the duality conditions (138), as explained above.

V. LOCAL EQUIVALENCE AND GLOBAL DIFFERENCE WITH THE STANDARD MAXWELL THEORY

As we have seen, any pair of dual maps generates a standard electromagnetic field. However, given a standard solution of Maxwell's equations in empty space, it is not true in general that there exist one pair of dual maps that generate this field. In this section we examine this question. First, we will prove that any radiation electromagnetic field is locally equal to an electromagnetic knot, and hence that the topological model is locally equivalent to the Maxwell standard theory, although they are nonequivalent from globally. Their difference relates to the behovior of the fields around the point at infinity. After that, we will examine more the difference more closely, showing the existence of what can be called a "hidden nonlinearity."

A. Local Equivalence

The electromagnetic knots satisfy a very important property. In a precise way, the following proposition holds true.

Proposition 2.   *Any standard radiation electromagnetic field in empty space with Faraday 2-form $\mathscr{F}^{st}$, regular in a bounded spacetime domain D, coincides locally with a knot around any point $P \in D$ in the following sense. There is a knot with 2-form $\mathscr{F}^{kn}$, such that $\mathscr{F}^{st} = \mathscr{F}^{kn}$ around P, except perhaps if P is in a zero measure set. The same property holds for $*\mathscr{F}^{st}$.*

This means that the difference between the set of the radiation solutions of the Maxwell equations and the set of the electromagnetic knots is not local but global. In other words: Radiation fields and knots are locally equal. A proof is the following.

Proof.   Let the Faraday 2-form of the standard radiation field $\mathscr{F}^{st}$ be expressed as (133), where p, q are two dimensionless functions of spacetime. We then define the functions η, δ as

$$\eta = \pi(p^2 + q^2), \qquad \delta = \frac{1}{2\pi} \arctan \frac{q}{p} \tag{141}$$

It is then clear that

$$\mathscr{F}^{st} = \sqrt{a} \, d\delta \wedge d\eta \tag{142}$$

so that η and δ give another election of the Clebsch variables of the standard field (they are obtained from p, q by a canonical transformation). If an electromagnetic knot is generated by the scalar $\phi = S \exp(i2\pi\gamma)$ through Eqs. (9)–(10), it is easy to show that

$$\mathscr{F}^{kn} = \sqrt{a} \, d\gamma \wedge d \frac{1}{(1 + S^2)} \tag{143}$$

This means that $\mathscr{F}^{st}$ will be a knot if there exist regular functions $S(\mathbf{r}), \gamma(\mathbf{r})$, one-valued at infinity, such that

$$\eta = \frac{1}{1 + S^2}, \qquad \delta = \gamma \tag{144}$$

The second equation poses no problem because δ was defined as a phase function. If $\eta < 1$, the solution for S is trivial, as the standard field with form $\mathscr{F}^{st}$ then becomes a knot. The same happens if η is bounded, say, if $\eta < A$, because we can then take as the Clebsch variables $\eta' = \eta/n'$, $\delta' = n'\delta$, where n' is an integer greater than A. Dropping the primes and entering the new Clebsch variables in (144), it is clear that there then exists a solution for S, γ.

Let us consider the case in which η is not bounded in D [but $F_{\mu\nu}^{st}$ is continuous and Eq. (142) is still valid]. Let Σ be the 3D set in which η diverges (a zero measure set). In general, $D - \Sigma$ consists of k connected open components D_j. Let $D_j^* \subset D_j$ be k open subsets in which η is bounded. In each one of them, we can define Clebsch variables η', δ', by the same method as before. It follows that the field is equal to a knot in each D_j^*. Now, the volume of $D - \cup D_j^*$ may be made as small as desired. This means that the magnetic field can be obtained by patching together those of the knots $\mathscr{F}_j^{kn}$, each one defined in the corresponding D_j^*, except for a set as small as required containing Σ. (Note that there is no problem if any D_j is not simply connected.) The same can be said of the dual to the Faraday 2-form $*\mathscr{F}^{st}$, which coincides with the corresponding 2-form of a knot, except perhaps in a zero measure set Σ'. This means that *any* radiation electromagnetic field coincides locally with an electromagnetic knot, except perhaps on a zero-measure set. In other words, standard radiation fields can be obtained by patching together electromagnetic knots generated by ϕ_j, θ_j, each one defined in a different domain, except at most on the zero-measure set $\Sigma \cup \Sigma'$. This ends the proof.

Traditionally, physics emphasizes the local properties. Indeed, many of its branches are based on partial differential equations, as happens, for instance, with continuum mechanics, field theory, or electromagnetism. In these cases, the corresponding basic equations are constructed by viewing the world locally, since these equations consist in relations between space (and time) derivatives of the coordinates. In consonance, most experiments make measurements in small, simply connected space regions and refer therefore also to local properties. (There are some exceptions; the Aharonov–Bohm effect is an interesting example.)

The local equivalence that we have just proved implies that the predictive contents of the Maxwell's theory and of this topological model are exactly the same when referred to local experiments, as most of them are. Accordingly, it is not possible to discern between the two by viewing locally. This is the operative meaning of local equivalence.

However, the fact that there is a difference of global character is very important and has interesting consequences. As we will see, it provides a topological structure. This is surprising and intriguing since it means that the linear Maxwell equations are compatible with the existence of topological constants of the motion, one of which is the electric charge. The topological model thus gives something more than Maxwell's theory: the quantization of the charge, as we will see in Section VIII.

It is convenient now to give examples of the expression of electromagnetic fields in the form (142) or equivalently (9)–(12). We now present three examples: the Coulomb potential, a plane wave, and a standing wave.

If $\phi = Pe^{i2\pi q}$, $\theta = Ve^{i2\pi u}$, it is easy to see that

$$p = \frac{1}{1 + P^2}, \qquad v = \frac{1}{1 + V^2}$$

where q and u are the other two Clebsch variables.

1. *Coulomb potential*, $\mathbf{E} = Q\mathbf{r}/(4\pi r^3)$, $\mathbf{B} = 0$. This field can be obtained from the scalars

$$\phi = \frac{ct}{r} \exp\left(iQ \frac{(c^2 t^2 + r^2)^2}{4c\sqrt{a} r^3 t} \log \frac{r}{r_0} \right), \qquad \theta = \tan \frac{\beta}{2} \exp\left(iQ \frac{\alpha}{\sqrt{a}} \right) \qquad (145)$$

where α, β are the azimuth and the polar angle and r_0 is any length. The Clebsch variables are

$$p = \frac{r^2}{(r^2 + c^2 t^2)}, \qquad q = Q \frac{(r^2 + c^2 t^2)^2}{8\pi c \sqrt{a} r^3 t} \log(r/r_0), \qquad v = \cos^2 \frac{\beta}{2}, \qquad u = \frac{Q\alpha}{2\pi\sqrt{a}}$$

As can be seen, both scalars are regular everywhere except at $r = 0$ and $r = \infty$.

2.   *Plane wave*, $\mathbf{E} = E_0(0, \sin\omega(x/c - t), 0)$, $\mathbf{B} = E_0(0, 0, \sin\omega(x/c - t))$. The two scalars and the corresponding Clebsch variables are

$$
\begin{aligned}
\phi &= \frac{1 + \cos\omega(x/c - t)}{\sin\omega(x/c - t)} \exp\left(i\frac{4\pi cE_0 y}{\sqrt{a\omega}}\right) \\
\theta &= \frac{1 + \cos\omega(x/c - t)}{\sin\omega(x/c - t)} \exp\left(i\frac{4\pi cE_0 z}{\sqrt{a\omega}}\right)
\end{aligned}
\tag{146}
$$

and $p = \frac{1}{2}(1 - \cos\omega(x/c - t))$, $q = (2cE_0 y/\sqrt{a\omega})$, $v = \frac{1}{2}(1 - \cos\omega(x/c - t))$, $u = (2cE_0 z/\sqrt{a\omega})$. It is seen that ϕ and θ do not represent smooth maps $S^3 \mapsto S^2$ because they are not well defined at infinity. However, there are smooth maps that coincide with them in any bounded domain and that are well defined at infinity. The fact that plane waves in all the space R^3 are not expressable as global knots is not a matter of concern, since a plane wave extending to all 3-space is not in fact a physical solution since it requires an infinite amount of energy.

3.   A *standing wave* given by

$$
\begin{aligned}
A_0 &= 0, \qquad A_1 = A_{01} \cos k_1 x \, \sin k_2 y \, \sin k_3 z \, \cos\omega t \\
A_2 &= A_{02} \sin k_1 x \, \cos k_2 y \, \sin k_3 z \, \cos\omega t \\
A_3 &= A_{03} \sin k_1 x \, \sin k_2 y \, \cos k_3 z \, \cos\omega t
\end{aligned}
\tag{147}
$$

which expresses one mode of a cubic cavity. The scalars that give this field can be taken as

$$
\phi = \sqrt{\frac{1-p}{p}} e^{i2\pi q}, \qquad \theta = \sqrt{\frac{1-v}{v}} e^{i2\pi u}
\tag{148}
$$

where the Clebsch variables are equal to

$$
p = \frac{1}{2}(1 + \sin k_1 x \, \sin k_2 y \, \sin k_3 z \cos\omega t), \qquad q = \sum_{i=1}^{3} \frac{2A_{0i}}{\sqrt{a k_i}} \log|\sin k_i x_i|
$$

$$
v = \frac{1}{2}(1 + \cos k_1 x \, \cos k_2 y \, \cos k_3 z \sin\omega t), \qquad u = \sum_{i=1}^{3} \frac{2(\mathbf{k} \times \mathbf{A_0})_i}{\sqrt{a\omega} k_i} \log|\cos k_i x_i|
$$

Note that the scalar field ϕ (resp. θ) is not well defined in the planes $k_i x_i = n_i \pi$ [resp. $k_i x_i = (n_i + \frac{1}{2})\pi$], where the n_i are integers, where q (resp. u) diverges. But there are scalars $\phi_{n_1 n_2 n_3}$, well defined and smooth in the finite domains

$n_1\pi < k_1 x < (n_1 + 1)\pi$, $n_2\pi < k_2 y < (n_2 + 1)\pi$, $n_3\pi < k_3 z < (n_3 + 1)\pi$, which generate the fields in each one of them (and similarly with θ). However, the electric and magnetic fields cannot be produced by a pair of smooth maps $S^3 \mapsto S^2$. As we stated before, the fields can be obtained by patching together knots defined in bounded domains. Locally, this electromagnetic wave coincides with a knot around any point (except for a zero measure set), but there is no knot coinciding with it throughout all the space R^3.

B. Global Difference

Example (2) given above is interesting. The standard plane wave is used very often even though it is in fact physically impossible—unless we can provide an infinite amount of energy to produce such a state. That it cannot be an electromagnetic knot is also clear. This is so because, in the case of the electromagnetic knots, only one magnetic line and one electric line passes through the point at infinity (because ϕ and θ are one-valued there). Quite to the contrary, for plane waves, an infinite number of lines go to infinity without coming back (the scalar field is not even defined at infinity). This illustrates the global difference between standard fields and electromagnetic knots. They cannot be differentiated locally, but they behave in quite different ways around infinity.

Example 1 above is considered in Section VIII to illustrate the topological quantization of the electric charge, which is one feature of the topological model.

VI. A HIDDEN NONLINEARITY

We have found a structure with two levels. At the deeper one, it is nonlinear since the scalars ϕ and θ obey highly nonlinear equations. However, the transformation $T: \sigma \to \mathscr{F}, *\mathscr{F}$ given by (9) and (12)

$$T: \sigma \to (\mathscr{F} = -a\phi^*\sigma, \quad *\mathscr{F} = a\theta^*\sigma) \tag{149}$$

where σ is the area 2-form in S^2, changes these nonlinear equation for ϕ and θ into the linear Maxwell's ones for $\mathscr{F}$, thus linearizing the theory. This is important; the Maxwell equations are the exact linearization of a nonlinear and topological theory (by change of variables, not by truncation!). The theory seems to be linear if the equation is assumed to be satisfied by the field $F_{\mu\nu}$, but it cannot be really linear since the topological quantization of the helicity imposes the nonlinear conditions

$$h_m = \int \mathbf{A} \cdot \mathbf{B}\, d^3 r = na, \qquad h_e = \int \mathbf{C} \cdot \mathbf{E}\, d^3 r = na \tag{150}$$

It is clear therefore that one cannot obtain another solution simply by multiplying **B** and **E** by a real number (or by adding two different solutions). A similar situation arises in the work by Evans [57].

We call this unexpected and curious property "hidden nonlinearity". It is due to the fact that the transformation T is not invertible, since there are solutions of Maxwell's equations for which $T^{-1}\mathscr{F}$ is not defined. In other words, in some cases there are no scalar fields ϕ, θ generating $\mathscr{F}$, which could be interpreted as maps $S^3 \mapsto S^2$ (as the three examples given above clearly show). As a consequence, although all the electromagnetic fields of the topological model obey the linear Maxwell equations, *they do not span the vector space of all the solutions, but form a nonlinear subset instead.* More precisely, we have seen that any standard electromagnetic field in empty space is locally equal to the addition of two electromagnetic knots, except for a zero measure set. However, the addition of all pairs of electromagnetic knots gives only a nonlinear subset of the set of all the standard electromagnetic fields. Some standard fields are lacking. This might appear disastrous, but the local equivalence shown earlier indicates that it is not a matter of concern.

Which standard electromagnetic fields must be excluded from the topological theory because they cannot be generated by a pair of dual maps? The fields to be excluded are those with helicities not verifying the equations (150) and also those for which the scalars do exist locally but do not behave well at infinity or are not of class C^1 and for which the Hopf index cannot be defined. Contrary to what it might seem, this is not necessarily a drawback of the model. In fact, it can be said that Maxwell's equations have too many solutions, since not all of them can be realized in nature and because some of them have energy, or momentum, that is infinite. Others are Coulomb or Liener–Wiechert potentials coupled to charges that are not integer multiples of the electron fundamental value e, or that would have been radiated by monopoles (if these particles do not exist), or have discontinuities in surfaces, meant to represent in a simple way changes of the field that are abrupt but continuous. Consequently, the fact that not all the standard solutions are included in the topological model is not necessarily a disadvantage.

In order to better understand the role of the hidden nonlinearity, let us examine two properties of the knots.

1. If $F_{\mu\nu}$ is a knot, all its integer multiples $nF_{\mu\nu}$ are also knots. It is easy to understand why. Let $\phi = S\exp(i2\pi s)$ and $\theta = Q\exp(i2\pi q)$. It is then a simple matter to see that $nF_{\mu\nu}$ and $n\,{}^*F_{\mu\nu}$ are generated by the scalars $\phi^{(n)} = S\exp(ni2\pi s)$ and $\theta^{(n)} = Q\exp(ni2\pi q)$, which are clearly defined if n is an integer. Note that the helicity of $nF_{\mu\nu}$ is equal to n^2 times that of $F_{\mu\nu}$.

2. If $F_{\mu\nu}$ is a knot and the scalars ϕ and θ never take the values 0 or ∞, then all $cF_{\mu\nu}$, where c is a real number, are also knots.

This is so because $cF_{\mu\nu}$ and $c\,{}^*F_{\mu\nu}$ are generated by $\phi^{(c)} = S\exp(ci2\pi s)$ and $\theta^{(c)} = Q\exp(ci2\pi q)$, respectively, which are clearly defined for any real c. Note that in this case, ϕ and θ are maps $S^3 \mapsto R^1 \times S^1$, which form only one homotopy class. Note also that the helicities vanish in this case.

This shows that there is still some linearity. In particular, there is a subset of knots that form a vector space and is therefore a linear sector of the model. It is the set of the knots with zero helicity or with unlinked lines. Note also that the theory is fully linear from the local point of view, as a consequence of the local equivalence with Maxwell's theory shown in Section V.A. By this we mean that the set of the electromagentic knots contains all the linear combinations of standard solutions around any point.

VII. TOPOLOGICAL QUANTIZATION OF ELECTROMAGNETIC HELICITY

As was shown in Section II, the magnetic and the electric helicities of any radiation electromagnetic field are equal. Moreover, in the case of the topological model, the helicities of the knots verify

$$h_m = \int \mathbf{A} \cdot \mathbf{B}\, d^3r = h_e = \int \mathbf{C} \cdot \mathbf{E}\, d^3r = na \tag{151}$$

where $\mathbf{B} = \nabla \times \mathbf{A}$, $\mathbf{E} = \nabla \times \mathbf{C}$, a is the normalizing constant of the model, and n is the common value of the Hopf indices of the two maps $\phi, \theta : S^3 \mapsto S^2$, which is related to the linking number of any pair of magnetic or electric lines, as explained in Section I.D.

Furthermore, it was shown in Section II.C that the semisum of the two helicities $\mathscr{H} = \frac{1}{2}(h_m + h_e) = na$, which we call the electromagnetic helicity, is a constant of the motion for any standard electromagnetic field in empty space:

$$\mathscr{H} = \hbar c \int d^3k(\bar{a}_R(\mathbf{k})a_R(\mathbf{k}) - \bar{a}_L(\mathbf{k})a_L(\mathbf{k})) \tag{152}$$

In the case of a knot, it follows that

$$n = \frac{\hbar c}{a} \int d^3k(\bar{a}_R(\mathbf{k})a_R(\mathbf{k}) - \bar{a}_L(\mathbf{k})a_L(\mathbf{k})) \tag{153}$$

In QED, a_R, a_L are taken to be annihilation operators (and $\bar{a}_R, \bar{a}_L$ creation operators) for photons, where the integral on the right-hand side of (152) and (153) is the operator for the difference between the numbers of right-handed and left-handed photons $N_R - N_L$. If the knots are classical, those Fourier transforms are functions, so that the integral in the right-hand side is the classical limit of this difference. Consequently, the value of $N_R - N_L$ for a knot is topologically quantized and takes the value $na/\hbar c$. (Note that this is true even if the knots are classical fields.) This suggests a criterion for the value of the normalizing constant. Taking $a = \hbar c$ (in natural units, this is $a = 1$ and in SU $a = \hbar c \epsilon_0$, where ϵ_0 being the permittivity of empty space), one has

$$n = N_R - N_L \tag{154}$$

Equation (154) relates, in a very simple and appealing way, two meanings of the term *helicity*, related to the wave and particle aspects of the field. At the left, the wave helicity is the Hopf index n, characterizing the way in which the force lines—either magnetic or electric—curl around one another (as explained before $n = \ell m^2$, where ℓ is the linking number and m the multiplicity of the map). At the right, the particle helicity is the difference between the numbers of right-handed and left-handed photons. This is certainly a nice property. It suggests that the electromagnetic knots are worthy of consideration. Note that this property gives a new interpretation of the number n. We know that it is a Hopf index. We see that it is furthermore the difference of the classical limit of the numbers of right-handed and left-handed photons.

All the electromagnetic knots verify the quantum conditions

$$h_m = h_e = n\hbar c, \qquad N_R - N_L = n \tag{155}$$

Note that the set of the electromagnetic knots contains some with very low energy, for which n is necessarily very small. Even if they can be defined as classical fields, the real system would have quantum behavior, since the action involved would be of the order of $\hbar$. On the other hand, there are states with n small and even zero, which have, however, macroscopic energy. They are those for which N_R, N_L are large. When n is large, the photon contents are high and the energy is macroscopic. These are the states for which the classical approximation is valid.

This suggest that the set of the electromagnetic knots give a classic limit with the right normalization.

VIII.   TOPOLOGICAL QUANTIZATION OF ELECTRIC AND MAGNETIC CHARGES

Quantization of the electric charge is one of the most important and intriguing laws of physics. However, the value of the fundamental charge is obtained through experiments, as all the efforts to predict it—or the fine-structure constant α—within a theoretical scheme have failed so far.

This important law is usually stated by saying that the electric charge of any particle is an integer multiple of a fundamental value e, the electron charge, whose value in the International System of Units is (SI units) $e = 1.6 \times 10^{-19}$ C. The Gauss theorem allows a different, although fully equivalent, statement of this property, namely, that the electric flux across any closed surface Σ that does not intersect any charge is always an integer multiple of e (we use the rationalized MKS system here). This can be written as

$$\int_{\Sigma} \omega = ne \qquad (156)$$

where ω is the 2-form $\mathbf{E} \cdot \mathbf{n} \, dS$, and $\mathbf{n}$ is a unit vector orthogonal to the surface, $\mathbf{E}$ is the electric field, and dS is the surface element. We could as well write (156) as

$$\int_{\Sigma} *\mathscr{F} = ne \qquad (157)$$

where $*\mathscr{F}$ is the dual to the Faraday 2-form $\mathscr{F} = \frac{1}{2} F_{\mu\nu} dx^{\mu} \wedge dx^{\nu}$. Stated in this way, the discretization of the charge is interesting because it shows a close similarity to the expression of the topological degree of a map. Assume that we have a regular map θ of Σ on a 2-sphere S^2 and let σ be the normalized area 2-form in S^2. It then happens that

$$\int_{\Sigma} \theta^* \sigma = n \qquad (158)$$

where $\theta^* \sigma$ is the pullback of σ and n an integer called the "degree of the map," which gives the number of times that S^2 is covered when one runs once through Σ (equal to the number of points in Σ in which θ takes any prescribed value). Note that θ^* in (158) indicates pullback by the map θ and must not be mistaken for the complex conjugate of θ, which will be written $\bar{\theta}$.

Comparison of (157) and (158) shows that there is a close formal similarity between the dual to the Faraday 2-form and the pullback of the area 2-form of a sphere S^2. It can be expressed in this way. Let an electromagnetic field be given, such that its form $*\mathscr{F}$ is regular except at the positions of some point charges. Let a map $\theta: R^3 \mapsto S^2$ also be given, which is regular except at some point singularities where its level curves converge or diverge. Then, Eqs. (157) and (158) are simultaneously satisfied for all the closed surfaces Σ that do not intersect any charge or singularity.

This means that the electric charge will be automatically and topologically discretized in a model in which these two forms—$*\mathscr{F}$ and $\theta^* \sigma$—are proportional; the fundamental charge is equal to the proportionality coefficient and the number of fundamental charges in a volume then have the meaning of a topological index.

This is exactly what happens in the topological model. Indeed, the dual to the Faraday 2-form is expressed in it as

$$*\mathscr{F} = \sqrt{a}\,\theta^* \sigma \tag{159}$$

where a is a normalizing constant with dimensions of action times velocity. Remember that the electric field is $\mathbf{E} = \sqrt{a}\,(2\pi i)^{-1}(1 + \bar{\theta}\theta)^{-2}\boldsymbol{\nabla}\bar{\theta} \times \boldsymbol{\nabla}\theta$; the electric lines are therefore the level curves of θ. The degree of the map $\Sigma \mapsto S^2$ induced by θ is given by (158); therefore

$$\int_\Sigma *\mathscr{F} = n\sqrt{a} \tag{160}$$

As this is equal to the charge Q inside Σ, it follows that $Q = n\sqrt{a}$, which implies that there is a fundamental charge $q_0 = \sqrt{a}$, where the degree n represents the number of fundamental charges inside Σ. This gives a topological interpretation of n, the number of fundamental charges inside any volume .

It is easy to understand that $n = 0$ if θ is regular in the interior of Σ. This is because each level curve of θ (i. e., each electric line) is labeled by its value along it—a complex number—and, in the regular case, any one of these lines enters into this interior as many times as it goes out of it. But assume that θ has a singularity at point P, from which the electric lines diverge or to which they converge. If Σ is a sphere around P, we can identify R^3 except P with $\Sigma \times R$, so that the induced map $\theta : \Sigma \mapsto S^2$ is regular. In this case, n need not vanish and is equal to the number of times that θ takes any prescribed complex value in Σ, with due account to the orientation. Otherwise stated, among the electric lines diverging from or converging to P, there are $|n|$ whose label is equal to any prescribed complex number.

This shows why the topological model embodies a topological quantization of the charge, because it entails the automatic verification of the equation (159). This mechanism for the quantization of the charge was first shown in Ref. 33 and developed later in Refs. 26,34, and 35. As the magnetic field is $\mathbf{B} = -\sqrt{a}\,(2\pi i)^{-1}(1 + \bar{\phi}\phi)^{-2}\boldsymbol{\nabla}\bar{\phi} \times \boldsymbol{\nabla}\phi$, the magnetic and electric lines are the level curves of ϕ and θ, respectively.

To better understand this discretization mechanism, let us take the case of a Coulomb potential [31,33], $\mathbf{E} = Q\mathbf{r}/(4\pi r^3)$, $\mathbf{B} = 0$. The corresponding scalar is then

$$\theta = \tan\left(\frac{\vartheta}{2}\right)\exp\left(i\,\frac{Q}{\sqrt{a}}\,\varphi\right) \tag{161}$$

where φ and ϑ are respectively the azimuth and the polar angle. The scalar (161) is clearly defined only if $Q = n\sqrt{a}$, where n is an integer. The lines diverging from the charge are labeled by the corresponding value of θ, so that there are $|n|$ lines going into or out of the singularity and having any prescribed complex number as their label. If $n = 1$, it turns out that $\theta = (x + iy)/(z + r)$.

This mechanism has a very curious aspect—it does not apply to the source but to the electromagnetic field itself. This is surprising; one would expect that the topology should operate by restricting the fields of the charged particles. However, in this model, the field that mediates the force is the one that is submitted to a topological condition. It must be emphasized furthermore that the maps $S^3 \mapsto S^2$, given by the two scalars ϕ, θ, are regular except for singularities at the position of point charges, either electric or magnetic (if the latter do exist). At these points, the level curves (i.e., the electric lines) either converge or diverge.

In the previous section, it was shown that the constant a must be equal to $\sqrt{\hbar c}$ in order to obtain the right quantization of the electromagnetic helicity. This implies that the topological model predicts that the fundamental charge, either electric or magnetic, has the value

$$q_0 = \sqrt{\hbar c} \tag{162}$$

(in the MKS system), which is about 3.3 times the electron charge. In SI units, this is $q_0 = \sqrt{\hbar c \epsilon_0} = 5.29 \times 10^{-19}$ C, and in natural units $q_0 = 1$. Note that this applies to both the electron charge and the hypothetical monopole charge. This property can be stated by saying that, in the topological model, the electromagnetic fields can be coupled only to point charges that are integer multiples of the fundamental charge $q_0 = \sqrt{\hbar c}$. Note that the same discretization mechanism would apply to the hypothetical magnetic charges (located at singularities of ϕ), and their fundamental values would also be $q_0 = \sqrt{\hbar c}$.

A. The Fine-Structure Constant at Infinite Energy Equal to $1/4\pi$?

As the topological model as presented here is classical, this value of q_0 must be interpreted as the fundamental bare charge, both electric and magnetic. The corresponding fine-structure constant is clearly $\alpha_0 = 1/4\pi$, which is certainly a nice number. We now argue that $1/4\pi$ is an appealing and interesting value for the unrenormalized fine-structure constant (i.e., neglecting the effect of the quantum vacuum). In that case, the topological model would describe the electromagnetic field at infinite energy.

The argument goes as follows. Let us combine this topological quantization of the charge with the appealing and plausible idea that, in the limit of very high energies, the interactions of charged particles could be determined by their bare charges (i.e., the value that their charges would have if they were not renormalized by the quantum vacuum; see, e.g., Section 11.8 of Ref. 66).

However, a warning is necessary. As the concept of bare charge is complex, it is convenient to speak instead of charge at a certain scale. To be precise and avoid confusion, when the expression "bare charge" is used here, it will be taken as equivalent and synonymous to "infinite energy limit of the charge" or, more correctly, "charge at infinite momentum transfer," defined as $e_\infty = \sqrt{4\pi\hbar c\alpha_\infty}$, where $\alpha_\infty = \lim \alpha(Q^2)$ when $Q^2 \to \infty$.

The possibility of a finite value for α_∞ is an intriguing idea worth studying. Indeed, it was discussed very early by Gell-Mann and Low in their classic and seminal paper "QED at small distances" [67], in which they showed that it is something to be seriously considered. However, they could not decide from their analysis whether e_∞ is finite or infinite. The standard QED statement that it is infinite was established later on the basis of perturbative calculations. Nevertheless, and contrary to an extended belief, the alternative presented by Gell-Mann and Low has not been really settled. It is still open, in spite of the many attempts to clarify this question.

The infinite energy charge e_∞ of an electron is partially screened by the sea of virtual pairs that are continuously being created and destroyed in empty space. It is hence said that it is renormalized. As the pairs are polarized, they generate a cloud of polarization charge near any charged particle, with the result that the observed value of the charge is smaller than e_∞. Moreover, the apparent electron charge increases as any probe goes deeper into the polarization cloud and is therefore less screened. This effect is difficult to measure, as it can be appreciated only at extremely short distances, but it has been observed indeed in experiments of electron–positron scattering at high energies [68]. In other words, the vacuum is dielectric. On the other hand, it is paramagnetic, since its effect on the magnetic field is due to the spin of the pairs. As a consequence, the hypothetical magnetic charge would be observed with a greater value at low energy than at very high energy, contrary to the electron charge.

The coinage "bare charge" is appropriate for e_∞, as it is easy to understand intuitively. When two electrons interact with very high momentum transfer, each one is located so deeply inside the polarization cloud around the other that no space is left between them to screen their charges, so that the bare values, namely, e_∞, interact directly. As unification is assumed to occur at very high energy, it is an appealing idea that $\alpha_\infty = \alpha_{GUT}$. Indeed, although this possibility is almost always neglected, it is certainly worth of careful consideration. [It is true that one could imagine that $\alpha(Q^2)$ has a plateau at the unification scale corresponding to a critical value smaller than α_∞, but we consider here the simpler situation in which that plateau does not exist.] This suggests that a unified theory could be a theory of bare particles (in the sense of neglecting the effect of the vacuum). If this were the case, nature would have provided us with a natural cutoff, in such a way that $\alpha_{GUT} = \alpha_\infty$ (where the subscript GUT denotes grand unified theory).

As a consequence of these considerations, it can be argued that the topological model implies the equalities $\alpha_{GUT} = \alpha_\infty = 1/4\pi$. The argument goes as follows.

1. The value of the fundamental charge implied by this topological quantization $e_0 = \sqrt{\hbar c}$ is in the right interval to verify $e_0 = e_\infty = g_\infty$, that is, to be equal to the common value of both the fundamental electric and magnetic infinite energy charges. This is so because, as the quantum vacuum is dielectric but paramagnetic, the following inequalities must be satisfied: $e < e_0 < g$, as they are indeed, since $e = 0.3028$, $e_0 = 1$, $g = e/2\alpha = 20.75$, in natural units. Note that it is impossible to have a complete symmetry between electricity and magnetism simultaneously at low and high energies. The lack of symmetry between the electron and the Dirac monopole charges would be due, in this view, to the vacuum polarization: according to the topological model, the electric and magnetic infinite energy charges are equal and verify $e_\infty g_\infty = e_0^2 = 1$, but they would be decreased and increased, respectively, by the sea of virtual pairs, until the electron and the monopole charge values verifying the Dirac relation $eg = 2\pi$ [14]. The qualitative picture seems nice and appealing.

2. Let us admit as a working hypothesis that two charged particles interact with their bare charges in the limit of very high energies (as explained above). There could be then a conflict between (a) a unified theory of electroweak and strong forces, in which $\alpha = \alpha_s$ at very high energies and (b) an infinite value of α_∞. This is so because unification implies that the curves of the running constants $\alpha(Q^2)$ and $\alpha_s(Q^2)$ must converge asymptotically to the same value α_{GUT}. It could be argued that, to have unification at a certain scale, it would suffice that these two curves be close in an energy interval, even if they cross and separate afterward. However, in that case, the unified theory would be just an approximate accident at certain energy interval. On the other hand, the assumption that both running constants go asymptotically to the same finite value α_{GUT} gives a much deeper meaning to the idea of unified theory, and is therefore much more appealing. In that case, e_∞ must be expected to be finite, and the equality $\alpha_{GUT} = \alpha_\infty$ must be satisfied.

3. The value $\alpha_0 = e_0^2/4\pi\hbar c = 1/4\pi = 0.0796$ for the infinite energy fine-structure constant α_∞ is thought-provoking and fitting, since α_{GUT} is believed to be in the interval $(0.05, 0.1)$. This reaffirms the assertion that the fundamental value of the charge given by the topological mechanism e_0 could be equal to e_∞, the infinite energy electron charge (and the infinite energy monopole charge also). It also supports the statement that α_{GUT} must be equal to α_0 and to $1/4\pi$. All this is certainly curious and intriguing since the topological mechanism for the quantization of the charge described here [26,33–35] is obtained simply by putting some topology in elementary classical low-energy electrodynamics [24–25,26,30,31].

We believe, therefore, that the following three ideas must be studied carefully: (1) the complete symmetry between electricity and magnetism at the level of the infinite energy charges, where both are equal to $\sqrt{\hbar c}$ and the symmetry is broken by the dielectric and paramagnetic quantum vacuum; (2) that the topological model on which the topological mechanism of quantization is based could give a theory of high-energy electromagnetism at the unification scale; and (3) that the value that it predicts for the fine-structure constant $\alpha_0 = 1/4\pi$ could be equal to the infinite energy limit α_∞ and also to α_{GUT}, the constant of the unified theory of strong and electroweak interactions.

In this way the three quantities (both the electric and the magnetic fine-structure constants at infinite momentum transfer and α_{GUT}) would be equal. Furhermore, there would be a complete symmetry between electricity, magnetism, and strong force at the level of bare particles (i.e., at $Q^2 = \infty$); this symmetry would be broken by the effect of the quantum vacuum.

IX. SUMMARY AND CONCLUSIONS

In this chapter we have presented a topological model of electromagnetism that was proposed by one of us (AFR) in 1989 [24,25]. It is based on the existence of a topological structure that underlies Maxwell's standard theory, in such a way that the Maxwell equations in empty space are the exact linearization (by change of variables, not by truncation) of some nonlinear equations with topological properties and constants of the motion. Although the model is classical, it embodies the topological quantizations of the helicity and the electric charge, which suggest that it clarifies the relationship between the classical and quantum aspects of the electromagnetism. Indeed, the model was developed in the spirit described by the Atiyah aphorism "Both topology and quantum physics go from the continuous to the discrete."

The main characteristics of the topological model are summarized as follows:

1. Its topological structure is induced by the topology of the force lines (both electric and magnetic). Indeed, it is based on the idea of electromagnetic knot, defined (in empty space) as a standard electromagnetic field in which any pair of magnetic lines and any pair of electric lines is a link. An electromagnetic knot is constructed by means of a pair of complex scalar fields ϕ, θ with only one value at infinity. The magnetic (resp. electric) lines are the level curves of ϕ (resp. θ). These scalars can be interpreted as giving two maps (termed *dual*) from the sphere S^3 to the sphere S^2, which are characterized by the common value of their Hopf indices n. The magnetic and electric helicities are $\int \mathbf{A} \cdot \mathbf{B} \, d^3r = \int \mathbf{C} \cdot \mathbf{E} \, d^3r = n$ (in natural units). An important feature is that the Faraday 2-form and its dual are the pullbacks of σ, the area 2-form in S^2, by the two scalars, so that $\mathscr{F} = -\phi^*\sigma, \ *\mathscr{F} = \theta^*\sigma$.

2. It is locally equivalent to Maxwell's standard theory in empty space (but globally disequivalent). This means that it cannot enter in conflict with Maxwell's theory in experiments of local nature.

3. The linear Maxwell equations appear in the model as the linearization by change of variables of nonlinear equations that refer to the scalars ϕ, θ. This introduces a subtle form of nonlinearity that we call "hidden nonlinearity." For this reason, the linearity of Maxwell's equations is compatible with the existence of topological constants of the motion.

4. One of these topological constants of the motion is the electromagnetic helicity, defined as the semisum of the magnetic and electric helicities, which is equal to the linking number of the force lines

$$ \mathscr{H} = \frac{1}{2} \int (\mathbf{A} \cdot \mathbf{B} + \mathbf{C} \cdot \mathbf{E}) \, d^3 r = n $$

Moreover, it turns out that $N_R - N_L = n$, where N_R, N_L are the classical expression of the number of right-handed and left-handed photons (i.e., obtained by substituting the Fourier transform functions $a_R(\mathbf{k})$, $a_L(\mathbf{k})$ for the quantum operators $a_{R,\mathbf{k}}$, $a_{L,\mathbf{k}}$). This establishes a nice relation between the wave and particle meaning of helicity (i.e. between the linking number of the force lines and the difference $N_R - N_L$ referring to the photonic content of the field). This suggests that the topological model could give the classical limit of the quantum theory with the right normalization.

5. Another topological constant is the electric charge, which is, moreover, topologically quantized; its fundamental value is $q_0 = \sqrt{\hbar c}$ in the rationalized MKS system ($q_0 = \sqrt{\hbar c \epsilon_0}$ in the SU system; $q_0 = 1$ in natural units). Furthermore, the number of fundamental charges inside a volume is equal to the degree of a map between two spheres S^2. It turns out that there are exactly $|m|$ electric lines going out from or coming into a point charge $q = m q_0$, for which ϕ is equal to any prescribed complex number (taking into account the orientation of the map).

The topological model is completely symmetric between electricity and magnetism, in the sense that it predicts that the fundamental hypothetical magnetic charge would also be q_0. Note that $q_0 = 3.3\,e$ and that the corresponding fine-structure constant is $\alpha_0 = 1/4\pi$. It is argued in Section VIII.A that q_0 could be interpreted as the bare electron and monopole charge. As the quantum vacuum is dielectric but paramagnetic, the observed electric charge must be smaller than q_0 (it is equal to $0.303\,q_0$), but the Dirac charge must be greater (it is equal to $20.75\,q_0$). This suggests that α_0 could be the fine-structure constant at infinite energy and, consequently, that the coupling constant of the grand unified theory could also be $\alpha_s = \alpha_0 = 1/4\pi$.

This is an indication that the topological model could give a theory of bare electromagnetism or, equivalently, of electromagnetism at infinite energy at the unification scale.

Our conclusion is that the topological model of electromagnetism is worth careful consideration.

Acknowledgments

We are grateful to Professors C. Aroca, M. V. Berry, D. Bouwmeester, A. Ibort, E. López, J. M. Montesinos, M. Soler (deceased), A. Tiemblo, and J. L. Vicent for discussions and encouragement.

References

1. E. Whittaker, *A History of the Theories of Aether and Electricity* (1910) (reprinted by Humanities Press Inc, New York, 1973).

2. J. C. Maxwell, *Trans. Cambridge. Phil. Soc.* **10**, 27–83 (1864).

3. J. C. Maxwell, in N. D. Niven (Ed.), *Scientific Papers of James Clerk Maxwell*, Cambridge, 1890 (Reprinted by Dover, New York, 1952).

4. M. Goldman, *The Demon in the Aether. The Life of James Clerk Maxwell*, Paul Harris, Edinburgh, 1983.

5. Lord Kelvin (then W. Thomson), *Trans. Roy. Soc. Edinburgh* **25**, 217–260 (1868).

6. P. G. Tait, *Scientific Papers*, Vol. 1, Cambridge Univ. Press, Cambridge, UK, 1911, pp. 136–150.

7. Th. Archibald, *Math. Mag.* **62**, 219 (1989)

8. M. Atiyah, *Q. J. Roy. Astron. Soc.* **29**, 287–299 (1988).

9. J. C. Maxwell, "Atomism," *Enciclopaedia Britannica*, II, 9th ed. 1875; also in J. M. Sánchez Ron (Ed.), *J. C. Maxwell, Scientific Papers*, CSIC, Madrid, 1998.

10. J. Z. Buchwald, "William Thomson (Lord Kelvin)," in C. Coulson (Ed.), *Dictionary of Scientific Biography*, Vols. 13–14, Scribner, New York, 1980.

11. Lord Kelvin (then W. Thomson), "Lecture to the Royal Society of Edinburgh," Feb. 18, 1967; *Proc. Roy. Soc. Edinburgh* **6**, 94 (1869); see also S. P. Thomson, *Life of Lord Kelvin*, Macmillan, London, 1910, p. 517.

12. M. Atiyah, *The Geometry and Physics of Knots,* Cambridge Univ. Press, 1990.

13. H. K. Moffatt, *Nature* **347**, 321 (1990).

14. P. A. M. Dirac, *Proc. Roy. Soc A.* **133**, 60 (1931).

15. A. M. Poliakov, *JETP Lett.* **20**, 194 (1974).

16. G. 't Hooft, *Nucl. Phys. B* **79**, 276 (1974).

17. Y. Aharonov and D. Bohm, *Phys. Rev.* **115**, 485 (1959).

18. T. H. Skyrme, *Proc. Roy. Soc. A* **260**, 127 (1961).

19. T. H. Skyrme, *Int. J. Mod. Phys. A* **3**, 2745–2751 (1988).

20. V. G. Makhankov, *The Skyrme Model*, Springer-Verlag, Berlin, 1993.

21. E. Witten, *Commun. Math. Phys.* **117**, 353 (1988).

22. E. Witten, *Commun. Math. Phys.* **121**, 351 (1989).

23. L. D. Faddeev and A. J. Niemi, *Nature* **387**, 58 (1997).

24. A. F. Rañada, *Lett. Math. Phys.* **18**, 97–106 (1989).

25. A. F. Rañada, *J. Phys. A: Math Gen.* **25**, 1621–1641 (1992).

26. A. F. Rañada, in M. Ferrero and A. van der Merwe (Eds.), *Fundamental Problems in Quantum Physics*, Kluwer, Dordrecht, 1995, pp. 267–277.

27. A. F. Rañada, *J. Phys. A: Math. Gen.* **23**, L815–L820 (1990).

28. A. F. Rañada, in V. G. Makhankov, V. K. Fedyanin, and O. K. Pashaev (Eds.), *Solitons and applications*, World Scientific, Singapore: 1990, pp. 180–194.

29. A. F. Rañada, in J. A. Ellison and H. Überall (Eds.), *Fundamental Problems in Classical and Quantum Dynamics*, Gordon & Breach, London, 1991, pp. 95–117.

30. A. F. Rañada and J. L. Trueba, *Phys. Lett. A* **202**, 337–342 (1995).

31. A. F. Rañada and J. L. Trueba, *Phys. Lett. A* **235**, 25–33 (1997).

32. J. L. Trueba, *Nudos Electromagnéticos* (Electromagnetic knots), Ph.D. Thesis, Complutense Univ. Madrid, 1997.

33. A. F. Rañada, *Ann. Fis.* (Madrid) *A* **87**, 55–59 (1991).

34. A. F. Rañada and J. L. Trueba, *Phys. Lett. B* **422**, 196–200 (1998).

35. A. F. Rañada, http://xxx.lanl.gov/list/hep-th/9904158 (1999).

36. R. Bott and L. W. Tu, *Differential forms in Algebraic Topology*, Springer, New York, 1982.

37. C. Nash and S. Sen, *Topology and Geometry for Physicists*, Academic Press, London, 1982.

38. H. Hopf, *Math. Annalen* **104** 637–665 (1931).

39. D. A. Nicole, *J. Phys. G: Nucl. Phys.* **4**, 1363 (1978).

40. A. Kundu and Yu. P. Rybakov, *J. Phys. A: Math. Gen.* **15**, 269 (1982).

41. A. Kundu, *Phys. Lett. B* **171**, 67 (1986).

42. H. K. Moffat, *Phil. Trans. Roy. Soc. A* **333**, 321–342 (1990).

43. J. H. C. Whitehead, *Proc. Natl. Acad. Sci.* (USA) **33**, 117–125 (1947).

44. H. K. Moffatt, *J. Fluid Mech.* **35**, 117–129 (1969).

45. L. Woltjer, *Proc. Natl. Acad. Sci.* (USA) **44**, 489 (1958).

46. J. J. Moreau, *C. R. Acad. Sci.* (Paris) **252**, 2810 (1961).

47. A. F. Rañada, *Eur. J. Phys.* **13**, 70 (1992).

48. J. L. Trueba and A. F. Rañada, *Eur. J. Phys.* **17**, 141 (1996).

49. G. E. Marsh, *Force-Free Magnetic Fields* World Scientific, Singapore, 1996.

50. M. A. Berger and G. B. Field, *J. Fluid Mech.* **147**, 133 (1984).

51. H. K. Moffatt and R. L. Ricca, *Proc. Roy. Soc. Lond.* **439A**, 411 (1992).

52. G. Calugareanu, *Rev. Math. Pure Appl.* **4**, 5 (1959).

53. V. I. Arnold, *Sel. Math. Sov.* **5**, 327 (1986).

54. M. W. Evans, *Found. Phys.* **24**, 1671 (1994).

55. M. W. Evans and J. P. Vigier (Eds.), *The Enigmatic Photon*, Vol. 1, Kluwer, Dordrecht, 1994.

56. M. W. Evans, J. P. Vigier, S. Roy, and S. Jeffers (Eds.), *The Enigmatic Photon*, Vol. 3, Kluwer, Dordrecht, 1996.

57. M. W. Evans, J. P. Vigier, S. Roy, and G. Hunter (Eds.), *The Enigmatic Photon*, Vol. 4, Kluwer, Dordrecht, 1997.

58. B. Hatfield, *Quantum Field Theory of Point Particles and Strings*, Addison-Wesley, Reading, MA, 1992.

59. G. Darboux, *Bull. Sci. Math.* (2), 6 (1882).

60. C. Godbillon, *Géometrie différentielle et méchanique analytique*, Hermann, Paris, 1969.

61. Y. Choquet-Bruhat, C. DeWitt-Morette, and M. Dillard-Bleick, *Analysis, Manifolds and Physics*, 2nd ed., North-Holland, Amsterdam, 1982.

62. S. Fritelli, S. Koshti, E. Newmann, and C. Rovelli, *Phys. Rev. D* **49**, 6883 (1994).

63. H. Lamb, *Hydrodynamics*, Dover, New York, 1932.

64. E. A. Kuznetsov and A. V. Mikhailov, *Phys. Lett. A* **77**, 37 (1980).

65. C. Misner, K. S. Thorne, and J. A. Wheeler, *Gravitation*, Freeman, San Francisco, 1973.

66. P. W. Milonni, *The Quantum Vacuum. An Introduction to Quantum Electrodynamics*, Academic, Boston, 1994.

67. M. Gell-Mann and F. Low, *Phys. Rev.* **95**, 1300–1312 (1954).

68. I. Levine, D. Koltick et al. (TOPAZ Collaboration), *Phys. Rev. Lett.* **78**, 424–427 (1997).

69. A. F. Rañada and J. L. Trueba, *Nature* **383**, 32 (1996).

70. A. F. Rañada, M. Soler, and J. L. Trueba, *Phys. Rev. E* **62**, 7181 (2000).

71. A. F. Rañada, M. Soler, and J. L. Trueba, in S. Singer (Ed.), *Balls of Fire. Recent Research in Ball Lighting*, Springer-Verlag, Berlin, in Press.

ELLIPSOIDS IN HOLOGRAPHY
AND RELATIVITY

NILS ABRAMSON

*Industrial Metrology and Optics, IIP, Royal Institute
of Technology, 11422 Stockholm, Sweden*

CONTENTS

A diagram, the holodiagram, which is based on a set of ellipses, was designed to simplify the making and evaluation of holograms. It was, however, soon found that this diagram could be used in many other fields of optics and, surprisingly, also in Einstein's theory of special relativity. Holography with ultrashort pulses "light-in-flight recording by holography" can produce slow-motion pictures of

Modern Nonlinear Optics, Part 3, Second Edition, Advances in Chemical Physics, Volume 119,
Edited by Myron W. Evans. Series Editors I. Prigogine and Stuart A. Rice.
ISBN 0-471-38932-3   © 2001 John Wiley & Sons, Inc.

light pulses. However, in such recordings a spherical light wave appears deformed into one of the ellipsoids of the holodiagram (the two focal points are the emitter respective the observer). The reason for this distortion is the limited velocity of the light used for observation in much the same way as the cause of the apparent deformations of fast-moving bodies described in special relativity. The main difference is that in holography the distance separating the two focal points of the ellipsoids is static while in relativity it is dynamic and caused by a high velocity of the observer. Using this new graphical approach to relativity, we find no reason for the Lorentz contraction; instead, we accept an elongation of the observation sphere.

I.   INTRODUCTION TO EINSTEIN'S SPECIAL RELATIVITY THEORY

Special relativity is based on Einstein's two postulates of 1905:

1. The same laws of electrodynamics and optics will be valid for all frames of reference for which the equations of mechanics hold good.
2. Light is always propagated in empty space with a definite velocity c which is independent of the state of motion of the emitting body.

In this chapter I would prefer to express these two postulates using the following statements:

1. If we are in a room that is totally isolated from the outside world, there is no experiment that can reveal a constant velocity of that room.
2. When we measure the speed of light (c) in vacuum, we always get the same result independent of any velocity of the observer and or of the source.

To make such strange effects possible, it is assumed that the velocity results in that time moves more slowly (so that seconds are longer), which is termed *relativistic time dilation*, and it was assumed (by Lorentz) that lengths (rulers) are shortened in the direction of travel (Lorentz contraction). The contraction effect was presented by Lorentz [1] to explain the Michelson–Morley experiments in 1881 and was later adapted by Einstein [2] in his famous special theory of relativity [2]. Time dilation is more "real" than the Lorentz contraction because it produces a permanent result, a lasting difference in the reading of a stationary and a traveling clock, while the Lorentz contraction is much more "apparent" as it produces no permanent result; there is a difference in length only as long as there is a difference in velocity of two rulers.

The Michelson–Morley interferometer compares the time of travel for light rays along just two perpendicular one-dimensional paths, while holography can be used to make that comparison for light rays in all three dimensions. Thus, there are reasons to believe that experiences from holography with ultrashort pulses could shine new light on the theory of relativity.

II. INTERSECTING MINKOWSKI LIGHTCONES

Let us start by studying the Minkowski diagram [3], which is based on one cone of illumination and one of observation. It was invented in 1908 to visualize relativistic relations between time and space. In Fig. 1. we see our modification of this diagram. The x and y axes represent two dimensions of our ordinary world, while the z axis represents time (t), multiplied by the speed of light (c), simply to make the scales of time and space of the same magnitude. Thus, in the x–ct coordinate system, the velocity of light is represented by a straight line at 45° to the ct axis. As all other possible velocities are lower than that of light, they are represented by straight lines inclined at an angle of less than 45° to the ct axis.

An ultrashort light pulse is emitted at A and slightly later an ultrafast detection is made at B. The separation in time and space between A and B could be either (1) *static* because A and B are fixed in space as in most holography (in which case we would refer to the one who makes the experiment the "rester" (2) *dynamic*, caused by an ultrahigh velocity (v) of the person

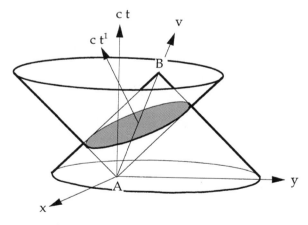

Figure 1. An ultrashort light pulse is emitted at A, which is the apex of a Minkowski lightcone. In our coordinate system x–y represent two axes of our ordinary world, while the third axis ct represents time. The widening of the cone upward represents the radius of the sphere of light as it increases with time. An ultrashort observation is later made at B, the apex of an inverted cone. The only way for light to be transmitted from A to B is by scattering objects placed where the two cones intersect. If the observer's velocity (v) is high, this intersection will be an ellipse that is inclined in relation to our stationary world.

(called the "traveler") who performs the experiment. In the original Minkowski diagram the distance A–B was zero, but to adapt the diagram to holography, we had to introduce a separation in both time and space. If there is a separation in time only, the world will be seen intersected by spheres, but when the separation in space is introduced, these spheres are transformed into ellipsoids.

We will first concentrate on the illumination cone. A spherical wavefront is emitted in all directions from a point source (A) and expands with the speed of light. In our chosen coordinate system, which is limited to only two space coordinates and one time coordinate, this phenomenon is represented by a cone with its apex at A, expanding in the direction of the positive time axis. The passing of time is represented by cross sections of the cone by planes parallel to the x–y plane at increasing ct values (Fig. 2). These intersections will, when projected down to the x–y plane, produce circles of increasing radius that in our 3D world represent the expanding spherical wavefront from the point source at A. As time increases, the circles around A expand, while those around B contract. The ellipses caused by the intersection of those circles stay fixed and unchanged as explained later, in Section IV.

Let us now more closely study the observation cone that intersects the illumination cone. If a point of illumination represents a point source of light, a point of observation represents a point sink of light, a point toward which spherical waves are shrinking. In the Minkowski diagram, it is represented by a cone that is inverted in relation to the lightcone, referred to as the observation cone (B), which like the light cone has a cone angle of 90° and the observer at the apex (B). Thus, an observer at B can see nothing outside this cone because of the limited speed of light (c). The only general way for light to pass from the illumination cone to the observation cone is by deflection, such as by scattering from matter that exists at the intersection of the two cones. The only exception is when the two cones just touch each other, which is the only case when light might pass directly from A to B.

The intersection of the two cones produce an ellipse that in three dimensions represents an ellipsoid of observation. In other words, it represents the traveler's surface of simultaneity that, to this traveler, who is situated at the apex (B) of the cone of course appears spherical. The traveler's time axis (ct^1) is at a different direction than the time axis of the rester's x–y plane (ct). Thus observations that appear to be simultaneous to the rester are not simultaneous to the traveler and vice versa, because time varies in a linear fashion along the line of travel.

To simplify the diagram of Figs. 1 and 2, we have drawn only two space dimensions and time. If all three space dimensions had been included the intersection of the two cones would represent the ellipsoid of observation, where the apices (A and B) of the two cones are the focal points. The situation will be the same regardless of whether the separation between A and B is static

MINKOWSKI LIGHT CONES

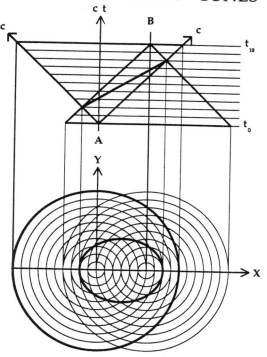

A = apex of light cone of illumination
B = apex of light cone of observation
Bold circle = true wave front at time t_{10}
Bold ellipse = apparent wave front at time t_{10}

Figure 2. The relation between the Minkowski diagram and the holodiagram designed for the creation and evaluation of holograms. The horizontal intersections of the two cones represent different points of time, and the two sets of circles formed by those intersections are identical to the two sets of circles originally used to produce the ellipses of the holodiagram as later described in Fig. 5.

or caused by an ultrahigh velocity. Thus, our concept of ellipsoids of observation applies just as well to the evaluation of apparent distortions at relativistic velocities as to radar, gated viewing, holographic interferometry, and holography with picosecond pulses [4]. As soon as there exists a separation between A and B, the spheres of observation are transformed into ellipsoids of observation.

In ordinary optics there usually exists perfect symmetry in relation to time. The light rays are the same regardless of whether times goes forward or backward. In our diagrams of Figs. 1 and 2, however, the situation is different. A change of A is observed after a delay, while a change of B is observed instantly, as understood by the following example. If a star (A) that is one lightyear away suddenly starts moving toward us, it will take one year before we see a Doppler blueshift, but if we (B) suddenly start moving toward the star, the blueshift is seen without delay. Thus, if A is moved, there is a delay until the cone of illumination and the ellipsoids of observation are changed, but a motion of B causes instantaneous change. On this the basis of this result, we will take a look at the "twin paradox" and explain it as a result of this asymmetry (Appendix A)

III. HOLOGRAPHIC USES OF THE HOLODIAGRAM

Both holography and the special theory of relativity are based on interferometry, which in turn depends on path lengths of light. Let such a pathlength be represented by a string. Fix the ends of the string with two nails on a blackboard and while keeping the string stretched draw a curve with a chalk. The result will be an ellipse with its two focal points at the nails (Fig. 3). Thus, we understand that if a light source is at one focal point (A) the pathlength for light to the other focal point (B) via any point on the ellipse will be constant. If we could draw a curve with chalk in three dimensions, the result would be a rotational symmetric ellipsoid still with the focal points A and B. Now shorten the string by the coherence length of a HeNe laser and produce another ellipse and finally, instead lengthen the string by another coherence length. If we place a HeNe laser at A, a hologram plate at B, and a reference mirror at the middle ellipse, we can record any object within the two outermost ellipses because the difference in pathlength for the light from the object and from the mirror will be within the coherence length. If the coherence length, or the pulse length, of the laser is very short, we will, on reconstruction of the hologram, see the object intersected by a thin ellipsoidal shell. This method, termed "light-in-flight recording by holography," can be used to observe the three-dimensional shape of either a wavefront (or pulsefront) or a real object [4].

If we instead produce a set of ellipses by frequently lengthening the string by the wavelength of light (λ), one interference fringe would form for every ellipsoid intersected by each point on an object as it moves between two exposures in holographic interferometry. On the basis of this idea, we produce the holodiagram [5]. The separation of the ellipses at a point C compared to their separation at the x axis is termed the k value, which depends on the angle ACB. As the peripherical angle on a circle is constant, the k value will be constant along arcs of circles passing through A and B as seen in Fig. 3. The moiré effect of two sets of ellipsoids visualize interference patterns in holographic interferometry [6].

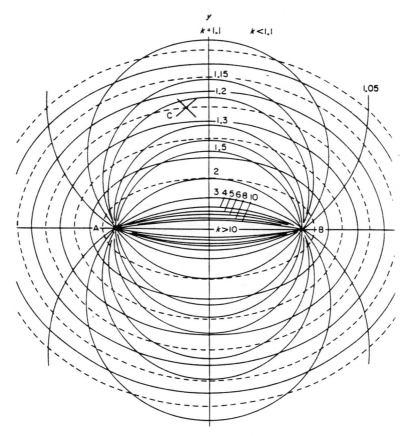

Figure 3. Let us assume that a string is fixed with one end at A and the other at B. Keeping the string stretched, a set of ellipses are drawn, and for each adjacent ellipse the string length is increased by a certain constant value (ΔL). The separation of the ellipses varies with a factor k that is constant along arcs of circles. In this holodiagram let A be the spatial filter in a holographic setup and B be the center of the hologram plate. If ΔL is the coherence length of light, the diagram can be used to optimize the use of a limited coherence in recording large objects. If ΔL instead represents the wavelength of light, the ellipses can be used to evaluate interference fringes in holographic interferometry or conventional interferometry with oblique illumination and observation. Finally, if ΔL represents a very short coherence length, or pulselength, the ellipses visualize the spherical wavefront from A as seen from B, deformed by the limited speed of the light used for the observation.

To make the diagram easier to study, we have painted every second area between the ellipses black as seen in Fig. 4. From this we can see how the thickness of these areas, which represent the k value, varies over the diagram. The k value is constant along arcs of circles that pass through A and B.

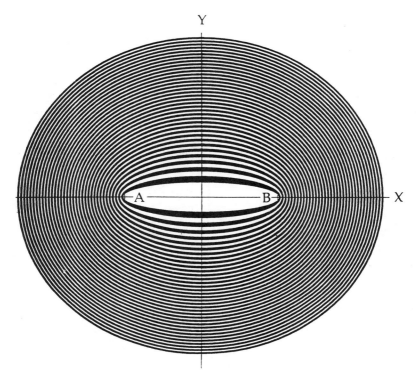

Figure 4. To make the diagram of Fig. 3 easier to study, we have painted every second area between the ellipses black. We see that the thickness of the areas varies with the k value throughout the diagram.

Therefore, in Fig. 3 we have printed the k value where these circles cross the Y axis. When the position in the holodiagram is known for a selected object point, then the sensitivity is known in both amplitude and direction. Thus, the holodiagram can be used to simplify the planning of the holographic setup and the evaluation of the displacement from the number of fringes.

IV. APPLICATION TO INTERFEROMETRY

If an object point is moved from one ellipsoid half way to an adjacent one, then the phase at B on the hologram plate will change by 180°. This means that the interference pattern at B moves half a fringe separation. If we make a double exposure with such an object motion in between the two exposures, then the fringes on the hologram plate will be displaced so that where there was darkness during the first exposure there will be brightness during the second and vice versa. For this reason the fringes on the hologram plate will be wiped out at B, and because there is no diffraction, the corresponding object point will appear dark during reconstruction. Consequently, if, for example, a long object is fixed to one end and the other end is moved so that it crosses five ellipsoids, the object will be covered by five fringes in the reconstructed holographic image.

A movement parallel to one ellipsoid will not change the string length and therefore causes no pathlength difference and consequently no fringes are formed on the reconstructed object image, while a movement perpendicular to the ellipsoids causes the most fringes. Later on we will show that the direction perpendicular to the ellipsoids will be on the surface of a set of hyperboloids. Thus, the hyperbolas represent the sensitivity direction, while the ellipses represent the direction of zero sensitivity, and the closiness of the ellipses is a measure of the sensitivity in that direction.

The separation between the ellipses varies over the diagram as seen in Fig. 3. Thus, along the x axis to the left of A and to the right of B the sensitivity is half the wavelength, and everywhere else it is lower. In between A and B it is almost zero, but as we move outward along the y axis, it increases until, at an infinite distance, it again becomes half the wavelength. We designate the separation as k times the wavelength. Thus, this k value is a *desensitizing factor* that represents a measure of how large a movement is necessary to produce one interference fringe.

Referring to Fig. 3 again, the displacement *in the sensitivity direction* can be calculated from the number of fringes on the reconstructed object in the following way: *Displacement is k times the number of fringes multiplied by half the wavelength.* A greater k value caused by more grazing incidence of the light rays works just *as if there was a longer wavelength or a redshift of the light.*

Let us study some more examples of this statement. When we look at a flat object that scatters light, such as a page in a book, it appears more and more mirror-like the more we tilt it, so that we look almost parallel to the surface. The reason is that, to see a surface, that surface must have a microscopic structure whose hills and valleys must be of a size comparable at least to the wavelength of light. If the structure is finer, we do not see the surface itself; instead, we see a mirror reflection, or, as it is also termed, a *specular* reflection. As we tilt the surface, the k value increases which produces the same result as if there had been an increase in the wavelength of the light, which is the same as a redshift.

Using the holodiagram, we have managed to lower the sensitivity of holographic interferometry so much that an object movement of 2 mm caused only two fringes. We also made an interferometer the "interferoscope" in which the sensitivity could be changed from 1 to 5 μm (micrometers) per fringe just by changing the k value [7]. In this case the k value was about 16; had it been unity, the sensitivity would have been half the wavelength or about 0.3 μm.

Another approach to this holodiagram is to draw one set of equidistant concentric circles centered at A and another set at B as seen in Fig. 5. A number

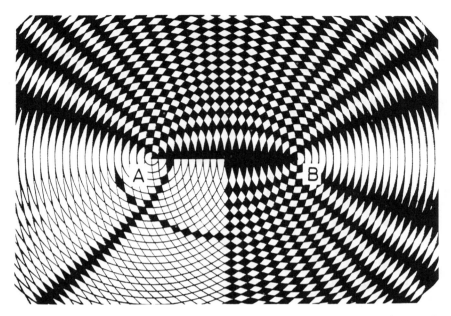

Figure 5. An alternative way to constract the holodiagram is to draw two sets of concentric circles where for each adjacent circle the radius is increased by a certain constant value. One set of ellipsoids and one set of hyperboloids are formed. If A and B are two sources of coherent light, or two points of observation, the ellipsoids move outward with a velocity *greater than the speed of light*, while the hyperboloids will be stationary. These hyperboloids represent the diffraction-limited resolution of a lens and are used for in-plane measurements in holography, moiré and speckle techniques. If A is a source of coherent light and B is a coherent point of observation, the hyperboloids move with a speed *greater than light* to the right while the ellipsoids will be stationary. These ellipsoids represent the interferometric limited resolution and are used for out-of-plane and 3D measurements.

of rhombs, or diamonds, are formed where the circles intersect. By painting every second rhomb black one set of ellipses and one set of hyperbolas are produced. The two diagonals of a rhomb are perpendicular, and thus the ellipses and the hyperbolas intersect at right angles. One diagonal of the rhombs represents the separation of the ellipses and the other, the separation of the hyperbolas. The radius from A and B are termed R_A (resp. R_B). The set of ellipses is represented by $R_A + R_B = $ constant and the set of hyperbolas, by $R_A - R_B = $ constant. If the separation of the concentric circles is 0.5λ, the separation of the ellipses (one diagonal of the rhomb) at a point C is

$$D_{\text{ell}} = \frac{0.5\lambda}{\cos \alpha} = k \cdot 0.5\lambda \tag{1}$$

while the separation of the hyperbolas (the other diagonal) is

$$D_{\text{hyp}} = \frac{0.5\lambda}{\sin \alpha} \tag{2}$$

where α is half the angle ACB of Fig. 3. The factor k of Eq. (1) is identical to the k of Fig. 3, and its value is

$$k = \frac{1}{\cos \alpha} \tag{3}$$

Interferometric measured displacement (d) is calculated $d = nk\,0.5\lambda$, where n is the number of interference fringes between the displaced point and a fixed point on the object. In holographic, or speckle, interferometry, A is a light source from which spherical waves radiate outward while B is a point of observation, or a light sink, toward which spherical waves move inward (see Fig. 6). In that case the hyperbolas will move with a *velocity greater than light* while the ellipses are

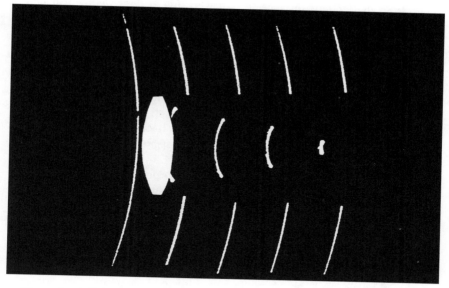

Figure 6. A light-in-flight recording of light focused by a lens. One single picosecond spherical pulse from a modelocked laser at A illuminated a white screen at an oblique angle. The screen was placed so that its normal passed trough the hologram plate at B. Part of the pulse was, after an appropriate delay, used as a reference beam at B. A cylindrical lens was fixed to the screen, and by multiple ephotographic exposures of the reconstructed image, the focusing effect of the lens was recorded.

stationary and Eq. (1) is used to find the displacement normal to the ellipsoids (out-of-plane displacement). If both A and B are light sources, the ellipsoids will move with a *velocity greater than light* while the hyperbolas (Young's fringes) are stationary, and the hyperbolas are used to evaluate the displacement normal to the hyperboloids (in-plane displacement). The situation will be similar if A and B both are points of observation while the light source could be anywhere. Finally, if A–B represents the diameter of a lens, the minimum separation of the hypebolas will represent the diffraction limited resolution of that lens [8] (except for a constant of 1.22).

V. LIGHT-IN-FLIGHT RECORDING BY HOLOGRAPHY

A hologram is recorded only if object light and reference light simultaneously illuminate the hologram plate. Thus, if the reference pulse is one picosecond (10^{-12} s = 0.3 mm) long, the reconstructed image of the object will be seen only where it is intersected by this light slice, the thickness of which depends on the length of the pulse. If the reference pulse illuminates the hologram plate at an angle from the left, for instance, it will work like a *light shutter* that with a *velocity greater than light* sweeps across the plate. Thus, what happens first to the object will be recorded farthest to the left on the plate, while what happens later will be farther to the right. If the hologram plate is studied from left to right, the reconstructed image functions like a movie that with picosecond resolution in slow motion shows the motion of the light pulse during, for example, 1 ns (nanosecond) (300 mm). This method, referred to as "light-in-flight (LIF) recording by holography," results in a frameless motion picture of the light as it is scattered by particles or any rough surface. It can be used to study the coherence function of pulses [9,10], and the 3D shape of wavefronts (Fig. 6) or of physical objects [11] (Fig. 7).

Figure 8 shows the holographic setup used to produce a LIF hologram of light reflected by a mirror. As the observer moves her eye, or a TV camera, from left to right behind the hologram plates he will see, as in a frameless motion picture, how the light pulse bounces off the mirror. Thus the method could be said to represent four-dimensional holography that can record the three dimensions of ordinary holography plus time.

If A and B are close together and if there are scattering particles, such as smoke in the air, while observing the reconstructed hologram the experimenter would find herself in the center of a spherical shell of light with a radius $R = 0.5\,ct$, where (t) is time interval between emission of light (at A) and recording (at B). If, however, A and B are separated she would find herself in the focal point B of an ellipsoid where A is the other focal point. The string length (referring to Fig. 3) is $R_A + R_B = ct$. The thickness of the ellipsoidal shell

Figure 7.   Light-in-flight recording of a set of spheres illuminated by a 3-ps laser pulse. The light source (*A*) and the point of observation at the hologram plate (*B*) were close together and far from the object. Thus, the intersecting ellipsoidal light slice can be approximated into a spheroidal light slice with a large radius and a thickness of 0.5 mm.

would be $k \times 0.5c \, \Delta t$, where Δt is pulselength (or coherence length) and k is the usual k value. If the distance $A - B$ were infinite, the observer at B would be in the focal point of a paraboloid. Thus, we have shown that a sphere of observation appears distorted into an ellipsoid and a flat surface of observation into a paraboloid and the sole reason for these distortions is the separation $A - B$.

In Fig. 9 we see at the top the ordinary holodiagram where the eccentricity of the ellipse is caused by the static separation of A and B. However, it is unimportant what the observer is doing during the time between emission and detection of light pulses. Thus, the observer could just as well be running from A to B so that the eccentricity of his observation ellipsoid is caused solely by the distance he has covered until he makes the observation. If his running speed is close to the velocity of light, this new dynamic holodiagram is identical to the ordinary, static holodiagram. This fact has inspired to the new graphical approach to special relativity, which will be explained in Section IX, but let us first study the development of Einstein's special relativity theory.

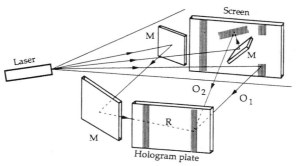

Exposure of Light-in-Flight

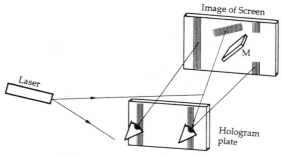

Reconstruction of Light-in-Flight

Figure 8. During exposure of a light-in-flight hologram only those parts are recorded for which the pathlengths of the reference beam and the object beam are equal. If during exposure you look through the left part of the hologram plate, you will see only a bright line on the left part of the screen. If you look through the right part of the hologram plate, you will see a bright line on the right part of the screen. You will also see the light reflected by the mirror on the screen because the pathlength of the reference beam R is equal to not only O_1 but also O_2. One could say that each part on the hologram plate records light of a certain age.

VI. THE MICHELSON–MORLEY EXPERIMENT

One way to understand special relativity is to see how time dilation and Lorentz' contraction of objects parallel to motion can be used to explain the null results of the Michelson–Morley [1] experiment, which was performed to measure the velocity of earth in relation to an assumed ether. The result was that the expected influence of such an ether on the velocity of light was not found. Let us now study this double-pass example, where one arm of a Michelson interferometer was perpendicular to the velocity of the earth's surface, while the other was parallel. When the interferometer is at rest, the lengths of both arms are identical.

Now, let me, the author, and you, the reader, be stationary in the stationary space and study the moving interferometer from there. The velocity of the interferometer from left to right is (v) while the speed of light is (c). The time

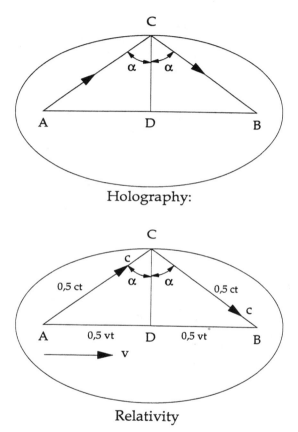

Holography:

Relativity

Figure 9. Two holodiagrams: (a) holography—the ordinary static holodiagram in which A is the light source, B is the point of observation (e.g., the center of the hologram plate), while C is an object for which the k value is $1/\cos \alpha$; (b) the dynamic holodiagram in which an experimenter emits a picosecond pulse at (A) and thereafter runs with a velocity close to the speed of light and makes a picosecond observation at B. The k value is as before:

$$k = \frac{1}{\cos \alpha} = \frac{ct}{\sqrt{c^2 t^2 - v^2 t^2}} = \frac{1}{\sqrt{1 - \dfrac{v^2}{c^2}}}$$

for light to travel along the perpendicular arm when the interferometer is stationary is t_0. When it is moving at velocity v, the pathlength becomes longer and therefore the travel time t_v becomes longer. However, this increase in traveling time of the light must not be observable by the traveling observer because that would be against the postulate (1) of special relativity. But why is this increase in traveling time not observed by the observer who is traveling with the interferometer? The accepted solution is that this increase in traveling time is rendered invisible because all clocks moving with the interferometer are delayed by a certain value so that each second becomes longer. A great number of experiments have supported this statement, and the slowing of time is termed *time dilation*. If the time dilation is $\frac{4}{3}$, a traveling clock will show 3 s when a stationary clock shows 4 s.

The delay along the parallel arm of the interferometer has to be exactly the same as that of the perpendicular arm; otherwise the difference in arriving time could be observed interferometrically and interpreted as a change in the speed of light, which would be against Einstein's postulates (1) and (2) . When the light is moving in the same direction as the interferometer, the travel time will be longer than when it is moving in the opposite direction. The total time will be longer than that of the perpendicular arm. Thus, judging from the time of flight, the parallel arm appears elongated in comparison to the perpendicular arm. In order to fulfill postulates (1) and (2), it was decided by Lorentz that this, elongation of the parallel arm that was apparent to the traveling observer was compensated for by introducing a corresponding assumed contraction, namely, the Lorentz contraction.

Thus we have found that time dilation has been proved by many experiments and, as it produces a permanent delay, there is no reason to doubt its existence. It is independent of the sign of the velocity of the interferometer and it produces the correct result for a single-pass measurement as well as for the demonstrated double pass. The slowing of time results in a longer wavelength of the light from the source traveling with the interferometer, consequently the number of waves in the perpendicular arm and thus the phase after a single or a double pass through that arm is independent of the velocity (v) of the interferometer. First we will just accept the time dilation as derived from the effect of the arm perpendicular to velocity and then solely the arm parallel to velocity.

The Lorentz contraction of the parallel arm is more complicated and cannot be measured directly as it is not permanent but disappears when the velocity (v) disappears. Thus, I find the discussion of assumed contractions or elongations of a moving object meaningless as they are, by definition, invisible. I look at them as only theoretical tools, and in the following section we will solely study how the stationary world appears deformed when studied and measured by an observer traveling with a velocity close to that of light.

VII. INTRODUCING SPHERES OF OBSERVATION TRANSFORMED INTO ELLIPSOIDS OF OBSERVATION

Let us compare static and dynamic separation of illumination and observation. A person is performing experiments based on gated viewing, which means that a short pulse of light (picosecond pulse) is emitted and, after a short time (e.g., 20 ns), she makes a high-speed recording with a picosecond exposure time. If the illumination point source (A) and the observation point (B) are close together, the experimenter will find herself surrounded by a luminous spherical shell with a radius of 3 m. This spherical shell can be seen only if something scatters the light, for instance, if the experiment is performed in a smoke-filled or dusty space. If there are large objects in the space, she will see these objects illuminated only in those places where they are intersected by the sphere. The experiment described can be used to map the space around the experimenter. This is identical to well-known radar methods. By changing the delay between emission and recording, intersections of differently sized spheres can be studied. In this way, the outside world is mapped in polar coordinates.

If the illumination point (A) and the observation point (B) are separated, the situation will be different. As the luminous sphere around A grows, the observer will see *nothing* until the true sphere reaches B. Then, she will find herself *inside an ellipsoidal luminous shell*. One focal point of the ellipsoid will be A; the other, B. By changing the delay between emission and recording, intersections of ellipsoids with different sizes, but identical focal points, can be studied. In this way the space around the experimenter can be mapped in bipolar coordinates. The experimenter should know the separation of A and B so that her mapping will be correct. If she erroneously believes the separation to be zero, she will misjudge the ellipsoids as spheres and make errors, especially in the measurement of lengths parallel to the line AB. She will also make angular errors because of the angular differences between points on the spherical and the ellipsoidal shells.

We shall take a closer look at the possibilities of applying the concept of the ellipsoids to visualize special relativity more generally. Our goal is to find a simple graphical way to predict the apparent distortions of objects that move at velocities close to that of light and to restore the true shape of an object from its relativistically distorted ultra-high-speed recording.

We have already described that, if the illumination point (A) is separated from the observation point (B), the gated viewing system produces recordings of intersections of ellipsoids having A and B as focal points. Now, let me, the author, and you, the reader, be stationary in a stationary space and study what a traveling experimenter (the traveler) will see of our stationary world when he travels past at relativistic velocity using picosecond illumination and observation.

Instead of an assumed contraction of fast moving objects, I have introduced the idea that the travelers' spheres of observation by the velocity are transformed into ellipsoids of observation. One advantage is that this new concept is easier to visualize and that it makes possible a simple graphic derivation of distortions of time and space caused by relativistic velocities. Another advantage is that it is mentally easier to accept a deformation of spheres of observation than a real deformation of rigid bodies depending on the velocity of the observer.

Our calculations refer to how a stationary observer (the rester) judges how a traveling observer (the traveler) judges the stationary world. We have restricted ourselves to this situation exclusively because it is convenient to visualize ourselves, you the reader and I the author, as stationary. When we are stationary, we find it to be a simple task to measure the true shape of a stationary object. We use optical instruments, measuring rods, or any other conventional measuring principle. We believe that we make no fundamental mistakes and thus accept our measurements as representing the true shape. No doubt the traveler has a much more difficult task. Thus we do not trust the traveler's results but refer to them as apparent shapes.

VIII.   THE PARADOX OF LIGHT SPHERES

In Fig. 10a we see a stationary car in the form of a cube. An experimenter emits an ultrashort light pulse from the center of the cube (A) and some nanoseconds later makes an ultrashort observation (B) from that same place. The true sphere of light emitted from (A) will reach all sides of the cube at the same time. Thus, while performing the observation from (*B*), the experimenter will simultaneously on all sides see bright points growing into circular rings of light. However, if the car is moving at velocity *v* close to *c*, the true sphere, as seen by a stationary observer, will not move with the car but remain stationary. The result will be that the sphere reaches point *E* on the side of the car earlier than, for instance, point *D* as seen in Fig. 10b. However, referring to Einstein's postulates 1 and 2, this fact must in some way be hidden to the traveling observer.

Figure 11 illustrates our explanation, which is that the traveler's sphere of observation is transformed into an ellipsoid of observation, as its focal points are the point of illumination (*A*) respective of the point of observation (*B*) separated by the velocity (*v*) as already described in Figs. 1, 2, and 9. The minor diameter of the ellipsoid is, however, unchanged and identical to the diameter of the sphere. The ellipsoid reaches *E* earlier than, for example, *D*, and therefore the different sides of the cube are observed at different points of time. However, to the traveler all the sides appear to be illuminated simultaneously, as by a sphere of light that touches all the sides at the same time. The reason why the ellipsoid to the traveler appears spherical is that *A* and *B* are focal points and thus $ACB = ADB = AEB = AFB$.

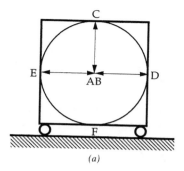

(a)

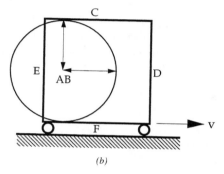

(b)

Figure 10. (a) Point A is the point source of light and B, the point of observation. To the traveling observer the sphere of light appears centered to the cubic car, independent of its velocity. Thus, she observes that all the sides are touched by this sphere simultaneously. (b) To the stationary observer the sphere of light appears fixed to his stationary world. Thus, the sphere will reach E earlier than D. This fact must be hidden from the traveling observer; otherwise he could measure the constant velocity of the car, which would be against Einsteins' special relativity theory.

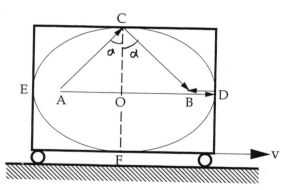

Figure 11. Our new explanation is that to the traveling observer, her sphere of observation is by the velocity transformed into an ellipsoid of observation. However, to the traveler this ellipsoid appears spherical because A and B are focal points of the ellipsoid and thus $ACB = ADB = AEB$. Thus, to her the car appears cubic because she observes that all sides of the car are touched by this sphere simultaneously, just as when the car was stationary. To the stationary observer this simultaneity is, however, not true.

The cubic car is shown elongated because time varies linearly along the car. Point E appears farther to the left because it was illuminated and thus recorded early when the whole moving car was farther to the left, while point D was illuminated and recorded later when the car had reached farther to the right. Thus the observations of the walls of the car appears simultaneous to the traveler but not to the stationary observer (compare to the inclined intersection of Figs. 1 and 2). As both the car and the ellipsoid of observation are elongated in the direction of travel, there is no need for an assumed contraction to make the velocity invisible to the traveler.

As time passes, the ellipsoid will grew and produce an ellipse on the ceiling of the car. Even this phenomenon will be invisible to the traveler because it has been mathematically proven that the intersection of an ellipsoid by a plane appears circular when observed from any focal point of the ellipsoid [12]. Further on, if the car had not been cubic but instead had consisted of a reflecting sphere with the light source in its center, the distortion into an ellipsoid would still be invisible. The traveler would find no difference in the reflected rays.

IX. TIME DILATION AND APPARENT LORENTZ CONTRACTION

Let us now see if this new idea about the observation ellipsoid produces the same results as derived from the Michelson interferometry experiment. We, who are stationary, the resters, understand that the traveler's observation ellipsoid has its focal points at A and B and that light with the speed of c travels ACB of Fig. 11. On the other hand, the traveler's observation sphere is perceived, by the traveler, as being centered at B, with the light simply having traveled with the speed of c in path ACB of Fig. 10a. Thus, the time dilation is the time t_v it takes for light to pass ACB divided by the time it takes to pass OCO (Fig. 11) where $OC = ct_0$, $CB = ct_v$, and $OB = vt_v$. Applying the Pythagorean theorem on the triangle OCB results in the accepted value of the time dilation [13]:

$$\text{Time dilation} = \frac{t_v}{t_0} = \frac{CB}{CO} = \frac{1}{\sqrt{1 - \dfrac{v^2}{c^2}}} = \frac{1}{\cos \alpha} = k \tag{4}$$

where k is the usual k value of the holodiagram. Stationary objects as measured by the observation ellipsoid, or by any measuring rod carried in the car, will *appear* to the traveler contracted by the inverted value of the major diameter ED to the minor diameter CF of his observation ellipsoid:

$$\text{Lorentz contraction} = \frac{CO}{CB} = \sqrt{1 - \frac{v^2}{c^2}} = \cos \alpha = \frac{1}{k} \tag{5}$$

This result is identical to the accepted value of the Lorentz contraction [14], but our graphic derivation shows that this is true only for objects that just pass by (Fig. 17). Objects in front appear elongated by OC/BD, while objects behind appear contracted by OC/BE.

When the traveler emits a laser beam in his direction OC, it will, in relation to the stationary world, have the direction AC, and its wavelength will be changed by the factor AC/OC. When looking in the direction OC, his line of sight will be changed to BC and the wavelength of light from the stationary world will be changed by the factor BC/OC (the same change will happen to his laser beam). This factor is well known as the "relativistic transversal Doppler shift." Finally, using our approach, we find that v can never exceed c because $v = c \sin \alpha$ (Fig. 9).

X. GRAPHICAL CALCULATIONS

Let us now examine the emitted light rays in more detail and assume, in another example, that the traveler who is moving from left to right directs a laser in the direction BH of Fig. 12. The direction of the beam in relation to the stationary world will then be AG. Point G is found by drawing a line parallel to the line of travel (the x axis) from the point (H) on the sphere of observation to the corresponding point (G) on the ellipsoid of observation. As the point of observation is identical in space and time in the two systems, the center of the sphere should coincide with the focal point of the ellipsoid of observation [15].

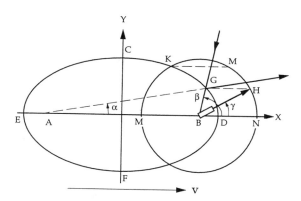

Figure 12. To the traveler an arbitrary point (G) of the stationary world appears to exist at H, which is found by drawing a line of constant Y value from G to the sphere. As the traveler directs the telescope in her direction BH, her line of sight in the rester's universe will be GB. As the traveler directs the laser in his direction BH, the direction of the laser beam will appear to be AG to the rester. The Lorentz contraction is MN/ED.

The angle of outgoing (emitted) light (α) and incoming (lines of sight) light (β) are calculated from Fig. 12:

$$\tan \alpha = \frac{\sin \gamma \sqrt{1 - \dfrac{v^2}{c^2}}}{\dfrac{v}{c} + \cos \lambda} \tag{6}$$

$$\tan \beta = \frac{\sin \gamma \sqrt{1 - \dfrac{v^2}{c^2}}}{-\dfrac{v}{c} + \cos \lambda} \tag{7}$$

These two equations, derived solely from Fig. 12, are identical to accepted relativistic equations (see, e.g., Ref. 16). From Fig. 13a it is easy to see that the

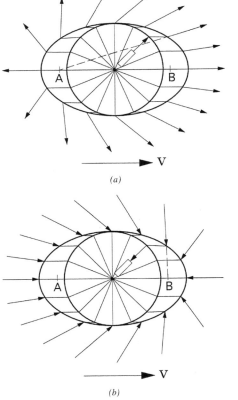

(a)

(b)

Figure 13. (a) The graphical method for finding the direction of light rays in Fig. 12 is independent of where along the x axis in which the circle is situated. Thus, for simplification it can just as well be placed in the center of the ellipse. Thus, we see how the outgoing laser beams are concentrated forward as by a positive lens. (b) Again using the method of Fig. 12, we find that the lines of sight are diverged backward so that the view forward appears demagnified as by a negative lens.

emitted rays are concentrated forward as if there was a focusing effect by a positive lens. This phenomenon, which was pointed out by Einstein, results in the light energy from a moving source appearing to be concentrated forward for two reasons. The light frequency is increased by the Doppler effect and the light rays are aberrated forward. This explains why the electron synchrotron radiation appears sharply peaked in intensity in the forward tangential direction of motion of the electrons. In the backward direction, we have the opposite effect. The light is defocused as if by a negative lens.

In Fig. 13b the line of sight is seen aberrated backward along a line through G from focal point B of Fig. 12 . The traveler is still moving to the right, and her direction of observation, the telescope axis, is in the direction BH. Thus, the stationary world around the traveler appears concentrated in the forward direction as if demagnified by a negative lens. In the backward direction we have the opposite effect. The stationary world appears magnified by a positive lens.

Let us again study Fig. 12 and calculate the Doppler ratio, which is the Doppler shifted wavelength divided by the original wavelength. The traveler observes the wavelength (λ_v) while measuring the true wavelength (λ_0) from the stationary world. Using some trigonometry, we get the following expression for the Doppler ratio

$$\frac{\lambda_v}{\lambda_0} = \frac{BG}{BH} = \frac{1 - \frac{v}{c}\cos\gamma}{\sqrt{1 - \frac{v^2}{c^2}}} = k\left(1 - \frac{v}{c}\cos\gamma\right) \tag{8}$$

where (λ_v) is the wavelength as seen by the observer who travels in relation to the light source and (λ_0) represents the wavelength as seen by the observer who is at rest in relation to the light source. This equation, derived solely from Fig. 12, is identical to accepted relativistic equations (see, e.g., Ref. 17).

Now let us study Fig. 12 again and seek the direction in which the traveler should experience zero Doppler shift. She should not look backward because in that direction there is a redshift. Nor should she look directly sideways because even then there is a redshift, the relativistic transverse redshift. As forward is the blueshift, she should look slightly forward. The way to find the zero Doppler shift for incoming light follows from Fig. 12 where K is the point of intersection for the traveler's sphere of observation as seen by the traveler, and the ellipsoid of observation as seen by the rester. Draw a line parallel to the x axis from K to the corresponding point on the sphere (M). Thus BM is the direction the traveler should look to see zero Doppler shift. The line BM is directed slightly forwards, and from that fact we understand that redshift has to be predominant in universe even if there were no expansion from a big bang but only a random velocity increasing with distance.

Now let us study Fig. 12 again and seek the direction (*l*) in which the traveler should experience zero Doppler shift. She should not look backward because in that direction there is a redshift. Nor should she look directly sideways because even then there is a redshift, the relativistic transverse redshift. As forward is the blueshift, she should look slightly forward. The way to find the zero Doppler shift for incoming light (*a*) is as follows:

Draw a line parallel to the *x* axis from *G* to the corresponding point on the sphere (*K*), where *G* is the point of intersection for the traveler's sphere of observation as seen by the traveler, and the ellipsoid of observation as seen by the rester. Thus *BK* at the angle γ is the direction the traveler should look to see zero Doppler shift. *BG* at the angle β represents that direction after the line of sight has been relativistically aberrated. The line *AL*, at the angle γ, represents the light rays that are emitted by the traveler and that the rester experiences as having zero Doppler shift.

All equations solely derived from Figs. 9 and 12 are identical to accepted relativistic equations (see, e.g., the references cited in Ref. 12).

XI. PREDOMINANT REDSHIFT EVEN WITHOUT THE BIG BANG

In Fig. 14, we see the distribution in space of redshift and blueshift as an object is traveling at different velocities. An object that emits light and moves to the right at a velocity (*v*) that is low compared to that of light (*c*) produces blueshifted light forward and redshifted light backward. The situation is the same for an observer traveling at the velocity (*v*); objects in front of him appear blueshifted while those behind him appear redshifted. These statements are based on the well-known Doppler effect. However, the closer the velocity is to the speed of light, the more of the light is redshifted and the less is blueshifted. This result can be derived either by our graphical method using the ellipses of the holodiagram (Fig. 12), or based on Einstein's statement that the time of clocks (or atoms) that travel fast in relation to the observer is slowed down [18,19].

Therefore, the fact that more distant stars are more redshifted (the Hubble effect) does not prove that they are moving away from us, only that they are moving at higher velocities than those stars that are closer by. Such a situation appear quite natural. Stars might move in a *random* way, but move faster the farther away they are. Perhaps because they are just like water molecules in a turbulent river, or perhaps even simpler, because they rotate in relation to a larger universe. In either case there would be a redshift that increases with distance. Thus, the expansion of the universe and the big bang are not directly proved by a redshift of distant stars.

These statements are still clearer with the help of the diagram in Fig. 15. The velocity away from the observer is named radial (v_R.), while the one perpendicular to that direction is v_T. Both velocities are expressed in fractions of the speed of light (c). Because c is the maximal velocity the diagram is limited by the circle:

$$(v_R)^2 + (v_T)^2 = c^2 \qquad (9)$$

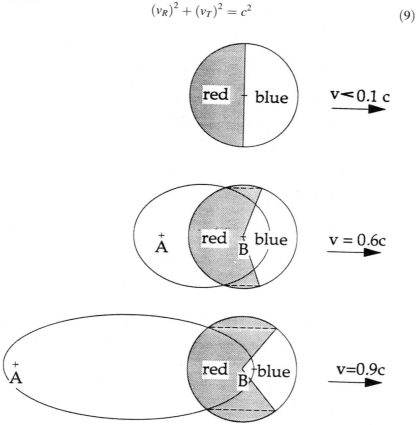

Figure 14. An object that emits light and moves to the right at a velocity v that is low compared to that of light c produces blueshifted light forward and redshifted light backward. The situation is the same for an observer traveling at the velocity v; objects in front of him appear blueshifted; those behind, redshifted. These conclusions are based on the Doppler effect. However, the closer the velocity is to the speed of light, the more of the light is redshifted and the less is blueshifted. This result can be derived either by our graphical method using the ellipses of the holodiagram as seen here or by equations based on Einstein's statement that the time of clocks (or atoms) that travel fast in relation to an observer are slowed down. Therefore the fact that more distant stars are more redshifted (the Hubble effect) does not prove that they are moving away from us, only that they are moving at higher velocities than those nearer. Thus the expansion of the universe and the big bang are not proved solely by a redshift of distant stars.

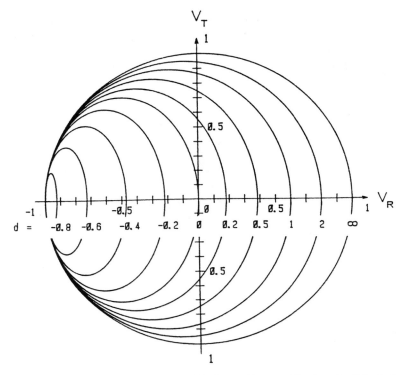

Figure 15. Visualization of the Doppler shift d caused by a combination of radial velocity V_R and transverse velocity V_T. From this diagram, we see that large redshifts (d is positive) can be caused even when there is a large velocity toward the observer (V_R is negative) if combined with a sufficiently large transverse velocity.

The Doppler shift (d) is calculated from the following equation:

$$d = \frac{(\gamma_v - \lambda_0)}{\lambda_0} = \frac{1 + \left(\dfrac{v_R}{c}\right)}{\sqrt{1 - \left(\dfrac{v_R^2 + v_T^2}{c^2}\right)}} \tag{10}$$

Thus the Doppler shift is the difference between the Doppler-shifted wavelength (λ) and the original wavelength (λ_0) divided by λ_0. The numerator is the classical Doppler redshift from a moving light source, while the denominator represents the red-shift caused by the relativistic time dilation resulting from the total velocity, which is independent of the direction of motion.

When the transverse velocity is zero, Fig. 15 shows that the Doppler shift (*d*) increases (redshift) with increasing radial velocity to the right, away from the observer. The Doppler shift is negative (blueshift) when the radial velocity is negative, to the left, toward the observer. When the radial velocity is zero, even the Doppler shift is zero.

However, when the transverse velocity differs from zero, everything becomes more complicated. Let us, for instance, study a Doppler shift of 0.5, which could be caused by a radial velocity of 0.38*c*. This could just as well have been caused by zero radial velocity combined with a transverse velocity of some 0.75*c*. It could even be caused by a radial velocity of −0.62*c* combined with *the same* transverse velocity of 0.75*c*. In the latter case, a motion *toward the observer*, which should result in a blueshift, results in a *redshift* when combined with a sufficiently large transverse velocity.

From the preceding statements, we understand that the true velocity of an object cannot be deduced from the redshift alone. For a given redshift there exists an upper limit to the radial velocity. It would be a great mistake to take for granted that this upper limit represents the actual velocity of the object. This cannot be determined without knowledge of the transverse velocity, and even then there might exist *two* possible radial velocities for *one* redshift, as demonstrated in previous example.

Thus, we have demonstrated, in the form of a diagram, the crosstalk between transverse velocity and a radial velocity measured by red shift. As we shall se in Section XII, the crosstalk between the radial velocity and the apparent transverse velocity is demonstrated by the use of another but similar diagram. By adding those two diagrams together we will finally find a method of evaluating the true velocity of the light source.

XII. TRANSFORMATION OF AN ORTHOGONAL COORDINATE SYSTEM

We shall now demonstrate how the diagram can be used for practical evaluation of the true shape of rigid bodies, whose images are relativistically distorted. The "true" shape is defined as the shape seen by an observer at rest in relation to the studied object. Again, let a traveling experimenter at high velocity pass through the stationary space (Fig. 16). She emits six picosecond light pulses with a constant time separation of *t*. After another time delay of *t*, she makes one single picosecond observation at *B*. Figure 16 shows how her spheres are transformed into ellipsoids. Let us look at one stationary straight line that is perpendicular to the direction of travel and see how it appears to the traveler. From every point at which the stationary line (*S–S*) of Fig. 16a is intersected by an ellipsoid a horizontal line is drawn to Fig. 16b until it intersects the corresponding sphere.

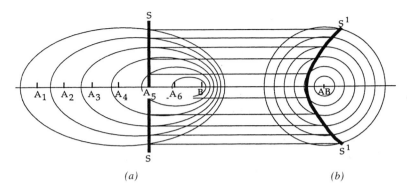

Figure 16. The traveler of Fig. 12 emits ultrashort light pulses at A_1, A_2, A_3, A_4, A_5, and A_6. Finally he makes one ultrashort observation at B. The vertical straight line S–S in the stationary world appears to the traveler to be distorted into the hyperbolic line S^1–S^1.

The curve connecting these intersections in Fig. 16b then represents the straight line of Fig. 16a as it appears distorted to the traveler.

In Fig. 17a a total stationary orthogonal coordinate system is shown, and in Fig. 17b we see the corresponding distorted image as observed by the traveler, who is represented by the small circle (i, o) passing from left to right. The

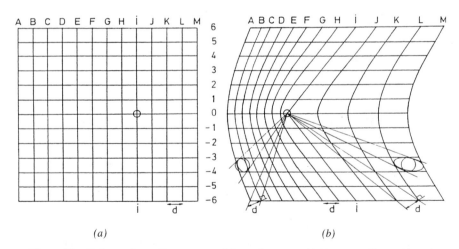

Figure 17. Orthogonal coordinate system of the stationary world (a) appears distorted into that of (b) to the traveler who exists at the small circle (i, o) and is moving to the right at a speed of $0.6c$. From the diagram we find that flat surfaces parallel to motion are not changed but those perpendicular to motion are transformed into hyperboloids, while the plane $(i$–$i)$ through the observer is transformed into a cone. The back side can be seen on all objects that have passed this cone. The separation of advancing hyperboloids is increased, while that of those moving away is decreased.

identical transformation would occur if the observer at *B* were stationary and instead the orthogonal coordinate system passed her with the constant speed of 0.6*c* from right to left.

All apparent displacements occur because different points on the object are studied at different points of time. During this time difference, the object has moved in relation to the observer, but only along its line of motion. Thus, flat *surfaces parallel to the direction of velocity are not changed* in respect to flatness, angle, or separation. From Fig. 17b, we find the following:

1. *Flat surfaces moving toward the observer* are transformed into hyperboloids that appear convex to the observer.

2. *Flat surfaces moving away from the observer* are transformed into similar hyperboloids that appear concave to the observer.

3. *The flat surface passing through the observer (i–i)* is transformed into a cone.

4. *The observer can see the back side* of all objects that have passed through the surface of the cone.

5. *The spacing of the surfaces moving toward the observer* is increased.

6. *The spacing of the surfaces moving away from the observer* is decreased.

7. Let the original spacing (*d*) rotate so that it is kept normal to the hyperboloid surface. It will then always occupy the same angle of view as the spacing of the hyperboloids. This confirms Terrell's statement, that, however, is true only for very small angles of view [20].

We have compared results from Fig. 17 with those of several other workers and found good agreement. Bhandari states that a vertical line moving at high speed assumes the shape of a hyperbola [21]. Mathews and Lakshmanan criticize the concept of relativistic rotation and introduce "the train paradox" [22]. When a fast-moving train is studied, should one imagine each boxcar to be rotated or the train as a whole rotated? What happens to the stationary rails? Finally, they conclude that the rotated appearance is not self consistent. We agree with this statement. The train is easily visualized in our Fig. 17b as one of the horizontal rows of deformed squares. From this, it is obvious that the distortion of the total train cannot be explained solely by rotation.

However, the statements by Terrell that objects appear rotated but nondistorted are verified in Fig. 17 when the studied objects subtend sufficiently small visual angles and changes in distances are neglected. Our diagram shows that the solid angle of sight of the separation of the hyperboloids varies as if the original separation (*d*) had rotated to keep it perpendicular to the hyperboloids. Thus, each infinitesimal original square that has been distorted into a diamond-like shape might, to the observer, appear to be rotated. Further, *small spheres are transformed into ellipsoid like shapes that, however, cover approximately the same solid angle of view as do the original spheres.* It is interesting to note that these two statements are equally true whether the object appears Lorentz contracted or expanded as seen in the diagram in Fig. 17.

Even if Terrell's statements in a limited way are verified by our diagram, it is so only when the observation is made by a camera. As soon as holography is included, it is easy to study variations in distances and then we understand that his statements are only approximations of the true story. Using holography shear and rotation can be distinguished so that the large-scale distortions as presented in Fig. 17 are observed.

Finally, Scott and van Driel show that stars appear moved toward the point of travel at increased speed [23]. It is also pointed out "although a sphere remains circular in outline, the apparent cross section may be grossly distorted and in some conditions the outside surface of the sphere appears concave." The last statement is verified by our Fig. 17 where a flat surface that has passed the observer appears concave.

We have up to now tested our graphical method of calculating relativistic apparent distortions by comparing it with accepted relativistic equations, and we have always found that it produces identical results. Let us do the same with Fig. 17, which was published in 1985 and compare it to Fig. 4.13b of Mook and Vargish's book [24] published in 1987, *Inside Relativity*. In their figure the orthogonal coordinate system is moving from left to right past a stationary observer, while in our Fig. 17 a stationary orthogonal coordinate system is studied by an observer moving from left to right. Therefore the two figures are turned in different directions.

There are two more differences; (1) the speed of our observer is $0.6c$, while the speed of their orthogonal system is $0.8c$; and (2) we have studied the apparent shape of flat surfaces as the observer is moving past, and through, those surfaces, while they have studied a grid painted on a flat screen that travels parallel to its surface at a distance of a tenth of the height of the grid. Therefore, in our figure, the flat surface passing through the observer is seen as a cone intersected along its axis producing a triangle with a sharp point at O. In their figure, however, the intersection does not pass through the axis, and we see another cone intersection that produces a hyperbola instead of a triangle. Apart from these difference in conditions, the diagrams are identical, which is quite impressive as their figure is produced by a computer using the accepted relativistic equations, while our figure is produced in a graphical way by using two pins, a string and a ruler.

Let us compare the apparent distortion of flat surfaces moving past an observer at increasing relativistic velocities. Figure 17b represents an orthogonal coordinate system moving from right to left at the speed of $0.6c$ past the stationary observer at B. As the velocity is increased and approaches that of light, the cone angle (i, o, i) approaches zero. Thus all approaching objects appear rotated through 90° so that their back side can be seen. The separation of advancing hyperboloids approaches infinity, while those moving away are transformed into paraboloids, and their separation along the line through the observer (O) approaches zero.

It is interesting to compare these results with the apparent shape of the only known flat surfaces that move with the speed of light, namely, flat wavefronts or flat sheets of light that are studied by its scattered light by using the light-in-flight technique. The cone angle is zero, and the back side of all visible wavefronts is seen. The separation of all approaching surfaces appears to be infinite. As the wavefronts pass by they are tilted through 45° instead of approaching 90°. The wavefronts that move away from the observer are transformed into paraboloids instead of hyperboloids. Their spacing appears to be half of their true value instead of zero.

The main difference between a set of flat solid surfaces moving perpendicular to their surfaces and flat wavefronts or flat sheets of light is that the former experience Lorentz contraction, which the light surfaces do not. We could expand this observation by stating that the Doppler shift is a sign and a measure of relativistic transformations. The light reflected from a moving solid surface is Doppler-shifted, but scattered light from a moving sheet of light is not. If wee look at an advancing object, it will appear most blueshifted when it approaches directly toward us. However, the color of a moving light sheet is independent of from what angle it is observed.

The apparent length of the cars of a passing train is equal to the true length only at the angle of sight of zero Doppler shift (close to $G–H$ in Fig. 17b). The apparent length of light pulses, on the other hand, is equal to the true length only at an angle of sight that is perpendicular to light propagation (d of Fig. 17b).

The rules for the apparent distortion of wavefronts or pulsefronts are much simpler than those of solid objects moving at relativistic velocities. In the following, we will repeat the three simple rules concerning the practical use of light-in-flight recordings for the study of the shape of wavefronts or stationary objects.

The curvature of a wavefront appears transformed into the curvature of a mirror surface shaped so that it would focus the total wavefront into the point of observation. The reason is that a focusing mirror reflects light in such a way that the total wavefront arrives to the focal point at one point of time. Thus, a small flat wavefront that passes by will appear tilted at 45°. A larger flat wavefront will not only appear tilted but will also be transformed into a paraboloid whose focal point is the point of observation. A spherical wavefront appears transformed into an ellipsoid, where one focal point is the point source of light (A) and the other is the point of observation (B). This configuration represents one of the ellipsoids of the holodiagram.

Light-in-flight recording by holography can be used to reveal the intersection of a light slice and a scattering surface. By positioning this surface in a special way, it is possible to produce a cross section of the apparent wavefront that is identical to a cross section of the true wavefront. Such an undistorted view of the true wavefront is formed if the scattering surface is a part of *a sphere the center of which is the point of observation*. A good approximation of this configuration is a flat surface at a large distance, parallel to the observer as seen in Fig. 8.1 of Ref. 4. There is another way to reach the same result, namely, to illuminate the object by light shining *toward A*, rather than by the light *from* a point source at A. In that case, the ellipsoids are changed into hyperboloids and the hyperboloid in the middle between A and B will be a flat surface.

XIII. CONCLUSION

In a new graphical way, we have solved optical problems found in the literature regarding special relativity. To arrive at these results, we used the accepted concept of the constancy of the speed of light and the following tools: a string and two nails for making ellipses and a ruler, pen, and paper to draw the diagrams.

No extensive mathematical knowledge or profound knowledge about relativity is needed to make and use the diagram. The technique is based on a slightly refined diagram, the holodiagram, which was initially designed for holography and conventional interferometry. Simplicity and visualization are the main advantages compared with application of the conventional equations presented by Einstein. The result is that for each velocity one diagram shows, in a concentrated form, the distribution in space of the following phenomena:

1. The Doppler shift, of which the transverse redshift is a special case
2. The aberration of light rays and lines of sight
3. The apparent rotation, which is found to be a part of the more general object distortion
4. The Lorentz contraction, which is found to be a special case of apparent expansions and contractions
5. The time dilation, which is found to be a special case of more general apparent speeding up and slowing down of time

Furthermore, we have found that the diagram can be used instantly after that the observer's velocity has changed but not when the changes its velocity object's has changed. From this fact, we conclude that the ellipses (ellipsoids) of Fig. 16 and the cone of the diagram of Fig. 17b move at infinite velocity with the observer but only with the speed of light with the observed object. *There is good reason to believe that this phenomenon produces the asymmetry needed to explain the twin paradox of Appendix A.*

Finally, we have shown that our method can be used to predict the relativistically distorted image or to restore the true image from it. The apparent distortion of a flat surface that approaches the speed of light is compared with the apparent distortion of a flat wavefront observed by holographic light-in-flight recording. It is found that these distortions are not identical but their main features are similar.

Inspired by the holographic uses of the holodiagram, we have introduced the new concept of spheres of observation that by velocity are transformed into ellipsoids of observation. In this way time dilations and apparent length contractions are explained as results of the eccentricity of ellipsoids. Our approach explains how a sphere of light can appear stationary in two frames of reference that move in relation to each other. It visualizes and simplifies in a graphical way the apparent distortions of time and space that are already generally accepted using the Lorentz transformations. When the ellipsoids are used to explain the null result of the Michelson–Morley experiment, there is no need to assume a real Lorentz contraction of rigid bodies, caused by a velocity of the observer. We have used our graphical concept to calculate, the time dilation, the apparent Lorentz, contraction, the transversal relativistic redshift, the relativistic aberration of light rays, and the apparent general distortion of objects. In all cases our results agree with those found in other publications [25]. However, using our approach we look at all those relativistic phenomena, except for time dilation, as caused solely by the influence of velocity on the measurement performed.

APPENDIX A: A MODIFICATION OF THE WELL-KNOWN TWIN PARADOX

Two spaceships, named the *Rester* and the *Traveler* and comanded by captains of those same names, both produce flashes of light at a frequency of exactly one flash per second. The ships are first close together, but then the traveler starts his ship's rockets and travels to a star one lightyear away at a constant speed that is one-tenth of the speed of light (c). After 10 years, the traveler turns and travels back to the rester again, where both captains compare experiences (Fig. 18).

The traveler says: I saw that the frequency of your flashes became lower as I traveled away from you, but after 10 years when I turned and traveled back to you their frequency became higher, so that when we met after 10 more years, I found that the number of your flashes and my flashes during the 20 years were exactly equal.

The rester says: I experienced exactly the same thing. The frequency of your flashes became lower as you traveled away from me, but when you turned and traveled back to me, their frequency became higher. *But, there was*

The modified Twin Paradox

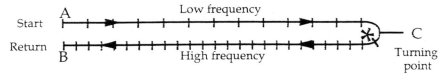

The separation as seen from the traveler

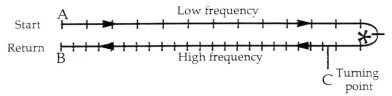

The separation as seen from the rester

Figure 18. Two rocket ships are at rest. Then one ship, the *Traveler*, starts its rockets and travels at the speed of 0.1*c* to a star one lightyear away, turns, and comes back. During the travel both ships produce flashes at exactly one flash per second. The traveler (the *Traveler's* captain) counts the flashes from the other ship, the *Rester*, and finds the frequency to be lower during the first 10 years and higher during the other 10. However, the captain of the other ship, the *Rester*, finds the *Traveler's* flash frequency to be lower during 11 years and higher during only 9 years. They have counted different numbers of flashes, as if the *Traveler's* time went slower.

one very important difference: I counted the flashes of lower frequency during 11 years and the higher frequency during only 9 years. Thus, it is *impossible* that the number of your flashes and my flashes during the 20 years were exactly equal. It appears as if I ought to have counted a lower number of your flashes.

The traveler: The reason why you experienced my turning no sooner than after 11 years is obvious; my turning point was one lightyear away from you.

The rester: The changes in frequency are well known as the *Doppler effect.* The reason is obvious—the longer the distance separating us, the more flashes that are flying in the space between us and that have not yet reached the observer. This phenomenon is, of course, symmetric and independent of who the sender is and who the observer is.

Thus, again, it is *impossible* that both the traveler and the rester counted the same number of flashes during the 20 years. But on the other hand it is also impossible that any flash simply disappeared. There is something fishy about this. Where did they go wrong in their assumptions?

If we believe in Einstein, the answer to this problem is simple. The statement that rester and traveler both counted the same number of flashes is incorrect; the number of flashes from the traveler not only *appears* lower to the rester but it *is* lower, because the time (the clock) of the traveler *is* slower. But does that really give the solution to the paradox? Well, that is certainly worth thinking about.

The ordinary version of the twin paradox is based on knowing Einstein's statement that the time of fast moving objects is slowed down. The question is: How do we know which one is moving if we have no references in empty space? I, however, want to point out that if the observer changes velocity, this immediately results in a change of the Doppler shift, but if the light source changes velocity there is a delay before the observer notices any Doppler shift. I think that this *delay between cause and effect* is of great importance in many fields of physics and has not received the recognition it deserves. It is the same as saying that if the position of point A of the holodiagram is changed, there will be a delay before the holodiagram has adjusted itself to the new situation, but if point B is moved, the holodiagram adjusts itself immediately.

APPENDIX B: FUNDAMENTAL CALCULATIONS BASED ON THE HOLODIAGRAM

If we define resolution as the shortest distance corresponding to the formation of one interference fringe, Fig. 5 represents the resolution limit of any possible optical system using the corresponding geometric configuration. The diffraction limited resolution and the "interference limited resolution" are represented by the separation of the hyperboloids and the ellipsoids, respectively.

In the following examples are presented where the resolution of Fig. 5 is compared to corresponding values quoted from different references in which other methods of calculation have been used. If there is a discrepancy, the value of the reference is noted.

1. A and B are two mutually coherent points of illumination. The hyperboloids represent interference surfaces in space. An object point passing through these surfaces will produce a beat signal. Any measuring system based on this configuration will have zero sensitivity parallel to the hyperboloids and maximal sensitivity normal to the hyperboloids (parallel to the ellipses). The resolution in this direction, is represented by the separation of the hyperboloids:

$$\text{Resolution} = \frac{0.5\lambda}{\sin\alpha}$$

The diagrams of Fig. 5 can be utilised for the understanding and the use of the following optical phenomena:

1.1. Conventional interferometry [26]

1.2. Young fringes [27]

1.3. Bragg angle [28]

1.4. Fresnel zone plate producing a virtual image (any cross section of the hyperboloids) [29]

1.5. Two-beam Doppler velocimeter [30]

1.6. Two-beam radar [31]

1.7. Ordinary and Lippman holography (A is object point, B is reference point) [32]

1.8. Hologram interferometry (fringes seen on the hologram plate when studied from the real image of an displaced object point) [33]; identical to the results of 2.6 and 3.4

1.9. Two-beam hologram interferometry (measurement of in-plane displacement) [34]

1.10. Objective speckles; speckle size $0.61\lambda/\sin\alpha$ [35]

1.11. Speckle photography for in-plane measurement: resolution: $0.61\lambda/\sin\alpha$. (A–B is the lens diameter) [36]

1.12. Two-beam speckle photography (measurement of in-plane displacement) [37]

1.13. Projected interference fringes (sensitivity direction is normal to object surface and the resolution corresponds to the movement of a surface point from one paraboloid to the adjacent) [38,39]

1.14. Focused spot size (A and B are two diametrical points on the lens); resolution $0.61\lambda/\sin\alpha$ [40]

1.15. Resolution of any optical system (necessary condition for resolution of an object with the size AB is that the lens of observation at least crosses one bright and one dark hyperboloid); identical to the results of 1.14 and 3.6

The moiré of two cross sections of the hyperboloids represents the difference of two interference patterns. If the two cross sections are identical but one is displaced, the moiré fringes also represent the loci of constant resolution in the displacement direction. If one focal point is fixed, and the other one is displaced, the resulting moiré pattern forms a new set of hyperboloids whose foci are the two positions of the displaced focal point. This new pattern is independent of the position of the fixed focal point (a rotation of one of the original spherical wavefronts of Fig. 5 produces no moiré effect). Therefore hologram interference fringes are independent of the position of the point source of the fixed reference beam.

2. *A is a point of illumination, B is a coherent point of observation.* B is the center of "spherical wavefronts of observation" and has been brought to coherence with A by receiving a direct reference beam. The ellipses represent imaginary interference surfaces in space. An object point passing through these surfaces will produce a beat signal. Any measuring system based on this configuration will have zero sensitivity parallel to the ellipsoids and maximal sensitivity normal to the ellipsoids (parallel to the hyperbolas). The resolution in this direction is represented by the separation of the ellipsoids:

$$\text{Resolution} = \frac{0.5\lambda}{\cos\alpha} = 0.5k\lambda$$

The diagrams of Fig. 3–5 can be utilized to understand and use the following phenomena:

2.1. Conventional interferometry and interferometry using oblique illumination and observation [41,42]

2.2. Bragg angle

2.3. Fresnel zone plate producing real image (any cross section of the ellipsoids) [53]

2.4. Holography (utilizing the coherence length, controlling the sensitivity to unwanted movements) [26,52]

2.5. Hologram interferometry (evaluating displacement, planning the resolution; one interference fringe is formed each time an ellipsoid is crossed by an object point) [26,43]

2.6. Hologram interferometry using two points of observation [44] (identical to the results of 1.9 and 3.4)

2.7. Hologram contouring using two frequency illuminations [45]

$$\text{Resolution} = \frac{\lambda}{\lambda_1 - \lambda_2} \cdot \frac{0.5\gamma}{\cos\alpha} = \frac{\lambda}{\lambda_1 - \lambda_2} \cdot 0.5k\lambda$$

2.8. Gated viewing (the ellipsoids represent surfaces of constant time delay): resolution $= ct\cos\alpha = kct$, where c is speed of light and t is pulsewidth

2.9. Doppler velocimeter [46]: $v = kf0.5\lambda$, where v is velocity and f is frequency

2.10. Doppler radar [47]: $v = kf0.5\lambda$

2.11. Relativistic Lorentz contraction $= 1/\cos a = 1/k$ (see Fig. 9)

2.12. Relativistic time dilation $= 1/k$

2.13. Relativistic transversal Doppler effect $= k$

The moiré effect of two cross sections of the ellipsoids (Fig. 4) represents the difference of two imaginary interference patterns. It also represents the interference fringes of hologram interferometry [48]. If the two cross sections are identical but one is displaced, the moiré fringes also represent the loci of constant resolution in the displacement direction. If one focal point of the ellipsoids is fixed and the other one is displaced, the resulting moiré pattern forms a set of hyperboloids whose foci are the two positions of the displaced focal point. This new pattern is independent of the position of the fixed focal point (a rotation of one of the original spherical wavefronts of Fig. 5 produces no moiré effect). Therefore, the movement of hologram interference fringes, when studied from different points of observation, is independent of the position of the fixed point source of illumination.

3. A and B are two mutually coherent points of observation. The information from A and B are brought together in a coherent way, such as by a semitransparent mirror or by the use of a reference beam. Points A and B could even be just two points on a lens. The hyperboloids represent imaginary interference surfaces in space. An object point passing through these surfaces will produce a beat signal. Any measuring system based on this configuration will have zero sensitivity parallel to the hyperboloids and maximal sensitivity, normal to the hyperboloids (parallel to the ellipses). The resolution in this direction is represented by the separation of the hyperboloids:

$$\text{Resolution} = \frac{0.5\lambda}{\sin \alpha}$$

The hyperbolas of Fig. 5 can be utilized to understand and use the following optical phenomena:

3.1. Doppler velocimeter using two points of observation [46].

3.2. Subjective speckles (A and B are two diametrical points on the observations lens). The number of subjective speckles seen on the object is equal to the number of objective speckles projected on to the lens; therefore 3.2 and 1.10 give identical results.

3.3. Speckle photography where the camera lens is blocked but for two diametrical holes (measurement of in-plane displacement) [49].

3.4. Hologram interferometry using two observations. The number of fringes passing an object point when the point of observation is moved from A to B is equal to the number of hyperboloids passed by that object point between the two exposures. This number is also equal to the number of hyperboloids of 1.8 seen between A and B. The results of 1.8 and 2.6 and 3.4 are identical.

3.5. Stellar interferometry (A and B are the two mirror systems in front of a telescope objective or the two antennas of a radiotelescope). The resolution for double star systems is $0.5\lambda/\sin\alpha$, for large star $0.61\lambda/\sin\alpha$ [50].

3.6. Diffraction-limited resolution of, for instance, a microscope (*A* and *B* are the two diametrical points on the objective lens). Maximal resolution is $0.61\lambda/\alpha$ [51].

Thus, we have studied a great number of different optical systems. The agreement between the accepted values of resolution and those found by the use of the "holodiagram" of Figs. 3 and 5 is very good. The discrepancies found, when a circular illuminating or observing area such as a lens is involved are caused by the fact that the total area of the lens is used, not only two diametrical points. To produce an image, some of the resolution has to be given up. The graphical approach of the holodiagram also agrees well with accepted relativistic equations.

The holodiagram appears to represent the fundamental resolution of any optical system and to verify the relativistic distortions of time and space. Thus, I

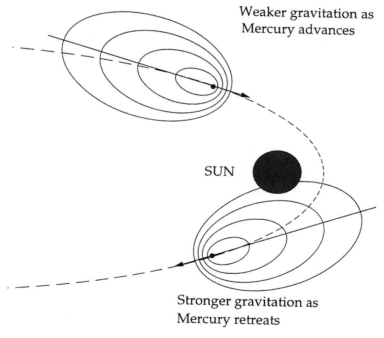

Figure 19. It is assumed here that the concept of the ellipsoids of observation apply to all fields moving with the velocity of light, such as electric or gravitational fields. Thus, precession of the perihelion of Mercury (the rotation of its elliptic orbit) can be explained by the asymmetry of the gravitational forces as the planet advances toward (resp. retreats from) the sun.

hope that it might become a useful educational tool that concentrates a large amount of information into one unifying method.

APPENDIX C: OTHER POSSIBLE APPLICATIONS

Up to now we have used the concept of ellipsoids of observation to explain apparent distortions of fast-moving objects as measured by light. However, these ellipsoids of observation apply just as well for observations using other fields that propagate with the speed of light, including electric or probably gravitational fields. Thus, the precession of the perihelion of Mercury [25] can be explained only by an asymmetry in the forces acting on the planet during its orbit. Such an asymmetry is visualized in Fig. 19, where, because of the assumed limited velocity of gravitation, the gravitational force varies as the planet advances toward (resp. retreats from) the sun. Using similar approaches, it is possible to develop our methods to visualize other relativistic effects such as kinetic energy and magnetic forces as functions of the eccentricity of ellipsoids.

References

1. H. A. Lorentz, "Michelsons' interference experiment" (1885), in *The Principle of Relativity*, Dover, New York, 1952, p. 7.

2. A. Einstein, "On the electrodynamics of moving bodies" (1905), in *The Principle of Relativity*, Dover, New York, 1952, pp. 37–65.

3. H. Minkowski, "Space and time" (1908), in *The Principle of Relativity*, Dover, New York, (1952), p. 75.

4. N. H. Abramson, *Light in Flight or the Holodiagram—the Columbi Egg of Optics*, SPIE Press, 1996, Vol. PM27.

5. N. H. Abramson, "The holo–diagram: A practical device for the making and evaluation of holograms" *Proc. Applications of Holography*, Strathclyde, Glasgow, 1968, pp. 45–55.

6. N. H. Abramson, *Nature* **231**, 65–67 (1971).

7. N. H. Abramson, *Optik* **1**, 56–71 (1969).

8. N. H. Abramson, *Appl. Opt.* **11**, 2562–2571 (1972).

9. D. I. Staselko, Y. N. Denisyuk, and A. G. Smirnow, *Opt. Spectrosc.* **26**, 413 (1969).

10. Y. N. Denisyuk, D. I. Staselko, and R. R. Herke, "On the effect of the time and spatial coherence of radiation source on the image produced by a hologram," *Proc. Applications of Holography*, Section 2, Besancon, 1970, pp. 1–8.

11. N. H. Abramson, *Appl. Opt.* **22**, 215–232 (1983).

12. Private communication with Donald Coxeter, Prof. Emeritus, Math. Dept. Univ. Toronto, Toronto, Ontario, Canada M5S 3G3.

13. P. A. Tipler, *College Physics*, Worth Publishers, 1987, p. 719.

14. P. A. Tipler, *College Physics*, Worth Publishers, 1987, p. 720.

15. N. H. Abramson, *Appl. Opt.* **24**, 3323–3329 (1985).

16. W. Rindler, *Special Relativity*, Wiley, New York, 1960, p. 49.

17. W. Rindler, *Special Relativity*, Wiley, New York, 1960, p. 47.

18. N. H. Abramson, *Spec. Sci. Technol.* **8**, 101–107 (1984).

19. Yu Hovsepyan, *Physics-Uspekhi* **41**(9), 941–944 (1998).

20. J. Terrell, *Phys. Rev.* **116**, 1041 (1959).

21. R. Bhandari, *Am. J. Phys.* **38**, 1200–1201 (1970).

22. P. M. Mathews and M. Laksmanan, *Nuovo Cimento* **12**, 168–180 (1972).

23. G. D. Scott and H. J. van Driel, *Am. J. Phys.* **38**, 971–977 (1970).

24. D. E. Mork and T. Vargish, *Inside Relativity*, Princeton Univ. Press, 1987.

25. A. Einstein, "The foundations of the general theory of relativity" (1916), in *The Principle of Relativity*, Dover, New York, 1952, pp. 111–164.

26. N. H. Abramson, *Appl. Opt.* **11**, 2562 (1972).

27. O. Bryngdahl, *J. Opt. Soc. Am.* **62**, 839 (1972).

28. N. H. Abramson, *Appl. Opt.* **8**, 1235 (1969).

29. M. Born and E. Wolf, *Principles of Optics*, Pergamon, New York, 1965, p. 261.

30. G. Freier, *University Physics*, Meredith, New York, 1965, p. 537.

31. M. Born and E. Wolf, *Principles of Optics*, Pergamon, New York, 1965, p. 371.

32. L. Lading, *Appl. Opt.* **10**, 1943 (1971).

33. W. M. Farmer, *Appl. Opt.* **11**, 770 (1972).

34. J. Collier, C. Burckhardt, and L. Lin, *Optical Holography*, Academic, New York, 1971.

35. Gates, *Opt. Technol.* **1**, 247 (1969).

36. J. Butters, in *The Engineering Uses of Holography*, Cambridge Univ. Press, Cambridge, UK, 1968, p. 163.

37. D. Gabor, *IBM J. Research* **14**, 509 (1970).

38. E. Archbold and A. Ennos, *Opt. Acta* **19**, 253 (1972).

39. J. Leendertz, *J. Phys. E* **3**, 214 (1972).

40. N. H. Abramson, *Laser Focus* **23**, 26 (1968).

41. R. Brooks and L. Heflinger, *Appl. Opt.* **8**, 935 (1969).

42. M. Born and E. Wolf, *Principles of Optics*, Pergamon, New York, 1965, p. 397.

43. N. Abramson, *Optik* **1**, 56 (1969).

44. J. D. Briers, *Appl. Opt.* **10**, 519 (1971).

45. K. Stetson and R. Powell, *J. Opt. Soc. Am.* **55**, 1694 (1965).

46. E. B. Aleksandrov and A. M. Bonch-Bruevich, *Sov. Phys.-Tech. Phys.* **12**, 258 (1967).

47. B. Hildebrand, *The Engineering Uses of Holography*, Cambridge Univ. Press, Cambridge, UK, 1968, p. 410.

48. R. M. Huflaker, *Appl. Opt.* **9**, 1026 (1970).

49. E. N. Leith, *Proc. IEEE* **59**, 1305 (1971).

50. N. Abramson, *Appl. Opt.* **10**, 2155 (1971).

51. D. E. Dufly, *Appl. Opt.* **11**, 1778 (1972).

52. M. Born and E. Wolf, *Principles of Optics*, Pergamon, New York, 1965, p. 275.

53. M. Born and E. Wolf, *Principles of Optics*, Pergamon, New York, 1965, p. 418.

54. R. Kurtz and H. Loh, *Appl. Opt.* **11**, 1998 (1972).

55. J. W. Y. Lit, *J. Opt. Soc. Am.* **62**, 491 (1972).

ASTROPHYSICS IN THE DARK: MACH'S PRINCIPLE LIGHTS THE WAY

D. F. ROSCOE

Department of Applied Mathematics, Sheffield University, Sheffield, United Kingdom

CONTENTS

Modern Nonlinear Optics, Part 3, Second Edition, Advances in Chemical Physics, Volume 119,
Edited by Myron W. Evans. Series Editors I. Prigogine and Stuart A. Rice.
ISBN 0-471-38932-3   © 2001 John Wiley & Sons, Inc.

I. INTRODUCTION

This chapter was originally planned as a review of the state of modern cosmology; however, on reflection, I decided that a more creative use of the opportunity to contribute to this volume would be to lay before the reader certain well-founded observational results that, at the very least, indicate that our cosmos is not quite as well understood as we are commonly inclined to believe.

Both pieces of evidence are in the mainstream literature, each is the subject of continuing work, and neither has had any impact to date on mainstream thinking.

After reviewing these observational results, I will argue that either alone presents modern cosmology with a potentially fatal crisis—beyond the ability of any patch-and-mend device to "save appearances." Accepting this, we go on to suggest that the root of the problem lies in the failure of modern gravitational theory to incorporate Mach's principle in a sufficiently fundamental way, and we briefly illustrate a way forward, to compensate for this deficit.

II. SOME BACKGROUND

A. General Comments

Both pieces of evidence concern spiral galaxies; the first concerns the nature of the cosmological redshift measured for such objects, while the second concerns the nature of spiral galaxy dynamics. To have a clear appreciation of the issues, some background knowledge is useful.

First, it is useful to know that a model spiral galaxy can be considered to consist of a central spherical bulge component, embedded in a rotating disk of stellar and gaseous material with the whole embedded in a spherical halo of very diffuse gas and "halo stars." It is conventionally believed that this latter component is much more massive than it appears, with the deficit made up of "dark matter."

Spiral galaxies rotate, with maximal rotation rates of, typically, 200 km/s reached at distances of typically 10 kpc from the galactic centers (kpc $\equiv$ kiloparsec, 1 parsec $\approx$ 3.25 lightyears). Consequently, for a galaxy seen edge-on, light from the receding arm will be Doppler-shifted to the red, while light from the approaching arm will be Doppler-shifted to the blue. The profile of any galaxy's rotation across its disk is called its *rotation curve*.

Additionally, galaxies appear to have individual motions relative to their local environment: the so-called peculiar motions. These motions are thought to be generated by local gravitational interactions, and are not thought to exceed a hundred or so km/s and give rise to additional true Doppler effects. Finally, the light from all galaxies has a global distance-dependent redshift component—the so-called cosmological redshift, which is usually quantified in units of velocity as if it were a true Doppler shift.

B. The Astrophysical Distance Scale

Briefly, astronomers have two basic methods for estimating distance scales for spiral galaxies that are independent of the observation of special standard candles such as supernovae.

The first of these is the widely known Hubble law, which relates the distance to an object to its measured redshift via the relationship $cz = Hd$, where c is the speed of light in the vacuum, z is the measured redshift of the object, and H is Hubble's constant—typically estimated to be about 75 $\mathrm{km\,s^{-1}\,Mpc^{-1}}$ where 1 Mpc $\approx$ 3.25 million lightyears.

The second is using the *Tully–Fisher* relationship, which provides a direct relationship between the absolute luminosity of a spiral galaxy and its maximum rotation speed. Since the maximum rotation speed can be directly estimated by observation, one can estimate the absolute luminosity; since we measure directly the *apparent* luminosity, the inverse-square law allows us to estimate the distance to the object concerned. For the present discussion, it is important to understand that, at *optical wavelengths*, estimates of maximum rotation speeds are generally *extrapolations* from rotation curve measurements.

C. Measuring Galactic Redshifts

In view of the foregoing, when astronomers are said to measure a galaxy's redshift, they begin by measuring a (very noisy) Doppler *profile* across the disk of the galaxy that contains the three redshift components. That component arising from the galaxy's own rotation is subtracted by some form of averaging process taken over the whole profile, leaving a measurement that consists of the required cosmological redshift together with an irreducible component arising from the galaxy's own peculiar motion.

It is crucial to understand that, even assuming a zero peculiar velocity, it is only rarely possible, even *in principle*, for this process to yield a cosmological redshift to better than 10 km/s accuracy—and, because astronomers have no particular need for highly accurate redshift determinations, the effort to obtain them is rarely made.

D. Measuring Galactic Rotation Curves

The rotation curve is calculated in two steps: (1) by subtracting the global redshift component (i.e., cosmological redshift + Doppler effect arising from peculiar motion) from the Doppler profile measured directly across the galaxy's disk and (2) by determining the actual dynamical centre of the galaxy.

The process of estimating the global redshift component and estimating the dynamical center is termed the process of "folding the rotation curve." Because of the very noisy nature of the data, this process is very far from trivial, especially if one is interested in accurate dynamical studies of spiral galaxies.

III. COSMOLOGICAL REDSHIFTS: ARE THEY QUANTIZED?

A. The Tifft Story

Around about 1980, William Tifft, a radio astronomer at the University of Arizona in Tucson, had the wild idea that, perhaps, the cosmological redshifts of galaxies had preferences for multiples of some basic unit. Subsequently, he looked and made two claims [1,2]:

- That the differential redshifts between galaxies in groups (obtained by subtracting redshifts in pairs) were quantized in steps of 72 km/s
- That the redshifts of galaxies measured with respect to our own galactic center were quantized in steps of 36 km/s

Initially, these claims raised quite a lot of interest—but it soon became apparent that the claimed effects were deeply problematical from the point of view of prevailing cosmology and, very conveniently, that Tifft's own statistical methods were very far from being robust. This latter fact made it very easy for the community to ignore a potentially very difficult problem for the status quo.

Subsequently, Tifft has formed increasingly complex hypotheses, claiming to see evidence for increasingly refined hierarchical systems of redshift quantization. Irrespective of whether there is anything substantive in his claims, Tifft failed to do the one absolutely necessary thing: perform a totally rigorous analysis of a single well-defined hypothesis that could withstand any criticism directed at it.

B. The Napier Story

The whole business would have probably faded away, forgotten for years, had the astronomer Bill Napier (then at the Royal Observatory, Edinburgh) not taken an interest in Tifft's claims around 1987. Napier was by then well known as the originator of "cometry catastrophy theory," according to which the long sequence of catastrophic species extinction, which is part of the geologic record, has arisen because of the cyclical motion of the solar system in and out of the galactic plane—with each passage through the plane bringing with it vastly increased risk of cometry collision (see Clube and Napier [3] for first report).

This work had given Napier considerable expertise in the analysis of phenomena that appear as potentially periodic, and it was this aspect of the Tifft claim that aroused his interest. Napier's personal view, then, was that the whole thing was probably nonsense and that the claimed periodicities would evaporate under rigorous investigation (private communications). Unlike Tifft, who simply set out to look for redshift quantization *at any periodicity*, Napier, and co-worker Guthrie, started with Tifft's specific claim that such a quantization existed with a period of 36 km/s—thus, he was in the quite different business of testing a specific well–defined hypothesis. For the sake of simplicity, I will not consider the equally important 72-km/s claim, since the story is essentially similar.

Napier began by using Monte Carlo methods to establish that an essential precondition for a rigorous analysis of the type proposed was the availability of a sufficiently large sample of redshifts, each with formal accuracy better than 5 km/s; anything less would result in even a real signal at ≈ 36 km/s being washed out by measurement errors.

Napier's co-worker, Guthrie, performed a very detailed literature search to assemble a sample of 97 redshift measurements of the required accuracy—taking care to reject any that had ever been used by Tifft in any of his claims. This sample formed the backbone of the subsequent Napier–Guthrie analysis.

Remember that the original claim was that the effect existed in redshift determinations that had been reduced to the frame of reference of our own galaxy's center. Since redshift determinations are routinely given in the solar frame of reference, this amounted to the need to correct the redshifts in the sample for the sun's motion with respect to the galactic center. At the time, 1989, the solar vector determinations resided inside a very large error box, and so Napier's analysis had a lot of slack associated with this part of it. Even so, it quickly became apparent that a very strong quantization effect emerged for estimated solar vectors anywhere inside the error box, at a periodicity of 37.6 km/s (against the claim for 36 km/s) [4–6]. Figure 1 shows the power spectrum arising from their analysis after redshift determinations have been corrected for the solar motion, using an estimated solar vector $V = 220$ km/s, $l = 95°, b = -12°$, where l is the galactic longitude and b is the galactic latitude. Extensive Monte Carlo simulations give a probability of $\approx 10^{-8}$ for a signal like that of Fig. 1 to have arisen by chance alone.

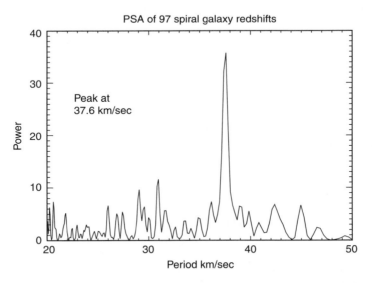

Figure 1. The power spectrum analysis of Guthrie and Napier's first sample of 97 redshifts. This peak power occurs when redshifts are corrected for the solar motion with respect to our galactic center.

Subsequent to this initial publication, the satellite Hipparchus has been launched, which has resulted in very refined conventional determinations of the solar vector error box; concurrently, Napier has reversed the Napier–Guthrie analysis, assuming the prior existence of the 37.6-km/s effect using it to obtain independent determinations of the solar vector error box. These lie wholly inside the Hipparchus error box determinations.

This analysis has been repeated on independent (although less accurate) samples (one of which was donated by one of the anonymous referees of the original publication) with similar results. Napier and Guthrie's parallel analysis of the claims for 72km/s for differential redshifts between galaxies in groups has been similarly successful, and has equally bizarre implications.

To summarize, Tifft's original claims have been strongly and independently substantiated by the Napier–Guthrie analysis; this latter analysis has appeared in the mainstream literature and stands increasingly secure as Hipparchus observations continue to tighten the solar error box. Any serious thought about these two effects soon convinces one that the implications for cosmology are profound—and very difficult to comprehend.

IV.   IS GALACTIC EVOLUTION CONFINED TO DISCRETE STATES?

This question arose in the course of a routine analysis of galaxy rotation curve data by myself (see Section II), which was originally driven by a prediction arising from an extension of the theory, to be briefly described in the latter sections of this chapter. This theoretical prediction was to the effect that, in an "idealized spiral galaxy" (i.e., one without any central bulge and with perfect rotational symmetry), then the circular velocity should have the general form $V = AR^\alpha$, where R is the radial distance from the center and (A, α) are parameters that vary from galaxy to galaxy. It is necessary to understand that, generally speaking, it is clear from real data that a simple power law *cannot* apply over the whole radial range of the typical spiral galaxy; at best, it can have any applicability at all only in the so-called optical disk (i.e., in that portion of the disk component of the galaxy that emits visible light). Astronomers generally treat rotation curves in their entirety, and describe them with very complicated phenomenologically derived functions, and ignoring the obvious fact that spiral galaxies are manifestly composed of at least three distinct dynamical regions: a spherically symmetric central part, a rotationally symmetric intermediate disky part, and an optically diffuse spherically symmetric outer part. The power law hypothesis arose from theoretical considerations of a purely disk mass distribution, and can therefore have a possible applicability only in the disk regions of spirals.

The original analysis (of the rotation curves of 900 individual galaxies) was based on the rationale that, if the data were viewed through the "filter" of the power law hypothesis, then, if the hypothesis was "good enough," unsuspected new relationships between the dynamical and the luminosity properties of spiral galaxies might be revealed. This turned out to be very much the case, and much that was new and interesting was uncovered and is now available in the literature [7].

However, this work led to the almost incidental discovery of an effect that, at the very least, reminds us that the cosmos is not quite as well understood as we like to believe, and that almost certainly indicates the need for a revolution in cosmology.

A.   A Numerical Coincidence

Given my complete initial ignorance of rotation curve data, and its typical forms, the story began with the decision to perform a practise minianalysis of the first rotation curve sample ever published—the 21 rotation curves published by Rubin et al. [8]—from the point of view of the power law hypothesis, $V = AR^\alpha$. Of the 21 rotation curves, 9 exhibited very strongly nonmonotonic behavior in their inner regions and were obviously poor candidates for any power law fit, and so were rejected on these purely subjective grounds. For each of the remaining 12 rotation curves, I computed the parameter pair $(\ln A, \alpha)$ by a simple regression procedure; the results for $\ln A$ (rounded to one decimal place) are tabulated in Table I. It was immediately clear that, after allowing for the rounding process, every singly value was within ± 0.15 of being an integer or half-integer value—a result that (ignoring its aposteriori nature) can be computed as being a 1 : 500 chance. Before continuing, it is necessary to clarify the fact that the $\ln A$ scale is ultimately determined by the galactic distance scale, which Rubin et al. fixed by using Hubble's law with $H = 50$ km s^{-1} Mpc^{-1}, the preferred value in the early 1980's. Consequently, had Rubin used a contemporary value (nearer to 80 km s^{-1} Mpc^{-1}), the integer/half-integer

TABLE I
Twelve RFT[a] 1980 Spirals

Galaxy	ln A	Galaxy	ln A
N3672	3.6	U3691	3.6
N3495	4.0	N4605	4.0
I0467	4.1	N0701	4.1
N1035	4.1	N4062	4.5
N2742	4.5	N4682	4.5
N7541	4.6	N4321	4.9

[a]Rubin–Ford–Thonnard.

structure would have given way to something else. The real point of interest is therefore the apparent regularity the ln A distribution manifested in Table I.

At the time, I considered this to be almost certainly a numerical coincidence, but one worth investigating once the primary task that I had in mind had been completed.

B.   Essential Data Reduction

It is necessary to introduce a little detail into the story. An essential step in the minianalysis referred to above was the decision to reject certain rotation curves, made on the basis of the subjective judgment that they were not "monotonic enough" in their innermost regions. In effect, I was rejecting those galaxies for which there appeared to be a particularly strong influence of the bulge on the disk dynamics.

While this subjective approach was perfectly justifiable for a small experimental analysis, it could play no part in the large analysis contemplated for which all decisions had to be made in an automatic "blackbox" fashion. We dealt with this problem by writing a piece of software that automatically cut out the innermost parts of rotation curves that were judged, according to objectively defined statistical criteria, to be unusually affected by the presence of the bulge. The effectiveness of this process was tested by means completely independent of the present considerations.

C.   Chasing the Coincidence

The primary task was to test the power law hypothesis, and to this end I had obtained a sample of 900 optical rotation curves, originally measured by Mathewson et al. [9], and *folded* (see Section II.D) by Persic and Salucci [10], two Italian astronomers.

For this sample, the galactic distance scale has been set by Mathewson, Ford, and Buchhorn (MFB) using the Tully–Fisher relationship, which sets distance scales by using an observed correlation between the maximum rotation velocity of a spiral and its absolute luminosity, and so is quite distinct from Hubble-based distance determinations. Even so, the Tully–Fisher method gives an absolute scale only after calibration, and the MFB calibration gave a scale that was statistically similar to a Hubble scale using $H = 85 \, \mathrm{km \, s^{-1} \, Mpc^{-1}}$.

With this information, the results of the primary analysis of the 900 rotation curves made it possible to recalibrate the Table I to give *specific* predictions for the existence of preferred ln A values in the folded MFB sample, and these predictions are given in Table II.

Table II represents a set of specific predictions about the ln A distribution for the 900 folded MFB rotation curves, and it is these that were to be tested against the MFB sample.

TABLE II
ln A Data

RFT Scale	Predicted Value with MFB Scale
3.5	3.81
4.0	4.22
4.5	4.63
5.0	5.04

D. The Results

We present the totality of our results. The computation of ln A for each galaxy requires the following:

- A measured rotation curve for the galaxy.
- The folding of each rotation curve (see Section II.D). This data reduction process is nontrivial, and different people have their own favored methods.
- An estimate of the maximal rotation velocity for the galaxy. Such estimates are problematical for optical data, such as those analyzed here, and different people have their own favored methods of estimation.
- A Tully–Fisher calibration for the sample to get the distance scale. These can vary between samples owing to the details of photometric methods used by astronomers.
- An automatic and predefined "blackbox" technique for removing the effects of the bulge on disk dynamics.

In the present case, we have three distinct samples obtained by two independent groups of astronomers, two distinct folding techniques, three distinct methods of estimating maximal rotation velocities, and one method for removing bulge effects that has been tested by means independent of any of the present considerations. As we shall see, the results are not affected by any of these variations.

E. The Mathewson–Ford–Buchhorn (MFB) Sample Folded by the Persic–Salucci Eyeball Method

Figure 2 gives the ln A frequency diagram for the rotation curves of 900 southern sky spirals, observed by the Australian astronomers Mathewson et al., (MFB) [9] using Australian telescopes at Siding Spring. MFB estimated maximum rotation speeds for each galaxy (for use in the Tully–Fisher distance relationship) using a subjective eyeball technique.

These rotation curves were folded by the Italian astronomers Persic and Salucci [10] using a case-by-case eyeball technique. The short vertical bars give the positions of the *predicted* peak centers given in Table II, and it is

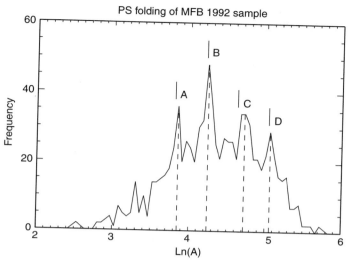

Figure 2. The MFB sample folded by the Persic–Salucci [10] eyeball method.

immediately clear that the predictions are strongly fulfilled on this sample. These initial results are now in the literature [11].

F. The MFB Sample Folded by the Roscoe Automatic Method

I considered the potential implications raised by the strongly positive result of the first test of the hypothesis of Table II to be so profound that tests on further samples became absolutely essential. Since the Italian folding method was (extremely) time-consuming, it was necessary to develop automatic methods for this part of the data reduction process; this turned out to be a nontrivial exercise, and the details of the method have been published [12].

For completeness, we show, in Fig. 3, the ln A frequency diagram for the rotation curves of the MFB sample (originally folded by Persic and Salucci [10]) folded by this automatic method; the vertical dotted lines indicate the peak centers of the Persic–Salucci solution. The clarity with which the peaks in this latter solution are reproduced in Fig. 3 indicates (1) that the peaks of Fig. 2 are not an artifact of the Persic–Salucci method; and (2) that the automatic algorithm works.

G.   The Mathewson–Ford Sample Folded by the Roscoe Automatic Method

The next sample of 1200+ rotation curves was observed by the Australians, Mathewson and Ford (MF) [13] (Buchhorn had discovered that astronomy pays

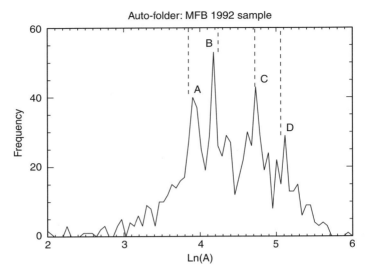

Figure 3.   The MFB sample folded by the automatic method of Roscoe; the vertical dotted lines indicate peak centers of Persic–Salucci [10] solution.

less than the bond markets), and was the first independent sample folded by the new software; the maximum rotation velocities of each galaxy were again estimated using a subjective eye-ball technique.

This sample is, on average, about 70% more distant that the MFB [9] sample so that, on average, only about 30% of the light is received at the telescope. For this reason, we expect a considerable attenuation of the signal.

The resulting ln A diagram is shown in Fig. 4, where the vertical dotted lines indicated the peak centers of the A, B, C, D peaks of Fig. 3. Notwithstanding the obvious signal attenuation (in comparison with Fig. 3), the coincidence of the peak structures is exact.

H.   The Courteau Sample Folded by the Roscoe Automatic Method

The results obtained from the first two samples analyzed (one of which was folded using two quite distinct methods) indicate that something profound has been uncovered—unless, perhaps, the observing astronomers [9] were somehow inadvertently introducing the signal into the sample. However, given the necessity of further reducing the data to remove bulge effects using our 'hole-cutting' technique, and the a priori ignorance on the part of MFB and MF of this future process, this seems to be an extremely remote possibility. Even so, given the profound nature of the claimed result, it is a possibility that must be accounted for. For this reason, we obtained the *only other available substantial*

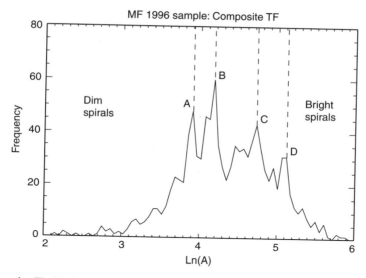

Figure 4. The Mathewson–Ford sample folded by the Roscoe automatic method; the vertical dotted lines indicate the peak centers of Fig. 3.

sample in existence, consisting of 305 northern sky spirals, and observed by the Canadian astronomer Courteau using U.S. telescopes at Lick Observatory and Las Palmas.

This sample was used by Courteau [14] in a study of systematic ways of defining maximum rotation speeds for spirals (remember that MFB and MF used subjective methods), and we present the ln A diagrams resulting from his stated best and worst methods of estimating these maximum speeds in Figs. 5 and 6, respectively. Except for the A peak (which is at the dim end of all the samples, and therefore suffers from small numbers), the peak positions are exactly reproduced.

I. Conclusions

We began with an explicit hypothesis made on the basis on the analysis of a very small sample (12 objects!), and a subsequent analysis of three other large samples has confirmed this hypothesis, in detail, and with a power that is virtually impossible to refute. But what can all this possibly mean?

In dimensionless form the power law $V = AR^{\alpha}$, which gave rise to the analysis in the first place, can be expressed as

$$\frac{V}{V_0} = \left(\frac{R}{R_0}\right)^{\alpha}$$

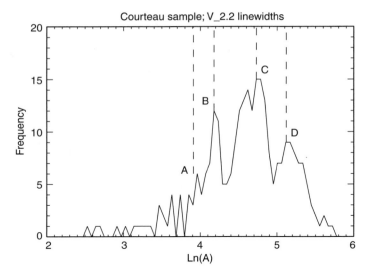

Figure 5. The Courteau sample with his best maximum rotation speed estimates; vertical dotted lines indicate peak centers in Fig. 3.

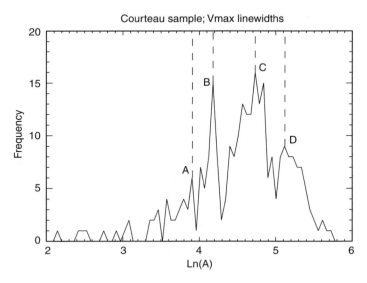

Figure 6. The Courteau sample with his worst maximum rotation speed estimates; vertical dotted lines indicate peak centers in Fig. 3.

so that $A \equiv (V_0/R_0)^\alpha$. A detailed analysis of the correlation between α and $\ln A$ [11] shows that V_0 and R_0 are each very strongly correlated with luminosity properties. Consequently, for absolute magnitude M (the astronomer's way of talking about absolute lumnosity) and surface brightness S, we can write

$$A \equiv F(M, S, \alpha)$$

which, since A appears to assume discrete values $k_1, k_2, \ldots$, implies

$$F(M, S, \alpha) = k_1, k_2, \ldots$$

Thus it appears that spiral galaxies are constrained to exist one of a set of discrete state planes in the three-dimensional (M, S, α) space. This then gives rise to one of two possibilities: either a spiral galaxy is "born" on one of these planes, and remains on this plane over its whole evolution; or a spiral galaxy remains on one of these planes for very long periods, with the possibility of transiting to other planes in very short periods of time.

There is currently no way of distinguishing between these possibilities, and neither is at all comprehendable from the point of view of any extant theory of galaxy formation—all of which are deeply embedded in the "standard model" of modern cosmology.

V.   A POSSIBLE THEORETICAL RESPONSE: MACH'S PRINCIPLE REVISITED

A.   Introduction

The findings of the previous sections indicate that astrophysics is a far less understood science than is generally believed—it may even turn out to be exciting again! But what could possibly constitute a rational theoretical response in the face of such phenomenology? We could try the mechanical approach, and explicitly try to formulate theories that addressed these phenomena directly; or we could step back, and ask if there is any way in which it could be argued that current theoretical perspectives fail to address identifiable fundamental issues. The approach that one chooses to take is, to a large extent, a matter of taste and (probably) prejudice. What is certainly true in the present case is that the "discrete state" phenomenon for spiral galaxies was discovered as a direct result of the theoretical prediction that circular velocities in "idealized" disk galaxies (i.e., spirals without bulges and with perfect rotational symmetry) should conform to the general power law $V_{rot} = AR^\alpha$, where (A, α) are two parameters that vary between objects; in turn, this prediction came from a theory that was constructed in the first instance to address what I saw as a fundamental failing of modern gravitation theory: its failure to encompass Mach's principle in a fundamental way.

This can be briefly explained as follows. Ordinary physical space is a metric 3-space, which means that it is a three-dimensional space within which we can perform measurements of distance and displacement. Very little thought convinces us that our concepts of a physical metric space are irreducibly connected to its matter content—that is, all our notions of distance and displacement are meaningless except insofar as they are defined as relations between objects. Similarly, all our concepts of physical time are irreducibly connected to the notion of material *process*. Consequently, it is impossible for us to conceive of physical models of "metric spacetime" without simultaneously imagining a universe of material and material process. From this, it seems clear to me that any theory that allows an internally self-consistent discussion of an *empty* metric spacetime is a deeply nonphysical theory. Since general relativity is exactly such a theory, it is fundamentally flawed, according to this view.

Since theories of gravitation are conventionally derived as point perturbations of some assumed inertial space (or flat spacetime, or equivalent), it follows that a prior condition to an understanding of gravitation is an understanding of the inertial space that is to be perturbed. The following section is devoted to the single problem of gaining such an understanding. As a means of clarifying the basic concepts involved, the development is restricted to its quasiclassical (i.e., nonrelativistic) form.

B. Historical Overview

The ideas underlying what is now known as Mach's principle can be traced to Berkeley [15,16], for which a good contemporary discussion can be found in Popper [17]. Berkeley's essential insight, formulated as a rejection of Newton's ideas of absolute space, was that the motion of any object had no meaning except insofar as that motion was referred to some other object, or set of objects. Mach ([18] reprint of 1883 German edition) went much further than Berkeley when he said "I have remained to the present day the only one who insists upon referring the law of inertia to the earth and, in the case of motions of great spatial and temporal extent, to the fixed stars." In this way, Mach formulated the idea that, ultimately, inertial frames should be defined with respect to the average rest frame of the visible universe.

It is a matter of history that Einstein was greatly influenced by Mach's ideas as expressed in the latter's *The Science of Mechanics* (see, e.g., Ref. 19) and believed that they were incorporated in his field equations as long as space was closed [20]. The modern general relativistic analysis gives detailed quantitative support to this latter view, showing how Mach's principle can be considered to arise as a consequence of the field equations when appropriate conditions are specified on an initial hypersurface in a closed evolving universe. In fact, in answer to Mach's question asking what would happen to inertia if mass were progressively removed from the universe, Lynden-Bell et al. [21] point out that, in a *closed* Friedmann universe, the maximum radius of this closed universe and the duration of its existence both shrink to zero as mass is progressively removed.

Thus, it is a matter of record that a satisfactory incorporation of Mach's principle within general relativity can be attained when the constraint of closure is imposed. However, there is still the point of view that, because general relativity allows solutions that give an internally consistent discussion of an *empty* inertial spacetime—whereas it is operationally impossible to define an inertial frame in the absence of matter—then the theory (general relativity) must have a non-fundamental basis at the classical level.

The present chapter attempts to lay the foundations of a theory of space, time, and material that addresses this perceived problem, and the main result is to show how a flat inertial space is irreducibly associated with a fractal $D = 2$ distribution of material. Furthermore, in the course of the development, fundamental insight is gained into the possible nature of "time."

C. Outline Analysis

We begin with a short review of Mach's principle, and draw from this the general conclusion that conceptions of an *empty* inertial spatiotemporal continuum are essentially nonphysical. The fact that we have apparently successful theories based exactly on such conceptions does not conflict with this statement—as long as we accept that, in such cases, the empty inertial spatiotemporal continuum is understood to be a metaphor for a deeper reality in which the metric (or inertial) properties of this spatiotemporal continuum are somehow projected out of an *unaccounted-for* universal distribution of material. For example, according to this view, the fact that general relativity admits an empty inertial spatiotemporal continuum as a special case (and was actually originally derived as a generalization of such a construct) implies that it is based on such a metaphor—and is therefore, according to this view, not sufficiently primitive to act as a basis from which fundamental theories of cosmology can be constructed.

By starting with a model universe consisting of objects that have no other properties except identity (and hence enumerability) existing in a formless continuum, we show how it is possible to project spatiotemporal metric properties from the objects onto the continuum. By considering idealized dynamical equilibrium conditions (which arise as a limiting case of a particular free parameter going to zero), we are then able to show how a globally inertial spatiotemporal continuum is necessarily identified with a material distribution that has a fractal dimension $D = 2$ in this projected space. This is a striking result since it bears a very close resemblance to the cosmic reality for the low-to-medium redshift regime.

However, this idealized limiting case material distribution is distinguished from an ordinary material distribution in the sense that the individual particles of which it is composed are each in a state of arbitrarily directed motion, but with equal-magnitude velocities for all particles—and in this sense is more like a quasiphoton gas distribution. For this reason, we interpret the distribution as a rudimentary representation of an inertial material vacuum, and present it as the appropriate physical background within which gravitational processes (as conventionally understood) can be described as point-source perturbations of an inertial spatiotemporal–material background. We briefly discuss how such processes can arise.

1.   Overview of the Nonrelativistic Formalism

In order to clarify the central arguments and to minimize conceptual problems in this initial development, we assume that the model universe is stationary in the sense that the overall statistical properties of the material distribution do not evolve in any way. Whilst this was intended merely as a simplifying assumption, it has the fundamental effect of making the development inherently nonrelativistic (in the sense that the system evolves within a curved metric 3-space, rather than being a geodesic structure within a spacetime continuum).

The latter consequence arises in the following way. Since the model universe is assumed to be stationary, there is no requirement to import a predetermined concept of "time" into the discussion at the beginning—although the qualitative notion of a generalized "temporal ordering" is assumed. The arguments used then lead to a formal model that allows the natural introduction of a generalized temporal ordering parameter, and this formal model is invariant with respect to any transformation of this latter parameter, which leaves the absolute ordering of events unchanged. This arbitrariness implies that the formal model is incomplete, and can be completed only by the imposition of an additional condition that constrains the temporal ordering parameter to be identifiable with some model of physical time. It is then found that such a model of physical time, defined in terms of "system process," arises automatically from the assumed isotropies within the system. In summary, the assumption of stationarity leads to the emergent concept of a physical "spatiotemporal continuum" that partitions into a metric 3-space together with a distinct model of physical time defined in terms of ordered material process in the metric 3-space. The fractal $D = 2$ inertial universe then arises as an idealized limiting case.

2.   Overview of the Relativistic Formalism

The relativistic formalism arises as a natural consequence of relaxing the constraint of a *stationary* universe. The formalism is not considered in any detail here but, briefly, its development can be described as follows. If the universe is not stationary, then it is evolving—and this implies the need for a predetermined concept of "time" to be included in the discussion at the outset. If this is defined in any of the ways that are, in practice, familiar to us then we can reasonably refer to it as "local process time." Arguments that exactly parallel those used in the stationary universe case considered in detail here then lead to a situation that is identical to that encountered in the Lagrangian formulation of general relativity—in that historical case, the equations of motion include a local coordinate time (which corresponds to our local process time) together with a global temporal ordering parameter, and the equations of motion are invariant with respect to any transformation of this latter parameter, which leaves the ordering of "spacetime" events unchanged. This implies that the equations of motion are incomplete—and the situation is resolved there by defining the global temporal ordering parameter to be "particle proper time." The solution we adopt for our evolving universe case is formally identical, so that everything is described in terms of a metric spacetime. By considering idealized dynamical equilibrium conditions, we are led to the concept of an inertial spacetime that is identical to the spacetime of special relativity—except that it is now irreducibly associated with a fractally distributed *relativistic* photon gas.

D.   Mach's Principle

Although most reading this chapter will have a general understanding of Mach's principle, its centrality to our argument makes a short review a worthwhile investment.

1.   Conventional Approach

Briefly, there are two kinds of mass: gravitational mass and inertial mass. *Gravitational mass* is what is measured on any kind of weighing machine (classically a pan balance, in which the mass to be measured is weighed against a collection of standard masses); *inertial mass* is what is measured in a collision experiment between the mass to be measured and a standard mass. In each case, the measured quantity is measured *relative* to some chosen standard, and therefore has no absolute significance.

The relevant facts about inertial mass are best explained first in the context of collisions between two smooth balls on a horizontal smooth surface viewed from a nonaccelerating frame of reference (the precise meaning of the term "nonaccelerating" is given shortly): Suppose we arrange for two balls, A and B, say, to be rolled along the same line at different speeds so that they collide, and then rebound (necessarily on the same line also), and that the *change* in the speeds of each ball is measured to be ΔV_A and ΔV_B, respectively. Then it is found that the ratio $\Delta V_A / \Delta V_B$ always has the same value *independently* of the initial speeds of the two balls. In other words, the calculated ratio appears to be a relative property of the balls, rather than being dependent on the initial conditions of the experiment. Now suppose that the experiment is repeated, but is now viewed from an accelerating frame of reference. It is now found that the ratio $\Delta V_A / \Delta V_B$ varies according to the initial speeds of the two balls.

The preceding paragraphs are clear, except in one respect: the notions of *accelerating* and *nonaccelerating* are undefined. This lack of rigor is usually rectified by *defining* the state of nonacceleration to be relative to the distant galaxies: specifically, by identifying that frame of reference that appears to be at rest with respect to the statistically averaged motion of distant galaxies, and then using this very special frame as a standard against which nonaccelerating motion is defined. It is then found that the ratio $\Delta V_A / \Delta V_B$ calculated in the collision experiments is constant in this class of frames; this ratio is termed the "relative inertial mass of the two balls," and the frames within which it can be measured (the nonaccelerating frames) are termed *inertial frames*.

The analysis described above makes it clear that there is some kind of relationship between the distant galaxies, and the idea of relative inertial mass - and the statement of the existence of such a relationship—is termed *Mach's principle*.

2.   *Alternative Approach*

Although the conventional approach outlined above contains all the essential components of Mach's principle, it does not focus on what, in our view, is *the* essential point about the principle: that it is *impossible* to define inertial frames in the absence of material. This fact is brought out most clearly in the following alternative approach.

Specifically, rather than define inertial frames with respect to the universal rest frame, we can define an inertial frame as any frame of reference within which the series of collision experiments discussed above yields the ratio $\Delta V_A / \Delta V_B$ to be a constant independently of the experiment's initial conditions. If this constant ratio is then termed the "relative inertial mass of the two balls," then the whole idea of the inertial frame and inertial mass is arrived at without any reference whatsoever to "distant galaxies"—and, in fact, is given a local context.

More significantly, this approach brings into the foreground *the* crucial point about Mach's principle: that it is *impossible* to define inertial frames in the absence of material. It is this argument that, in our view, renders general relativity—which allows an internally consistent discussion of empty inertial spacetime—into a nonfundamental theory.

E.   A Qualitative Description of the New Approach

We have argued that the fundamental significance of Mach's principle arises from its implication of the impossibility of defining inertial frames in the absence of material; or, as a generalization, we can say that it is *impossible* to conceive of a physical spatiotemporal continuum in the absence of material. It follows from this that, if we are to arrive at a consistent and fundamental implementation of Mach's principle, then we need a theory of the world according to which (roughly speaking) notions of the spatiotemporal continuum are somehow projected out of primary relationships between objects. In other words, we require a theory in which notions of metrical space and time are to be considered as metaphors for these primary relationships. Our starting point is to consider the calibration of a radial measure which conforms to these ideas.

Consider the following perfectly conventional procedure that assumes that we "know" what is meant by a given radial displacement, say, R. On a sufficeintly large scale (e.g., $> 10^8$ lightyears), we can reasonably assume that it is possible to write down a relationship describing the amount of mass contained within a given spherical volume, for instance

$$M = U(R) \qquad (1)$$

where U is, in principle, determinable. Of course, a classical description of this type ignores the discrete nature of real material; however, overlooking this point, such a description is completely conventional and unremarkable. Because M obviously increases as R increases, then U is said to be monotonic, with the consequence that the above relationship can be inverted to give

$$R = G(M) \qquad (2)$$

which, because (1) is unremarkable, is also unremarkable.

In the conventional view, (1) is logically prior to (2); however, it is perfectly possible to reverse the logical priority of (1) and (2) so that, in effect, we can choose to *define* the radial measure in terms of (2) rather than assume that it is known by some independent means. If this is done, then, we have immediately, made it impossible to conceive of radial measure in the absence of material. With this as a starting point, we are able to construct a completely Machian cosmology in a way outlined in the following sections.

F. A Discrete Model Universe

The model universe is intended as an idealization of our actual universe, and is defined as follows:

- It consists of an infinity of identical, but labeled, discrete material particles that are primitive, possessing no other properties beyond being countable.
- "Time" is to be understood, in a qualitative way, as a measure of process or ordered change in the model universe.
- There is at least one origin about which the distribution of material particles is statistically isotropic—meaning that the results of sampling

along arbitrary lines of sight over sufficiently long characteristic "times" are independent of the directions of lines of sight.

- The distribution of material is statistically stationary—meaning that the results of sampling along arbitrary lines of sight over sufficiently long characteristic times are independent of sampling epoch.

Although concepts of invariant spatiotemporal measurement are implicitly assumed to exist in this model universe, we make no apriori assumptions about their quantitative definition, but require that such definitions arise naturally from the structure of the model universe and from the following analysis.

1.   The Invariant Calibration of a Radial Coordinate in Terms of Counting Primitive Objects

In Eq. (2), we have already introduced, in a qualitative way, the idea that the radial magnitude of a given sphere can be *defined* in terms of the amount of material contained within that sphere and, in this section, we seek to make this idea more rigorous. To this end, we note that the most primitive invariant that can be conceived is that based on the counting of objects in a countable set, and we show how this fundamental idea can be used to define the concept of invariant distance in the model universe.

The isotropy properties assumed for the model universe imply that it is statistically spherically symmetric about the chosen origin. If, for the sake of simplicity, it is assumed that the characteristic sampling times over which the assumed statistical isotropies become exact are infinitesimal, then the idea of statistical spherical symmetry, gives way to the idea of exact spherical symmetry thereby allowing the idea of some kind of rotationally invariant radial coordinate to exist. As a first step toward defining such an idea, suppose only that the means exists to define a succession of nested spheres, $S_1 \subset S_2 \subset \cdots \subset S_p$, about the chosen origin; since the model universe with infinitesimal characteristic sampling times is stationary, then the flux of particles across the spheres is such that these spheres will always contain fixed numbers of particles, say $N_1, N_2, \ldots, N_p$, respectively.

Since the only invariant quantity associated with any given sphere, say S, is the *number* of material particles contained within it, such as N, then the only way to associate an invariant radial coordinate, say, r with S is to *define* it according to $r = r_0 f(N)$, where r_0 is a fixed scale constant having units of "length" and the function f is restricted by the requirements $f(N_a) > f(N_b)$ whenever $N_a > N_b$, $f(N) > 0$ for all $N > 0$, and $f(0) = 0$. To summarize, an invariant calibration of a radial coordinate in the model universe is given by $r = r_0 f(N)$ where

- $f(N_a) > f(N_b)$ whenever $N_a > N_b$.
- $f(N) > 0$ for all $N > 0$ and $f(0) = 0$.

Once a radial coordinate has been invariantly calibrated, it is a matter of routine to define a rectangular coordinate system based on this radial calibration; this is taken as done for the remainder of this chapter.

2. The Mass Model

At this stage, since no notion of *inertial frame* has been introduced, the idea of *inertial mass* cannot be defined. However, we have assumed the model universe to be composed of a countable infinity of labeled—but otherwise indistinguishable—material particles so that we can associate with each individual particle a property called *mass* that quantifies the amount of material in the particle, and is represented by a scale constant, say, m_0, having units of mass.

The radial parameter about any point is *defined* by $r = r_0 f(N)$; since this function is constrained to be monotonic, its inverse exists so that, by definition, $N = f^{-1}(r/r_0)$. Suppose that we now introduce the scale constant m_0; then $N m_0 = m_0 f^{-1}(r/r_0) \equiv M(r)$ can be *interpreted* as quantifying the total amount of material inside a sphere of radius r centered on the assumed origin. Although $r = r_0 f(N)$ and $M(r) = N m_0$ are equivalent, the development that follows is based on using $M(r)$ as a description of the mass distribution given as a function of an invariant radial distance parameter, r, of undefined calibration.

It is clear from the foregoing discussion that r is defined as a necessarily discrete parameter. However, to enable the use of familiar techniques, it will hereafter be supposed that r represents a continuum—with the understanding that a fully consistent treatment will require the use of discrete mathematics throughout.

G.   The Absolute Magnitudes of Arbitrary Displacements in the Model Universe

We have so far defined, in general terms, an invariant radial coordinate calibration procedure in terms of the radial distribution of material valid from the assumed origin, and have noted that such a procedure allows a routine definition of orthogonal coordinate axes. Whilst this process has provided a means of describing arbitrary displacements relative to the global material distribution, it does not provide the means by which an invariant *magnitude* can be assigned to such displacements—that is, there is no metric defined for the model universe. In the following, we show how the notion of "metric" can be considered to be projected from the mass distribution.

1.   Change in Perspective as a General Indicator of Displacement in a Material Universe

In order to understand how the notion of metric can be defined, we begin by noting the following empirical circumstances from our familiar world:

- In reality, one (an observer) recognizes the fact of a spatial displacement by reference to one's changed perspective of one's material universe.
- The same observer can judge the magnitude of a displacement in terms of the magnitude of the changes in the perspective of the material distribution arising from the displacement.

These circumstances suggest the possibility of using the concept of "perspective change" in the model universe as a means of associating absolute magnitudes to coordinate displacements. However, before this can be done, we must first give a quantitative meaning to the notion of perspective in the model universe.

2.   Perspective in the Model Universe

In general terms, a "perspective" implies the existence of an observed object plus a particular angle of view onto the object. If, in the context of the mass-model, $M(r)$, the observed object is considered defined by the specification of a constant-mass surface (r = constant), then, subject to the magnitude of the normal gradient vector, ∇M, as a monotonic function of r, total perspective information is precisely carried by the normal gradient vector itself. To see this, we note that the assumed monotonicity of the magnitude of ∇M means that it is in a $1:1$ relation with r; consequently, this magnitude defines *which* constant-mass surface is observed. Simultaneously, the direction of ∇M, which is always radial, defines an angle of view onto this constant-mass surface.

So, to summarize, an observer's perspective of the mass model, $M(r)$, can be considered defined by the normal gradient vector, $\mathbf{n} \equiv \nabla M$, at the observer's position.

3.   Change in Perspective in the Model Universe

We now consider the change in perspective arising from an infinitesimal change in coordinate position. Defining the components of the normal gradient vector (the perspective) as $n_a \equiv \nabla_a M, a = 1, 2, 3$, then the *change* in perspective for a coordinate displacement $d\mathbf{r} \equiv (dx^1, dx^2, dx^3)$ is given by

$$dn_a = \nabla_j(\nabla_a M)dx^j \equiv g_{ja}dx^j, \qquad g_{ab} \equiv \nabla_a \nabla_b M \qquad (3)$$

for which it is assumed that the geometrical connections required to give this latter expression an unambiguous meaning will be defined in due course. Given that g_{ab} is non-singular, we now note that (3) provides a $1:1$ relationship between the contravariant vector dx^a (defining change in the observer's coordinate position) and the covariant vector dn_a (defining the corresponding change in the observer's perspective). It follows that we can define dn_a as the covariant form of dx^a, so that g_{ab} automatically becomes the mass model metric tensor. The scalar product $dS^2 \equiv dn_i dx^i$ is then the absolute magnitude of the coordinate displacement, dx^a, defined relative to the change in perspective arising from the coordinate displacement.

The units of dS^2 are easily seen to be those of *mass* only and so, in order to make them those of length^2—as dimensional consistency requires—we define the working invariant as $ds^2 \equiv (2r_0^2/m_0)dS^2$, where r_0 and m_0 are scaling constants for the distance and mass scales, respectively, and the numerical factor has been introduced for later convenience.

Finally, if we want

$$ds^2 \equiv \left(\frac{r_0^2}{2m_0}\right)dn_i dx^i \equiv \left(\frac{r_0^2}{2m_0}\right)g_{ij}dx^i dx^j \tag{4}$$

to behave sensibly in the sense that $ds^2 = 0$ only when $d\mathbf{r} = 0$, we must replace the condition of nonsingularity of g_{ab} by the condition that it is strictly positive (or negative) definite; in the physical context of the present problem, this will be considered to be a self-evident requirement.

4. The Connection Coefficients

We have assumed that the geometric connection coefficients can be defined in some sensible way. To do this, we simply note that, in order to define conservation laws (i.e., to do physics) in a Riemannian space, it is necessary to be have a generalized form of Gauss' divergence theorem in the space. This is certainly possible when the connections are defined to be the metrical connections, but it is by no means clear that it is ever possible otherwise. Consequently, the connections are assumed to be metrical and so g_{ab}, given in (3), can be written explicitly as

$$g_{ab} \equiv \nabla_a \nabla_b M \equiv \frac{\partial^2 M}{\partial x^a \partial x^b} - \Gamma_{ab}^k \frac{\partial M}{\partial x^k} \tag{5}$$

where Γ_{ab}^k are the Christoffel symbols, and given by

$$\Gamma_{ab}^k = \frac{1}{2}g^{kj}\left(\frac{\partial g_{bj}}{\partial x^a} + \frac{\partial g_{ja}}{\partial x^b} - \frac{\partial g_{ab}}{\partial x^j}\right)$$

H. The Metric Tensor Given in Terms of the Mass Model

It can be shown how, for an arbitrarily defined mass model, $M(r)$, (5) can be exactly resolved to give an explicit form for g_{ab} in terms of such a general $M(r)$. Defining

$$\mathbf{r} \equiv (x^1, x^2, x^3), \qquad \Phi \equiv \frac{1}{2}\langle \mathbf{r}|\mathbf{r}\rangle \quad \text{and} \quad M' \equiv \frac{dM}{d\Phi}$$

where $\langle \cdot|\cdot \rangle$ denotes a scalar product, it is found that

$$g_{ab} = A\delta_{ab} + Bx^a x^b \tag{6}$$

where

$$A \equiv \frac{d_0 M + m_1}{\Phi}, \qquad B \equiv -\frac{A}{2\Phi} + \frac{d_0 M' M'}{2A\Phi}$$

for arbitrary constants d_0 and m_1, where, as inspection of the structure of these expressions for A and B shows, d_0 is dimensionless and m_1 has dimensions of mass. Noting that M always occurs in the form $d_0 M + m_1$, it is convenient to write $\mathscr{M} \equiv d_0 M + m_1$, and to write A and B as

$$A \equiv \frac{\mathscr{M}}{\Phi}, \qquad B \equiv -\left(\frac{\mathscr{M}}{2\Phi^2} - \frac{\mathscr{M}' \mathscr{M}'}{2d_0 \mathscr{M}}\right) \tag{7}$$

I. Geodesic Distance Determined in Terms of Matter Distribution

We calibrate the radial displacement parameter so that it coincides with the geodesic radius, and find the remarkable result that, on sufficiently large scales, the calibrated radius of a sphere centered on the chosen origin in the model universe then varies as the square root of the mass contained within the sphere.

Using (6) and (7) in (4), and after using $x^i \, dx^i \equiv r \, dr$ and $\Phi \equiv r^2/2$, we find, the following for an arbitrary displacement:

$$ds^2 = \left(\frac{r_0^2}{2m_0}\right)\left\{\frac{\mathscr{M}}{\Phi} dx^i dx^i - \Phi\left(\frac{\mathscr{M}}{\Phi^2} - \frac{\mathscr{M}' \mathscr{M}'}{d_0 \mathscr{M}}\right) dr^2\right\}$$

Now suppose that the displacement is purely radial; in this case, we find

$$ds^2 = \left(\frac{r_0^2}{2m_0}\right)\left\{\Phi\left(\frac{\mathscr{M}' \mathscr{M}'}{d_0 \mathscr{M}}\right) dr^2\right\}$$

Use of $\mathscr{M}' \equiv d\mathscr{M}/d\Phi$ reduces this latter relationship to

$$ds^2 = \frac{r_0^2}{d_0 m_0}\left(d\sqrt{\mathscr{M}}\right)^2 \rightarrow ds = \frac{r_0}{\sqrt{d_0 m_0}} d\sqrt{\mathscr{M}}$$

which defines the invariant magnitude of an infinitesimal *radial* displacement purely in terms of $\mathscr{M} \equiv d_0 M + m_1$, which represents the mass model. From this, we easily see that if we make the association $r \equiv s$ (which we can, since r is so far uncalibrated) so that the radial coordinate r effectively coincides with the geodesic distance, then geodesic radial displacement from the chosen coordinate origin is *defined* by

$$r = \frac{r_0}{\sqrt{d_0 m_0}}\left(\sqrt{\mathscr{M}} - \sqrt{\mathscr{M}_0}\right)$$

where $\mathcal{M}_0$ is the value of $\mathcal{M}$ at $r = 0$; the significance of this result lies in the fact that it says the perception of physical displacement is *created* by the matter distribution.

For convenience, this result is restated as follows. Using $\mathcal{M} \equiv d_0 M + m_1$ and noting that $M(r = 0) = 0$ necessarily, then $\mathcal{M}_0 = m_1$ from which the preceding result can be equivalently arranged as

$$\mathcal{M} = \left[\frac{\sqrt{d_0 m_0}}{r_0} r + \sqrt{m_1} \right]^2 \tag{8}$$

Using $\mathcal{M} \equiv d_0 M + m_1$ again, the mass distribution function can be expressed in terms of the invariant radial displacement as

$$M = m_0 \left(\frac{r}{r_0} \right)^2 + 2\sqrt{\frac{m_0 m_1}{d_0}} \left(\frac{r}{r_0} \right) \tag{9}$$

which, for large r, can be approximated as $M \approx m_0 (r/r_0)^2$. In other words, on a sufficently large scale, radial distance varies as the square root of mass from the chosen origin—or, equivalently, the mass varies as r^2. As a consequence of this, M/r^2 is a global constant on a large enough scale and has the limiting value m_0/r_0^2; for the remainder of this paper the notation $g_0 \equiv m_0/r_0^2$ is employed.

J. A Qualitative Discussion of the Temporal Dimension

So far, the concept of time has entered the discussion only in the form of the qualitative definition given in Section V.F; it has not entered in any quantitative way, and, until it does, there can be no discussion of dynamical processes.

Since, in its most general definition, time is a parameter that orders change within a system, a necessary prerequisite for its quantitative definition in the model universe is a notion of change within that universe, and the only kind of change that can be defined in such a simple place as the model universe is that of internal change arising from the spatial displacement of particles. Furthermore, since the system is populated solely by primitive particles that possess only the property of enumerability (and hence quantification in terms of the *amount* of material present), then, in effect, all change is gravitational change. This fact is incorporated into the cosmology to be derived by constraining all particle displacements to satisfy the "weak equivalence principle." We are then led to a Lagrangian description of particle motions in which the Lagrange density is degree zero in its temporal ordering parameter. From this, it follows that the corresponding Euler–Lagrange equations form an *incomplete* set.

The origin of this problem traces back to the fact that, because the Lagrangian density is degree zero in the temporal ordering parameter, it is then invariant with respect to any transformation of this parameter that preserves the ordering.

This implies that, in general, temporal ordering parameters cannot be identified directly with physical time—they merely share one essential characteristic. This situation is identical to that encountered in the Lagrangian formulation of general relativity; there, the situation is resolved by defining the concept of "particle proper time." In the present case, this is not an option because the notion of particle proper time involves the prior definition of a system of observer's clocks—so that some notion of clock time is factored into the prior assumptions on which general relativity is built.

In the present case, it turns out that the isotropies already imposed on the system conspire to provide an automatic resolution of the problem that is consistent with the already assumed interpretation of time as a measure of ordered change in the model universe. To be specific, it turns out that the elapsed time associated with any given particle displacement is proportional, via a scalar field, to the invariant spatial measure attached to that displacement. Thus, physical time is defined directly in terms of the invariant measures of *process* with the model universe.

K. Dynamical Constraints in the Model Universe

First, and as already noted, the model universe is populated exclusively by primitive particles that possess solely the property of enumeration, and hence quantification. Consequently, all motions in the model universe are effectively gravitational, and we model this circumstance by constraining all such motions to satisfy the weak equivalence principle, by which we mean that the trajectory of a body is independent of its internal constitution. This constraint can be expressed as follows:

Constraint 1.   *Particle trajectories are independent of the specific mass values of the particles concerned.*

Second, given the isotropy conditions imposed on the model universe from the chosen origin, symmetry arguments lead to the conclusion that the net action of the whole universe of particles acting on any given single particle is such that any net acceleration of the particle must always appear to be directed through the coordinate origin. Note that this conclusion is *independent* of any notions of retarded or instantaneous action. This constraint can then be stated as follows:

Constraint 2.   *Any acceleration of any given material particle must necessarily be along the line connecting the particular particle to the coordinate origin.*

L. Gravitational Trajectories

Suppose that p and q are two arbitrarily chosen point coordinates on the trajectory of the chosen particle, and suppose that (4) is integrated between these points to give the scalar invariant

$$I(p, q) = \int_p^q \left(\frac{1}{\sqrt{2g_0}} \right) \sqrt{dn_i dx^i} \equiv \int_p^q \left(\frac{1}{\sqrt{2g_0}} \right) \sqrt{g_{ij} dx^i dx^j} \tag{10}$$

Then, in accordance with the foregoing interpretation, $I(p, q)$ gives a scalar record of how the particle has moved between p and q defined with respect to the particle's continually changing relationship with the mass model, $M(r)$.

Now suppose that $I(p, q)$ is minimized with respect to choice of the trajectory connecting p and q; this minimizing trajectory can then be interpreted as a geodesic in the Riemannian space that has g_{ab} as its metric tensor. Given that g_{ab} is defined in terms of the mass model $M(r)$—the existence of which is independent of any notion of inertial mass, then the existence of the metric space, and of geodesic curves within it, is likewise explicitly independent of any concept of inertial mass. It follows that the identification of the particle trajectory r with these geodesics means that particle trajectories are similarly independent of any concept of inertial mass, and can be considered as the modeling step defining that general subclass of trajectories that conform to that characteristic phenomenology of gravitation defined by Constraint 1 (in Section V.K).

M. The Equations of Motion

While the mass distribution, represented by $\mathcal{M}$, has been explicitly determined in terms of the geodesic distance at (8), it is convenient to develop the theory in terms of unspecified $\mathcal{M}$.

The geodesic equations in the space with the metric tensor (6) can be obtained, in the usual way, by defining the Lagrangian density

$$\mathcal{L} \equiv \left(\frac{1}{\sqrt{2g_0}} \right) \sqrt{g_{ij} \dot{x}^i \dot{x}^j} = \left(\frac{1}{\sqrt{2g_0}} \right) (A \langle \dot{r} | \dot{r} \rangle + B \dot{\Phi}^2)^{1/2} \tag{11}$$

where $\dot{x}^i \equiv dx^i/dt$, and so on, and writing down the Euler–Lagrange equations

$$2A\ddot{r} + \left(2A'\dot{\Phi} - 2\frac{\dot{\mathcal{L}}}{\mathcal{L}} A \right) \dot{r} + \left(B'\dot{\Phi}^2 + 2B\ddot{\Phi} - A'\langle \dot{r} | \dot{r} \rangle - 2\frac{\dot{\mathcal{L}}}{\mathcal{L}} B\dot{\Phi} \right) r = 0 \tag{12}$$

where $\dot{r} \equiv dr/dt$ and $A' \equiv dA/d\Phi$, and so forth. By identifying particle trajectories with geodesic curves, this equation is now interpreted as the equation of motion, referred to the chosen origin, of a single particle satisfying Constraint 1.

However, noting that the variational principle, Eq. (10), is of order zero in its temporal ordering parameter, we can conclude that the principle is invariant with respect to arbitrary transformations of this parameter; in turn, this means that the temporal ordering parameter cannot be identified with physical time.

This problem manifests itself formally in the statement that the equations of motion (12) do not form a complete set, so that it becomes necessary to specify some extra condition to close the system.

A similar circumstance arises in general relativity theory when the equations of motion are derived from an action integral that is formally identical to (10). In that case, the system is closed by specifying the arbitrary time parameter to be the "proper time," so that

$$d\tau = \mathcal{L}(x^j, dx^j) \rightarrow \mathcal{L}\left(x^j, \frac{dx^j}{d\tau}\right) = 1 \qquad (13)$$

which is then considered as the necessary extra condition required to close the system. In the present circumstance, we are rescued by the, as yet, unused Constraint 2.

N. The Quantitative Definition of Physical Time

1. Completion of Equations of Motion

Consider Constraint 2, which states that any particle accelerations must necessarily be directed through the coordinate origin. This latter condition simply means that the equations of motion must have the general structure $\ddot{\mathbf{r}} = G(t, \mathbf{r}, \dot{\mathbf{r}})\mathbf{r}$ for scalar function $G(t, \mathbf{r}, \dot{\mathbf{r}})$. In other words, (12) satisfies Constraint 2 if the coefficient of $\dot{\mathbf{r}}$ is zero, so that

$$\left(2A'\dot{\Phi} - 2\frac{\dot{\mathcal{L}}}{\mathcal{L}}A\right) = 0 \rightarrow \frac{A'}{A}\dot{\Phi} = \frac{\dot{\mathcal{L}}}{\mathcal{L}} \rightarrow \mathcal{L} = k_0 A \qquad (14)$$

for arbitrary constant k_0, which is necessarily positive since $A > 0$ and $\mathcal{L} > 0$. The condition (14), which guarantees Constraint 2, can be considered as the condition required to close the incomplete set (12), and is directly analogous to (13), the condition that defines "proper time" in general relativity.

2. Physical Time Defined Quantitatively as Process

Equation (14) can be considered as the equation that removes the preexisting arbitrariness in the time parameter by *defining* physical time; from (14) and (11) we have

$$\mathcal{L}^2 = k_0^2 A^2 \rightarrow A\langle\dot{\mathbf{r}}|\dot{\mathbf{r}}\rangle + B\dot{\Phi}^2 = 2g_0 k_0^2 A^2 \rightarrow g_{ij}\dot{x}^i\dot{x}^j = 2g_0 k_0^2 A^2 \qquad (15)$$

so that, in explicit terms, physical time is *defined* by the relation

$$dt^2 = \left(\frac{1}{2g_0 k_0^2 A^2}\right)g_{ij}dx^i dx^j \qquad (16)$$

In short, the elapsing of time is given a direct physical interpretation in terms of the process of *displacement* in the model universe.

Finally, noting that, by (16), the dimensions of k_0^2 are those of $L^6/[T^2 \times M^2]$, then the fact that $g_0 \equiv m_0/r_0^2$ (see Section V.I) suggests the change of notation $k_0^2 \propto v_0^2/g_0^2$, where v_0 is a constant having the dimensions (but not the interpretation) of velocity. So, as a means of making the dimensions that appear in the development more transparent, it is found convenient to use the particular replacement $k_0^2 \equiv v_0^2/(4d_0^2 g_0^2)$, where d_0 is the dimensionless global constant introduced in Section V.H. With this replacement, the *definition* of physical time, given at (16), becomes

$$dt^2 = \left(\frac{4d_0^2 g_0}{v_0^2 A^2}\right) g_{ij} dx^i dx^j \qquad (17)$$

since, as is easily seen from the definition of g_{ab} given in Section V.H, $g_{ij} dx^i dx^j$ is necessarily finite and nonzero for a nontrivial displacement $d\mathbf{r}$.

3.   The Necessity of $v_0^2 \neq 0$

Equation (17) provides a definition of physical time in terms of basic process (displacement) in the model universe. Since the parameter v_0^2 occurs nowhere other than in its explicit position in (17), it is clear that setting $v_0^2 = 0$ is equivalent to physical time becoming undefined. Therefore, of necessity, $v_0^2 \neq 0$.

O.   The Cosmological Potential

The model is most conveniently interpreted when expressed in potential terms, and so, in the following, paragraphs, we show how this is done.

1.   The Equations of Motion: Potential Form

From Section V.N, when (14) is used in (12), there results

$$2A\ddot{\mathbf{r}} + \left(B'\dot{\Phi}^2 + 2B\ddot{\Phi} - A'\langle\dot{\mathbf{r}}|\dot{\mathbf{r}}\rangle - 2\frac{A'}{A}B\dot{\Phi}^2\right)\mathbf{r} = 0 \qquad (18)$$

Suppose we define a function V according to $V \equiv C_0 - \langle\dot{\mathbf{r}}|\dot{\mathbf{r}}\rangle/2$, for some arbitrary constant C_0; then, by (15)

$$V \equiv C_0 - \frac{1}{2}\langle\dot{\mathbf{r}}|\dot{\mathbf{r}}\rangle = C_0 - \frac{v_0^2}{4d_0^2 g_0}A + \frac{B}{2A}\dot{\Phi}^2 \qquad (19)$$

where A and B are defined in (7). With the unit vector $\hat{\mathbf{r}}$, this function can be used to express (18) in the potential form

$$\ddot{\mathbf{r}} = -\frac{dV}{dr}\hat{\mathbf{r}} \qquad (20)$$

so that V is a potential function and C_0 is the arbitrary constant usually associated with a potential function.

2.   The Potential Function, V, as a Function of r

From (19), we have

$$2C_0 - 2V = \dot{r}^2 + r^2\dot{\theta}^2 = \frac{v_0^2}{2d_0^2 g_0}A - \frac{B}{A}r^2\dot{r}^2$$

so that V is effectively given in terms of r and $\dot{r}$. In order to clarify things further, we now eliminate the explicit appearance of $\dot{r}$. Since all forces are central, angular momentum is conserved; consequently, after using conserved angular momentum, h, and the definitions of A, B, and $\mathcal{M}$ given in Section V.H, the foregoing equations can be written as

$$2C_0 - 2V = \dot{r}^2 + r^2\dot{\theta}^2 = v_0^2 + \frac{4v_0^2}{r}\sqrt{\frac{m_1}{d_0 g_0}} + \frac{d_0 - 1}{r^2}\left(\frac{6m_1 v_0^2}{d_0^2 g_0} - h^2\right)$$

$$+ \frac{2}{r^3}\sqrt{\frac{d_0 m_1}{g_0}}\left(\frac{2m_1 v_0^2}{d_0^2 g_0} - h^2\right) + \frac{1}{r^4}\frac{m_1}{g_0}\left(\frac{m_1 v_0^2}{d_0^2 g_0} - h^2\right)$$

$$(21)$$

so that $V(r)$ is effectively given by the right-hand side of (21).

P.   A Discussion of the Potential Function

It is clear from (24) that m_1 plays the role of the mass of the central source that generates the potential, V. A detailed analysis of the behavior of V shows that there are two distinct classes of solution depending on the free parameters of the system:

- A constant potential universe within which all points are dynamically indistinguishable; this corresponds to an inertial material universe, and arises in the case $m_1 = 0, d_0 = 1$.
- All other possibilities give rise to a "distinguished origin" universe in which either.

  There is a singularity at the center, $r = 0$.

  There is no singularity at $r = 0$ and, instead, the origin is the center of a nontrivial sphere of radius $R_{min} > 0$, which acts as an impervious boundary between the exterior universe and the potential source. In effect, this sphere provides the source with a nontrivial spatial extension so that the classical notion of the massive point source is avoided.

Of these possibilities, the constant potential universe is the one that suits the needs of a realistic cosmology, and this possibility is discussed in detail in the following sections.

However, of the two cases in the distinguished origin universe, the no-singularity case offers the interesting possibility of being able to model the gravitational effects created by a central massive source, but without the non-physical singularity at the origin. This case is mentioned here for future reference.

Q.   The Fractal $D=2$ Inertial Universe

Reference to Eq. (21) shows that the parameter choice $m_1 = 0$ and $d_0 = 1$ makes the potential function constant everywhere, while Eq. (9) shows how, for this case, universal matter in an equilibrium universe is necessarily distributed as an exact fractal with $D = 2$. Thus, the fractal $D = 2$ material universe is necessarily a globally inertial equilibrium universe.

Given that gravitational phenomena are usually considered to arise as mass-driven perturbations of flat inertial backgrounds, the foregoing result—to the effect that the inertial background is necessarily associated with a nontrivial fractal matter distribution—must necessarily give rise to completely new perspectives about the nature and properties of gravitational phenomena. However, as we show in Section V.Q.1, the kinematics in this inertial universe is unusual, and suggests that the inertial material distribution is more properly interpreted as a material vacuum out of which (presumably) we can consider ordinary material to condense in some fashion.

1.   The Quasiphoton Fractal Gas

For the case $m_1 = 0$, $d_0 = 1$, the definition M in (9) together with the definitions of A and B in Section V.H. give

$$A = \frac{2m_0}{r_0^2}, \qquad B = 0$$

so that, by (19) (remembering that $g_0 \equiv m_0/r_0^2$), we have

$$\langle \dot{\mathbf{r}} | \dot{\mathbf{r}} \rangle = v_0^2 \tag{22}$$

for all displacements in the model universe. It is (almost) natural to assume that the constant v_0^2 in (22) simply refers to the constant velocity of any given particle, and likewise to assume that this can differ between particles. However, each of these assumptions would be wrong since—as we now show—v_0^2 is (1) more properly interpreted as a conversion factor from spatial to temporal units and, (2) a *global* constant that applies equally to all particles.

To understand these points, we begin by noting that (22) is a special case of (15) and so, by (16), can be equivalently written as

$$dt^2 = \frac{1}{v_0^2} \langle d\mathbf{r} | d\mathbf{r} \rangle \tag{23}$$

which, by the considerations of Section V.N.2, we recognize as the *definition* of the elapsed time experienced by any particle undergoing a spatial displacement $d\mathbf{r}$ in the model inertial universe. Since this universe is isotropic about all points, there is nothing that can distinguish between two separated particles (other than their separateness) undergoing displacements of equal magnitudes; consequently, each must be considered to have experienced equal elapsed times. It follows from this that v_0^2 is not to be considered as a locally defined particle velocity, but is a *globally* defined constant that has the effect of converting between spatial and temporal units of measurement.

We now see that the model inertial universe, with (23) as a global relationship, bears a close formal resemblance to a universe filled purely with Einsteinien photons—the difference is, of course, that the particles in the model inertial universe are assumed to be countable and to have mass properties. This formal resemblance means that the model inertial universe can be likened to a quasiphoton fractal gas universe.

R. A Quasifractal Mass Distribution Law, $M \approx r^2$: The Evidence

A basic assumption of the *standard model* of modern cosmology is that, on some scale, the universe is homogeneous; however, in early responses to suspicions that the accruing data were more consistent with Charlier's conceptions of an hierarchical universe [22–24] than with the requirements of the *standard model*, De Vaucouleurs [25] showed that, within wide limits, the available data satisfied a mass distribution law $M \approx r^{1.3}$, while Peebles [26] found $M \approx r^{1.23}$. The situation, from the point of view of the *standard model*, has continued to deteriorate with the growth of the database to the point that "the scale of the largest inhomogeneities discovered to date) is comparable with the extent of the surveys, so that the largest known structures are limited by the boundaries of the survey in which they are detected" [27].

For example, several redshift surveys, such as those performed by Huchra et al. [28], Giovanelli et al. [29], De Lapparent et al. [30], Broadhurst et al. [317], Da Costa et al. [32] and Vettolani et al. [33], have discovered massive structures such as sheets, filaments, superclusters, and voids, and show that large structures are common features of the observable universe; the most significant conclusion to be drawn from all of these surveys is that the scale of the largest inhomogeneities observed is comparable with the spatial extent of the surveys themselves.

More recently several quantitative analyses of both pencil-beam and wide-angle surveys of galaxy distributions have been performed; three examples are given by Joyce et al. [34], who analyzed the CfA2-South catalog to find fractal behavior with $D = 1.9 \pm 0.1$; Labini and Montuori [35] analyzed the APM-Stromlo survey to find fractal behavior with $D = 2.1 \pm 0.1$, while Labini et al. [36] analyzed the Perseus–Pisces survey to find fractal behavior with $D = 2.0 \pm 0.1$. There are many other papers of this nature in the literature, all supporting the view that, out to medium depth, at least, galaxy distributions appear to be fractal with $D \approx 2$.

This latter view is now widely accepted (see e.g., Wu et al. [37]), and the open question has become whether there is a transition to homogeneity on some sufficiently large scale. For example, Scaramella et al. [38] analyze the ESO Slice Project redshift survey, while Martinez et al. [39] analyze the Perseus–Pisces, the APM-Stromlo, and the 1.2-Jy IRAS redshift surveys, with both groups finding evidence for a crossover to homogeneity at large scales. In response, the Scaramella et al. analysis has been criticized on various grounds by Joyce et al. [40].

So, to date, evidence that galaxy distributions are fractal with $D \approx 2$ on small to medium scales is widely accepted, but there is a lively open debate over the existence, or otherwise, of a crossover to homogeneity on large scales.

To summarize, there is considerable debate centered around the question of whether the material in the universe is distributed fractally, with supporters of the bigbang picture arguing that, basically, it is not, while the supporters of the fractal picture argue that it is with the weight of evidence supporting $D \approx 2$. This latter position corresponds exactly to the picture predicted by the present approach.

S. Conclusions

The main result arising from the present stationary universe analysis is that a perfectly inertial universe, which arises as an *idealized* limiting case, necessarily consists of a fractal, $D = 2$, distribution of material. This result is to be compared with the real universe, which approximates very closely perfectly inertial conditions on even quite small scales, and that appears to be fractal with $D \approx 2$ on the medium scale.

Since gravitational phenomena are conventionally considered to arise as mass-driven perturbations of a flat inertial background, the main result of the analysis—that the flat inertial background is irreducibly associated with a nontrivial fractal distribution of material—must necessarily lead to novel insights into the nature and causes of gravitational phenomena.

The material background has the structure of a quasi–photon gas in the sense that its individual particles move in arbitrary directions but with identical velocity magnitudes. For this reason, we interpret the material inertial background as a form of material vacuum so that, ultimately, on the proposed view, gravitational phenomena are to be seen as disturbances of a material vacuum and the present analysis is to be interpreted in terms of a rudimentary vacuum physics.

VI. OVERALL CONCLUSIONS

In the first sections of this chapter, we described two distinct forms of phenomenology that are both extremely difficult to comprehend from the perspective of conventional astrophysics and cosmology; any serious consideration of either soon leads one to the conclusion that some form of new physics is probably indicated. But, having arrived at such a tentative conclusion, it is extremely difficult to imagine what form of new physics might lead to such phenomena.

However, in the present case, the second of these two phenomena was discovered *directly* as a consequence of believing that Mach's principle is not incorporated into modern gravitation theory in any fundamental way, and replacing this with a theory that addresses this problem directly. This resulting theory gives completely new quantitative insight into the nature of "time as material process" and leads to what can be described only as a rudimentary form of vacuum physics. This theory associates inertial spacetime directly with a material vacuum that possesses a fractal dimension of 2—a result that chimes perfectly with modern galaxy surveys out to medium distances, especially if one imagines that, somehow, galaxies condense out of the material vacuum.

In conclusion, there is good reasons to believe that the rudimentary vacuum physics described in this chapter represents a potentially sound foundation for the study of (at least) the "discrete dynamical states" phenomenology. Given that it stretches incredulity to believe that two sets of new physics are indicated by the two phenomenologies described, it also seems quite possible that this vacuum phsyics has the potential to address the "quantized redshift" phenomenology as well.

References

1. W. G., Tifft, ApJ **257**, 442 (1982).

2. W. G. Tifft and W. J. Cocke, ApJ **287**, 492 (1984).

3. S. V. M. Clube and W. M. Napier, *Nature* **303**, 10 (1983).

4. B. N. G. Guthrie and W. M. Napier, MNRAS **243**, 431 (1990).

5. B. N. G. Guthrie and W. M. Napier, MNRAS **253**, 533 (1991).

6. B. N. G. Guthrie and W. M. Napier, A&A **310**, 353 (1996).

7. D. F. Roscoe, A&A **343**, 788–800 (1999).

8. V. C. Rubin, W. K. Ford, and N. Thonard, ApJ **238**, 471 (1980).

9. D. S. Mathewson, V. L. Ford, and M. Buchhorn, ApJS **81**, 413 (1992).

10. M. Persic and P. Salucci, ApJS **99**, 501 (1995).

11. D. F. Roscoe, A&A **343**, 697–704 (1999).

12. D. F. Roscoe, A&AS **140**, 247–260 (1999).

13. D. S. Mathewson and V. L. Ford, ApJS **107**, 97 (1996).

14. S. Courteau, AJ, **114**(6), 2402–2427 (1997).

15. G. Berkeley, *Principles of Human Knowledge*, 1710.

16. G. Berkeley, *De Motu*, 1721.

17. K. R. Popper, *J. Phil. Sci.* **4**, 26 (1953).

18. E. Mach, *The Science of Mechanics—a critical and Historical Account of Its Development*, Open Court, La Salle, 1960.

19. A. Pais, in *The Science and Life of Albert Einstein*, Oxford Univ. Press, 1982.

20. A. Einstein, *The Meaning of Relativity*, 3rd ed, Princeton Univ. Press, 1950.

21. D. Lynden-Bell, J. Katz, and J. Bicak, MNRAS **272**, 150 (1995).

22. C. V. L. Charlier, *Astron. Fysik* **4**, 1 (1908).

23. C. V. L. Charlier, *Ark. Mat. Astron. Physik* **16**, 1 (1922).

24. C. V. L. Charlier, PASP **37**, 177 (1924).

25. G. De Vaucouleurs, *Science* **167**, 1203 (1970).

26. P. J. E. Peebles, *The Large Scale Structure of the Universe*, Princeton Univ. Press, Princeton, NJ, 1980.

27. Yu. V. Barysev, F. Sylos Labini, M. Montuori, and L. Pietronero, *Vistas Astron.* **38**, 419 (1995).

28. J. Huchra, M. Davis, D. Latham, and J. Tonry, ApJS **52**, 89 (1983).

29. R. Giovanelli, M. P. Haynes, and G. L. Chincarini, ApJ **300**, 77 (1986).

30. V. De Lapparent, M. J. Geller, and J. P. Huchra, ApJ **332**, 44 (1988).

31. T. J. Broadhurst, R. S. Ellis, D. C. Koo, and A. S. Szalay, *Nature* **343**, 726 (1990).

32. L. N. Da Costa, M. J. Geller, P. S. Pellegrini, D. W. Latham, A. P. Fairall, R. O. Marzke, C. N. A. Willmer, J. P. Huchra, J. H. Calderon, M. Ramella, and M. J. Kurtz, ApJ **424**, L1 (1994).

33. G. Vettolani et al., *Proc. Schloss Rindberg Workshop: Studying the Universe with Clusters of Galaxies*, 1994.

34. M. Joyce, M. Montuori, and F. S. Labini, ApJ **514**, L5 (1999).

35. F. S. Labini and M. Montuori, A&A **331**, 809 (1998).

36. F. S. Labini M. Montuori, and L. Pietronero, *Phys. Lett.* **293**, 62 (1998).

37. K. K. S. Wu, O. Lahav, and M. J. Rees, *Nature* **397**, 225 (1999).

38. R. Scaramella, L. Guzzo, G. Zamorani, E. Zucca, C. Balkowski, A. Blanchard, A. Cappi, V. Cayatte, G. Chincarini, C. Collins, A. Fiorani, D. Maccagni, H. MacGillivray, S. Maurogordato, R. Merighi, M. Mignoli, D. Proust, M. Ramella, G. M. Stirpe, and G. Vettolani, A&A **334**, 404 (1998).

39. V. J. Martinez, M. J. PonsBorderia, R. A. Moyeed, and M. J. Graham, MNRAS **298**, 1212 (1998).

40. M. Joyce, M. Montuori, F. S. Labini, and L. Pietronero, A&A **344**, 387 (1999).

A SEMICLASSICAL MODEL OF THE PHOTON BASED ON OBJECTIVE REALITY AND CONTAINING LONGITUDINAL FIELD COMPONENTS

HÉCTOR A. MÚNERA

Department of Physics, Universidad Nacional de Colombia, Bogotá, Colombia

CONTENTS

Modern Nonlinear Optics, Part 3, Second Edition, Advances in Chemical Physics, Volume 119,
Edited by Myron W. Evans. Series Editors I. Prigogine and Stuart A. Rice.
ISBN 0-471-38932-3 © 2001 John Wiley & Sons, Inc.

This chapter starts with a revision, from the viewpoint of objective reality, of some physical, logical, and conceptual inconsistencies in the description of the photon in free space. Then, in the context of a four-dimensional ether, we introduce the novel concept of dynamic rest mass as a signed flow of ether fluid. Here, particles (antiparticles) are extended objects formed by a momentum flow along the positive (negative) direction of the normal to the 3D surface of Dirac's sea of energy. Therefrom, the photon is modeled as a semiclassical particle–antiparticle doublet, which can explain the meaning of frequency and rest mass of photon. For the photon's ground state, it predicts two values of spin (± 1) and de Broglie's energy equation. In the excited state, the photon has three possible values of spin: 0, ± 2. It also leads to solutions of Maxwell's equations containing both advanced and retarded components. In the near field there are longitudinal field components that disappear in the far field. In this sense, Maxwell's equations are identified as a limiting case at macroscopic distances.

I. INTRODUCTION

The possibility that the propagation of light could have a dual nature arose in the seventeenth century as a controversy between Newton and Huygens: corpuscular versus undulatory descriptions, respectively. When Maxwell's electromagnetic theory was developed in the nineteenth century, the matter seemed settled in favor of the proponents of wave-like electromagnetic phenomena.

However, right at the beginning of the twentieth century, Planck and Einstein again introduced the corpuscular view with the notion of the photon. The energy E of such a particle is given by de Broglie's relation

$$E = \hbar\omega = h\nu \tag{1}$$

where, as usual, $\hbar = h/2\pi$ are the reduced and nonreduced constants of Planck, and $\omega = 2\pi\nu$ are frequencies associated with the photon in radians per second and cycles per second respectively.

A hundred years after birth, the inner structure of the photon remains a mistery. This is particularly true when one searches for an objective reality representation. In this chapter, the focus is on the nature of the photon and the propagation of electromagnetic radiation in free space. Questions arise in at least three different areas: the rest mass, the velocity, and the solutions of

Maxwell's and wave equations in free space. As discussed elsewhere in the book, one can also question the validity of the conventional representation itself; in this chapter, however, we will keep as close as possible to Maxwell's equations. Without any pretension for completeness, some of the issues are as follows.

A. Questions Related to the Rest Mass of the Photon

There is an *apparent* incompatibility between $E_0 = 0$, and other properties of the photon. For instance

1. *The Frequency.* What is *the thing* that oscillates with frequency ν ? Clearly, it cannot be an entity that preexists at rest. Hence, E in Eq.(1) must relate to an entity in a state of oscillatory motion that disappears when motion ends. Then, it seems that, from this viewpoint, the photon behaves as a wave.

2. *The Duality.* Hence, if the photon is a wave, what is the origin of the particle-like behavior?

3. *The Spin.* How can we reconcile spin $s = 1$ and $E_0 = 0$? Spin is a constant, identical for all photons of arbitrary energy E. Hence, it is independent of energy and, therefore, it does not depend of the state of motion characterized by ω. Also, evidently, spin cannot be a property of a nonexistent rest mass. Therefore, spin is associated with what?

One possible solution is to relax the condition that rest mass is zero exactly, and allow for a tiny rest mass. Many distinguished physicists have explored this alternative, including Einstein, de Broglie, Schrödinger, and Vigier [1–8]. For additional bibliography, see Ref. 8, where Vigier explores the idea anew in the context of his interpretation of the nonnull results of Michelson and Morley; Vigier suggests a value $m_0 \sim 10^{-65}$ g. (See pp. 275–291 of Jeffers et al. [9] for a compilation of Vigier's quantum-mechanical papers.)

A weaker alternative would be to admit $E_0 = 0$ in a preferred frame of reference Σ, but to allow for a local nonzero rest mass $m_0 \sim 10^{-34}$ g as an artifact of the total motion of the earth relative to Σ [10]. However, this value is too high compared to the limits set to the photon mass, typically in the range $m_0 \sim 10^{-52}–10^{-43}$ g [11]. At any rate, there are two implications of a nonzero photonic mass:

1. Locally, the speed of light is not a constant over all frequencies.

2. An absolute inertial frame of reference Σ must be reintroduced. In plain words, introduce a modern version of the ether.

In the present writing, we propose a *novel alternative*: rest mass is not absence of mass but neutrality of momentum flux within a well-defined spatial region (see

Section IV). In this way, the properties of photon and electromagnetic radiation naturally arise from flow of momentum in Euclidean space.

B.   Questions Related to the Velocity of the Photon

There are at least four different velocities associated with photons:

- Particle velocity c
- Phase velocity in the wave representation v_{ph}
- Group velocity in the wave representation v_{gr}
- Velocity of momentum and energy transport v_{en}

To describe the transport of sound, Lord Rayleigh [12] introduced the concepts of phase and group velocities. Since there is a fluid for the propagation of sound, there is no particular difficulty in understanding the various physical processes arising therein [13]. The same concepts v_{ph} and v_{gr} were applied to light by the same Lord Rayleigh [14] and Gouy [15] and later on by Lamb [16] and [17]. For propagation of electromagnetic waves in dispersive media, there is no particular difficulty in building a physical picture for the underlying processes. For completeness, it is noted that negative group velocities were theoretically predicted at the beginning of the twentieth century. Lamb [16, p. 479] noted that "It is hardly to be expected that the notion of a negative group-velocity will have any very important physical application."

A paper by Mitchell and Chiao [18] reports some experimental evidence indicating the physical existence of negative group delays, but there are some challenges to the theoretical interpretation [19].

The situation with photon propagation in free space is quite diferent. If vacuum is equated to absence of a fluid, what is the support for the waves? Of course, particle-like propagation solves the problem, but it (strictly) invalidates Maxwell's equations in vacuum. There is a positive aspect. Since vacuum is nondispersive, all velocities have the same magnitude.

Let us concentrate on the particle aspect only. The main issue is to identify an space (three- or four-dimensional?) where photons propagate with constant speed c. Einstein's second postulate of the special theory of relativity (STR) requires the speed of light in free space to be the same for all inertial observers. This postulate is conventionally interpreted as implying the non-existence of a preferred frame Σ. As discussed in section II, the exactly opposite view will be adopted here.

There is another curiosity related to the speed of photons in STR. Long ago, it was noted [20–22] that the second postulate of STR (speed of light invariance) may be derived from a pair of more fundamental assumptions: the principle of relativity for inertial observers plus the principle of isotropy of spacetime. More recent work along the same lines [23–25] implies that the parallel addition theorem of STR becomes

$$V = \frac{v_1 + v_2}{1 + K^{-2}v_1v_2} \tag{2}$$

where K is an arbitrary constant with dimensions of speed [25]. Equation (2) suggests that there is no need for the speed of light to appear in STR, and that another more fundamental constant speed could play the role.

Photon speed also appears in the most popular equation of physics:

$$E = mc^2 \tag{3}$$

Of course, Eq. (3) is valid for *any particle*. The question is: Why is the speed of the photon there? One may conjecture with DiMarzio [26] that there is a more fundamental meaning for c. In this context, Múnera [27] explored the possibility of deriving the main predictions of STR from Newton's theory plus a postulate of mass–energy equivalence: $E = mK^2$. The value of the unknown constant K was obtained from the acceleration of electrons [28]. The numerical value is c within the limits of accuracy of the (large) experimental error.

C. Questions Related to Solutions of Electromagnetic Equations in vacuo

According to the conventional view, charge density ρ and electric current density **J** do not exist in free space. Hence, there are no sources for the electric and magnetic fields. However, both Maxwell's equations and the homogeneous wave equations have nontrivial solutions for the fields **E** and **B**. Then, what is the origin of the electromagnetic field? There is a possible solution. After being produced, fields **E** and **B** have existence independently of the source. This interpretation implicitly requires an underlying substance, or at least a 3D space, where the fields **E** and **B** linger.

A second question concerns the existence of longitudinal components of the magnetic field. Maxwell's equations in free space are (completely??) equivalent to two homogeneous *uncoupled* wave equations for the vector fields **E** and **B**. The uncoupled wave equations admit longitudinal components for both fields **E** and **B**. However, longitudinal components are prohibited in the conventional interpretation of Maxwell's equations.

A third similar question arises in the potential representation of fields **E** and **B**. Conventionally, a magnetic scalar potential is not included as part of the solution of Maxwell's equations. However, there is no a priori prohibition for the existence of such solution within a general formulation. Sections III.A–III.E consider previous issues in some detail.

Another question is related to the symmetrization of Maxwell's equations. Dirac asked himself whether there existed magnetic monopoles, and proposed inclusion of a magnetic source to make Maxwell equations symmetric.

Section III. F exhibits a different route to symmetrize the equations, without the introduction of magnetic monopoles.

Finally, both the Maxwell equations and the wave equations admit solutions in terms of retarded and advanced potentials. Long ago, Conway [29] recognized the existence of such solutions in his study of the electron. Referring to *"the convergence,* of the disturbances" Conway said that "the medium now contains the *future* history of the motion of the electron" (Ref. 29, p. 160, emphasis in original).

Obviously, such interpetation led to disregard advanced solutions as nonphysical. For instance, Ritz [30] and Tetrode [31] considered that the mathematical existence of advanced solutions was a major weakness of Maxwell's equations. An attempt to provide a physical basis for advanced potentials is due to Lewis, who proposed focusing on the process of propagation from an emitter to an absorber far away from the emitter [32]. This concept also appears in the work of Wheeler and Feynman [33]. However, such model constitutes another form of causality violation. Lewis [32, p. 25] himself stated: "I shall not attempt to conceal the conflict between these views and common sense."

D.   A Model for the Photon

From a pragmatic viewpoint, there is no need for a model of the photon. One may be content with a description of the particle based entirely on the equations that it obeys. This is a very respectable scientific stance. There is another equally respectable scientific position—try to understand the mathematical equations in relation to a physical model. In previous paragraph we mentioned the attempts of several investigators [30–33]. More recent trials are those of Warburton [34], Fox [35], Scully and Sargent [36], Hunter and Wadlinger [37,38], Evans and Vigier [39], Barbosa and González [40], and Lehnert [41]. For additional contemporary models see Hunter et al. [42].

This chapter describes the programme of the present author to develop yet another representation for the photon in a semiclassical setup. Section II discusses the concept of a preferred frame, and briefly evaluates the empirical evidence against it. Section III discusses some properties of Maxwell's equations that shaped our model. Section IV presents a four-dimensional ether, which leads to a photon model in Section V. The model is based on a rotating doublet, and contains retarded and advanced potentials in a setup that hopefully avoids the pitfalls of previous attempts. A closing section, Section VI, summarizes the main findings.

II.   THE EXISTENCE OF A PREFERRED FRAME

A.   The Meaning of a Constant Speed of Light

Let us concentrate here on the photon as a particle only. The main task is to identify the family of frames where the photon propagates with a constant speed c.[1] In his early papers, Einstein accepted the notion of a preferred frame both for the special and the general theories of relativity. In his words [43, p. 17]:

Newton might no less well have called his absolute space 'Aether'; what is essential is merely that besides observable objects, another thing, which is not perceptible, must be looked upon as real, to enable acceleration or rotation to be looked upon as something real.

Einstein's second postulate requires the speed of electromagnetic radiation in free space to be the same for all *inertial observers*. The special theory of relativity (STR) is conventionally interpreted as a prohibition for the existence of a preferred frame Σ.

However, the operational identification of inertial observers without a preferred frame is plagued with difficulties, as a cursory examination of textbooks in classical and relativistic mechanics will immediately show. A possible solution is to identify inertial observers with the class of observers in free fall in arbitrary gravitational fields [44]. But this is a direct link to the frame where gravitation exists, which is the very same 'Aether' acknowledged by Einstein himself.

In the spirit of Lorentz [45], de Broglie, and Vigier, let us postulate the existence of a preferred frame Σ. Operationally, Σ may be identified with the frame of cosmic background radiation (CBR), whose isotropic thermal nature was established by measurements during the COBE-FIRAS project [46]. Then, the principle of relativity simply states that all frames that are not accelerated relative to Σ, are equivalent to it.

More formally, consider any frame S with an observer at the origin, and let the acceleration of the origin relative to Σ be $\mathbf{a}_0^\Sigma = 0$. Let S^Σ be the class of inertial frames equivalent to Σ:

$$S^\Sigma = \{S | \mathbf{a}_0^\Sigma = 0\} \tag{4}$$

Then, the principle of relativity simply states that all frames belonging to S^Σ are equivalent. Hence, in this limited context, Einstein's second postulate reduces to the statement that speed of electromagnetic radiation is a constant c in S^Σ.

More generally, consider the class of all frames S_i whose origin has the same acceleration $\mathbf{a}_i$ relative to Σ:

$$S_i = S(\mathbf{a}_i) = \{S | \mathbf{a}_0^\Sigma = \mathbf{a}_i\} \tag{5}$$

Clearly, any two frames belonging to some S_i are in inertial relation (i.e., either at relative rest, or in relative uniform motion). An example is provided by two frames in free fall in a region of constant gravitational field [assuming, of course,

---

[1]Often, following the optical tradition, we will refer to the speed of electromagnetic radiation in general as the *speed of light*.

that gravitation is defined with respect to the preferred frame (see quotation [43] at the beginning of this section.).

According to the principle of relativity, the speed of light is a constant c_i in any set S_i. Up to this point our interpretation and the conventional interpretation of STR coincide. The difference arises when the STR requires that

$$c_i = c, \ \forall i \tag{6}$$

We will not pursue this issue any further here, it is simply noted that condition (6) is much stronger than our assumption of a constant c in S^Σ [Eq. (4)].

B. The Empirical Evidence

Let us turn to the often forgotten, but all important question of emipirical verification. According to current information, our solar system moves relative to CBR with speed $v_S \sim 10^{-3} c$ [46–48]. Diurnal and annual rotation of the earth lead to local anisotropies that were documented long ago in different contexts. For instance, Esclangon [49] measured an anisotropic effect for the propagation of light along two perpendicular axes: northwest and northeast. The time difference for propagation along the two perpendicular directions *depended of sidereal time*, thus suggesting an absolute motion of the earth.

Now then, is there any *direct* evidence for the nonexistence of Σ? According to the present author's knowledge, the only available evidence is the *claimed* nullresult of the Michelson–Morley type of experiments.

Michelson and Morley [50] used an interferometer to measure the speed of light along two orthogonal directions: parallel and perpendicular to the earth's orbital speed. They found that the speeds differed by a value somewhere in the range between 5 and 7.5 km/s. Michelson and Morley were extremely surprised because they expected to observe a difference of 30 km/s. At that time they had no plausible explanation for their empirical observation and decided to *interpret* the outcome of the experiment as a null result: no difference in speed along both direction (apparently, the reason for this choice was that Fresnel's theory predicted no difference).

Of course, such interpretation nicely fitted with Einstein's second postulate, proposed 18 years later. Tolman [51, p. 27] explicitly said:

> In support of this principle is the general fact that no "ether drift" has ever been detected, but, especially, the conclusive experiments of Michelson and Morley, and Trouton and Noble, in which, a motion through the ether, of the earth in its path around the sun would certainly have been detected.

Eventually, along with the success of relativity theory, the incorrect interpretation (i.e., that the outcome of the experiment was a nullresult) became entrenched in mainstream physics.

At the beginning of the twentieth century, there were several isolated voices claiming for a revision of the Michelson–Morley interpretation. Hicks [52] performed a theoretical analysis of the Michelson–Morley experiment and concluded that data were consistent with a somewhat larger magnitude of the difference of speeds. More importantly, he noted that the data followed a periodic curve proportional to $\cos 2\theta$, where angle θ refers to a rotation of the interferometer relative to the presumed direction of orbital velocity. The functional dependence present in the results is of the form to be expected if there existed Σ.

The most important critic of Michelson and Morley's interpetation was, no doubt, Dayton C. Miller. He was a collaborator of Morley in the work that followed the initial experiments. Miller applied some of the corrections suggested by Hicks [52] to the results of Michelson-Morley experiment. Miller reports that, after the corrections, the difference of speeds measured in the original experiment were 8.8 km/s for the noon observations and 8.0 km/s for the evening observations [53, p. 207]; clearly, *nonnull results*.

After Morley's retirement, Miller continued a lonely quest for more than 20 years. He repeated the experiment many times at Cleveland and at Mount Wilson, and typically found a nonnull difference of speed around 10 km/s [53–55]. More importantly, he carried out measurements during a whole calendar day, spaced at intervals of three months. He identified seasonal variations both in the magnitude of the difference of speeds, and in the shape of the daily curve [53]. He ascribed the seasonal variations to a motion of the solar system of 208 km/s relative to the fixed stars. This velocity is of the same order of magnitude as the currently accepted $v_S \sim 10^{-3}c$!! However, his argumentation was not clear enough at that time.

Shankland and collaborators [56] thoroughly reviewed Miller's results, and applied formal statistical tests to Miller's data to conclude that [56, p. 171] "there can be little doubt that statistical fluctuations alone cannot account for the periodic fringe-shifts observed by Miller." To any outsider, this remark highly commends the experimental quality of Miller's work. However, regarding the curves depicting seasonal variations, Shankland et al. also noted that, according to their (Shankland's) theory [56, p. 172] "the four curves should have a common maximum (or minimum) . . . only the amplitude may be different at different epochs."

Hence, they concluded that Miller's experimentally observed seasonal variations were simple experimental artifacts!! Evidently, Shankland et al. [56] did not grasp the full meaning of Miller's suggestion that the sun was in motion relative to the fixed stars.

The conclusions of the paper by Shankland and collaborators [56] exemplify an obnoxious practice. Empirical evidence is used in an unconventional manner (to say the least). Indeed, experiments are typically carried out to check a theory. If evidence opposes theory, then the latter is suspect. On the contrary, Shankland et al. did the exact opposite: since evidence opposed their theory, they disregarded the evidence, not the theory.

Several authors have independently revisited Miller's work. Vigier's [7] interpretation was mentioned in Section I. In 1988 the present author began a revision of *all experiments of the Michelson–Morley type* (M-M experiments) reported in the literature. The review is published as Múnera [57].

M-M experiments typically yielded finite (nonzero) differences of speed along two perpendicular positions of the interferometer's reference arm. Such difference is consistently lower than the value to be expected from orbital motion alone (30 km/s), within the naive conventional approximation of not taking into account diurnal variations due to earth rotation. With the exception of Miller, all authors consistently interpreted their observations as nullresults.

A most surprising finding in our review was that no significant effort was made by the experimenters to try and find a theory closer to the empirical observations. For instance, solar motion $v_S \sim 10^{-3}c$ is not included in the analysis. The only exception is Miller [53], who used his observations to derive such a value for solar motion. This criticism is particularly valid for the more recent experiments.

Múnera [57] took into account both earth rotation and orbital motion, as a function of the local latitude and longitude. Prediction of the variation of speed difference as function of time of day are given in Múnera [57] for the locations of Miller's experiment. The qualitative shape of the variations is of the same sort observed by Miller in the 1930s. However, the magnitudes are not correct because solar motion was not included.[2] Selleri [58] allowed for small violations of Lorentz invariance; a correction factor around 10^{-3} reproduces Miller's observations. Also independently, Allais [59] revisited Miller's work. He argues that Miller's seasonal variations are strong proof for a local anisotropy of space.

In summary, the only *direct* evidence against the existence of a preferred frame Σ is the interpretation of Michelson–Morley experiments as being a nullresult. To put it mildly, this evidence is fairly weak. On the contrary, there is mounting evidence for the existence of local anisotropies [49,59,60], which can be interpreted as motion of the earth relative to Σ. Additionally, a replication of Faraday induction experiments with a rotating permanent magnet yielded a positive outcome [61]. Such results may be interpreted as an indication of the existence of absolute motion, and hence of Σ. As usual, the final referee will be empirical evidence. Hence, there is a pressing need to carry out new

---

[2]Solar motion was not included because our intent was to make predictions within the same assumptions used in the original papers: orbital and rotational motion only.

experiments of the Michelson–Morley type using high-resolution modern equipment to check several competing new interpretations.

III. SOLUTIONS OF MAXWELL'S EQUATIONS IN FREE SPACE

A. Maxwell's Equations for Ether

Let us picture the *"vacuum"* as a fluid filling the preferred frame Σ. As usual in electromagnetic theory, any material substance is described by three parameters [62, Chap. 1]:[3]

- The dielectric constant, or permittivity, ε
- The magnetic permeability, μ
- The specific conductivity, σ [units: s^{-1} (reciprocal seconds)]

For the ether it is *postulated* here that

$$\varepsilon = 1, \qquad \mu = 1, \qquad \sigma = 0 \tag{7}$$

Then, Maxwell's equations for any arbitrary material medium reduce to the system of Maxwell equations (MEs) for the vacuum (see any standard source, such as Ref. 62,63,64, or 65). In CGS units, MEs are [66]:

$$\nabla \times \mathbf{E} = -\frac{\partial \mathbf{B}}{\partial u} \tag{8}$$

$$\nabla \times \mathbf{B} = +\frac{\partial \mathbf{E}}{\partial u} + \frac{4\pi}{c}\mathbf{J} \tag{9}$$

$$\nabla \cdot \mathbf{E} = 4\pi\rho_e \tag{10}$$

$$\nabla \cdot \mathbf{B} = 0 \tag{11}$$

where time is expressed as a length u^4

$$u = ct \tag{12}$$

and $\mathbf{E}$, $\mathbf{B}$ are the electric and magnetic fields in vacuum, respectively (units: dyne esu^{-1} = esu cm^{-2}), ρ_e is charge density (units: esu cm^{-3}), and $\mathbf{J}$ is current density (units: esu s^{-1} cm^{-2}).

The set of four equations may be divided into:

---

[3]In the rest of this chapter, references to the authoritative book of Born and Wolf will be [62] (reference number 62 in brackets) followed by page or section number(s) in that book.

[4]In my earlier papers time was represented by w. A different symbol is used here to avoid confusion with photon's omega ω.

- Two induction expressions: Fraday [Eq. (8)] and Ampère [Eq. (9)]
- Two source conditions: electric [Eq. (10)] and (absence of) magnetic source [Eq. (11)]

Consider a region of Σ in our neighbourhood (this is our three-dimensional space), where there exists an electromagnetic field $\mathbf{E}, \mathbf{B}$ and an associated current $\mathbf{J}$. Let these three vector quantities obey Ampère's Eq. (9). Operate with $\nabla\cdot$ on (10) and substitute Eq. (10) to get the charge continuity condition [62, p. 2]:

$$c\frac{\partial \rho_e}{\partial u} + \nabla \cdot \mathbf{J} = 0 \qquad (13)$$

The derivation of Eq. (13) means that the equation of continuity is a mere mathematical consequence of only two of Maxwell's equations; that is, the condition of continuity does not add additional physical information to Maxwell's equations.

If the distribution of charge in a certain region of space is time-independent, $\rho_e = \rho(\mathbf{r})$, then Eq. (13) reduces to

$$\nabla \cdot \mathbf{J} = 0 \qquad (14)$$

As noted elsewhere [67], Eq. (14) means that the continuity condition does not prohibit the existence of an electromagnetic current density $\mathbf{J}$ in free space. It is stressed that Eq. (14) is a mathematical prediction of Maxwell's equations, completely independent of any interpretation.

On the interpretational side, at least three different mechanisms may lead to a current density $\mathbf{J}$:

- Motion of electric charges in the medium (superscript m) leading to a convection current density $\mathbf{J}_e = \rho_e^m \mathbf{v}_m$
- Resistive dissipation in the medium producing a conduction current density $\mathbf{J}_c = \sigma\mathbf{E}$
- Nonresistive redistribution of energy within the electromagnetic field $\mathbf{E}, \mathbf{B}$, described by Ampère's Eq. (9) and leading to an electromagnetic displacement current $\mathbf{J}_d$

Then

$$\mathbf{J} = \mathbf{J}_e + \mathbf{J}_c + \mathbf{J}_d \qquad (15)$$

The first two terms on the right-hand side of Eq. (15) are conventionally ascribed to dispersive media [62, p. 9], while the third term is the displacement current density $\mathbf{J}_d$ [66, Chap. 9]. The latter may be easily observed in material media (air); see, for instance, Carver and Rajhel [68] and Bartlett and Corle [69]. It is assumed here that $\mathbf{J}_d$ also exists in a vacuum with $\sigma = 0$. Therefore, in free space without charges, $\mathbf{J}_e = \mathbf{J}_c = 0$, Ampère's law leads to

$$\mathbf{J}_d = \frac{c}{4\pi}\left(\nabla \times \mathbf{B} - \frac{\partial \mathbf{E}}{\partial u}\right) \tag{16}$$

However, the interpetation of $\mathbf{J}_d$ is open to considerable controversy, as a quick reference to conventional journals indicates; see, for instance, Warburton [34], French and Tessman [70], Rosser [71], and references cited therein.

Some extensions of MEs indentify free space with a medium having $\sigma \neq 0$ [72]; in such cases there is a dispersive loss of energy. If one wishes to maintain a relativistic theory, previous fact introduces additional complications. To correct such new difficulties, Lehnert [41,73] postulated additional sources in vacuo ($\nabla \cdot \mathbf{E} \neq 0$ when $\sigma \neq 0$).

Let us turn now to different aspects of the solutions of Maxwell's equations with $\sigma = 0$.

B. Source of Electromagnetic Field in Free Space

Consider propagation of photons in vacuum. The first issue is to determine the meaning of such propagation. It seems fairly clear that Maxwell's equations describe the propagation of electromagnetic field $\mathbf{E}, \mathbf{B}$ in general. However, as mentioned in the introduction, what is the mechanism for undulatory propagation in vacuum?

To understand propagation of photon, it is necessary to define the photon. There are many possibilities, including the following:

1. The photon and the electromagnetic wave are different manifestations of the same reality (particle–wave duality).
2. The photon is a different entity that is guided by the electromagnetic wave (de Broglie and Vigier; see references at the end of this chapter).
3. The photon is not a particle, but a process. A prime example is the absorber model of Lewis [32] and Wheeler and Feynman [33]; for a revival of the same idea, see Whitney [74]. For a related concept with emphasis on path, see Ryff [75].
4. The photon and the electromagnetic field are different entities. For instance, Ritz [30] proposed an emission theory. In Section V we will try a similar dichotomy, but maintaining a constant speed of signal relative to Σ.

For the time being, let us consider the conventional view: wave–particle duality. Then, propogation of photon is the same as propagation of electromagnetic field $\mathbf{E}, \mathbf{B}$. In free space the charge density is null everywhere, except possibly at the source. The photon is chargeless; hence, if Maxwell's equations are applicable to a photon in vacuum, $\rho_e = 0$ everywhere. This leads to some contradiction.

There is a cause–effect relationship between electric charge density and electric field, represented by Eq. (10). Since $\rho_e = 0$, it should follow that $\mathbf{E} = 0$. Such trivial solution, however, cannot possibly represent a photon. There is another alternative. Induction Eqs. (8) and (9) relate to $\mathbf{E}$ and $\mathbf{B}$ so that, if $\mathbf{B}$ were an independent variable, variations of magnetic field could, in principle, induce an electric field. However, magnetic field $\mathbf{B}$ is conventionally ascribed to moving charges [66]. Again, $\rho_e = 0$ forbids $\mathbf{B}$, and a fortiori $\mathbf{E}$. It seems that there is some violation of causality: an electromagnetic field represented by $\mathbf{E}$ and $\mathbf{B}$ (effect) without a source ρ_e (cause).

Considerations of this sort led us to suggest that, in order to avoid violations of causality within the wave–particle duality, there are two possible interpretations of Maxwell's equations [76]:

- *Alternative 1*. Assume that electromagnetic free field is a primitive concept.
- *Alternative 2*. Admit that there are hidden charge doublets at the photon source.

The first route leads to a model of the physical world where the concept of particle is derived, while the second assumption implies the opposite view. Long ago, Bateman [77, pp. 9–10] reached a conclusion similar to our second alternative: "Since the electric force is ultimately at right angles to the radius there is no total charge associated with the singularity, for the charge is equal to the surface-integral of the normal electric force over a large sphere concentric with the origin and this integral is evidently zero. We are consequently justified in regarding the singularity as a *doublet* and in fact as a simple electric doublet of varying moment as indicated by the way in which the electric and magnetic forces become infinite" (emphasis added).

More recently, McLennan [78] also analysed the meaning of MEs, in particular the implications of Gauss' theorem in the context of $\rho_e = 0$. He concluded, however, that there should exist two different sets of MEs: one for the field, another for the source. This view is reminiscent of some remarks made earlier by Warburton [34].

From a completely different viewpoint, Mannheim [79, p. 913] proposed a theory to quantize relativistic fermions using classical coordinates. In the conclusions he suggests that "gauge fields may not be fundamental at all but may be fermion composites."

Section IV describes the aether as a four-dimensional fluid; this is equivalent to assigning objective reality to the field. Thence, in section 5 the photon is modelled as a charge doublet, that acts as the source. In the context of this section, our model contains elements of both Bateman (the doublet) and McLennan (the differentiation between field and source).

Summarizing, in the conventional wave-particle view, the condition $\rho_e = 0$ in free space may be described as being charge-neutral, rather than charge-free.

C. The Meaning of Current Density in Free Space

The conventional view that $\mathbf{J} = 0$ when $\rho_e = 0$ is not a result of MEs. As noted in Section III.A, MEs only lead to Eq. (14), which does not mean *a fortiori* that $\mathbf{J} = 0$.

To see this, let us introduce some definitions first. The Poynting vector $\mathbf{G}$ represents energy flux density (units: $\mathrm{esu}^2\,\mathrm{cm}^{-3}\,\mathrm{s}^{-1} = \mathrm{erg\,cm^2\,s^{-1}}$). It is a capability of the electromagnetic field to perform work defined as

$$\mathbf{G} = \frac{c}{4\pi}\mathbf{E} \times \mathbf{B} \tag{17}$$

The internal energy transfer along the electric field is Q_I (units: $\mathrm{erg\,s^{-1}}$) given by

$$Q_I = \int_V \mathbf{E} \cdot \mathbf{J}\, dV \tag{18}$$

The electric and magnetic energy densities w_e, w_m in free space (units: $\mathrm{erg\,cm^{-3}}$) are defined as

$$w_e = \frac{1}{4\pi}\int_V \mathbf{E}\cdot d\mathbf{E} = \frac{E^2}{8\pi}, \qquad w_m = \int_V \mathbf{B}\cdot d\mathbf{B} = \frac{B^2}{8\pi} \tag{19}$$

The electromagnetic energy associated with a volume V is W (units: erg):

$$W = \int_V (w_e + w_m)\, dV = \frac{1}{8\pi}\int (E^2 + B^2)\, dV \tag{20}$$

A conventional interpretation is "that W represents the total energy contained within the volume" [62, p.8]. McLennan [78b] challenges this interpretation proposing that, instead, W is a potential energy. Along the same line of thought, long ago Ritz [30] identified w_e with potential energy and w_m with kinetic energy [30, pp. 157–158].

Independently of interpretation, MEs directly lead to [62, Sec. 1.1.4]

$$\frac{dW}{dt} + Q_I + \oint_S \mathbf{G}\cdot \mathbf{n}\, dS = \frac{dW}{dt} + Q_I + Q_P = 0 \tag{21}$$

where $\mathbf{n}$ is a unit vector[5] orthogonal to the surface S that bounds the integration volume V, and the energy Q_P propagated by the Poynting vector is implicitly defined.

---

[5]Unit vectors are represented by lowercase boldface characters. If there is a risk of confusion, an additional caret is used.

Equation (21) implies that time variations of energy, inside any V, manifest in two different forms. It is pointed out that, conventionally, only the first one is allowed:

- Transport Q_P along the direction of propagation $\mathbf{g}$ (this unit vector is parallel to the Poynting vector)
- Internal rearrangement of energy Q_I

If one adopts McLennan's [78b] interpretation, then Eq. (21) is a realization of a standard theorem of Newtonian mechanics: conservation of total energy = conservation of kinetic plus potential energy (see, e.g., Chap. 4 of Kleppner and Kolenkow, [80]). The reason is simple: Coulomb electric force is central, then work is path independent, and total energy is function of position only. The time derivative of total energy is of course zero, as in Eq. (21). In this interpretation Q_P and Q_I are manifestations of kinetic energy.

The differential form of Eq. (21) is

$$\frac{\partial(w_e + w_m)}{\partial t} = -\mathbf{E} \cdot \mathbf{J} - \nabla \cdot \mathbf{G} \tag{22}$$

This equation simply states that variations of energy density at a given spacetime point are due to transport along the Poynting vector plus transport along electric field. Clearly, when $\mathbf{J} = 0$ all transport is along $\mathbf{g}$ [62, Eq. 43, p. 10].

D. Equivalence of Maxwell's and Wave Equations

It is easy to decouple $\mathbf{E}$ and $\mathbf{B}$ in Maxwell's equations, thus obtaining two vector wave equations. Operate with $\nabla \times$ on Eq. (8), and substitute Eqs. (9) and (10) to get

$$\Box \mathbf{E} = \nabla^2 \mathbf{E} - \frac{\partial^2 \mathbf{E}}{\partial u^2} = -\frac{4\pi}{c} \left(c\nabla\rho_e - \frac{\partial \mathbf{J}}{\partial u} \right) \tag{23}$$

where the D'Alembertian operator is defined as (units: cm^{-2})

$$\Box = \nabla^2 - \frac{\partial^2}{\partial u^2} \tag{24}$$

Likewise, to obtain a wave equation for magnetic field, operate with $\nabla \times$ on Eq. (9), and substitute Eqs. (8) and (11) to get

$$\Box \mathbf{B} = \nabla^2 \mathbf{B} - \frac{\partial^2 \mathbf{B}}{\partial u^2} = -\frac{4\pi}{c} (c\nabla\rho_e + \nabla \times \mathbf{J}) \tag{25}$$

Again, as in Section III. B, Eqs. (23) and (25) clearly depict the cause–effect relationship between the fields $\mathbf{E}$, $\mathbf{B}$ and the source $\rho_e \neq 0$. If field is a primitive concept, then the left-hand side is the cause, whereas the right-hand side is the cause if charge is primitive.

In free space with $\rho_e = 0$, then $\mathbf{J}_e = 0$. Expressions (23) and (25) reduce to

$$\Box \mathbf{E} = \nabla^2 \mathbf{E} - \frac{\partial^2 \mathbf{E}}{\partial u^2} = -\frac{4\pi}{c}\left(\frac{\partial \mathbf{J}_d}{\partial u}\right) \tag{26}$$

$$\Box \mathbf{B} = \nabla^2 \mathbf{B} - \frac{\partial^2 \mathbf{B}}{\partial u^2} = -\frac{4\pi}{c}(\nabla \times \mathbf{J}_d) \tag{27}$$

If Eq. (16) is substituted into (27), then an identity follows, which suggests that Eq. (27) is not an independent condition. Of course, when $\mathbf{J}_d = 0$ the conventional homogeneous wave equations obtain

$$\Box \mathbf{E} = 0 \tag{28}$$
$$\Box \mathbf{B} = 0 \tag{29}$$

As in Section III.B, the question of causality immediately arises. If there is no source, why are there nontrivial solutions for the fields $\mathbf{E}, \mathbf{B}$? Again, the alternatives discussed in Section III.B may provide an answer. Another standpoint is to assume that Eqs. (28) and (29) describe an independent reality. Then, after being produced, fields $\mathbf{E}, \mathbf{B}$ exist on their own, quite independently of the continued existence of the source. Of course, the properties of the fields depend of the source at the moment of emission.

There is a curiosity here. The process to decouple the Maxwellian fields $\mathbf{E}, \mathbf{B}$ that was explained above is completely algebraic. There are no additional physical concepts introduced in the process. Therefore, Eqs. (28) and (29) should be completely equivalent to Maxwell equations. That is, *the set of solutions to the wave equations* represented by (26) and (27), or by (28) and (29), *should be the same as the set of solutions to Maxwell's equations.* However, as noted elsewhere by Múnera and Guzmán [81], this is not the case. A possible explanation may be the nature of Eq. (27): it is an identity.

Let us illustrate previous claim with an elementary example for $\rho = 0$. In Cartesian coordinates, let

$$\mathbf{E} = E_x \mathbf{i} + E_y \mathbf{j} + E_z \mathbf{k}, \quad E_y = E_z = 0, \quad E_x = A^E \sin[k(z-u)] \tag{30}$$
$$\mathbf{B} = B_x \mathbf{i} + B_y \mathbf{j} + B_z \mathbf{k}, \quad B_y = B_z = 0, \quad B_x = A^B \sin[k(z-u)] \tag{31}$$

Evidently, fields $\mathbf{E}, \mathbf{B}$ are a solution of wave equations (28) and (29), respectively. However, fields $\mathbf{E}, \mathbf{B}$ are only *partially* consistent with Maxwell equations, as follows:

- Equation (30) is directly consistent with the electric source Eq. (10).
- Equation (31) is directly consistent with the magnetic source condition (11).
- Equations (30) and (31) are consistent with Ampère's Eq. (9), interpreted as a definition for $\mathbf{J}_d$ (Eq. 16). In this sense, any arbitrary pair of vectors is a solution of Ampère's equation.
- However, Eqs. (30) and (31) are not consistent with Faraday's Eq. (8). Indeed, a direct substitution shows that

$$\nabla \times \mathbf{E} = +kA^E \cos\left[k(z-u)\right]\mathbf{j} \neq -\frac{\partial \mathbf{B}}{\partial u} = +kA^B \cos\left[k(z-u)\right]\mathbf{i} \qquad (32)$$

Previous example demonstrates in a straightforward manner that the uncoupled wave equations (28) and (29) are not completely equivalent to Maxwell's equations. We have no explanation for this fact, other than the (possible??) lack of independence of Eq. (27) noted above.

This finding may be related to similar remarks of Ritz [30] regarding Lorentz electron theory [45]. Ritz concluded that the solutions to the wave equations were more fundamental than Maxwell's equations. In his words [30, p. 172]: "on voit qu'en dernière analyse *c'est la formule des actions élémentaires, et non le système de equations aux dérivées partielles, qui est l'expression exacte et complète de la théorie de Lorentz*" (emphasis in original).

Conventional electromagnetic theory is fully aware of this difficulty, but no attention is paid to the inconsistency. Pragmatically, Jackson simply notes that solutions to the wave equations must also satisfy Maxwell's equations [63, Chap. 7, p. 198], and go on to use Faraday's Eq. (8) as a coupling condition for the two wave equations. We will return to this point in Section V.

In the present simple example, Eq. (32) immediately suggests a valid solution; namely, that the magnetic field must lie along the y axis, thus leading to the well-known orthogonality between the electric and magnetic fields. A bona fide solution for Maxwell's equations is then provided by the electric field of Eq. (30), and

$$\mathbf{B} = B_x\mathbf{i} + B_y\mathbf{j} + B_z\mathbf{k}, \qquad B_x = B_z = 0, \quad B_y = A^B \sin\left[k(z-u)\right] \qquad (33)$$

Equations (30) and (31) are a solution of Faraday's equation provided that $A^E = A^B = A$. Then, in a single stroke, Faraday's condition achieves two different things: orthogonality and equal amplitude of fields $\mathbf{E}$ and $\mathbf{B}$.

Summarizing the discussion in this section. It seems as if the entire physical information about the behavior of the electromagnetic field were contained in Faraday's equation. The other three equations play a minor role: definitions of current density, electric source, and absence of magnetic source.

E. Longitudinal Components of Magnetic Field

In a long series of publications, Evans and Vigier [39] and Evans et al. [82] have suggested the existence of a non-Maxwellian longitudinal component of magnetic field. Here we want to explore a related problem: the sense in which longitudinal components of magnetic field may exist within the realm of the conventional Maxwellian theory, in the extended sense of Eq. (16).

Let us consider the propagation of electromagnetic waves with both fields nonzero: $\mathbf{E} \neq 0$ and $\mathbf{B} \neq 0$. As usual, propagation is parallel to the Poynting vector $\mathbf{G}$, defined in Eq. (17). Evidently, by definition, vector $\mathbf{G}$ is perpendicular to both fields $\mathbf{E}$ and $\mathbf{B}$. Hence, there cannot exist components of the magnetic field $\mathbf{B}$ parallel to the *instantaneous* direction of propagation $\mathbf{G}$.

In order to determine the direction of propagation of some arbitrary wave, the observer must make a measurement or observation during a finite period of time T_M, at some three-dimensional location, say, a small volume at some position $\Delta V(\mathbf{r})$. The result of the measurement will be some deposition of energy $\Delta W(\mathbf{r})$ within $\Delta V(\mathbf{r})$ given by [recall Eq. 21)]

$$\Delta W = -\int_0^{T_M} \frac{dW}{dt} dt = \int_0^{T_M} (Q_I + Q_P)\, dt = \int_0^{T_M} Q_I\, dt + \int_0^{T_M} \oint_S \mathbf{G} \cdot \mathbf{n}\, dS\, dt \quad (34)$$

The average energy deposited per unit volume during the observation time is obtained from Eq. (22) as

$$\frac{\Delta W}{\Delta T} = \left\langle -\frac{\partial(w_e + w_m)}{\partial t} \right\rangle = \frac{1}{T_M} \left(\int_0^{T_M} \mathbf{E} \cdot \mathbf{J}\, dt + \int_0^{T_{Ms}} \nabla \cdot \mathbf{G}\, dt \right) \quad (35)$$

Equations (34) and (35) show that the result of any actual measurement will strongly depend of the direction of propagation $\mathbf{G}$, that is, of the surface through which the electromagnetic radiation enters the detector. As a first approximation, let us concentrate on the average direction of propagation $\langle \mathbf{G} \rangle$:

$$\langle \mathbf{G} \rangle = \frac{1}{T_M} \int_0^{T_M} \mathbf{G}\, dt = \left(\frac{1}{T_M} \int_0^{T_M} G(t)\, dt \right) \mathbf{g} = \langle G \rangle \mathbf{g} \quad (36)$$

In the conventional solution of MEs as plane electromagnetic waves, the direction of propagation $\mathbf{G}$ is always perpendicular to the plane; hence, it is time-independent. Let an external observer of the electromagnetic wave define the z-axis as parallel to $\mathbf{G}$. Then, $\mathbf{G} = G(t)\mathbf{k}$, and $\langle \mathbf{G} \rangle = \langle G \rangle \mathbf{k}$.

Consider a non-plane-wave solution of Maxwell's equations, whose direction of propagation varies with respect to the z axis. In general it holds that

$$\langle \mathbf{G} \rangle = \langle G_x \rangle \mathbf{i} + \langle G_y \rangle \mathbf{j} + \langle G_z \rangle \mathbf{k} \quad (37)$$

The *instantaneous* direction of propagation is, of course, perpendicular to the plane defined by the *instantaneous* electromagnetic fields **E** and **B**. But this time-dependent direction need not be parallel to the z axis, physically defined as the direction for the average propagation of energy. Let us illustrate the point with variations of the same simple example of previous section, for additional details see Múnera and Guzmán [67].

Example 1. Consider the following fields **E** and **B**, which are solutions of Maxwell's equations for $\rho_e = 0$:

$$\mathbf{E} = E_x\mathbf{i} + E_y\mathbf{j} + E_z\mathbf{k}, \qquad E_y = E_z = 0, \qquad E_x = A\sin[k(z - u)] \quad (38)$$

$$\mathbf{B} = B_x\mathbf{i} + B_y\mathbf{j} + B_z\mathbf{k}, \qquad B_x = B_z = 0, \qquad B_y = A\sin[k(z - u)] \quad (39)$$

This is a monochromatic linearly polarized wave with electric field vibrating in the x–z plane, and the magnetic field vibrating in the y–z plane. We have adopted the practice of explicitly identifying the plane of vibration [62, p. 29]. Current density associated with **E** and **B** is given by Eq. (16) as $\mathbf{J} = 0$. It is stressed that $\mathbf{J} = 0$ is obtained here from the fields, whereas the conventional approach is to assume the current to be zero on the grounds that $\rho_e = 0$.

The Poynting vector and its time-average are [Eqs. (17) and (26), respectively]:

$$\mathbf{G} = \frac{cA^2}{4\pi}\sin^2(k(z - u))\mathbf{k} \tag{40}$$

$$\langle\mathbf{G}\rangle = \frac{cA^2}{8\pi}\mathbf{k} \tag{41}$$

where the average is taken over an integer number of cycles m

$$T_M = \frac{2\pi m}{kc} = \frac{2\pi m}{\omega} \tag{42}$$

Clearly, there are no longitudinal components in a plane wave.

Example 2. Let us consider a variation of Example 1. The plane wave of previous example is perturbed with the addition of a small longitudinal component of magnetic field to get

$$\mathbf{E} = E_x\mathbf{i} + E_y\mathbf{j} + E_z\mathbf{k}, \qquad E_y = E_z = 0, \qquad E_x = A\sin[k(z - u)] - B\sin[k_L(y - u)] \tag{43}$$

$$\mathbf{B} = B_x\mathbf{i} + B_y\mathbf{j} + B_z\mathbf{k} \qquad B_x = 0, \qquad B_y = A\sin[k(z - u)], \qquad B_z = B\sin[k_L(y - u)] \tag{44}$$

Again, Eq. (16) yields $\mathbf{J} = 0$. The Poynting vector $\mathbf{G}$ has components

$$G_x = 0, \qquad G_y = \frac{c}{4\pi}(AB\sin[k(z-u)]\sin[k_L(y-u)] - B^2\sin^2[k_L(y-u)]$$
$$G_z = \frac{c}{4\pi}(A^2\sin^2[k(z-u)] - AB\sin[k(z-u)]\sin[k_L(y-u)]) \tag{45}$$

This example is a nonplanar linearly polarized wave. The direction of vibration of the electric field is still along the x axis, while the magnetic field and the Poynting vector are both contained on the y–z plane. The instantaneous direction of magnetic field is along angle θ given by

$$\tan\theta = \frac{B_y}{B_z} \tag{46}$$

where the angle is measured from the z axis (direction of the unperturbed wave). Let us take the average of Eqs. (45) during T_M. The integration leads to particularly simple results when the ratio $R = (k_L/k) = (\omega_L/\omega)$ is rational; on the contrary, when R is not rational, it is not possible to find a time of integration such that the longitudinal magnetic components disappear. In the rational case, the observation period is chosen such that

$$T_M = \frac{2\pi m}{kc} = \frac{2\pi m}{\omega} = \frac{2\pi n}{k_L c} = \frac{2\pi n}{\omega_L} \tag{47}$$

where n, m are arbitrary integers.[6] The results are

$$\langle G_x \rangle = 0, \qquad \langle G_y \rangle = -\frac{cB^2}{8\pi}, \qquad \langle G_z \rangle = \frac{cA^2}{8\pi} \tag{48}$$

When R is rational, there is a surprising find for a linearly polarized wave; specifically, the average energy along the direction of propagation (the z axis) is the same for the unperturbed [Eq. (41)] and the perturbed cases [Eq. (48)]. This is important because a direct measurement of intensity cannot distinguish between the two physically different situations.

On the other hand, the average propagation of energy along the y axis is quite different: zero in the plane case (Example 1), and negative in the nonplanar Example 2. This means that the wave absorbs energy from the surroundings. As

---

[6]This is a sort of quantization of frequencies [an unexpected connection between classical MEs and quantum mechanics (QM)]. Vigier [8, p. 14] mentions another instance of a Maxwellian connection to QM.

expected, the energy intensity ratio depends of the amplitudes of the waves:

$$\frac{\langle G_y \rangle}{\langle G_z \rangle} = -\frac{B^2}{A^2} \tag{49}$$

Assuming that physical polarized light is closer to Example 2 than to Example 1, then one would expect to see transfer of energy in a direction perpendicular to the propagation of a finite light beam. In a recent experiment a transversal light current was induced by a magnetic field [83]. As a wild conjecture, the classical mechanism contained in Example 2 might be at work there.

Example 3. Consider now a variation of Example 2: addition of scalar magnetic potential. Consider a generic magnetic potential of the form

$$\Phi^m(\mathbf{r}, u) = \Phi_0 M(\mathbf{r}) e^{-H_0 u} \tag{50}$$

where Φ_0 is a reference potential (units: erg esu^{-1}) and H_0 is a constant (units: cm^{-1}), and $M(\mathbf{r})$ is a solution of the dimensionless wave equation $\nabla^2 M(\mathbf{r}) = 0$. Let us apply a gauge transformation to the solution of MEs in Example 2 as $\mathbf{B} \rightarrow \mathbf{B} + \nabla \Phi^m(\mathbf{r}, u)$. A new solution of Maxwell's equations is then

$$E_x = A \sin[k(z - u)] - B \sin[k_L(y - u)] + H_0 \Phi_0 e^{-H_0 u} \int \frac{\partial M}{\partial y} \, dz - axF_1(u)$$

$$E_y = -H_0 \Phi_0 e^{-H_0 u} \int \frac{\partial M}{\partial x} \, dz + ayF_1(u) \tag{51}$$

$$E_z = F_2(u)$$

$$B_x = \Phi_0 e^{-H_0 u} \frac{\partial M}{\partial x}$$

$$B_y = A \sin[k(z - u)] + \Phi_0 e^{-H_0 u} \frac{\partial M}{\partial x} \tag{52}$$

$$B_z = B \sin[k_L(y - u)] + \Phi_0 e^{-H_0 u} \frac{\partial M}{\partial z}$$

where $a, F_1(u), F_2(u)$ are a real constant and two arbitrary functions of time, to be determined from boundary conditions. Equation (16) yields $\mathbf{J} \neq 0$, with components

$$J_x = \frac{c}{4\pi} \left(H_0^2 \Phi_0 e^{-H_0 u} \int \frac{\partial M}{\partial y} \, dz + ax \frac{dF_1(u)}{du} \right)$$

$$J_y = \frac{c}{4\pi} \left(H_0^2 \Phi_0 e^{-H_0 u} \int \frac{\partial M}{\partial x} \, dz + ay \frac{dF_1(u)}{du} \right) \tag{53}$$

$$J_z = -\frac{c}{4\pi} \frac{dF_2(u)}{du}$$

The energy transported by the field is given by Eq. (17), and differs from the standard equation in the presence of the term $\mathbf{E} \cdot \mathbf{J} \neq 0$. According to sign, energy may be extracted from (lost to) Dirac's sea of energy. The explicit average components of the Poynting vector are given in Múnera and Guzmán [67]. It may be expected that, if the magnetic potential exists, both $H_0 \rightarrow 0$ and $\Phi_0 \rightarrow 0$. Hence, very precise measurements will be required to detect its contribution to total energy flux.

F. Symmetrization of Maxwell's Equations

Many people in the past have wondered why each pair (8)–(9) and (10)–(11) in Maxwell's equations do not have *exactly the same* structure. For instance, to make Eq. (11) exactly alike (10), then, according to Dirac's [84] suggestion, the existence of magnetic sources. Despite the fact that monopoles have never been convincingly observed, the subject is still alive. Zeleny [85] derives magnetic monopoles by assuming that "the field mediating the electromagnetic interaction shall be the antisymmetric tensor field," while Adawi [86] connects them to special relativity.

Múnera and Guzmán [87] tackled the question from a different angle and without the introduction of new sources. The starting point is fairly simple:

1. Acknowledge that the pair of electromagnetic fields $\mathbf{E}, \mathbf{B}$ represent some physical reality.
2. Assume that the pair $\mathbf{E}, \mathbf{B}$ can be handled as any pair of vectors, regardless of the axial symmetry of $\mathbf{B}$.
3. Obtain new vectors $\mathbf{P}, \mathbf{N}$ as a linear combination of the electromagnetic pair $\mathbf{E}, \mathbf{B}$. Consequently, the new vectors (defined below) must have physical nature:

$$\mathbf{P} = \mathbf{E} + \mathbf{B} \tag{54}$$

$$\mathbf{N} = \mathbf{B} - \mathbf{E} \tag{55}$$

From Eqs. (54) and (55) it follows that

$$\mathbf{E} = \frac{\mathbf{P} - \mathbf{N}}{2} \tag{56}$$

$$\mathbf{B} = \frac{\mathbf{P} + \mathbf{N}}{2} \tag{57}$$

Direct substitution of (56) and (57) into MEs (8)–(11) easily leads to two *symmetric* induction equations plus two *symmetric* source equations:

$$\nabla \times \mathbf{P} = -\frac{\partial \mathbf{N}}{\partial u} + \frac{4\pi \mathbf{J}}{c} \tag{58}$$

$$\nabla \times \mathbf{N} = +\frac{\partial \mathbf{P}}{\partial u} + \frac{4\pi \mathbf{J}}{c} \tag{59}$$

$$\nabla \cdot \mathbf{P} = +4\pi \rho_e \tag{60}$$

$$\nabla \cdot \mathbf{N} = -4\pi \rho_e \tag{61}$$

There is a clear symmetry. In particular there are *electrical* sources for both fields **P**, **N**. There is a simple change of sign in the source, but monopoles do not arise However, there is no such sign difference for the current density **J**. There are two equations of continuity, one for each field **P**, **N**, while there is only one in the unsymmetrized version of Maxwell's equations.

Since the derivation is completely algebraic, the symmetrized set of equations (58)–(61) should be identical in every respect to the conventional MEs (8)–(11). Surprisingly, there are some slight differences as discussed in Múnera and Guzmán [87]. One of them is related to the Coulomb gauge, as follows.

Let us express Eqs. (58)–(61) in terms of potentials, rather than fields. Toward that end, let us invoke a general result from field theory (see, e.g., Kellogg [88], p. 76): Any vector field **F**(**r**,w), sufficiently differentiable, is the sum of a gradient and a curl. Then, fields **P**, **N** are given by

$$\mathbf{P} = \nabla \times \mathbf{A}^P - \nabla U^P \tag{62}$$

$$\mathbf{N} = \nabla \times \mathbf{A}^N - \nabla U^N \tag{63}$$

Here $\mathbf{A}^P$, $\mathbf{A}^N$ are the individual vector potentials, and U^P, U^N are arbitrary scalar potentials. Since source equations for **P**, **N** are nonsolenoidal, there is no doubt regarding the presence of the gradients of the arbitrary functions U^P, U^N. Note that the magnetic scalar potential U^B associated with **B** is not ignored.[7]

Substitute now definitions (62) and (63) into the symmetrized Maxwell's equations to get

$$\nabla \times \left(\nabla \times \mathbf{A}^P + \frac{\partial \mathbf{A}^N}{\partial u} \right) = +\frac{\partial (\nabla U^N)}{\partial u} + \frac{4\pi \mathbf{J}}{c} \tag{64}$$

$$\nabla \times \left(\nabla \times \mathbf{A}^N - \frac{\partial \mathbf{A}^P}{\partial u} \right) = -\frac{\partial (\nabla U^P)}{\partial u} + \frac{4\pi \mathbf{J}}{c} \tag{65}$$

$$\nabla^2 U^P = -4\pi \rho_e \tag{66}$$

$$\nabla^2 U^N = +4\pi \rho_e \tag{67}$$

---

[7]The conventional practice is to ignore it on the grounds that the source is solenoidal.

Source expressions (66) and (67) are Poisson equations that were obtained directly from Maxwell's equations *without invoking the Coulomb gauge condition* $\nabla \cdot \mathbf{A}^B = 0$. However, according to standard textbooks, for instance, Chap. 6 of Jackson [63], the Coulomb gauge condition is required in order to impose *transversality on* $\mathbf{A}^B$. Therefore, we are led to a dichotomy: fields $\mathbf{P}, \mathbf{N}$ are either (1) free from the transversality constraint or (2) have the transversality trait built in. Either case is a surprise, because there was no additional physics involved in the derivation of the symmetrized set. In our original paper [87], no special meaning was attached to vector fields $\mathbf{P}, \mathbf{N}$. There is now a suggestion, presented in Section V of this writing.

IV. A FOUR-DIMENSIONAL ETHER

The idea that the modern "vacuum" (=ether in this writing) is of hydrodynamic nature is a recurrent one. Dirac [89] acknowledged its plausibility, for an elaboration, see Cufaro-Petroni and Vigier [90], and for a more recent summary, Chebotarev [91].

Other examples of ether are a superfluid of particle–antiparticle pairs [92], a fluid of "stuff" particles [26], and a variety of fluids [93–97]. From such fluids, electrodynamic and particle models easily follow; see Thomson [98], Hofer [99], Marmanis [100], and Dmitriyev [101].

The present author has proposed a four-dimensional (4D) hydrodynamic model that allows for a variable component of the 4-velocity along the time axis [102, 103]. The model leads to a 4D force as the gradient of the 4-pressure; the 3D-electromagnetic force is a particular case [104].

A. A Four-Dimensional Equation of Motion

Let us assume the existence of a four-dimensional (4D) flat Euclidean space $\Sigma = (u, x, y, z)$, where the time dimension $u = v_u t$ behaves exactly the same as the three spatial dimensions [102, 104]. Further, let Σ be filled with a fluid of preons (=tiny particles of mass m and Planck length dimensions). These particles are in continual motion with speed $\mathscr{V} = (v_u, v_x, v_y, v_z) = (v_u, \mathbf{V})$. No a priori limits are set on the speed v_u of preons along the u-axis.[8]

Note that the limitations of the special theory of relativity (STR), when applicable, refer to $V = |\mathbf{V}| = (v_x^2 + v_y^2 + v_z^2)^{1/2}$, which is the speed of particles in our 3D world. However, v_u is the projection of the 4D velocity $\mathscr{V}$ onto the u axis, which is not a spatial speed. Here, we extend the notion of absolute space to 4D $(= R^4)$, whereas the spacetime of STR is (ct, x, y, z), specifically, $R^{1,3}$.

---

[8]4D concepts and vectors are represented by either calligraphic or Greek uppercase letters, while 3D vectors are in the usual boldface.

Motion of individual preons in Σ is governed by a 4D equation of motion, given by the following matrix expression [102]:

$$\partial_\mu(\rho\mathscr{V}\mathscr{V}) = -\partial_\mu\tau_{4x4} - \partial_\mu P \qquad (68)$$

where $\rho = nm$ is the preonic fluid mass density, n is the number of preons per unit 3D volume, the vector operator $\partial_\mu = (\partial_u, \nabla), \partial_u = \partial/\partial u$ is a 4D gradient, the 4D stress tensor τ_{4x4} is a 4×4 matrix, $P = P(u, x, y, z)$ is the pressure generated by the preonic fluid, and the Greek index $\mu = (u, x, y, z)$. The energy-momentum tensor $\rho\mathscr{V}\mathscr{V}$ results from the dyadic product $\mathscr{V}\mathscr{V}$.

Now consider an arbitrary 3D hypersurface formed by a projection of the 4D universe onto the u axis, say, $u = u_0 = v_u^0 t_0$ (see Fig. 1). The plane $u - \mathbf{r}$ may be interpreted in two complementary ways:

Interpretation 1 (Fig. 1a). At an arbitrary time t_0 (say, the present), the line $u = u_0$ divides the plane into three classes of particles:

- Preons moving with $v_u > v_u^0$ (upper region)
- Preons moving with $v_u < v_u^0$ (lower region)
- Preons moving with $v_u = v_u^0$ (on the horizontal line)

Interpretation 2 (Fig. 1b). For the class of preons moving with $v_u = v_u^0$, the line $u = u_0$ divides the plane into three periods of time:

- The future for $t > t_0$ (upper half-plane)
- The past for $t < t_0$ (lower half-plane)
- The present $t = t_0$ (on the line)

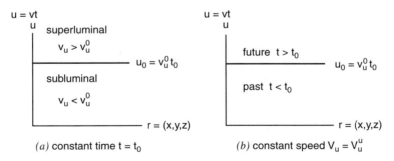

Figure 1. Four-dimensional representation of universe as a $u - \mathbf{r}$ diagram; (a) constant time $t = t_0$; (b) constant speed $v_u = v_u^0$. The projection on the u axis is a 3D hypersurface. This horizontal line partitions the universe into two half-spaces. (a) For a given time t_0 (say, the present) the upper (resp. lower) space corresponds to universes with higher (resp. lower) speeds on the u axis. (b) For a given $u = u_0$, the upper (resp. lower) space corresponds to the future (resp. past). See discussion in the text.

The conventional worldlines of STR and the space underlying Feynman diagrams belong to interpretation 2 with v_u^0 unspecified.

Let us postulate that we live in a 3D hypersurface that slides along the u axis with speed $v_u^0 = c_a$, where the u axis coincides with the arrow of time. The 4-velocity is then a (row or column) vector $\mathscr{V}_a = (\pm c_a, v_x, v_y, v_z)$. The plus (resp. minus) sign corresponds to the speed of preons that enter (resp. leave) our 3D world, parallel (resp. antiparallel) to the *time arrow*. It will be seen below that this constant c_a is the one that enters Einstein's mass–energy equation, and corresponds to the speed of our 3D world along the time axis (interpretation 2 in Fig. 1). The speed of electromagnetic radiation in free space is a different constant c. The value of the latter may be either identical or numerically close to c_a, depending of whether one adopts a relativistic or an emission theory for photons, respectively (see Section V).

The meaning of the $u - \mathbf{r}$ plane under interpretation 1 can now be rephrased as follows. At any arbitrary time t_0 (say, the present), our 3D world separates superluminal from subluminal preons. Furthermore, as seen in Fig. 1, there is a continuous exchange of preons between our hypersurface and the two half-spaces above and below.

For events inside our hypersurface, Eq. (68) reduces to

$$\partial_\mu(\rho \mathscr{V}_a \mathscr{V}_a) = -\partial_\mu \tau_{4x4} - \partial_\mu P \qquad (69)$$

The stress tensor τ_{4x4} is formed by the conventional 3×3 viscosity matrix τ_{3x3} associated with the 3D spatial dimensions, and by the elements associated with the u-dimension given by

$$\tau_{uj} = \tau_{ju} = \frac{S_j}{c_a}, \qquad j = (x, y, z) \qquad (70)$$

$$\partial_u \tau_{uu} = \frac{1}{c_a} \frac{\partial \tau_{uu}}{\partial t} = -\frac{1}{c_a} \sum_{all\pm} S_u^\pm \delta(\mathbf{r} - \mathbf{r}^\pm) = -c_a \sum_{all\pm} q^\pm \delta(\mathbf{r} - \mathbf{r}^\pm) \qquad (71)$$

where $\mathbf{S} = (S_x, S_y, S_z)$ is a (displacement) energy flux density along axes x, y, z with dimensions of energy per unit time per unit area, the source/sink $S_u^\pm = q^\pm c_a^2$ is a concentrated energy flow along the u axis with dimensions of energy per unit time, the preonic flow $q^\pm$ has dimensions of mass per unit time, and $\delta(\mathbf{r} - \mathbf{r}^\pm)$ is Dirac's tridimensional delta function (dimensions: L^{-3}, $L =$ length) representing the position of the source/sink.

Equation (70) may be interpreted as a transfer of energy (by displacement) from the u axis into the spatial axes (or the other way around), whereas Eq. (71) is a flow of momentum along the u axis. It follows that the 4D source $\mathscr{S} = (S_u^\pm, \mathbf{S})$ represents a *convective* transfer of momentum and energy, that is mediated by preons flowing from one region of the 4D fluid into another. Therefore, there is conservation of preons in the whole 4D universe; hence, conservation of energy and momentum also follow.

It is mentioned in passing that other fluid theories contain expressions similar to our Eq. (69) (e.g., Eq. 3 in Ref. 97). However, our approach is fundamentally different because we allow for interaction between our world and other regions of Σ where preons move with $v_u \neq c_a$, as described by the more general Eq. (68). This interaction gives rise to the 4D source $\mathscr{S} = (S_u^{\pm}, \mathbf{S})$ described by Eqs. (70) and (71).[9]

By analogy with the standard 3D case, the 4D preonic fluid exerts force, and performs work along the *four* dimensions u, x, y, z, via its hydrodynamic pressure P. Thence, P is interpreted as potential energy per unit volume. The force density associated with the preons contained in a unit 3D volume is

$$\mathscr{F} = -\partial_\mu P \tag{72}$$

Summarizing, the preonic fluid was described by the general equation of motion (68), that we interpret as a *unified field equation* (UFE) representing all forces. Our expression is completely based on conventional hydrodynamics. No sources were included here; pressure is a result of the motion of preons. Consequently our model differs of Einstein's general field equation in regard to the sources. The possibility that gravitational tensor sources could be substituted for something simpler has been noted previously [106, p. 97]. For additional details, see Múnera [102,104].

B. Electromagnetic Force

In the spirit of effective field theories, it is expected that Maxwell equations should be a special case of UFE, valid in restricted regions of Σ. It has been shown elsewhere [104] that the UFE reduces to Maxwell's equations when the following three conditions hold:

1. Preons have $v_u = c_a$ (our 3D-world).
2. ρ is a constant, then $\partial_\mu \rho = 0$.
3. The preonic flow is inviscid: $\tau_{3\times3} = 0$.

The first row in Eq. (69) corresponds to the u axis. Substituting conditions 2 and 3 above, this leads to the scalar equation

$$\partial_u \rho_e + \frac{1}{c_a} \nabla \cdot \mathbf{J}_d = -k_e(\rho c_a \nabla \cdot \mathbf{v} + \partial_u P) \tag{73}$$

---

[9]Readers who are uncomfortable with the notion of a fourth dimension may refer to an alternative interpretation. The source term may be viewed as a transfer of energy and momentum from one region of 3D space to another, under the assumption that our 3D space may be topologically disconnected, as in Gribov [105].

where

$$\mathbf{J}_d = k_e \mathbf{S} \tag{74}$$

$$I = -k_e \sum_{all\pm} S_u^{\pm} \tag{75}$$

$$\partial_u \rho_e = -\frac{k_e}{c_a} \sum_{all\pm} S_u^{\pm} \delta(\mathbf{r} - \mathbf{r}_{\pm}) \tag{76}$$

The dimensional constant k_e has units of charge per unit energy. Hence, $I, \mathbf{J}_d$ are proportional to energy flow along the u axis, and flow of energy into (from) our 3D world from (resp. into) the u axis. In this sense, electric current and charge density are simple *auxiliary* 3D concepts associated with the 4D energy-momentum source $\mathscr{S} = (S_u^{\pm}, \mathbf{S})$. This result is reminiscent of some opinion of Warburton [34], who claims that the displacement current is not a fundamental concept.

The definition of electric charge density in Eq. (76) agrees with our opinion that $\rho_e = 0$ in Maxwell's equations represents charge neutrality (see Section III); the simplest case is $S_u^+ + S_u^- = 0$. Also note that $\mathbf{J}_d$ defined by Eq. (74) is independent of ρ_e thus allowing for the existence of a displacement current in the absence of electric charge, as also discussed in Section III.

Equation (73) is a generalization of the continuity equation for electric charge. Indeed, the left-hand side is the standard Eq. (13), which obtains when

$$\rho c_a \nabla \cdot \mathbf{v} + \partial_u P = 0 \Rightarrow \mathscr{F}_u = -\partial_u P = \rho c_a \nabla \cdot \mathbf{v} \tag{77}$$

which means that there exists a force density $\mathscr{F}_u$ along the u axis when $\nabla \cdot \mathbf{v} \neq 0$. Therefore, compressibility of the preonic fluid leads to the creation of sources and sinks. It is noted that most conventional fluid models in the literature stay within the boundaries of incompressible fluids.

The magnetic vector potential $\mathbf{A}^B$ is identified with the convective transport of momentum by individual preons:

$$\mathbf{A}^B = -\frac{nmc_a \mathbf{v}}{K_e} = -\frac{\rho c_a \mathbf{v}}{K_e} \tag{78}$$

where K_e is a dimensional constant with dimensions of charge density. From this definition it follows that

$$\mathbf{B} = -\frac{\rho c_a \nabla \times \mathbf{v}}{K_e} \tag{79}$$

$$\mathbf{E} = \frac{\rho c_a \partial_u \mathbf{v}}{K_e} + \nabla U^E \tag{80}$$

Definitions for electromagnetic field in Eqs. (79) and (80) are similar to Hofer's [99]. There is a difference: we start from an equation of motion for a 4D ether, while Hofer starts from a wave equation for 3D momentum density (his eq. 16). Our **B** is also similar to Marmanis [100], but his **E** is quite different.

After substitution of Eqs. (74)–(77) into the spatial part of Eq. (69) we get

$$\rho c_a \partial_u \mathbf{v} + \frac{\rho}{2} \nabla v^2 - \rho \mathbf{v} \times (\nabla \times \mathbf{v}) + \frac{\mathbf{v}}{c_a} \mathscr{F}_u + \frac{1}{c_a k_e} \partial_u \mathbf{J}_d = -\nabla P \qquad (81)$$

Further substitution of Eqs. (77)–(80) leads to the 3D force density

$$\mathbf{F} = -\nabla P = K_e \mathbf{E} + K_e \frac{\mathbf{v}}{c_a} \times \mathbf{B} + \frac{\mathbf{v}}{c_a} \mathscr{F}_u + \frac{1}{c_a k_e} \partial_u \mathbf{J}_d \qquad (82)$$

Note that the spatial components of the equation of motion (69) directly represent force density in our 3D world. Also note that the right-hand side of Eq. (82) is independent of the explicit values of constants K_e and k_e (as expected because charge is not a fundamental concept here).

The first two terms on the right-hand side are the Coulomb and the Lorentz forces. There are two additional terms in Eq. (82):

1. A *displacement* induction force density

$$\mathbf{F}_d = \frac{1}{c_a k_e} \partial_u \mathbf{J}_d = \frac{1}{c_a} \partial_u \mathbf{S} \qquad (83)$$

   produced by temporal variations of the displacement energy flux **S**. As noted before, this flux is independent of the existence of electric charge density.

2. A force associated with regions of compressible preonic fluid:

$$\mathbf{F}_C = \frac{\mathbf{v}}{c_a} \mathscr{F}_u = -\frac{K_e \mathscr{F}_u}{\rho c_a^2} \mathbf{A}^B = \rho \mathbf{v} \nabla \cdot \mathbf{v} \qquad (84)$$

As announced, both force terms are independent of the auxiliary constants K_e and k_e. Since $\mathbf{F}_C$ appears in regions of compressible fluid, it probably is associated with variable preonic density ρ. Then, it could be concluded at first sight that, strictly speaking, Maxwell's equations are not applicable in the presence of $\mathbf{F}_C$. However, if the photon is associated with a source-sink pair, as in Section V, it can still hold that density is a constant on the average.

C. Particles as Solitons in 4D Ether

In Eq. (72) there is a component of force along the u dimension, leading to the appeareance of sources and sinks in our hypersurface, as follows:

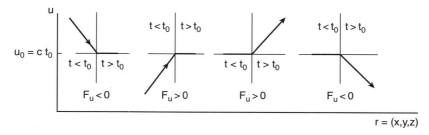

Figure 2. The four mechanisms for producing sources and sinks (see the text).

Sources S_u^+ are produced by the fourth component of force, which acts on preons outside our hypersurface, via two mechanisms (Fig. 2, left side):

- Preons moving with $v_u > c_a$ are decelerated to enter our world at $t = t_0$ with $v_u = c_a$.
- Preons moving with $v_u < c_a$ are accelerated to enter our world at $t = t_0$ with $v_u = c_a$.

Sinks S_u^- are produced by the fourth component of force, which acts on preons in our hypersurface, where they move with $v_u = c_a$ at $t < t_0$. There are two mechanisms (Fig. 2, right side):

- Preons are accelerated and leave our world at $t = t_0$ with $v_u > c_a$.
- Preons are decelerated and leave our world at $t = t_0$ with $v_u < c_a$.

In the representation advanced above, our 3D world is bounded by a hypersurface, whose normal points into our world. This is interpreted as the surface of Dirac's sea of energy momentum. Sources and sinks correspond to punctures on the hypersurface driven by $\mathscr{F}_u$, identified with particles and antiparticles, respectively. In this way, particles and antiparticles become solitons in the 4D ether.

The mass of a particle (resp. antiparticle) is then proportional to the preonic mass flow into (resp. out of) our 3D-world, which carry a momentum flux $q^{\pm}c_a$. Particles (resp. antiparticles) are solitons of steady flow, whose rest mass $M_0^{\pm}$ is the result of a transfer of energy from (resp. into) the u axis during the duration T_m of a measurement inside a 3D volume whose size corresponds to the volume of the particle. Then

$$M_0^{\pm}c_a^2 = q^{\pm}c_a^2 T_m \tag{85}$$

Notice that this generic model for particles immediately solves one of the greatest difficulties in quantum mechanics: the infinities associated with electric and gravitational energy. Indeed, momentum flows at a finite rate, but the total amount of energy transferred by the field depends of the interaction (or observation) time.

As briefly recollected by Ohanian [107], the mechanical origin of spin was mentioned as a possibility at the beginning of the twentieth century. Quantum theory adopted a *point* model for particles, which completely closed the door to a mechanical interpretation of spin. Corben [108–111] tried to develop a relativistic composite model for particles, where the basic components were punctual, but allowing for a separation between the center of mass and the center of charge. Corben argued that one of the components could have negative mass.

Other, more recent, attempts to develop nonconventional particle models are those of Vigier [8], who proposed an *extended* model for the electron, and Costella et al. [112] with a classical representation for antiparticles.

It appears that a mechanical model for spin must start from extensive particles. It is expected that the 4D solitons will exhibit vorticity in many instances. Let the moment of inertia I associated with a particle be

$$I = M_0^{\pm} r_g^2 \tag{86}$$

where the radius of gyration r_g is a property of the *extended* vortices that form the soliton. Detailed models are currently under development and will be published at a later date.

An electron model is needed for the model of the photon in Section V. As a first approximation, let the electron (resp. positron) be a thin disk of mass m_e and radius r_e rotating at an average angular velocity ω_e. The angular momentum of the disk is then

$$s = I\omega_e = \frac{m_e r_e^2}{2} \omega_e = \pm \frac{\hbar}{2} \tag{87}$$

Let the magnetic moment μ be associated with the vortex radius r_e then

$$\mu = \frac{\pi r_e^2 e v}{c} = \frac{e r_e^2 \omega_e}{2c} \Rightarrow \frac{\mu}{s} = \frac{e}{m_e c} \tag{88}$$

Note that this extended semiclassical model leads to the correct ratio of magnetic moment to spin [Eq. (88)]. Failure of classical theory to account for the correct ratio is one of the main arguments in favor of a quantum model. However, it is noted that Eq. (88) results from a first approximation to a definitive model of a vortex.

V. A CHARGE-NEUTRAL/MASS-NEUTRAL PHOTON

A. Historical Introduction

The idea that the photon may be a composite particle is not new. Long ago de Broglie [113] (see also de Broglie's treatise on light and matter [114]) suggested that the photon was a composite state of a neutrino–antineutrino pair; such a pair, however, did not obey Bose statistics. To avoid this difficulty Jordan [115] introduced neutrinos with different momenta. Over the years, additional adjustments were made by other investigators [116–121]. Since the photon restmass is zero, or very small [7], neutrinos are chosen as its components. However, one would naively expect that if the photon is a composite particle it may under some conditions decay or be separated into its components. Indeed, photon pair production leads to an electron–positron pair, but not to a neutrino–antineutrino one. However, from the viewpoint of total rest mass, an electron plus a positron cannot be the components of a low energy photon. There is a clear difficulty for composite models along this line of thought.

In a later paper the same de Broglie [3] conjectured that photons may contain two Dirac corpuscles. The idea that dipoles are related to the photon has been around for a while, for example, see Bateman [77], Warburton [34], Hunter and Wadlinger [37,38], McLennan [78], and Barbosa and González [40].

Spin is typically treated as a quantum phenomenon; an easily accesible and readable account is given by Ohanian [107]. However, the possibility that spin may be a phenomenon with classical overtones has been a recurrent one [79,107–111,122–124]. The connection between the classical polarization of light and quantum mechanics was noted long ago by Fano [125], while the connection between polarization and Clifford algebra for spinors was noted more recently [126]. Finally, some philosophers have suggested that spin is a mere property of space [127].

The development of our photon model started from the interpretative difficulties discussed in Section III. In particular, the possibility that zero charge in vacuum may be interpreted as neutrality of charge almost everywhere, rather than as complete absence of charge [76]. The symmetrization of Maxwell's equations in Section III.E hinted at the presence of two charges of different sign. Pair production and pair annihilation processes further hint that the constituents of the photon are a positron–electron pair, but there was always the nagging question of rest mass. Finally, the notion of a 4D ether led to the concept of a dynamic mass: a flow of ether fluid across the projection of our world onto the u axis. In this way the photon may be modeled as a source–sink pair (into and out of our 3D world), having a zero net mass flow, thus accounting for a photon zero rest mass. Such a model is strongly reminiscent of Newtonian mechanics. The concept, however, is not completely new if one recalls that Feynman was able to derive Maxwell's equations from Newton's equations of motion plus the commutation relations [128].

Ritz' [30] remarkable paper[10] derives the retarded electromagnetic potential from the electrostatic field of the electron, as in Section V.C below. McLennan's [78] suggestion that the photon is a rotating dipole is analogous to the model to be developed in Section V.B. The ellipsoidal shape of Hunter and Wadlinger [37] also appears in our model. However, to the present author's knowledge, models in the literature do not explicitly derive the electric and magnetic fields as a combination of the elementary electrostatic fields of the particle–antiparticle pair; put in a different way, models do not have a source of advanced potential in the same spatial region as the source of retarded potential.

Overall, the most significant differences between our approach and other photon models known to the present author are

1. The source of electromagnetic field is explicitly identified as a positron–electron pair.

2. The source of advanced potentials is at the photon itself.

3. Spin of photon is connected to both orbital and rotational motion.

4. Momentum of the photon is generated in the plane of rotation.

5. Mass is a dynamic concept.

6. Rest mass is identified with mass neutrality.

7. Potential energy in the electromagnetic field is a result of linear momentum transport.

B.   The Photon as a Soliton Doublet

Let the photon be a 4D soliton doublet, which manifests in our world as an electron–positron pair, orbiting the common center of mass at distance r_γ with angular velocity ω_γ. There is an incoming momentum flux $q^+ c_a$ and a canceling outgoing flow of equal magnitude $q^- c_a$, so that net momentum flux across the soliton *doublet*[11] is zero. The word *dipole* was not used to stress the fact that the pair is formed by a particle and an antiparticle, having zero net mass.

The model is similar to a positronium atom, so that nonrelativistic quantum behavior of the photon may be obtained from conventional quantum mechanics. Relativistic predictions require application of Dirac's theory; see, for instance, Chaps. XI-XII of the standard textbook by Dirac [129] himself. As a first

---

[10]Ritz paper is remarkable in every sense; it is 130 pages long. We saw Ritz' paper while in the last stage of preparation of this chapter, and learned that we had rediscovered two concepts that he used in exactly the same way: the ballistic fluid and the electrostatic field of the electron.

[11]The sense of word "doublet" here is related to a quantum mechanical doublets: two states of a particle.

approximation, this chapter uses a semiclassical model, reminiscent of Bohr's atom.

Rotation of the positron–electron pair is produced by the mutual Coulomb central force. Hence, the torque is zero.[12] It follows that motion is in a plane perpendicular to orbital angular momentum **L**. Furthermore, the magnitude of **L** *is a constant*.

Previous results considerably simplify the analysis of motion of the doublet in the preferred frame Σ, which is an inertial system by definition. Let the z axis be perpendicular to the plane of motion at the center of mass of the electron–positron pair and. Vector **L** lies along the z axis, and the magnitude is

$$L_z = I_\gamma \omega_\gamma = 2m_e r_\gamma^2 \omega_\gamma \qquad (89)$$

where I_γ is the moment of inertia of the source–sink pair in the nonrelativistic region. Note that both electron and positron contribute to I_γ because the whole preonic mass of the pair still is on the 3D surface of our world (i.e., the preonic mass of the antiparticle has not yet gone into the u axis).

Since motion takes place on the x–y plane, total photon spin is given by

$$\mathbf{J} = \mathbf{S} + \mathbf{L} = (I_e^+ \omega_e^+ + I_e^- \omega_e^-)\mathbf{k} + I_\gamma \omega_\gamma \mathbf{k} = n\hbar\mathbf{k} \qquad (90)$$

where the first term on the right-hand side represents rotational motion of electron and positron around their internal axes, given by Eqs. (86) and (87). All angular velocities are positive when rotation is counterclockwise, and negative in the opposite sense.

Now, L_z is a constant because of central forces, and electron/positron spin is also a constant given by Eq. (87). Hence the magnitude of photon spin is also a constant, which has been equated to $n\hbar$ in Eq. (90). Parameter n allows for different levels of energy of the system. Quantization is introduced as in Bohr's atom assuming that orbits are stable for values of energy corresponding to integer values of n.

In photon ground state, spins of the electron–positron pair form a singlet such that

$$S = s^+ + s^- = \frac{\hbar}{2} - \frac{\hbar}{2} = 0 \qquad (91)$$

Then, photon spin in ground state is

$$\mathbf{J} = \mathbf{S} + \mathbf{L} = L_z \mathbf{k} = 2m_e r_\gamma^2 \omega_\gamma = \hbar\mathbf{k} \qquad (92)$$

---

[12]This is a well-known result from Newtonian mechanics; see, for instance, Chap. 6 of Kleppner and Kolenkow [80].

for counterclockwise orbital rotation. For clockwise rotation the opposite sign holds. Then, this model naturally predicts different photon helicity.

In some excited state, either the electron or the positron may flip-spin, producing $S = \pm\hbar$. This would lead to $J = \pm 2\hbar, 0$. Of course, quantum mechanically all values $-2, -1, 0, +1, +2$ are allowed.

Although L_z is a constant, there is no limitation on the values of ω_γ and r_γ. This means that even for a fixed value of n the orbit need not be circular. In ground state $L_z = 2m_e r_\gamma^2 \omega_\gamma = \hbar$, so that both ω_γ and r_γ may vary, producing an elliptical motion. Of course, as seen below, the value of energy will vary between two extremes associated with the major and minor axes of the ellipse.

The kinetic energy of orbital motion is

$$K_{\text{orb}} = \frac{I_\gamma \omega_\gamma^2}{2} = \frac{L_z \omega_\gamma}{2} = \frac{n\hbar\omega_\gamma}{2} \tag{93}$$

The kinetic energy of rotation of the electron around its proper axis equals the kinetic energy of the positron. The total rotational energy is then

$$K_{\text{rot}} = 2\frac{I_e \omega_e^2}{2} = s\omega_e = \frac{\hbar\omega_e}{2} \tag{94}$$

The total kinetic energy of the photon is

$$K_\gamma = K_{\text{orb}} + K_{\text{rot}} = \frac{n\hbar\omega_\gamma}{2} + \frac{\hbar\omega_e}{2} = \frac{\hbar}{2}(n\omega_\gamma + \omega_e) \tag{95}$$

For low-energy photons, the rotational and orbital motions occupy the same region in 3D space, where the same preonic fluid participates in both motions. Then, as a first approximation, it is assumed that $\omega_\gamma = \omega_e = \omega$.[13] Substituting in (95), the total kinetic energy of a photon in state n is

$$K_n = \frac{(n+1)\hbar\omega}{2} \tag{96}$$

The ground-state for the photon occurs for $n = 1$ with energy

$$K_1 = \hbar\omega = h\nu \tag{97}$$

which is de Broglie's famous expression. It is noted that in the relativistic energy equation, total energy is the sum of rest mass energy, plus energy carried by

---

[13] An alternative rationalization: a resonant orbital and rotational motion.

linear momentum (i.e., kinetic energy). In the photon, all energy is kinetic. This explains why potential energy was not considered.

For photon ground state, it is easily checked that rotational motion and orbital motion occur in the same 3D region. From from Eqs. (92), (86), and (87) we obtain

$$r_\gamma = r_g \qquad (98)$$

A 2D pictorial representation of the photon model is shown in Fig. 3a. For macroscopic fluids, there are three-dimensional representations of sink–source pairs in Brandt and Schneider [130].

As a numerical example, consider an X ray with $v = 10^{18}$ Hz. The tangential speed of rotation of the preonic fluid is about $v \approx 0.1c$ and the radii in Eq. (98) are $r_\gamma = r_g \approx 10^{-12}$ m. For a microwave radiation of $v = 10^{10}$ Hz the tangential speed of rotation of the preonic fluid is about $v \approx 10^{-5}c$ and the radii in Eq. (98) are $r_\gamma = r_g \approx 10^{-8}$ m. The low values of tangential speed justify the use of the nonrelativistic mass in the moment of inertia [recall Eqs. (86) and (87) in Section IV, and all previous equations in this section).

Finally, let us consider elliptical motion for $n = 1$. Angular velocity in Eq. (97) is given by

$$\omega = \frac{\omega_m + \omega_M}{2}, \qquad \Delta\omega = \frac{\omega_m - \omega_M}{2} \qquad (99)$$

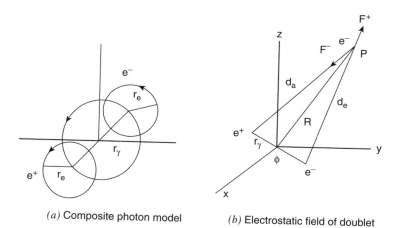

(a) Composite photon model   (b) Electrostatic field of doublet

Figure 3. The photon as a rotating doublet: (a) composite photon model—extended electron–positron pair rotating in x–y plane; (b) electrostatic field of doublet—electrostatic force on a test particle at rest.

where ω_m, ω_M correspond to angular velocities at the minor and major axes respectively, calculated with the help of Eq. (92). There are small variations in energy K_n that manifest as a spread in frequency $\Delta\omega$, neatly leading to wavepackets. Similar variations of ω_γ are expected within each state $n = j, j = 1, 2, \ldots$. This subject will be treated in detail elsewhere.

In summary, the photon has been modeled as a doublet in rotation in the preferred frame Σ. Spin and energy have been obtained from a semiclassical analysis. Polarization corresponds to a fixed direction of vector $\mathbf{L}$ in Σ. In a nonpolarized photon vector $\mathbf{L}$ has a time-dependent direction. A particular case of nonpolarization is the ellipsoid, as in Hunter and Wadlinger [37]. Our Eq. (96) allows for the existence of multiphotons that vary in steps of half the ground-state photon energy; such prediction differs of the prediction of Hunter and Wadlinger [37]. Photons in motion with respect to Σ will be considered elsewhere. The photon is the source of the electromagnetic field, as explained next.

C. Electrostatic Force of a Rotating Doublet

Consider an electron at rest in the preferred frame Σ. The force on a stationary (negative) test particle is given by Coulomb's law

$$\mathbf{F}^+ = F^+\hat{\mathbf{r}} = F_x^+\mathbf{i} + F_Y^+\mathbf{j} + F_Z^+\mathbf{k}, \qquad F_x^+ = \frac{x}{d_e}F^+, \qquad F_y^+ = \frac{y}{d_e}F^+$$

$$F_z^+ = \frac{z}{d_e}F^+ \tag{100}$$

$$F^+ = \frac{e^2}{d_e^2} = \frac{q^+ c A_{\text{test}}}{4\pi d_e^2} \tag{101}$$

where the plus sign refers to the source and A_{test} is the area associated with the test particle. The first part of Eq. (101) is the standard expression, which may be used by readers who prefer to avoid interpretational aspects.

The second part of Eq. (101) is based on the 4D model of Section IV. The electron is an isotropic 3D source that emits a momentum flow q^+c at time t_e. The preonic fluid[14] propagates in straight line with constant speed c from the point of emission to a spatially separated point P located at distance d_e at the moment of emission. The preonic fluid carries momentum, which materializes as force during interaction with an obstacle, say, a test charge. To be specific, let the test particle be an electron of effective radius r_e, then $A_{\text{test}} = \pi r_e^2$. From

---

[14]This is equivalent to the flow of "fictive" particles in Ritz [30]. A summary of the latter is presented by [35].

Eq. (101), the preonic mass flow $q^{\pm}$ associated with the electron (resp. positron) mass, in CGS units (g s^{-1}) is

$$q^+ = +q = \frac{4e^2}{cr_e^2}, \qquad q^- = -q \tag{102}$$

Equation (101) allows for different theories, according to the value of the speed of propagation:

1. Einsteinian relativity when c is independent of motion of emitter and energy of photon. Observers accelerated relative to Σ[15] will perceive the same constant c provided that source and observer be in inertial relation, that is, either at relative rest or in relative constant motion. Presumably, $c = c_a$.

2. Lorentzian relativity where c is a constant in Σ, independently of motion of emitter, and energy of photon. Then, $c = c_a$. Observers in motion relative to Σ will perceive speeds of propagation different from c_a.

3. Modified Lorentzian relativity when c is independent of motion of emitter, but depends of energy of photon. For instance, $c = (c_a^2 - r_\gamma^2 \omega_\gamma^2)^{1/2}$. Observers in motion relative to Σ will measure speeds of propagation different from c.

4. Emission theory (see the paper by Cyrenika [131] for the principles of emission theory) when c depends of photon energy and speed of the emitter. This is the case of Ritz [30] and other emission theories reviewed by Fox [35].

The photon model here refers to a photon at rest in Σ. The four theories just mentioned are compatible with Eq. (101). Detailed predictions of each theory are different, so that crucial tests may be designed and carried out. For instance, the Michelson–Morley experiment is conventionally interpreted as a demonstration of Einsteinian relativity, but the evidence is not convincing, as discussed in Section II. Another example, to discriminate between relativistic theories (1) and (2) and emission theories (3) and (4), it is necessary to measure with high precision the velocity of photons with energy higher than 100 keV.

The ballistic preonic fluid reaches the stationary observation point P with some delay at time t_P:

$$t_P = t_e + \frac{d_e}{c} \tag{103}$$

Clearly, electrostatic force is a retarded concept. Of course, if source is at rest, force does not change over time, so that the force field appears as static, as if it were time-independent.

---

[15]Actually, the preferred frame is undefined in special theory of relativity.

Now consider a change in the origin of coordinates. The source is still on the x–y plane but located at distance r_γ from the origin at azimuthal angle φ. The electrostatic force at any arbitrary $P = (x, y, z)$ is immediately obtained from Eq. (100) with appropriate cosine angles (see Fig. 3b):

$$F_x^+ = \frac{x - r_\gamma \cos \varphi}{d_e} F^+, \qquad F_y^+ = \frac{y - r_\gamma \sin \varphi}{d_e} F^+, \qquad F_z^+ = \frac{z}{d_e} F^+ \qquad (104)$$

Now consider a sink (i.e., positron) on the x–y plane, located at distance r_γ from the origin at azimuthal angle $\varphi + \pi$. There is an attractive electrostatic force at any arbitrary $P = (x, y, z)$ given by

$$F_x^- = \frac{x + r_\gamma \cos \varphi}{d_a} F^-, \qquad F_y^- = \frac{y + r_\gamma \sin \varphi}{d_a} F^-, \qquad F_z^- = \frac{z}{d_a} F^- \qquad (105)$$

$$F^- = -\frac{e^2}{d_a^2} = \frac{q^- c A_{\text{test}}}{4\pi d_a^2} \qquad (106)$$

where the minus sign refers to the sink and d_a is the distance from sink to the stationary P at the moment of absorption. In this ballistic model, preons flow past P at time t_P in the direction of the sink. Preons arrive to the sink at some later time t_a where they are absorbed:

$$t_a = t_P + \frac{d_a}{c} \qquad (107)$$

Summarizing the previous discussion, individual forces are active along the respective rays. If test charge is negative, electron force is repulsive, whereas positron force is attractive. Also, distance d_e is evaluated at the time of emission t_e, while d_a is evaluated at the moment of absorption. Net force is the vector addition.

Up to this point the electron and positron forces were treated as independent. Now consider a doublet in a photon. Let τ be the proper time of the photon. Preons absorbed at τ were at P at an earlier time $t_P = \tau - (d_a/c)$, while preons emitted at τ arrive to P at a later time $t_P = \tau + (d_e/c)$. However, from the viewpoint of observer P the interest is in determining force at an arbitrary time t_P. This is the combined effect of two flows of fluid:

- Preons en route to the sink, where they will arrive at a later time: $\tau = t_a = t_P + (d_a/c)$
- Preons coming from the source, emitted earlier at $\tau = t_e = t_P - (d_e/c)$

These two processes constitute a straightforward, causal explanation of advanced and retarded forces and potentials. Net effect at P is obtained by vector addition.

If the doublet rotates, there is still emission and absorption. The same expressions are still valid, provided that there is independence among time intervals. In other words, that emission and absorption processes at time $\tau + d\tau$ be independent of previous processes in the doublet at time τ. Such assumption is fairly weak.

Consider the simplest case: a doublet rotating with constant angular velocity ω on the x–y plane. Let $\tau = 0$ at $\varphi = 0$. Then, $\varphi = \omega\tau$. If proper time is not measured from the x axis, there is an additional phase angle (with an appropriate sign): $\varphi = \omega\tau + \varphi_0$. Substitute in Eqs. (104) and (105), and introduce the observer's time to get

$$F_x^+ = \frac{x - r_\gamma\cos(\omega t_P - kd_e)}{d_e}\,F^+, \qquad F_y^+ = \frac{y - r_\gamma\sin(\omega t_P - kd_e)}{d_e}\,F^+$$

$$F_z^+ = \frac{z}{d_e}F^+ \tag{108}$$

$$F_x^- = \frac{x + r_\gamma\cos(\omega t_P + kd_a)}{d_a}\,F^-, \qquad F_y^- = \frac{y + r_\gamma\sin(\omega t_P + kd_a)}{d_a}\,F^-$$

$$F_z^- = \frac{z}{d_a}F^- \tag{109}$$

where, as usual, the wavenumber in vacuo is

$$k = \frac{\omega}{c} \tag{110}$$

Explicit expressions for the distances contain trascendental expressions:

$$d_e^2 = \left(x - r_\gamma\cos\left(\omega t_P - kd_e\right)\right)^2 + \left(y - r_\gamma\sin\left(\omega t_P - kd_e\right)\right)^2 + z^2$$

$$d_e^2 = R^2 + r_\gamma^2 - 2r_\gamma(x\cos\left(\omega t_P - kd_e\right) + y\sin\left(\omega t_P - kd_e\right)) \tag{111}$$

$$d_a^2 = \left(x + r_\gamma\cos\left(\omega t_P + kd_a\right)\right)^2 + \left(y + r_\gamma\sin\left(\omega t_P + kd_a\right)\right)^2 + z^2$$

$$d_a^2 = R^2 + r_\gamma^2 + 2r_\gamma(x\cos\left(\omega t_P + kd_a\right) + y\sin\left(\omega t_P + kd_a\right)) \tag{112}$$

$$R^2 = x^2 + y^2 + z^2$$

Evidently, both forces change direction with time as the individual rays associated with the source and the sink rotate. Also, the wavenumber, Eq. (110), formally is the conventional expression; of course, k describes propagation of the electromagnetic wave. However, the origin of angular velocity ω is the rotation of the source.

It is noteworthy that the forcefield of the doublet was calculated as a mere electrostatic force of the components. Of course, there is acceleration associated with the rotation of the doublet, but no explicit allowance was made for such fact. Only time delays and geometrical positions are involved in our calculation. In this sense, Eq. (108) is a straightforward way of obtaining the forcefield of an accelerated electron, calculated in a different approach by Conway [29].

There are three regions associated with Eqs. (108) and (109):

- *Inner region*, inside the photon, defined by $R < r_\gamma$.
- *Near-field region*, when distance is of order of magnitude of photon dimensions, $R \approx r_\gamma$.
- *Far-field region or radiation zone*, defined by $R \gg r_\gamma$. At macroscopic distances, the longitudinal component of force is negligible, so that the radiation field is almost plane (i.e., perpendicular to the z-axis in Fig. 3).

The boundaries between the regions are frequency-dependent, as shown by the numerical values of Section V.B. Also note that, in all three regions, there is no longitudinal component of the net force along the z-axis. Strictly speaking, there are longitudinal net forces elsewhere.

Finally, Eqs. (108) and (109) may be easily adapted to emission by atomic transitions. For the hydrogen atom, one only needs to substitute reduced masses as appropriate. The same is true, as a first approximation, for more complex atoms, where an electron undergoing a transition sees the rest of the atom as a positive charge. Computer animations of such transitions have been independently produced by Barbosa and González [40].

D. Symmetric Electromagnetic Fields of the Doublet

At macroscopic distances, within some beam area around the z axis, the field of forces $\mathbf{F}^+, \mathbf{F}^-$ is plane FAPP.[16] However, forces $\mathbf{F}^+, \mathbf{F}^-$ are not orthogonal in general; so, they cannot represent the conventional electric and magnetic fields.

Here we develop an alternative based on the symmetrized fields $\mathbf{P}, \mathbf{N}$ discussed in Section III.F. Let the Maxwell-like symmetric fields $\mathbf{P}, \mathbf{N}$ be defined as

$$\mathbf{P} = -\frac{\mathbf{F}^+}{e}, \qquad \mathbf{N} = -\frac{\mathbf{F}^-}{e} \tag{113}$$

where the forces are given by Eqs. (108) and (109). Sources and currents associated with fields $\mathbf{P}, \mathbf{N}$ are implicitly defined by

$$\nabla \times \mathbf{P} = -\frac{\partial \mathbf{N}}{\partial u} - \frac{4\pi \mathbf{J}^N}{c} \tag{114}$$

$$\nabla \times \mathbf{N} = +\frac{\partial \mathbf{P}}{\partial u} + \frac{4\pi \mathbf{J}^P}{c} \tag{115}$$

$$\nabla \cdot \mathbf{P} = 4\pi \rho^P \tag{116}$$

$$\nabla \cdot \mathbf{N} = 4\pi \rho^N \tag{117}$$

[16]Plane FAPP = plane for all practical purposes.

Two identical continuity equations immediately obtain:

$$\nabla \cdot \mathbf{J}^X + c \frac{\partial \rho^X}{\partial u} = 0, \qquad X = P, N \tag{118}$$

Explicit values for currents and charge densities may be calculated from the corresponding defining expressions, plus Eqs. (113), and (108)–(110). For instance, charge density ρ^X follows from

$$\nabla \cdot \mathbf{P} = -2Cr_\gamma k \frac{x \sin(\omega t_P - kd_e) - y \cos(\omega t_P - kd_e)}{d_e^3 D_e} \tag{119}$$

$$\nabla \cdot \mathbf{N} = +2Cr_\gamma k \frac{x \sin(\omega t_P - kd_a) - y \cos(\omega t_P - kd_a)}{d_a^3 D_a} \tag{120}$$

$$D_e = d_e + r_\gamma k(x \sin(\omega t_P - kd_e) - y \cos(\omega t_P - kd_e))$$
$$D_a = d_a + r_\gamma k(x \sin(\omega t_P + kd_a) - y \cos(\omega t_P + kd_a)) \tag{121}$$

$$C = \frac{q^+ c A_{\text{test}}}{4\pi e} = \frac{q c r_e^2}{4e} = e \tag{122}$$

Constant C was evaluated noticing that $q^+ = +q, q^- = -q$ and using Eq. (102). Fields $\mathbf{P}, \mathbf{N}$ are not orthogonal in general. Their magnitudes are

$$|\mathbf{P}| = \frac{C}{d_e^2} = \frac{e}{d_e^2}, \qquad |\mathbf{N}| = \frac{C}{d_a^2} = \frac{e}{d_a^2} \tag{123}$$

From Eqs. (111) and (112), in the far field the magnitudes of $\mathbf{P}, \mathbf{N}$ are equal FAPP:

$$|\mathbf{P}| \cong |\mathbf{N}| \cong \frac{C}{R^2} \cong \frac{e}{R^2} \tag{124}$$

This last result leads to the conventional orthogonality of the electromagnetic field, as seen in next section.

E. Extended Maxwell Equations

Let us now define the conventional electromagnetic field as

$$\mathbf{E} = \frac{\mathbf{P} - \mathbf{N}}{2}, \qquad \mathbf{J}^E = \frac{\mathbf{J}^P - \mathbf{J}^N}{2}, \qquad \rho^E = \frac{\rho^P - \rho^N}{2} \tag{125}$$

$$\mathbf{B} = \frac{\mathbf{P} + \mathbf{N}}{2}, \qquad \mathbf{J}^B = \frac{\mathbf{J}^P + \mathbf{J}^N}{2}, \qquad \rho^B = \frac{\rho^P + \rho^N}{2} \tag{126}$$

Simple manipulation of Eqs. (114)–(117) leads to

$$\nabla \times \mathbf{E} = -\frac{\partial \mathbf{B}}{\partial u} - \frac{4\pi \mathbf{J}^B}{c} \tag{127}$$

$$\nabla \times \mathbf{B} = +\frac{\partial \mathbf{E}}{\partial u} + \frac{4\pi \mathbf{J}^E}{c} \tag{128}$$

$$\nabla \cdot \mathbf{E} = 4\pi \rho^E \tag{129}$$

$$\nabla \cdot \mathbf{B} = 4\pi \rho^B \tag{130}$$

There are also two identical continuity equations:

$$\nabla \cdot \mathbf{J}^X + c\frac{\partial \rho^X}{\partial u} = 0, \qquad X = E, B \tag{131}$$

Explicit values for currents and charge densities may be calculated from the corresponding defining expressions. Charge densities are as follows:

$$\rho^E = -\frac{Cr_\gamma k}{4\pi}\left(\frac{x\sin(\omega t_P - kd_e) - y\cos(\omega t_P - kd_e)}{d_e^3 D_e}\right.$$
$$\left. + \frac{x\sin(\omega t_P + kd_a) - y\cos(\omega t_P + kd_a)}{d_a^3 D_a}\right) \tag{132}$$

$$\rho^B = -\frac{Cr_\gamma k}{4\pi}\left(\frac{x\sin(\omega t_P - kd_e) - y\cos(\omega t_P - kd_e)}{d_e^3 D_e}\right.$$
$$\left. - \frac{x\sin(\omega t_P + kd_a) - y\cos(\omega t_P + kd_a)}{d_a^3 D_a}\right) \tag{133}$$

In the far field it holds that $d_e \cong d_a \cong D_e \cong D_a \cong R$; then

$$\rho^E = -\frac{er_\gamma k}{2\pi R^4}\cos(kR)(x\sin(\omega t_P) - y\cos(\omega t_P)) \tag{134}$$

$$\rho^B = +\frac{er_\gamma k}{2\pi R^4}\sin(kR)(x\cos(\omega t_P) + y\sin(\omega t_P)) \tag{135}$$

On the z axis both charge densities are zero, but $\mathbf{E}, \mathbf{B}$ fields are non solenoidal elsewhere. The last two equations contain both advanced and retarded potentials. These expressions may be constrasted with conventional results for electric dipoles containing retarded potentials only; see, for instance, Panofsky and Phillips [65, Chap. 14], and Born and Wolf [62, pp. 84–87].

Fields $\mathbf{E}, \mathbf{B}$ are orthogonal if

$$\mathbf{E} \cdot \mathbf{B} = 0 \Rightarrow P^2 - N^2 = 0 \tag{136}$$

As noted in Eq. (124), the condition is valid FAPP in the far field. However, in the inner and the near-field regions, fields $\mathbf{E}, \mathbf{B}$ are not orthogonal, except along the z-axis.

In summary, the photon model proposed leads to an extended symmetric set of Maxwell's equations, that contains a magnetic source and a magnetic current, both of electric origin. Conventional Maxwell's equations appear as a limiting case in far-field with both $\mathbf{J}^B = 0, \rho^B = 0$.

VI.   CONCLUDING REMARKS

The evidence against the existence of a preferred frame Σ was briefly reviewed in Section II. It appears that there is no strong evidence against Σ. On the contrary, there is mounting evidence on the existence of local anisotropies [59] that may be interpreted as supporting the existence of Σ. Our own analysis of all experiments of the Michelson–Morley type supports Σ, rather than Einstein's second postulate [57].

In Section III we reviewed our own work on the solutions of Maxwell's equations, which hint to the existence of non-conventional magnetic scalar potentials in free space. The symmetrized set of Maxwell's equations [87] suggests the existence of two novel electromagnetic fields $\mathbf{P}, \mathbf{N}$, that lead to the conventional fields $\mathbf{E}, \mathbf{B}$.

Section IV reviews our more recently developed 4D ether model [102–104], which is based on the premise of the existence of Σ. Rest mass is associated to a flow of primordial fluid (preons). This novel dynamic concept of mass solves at once several longstanding difficulties; two of them are (1) the infinities associated with electric and gravitational fields and (2) the stability of orbits under Coulomb attraction. Indeed, there is a permanent flow of momentum across a particle (source); the momentum flux is occasionally tapped by interaction with a (test) particle. Such process does not change the total momentum flux available at the source; hence, there is no loss of potential energy as in the conventional interpretation. The total momentum that crosses a source is, of course, infinite in an infinite time, but the source is always finite.

In Section V, the photon was described as an electron–positron pair in rotation in Σ. In the 4D ether, antiparticles are dynamic sinks, so that an electron–positron pair has a zero net momentum flux, thus explaining the photon rest mass. The photon is then a composite charge-neutral and mass-neutral entity.

The primordial fluid propagates with a constant speed c in Σ, originating the Coulombian forcefield of the individual particles that constitute the photon. An observer at rest in Σ sees a rotating field as the electron–positron pair rotates. Photon spin and de Broglie's energy relation correspond to the ground state of the composite particle.

The electromagnetic field of the composite photon contains advanced and retarded components, without any causality breach. The forcefield of doublet is described by the *symmetric* Maxwell's equations [87b]. Three different regions appear in the forcefield: inner, near-field, and far-field. Longitudinal components of force are always present. However, in the far field, they dissapear for practical purposes. In this sense, the equations developed contain the standard set as a limiting case.

From the symmetric set, an extended set of Maxwell equations was exhibited in Section V.E. This set contains currents and sources for both fields **E**, **B**. The old conjecture of Dirac's is vindicated, but the origin of charge density is always electric (i.e., no magnetic monopole). Standard Maxwell's equations are a limiting case in far field.

Falaco solitons were reported [132] as pairs of solitons that exist on the surface of a fluid (water), and are interconnected through the third spatial dimension. Our model for the photon is a pair of 3D solitons interconnected through the fourth dimension.

Theoretical issues to be pursued at some future time are

- Calculation of forcefield when the observer is in motion relative to Σ.
- Calculation of forcefield in the (Lorentzian) relativistic case. The equations given here are applicable up to about 100 keV photons.
- Connection between the equation of motion describing our 4D ether and the Bohm–Vigier [133] relativistic hydrodynamics.

As always in science, empirical test is required to validate any theory. Some testable matters are

- Experiments of the Michelson–Morley type may help confirm the existence of Σ, thus disproving Einsteinian relativity. Or, the other way around.
- Measurement of speed of propagation of energetic photons (1 MeV and above) may confirm whether c (in Σ or elsewhere) is frequency-dependent. A revision of astrophysical data may be useful here, such as comparison of speed of propagation of neutrinos and energetic gamma rays in free space.
- Emission of light in external magnetic fields may be reinterpreted under the photon model proposed here.
- Spin and velocity of photons from pair production and bremsstrahlung[17] may help decide between emission theories and (Lorentzian or Einsteinian) relativistic theories.

---

[17]Radiation emitted by accelerated charged particles.

Acknowledgments

The author is indebted to the Editor for his kind invitation to contribute this chapter. Participants at two colloquia and conversations with colleagues of the Department of Physics at National University, Bogotá, Colombia helped clarify ideas and the form of presentation. Particular mention is due to E. Alfonso, G. Arenas, E. Barbosa, C. Quimbay, and V. Tapia; the latter also directed my attention to relevant literature, in particular Corben's papers. All mistakes and misunderstandings are mine.

References

1. A. Einstein, *Ann. Phys.* (Leipzig) **7**, 132 (1905).

2. A. Einstein, *Ann. Phys.* (Leipzig) **18**, 121 (1917).

3. L. de Broglie, *La Mechanique Ondulatoire du Photon, Vol. 1, Une Nouvelle Théorie de la Lumière*, Hermann, Paris, 1940, pp. 121–165.

4. I. Bass and E. Schrödinger, *Proc. Roy. Soc. Lond.* **232A**, 1 (1955).

5. L. de Broglie and J.-P. Vigier, *Phys. Rev. Lett.* **28**, 1001 (1972).

6. J. P. Narlikar et al., *Phys. Lett. A* **154**, 203 (1991).

7. J.-P. Vigier, *Apeiron* **4**(2–3), 71–76 (1997).

8. J.-P. Vigier, in G. Hunter, S. Jeffers, and J.-P. Vigier (Eds.), *Causality and Locality in Modern Physics*, Kluwer, Dordrecht, 1998, pp. 1–22.

9. S. Jeffers, B. Lehnert, N. Abramson, and L. Chebotarev (Eds.), *Jean-Pierre Vigier and the Stochastic Interpretation of Quantum Mechanics*, Apeiron, Montreal, 2000.

10. H. A. Múnera, *Apeiron* **4**(2–3), 77–79 (1997).

11. T. Villela, N. Figueiredo, and C. A. Wuensche, *Photon Mass Inferred from Cosmic Microwave Background Radiation: Maxwell's Equations in three-Dimensional Space,"* Instituto Nacional de Pesquisas Espaciais, Brazil, circa 1994, pp. 65–73.

12. J. W. S. (Lord) Rayleigh, *Proc. Lond. Math. Soc.* **9**, 21–26 (1877).

13. J. W. S. (Lord) Rayleigh, *The Theory of Sound*, 1884; 2nd ed., Dover Publications, 1945.

14. J. W. S. (Lord) Rayleigh, (a) *Nature* **24**, 382–383 (1881); (b) ibid. **25**, 52 (1881).

15. M. Gouy (a) *C. R. Acad. Sci.* (Paris) (Nov. 29, 1880); (b) ibid. (Jan. 3, 1881); (c) *Ann. Chem. Phys. Ser. 6* **16**, 262–288 (1889).

16. H. Lamb, *Proc. Lond. Math. Soc. Ser. 2* **1**, 473–479 (1994).

17. L. Brillouin, *Wave Propagation and Group Velocity*, Academic, New York, 1960.

18. M. W. Mitchell and R. Y. Chiao, *Am. J. Phys.* **66**(1), 14–19 (1998).

19. A. E. Chubykalo and V. V. Dvoeglazov, *Apeiron* **5**(3–4), 255–256 (1998).

20. E. Hahn, *Arch. Math. Phys.* **21**, 1 (1913).

21. P. Frank and H. Rothe, *Ann der Phys.* **34**, 825 (1911).

22. W. V. Ignatowski, *Phys. Z.* **21**, 972 (1910).

23. Y. P. Terletskii, *Paradoxes in the Theory of Relativity,* Plenum, New York, 1968.

24. E. Recami, *Found. Phys.* **8**(5–6), 329–340 (1978).

25. N. D. Mermin, *Am. J. Phys.* **52**(2), 119–124 (1984); ibid. **52**(11), 967 (1984).

26. E. A. Di Marzio (a) *Found. Phys.* **7**, 511–528 (1977); (b) ibid., 885–905 (1977).

27. H. A. Múnera, *Phys. Essays* **6**(2), 173–180 (1993).

28. W. Bertozzi, *Am. J. Phys.* **32**, 551–555 (1964).

29. A. W. Conway, *Proc. Lond. Math. Soc. Ser. 2* **1**, 154–165 (1904).

30. W. Ritz, *Ann. Chim. Phys. Ser. 8* **13**, 145–275 (1908).

31. H. Tetrode, *Z. Phys.* **10**, 317 (1922).

32. G. N. Lewis, *Proc. Nat. Acad. Sci.* (USA) **12**, 22–29 (1926).

33. J. A. Wheeler and R. P. Feynman, *Revs. Mod. Phys.* (a) **17**(2–3), 157–181 (1945) (b) ibid. **21**, 425 (1949).

34. F. W. Warburton, *Am. J. Phys.* **22**, 299–305 (1954).

35. J. G. Fox, *Am. J. Phys.* **33**(1), 1–17 (1965).

36. M. O. Scully and M. Sargent III, *Phys. Today* 38–47 (March 1972).

37. G. Hunter and R. L. P. Wadlinger, in *Proc. Quantum Uncertainties NATO Workshop, NATO ASI Series B*, 1986, Vol. 162, pp. 331–343.

38. G. Hunter and R. L. P. Wadlinger, *Phys. Essays* **2**(2), 154 (1989).

39. M. Evans and J.-P. Vigier, *The Enigmatic Photon* (a) Vol. 1; *The Field B*$^{(3)}$; (b) Vol. 2, *Non Abelian Electrodynamics*, Kluwer, Dordrecht, Vol. 1, 1994; Vol. 2, 1995.

40. E. Barbosa and F. González, *Antiphoton and the Probabilistic Interpretation of Quantum Mechanics*, thesis, Univ. Nacional, Bogotá, 1995; presented at Colombian Congress of Physics, Bogotá, 1999.

41. B. Lehnert, in V. V. Dvoeglazov (Ed.), *Photon and Old Problems in Light of New Ideas*, Nova Science Publishers, 2000.

42. G. Hunter, S. Jeffers, and J.-P. Vigier (Eds.), *Causality and Locality in Modern Physics*, Kluwer, Dordrecht, 1998.

43. A. Einstein, "Ether and the theory of relativity," address at Univ. Leyden, 1920, reprinted in *Sidelights on Relativity*, Dover, New York, 1983.

44. E. F. Taylor and J. A. Wheeler, *Spacetime Physics Introduction to Special Relativity*, 2nd ed., Freeman, New York, 1992.

45. H. A. Lorentz, *Arch. Néerlandaises Sci. exactes naturelles* **21**, 103–176 (1887).

46. S. L. Glashow, *Phys. Lett. B* **430**, 54–56 (1998).

47. E. K. Conklin, *Nature* **222**, 971–972 (1969).

48. P. S. Henry, *Nature* **231**, 516–518 (1971).

49. E. Esclangon, (a) *C. R. Acad. Sci.* (Paris) **182**, 921–923 (1926); (b) ibid. **183**, 116–118 (1926).

50. A. A. Michelson and E. W. Morley, *Philos. Mag., Ser. 5* **24**(151), 449–463 (1887).

51. R. C. Tolman, *Phys. Rev.* **31**, 26–40 (1910).

52. W. M. Hicks, *Philos. Mag. Ser. 6* **3**, 9–42 (1902).

53. D. C. Miller, *Revs. Mod. Phys.* **5**, 203–242 (1933).

54. D. C. Miller, *Proc. Nat. Acad. Sci.* (USA) **11**, 306–314 (1925).

55. D. C. Miller, *Astrophys. J.* **68**, 352–367 (1928).

56. R. S. Shankland, S. W. McCuskey, F. C. Leone, and G. Kuerti, *Revs. Mod. Phys.* **27**(2), 167–178 (1955).

57. H. A. Múnera, *Apeiron* **5**(1–2), 37–54 (1998).

58. F. Selleri, in M. Duffy and M. Wegener (Eds.), *Recent Advances in Relativity Theory*, Hadronic Press, 2000.

59. M. Allais, (a) *L'Anisotropie de l'Espace—les données de l'expérience*, Éditions Clément Juglar, Paris, 1997; (b) *C. R. Acad. Sci. Ser.IIb* (Paris) **327**, 1405–1410 (1999); (c) ibid. **327**, 1411–1419 (1999).

60. Yu. A. Baurov, *Phys. Lett. A* **181**, 283–288 (1993).

61. M. J. Crooks, D. B. Litvin, P. W. Matthews, R. Macaulay, and J. Shaw, *Am. J. Phys.* **46**(7), 729–731(1978).

62. M. Born and E. Wolf, *Principles of Optics—Electromagnetic Theory of Propagation, Interference and Diffraction of Light*, 7th expanded ed., Cambridge Univ. Press, 1999.

63. J. D. Jackson, *Electrodinámica Clásica* (Spanish transl. of *Classical Electrodynamics*), Editorial Alhambra, Madrid, 1966.

64. A. Sommerfeld, *Electrodynamics, Lectures on Theoretical Physics*, Vol. III, Academic, New York, 1964.

65. W. K. H. Panofsky and M. Phillips, *Classical Electricity and Magnetism*, 2nd ed., Addison-Wesley, Reading, MA, 1962.

66. E. M. Purcell, *Electricity and Magnetism*, Berkeley Physics Course, 2nd ed., Vol. 2, McGraw-Hill, 1985, Chap. 9.

67. H. A. Múnera and O. Guzmán, *Apeiron* **7**(1–2), 59–66 (2000).

68. T. R. Carver and J. Rajhel, *Am. J. Phys.* **42**(3), 246–249 (1974).

69. D. F. Bartlett and T. R. Corle, *Phys. Rev. Lett.* **55**(1), 59–62 (1985).

70. A. P. French and J. R. Tessman, *Am. J. Phys.* **31**, 201–204 (1963).

71. W. G. V. Rosser, *Am. J. Phys.* **44**(12), 1221–1223 (1976).

72. S. Roy and M. Roy, in S. Jeffers, S. Roy, J.-P. Vigier, and G. Hunter (Eds.), *The Present Status of the Quantum Theory of Light*, Kluwer, Dordrecht, 1997.

73. B. Lehnert, S. Roy, and A. Deb, *Apeiron* **7**(1–2), 53–58 (2000).

74. C. K. Whitney, in G. Hunter, S. Jeffers, and J.-P. Vigier (Eds.), *Causality and Locality in Modern Physics*, Kluwer, Dordrecht, 1998, pp. 219–226.

75. L. C. Ryff, in G. Hunter, S. Jeffers, and J.-P. Vigier (Eds.), *Causality and Locality in Modern Physics*, Kluwer, Dordrecht, 1998, pp. 203–207.

76. A. E. Chubykalo, H. A. Múnera, and R. Smirnov-Rueda, *Found. Phys. Lett.* **11**(6), 573–584 (1998).

77. H. Bateman, *The Mathematical Analysis of Electrical & Optical Wave-Motion*, Cambridge Univ. Press, 1914; reprinted by Dover, New York, 1955.

78. D. E. McLennan, (a) *Phys. Essays* **1**(3), 171–175 (1988); (b) ibid. **1**(3), 179–183 (1988); (c) ibid. **1**(3), 285–288 (1988); (d) ibid. **2**(1), 51–54 (1989).

79. P. D. Mannheim, *Phys. Rev. D* **32**(4), 898–913 (1985).

80. D. Kleppner and R. J. Kolenkow, *An Introduction to Mechanics*, McGraw-Hill, 1978.

81. H. A. Múnera and O. Guzmán, *Apeiron* **4**(2–3), 63–70 (1997).

82. M. W. Evans, J.-P. Vigier, S. Roy, and S. Jeffers, *The Enigmatic Photon*, Vol. 3, *Theory and Practice of the $B^{(3)}$ Field*, Kluwer, Dordrecht, 1996.

83. G. L. J. A. Rikken and B. A. van Tiggelen, *Nature* **381**, 54–55 (1986).

84. P. A. M. Dirac, *Proc. Roy. Soc. Lond. Ser. A* **133**, 60–72 (1931).

85. W. B. Zeleny, *Am. J. Phys.* **59**(5), 412–415 (1991).

86. I. Adawi, *Am. J. Phys.* **59**(5), 410–412 (1991).

87. H. A. Múnera and O. Guzmán, *Mod. Phys. Lett. A* **12**(28), 2089–2101 (1997).

88. O. D. Kellogg, *Foundations of Potential Theory*, Dover, New York, (1953).

89. P. A. M. Dirac, *Nature* (Lond.) **169**, 702 (1952).

90. N. Cufaro-Petroni and J. P. Vigier, *Found. Phys.* **13**(2), 253–285(1983).

91. L. V. Chebotarev, in S. Jeffers, B. Lehnert, N. Abramson, and L. Chebotarev (Eds.), *Jean-Pierre Vigier and the Stochastic Interpretation of Quantum Mechanics*, Apeiron, Montreal, 2000, pp. 1–17.

92. K. P. Sinha, C. Sivaram, and E. C. G. Sudarshan, *Found. Physics* **6**(1), 65–70 (1976).

93. B. G. Wallace, *Found. Phys.* **3**(3), 381–388 (1973).

94. V. Shekhawat, *Found. Physics* **6**, 221–235 (1976).

95. A. Widom and Y. N. Srivastava, *Mod. Phys. Lett. B* **4**, 1–8 (1990).

96. F. Winterberg, *Z. Naturforsch.* **52a**, 185 (1997).

97. M. Ribaric and L. Sustersik, *Transport Theoretic Extensions of Quantum Field Theories*, E-print archive: hep-th/9710220 (Oct. 97) *Framework for a Theory that Underlies the Standard Model*, LANL electronic file hep-th/9810138 (Oct. 1998).

98. J. J. Thomson, *Phil. Mag. Ser. 7* **12**, 1057–1063 (1931).

99. W. A. Hofer, *Physica A* **256**, 178–196 (1998).

100. H. Marmanis, *Phys. Fluids* **10**(6), 1428–37 (1998); "Erratum," ibid. **10**(11), 3031 (1998).

101. V. P. Dmitriyev, *Galilean Electrodyn.* **10**(5), 95–99 (1999).

102. H. A. Múnera, in A. E. Chubykalo, V. V. Dvoeglazov, D. J. Ernst, V. G. Kadyshevsky, and Y. S. Kim (Eds.), *Proceedings of the International Workshop Lorentz Group, CPT and Neutrinos*, World Scientific, Singapore, 2000, pp. 425–433.

103. H. A. Múnera, "The photon as a charge-neutral and mass-neutral composite particle. Part I. The qualitative model," paper presented at *Gravitation and Cosmology: From the Hubble Radius to the Planck Scale, Vigier 2000 Symposium*, University of California, Berkeley, USA (August 21–25, 2000) (to be published in the Proceedings edited by R. L. Amoroso, G. Hunter, M. Kafatos, and J.-P. Vigier).

104. H. A. Múnera, *Apeiron* **7**(1–2), 67–75 (2000).

105. V. N. Gribov, *Nucl. Phys. B* **139**, 1–19 (1978).

106. H. C. Ohanian, *Gravitation and Spacetime*, Norton, New York, 1976.

107. H. C. Ohanian, *Am. J. Phys.* **54**(6), 500–505 (1986).

108. H. C. Corben, *Phys. Rev.* **121**(6), 1833–1839 (1961).

109. H. C. Corben, *Nuovo Cimento* **20**(3), 529–541 (1961).

110. H. C. Corben, *Am. J. Phys.* **61**(6), 19–29 (1993).

111. H. C. Corben, *Int. J. Theor. Phys.* **34**(1), 19–29 (1995).

112. J. P. Costella, B. H. J. McKellar, and A. A. Rawlison, *Am. J. Phys.* **65**(9), 835–841(1997).

113. L. de Broglie, *C. R. Acad. Sci.* (Paris) **195**, 862 (1932); ibid. **199**, 813 (1934).

114. L. de Broglie, *Matter and Light, The New Physics*, Dover, New York, 1939, p. 140.

115. P. Jordan, *Z. Physik* **93**, 464 (1935).

116. P. Kronig, *Physica* **3**, 1120 (1936).

117. M. H. L. Pryce, *Proc. Roy. Soc. Lond.* **165**, 247 (1938).

118. I. M. Barbour, A. Bietti, and B. F. Toushek, *Nuovo Cimento* **28**, 453 (1963).

119. B. Ferreti, *Nuovo Cimento* **28**, 265 (1964).

120. W. A. Perkins, *Phys. Rev. B* **137**, 1291 (1965).

121. P. Bandyopadhyay and P. R. Chaudhuri, *Phys. Rev. D* **3**(6), 1378–1381 (1971).

122. K. Rafanelli and R. Schiller, *Phys. Rev.* **135**(1B), B279–B281 (1964).

123. J. Stachel and J. Plebanski, *J. Math. Phys.* **18**(12), 2368–2374 (1977).

124. F. H. Gaioli and E. T. García-Alvarez, *Found. Phys.* **28**(10), 1539–1550 (1998).

125. U. Fano, *Phys. Rev.* **93**(1), 121–123 (1954).

126. W. E. Baylis, J. Bonenfant, J. Derbyshire, and J. Hurschilt, *Am. J. Phys.* **61**(6), 534–545 (1993).

127. R. Weingard and G. Smith, *Synthese* **50**, 213–231 (1982).

128. F. J. Dyson, *Am. J. Phys.* **58**(3), 209–211 (1990).

129. P. A. M. Dirac, *The Principles of Quantum Mechanics*, Clarendon Press, Oxford, 1958.

130. S. Brandt and H. Schneider, *Am. J. Phys.* **44**(12), 1160–1171 (1976).

131. A. A. Cyrenika, *Apeiron* **7**(1–2), 89–106 (2000).

132. R. M. Kiehn, paper presented at Gravitation and Cosmology: From the Hubble Radius to the Planck Scale, Vigier 2000 Symp. Univ. of California, Berkeley, USA (Aug. 21–25, 2000) (to be published in Proceedings edited by R. L. Amoroso, G. Hunter, M. Kafatos, and J.-P. Vigier).

133. D. Bohm and J.-P. Vigier, *Phys. Rev.* **109**(6), 1882–1891 (1958).

SIGNIFICANCE OF THE SAGNAC EFFECT: BEYOND THE CONTEMPORARY PHYSICS

PAL R. MOLNAR AND MILAN MESZAROS

The Alpha Foundation, Institute of Physics, Budapest, Hungary

CONTENTS

I. THE PARADIGM

A. Analytical Viewpoint

During the historical development, the notions of electrodynamics and the theory of light have become complicated complexes of concepts [1]. And what is more, nowadays they are incomplete, or in the worst case wholly confusing. The laws

Modern Nonlinear Optics, Part 3, Second Edition, Advances in Chemical Physics, Volume 119,
Edited by Myron W. Evans. Series Editors I. Prigogine and Stuart A. Rice.
ISBN 0-471-38932-3 © 2001 John Wiley & Sons, Inc.

of electrodynamics in their present form are not valid in rotating and deforming systems in general [2]. These turbulent notion complexes—which are inadequate for the inner connections, as verified by experiments, measurement results, and certain electrodynamical states and processes—have to be broken open, disintegrated, and then disjoined. Henceforth, we must search for those genuine, pure, and simple electrodynamical ideas that can be joined in an imminent natural and adequate manner. Consequently, progress can be achieved only by careful analysis.

Some of the unsolved problems in contemporary electrodynamics draw attention to deeper (more profound) evidence, new ideas and new theories or equations. The aim of this historical introduction is to find the deeper evidence and new basic concepts and connections. The guiding principle is the investigation of light propagation.

B. Profound Evidence and Connections

The childhood of optics was in ancient religious Egypt. The first survived written relics of the optics originates from antique Greek science. Euclid was regarded as one of the founders of geometric optics because of his books on optics and catoptrics (catoptric light, reflected from a mirror).

The geometric description of the light propagation and the kinetics description of motion were closely correlated in the history of science. Among the main evidence of classical Newtonian mechanics is Euclidean geometry based on optical effects. In Newtonian physics, space has an affine structure but time is absolute. The basic idea is the inertial system, and the relations are the linear force laws. The affine structure allows linear transformations in space between the inertial coordinate systems, but not in time. This is the Galilean transformation:

$$x' = x + x_0 + vt, \qquad t' = t + t_0 \tag{1}$$

This is a law of choice for any motion equation.

The revolution in physics at the end of nineteenth century was determined by the new properties of light propagation and heat radiation. However, there remain many unsolved problems in these fields [2].

The laws of sound propagation in different media include the concept of ether, which is the hypothetical bearing substance of light and electromagnetic waves.

II. HISTORICAL OVERVIEW

A. The Main Experiments

The first measurement for the determination of velocity of sound was made by Mersenne in 1636. In 1687 Newton gave a rough formula for the velocity of sound. It was further developed by Laplace in 1816, based on the adiabatic changes of states for gases. In 1866 Kundt constructed the so-called Kundt tube, which can determine the velocity of sound in liquids and solid materials. He found that the velocity of sound grows because of the solidity of bearer materials. In the framework of classical mechanics, this observation inspired the notion that ether is an extremely solid substance.

The first attempt to determine the speed of light was made by Galileo in 1641. Descartes assumed an infinite speed of light based on the unsuccessful Galilean measurement.

In 1676, after 20 years of observation of the motion of Jupiter's Io moon, Römer published his result about the speed of light, which was calculated as $c = 220,000$ km/s [3].

In 1727 Bradley performed a much more precise experiment to determine the speed of light. His measurements were based on the aberration of stars, and the results of these measurements closely approximated today's values.

Arago was the first to measure the speed of light under laboratory conditions [4]. This measurement gave the Bradley's value for the speed of light. In 1850 Arago's followers Foucault [5], and Fizeau [6] proved that the speed of light is higher in air than in liquid. These measurements closed down the old debate in the spirit of the wave nature of light. In that time this seemed to verify the concept of ether as the bearing substance of light.

The first experimental investigation for the magnitude of change in light speed in moving media was made by Fizeau in 1851 [7]. His experiment proved that the velocity of the propagation is greater in the direction of motion of the medium than in the opposite direction; that is, the light is carried along with the moving medium. This theory was developed and confirmed by Michelson and Morley in 1886. In 1926 Michelson developed the Foucault's rotating-mirror experiment. The result of Michelson's experiment [8] is $c = 2.99769 \times 10^8 \pm 4 \times 10^5$ m/s [where c is (longitudinal) speed of light].

B. The Turning Point: Michelson–Morley Experiment

In 1867 Maxwell published his book on electromagnetism [9]. Maxwell's work has a basic importance, not only in the electromagnetism but also in optics. It also provided a common frame of reference for the propagation of electromagnetic and light waves.

The Maxwell equations are valid only in the unique inertial coordinate system, but they are not invariant for the Galilean transformation (1). This means that the Maxwell equations do not satisfy the requirements of classical equation of motion. This problem was apparently solved by the introduction of the concept of ether, the bearing substance of light. The challenge was to determine ether as the unique inertial system, or earth's motion in this ether.

Maxwell in another work [10] raised the question as to whether the translation motion of the earth relative to the ether can be observed experimentally.

An electromagnetic inertial system could be found by measurement, which could be used in astronomical calculations as well. Furthermore, space must be provided for formulating an equation of motion that is less rigorous than that used in Galilean relativity theory.

Numerous unsuccessful measurements were made to determine the motion of earth in the ether. These measurements were not able to give results compatible within the framework of classical Newtonian mechanics, even though that the earth has an orbital velocity $v_o \sim 30{,}000$ m/s (where v_o is velocity of the earth to the ether). In 1887 Michelson and Morley also determined the earth's orbital velocity by their precision interferometer [11]. The updated arrangement of Michelson–Morley experiment (M-M experiment) can be seen in Fig. 1.

According to classical mechanics, the traveling times of light T for the arms d_1 and d_2 can be given as Follows:

$$T_{OAO'} = \frac{2d_1}{c}\frac{1}{1-(v^2/c^2)}, \qquad T_{OBO'} = \frac{2d_2}{c}\frac{1}{\sqrt{1-(v^2/c^2)}} \qquad (2)$$

Fitting the length of interferometer's arms—according to the zero difference of traveling times (zero interference)—it is given that $\Delta T = T_{OBO'} - T_{OAO'} = 0$. Then the lengths of two arms can be determined exactly:

$$d_1 = d_2\sqrt{1-\left(\frac{v^2}{c^2}\right)} \qquad (3)$$

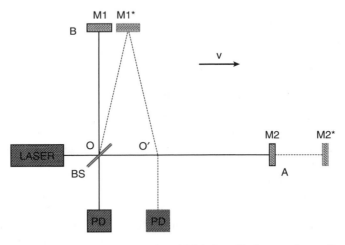

Figure 1. An up-to-date arrangement the of Michelson–Morley experiment. Here LASER means the source of light, BS means beamsplitter, M1 and M2 are mirrors on the end of arms, PD is the phase detector (interferometer), and v is the earth's orbital velocity, which is regarded as the inertial motion for short time periods.

According to classical physics, the difference of traveling times ΔT^* and the interference picture must be changed, turned around the instrument with 90°:

$$\Delta T^* = T^*_{OBD'} - T^*_{OAO'} = \frac{2}{c(1 - (v^2/c^2))} \left\{ d_2 - d_2 \sqrt{1 - \left(\frac{v^2}{c^2}\right)} \right\} \quad (4)$$

Substituting Eq. (3) into Eq. (4) and arranging, the traveling time difference for $v^2 \ll c^2$ is

$$\Delta T^* = \frac{2d_2c}{c^2 - v^2} \frac{v^2}{c^2} \quad (5)$$

Their experiments proved that the travelling-times differences did not change along the two arms $\Delta T^* = 0$ for any turning round of instrument. In other words, there was no change in phase relations or interference fringes. Thus, one might suppose that the solar system moved relative to the ether possessing a velocity that coincided with that of the orbital velocity of the earth, and, by coincidence, the experiment was carried out during a period when the earth was moving relative to the sun in the same direction as the ether. This experiments essentially contradict classical Newtonian mechanics. The Michelson–Morley measurements, which resulted in a negative outcome, have had one of the most remarkable influences on the development of twentieth-century physics. A modern setup can be seen in Fig. 2.

C. The Sagnac-Type Experiments

The earth's rotation around its axis can be seen from the apparent motion of the stars. The rotation can also be observed by mechanical experiments carried out on the surface of the earth, that is, with the help of Foucault's pendulum, or by observing of the motion of a rapidly rotating gyroscope. It is important that the rotation of the earth can also be observed by closed optical experiments.

This effect was first demonstrated in 1911 by Harress and in 1913 by Sagnac, so it is now often called the *Sagnac effect*. Sagnac determined a rotation by a closed optical instrument [12]. Sagnac also fixed an interferometer onto a rotating disc. A flowchart of the basic arrangement of the essential features in the Sagnac experiment is shown in Fig. 3.

It is clear that the rotation occurs relative to the carrier of electromagnetic waves; this is the observed rotation relative to the ether.

This measurement was improved by Michelson and Gale in 1925 using the earth instead of rotating disk [13].

In 1926 the Michelson–Gale experiment was confirmed by Pogány [14], who determined the surface velocity of the rotating earth by a closed optical

Figure 2.   An up-to-date setting of a M-M-type experiment.

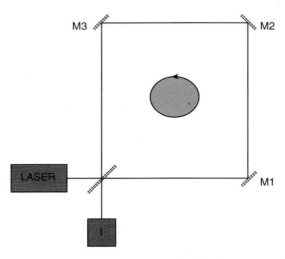

Figure 3.   Arrangement of Sagnac the experiment. Here, LASER represents the source of light, the first mirror is a beamsplitter, M1–M3 are mirrors on the end of arms, and I represents the interferometer.

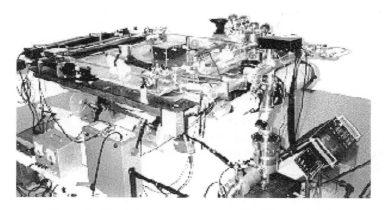

Figure 4. The CI laser-gyroscope arranged by Bilger et al. [15].

instrument, $V_R \sim 300$ m/s, in Budapest's latitude. Because of its precision, this experiment it is used in some military applications, such as in laser gyroscope techniques. It is also commonly used today in guidance and navigation systems for airlines, nautical ships, spacecraft, and in many other applications. A laser gyroscope is shown in Fig. 4.

Because of the incredible precision of interferometric techniques, this measured velocity is altogether one percent of the earth's circumference velocity derived from the orbital motion. Very-long-baseline interferometry (VLBI)—which is an exhaustively improved *Pogány* experiment—can detect $\Delta\omega \approx 10^{-9}$ in the earth's rotation.

Sagnac-type experiments are versatile and more accurate than the M-M-type experiments, which cannot detect rotation. Sagnac-type experiments demonstrated that the caused phase shift is proportional to the angular velocity ω and the measure of the enclosed surface S in a rotating system.

III. ANALYSIS OF MICHELSON–MORLEY EXPERIMENT

A. The Least-Arbitrariness Principle: The Necessary Hidden Variables

In order to explain the negative result of the M-M-type experiments, a whole series of hypotheses were proposed, all of which were eventually found to be untenable. This first explanation consists in the assumption that the ether at the earth's surface is carried along by the earth, adhering to the earth like the earth's atmosphere. This explanation became very improbable in the light of Fizeau's experiment on light propagation in media with motion. This experiment suggested that the ether is not carried along or at most is only partially carried along by moving medium [7,8].

Numerous researchers tried to determine the velocity of the earth motion to the ether by electromagnetic and optical methods. These experiments predicted that the earth with the experimental instruments always are standing in (or moving along with) the ether, which really is a tenacious contradiction of contemporary physics.

The physicists tried to solve this profound problem by the *principle of least arbitrariness or* a fortiori [2c]. This principle means the optimum relation among the introduced hidden variables, which are necessary to description of the phenomena. (This maxim is well known and accepted in the scientific community as (*Occam's razor.*)

B.   The Lorentz Interpretation of M-M Experiments

Lorentz [16] and his colleagues introduced a hidden variable: the *contraction form factor* $\beta = (1 - v^2/c^2)^{1/2}$ in Eq. (3). In the case of $d_1 = d_2$, Eq. (3) provides a simple solution of this contradiction. In Eq. (5) the difference in traveling times can be eliminated if, for example, d depends on the velocity only:

$$d^* = d\beta \qquad (6)$$

(where β is the contraction from factor).

Of course, in Eq. (6) the contraction form factor β is valid only in the arm that is parallel to the velocity vector. Equation (6) was interpreted by Lorentz and Fitz-Gerald as a real contraction [17]. It is important to see that in Eq. (6) the hidden parameter β is only one possible solution for the contradiction, but the result of the M-M experiment allows numerous other solutions based on the inner properties and features of the light. The M-M experiment destroyed the world picture of classical physics, and it required a new physical system of paradigms. Thus, for example, the applicability of Galilean relativity principle was rendered invalid.

One of the most important requirements for an axiomatic theory is to determine the validity-round of the laws, and to verify of the self-consistency in the theory. The M-M experiment proved that the prediction of the classical physics was not valid for light propagation, or rather, for Maxwell's theory of electromagnetism. This is an applicability limit of Newtonian physics. Beyond this limit, Newtonian physics becomes incomplete.

Lorentz, Fitz-Gerald, and others were able to formally explain the lack of changing in interference fringes [1] using a hidden variable that is essentially the quotient of the theoretical and the measured results. This method, combined with the least-arbitrariness principle, obtained the optimal hidden parameter, which was satisfied by the experiment. The operator of the optimal hidden parameters used in the description of the M-M experiment is the generalized form factor, the so-called Lorentz transformation. Lorentz believed a fortiori that this operator functions in connection with the ether's wind, and that this wind is the actual cause of the assumed bodies' contractions. The merit of the Lorentz transformation is the verification for the invariance in the Maxwell equation. However, one disadvantage of the Lorentz interpretation is that the contraction is independent of the material properties of bodies.

C. The Einstein Interpretation of M-M Experiments

Einstein created a tabula rasa in his 1905 paper titled "On the electrodynamics of moving bodies" [18]. He rejected the paradigm of ether as well as the classical concepts of space and time, and founded a new physics by the exclusion of inner forces called the *special relativity theory*. He stated two axioms: (1) the principle of relativity and (2) the homogeneous and isotropic propagation of light in any inertial coordinate system of the vacuum. The homogeneous isotropic light propagation can be satisfied by the Lorentz-contracted spacetime. Of course, without the concept of physical ether, the ether wind theory is meaningless. Einstein refused the material explanation of Lorentz and Fitz-Gerald, but kept the contraction form factor β without another material interpretation. It is clear that the nonmaterial interpretation given by Einstein is high-handed, but it is still questionable that it is the least arbitrary.

It is well known that Einstein's interpretation for the Michelson–Morley-type experiments was self-consistent in mathematical sense, although he lost the genuine concepts and the traditional a priori and anthropic relations of space and time forever. With this step the science left its childhood or rather, lost its innocence. In this way Einstein created the opportunity for any extravagant interpretations of strange experiments, and so any other physical concepts, for example, the propagation theory, became illusory.

D. Interferometers: Standing-Wave Systems

As it was confirmed that the notions of electrodynamics and the theory of light propagation have become complicated complexes of concepts and they are wholly confusing. These inadequate notion complexes have to be broken open, disintegrated, then disjoined.

Let us study the M-M- and the Sagnac-type experiments without any preconceptions. We can then see that the interferometers are unable to measure the traveling times; they can measure only the interference fringes of standing waves. This means that description of the M-M experiment allows the use of the wavelengths and phases, but not the traveling times and the speed of propagation. In a strict sense, the Michelson–type the interferometers are unable to measure the velocity of propagation and traveling times in the arms. Specifically to measure traveling times, it an exact optical distance measurement theory and method would be necessary. (In connection with the restrictions of the least-arbitrariness principle in the geometric optics, the principle of least action can give the path of light as the distance.)

The fine distinction between traveling times and the shift in interference fringes may not be clear from the point of view of Newtonian mechanics, which predicts both to be changing. Finally, classical physics and the geometric optics are refuted or restricted by experience, notwithstanding the fact that these are self-consistent theories in their own right.

IV.   ANALYSIS OF SAGNAC-TYPE EXPERIMENTS

A.   The Classical Arrangements

Consider a disk of radius R rotating with an angular velocity ω around its axis [1,12–14]. Suppose a large number of mirrors n are arranged on its periphery in such a way that a light signal starting, say, from a point A of the periphery is guided along a path very nearly coinciding with the edge of the disk. If the disk is at rest, a signal starting at time $t=0$ from a point A on the periphery arrives back into A at a time

$$T = \frac{2\pi R}{c} \tag{7}$$

However, if the disk is rotating with a circumference velocity $v_R = \omega R$ and the light signal is moving in the direction of rotation, then, at time $T = 2\pi R/c$, it will reach a point A_0 located at the location that A had left at $t = 0$. The signal has to catch up to point A, which is moving away; the signal will reach this location at a later time T_+, so that $cT_+ = 2\pi R + v_R T_+$; therefore

$$T_+ = \frac{2\pi R}{c - v_R} > T \tag{8}$$

(where v_R is circumference velocity).

Now suppose that the light moves relative to the edge of the disk c_+ -according to classical physics and according to Eq. (8), in the direction of velocity

$$c_+ := c - v_R \tag{9}$$

(where c_+ is speed of light in the direction of velocity).

Suppose that the velocity of the beam is relative to the disk but that we have calculated the traveling time only and that the signal starting from A must again catch up with point A, which is moving away.

If the light signal moves in the opposite direction, it reaches A sooner that at $t = T$ as point A moves then toward the signal. In this case we find for the time at which the signal reaches A

$$T^- = \frac{2\pi R}{c + v_R} < T \qquad (10)$$

or we may assume that the speed of light traveling in the opposite direction is velocity c_-:

$$c_- := c + v_R \qquad (11)$$

(c_- is speed of light opposite the velocity).

In the boundary transition ($n \rightarrow \infty$), the polygon—constructed by the mirrors—becomes a circle with radius R, and the difference of the times needed to circle around the disk in the opposite direction is thus

$$\Delta T = T_+ - T_- = 2\pi R \left(\frac{1}{c - v_R} - \frac{1}{c + v_R} \right) = \frac{4\pi R v_R}{c^2 - v_R^2} \cong \frac{4S\omega}{c^2} \qquad (12)$$

where $S = \pi R^2$ is the area of the disc circled round by the beams and ω is angular velocity.

Of course, according to the Section III.D, this calculation should really be carried out at wavelengths $\lambda - s$ instead of traveling times $T - s$. The Sagnac-type experiments are also standing-wave systems. Then the magnitude of shift of the interference fringes with the above ω

$$\Delta\lambda = \lambda_+ - \lambda_- = 2\pi R \left(\frac{c}{c_+} - \frac{c}{c_-} \right) \cong \frac{4S\omega}{c} \qquad (13)$$

which has been confirmed by experiments [12–14] without any doubt.

Naturally this coincidence does not mean that the geometric optics added to the classical physics could be used for the exact description of the light propagation since the Michelson–Morley experiment refuted its validity forever. It is evident that there are possible new mathematical definitions for c_+ and c_- instead of the ordinary speed addition rule of the classical physics seen in Eqs. (9) and (11). These can be compatible with the experimental results as well.

B. The Relativistic Calculation

The major absurdity of the result of the Sagnac-type experiments is that the calculation was carried out by the geometric optics exclusively. Of course, the calculation should carried out using the special relativity theory exhaustively.

The validity of a physical theory depends on, among other things, the certainty and completeness by which the theory is ordered to the totality of experiences [2c]. Consequently, the special relativity theory must also be confronted with observation and experiment carried out on the physical system examined. In any given case one has to clarify the mutuality of the special relativity and the Sagnac effect. In this case, the second postulate of the special relativity theory must be satisfied; that is, the speed of light must be the same in every direction

$$c_+ = c_- = c \qquad (14)$$

by definition. Substituting Eq. (14) into Eq. (13) a zero shift of interference fringes, we obtain. $\Delta\lambda = \lambda_+ - \lambda_- = 0$, which is contrary to the experiments.

This means that the special relativity theory does not predict any shift of interference fringes that is contrary to the experiments. The standing-wave approach of Sagnac-type experiments allows a freedom in the definition of c_+ and c_- instead of Eqs. (9) and (11), but the second postulate of the special relativity theory is out of this range.

Of course, the Sagnac-type experiments were not made in a perfect inertial systems. The earth's orbital motion around the sun is also a noninertial system. But the circumference velocities in both cases are extremely low, $v/c \ll 1$, and—in the first approximation—these frames of reference are almost inertial systems.

The Sagnac-type experiments proved that the circumference velocity can be detected by purely and closed optical instruments as well. The circumference velocity of the rotating earth, $v_R \sim 300$ m/s, is extremely low to the earth's orbital velocity, which is also a circumference velocity, with $v_o \propto 100 \times v_R$. In both cases, Michelson–Morley and Sagnac wanted to determine the circumference velocities. The M-M experiments were unable to determine the earth's orbital circumference velocity, but the Sagnac experiment determined the rotating earth's circumference velocity. On the basis of the Michelson–Morley-type experiments, Einstein postulated the constancy of the speed of light, so the results of the Sagnac-type experiments—with different speeds of light—contradict the special relativity theory.

In a strict sense, the classical Newtonian mechanics and the Maxwell's theory of electromagnetism are not compatible. The M-M-type experiments refuted the geometric optics completed by classical mechanics. In classical mechanics the inertial system was a basic concept, and the equation of motion must be invariant to the Galilean transformation Eq. (1). After the M-M experiments, Eq. (1) and so any equations of motion became invalid. Einstein realized that only the Maxwell equations are invariant for the Lorentz transformation. Therefore he believed that they are the authentic equations of motion, and so he created new concepts for the space, time, inertia, and so on. Within this framework the Lorentz transformation is the law of choice for the equation of motion. Sagnac's result proved that Einstein's method contradicts experience. Besides, on a deeper level it is proved that Maxwell's equations are not applicable for the slowly rotating systems. So, in an authentic theory of light, Maxwell's equations must be changed to allow for a description of rotating and deforming systems [19,20].

C. The Incompleteness of the Theory of Light

The classical theory of light—consisting in the complexes of concepts such as light propagation and interference—employs geometric optics added to classical physics and the Maxwell theory of electromagnetism. These turbulent notion complexes suffered from logical inconsistencies. [For example, the Maxwell equations are not invariant to the Galilean transformations (1) since those are not equation of motion in the mechanical sense.] This conceptual conglomeration was broken open by the Michelson–Morley-type experiments. In the present case, the incompleteness of classical light theory means that it cannot describe and explain the M-M-type experiments within the frame of the theory. For a complete, accurate description and explanation, a new theory was needed. Einstein believed the new theory to be nonclassical, and so he created the special relativity theory. The relativistic theory of light is similar in composition to the classical one, except that classical mechanics is changed to the relativistic mechanics. The relativistic theory of light—beside the explanation of the M-M-type experiments—was free from the logical problem of the classical light theory described above.

Eight years later Sagnac made a crucial experiment. The Sagnac-type experiments are broken open the complexes of concepts of relativistic light theory. Thus it became an incomplete theory since its prediction of the shift of interference is $\Delta\lambda = \lambda_+ - \lambda_- = 0$, contrary to the Sagnac-type experiments.

We need to find a complete theory of light based on more profound evidence, new basic concepts, and authentic connections.

V. SUMMARY

The complete theory of light should describe and explain the totality of experiences, that is, the M-M- and Sagnac-type experiments simultaneously.

In the spirit of the standing-wave picture of Sagnac-type experiments, this theory needs to recalculate the result of the Michelson-Morley experiment as well. In the M-M experiment there is a new unknown hidden parameter c_p, which denotes the speed of light in the direction perpendicular to the earth's velocity. The traveled path of light in the perpendicular arm $\lambda_p := 2Tc_p$ [dim$\{\lambda\}$ = meter]. [where c_p is speed of light perpendicular to the velocity (transversal light speed)]. The difference of the paths traveled in the interferometer is

$$\delta\lambda^* = \lambda_+^* + \lambda_-^* - 2\lambda_p^* = d_2\left(\frac{c}{c_+} + \frac{c}{c_-}\right) - 2d_1\frac{c}{c_p} \tag{15}$$

It can be seen that the second postulate of special relativity theory [Eq. (14)] leads to the form

$$c_+ = c_- = c_p = c \tag{16}$$

Substituting Eq. (16) into Eq. (15), we obtain a zero interference change, corresponding to with the M-M experiment. The M-M experiments are only a limited part of the totality of experiences.

The Michelson–Morley- and Sagnac-type experiments give only two independent equations—Eqs. (13) and (15)—for three unknown hidden parameters c_+, c_-, and c_p. In the present case the incompleteness means that there are three unknown parameters for two equations. A third equation is needed in the form of a crucial experiment for the unique solutions. (Of course, this crucial experiment must be independent of the M-M- and Sagnac-type experiments.) In this manner we will be able to develop an authentic nonquantized (complete) theory of light.

After the frequent metaphysical optimism of a century ago, we again return to the fundamental questions.

References

1. L. Jánossy, *Theory of Relativity Based on Physical Reality*, Academic Press, Budapest, 1971.

2. (a) M. Meszaros and P. Molnar, *Ann. Physik* **46**, 153 (1989); (b) P. Molnar and M. Meszaros, *Ann. Physik* **46**, 381 (1989); (c) M. Meszaros and P. Molnar, *Phys. Essays* **3**, 284 (1990) and ibid. **5**, 463 (1992); (d) P. R. Molnar, T. Borbely, and B. Fajszi, in M.W. Evans et al. (Eds.), *The Enigmatic Photon, Vol. 4: Developments*, Kluwer, Dordrecht, 1997, p. 205; (e) M. Meszaros, idem., p. 147.

3. R. Römer, *Mem. Acad. Sci.* (Paris) 1675; see also C. Ramsauer, *Grudversuche der Physik in historischer Darstellung* I, Springer-Verlag, 1953.

4. F. Arago, *Compt. Rend. Hebd.* **7**, 954 (1838); ibid. **30**, 489 (1850); ibid. **55**, 792 (1862).

5. L. Foucault, *Compt. Rend. Hebd.* **30**, 551 (1850); ibid. **55**, 501 (1862).

6. H. Fizeau, *Compt. Rend. Hebd.* **29**, 90 (1849); *Ann Phys.* **79**, 167 (1850).

7. H. Fizeau, *Compt. Rend. Hebd.* **33**, 349 (1851).

8. A. A. Michelson, *Sill. J.* **15**, 394 (1878); ibid. **18**, 390 (1879); ibid., *Nature* **21**, 94, 120 (1880); ibid., *Naut. Alm.* 235 (1885); A. A. Michelson and E.W. Morley, *Am. J. Sci.* **31**, 377 (1886); A. A. Michelson, *Astrophys. J.* **37**, 190 (1913); ibid. **60**, 256 (1924); ibid. **65**, 1 (1927).

9. J. C. Maxwell, *Treatise on Electricity and Magnetism* I–II, Dover, New York, 1954.

10. J. C. Maxwell, *Nature* **21**, 314 (1879–1880); idem, "Ether" in *The Enciclopedia Britannica*, 9th ed. (1875–1889); reprinted in W. D. Niven (Ed.), *The Scientific Papers of James Clark Maxwell*, (Dover, New York, 1965).

11. A. A. Michelson and E.W. Morley, *Am. J. Sci.* **34**, 333 (1887); reprinted in L. Pearce Williams (Ed.), *Relativity Theory: Its Origin and Impact on Modern Thought*, Wiley, New York, 1968.

12. G. Sagnac, *Compt. Rend.* **157**, 708 (1913); *J. Phys.* **5**, 177 (1914).

13. A. A. Michelson and H. G. Gale, *Astrophys. J.* **61**, 140 (1925).

14. B. Pogány, *Ann. Phys.* **80**, 217 (1926); **85**, 244 (1928); *Naturwissenschaften* **15**, 177 (1927).

15. H. R. Bilger, G. Stedman, U. Schreiber, and J. Schulte, in *proc. 13th Working Meeting on European VLBI for Geodesy and Astrometry*, Viechtach, 1999; an interview by F. Su, "World's largest ring laser gyro will measure earth's rotation," http://www.spie.org/web/oer/september/sep96/gyro.

16. H. A. Lorentz, *Zittungsverlagen Akad. van Wettenschappen* **1**, 74 (Nov. 26, 1892); *Versuch einer Theorie der elektrischen und optischen Erscheinungen in bewegten Körpern*, Brill, Leiden, 1895; *Proc. Acad. Sci.* (Amsterdam) (Engl. version) **6**, 809 (1904) (The third reference and a translated excerpt form the second are available *in The Principle of Relativity*, Ref. 1.; idem., *The Theory of Electrons*, Leipzig, 1909).

17. G. F. Fitz-Gerald, quoted by O. Lodge, *Nature* **46**, 165 (1892); see also O. Lodge, *Phil. Trans. Roy. Soc.* **184A** (1893).

18. A. Einstein, *Annal. Phys.* **17**, 891 (1905).

19. M. W. Evans et al. (Eds.), *The Enigmatic Photon*, Vol. 4: *Developments*, (Kluwer, Dordrecht, 1997) (a whole series of AIAS papers on nonconventional electromagnetism can be found at http://www.ott.doe.gov/electromagnetic/).

20. M. Meszaros, *Int. J. Theor. Phys.* **33**, 1035 (1994).

NON-ABELIAN ELECTRODYNAMICS: PROGRESS AND PROBLEMS

LAWRENCE B. CROWELL

The Alpha Foundation, Institute of Physics, Budapest, Hungary

CONTENTS

I. INTRODUCTION

Non-Abelian electrodynamics is an interesting proposal that electrodynamics has a more general gauge structure. The basis advanced by Barrett, Harmuth, and Evans proposes that electrodynamics has a more complex structure than one

Modern Nonlinear Optics, Part 3, Second Edition, Advances in Chemical Physics, Volume 119,
Edited by Myron W. Evans. Series Editors I. Prigogine and Stuart A. Rice.
ISBN 0-471-38932-3 © 2001 John Wiley & Sons, Inc.

described by the $U(1)$ gauge group. Initially it was advanced as an extension of electrodynamics for the derivation of solutions to Maxwell's equations. Later Evans suggested that this extension may have physical implications. The principal implication is the existence of the $\mathbf{B}^{(3)}$ magnetic field. This field emerges from the commutator of gauge potentials in the Yang–Mills equations. This field is written most often as

$$\mathbf{B}^{(3)} = \frac{ie}{\hbar}\mathbf{A}^{(1)} \times \mathbf{A}^{(2)} \tag{1}$$

The particular gauge potentials are orthogonal to each other by complex conjugation $\mathbf{A}^{(2)} = \mathbf{A}^{(1)*}$, with

$$\mathbf{A}^{(1)} = \mathbf{i}A_x + \mathbf{i}\mathbf{j}A_y \tag{2}$$

The definition of the $\mathbf{B}^{(3)}$ field in this manner illustrates that the internal index associated with the extended gauge group is identified with coordinates that are orthogonal to the direction of propagation of the electromagnetic field. This has various implications, which, if interpreted classically, mean that the Stokes parameters of an electromagnetic field determine this field.

Few of the claims for classical effects in non-Abelian electrodynamics have been conclusively demonstrated, but it may account for the Sagnac effect in interferometry. This is a pure phase effect associated with rotating inter-ferometers, which can be predicted according to non-Abelian electrodynamics. However, as a purely phase effect this does not mean that the magnetic field, called the $\mathbf{B}^{(3)}$ field, has been directly measured as yet. If one regards the electrodynamic field as quantum-mechanically composed of harmonic oscillator states, this $\mathbf{B}^{(3)}$ field has not been demonstrated to have eigenstates according to $\langle \mathbf{B}^{(3)} \cdot \mathbf{B}^{(3)} \rangle$. This does not mean that non-Abelian electrodynamics is false, but rather suggests that the eigenstates of this 3-magnetic field do not present themselves readily. Even if this magnetic field were absent, there would still exist subtle phase effects, which would manifest themselves in the interaction of electromagnetic fields and media. Classically, it appears as if this field should appear with a field strength that depends on $1/\omega^2$, which means that for long wavelength electromagnetic fields it should be quite large. If this field did occur readily, it would be abundantly present in a microwave beam with a coherent polarization, which would have certainly been discovered sometime during or shortly after World War II with the rapid development of radar. One of the purposes of this chapter is to address this matter.

There have been reports of the inverse Faraday effect that are predicted by non-Abelian electrodynamics. However, these reports are comparatively dated, and no updated results appear to have been reported. In 1998 the Varian

Corporation attempted to measure the $\mathbf{B}^{(3)}$ field. However, the results were null, and an inconclusive direct measurements of the $\mathbf{B}^{(3)}$ field still remains elusive. On the theoretical front non-Abelian electrodynamics remains controversial and not widely upheld. Some objections are not entirely reasonable. On the other hand, Waldyr Rodriques objected to certain assumptions, proposed by M. W. Evans, that relates coefficients in Whittaker's 1904 paper on electrodynamics to the putative existence of longitudinal modes in non-Abelian electromagnetic waves in vacuum. Rodrigues' objections appear reasonable. However, this response was quite forceful and direct, and resulted in his refusal to consider anything involving non-Abelian electrodynamics.

In order to address this question, it is requisite that the quantum field theoretic aspects of non-Abelian electrodynamics must be considered. In fact, theoretical reasons for the apparent paucity of the $\mathbf{B}^{(3)}$ field were discovered through examination of non-Abelian electrodynamics at high energy and its unification with the weak interactions. Non-Abelian electrodynamics at high energy should emerge from an $SU(2)$ gauge field theory, and this appears to have an elegant duality with the weak interactions with which it is unified within the TeV (teraelectronvolt) range in high-energy physics. This lead to the prospect that if this putative theory were true, then there should exist an additional Z-like boson, referred to as Z_γ which should appear with a mass in the TeV range. The additional degrees of freedom in the field define the $U(1)$ gauge theory plus various non-Lagrangian symmetries. These curious non-Lagrangian symmetries emerge from a Lagrangian that vanishes, and thus have no action or dynamics. It is here that the $\mathbf{B}^{(3)}$ field exists, or rather nonexists. This also implies that the field is a vacuum effect that induces squeezed states and other nonlinear effects. The definition of the $\mathbf{B}^{(3)}$ according to Eqs. (1) and (2) suggests that this is the case, as it is determined by the orthogonal polarization directions of an electromagnetic field. The apparent immeasurablity of the $\mathbf{B}^{(3)}$ field suggests that this field is a pseudofield that has subtle effects.

These results are somewhat at odds with the classical ideas of Evans; however, the quantum field theoretic calculations performed in this section lead to the conclusion that $\langle \mathbf{B}^{(3)} \cdot \mathbf{B}^{(3)} \rangle = 0$. This means that this 3-field simply is not a classical effect, and that the classical calculations may be valid only as formalistic tools. If $\langle \mathbf{B}^{(3)} \cdot \mathbf{B}^{(3)} \rangle$ is demonstrated to exist, then this implies that additional physics is involved that either generalizes the quantum-mechanical results, or new physics that is involved with the quantum–classical transition.

Equally of interest in the prospects this may have for high-energy physics. An extended $SU(2) \times SU(2)$ standard model has features of gauge field duality proposed by Montenen and Olive. This theory further embeds into an $SO(10)$ grand unification scheme that includes the $SU(3)$ gauge field for the strong interactions. Also, since this field predicts the existence of additional Z bosons, this also has an influence on the gauge hierarchy problem. Within the

$SU(2) \times SU(2)$ extended standard model the two gauge groups are chiral with opposite handedness. At low energy the weak angle mixes the two chrial fields so that one field theory is chiral while the other is vector. The weak interactions are the chiral fields, while electromagnetism on the physical vacuum is vectorial. The result of this transition to low energy is the production of a massive $\mathbf{A}^{(3)}$ field. This massive field is identified as the Z_γ boson. This massive neutral boson has been recently suggested through deviations in neutrino production at LEP1. Yet, this type of theory can be tested or falsified only in the multi-TeV range in energy. Only until the Z_γ is directly produced and its existence is deduced through its decay products can it be determined whether this dual theory of electroweak interactions is acceptable.

This section will be broken into a number of discussions. The first will be on a naive $SU(2) \times SU(2)$ extended standard model, followed by a more general chiral theory and a discussion on the lack of Lagrangian dynamics associated with the $\mathbf{B}^{(3)}$ field. This will be followed by an examination of non-Abelian QED at nonrelativistic energies and then at relativistic energies. It will conclude with a discussion of a putative $SO(10)$ gauge unification that includes the strong interactions.

II.   THE $SU(2) \times SU(2)$ EXTENDED STANDARD MODEL

If we are to consider the prospects for non-Abelian electrodynamics, it is best for aesthetic reasons to consider what this implies for an $SU(2) \times SU(2)$ extended standard model of electroweak interactions. This model has the pleasing quality of gauge duality, and it can be examined to determine whether there are inconsistencies with high-energy physics data at the TeV range in energy. It also may indicate something of the appearance of electrodynamics at low energy. If this is consistent with the abundant data, then the model is at least tentative. It will be demonstrated that this leads to certain conclusions about the ontological status of the $\mathbf{B}^{(3)}$ field.

Consider an extended standard model to determine what form the electromagnetic and weak interactions assume on the physical vacuum defined by the Higgs mechanism. Such a theory would then be $SU(2) \times SU(2)$. We will at first consider such a theory with one Higgs field. The covariant derivative will then be

$$\mathscr{D}_\mu = \partial_\mu + ig'\sigma \cdot \mathbf{A}_\mu + ig\tau \cdot \mathbf{b}_\mu \tag{3}$$

where σ and τ are the generators for the two $SU(2)$ gauge fields represented as Pauli matrices and $\mathbf{A}, \mathbf{b}$ are the gauge connections defined on the two $SU(2)$ principal bundles. There is an additional Lagrangian for the ϕ^4 scalar field [8]:

$$\mathscr{L}_\phi = \frac{1}{2}|\mathscr{D}_\mu(\phi)|^2 - \frac{1}{2}\mu^2|\phi|^2 + \frac{1}{4}|\lambda|(|\phi|^2)^2 \tag{4}$$

The expectation value for the scalar field is then

$$\langle \phi_0 \rangle = \left(0, \frac{v}{\sqrt{2}} \right) \tag{5}$$

for $v = \sqrt{-\mu^2/\lambda}$. At this point the generators for the theory on the broken vacuum are

$$\langle \phi_0 \rangle \sigma_1 = \left(\frac{v}{\sqrt{2}}, 0 \right)$$

$$\langle \phi_0 \rangle \sigma_2 = \left(i \frac{v}{\sqrt{2}}, 0 \right) \tag{6}$$

$$\langle \phi_0 \rangle \sigma_3 = \left(0, -\frac{v}{\sqrt{2}} \right)$$

These hold similarly for the generators of the other $SU(2)$ sector of the theory. There is a formula for the hypercharge, due to Nishijima, that when applied directly, would lead to an electric charge:

$$Q\langle \phi_0 \rangle = \frac{1}{2} \langle \phi_0 \rangle (\sigma_3 + \tau_1) = \left(0, -\frac{v}{\sqrt{2}}, 0, \frac{v}{\sqrt{2}} \right) \tag{7}$$

This would mean that there are two photons that carry a $\pm$ charge, respectively. We are obviously treating the hypercharge incorrectly. It is then proposed that the equation for hypercharge be modified as

$$Q\langle \phi_0 \rangle = \frac{1}{2} \langle \phi_0 \rangle (\mathbf{n}_2 \cdot \tau_3 + \mathbf{n}_1 \cdot \sigma_1) = 0 \tag{8}$$

where the vectors $\mathbf{n}_1$ and $\mathbf{n}_2$ are unit vectors on the doublet defined by the two eigenstates of the vacuum. This projection onto σ_1 and τ_3 is an ad hoc change to the theory that is required since we are using a single Higgs field on both bundles on both $SU(2)$ connections. This condition, an artifact of using one Higgs field, will be relaxed later. Now the generators of the theory have a broken symmetry on the physical vacuum. Therefore the photon is defined according to the σ_1 generator in one $SU(2)$ sector of the theory, while the charged neutral current of the weak interaction is defined on the τ_3 generator.

We now consider the role of the ϕ^4 scalar field with the basic Lagrangian containing the follwoing electroweak Lagrangians:

$$\mathcal{L} = -\frac{1}{4} F^a_{\mu\nu} F^{a\mu\nu} - \frac{1}{4} G^a_{\mu\nu} G^{a\mu\nu} + |\mathcal{D}_\mu \phi|^2 - \frac{1}{2} \mu^2 |\phi|^2 + \frac{1}{4} \lambda (|\phi|^2)^2 \tag{9}$$

Here $G^a_{\mu\nu}$ and $F^a_{\mu\nu}$ are elements of the field strength tensors for the two $SU(2)$ principal bundles. So far the theory is entirely parallel to the basic standard model of electroweak intereactions. In further work the Dirac and Yukawa Lagrangians that couple the Higgs field to the leptons and quarks will be included. It will then be pointed out how this will modify the $\mathbf{B}^{(3)}$ field. The ϕ^4 field may be written according to a small displacement in the vacuum energy:

$$\phi' = \phi + \langle\phi_0\rangle \simeq \frac{(v + \xi + i\chi)}{\sqrt{2}} \tag{10}$$

The fields ξ and χ are orthogonal components in the complex phase plane for the oscillations due to the small displacement of the scalar field. The small displacement of the scalar field is then completely characterized. The scalar field Lagrangian then becomes

$$\mathcal{L}_\phi = \frac{1}{2}\left(\partial_\mu\xi\partial^\mu\xi - 2\mu^2\xi^2\right) + \frac{1}{2}v^2\left(g'\mathbf{A}_\mu + g\mathbf{b}_\mu\right.$$
$$\left. + \left(\frac{1}{gv} + \frac{1}{g'v}\right)\partial_\mu\chi\right)\cdot\left(g'\mathbf{A}^\mu + g\mathbf{b}^\mu + \left(\frac{1}{gv} + \frac{1}{g'v}\right)\partial^\mu\chi\right) \tag{11}$$

The Lie algebraic indices are implied. The Higgs field is described by the harmonic oscillator equation where the field has the mass $M_H \simeq 1.0$ TeV/c^2. On the physical vacuum the gauge fields are

$$g'\mathbf{A}_\mu + g\mathbf{b}_\mu \rightarrow g'\mathbf{A}_\mu + g\mathbf{b}_\mu + \frac{1}{gv}\partial_\mu\chi = g'\mathbf{A}'_\mu + g\mathbf{b}'_\mu \tag{12}$$

which corresponds to a phase rotation induced by the transition of the vacuum to the physical vacuum. Let us now break the Lagrangian, now expanded about the minimum of the scalar potential, out into components:

$$\mathcal{L}_\phi = \frac{1}{2}\left(\partial_\mu\xi\partial^\mu\xi - 2\mu^2\xi^2\right) + \frac{1}{8}v^2 \times \left(g'^2|\mathbf{B}^3|^3 + g'^2(|\mathbf{W}^+|^2 + |\mathbf{W}^-|^2)\right.$$
$$\left. + g^2|\mathbf{A}^1|^2 + g^2|\mathbf{A}^3 + i|\mathbf{A}^2|^2\right) \tag{13}$$

where we have identified the charged weak gauge fields as

$$W^\pm_\mu = \frac{1}{\sqrt{2}}\left(\mathbf{b}^1_\mu \pm i\mathbf{b}^2_\mu\right) \tag{14}$$

The masses of these two fields are then $gv/2$. From what is left, we are forced to define the fields

$$A_\mu = \frac{1}{\sqrt{g^2 + g'^2}} (gA_\mu^3 + g'\mathbf{b}3_\mu - gA_\mu^1) \tag{15a}$$

$$Z_\mu^0 = \frac{1}{\sqrt{g^2 + g'^2}} (g'A_\mu^3 + g\mathbf{b}3_\mu + g'A_\mu^1) \tag{15b}$$

In order to make this consistent with the $SU(2) \times U(1)$ electroweak interaction [4], theory we initially require that $A^3{}_\mu = 0$ everywhere on scales larger than at unification. If this were nonzero, then Z_0 would have a larger mass or there would be an additional massive boson along with the Z_0 neutral boson. The first case is not been observed, and the second case is to be determined. This assumption, while ad hoc at this point, is made to restrict this gauge freedom and will be relaxed later in a more complete discussion of the 3-photon. This condition is relaxed in the following discussion or chiral and vector fields. This leads to the standard result that the mass of the photon is zero and that the mass of the Z_0 particle is

$$M_{Z^0} = \sqrt{g^2 + g'^2}\,\frac{v}{2} = \sqrt{1 + \left(\frac{g'}{g}\right)^2}\, M_W \tag{16}$$

The weak angles are defined trigonometrically by the terms $g/(g^2 + g'^2)$ and $g'/(g^2 + g'^2)$. This means that the field strength tensor $F_{\mu\nu}^3$ satisfies

$$F_{\mu\nu}^3 = \partial_\nu A_\mu^3 - \partial_\mu A_\nu^3 - i\frac{e}{\hbar}[A_\nu^1, A_\mu^2]$$
$$= -i\frac{e}{\hbar}[A_\nu^1, A_\mu^2] \tag{17}$$

and further implies that the third component of the magnetic field in the $SU(2)$ sector is

$$B_j^3 = \epsilon_j{}^{\mu\nu} F_{\mu\nu}^3$$
$$= -i\frac{e}{\hbar}(\mathbf{A}^1 \times \mathbf{A}^2)_j \tag{18}$$

This is the form of the $\mathbf{B}^{(3)}$ magnetic field. This also implies that the $\mathbf{E}^{(3)}$ electric field is then

$$\mathbf{E}^3{}_j = \epsilon_j{}^{0\mu} F_{0\mu}^3 = -i\frac{e}{\hbar}(\mathbf{A}^1 \times \mathbf{A}^2)_j \tag{19}$$

This demonstrates that $\mathbf{E}^{(3)} = \mathbf{B}^{(3)}$ in naturalized units. This is leads to the suspicion that the $B(3)$ field is zero.

The duality between these electric and magnetic field means that the Lagrangian vanishes. The vanishing of this Lagrangian on symmetry principles means that no dynamics can be determined. This would indicate that this particular model simply reproduces $U(1)$ electrodynamics, plus additional non-Lagrangian symmetries. Within this picture it appears as if the $\mathbf{B}^{(3)} = 0$, and that it simply represents the occurrence of various non-Lagrangian symmetries, but where there are no dynamics for the $\mathbf{B}^{(3)}$ field.

This result is a curious and troubling one for the prospect that there can be a classical $\mathbf{B}^{(3)}$ field that has real dynamics. This would imply that non-Abelian symmetry is determined by a Lagrangian of the form $\frac{1}{2}(E^{3^2} - B^{3^2})$, where this is automatically zero by duality. However, if this were the case, we would still have non-Abelian symmetry as a nonLagrangian symmetry. This strongly supports the possibility that the electrodynamic vacuum will continue to exhibit non-Abelian symmetries, such as squeezed states, even if we impose $\mathbf{E}^3 = \mathbf{B}^3 = 0$.

However, it can be suggested that the $\mathbf{B}^3 = \mathbf{E}^{(3)}$ field duality is broken when we consider the Lagrangian for the 3-field with the massive $A^{(3)}$ field introduced as

$$\mathcal{L} = \frac{1}{2}F^3_{\mu\nu}F^{3\mu\nu} + \frac{1}{2}\mu A^3{}_\mu A^{3\mu} - \left(\frac{1}{c}\right)j^3{}_\mu A^{3\mu}$$

$$\mathcal{L} = \frac{1}{2}(E^{3^2} - B^{3^2}) + \frac{1}{2}\mu A^3{}_\mu A^{3\mu} - \left(\frac{1}{c}\right)j^3_\mu A^{3\mu} \tag{20}$$

The middle term is a Proca Lagrangian for a massive photon. Here the mass of this photon is assumed to be larger than the masses of the $W^\pm$ and W^0 bosons. The current $j^3{}_\mu$ is determined by the charged fermions with masses given by the Yukawa interactions with the Higgs field. These are yet to be explored. Now consider the term in the Euler–Lagrange equation

$$\frac{\partial L}{\partial D^\mu A^{3\nu}} = [A_\mu, A_\nu] \tag{21a}$$

with covariant derivatives that enter into the Euler–Lagrange equation as

$$D_\mu A^{3\nu} = \partial_\mu A^{3\nu} + i\left(\frac{e}{\hbar}\right)\epsilon^{3ab}[A^a_\mu, A^b_\nu] \tag{21b}$$

and the subsequent setting of $A^3 \to 0$. Then the full Euler–Lagrange equation

$$D^\mu \frac{\partial L}{\partial(D^\mu A^{3\nu})} - \frac{\partial L}{\partial A^{3\nu}} = 0 \tag{22}$$

is

$$\nabla \times \mathbf{B}^3 + \mu^2 \mathbf{A}^3 - \mathbf{j}^3 = \frac{\partial \mathbf{E}^3}{\partial t} \tag{23}$$

which is just a form of the Faraday–Maxwell equation. However, the Hodge–star dual of this equation, the Maxwell equation, does not contain the current term,

$$\nabla \times \mathbf{E}^3 + \mu^2 \mathbf{A}^3 = -\frac{\partial \mathbf{B}^3}{\partial t} \tag{24}$$

The nonvanishing $A^{(3)}$ field at high energy will then break the duality between the $\mathbf{E}^{(3)}$ and $\mathbf{B}^{(3)}$ fields.

There is rub to this construction. This Proca equation is really only applicable on a scale that approaches high-energy physics where the $\mathbf{A}^{(3)}$ boson has appreciable influence. This will be only at a range of 10^{-17} cm. On the scale of atomic physics 10^{-8} cm, where quantum optics is applicable, this influence will be insignificant. In effect on a scale where the $\mathbf{A}^{(3)}$ does not exist, as it has decayed into pion pairs, the duality is established and there is no Lagrangian for the $\mathbf{B}^{(3)}$ field. This puts us back to square one, where we must consider non-Abelian electrodynamics as effectively $U(1)$ electrodynamics plus additional nonLagrangian and nonHamiltonian symmetries.

It has been demonstrated that there is an $SU(2) \times SU(2)$ electroweak theory that gives rise to the Z_0, $W^{\pm}$ gauge vector bosons plus electromagnetism with the photon theory with the cyclic condition for the $\mathbf{B}^{(3)}$ field. What has not been worked out are the implications for quark and lepton masses by inclusion of Yukawa coupling Lagrangians. However, that sector of the theory has little bearing on this examination of the electromagnetic theory, with $A^3 = 0$, that emerges from the $SU(2) \times SU(2)$ gauge theory. We now have a theory for electromagnetism on the physical vacuum that is

$$\mathscr{L} = -(1/4)F^{\mu\nu}F_{\mu\nu} - (1/4)G^{a\mu\nu}G^a_{\mu\nu} + \frac{1}{2}((\mathbf{E}^3)^2 - (\mathbf{B}^3)^2)$$

$$+ M_0|Z_0|^2 + M_w|W^{\pm}|^2 + \frac{1}{2}(|\partial\xi|^2 - 2\mu^2|\xi|^2)$$

$$+ \text{Dirac Lagrangians} + \text{Yukawa [Fermi–Higgs]} \tag{25}$$

where $F_{\mu\nu}$ and $G^a_{\mu\nu}$ are the field tensor components for standard electromagnetic and weak interaction fields, and the cyclic electric and magnetic fields define the Lagrangian in the third term. The occurrence of the massive Z_0 and $W^{\pm}$ particles obviously breaks the gauge symmetry of the $SU(2)$ weak interaction.

III. THE THEORY AND ITS PROBLEMS AND THEIR REMEDIES

So here we have constructed, in some ways rather artificially, an $SU(2) \times SU(2)$ gauge theory that is able to reproduce the standard model $U(1) \times SU(2)$ with the additional cyclic magnetic field given by Eq. (19). However, we are left with two uncomfortable conditions imposed on the theory to make this work. The first is that the electric charge of gauge bosons is treated in an ad hoc fashion so that we do not have photons $\mathbf{A}^1$ and $\mathbf{A}^2$ that carry a unit of electric charges ± 1. The second problem is that we have, by hand, eliminated the $\mathbf{A}^{(3)}$ vector potential. If this were nonzero, we would have the following gauge potential:

$$\omega_\mu'^3 = \frac{g}{\sqrt{g^2 + g'^2}} A_\mu^3 \tag{26}$$

This field would have a mass equal to $\sqrt{g^2 + g'^2}v/2$ and would then contribute a large decay signal at the same scattering transverse momenta where the Z_0 is seen.

The problem is that we have a theory with two $SU(2)$ algebras that both act on the same Fermi spinor fields. We further are using one Higgs field to compute the vacuum expectation values for both fields. The obvious thing to do is to first consider that each $SU(2)$ acts on a separate spinor field's doublets. Next the theory demands that we consider that there be two Higgs fields that compute separate physical vacuums for each $SU(2)$ sector independently. This means that the two Higgs fields will give 2×2 vacuum expectations, which may be considered to be diagonal. If two entries in each of these matrices are equal then we conclude that the resulting massive fermion in each of the two spinor doublets are the same field. Further, if the spinor in one doublet assumes a very large mass then at low energies this doublet will appear as a singlet and the gauge theory that acts on it will be $O(3)_b$ with the algebra of singlets

$$\mathbf{e_i} = \epsilon_{ijk}[\mathbf{e}_j, \mathbf{e}_k] \tag{27}$$

This will leave a theory on the physical vacuum that involves transformations on a singlet according to a broken $O(3)_b$ gauge theory, and transformations on a doublet according to a broken $SU(2)$ gauge theory. The broken $O(3)_b$ gauge theory reflects the occurrence of a very massive $\mathbf{A}^{(3)}$ photon, but massless $\mathbf{A}^1$ and $\mathbf{A}^2$ fields. This broken $O(3)_b$ gauge theory then reduces to electromagnetism with the cyclicity condition. The broken $SU(2)$ theory reflects the occurrence of massive charged and neutral weakly interacting bosons. Further, since the Lagrangian for the 3-fields is zero, this would further imply that the electromagnetic gauge theory is $U(1)$. This would mean that the electromagnetic field singlets will not obey the algebra given, in equation 27.

To take this theory further would be to embed it into an $SU(4)$ gauge theory. The gauge potentials are described by 4×4 traceless Hermitian matrices and the Dirac spinor has 16 components. The neutrality of the photon is then given by the sum over charges, which vanishes by the tracelessness of the theory. The Higgs field is described by a 4×4 matrix of entries.

It is concluded within the "toy model" above that the $\mathbf{B}^{(3)}$ field, or more likely a pseudofield, is consistent with an extended $SU(2) \times SU(2)$ model of electroweak interactions. A more complete formalism of the $SU(2) \times SU(2)$ theory with fermion masses will yield more general results. A direct measurement of $\mathbf{B}^{(3)}$ should have a major impact on the future of unified field theory and superstring theories. The first such measurement was reported in Ref. 14, (see also Refs. 6 and 7).

IV. CHIRAL AND VECTOR FIELDS IN $SU(2) \times SU(2)$ ELECTROWEAK FIELD

The cyclic theory of electromagnetism has been demonstrated to be consistent with a $SU(2) \times SU(2)$ theory of electroweak unification [15]. It has been demonstrated that if we set $\mathbf{A}^3 = 0$ on the physical vacuum that a cyclic theory of electromagnetism is derived. This theory contains longitudinal $\mathbf{E}^{(3)}$ and $\mathbf{B}^{(3)}$ fields that are dual $\mathbf{E}^3 = \mathbf{B}^{(3)}$, but where this duality is broken by current interactions. By setting $\mathbf{A}^3 = 0$ the transverse 3-modes of the theory have been completely eliminated by this arbitrary restriction of this gauge freedom. The elimination of these transverse 3-modes guarantees that photons are entirely defined by the $\sigma^{1,2}$ generators of the $SU(2)$ theory of electrodynamics. Since the field defined by the $\sigma^{(3)}$ generators are longitudinal this means they are irrotational $\nabla \times \mathbf{E}^3 = \nabla \times \mathbf{B}^3 = 0$ and thus time independent. According to Maxwell's equations, this means that there are no electromagnetic waves or photons associated with this field.

V. AXIAL VECTOR $SU(2) \times SU(2)$ FIELDS: A FIRST LOOK

To start we examine a putative model of a chiral vector model at low energies to determine what sorts of processes may be involved with the broken symmetry of such a model. We start by naively considering a chiral vector model to see what sorts of structure may emerge at low energy without explicit consideration of the Higgs mechanism. The field theory starts out as a twisted bundle of two chiral groups $SU(2) \times SU(2)$ and emerges as a theory that is an axial-vector theory at low energy. We consider initially the situation where the theory is an axial vector theory at low energy. We then consider the situation where there is a breakdown of chiral symmetry. This is then used to set up the more complete situation that involves the breakdown of the chiral theory at high energy into an axial vector theory at low energy.

Now we relax the condition that $\mathbf{A}^3 = 0$. This statement would physically mean that the current for this gauge boson is highly nonconserved with a very large mass so that the interaction scale is far smaller than the scale for the cyclic electromagnetic field. In relaxing this condition we will find that we still have a violation of current conservation.

With $\mathbf{A}^3 \neq 0$ we have the following fields [15]:

$$A'^1_\mu = \frac{1}{\sqrt{g^2 + g'^2}}(gA^3_\mu + g'b^3_\mu - gA^1_\mu)$$

$$Z^0_\mu = \frac{1}{\sqrt{g^2 + g'^2}}(gb^3_\mu + g'A^1_\mu) \tag{28}$$

$$\omega^3_\mu = \frac{g'}{\sqrt{g^2 + g'^2}}A^3_\mu$$

One purpose here is to examine the ω^3_μ connection; which will have a chiral component. This at first implies that the $\mathbf{B}^{(3)}$ field is partly chiral, or that it is mixed with the chiral component of the other $SU(2)$ chiral field in some manner to remove its chirality.

The theory of $SU(2)$ electromagnetism, at high energy, is very similar to the theory of weak interactions in its formal structure. Further, it has implications for the theory of leptons. The electromagnetic interaction acts upon a doublet, where this doublet is most often treated as an element of a Fermi doublet of charged leptons and their neutrinos in the $SU(2)$ theory of weak interactions.

Following in analogy with the theory of weak interactions we let ψ be a doublet that describes an electron according to the 1 field and the 3 field. Here we illustrate the sort of physics that would occur with a chiral theory. We start with the free particle Dirac Lagrangian and let the differential become gauge covariant,

$$\mathcal{L} = \bar{\psi}(i\gamma^\mu \mathcal{D}_\mu - m)\psi$$

$$= \bar{\psi}(i\gamma^\mu \partial_\mu - m)\psi - gA^b_\mu\bar{\psi}\gamma^\mu\sigma_b\psi$$

$$= \mathcal{L}_{\text{free}} + A^b_\mu J^\mu_b \tag{29}$$

where $\bar{\psi} = \psi^\dagger\gamma_4$. From here we decompose the current J^b_μ into vector and chiral components:

$$J^b_\mu = \psi^\dagger\gamma_4\gamma_\mu(1 + \gamma_5)\sigma^3\psi = V^b_\mu + \chi^b_\mu \tag{30}$$

This is analogous to the current algebra for the weak and electromagnetic interactions between fermions. We have the two vector current operators [16]

$$V_\mu^a = \frac{i}{2} \bar{\psi} \gamma_\mu \sigma^a, \psi \tag{31}$$

and the two axial vector current operators:

$$\chi_\mu^b = \frac{i}{2} \bar{\psi} \gamma_\mu \gamma_5 \tau^b \psi \tag{32}$$

Here $\gamma_5 = i\gamma_1\gamma_2\gamma_3\gamma_4$, and τ^b are Pauli matrices. These define an algebra of equal time commutators:

$$[V_4^a, V_\mu^b] = it^{abc} V_\mu^c$$
$$[V_4^a, \chi_\mu^b] = -it^{abc} \chi_\mu^b \tag{33}$$

If we set $\mu = 4$, we then have the algebra

$$[V_4^a, V_4^b] = it^{abc} V_4^c$$

and

$$[V_4^a, \chi_4^b] = -it^{abc} \chi_4^c \tag{34}$$

If we set

$$Q_\pm^a = \frac{1}{2}(V_4^a \pm \chi_4^a) \tag{35}$$

we then have the algebra

$$[Q_+^a, Q_+^b] = it^{abc} Q_+^c$$
$$[Q_-^a, Q_-^b] = it^{abc} Q_-^c \tag{36}$$
$$[Q_+^a, Q_-^b] = 0$$

This can be seen to define the $SU(2) \times SU(2)$ algebra.

The action of the parity operator on V_4^b and χ_4^b due to the presence of γ_5 in the axial vector current.

$$PV_4^b P^\dagger = V_4^b$$
$$P\chi_4^b P^\dagger = -\chi_4^b \tag{37}$$

As such, one $SU(2)$ differs from the other by the action of the parity operator and the total group is the chiral group $SU(2) \times SU(2)_P$. This illustrates the sort of current that exists with chrial gauge theory, and what will exist with a right and left handed chrial $SU(2) \times SU(2)$ theory.

We have, at low energy, half vector and half chiral vector theory $SU(2) \times SU(2)_P$. On the physical vacuum, we have the vector gauge theory described by $\mathbf{A}^1 = \mathbf{A}^{2*}$ and $\mathbf{B}^3 = \nabla \times \mathbf{A}^3 + (ie/\hbar)\mathbf{A}^1 \times \mathbf{A}^2$ and the theory of weak interactions with matrix elements of the form $\bar{v}\gamma_\mu(1 - \gamma_5)e$ and are thus half vector and chiral on the level of elements of the left- and right-handed components of doublets. We then demand that on the physical vacuum we must have a mixture of vector and chiral gauge connections, within both the electromagnetic and weak interactions, due to the breakdown of symmetry. This will mean that the gauge potential $\mathbf{A}^{(3)}$ will be massive and short-ranged.

One occurrence is a violation of the conservation of the axial vector current. We have that the 1 and 2 currents are conserved and invariant. On the high-energy vacuum we expect that currents should obey

$$\partial^\mu J_\mu^b = 0 \tag{38}$$

where $b \in \{1, 2\}$, which are absolutely conserved currents. However, for the A_μ^3 fields we have the nonconserved current equation [17]

$$\partial^\mu J_\mu^3 = im_\psi \psi^\dagger \gamma_4 \gamma_5 \sigma^3 \psi \tag{39}$$

where inhomogeneous terms correspond to the quark–antiquark and lepton–antilepton pairs that are formed from the decay of these particles. This breaks the chiral symmetry of the theory. Then this current's action on the physical vacuum is such that when projected on a massive eigenstates for the 3-photon with transverse modes

$$\langle 0|\partial^\mu J_\mu^3|X_k\rangle = \left(\frac{m^2}{\sqrt{\omega(k)\omega(k')}}\right)\langle X_{k'}|X_k\rangle e^{ikx} \tag{40}$$

The mass of the chiral $\{1, 2\}$-bosons will then vanish, while the mass of the chiral 3-boson will be m. So rather than strictly setting $\mathbf{A}^3 = 0$, it is a separate chiral gauge field that obeys axial vector nonconservation and only occurs at short ranges.

Now that we have an idea of what nature may look like on the physical vacuum, we need to examine how we in fact can have symmetry breaking and an $SU(2) \times SU(2)_P$ gauge theory that gives rise to some of the requirements of $O(3)_b$ electromagnetism mentioned above. A mixing of the two chiral $SU(2)$ bundles at low energy will effect the production of vector gauge bosons for the electromagnetic interaction. It is apparent that we need to invoke the mixing of two chiral gauge bosons in such a manner as to produce a vector theory of electromagnetism at low energy with a broken chiral theory of weak interactions.

VI. CHIRAL AND VECTOR GAUGE THEORY ON THE PHYSICAL VACUUM FROM A GAUGE THEORY WITH A CHIRAL TWISTED BUNDLE

The $SU(2) \times SU(2)$ theory should mimic the standard model with the addition of the $\mathbf{B}^3 = (e/\hbar)\mathbf{A}^1 \times \mathbf{A}^2$ field at low energies. This means that we demand that a field theory that is completely chiral at high energy becomes a field theory that is vector and chiral in separate sectors on the physical vacuum of low energies. This means that a field theory that is chiral at high energy will combine with the other chiral field in the twisted bundle to produce a vector field plus a broken chiral field at low energy. Generally this means that a field theory that has two chiral bundles at high energies can become vector and chiral within various independent fields that are decoupled on physical vacuum at low energies.

We consider a toy model where there are two fermion fields ψ and χ, where each of these fields consists of the two component right- and left-handed fields R_ψ, L_ψ and R_χ, L_χ. These Fermi doublets have the masses m_1 and m_2. We then have the two gauge potentials A_μ and B_μ that interact respectively with the ψ and χ fields. In general, with more Fermi fields, this situation becomes more complex, where these two Fermi fields are degeneracies that split into the multiplet of fermions known. In this situation there are four possible masses for these fields on the physical vacuum. These masses occur from Yukawa couplings with the Higgs field on the physical vacuum. These will give Lagrangians terms of the form $Y_\phi R_\psi^\dagger \phi L_\chi + H.C.$ and $Y_\eta L_\psi^\dagger \eta R_\chi + H.C.$, where we now have a two-component ϕ^4 field for the Higgs mechanism. These two components assume the minimal expectation values $\langle \phi_0 \rangle$ and $\langle \eta_0 \rangle$ on the physical vacuum. We then have the Lagrangian [18]

$$\mathscr{L} = \bar{\psi}(i\gamma^\mu(\partial_\mu + igA_\mu) - m_1)\psi + \bar{\chi}(i\gamma^\mu(\partial_\mu + igB_\mu) - m_2)\chi$$
$$- Y_\phi R_\psi^\dagger \phi L_\chi + H.C. - Y_\eta L_\psi^\dagger \eta R_\chi + H.C. \tag{41}$$

(where $H.C.$ = higher contributions), which can be further broken into the left and right two-component spinors:

$$\mathscr{L} = R_\psi^\dagger i\sigma^\mu(\partial_\mu + igA_\mu)R_\psi + L_\psi^\dagger i\sigma^\mu(\partial_\mu + igA_\mu)L_\psi$$
$$+ R_\chi^\dagger i\sigma^\mu(\partial_\mu + igB_\mu)R_\chi + L_\chi^\dagger i\sigma^\mu(\partial_\mu + igB_\mu)L_\chi$$
$$- m_1 R_\psi^\dagger L_\psi - m_1 L_\psi^\dagger R_\psi - m_2 R_\chi^\dagger L_\chi - m_2 L_\chi^\dagger R_\chi$$
$$- Y_\phi R_\psi^\dagger \phi L_\chi + Y_\phi^* L_\chi^\dagger \phi^* R_\psi - Y_\eta L_\psi^\dagger \eta R_\chi + Y_\eta^* R_\chi^\dagger \eta^* L_\psi \tag{42}$$

The gauge potentials A_μ and B_μ are 2×2 Hermitian traceless matrices and the Higgs fields ϕ and χ are also 2×2 matrices. These expectations are real valued, and so we then expect that the non-zero contributions of the Higgs field on the physical vacuum are given by the diagonal matrix entries [18]

$$\langle \phi \rangle = \begin{pmatrix} \langle \phi^1 \rangle & 0 \\ 0 & \langle \phi^2 \rangle \end{pmatrix} \qquad \langle \chi \rangle = \begin{pmatrix} \langle \chi^1 \rangle & 0 \\ 0 & \langle \chi^2 \rangle \end{pmatrix} \tag{43}$$

In a 1999 paper, these issues were not discussed [15]. There this matrix was proportional to the identity matrix and the matrix nature of the Higgs field was conveniently ignored. This means that the $SU(2) \times SU(2)$ electroweak theory shares certain generic features with the $SU(2) \times U(1)$ theory. The values of the vacuum expectations are such that at high energy the left-handed fields R_χ and the right-handed doublet field L_ψ couple to the $SU(2)$ vector boson field B_μ, while at low energy the theory is one with a left-handed $SU(2)$ doublet R_ψ that interacts with the right-handed doublet L_χ through the massive gauge fields A_μ. Then the mass terms from the Yukawa coupling Lagrangians will then give

$$m' = Y_\eta \langle \chi^1 \rangle \gg m'' = Y_\eta \langle \chi^2 \rangle \gg \tag{44}$$

$$m''' = Y_\phi \langle \phi^1 \rangle \gg m'''' = Y_\phi \langle \phi^2 \rangle \tag{45}$$

Further, if the $SU(2)$ theory for B_μ potentials are right-handed chiral and the $SU(2)$ theory for A_μ potentials are left-handed chiral, then we see that a chiral theory at high energies can become a vector theory at low energies. The converse may also be true in another model.

In the switch between chirality and vectorality at different energies, there is an element of broken gauge symmetry. So far we would have a theory of a broken gauge theory at low energy. However, there is a way to express this idea so that at low energy we have a gauge theory accompanied by a broken gauge symmetry. To illustrate this let us assume we have a simple Lagrangian that couples the left-handed fields ψ_l to the right-handed boson A_μ and the right-handed fields ψ_r to the left handed boson B_μ

$$\mathscr{L} = \bar{\psi}_l(i\gamma^\mu(\partial_\mu + igA_\mu) - m_1)\psi_l + \bar{\psi}_r(i\gamma^\mu(\partial_\mu + igB_\mu) - m_2)\psi_r$$
$$- Y_\phi \psi_l^\dagger \phi \psi_r - Y_\phi^* \psi_r^\dagger \eta \psi_l \tag{46}$$

If the coupling constant Y_ϕ is comparable to the coupling constant g, then the Fermi expectation energies of the Fermions occur at the mean value for the Higgs field $\langle \phi_0 \rangle$. In this case the vacuum expectation of the vacuum is proportional to the identity matrix. This means that the masses acquired by the right chiral plus left chiral gauge bosons $A_\mu + B_\mu$ are zero, while the left chiral minus right chiral gauge bosons $A_\mu - B_\mu$ acquires masses approximately $Y_\phi \langle \phi_0 \rangle$. The theory at low energies is a theory with an unbroken vector gauge theory plus a broken chiral gauge theory [18]. It is also the case that we demand that the charges of the two chiral fields $A^{1,2}, B^{1,2}$ that add are opposite so that the resulting vector gauge bosons are chargeless.

Just as we have gauge theories that can change their vector and chiral character, so also do the doublets of the theory. In so doing this will give rise to the doublets of leptons and quarks plus doublets of very massive fermions. These massive fermions should be observable in the multi-TeV range of energy.

VII. THE OCCURRENCE OF $O(3)_b$ ELECTRODYNAMICS ON THE PHYSICAL VACUUM

The two parts of the twisted bundle are copies of $SU(2)$ with doublet fermion structures. However, one of the fermions has the extremely large mass, $m' = Y_\eta \langle \chi^1 \rangle$, which is presumed to be unstable and is not observed at low energies. So one sector of the twisted bundle is left with the same abelian structure, but with a singlet fermion. This means that the $SU(2)$ gauge theory will be defined by the algebra over the basis elements $\hat{e}_i, i \in \{1, 2, 3\}$:

$$[\hat{e}_i, \hat{e}_j] = i\epsilon_{ijk}\hat{e}_k \tag{47}$$

We also need to examine the photon masses. We define the Higgs field by a small expansion around the vacuum expectations $\eta^1 = \xi^1 + \langle \eta_0^1 \rangle$ and $\eta^2 = \xi^2 + \langle \eta_0^2 \rangle$. The contraction of the generators σ^1 and σ^2 with the Higgs field matrix and right and left fields gives

$$\sigma^1 \cdot \eta R + \sigma^2 \cdot \eta L = 0 \tag{48}$$

which confirms that the charges of the $\mathbf{A}^1$ and $\mathbf{A}^2$ fields are zero. These fields on the low-energy vacuum can be thought of as massless fields composed of two gauge bosons, with masses $\sqrt{m' + m''} \gg M_Z$ and with opposite charges. These electrically charged fields can be thought of as $\mathbf{A}^\pm = \mathbf{A}^1 \pm \mathbf{A}^2$. These particles cancel each other and gives rise to massless vector photon gauge fields. The field $\mathbf{A}^{(3)}$ also has this mass. This massive field is also unstable and decays into particle pairs.

With the action of the more massive Higgs field we are left with the gauge theory $SU(2) \times O(3)$, where the first gauge group acts on doublets and the last gauge group acts on singlets. Further on a lower-energy scale, or equivalently sufficiently long timescales, the field $\mathbf{A}^{(3)}$ has decayed and vanished. At this scale the second gauge group is then represented by $O(3)_p$ meaning a partial group. This group describes Maxwell's equations along with the definition of the field $\mathbf{A}^1 \times \mathbf{A}^2$.

From this point we can then treat the action of the second Higgs field on this group in a manner described in Ref. 15. If we set the second Higgs field to have zero-vacuum expectation $\langle \phi^2 \rangle = 0$, then the symmetry breaking mechanism effectively collapses to this formalism, which is similar to the standard $SU(2) \times U(1)$ model Higgs mechanism. We can the arrive at a vector electromagnetic gauge theory $O(3)_p$, where p stands for partial, and a broken chiral $SU(2)$ weak interaction theory. The mass of the vector boson sector is in the $\mathbf{A}^{(3)}$ boson plus the $W^\pm$ and Z^0 particles.

VIII. THE $SU(4)$ MODEL

It is possible to consider the two $SU(2)$ group theories as being represented as the block diagonals of the larger $SU(4)$ gauge theory. The Lagrangian density for the system is then

$$\mathcal{L} = \bar{\psi}(i\gamma^\mu(\partial_\mu + igA_\mu) - m_1)\psi - Y\bar{\psi}\phi\psi \qquad (49)$$

The gauge potentials A_μ now have 4×4 traceless representations. The scalar field theory that describes the vacuum will now satisfy field equations that involve all 16 components of the gauge potential. By selectively coupling these fields to the fermions, it should be possible to formulate a theory that recovers a low-energy theory that is the standard model with the $O(3)_p$ gauge theory of electromagnetism.

What has been presented is an outline of an $SU(2) \times SU(2)$ electroweak theory that can give rise to the non-Abelian $O(3)_b$ theory of quantum electrodynamics on the physical vacuum. The details of the fermions and their masses has yet to be worked through, as well as the mass of the $\mathbf{A}^{(3)}$ boson. This vector boson as well as the additional fermions should be observable within the 10-Tev range of energy. This may be accessible by the CERN Large Hadron Collider in the near future.

The principal purpose here has been to demonstrate what sort of electroweak interaction physics may be required for the existence of an $O(3)_b$ theory of quantum electrodynamics on the low-energy physical vacuum. This demonstrates that an extended standard model of electroweak interactions can support such a theory with the addition of new physics at high energy.

IX. DUALITY IN GRAND UNIFIED FIELD THEORY, AND LEP1 DATA

The preceding construction indicates that the electromagnetic and weak interactions may be dual-field theories. If the preceding construction is experimentally verified, then this would be the first empirical indication that the universe is indeed dual according to a theory along the lines of Olive–Montenen [11–13]. Within this theory there are coupling constants that have inverse relationships, or convergences at high energy, so that one field is weak and the other is strong at low energy. In this case the electromagnetic field is comparatively strong, but not when compared to the nuclear force, and the other is very weak. It may be that both field theories have coupling constants that are both lowered and diverge at low energy within a grand unified theory (GUT). The examination of this electroweak theory within such a construction has not been been done. Nonetheless, the experimental finding of the $\mathbf{A}^{(3)}$ would bring a tremendous change in our views on the foundations of physics.

It was recently suggested by Erler and Langacker [19] that an anomaly in Z decay widths points to the existence of Z' bosons. These are predicted to exist with a mass estimate of 812 GeV$^{+339}_{-152}$ within an $SO(10)$ GUT model and a Higgs mass posited at 145 GeV$^{+103}_{-61}$. This suggests that a massive neutral boson predicted by grand unified theories has been detected. Further, variants of string theories predict the existence of a large number of these neutral massive bosons.

Analyses of the hadronic peak cross section data obtained at LEP1 [20] implies a small amount of missing invisible width in Z decays. These data imply an effective number of massless neutrinos, $N = 2.985 \pm 0.008$, which is below the prediction of 3 standard neutrinos by the standard model of electroweak interactions. The weak charge Q_W in atomic parity violation can be interpreted as a measurement of the S parameter. This indicates a new $Q_W = -72.06 \pm 0.44$ is found to be above the standard model prediction. This effect is interpreted as due to the occurrence of the Z' particle, which will be refered to as the Z_γ particle.

$SO(10)$ has the six roots $\alpha^i, i = 1, \ldots, 6$. The angle between the connected roots are all $120°$, where the roots α^3, α^4 are connected to each other and two other roots. The Dynkin diagram is illustrated below:

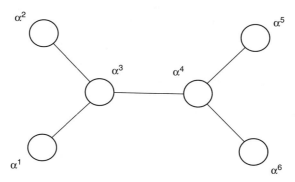

Figure 1.   Extended Dynkin diagram for $SO(10)$.

The decomposition of $SO(10) \rightarrow SU(5) \times U(1)$ is performed by removing the circles representing the roots $\alpha^{1,2,5,6}$ connected by a single branch. The remaining connected graph describes the $SU(5)$ group. However, by removing the circle α^4 connected by three branches forces $SO(10)$ to decompose into $SU(2) \times SU(2) \times SU(4)$. Here we have an $SU(2)$ and a mirror $SU(2)$ that describe opposite-handed chiral gauge fields, plus an $SU(4)$ gauge field. The chiral fields are precisely the sort of electroweak structure above and proposed in Ref. 15. Presumably since $SU(4)$ can be represented by a 4 that is $3 \oplus 1$ and $\bar{4}$ as $\bar{3} \oplus 1$, we can decompose this into $SU(3) \times U(1)$. Further, the neutrino short fall is a signature of the opposite chiralities of the two "mirrored" SU(2) gauge fields [15].

The $SU(2) \times SU(2) \rightarrow SU(2) \times O(3)_b$ predicts the occurrence of a massive photon. So it is possible that these data could corroborate the extended standard model that expands the electromagnetic sector of the theory. What we really understand empirically is QCD and electroweak standard model, and we may have some idea about quantum gravity for at least we do have general relativity and quantum mechanics. This leads to the strange situation where we have reasonable data on low-TeV range physics and potential ideas about quantum gravity at 10^{19} GeV, with a void of greater ignorance in between. However, these data and analyses suggest theoretical information about GUTs and cast some light on this energy region.

These experimental data do suggest that non-Abelian electrodynamics is tentatively a valid theory, at least as an extended theory that predicts nonHamiltonian vacuum symmetries. It also suggests that at high energy, electrodynamics and the weak interactions are dual-field theories. This duality would then exist at energies that may be probed in the TeV range of energy. In order to completely verify that this is the case experiments at the TeV range need to be performed where the Z_γ and Higgs boson can be directly produced.

This leaves open the question about the nuclear interaction. It is tempting to conjecture that there is a dual field theory to the $SU(3)$ nuclear interaction or quantum chromodynamics (QCD). It is easy to presume that such a construction would proceed in a manner outlined above with the chiral $SU(2) \times SU(2)$ electroweak field theory. This would then imply that there exists an additional weak field in nature. If the field theory is similar in construction, then there may exist some massive particle with weak coupling. It would then be tempting to pursue calculations to predict the existence of such particles. However, it must be stressed that this is rather speculative and has speculative implications for the foundations of physics.

It is tempting to think that there may be a generalized $SU(3) \times SU(3)$ type of theory for the strong interactions. As in the abovementiond $SO(10)$ theory, we see that the nuclear interactions are embedded in an $SU(4)$ theory. This would mean that there exist chiral colored gluons associated with QCD. This can most easily be seen if the $U(1)$ group associated with QCD according to $SU(4) = SU(3) \times U(1)$. The $U(1)$ group describes local phase changes according to

$$\psi \rightarrow e^{i\phi}\psi \tag{50}$$

We may assign this $U(1)$ group to a chiral transformation, similar to a G parity operator, according to

$$\psi \rightarrow e^{i\phi\gamma_5}\psi \tag{51}$$

The Dirac Lagrangian would then assume the form

$$\mathscr{L} = \frac{1}{2}(\bar{\psi}(1 + \gamma_5)\gamma_\mu\partial^\mu\psi) \tag{52}$$

where at high energies, before the Higgs field has assigned masses to the quarks through Yukawa couplings, the QCD sector would be chiral invariant. Once the quarks acquire mass, there is chiral breaking. One may then have a field where the dominant amplitudes favor vector gluons, but where there is a small chromochiral amplitude. This would also mean that quarks would exhibit a small chiral breaking. Further, if the coupling constants for the chiral component of the chromofield are very weak then we have in effect a duality within QCD.

It is then apparent that the extension of electromagnetism to a higher symmetry group, such as $SU(2)$ at higher energy will have implications for the spectra of elementary particles at high energy. In this way, even if electromagnetism at low energy fails to demonstrate a $B(3)$ field, the predictions of an extended electromagnetism may either be demonstrated or the theory falsified.

X. QUANTUM ELECTRODYNAMICS

In this section we discuss the nonrelativistic $O(3)_b$ quantum electrodynamics. This discussion covers the basic physics of $U(1)$ electrodynamics and leads into a discussion of nonrelativistic $O(3)_b$ quantum electrodynamics. This discussion will introduce the quantum picture of the interaction between a fermion and the electromagnetic field with the $\mathbf{B}^{(3)}$ magnetic field. Here it is demonstrated that the existence of the $\mathbf{B}^{(3)}$ field implies photon–photon interactions. In nonrelativistic quantum electrodynamics this leads to nonlinear wave equations. Some presentation is given on relativistic quantum electrodynamics and the occurrence of Feynman diagrams that emerge from the $\mathbf{B}^{(3)}$ are demonstrated to lead to new subtle corrections. Numerical results with the interaction of a fermion, identical in form to a 2-state atom, with photons in a cavity are discussed. This concludes with a demonstration of the Lamb shift and renormalizability.

One of the oldest subjects of physical science is electrodynamics. The study has its early origins in the study of optics by Willebrord Snellius and the studies of magnetism by William Gilbert in the sixteenth century [1]. It took nearly three centuries for the theory of classical electromagnetism to reach fruition with Maxwell [2]. This grand synthesis at first appeared to solve the most fundamental questions of the day, but an historical retrospective shows that it posed as many questions as it solved. The resolutions to these problems were found in the theory of special relativity and in quantum theory. The first of these was an answer to the problem of what is the speed of an electromagnetic wave on any given reference frame, and the second was a resolution to the blackbody radiation problem. The latter solution advanced by Planck assumed that light existed in discrete packets of energy that were emitted and absorbed [3]. This initiated the study of the interaction between quantized electromagnetic waves and matter with discrete quantized energy levels. This theory is called *quantum electrodynamics.*

The formalism of quantum electrodynamics may appear arcane to the uninitiated, but in reality it is based on rather simple concepts. The first of these is that the radiation field is described by a set of harmonic oscillators. The harmonic oscillator is essentially a spring loaded with a mass or a pendulum that swings through a small angle. The pendulum has an old history with physics that began with Galileo. Early in the formalism of quantum mechanics this was a system examined and quantized. An analysis with the Schrödinger wave equation leads to some complexities with recurrence relations and Hermite polynomials. However, with the Heisenberg formalism the quantum theory of the harmonics oscillator reduces to a simple model with evenly spaced states that have an associated energy $\left(n + \frac{1}{2}\right)\hbar\omega$. Here the number n corresponds to the number of photons with angular frequency $\omega = ck$ in the system. For $n = 0$, we see that the absence of photons predicts that there is still an energy associated with the vacuum. This nonzero value for the ground state of the harmonic oscillator has been a source of controversy as well as profound physical insight. A second assumption that is often made is that these photons exist within a cavity. This allows for a simplification of the meaning to counting modes. The third concept is that atoms that interact with these photons also have energy levels. The simplest example would be atoms with two states. Here an atom that absorbs a photon can only do so by changing its internal state from the lower state to the excited state, and an atom can emit a photon only by changing its internal state from the excited state to the lower state. These atomic interactions with the electromagnetic field will change the photon number by ± 1.

How does one proceed to take the classical theory of electromagnetism, or Maxwell's equations, and cast them in a quantum mechanical context? It is best to start with the definitions of the electric and magnetic fields

$$\mathbf{E} = -\nabla\phi - \frac{1}{c}\frac{\partial\mathbf{A}}{\partial t} \tag{53}$$

and

$$\mathbf{B} = \nabla \times \mathbf{A} \qquad (54)$$

The quantity $\mathbf{A}$ appears in these equations and is the vector potential of electromagnetic theory. In a very elementary discussion of the static electric field we are introduced to the theory of Coulomb. It is demonstrated that the electric field can be written as the gradient of a scalar potential $\mathbf{E} = -\nabla\phi, \phi = \kappa q/r$. It is also demonstrated that the addition of a constant term to this potential leaves the electric field invariant. Where you choose to set the potential to zero is purely arbitrary. In order to describe a time-varying electric field a time dependent vector potential must be introduced $\mathbf{A}$. If one takes any scalar function χ and uses it in the substitutions

$$\mathbf{A}' = \mathbf{A} - \kappa\nabla\chi, \qquad \kappa = \text{constant} \qquad (55)$$

$$\phi' = \phi + \gamma\frac{\partial\chi}{\partial t}, \qquad \gamma = \text{constant} \qquad (56)$$

it is easy to demonstrate that the electric and magnetic fields are left invariant. This means that the analyst can choose the form of the vector potential in an arbitrary fashion. This is defined as a choice of gauge that is described by either writing an explicit form for the vector potential or by writing an auxiliary differential equation. As an example we may then choose

$$\mathbf{A}(\mathbf{r}, t) = A_0\mathbf{e}\exp\left(i(\mathbf{k} \cdot \mathbf{r} - \omega t)\right)$$
$$\phi = 0 \qquad (57)$$

which is equivalent to stating that $\nabla \cdot \mathbf{A} = 0$. It is then fitting that the Maxwell's equations are presented, as they are invariant under all possible gauge transformations

$$\nabla \times \mathbf{H} = \mathbf{j} + \frac{\partial\mathbf{D}}{\partial t}, \qquad \nabla \times \mathbf{E} = -\frac{\partial\mathbf{B}}{\partial t}$$

$$\nabla \cdot \mathbf{D} = \rho, \qquad \nabla \cdot \mathbf{B} = \mathbf{0} \qquad (58)$$

$$\mathbf{D} = \epsilon_0\mathbf{E} + \mathbf{P} = \epsilon\mathbf{E}, \qquad \mathbf{B} = \mu_0(\mathbf{H} + \mathbf{M}) = \mu\mathbf{H}$$

The connection to quantum theory is made with the recognition that this transformation changes the phase of a wave function of a particle that interacts with the electromagnetic field:

$$\psi \rightarrow e^{-i\chi}\psi \qquad (59)$$

The equation that describes the interaction of a nonrelativistic electron with the electromagnetic field is the Pauli equation

$$i\hbar\frac{\partial\psi}{\partial t} = \frac{1}{2m}\left(\sigma\cdot\left(\mathbf{p}-\frac{e}{c}\mathbf{A}\right)\right)\left(\sigma\cdot\left(\mathbf{p}-\frac{e}{c}\mathbf{A}\right)\right)\psi + e\phi\psi \tag{60}$$

where the $|\mathbf{A}|^2$ potential term is dropped in $U(1)$ electrodynamics. Now consider this equation under the phase shift $\psi \to e^{-i\chi}\psi$:

$$\begin{aligned}\nabla\psi &= \nabla(e^{-i\chi}\psi) = e^{-i\chi}\nabla\psi - i\nabla\chi e^{i\chi}\psi \\ &= e^{-i\chi}(\nabla - i\nabla\chi)\psi\end{aligned} \tag{61}$$

This means that the generalized momentum operator is

$$\left(\sigma\cdot\left(\mathbf{p}-\frac{e}{c}\mathbf{A}\right)\right)\psi' \to e^{i\chi}\left(\sigma\cdot\left(\mathbf{p}-\hbar\nabla\chi-\frac{e}{c}\mathbf{A}\right)\right)\psi \tag{62}$$

which recovers the preceding gauge transformations for $\mathbf{A}$ as $\mathbf{A} \to \mathbf{A} - (e/c)\nabla\chi$, rendering the quantity in Eq. (62) is gauge-invariant.

We have our first connection between quantum mechanics and electromagnetism—a local phase shift in a wavefunction is coexistent with a local gauge transformation in the vector and scalar potential for the electromagnetic field. So far nothing has been changed with the formal description of the electric and magnetic field. This is good news, for this means that the electromagnetic field can be described by the classical equations of Maxwell. This can be stated that the probability amplitude for the absorption or emission of a photon by an atom is equal to the amplitude given by the absorption and emission of an electromagnetic wave described by the classical electrodynamics of Maxwell's equations. This statement must be accompanied by the stipulation that the classical wave is normalized. Then energy density of the wave is $\hbar\omega$ times the probability per unit volume for the occurrence of the photon, and the classical wave is broken into two complex components $e^{-i\omega t}$ and $e^{i\omega t}$ that represent the phase of an absorbed and emitted photon. These phases will, by the first stipulation, be multiplied by the appropriate probability amplitudes for absorption and emission. This sets us up for an examination of the semiclassical theory of radiation and its interaction with quantized atoms.

We know that the electromagnetic field is described within a box. This means that the number of states per unit volume is dependent on the number of discrete modes per volume $|\mathbf{k}|^2(2\pi)^3\Delta|\mathbf{k}|$. This can easily be carried over to the continuous version if we let the wall of the cavity separate to arbitrary distances. The density of states is then $k^2(2\pi c)^3(dk\,d\Omega)/\hbar d\omega$. This describes the density of states that are available for an atom to interact with. We then have that if we have an atom in the state ψ_i that it may then enter into the state ψ_f, with respective energies E_i and E_f. The probability per increment in time is proportional to the transition probability for this event times the density of states. Assume that the time over which this transition occurs is far larger than the periodicity of the electromagnetic field. The transition probability is then proportional to the modulus square of the vector potential when averaged over many periods of the field. This then gives the Fermi Golden Rule [4]:

$$\frac{\Delta}{\Delta t}\text{Prob}_{\psi_i \to \psi_f} = \frac{2\pi}{\hbar^2}|A|^2 \left(\frac{\omega^2}{(2\pi c)^3}\right)d\Omega \tag{63}$$

This process is illustrated as

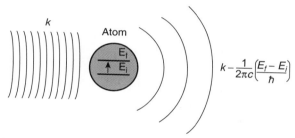

Figure 2.   The interaction between an electromagnetic field and a twostate atom.

All we need to do is to estimate the average of the potential. To do this, the form of the electric and magnetic fields are used in the normalized energy density of the electromagnetic field

$$E = \hbar\omega = \frac{1}{8\pi}(|\mathbf{E}|^2 + |\mathbf{B}|^2) = \frac{1}{8\pi}\frac{A^2\omega^2}{c^2} \tag{64}$$

which gives the averaged potential as $A = \sqrt{8\pi\hbar c^2/\omega}$ this gives us the transition probability per unit time

$$\frac{\Delta}{\Delta t}\text{Prob}_{\psi_i \to \psi_f} = \frac{2}{\pi}\frac{e^2\omega}{\hbar c} \tag{65}$$

This gives us an order-of-magnitude estimate for this transition. It assumes that the potential is absorbed or emitted with no regard to its components $e^{i\omega t}$ and $e^{-i\omega t}$. As such, this can be regarded as only a rather crude estimate. However, we are beginning to make progress in our understanding of how electromagnetic fields interact quantum-mechanically with atoms.

Returning to Eq. (2), we express this according to the matrix element U_{ij} that will be determined explicitly:

$$\frac{\Delta}{\Delta t}\text{Prob}_{\psi_i \to \psi_f} = \frac{2\pi}{\hbar^2} \frac{|U_{if}|^2 \omega^2}{(2\pi c)^3 d\Omega} \tag{66}$$

This matrix element is the expectation of a time-dependent perturbative or interaction Hamiltonian, $V = e^{i\omega t} U(\mathbf{r})$;

$$U_{fi} = \int d^3r \psi_f^* V \psi_i \tag{67}$$

Since the Pauli–Schrödinger equation is of the form $i\hbar \frac{\partial \psi}{\partial t} = H\psi$, we may write the wave functions as $\psi_{if} = e^{iE_{if}t/\hbar}\psi(0)_{if}$. We then have the transition matrix element written as,

$$U_{fi} = \int d^3r \psi(0)_f^* U(\mathbf{r})\psi(0)_i \exp\left(i\frac{(E_f - E_i)t}{\hbar} + i\omega t\right) \tag{68}$$

The initial and final states of the system are $E_i - \hbar\omega$ and E_f. We expect that the interaction occurs where $E_i - \hbar\omega = E_f$. This means that we may set the phase equal to zero and interactions that are slightly off resonant are ignored, and

$$U_{fi} = \int d^3r \psi(0)_f^* U(\mathbf{r})\psi(0)_i \tag{69}$$

The interaction Hamiltonian can be extracted from the Pauli Hamiltonian plus a dipole interaction Hamiltonian

$$
\begin{aligned}
H &= \frac{1}{2m}\left(\mathbf{p} - \frac{e}{c}\mathbf{A}\right)^2 - \frac{e\hbar}{2mc}\boldsymbol{\sigma}\cdot\nabla\mathbf{A} \\
&= \frac{1}{2m}p^2 - \frac{e}{2mc}(\mathbf{p}\cdot\mathbf{A} + \mathbf{A}\cdot\mathbf{p}) - \frac{e\hbar}{2mc}\boldsymbol{\sigma}\cdot\nabla\mathbf{A} \\
&\quad + \frac{e^2}{2mc^2}\mathbf{A}\cdot\mathbf{A} + \frac{e\hbar}{2mc}\boldsymbol{\sigma}\cdot\nabla\mathbf{A}
\end{aligned}
\tag{70}
$$

The second and third terms are the interaction terms that couple the atom, here modeled as a two-state system with Pauli matrices, to the electromagnetic field. We consider the momentum to be the operator $\mathbf{p} = \frac{\hbar}{i}\nabla$ and consider this operator as not only operating on the vector potential but on the wavefunction. Hence we find that

$$
\begin{aligned}
\nabla \times \mathbf{A} &= i\mathbf{k} \times \mathbf{e}A e^{i\mathbf{k}\cdot\mathbf{r} - i\omega t} \\
\mathbf{p}\cdot\mathbf{A} &= \mathbf{A}\cdot(\mathbf{p} + \hbar\mathbf{k})
\end{aligned}
\tag{71}
$$

This leads to a more complete form of the interaction Hamiltonian

$$U_{fi} = -\frac{e}{2mc}A\int d^3r\phi_f^*(0)\left(\mathbf{p}\cdot\mathbf{e}+\mathbf{e}\cdot\mathbf{p}-i\hbar\sigma\cdot(\mathbf{k}\times\mathbf{e})e^{i\mathbf{k}\cdot\mathbf{r}}\right)\psi_i(0) \qquad (72)$$

This result is an exact expression for the transition matrix element. Physically we have a dipole interaction with the vector potential and a dipole interaction with the magnetic field modulated by a phase factor. The problem is that this integral is difficult to compute. An approximation can be invoked. The wavevector has a magnitude equal to $1/\lambda$. The position $\mathbf{r}$ is set to the position of an atom and is on the order of the radius of that atom. Thus $\mathbf{K}\cdot\mathbf{r}\simeq a/\lambda$. So if the wavelength of the radiation is much larger than the radius of the atom, which is the case with optical radiation, we may then invoke the approximation $e^{i\mathbf{k}\cdot\mathbf{r}}\simeq 1+i\mathbf{k}\cdot\mathbf{r}$. This is commonly known as the *Born approximation*. This first-order term under this approximation is also seen to vanish in the first two terms as it multiplies the term $\mathbf{p}\cdot\mathbf{e}$. A further simplification occurs, since the term $\sigma\cdot(\mathbf{k}\times\mathbf{e})$ has only diagonal entries, and our transition matrix evaluates these over orthogonal states. Hence, the last term vanishes. We are then left with the simplified variant of the transition matrix:

$$\begin{aligned}U_{fi} &= -\frac{e}{mc}A\int d^3r\phi_f^*(0)\mathbf{p}\cdot\mathbf{e}\phi_i(0)\\ &= -\frac{e}{mc}A\langle f|\mathbf{p}|i\rangle\cdot\mathbf{e}\end{aligned} \qquad (73)$$

The element $\mathbf{p}_{fi}\cdot\mathbf{e}=|\mathbf{p}_{fi}|\cos(\theta)$, where θ is the angle between these two vectors. However this angle is $\pi/2$ different from the coordinate angle evaluated in $d^3r=r^2drd\sin(\theta)d\phi$, so we set $\theta\rightarrow\theta+\pi/2$. This means that the transition probability per unit time assumes the form

$$U_{if} = -\frac{1}{\hbar}\left(\frac{e}{mc}\right)^2 A\int_0^\pi\int_0^{2\pi}|\mathbf{p}_{fi}\cdot\mathbf{e}|^2\sin^3(\theta)r^2\,d\theta\,d\phi \qquad (74)$$

Recognizing that $\mathbf{p}_{fi}=im\omega\mathbf{r}_{fi}$ and performing, the integration, we find that

$$\frac{\Delta}{\Delta t}P_{i\rightarrow j} = \frac{4}{3}\frac{e^2}{\hbar c}\frac{\omega^3}{c^2}|r_{fi}|^2 \qquad (75)$$

As a final side note, the term $\alpha=\frac{e^2}{\hbar c}\simeq\frac{1}{137}$ is the fine-structure constant for the electromagnetic interaction. This is a dimensionless quantity that gives the interaction strength between photons and charged particles.

So far we have the transition probability per unit time. What is measured is the transition probability over a given time as measured from a statistical ensemble of identical systems. A quantum operator, O_{op} evolves in time according to the Schrödinger equation

$$i\hbar \frac{\partial O_{op}}{\partial t} = [H, O_{op}] \tag{76}$$

For the matrix U_{ij} defined at a time t, we have the solution to the Schrödinger equation with this initial condition;

$$a_{ij}(t) = -\frac{i}{\hbar} \int_0^T e^{iE_j t/\hbar} U_{ij}(t) e^{-iE_i t/\hbar} \, dt \tag{77}$$

The use of the symbol a_{ij} is to indicate that this represents the absorption of a photon by an atom. Further, the matrix $U_{ij}(t) = e^{-i\omega t} U_{ij}(0)$, and when placed substituted into Eq. (77), we arrive at an expression for $a_{ij}(t)$. Now, when $a_{ij}(t)$ is multiplied by its complex conjugate, we have

$$|a_{ij}|^2 = \frac{4 \sin^2(\Delta T/2\hbar)}{\Delta^2} |U_{ij}(0)|^2, \qquad \Delta = E_i - E_j - \hbar\omega \tag{78}$$

This gives the probability for the absorption of a photon with a frequency ω traveling along a particular angle pair in spherical coordinates. This must then be integrated over by the solid angle $d\Omega$ and evaluated.

So far considerable progress has been made. We have a fairly reasonable understanding of how the electromagnetic field interacts with an atom, and have in hand an expression that gives the transition probability for the absorption and emission of a photon by an atom. This expression has been demonstrated to be remarkably accurate in its description of the interaction of light with atomic structure. Additional features may be included to account for the permutation symmetry of various photons that interact with an atom. Explicit consideration may also be given for the probability that the atoms may also emit a photon once in the excited state. These considerations can be found in many textbooks on quantum electrodynamics.

What has been presented here is a semiclassical theory of $U(1)$ quantum electrodynamics. Here the electromagnetic field is treated in a purely classical manner, but where the electromagnetic potential has been normalized to include one photon per some unit volume. Here the absorption and emission of a photon is treated in a purely perturbative manner. Further, the field normalization is done so that each unit volume contains the equivalent of n photons and that the energy is computed accordingly. However, this is not a complete theory, for it is known that the transition probability is proportional to $n + 1$. So the semi-classical theory is only appropriate when the number of photons is comparatively large.

A.   The Physical Basis for Non-Abelian Electrodynamics

An initial study of electrodynamics was the practical art of optics and glass-making. Through the middle ages, optics was a substudy of glassmaking, and was done by artisans who learned through practical experience. The subject reached it first measure of academic importance with Willebrord Snellius (1591–1626). He spent years working on the principles of optics involved with the process of vision; apparently the need for corrective eyewear was a growing market, and somebody had to find a complete understanding of how optics could assist the physician. In his treatise *Di Optrice* he laid down the first law of optical refraction. He recognized that the angle of incidence, with respect to the normal of a material surface, that a light ray hit a medium was related to the angle at which that light ray went through the transparent medium. So the paths of light outside and outside the glass with respect to the normal were related to each other by a constant later called the index of refraction. This ushered in the law of sines. He further went on to derive equations for curved thin lenses, based on this principle that were able to determine the position at which an image would form. This is the elementary lens maker's formula learned in first-year physics. This physics was extended by Newton when he demonstrated that the index of refraction may have a dependency on the color of light. In this manner light could be split by a prism.

The theory of light reached its second step forward with Huygens, who demonstrated that light was a wave that obeyed various diffractive properties [5]. Of course, there later came Faraday and then Maxwell, who brought in the complete theory of classical electromagnetism. The wave aspect of light tended to eclipse the older geometric optical view of light intellectually. However, the art of geometric optics grew into a very refined art. Before the advent of computers, it required dozens or hundreds of human "computers" to complete the calculations required to characterize a particular optical system of lenses. The issue of refractive optics appeared to be in a sort of state of completion and was a matter of "simple calculation" that could be done by a machine.

Reality is not so simple. Suppose that the index of refraction depends on the intensity of the light, or in a modern setting the electromagnetic fields, that pass through it. Suddenly we are confronted with having to revise our notion of the index of refraction; it is not necessarily a constant. Snellius had to compute the paths of rays that passed through a thin lens by considering the geometry in the curvature of a lens. Today nonlinear optics is a study that has to consider the variable index of refraction that was dependent on the field strengths of the optical radiation being transmitted. This has become an important issue in the modern world. Optical fibers that transmit information as pulses of light are developed to transmit shorter pulses so that the date transmission rate can be increased. An optical fiber with a constant index of refraction has serious limitations. The radiation transmitted will reflect off the sides of the fiber, but at various angles. There will then be a spread in the optical pulse as it travels down the optical fiber since various photons will be reflected at slightly different angles. However, an optical fiber that has an index of refraction that is dependent on the field strength will tend to "bunch" these photons into a single stream and thus eliminate this unfortunate problem.

The laws of electromagnetism are based on the theory of gauge fields. The electromagnetic vector potential defines components of a gauge connection 1-form. This gauge connection defines a field strength 2-form:

$$\mathbf{dA} = \mathbf{F} \qquad (79)$$

In general this emerges because the differential operator $\mathbf{d}$ is gauge-covariant when it acts on a section of the bundle, or physically when it acts on a wave function $\mathbf{d} \rightarrow \mathbf{d} + q\mathbf{A}$. The application of this covariant differential operator twice on a function gives

$$(\mathbf{d} + q\mathbf{A}) \wedge (\mathbf{d} + q\mathbf{A})\psi = \mathscr{D} \wedge \mathscr{D}\psi$$
$$= q(\mathbf{dA} + q\mathbf{A} \wedge \mathbf{A})\psi \qquad (80)$$

If the gauge connection is Abelian, then the term $e\mathbf{A} \wedge \mathbf{A}$ vanishes by the antisymmetry of the wedge product. This means that $\mathscr{D}^2\psi = q\mathbf{dA}\psi$. This is an example of an Abelian gauge theory, defined according to that vanishing of commutators between gauge potentials.

In general gauge theories are such that there is more that one particular gauge potential or connection coefficient $\mathbf{A^a}$, where a is an index that spans a Lie algebra, such as $SU(2)$ and $SU(3)$, so that $q\mathbf{A}^a \wedge \mathbf{A}^b$ is in general nonvanishing. The gauge theories for the weak and nuclear interactions are such non-Abelian gauge theories. Physically the occurrence of these antisymmetric terms means that the gauge vector boson, the analog of the photon, carries a charge associated with the field sources. This causes the field lines, analogous to the electric and magnetic field lines, to attract each other. Thus the field lines between two particles, that are themselves sources of the field, tend to clump into a tube-like structure. If the coupling constant, the term analogous to the electric charge, is very large, this tube becomes a very tightly bound structure. In the case of quantum chromodynamics (QCD), mesons consist of two quarks as sources of the field lines in such a flux tube of field lines, and baryons consist of three quarks that sit in a bubble or bag of such self confined field lines.

It is, in general, difficult to obtain real solutions from such field theories. These difficulties have two sources. The first is that in QCD you have three quarks in the bubble, and such 3-body problems are not exactly solvable. This is further compounded by the fact that the virtual quanta are themselves carriers of the various charges and so one essentially has a many-body problem as one computes higher-order perturbative Feynman diagrams. The second is that if the coupling constant is strong then the perturbation terms in the expansion contribute equally to all orders. This means that in general one has to compute an infinite number of such perturbation terms to determine anything about the theory. Fortunately, in the case of QCD a process called *quark antiscreening* implies that at sufficiently high energies the quarks behave more freely as the coupling constant is renormalized to a smaller value and this problem is ameliorated. This does mean that nobody knows precisely how to compute the problem of a proton in free space with no interactions with other particles. Lattice gauge methods have been written as algorithms and run on computers and approximate answers have been garnered.

Electromagnetism is considered to be an Abelian gauge theory. This is most often expressed according to Maxwell's equations. This theory is remarkably successful, but is called into question when one has nonlinear optical and electromagnetic systems. This occurs when electric permitivities are themselves a function of the electric field. So this term, most often treated as a constant, contributes some term that is a function of the electric field to some power greater than one. It is standard to consider these effects as phenomenology associated with atoms within the medium. However, one can view the occurrence of these atoms as effectively changing the electromagnetic vacuum, and so this physics is ultimately electromagnetic. These nonlinear terms then have the appearance as the magnitude of the elements of the 2-form $q\mathbf{A} \wedge \mathbf{A}$.

This suggests that electromagnetism may in fact have a deeper non-Abelian structure. In what follows it is assumed that the $\mathbf{B}^{(3)}$ field exists. It is likely that the $\mathbf{B}^{(3)}$ field exists only as a manifestation of nonlinear optics. This is an aspect of non-Abelian electrodynamics that has been quite under studied. Later, a discussion of squeezed state operators in connection to non-Abelian electrodynamics is mentioned. However, its role in nonlinear optics is an open topic for work.

An illustration of this fact comes from the nonlinear Schrödinger equation. This equation describes an electromagnetic wave in a nonlinear medium, where the dispersive effects of the wave in that medium are compensated for by a refocusing property of that nonlinear medium. The result is that this electromagnetic wave is a soliton. Suppose that we have a Fabry–Perot cavity of infinite extend in the x direction that is pumped with a laser [6,7]. The modes allowed in that cavity can be expanded in a Fourier series as follows:

$$\mathbf{E}(x, y, z, t) = \sum_{m,n} \mathscr{E}(x, t)\phi_{mn}(y, z)e^{-i\omega_0 t} + H.C. \tag{81}$$

The fundamental wave equation to emerge from Maxwell's equation is

$$\left(\partial_x^2 - \frac{1}{c^2}\partial_t^2\right)\mathbf{E}(x,t) = \frac{1}{c^2}\partial_t\mathbf{P}(x,t) \tag{82}$$

If we input the mode expansion into this wave equation we arrive at the wave equation

$$i\partial_t\mathscr{E} = -\frac{c}{2k_0}\partial_x^2\mathscr{E} - \omega_0\mathbf{P}(x,t)$$

$$+ \text{ inhomogenous driving and dissipation terms} \tag{83}$$

We will ignore these inhomogenous terms. The polarization vector is going to have contributions from the linear electric susceptibility and the nonlinear electric susceptibility due to the nonlinear response of the atoms:

$$\mathbf{P} = \chi_l\mathscr{E}(x,t) + \chi_{nl}|\mathscr{E}|^2\mathscr{E} \tag{84}$$

With an appropriate redefinition of constants we arrive at the following wave equation for the propagation of field $\mathscr{E}$:

$$i\partial_t\mathscr{E} = -\frac{c}{2k_0}\partial_x^2\mathscr{E} - \omega_0 n|\mathscr{E}|^2\mathscr{E} \tag{85}$$

The solution to this cubic Schrödinger equation is $\mathscr{E} = \mathscr{E}_0 \operatorname{sech}(\kappa x)e^{i\omega t}$, where $2\kappa = k_0\sqrt{n|\mathscr{E}|^2}$, which is a soliton wave.

It is noted that the derivation of this equation involves the phenomenological concept of the nonlinear response of the atoms. This equation is derived on the basis of the standard Abelian theory of electromagnetism, which is linear, and where the nonlinearity obtains by imposing nonlinear material responses. The physical underpinnings of these nonlinearities are not completely described. This soliton wave corresponds to diphotons, or photon bunches.

It is then advanced that electromagnetism is expanded into a theory with three vector potentials and the conjugate product that determines an additional magnetic field,

$$\mathbf{B}^{(3)} = \frac{ie}{\hbar}\mathbf{A}^{(1)} \times \mathbf{A}^{(2)} \tag{86}$$

where $\mathbf{A}^{(1)}$ is the complex vector potential field and $\mathbf{A}^{(2)} = \mathbf{A}^{(1)*}$ of the electromagnetic field. This additional magnetic field $\mathbf{B}^{(3)}$ has been described through the physics of fermion resonance, and with empirical evidence for this magnetic field as given by the optical conjugate product $\mathbf{A}^{(1)} \times \mathbf{A}^{(2)}$ [8]. This magnetic field may enter into Dirac's theory of the electron so that the interaction of a fermion with this field is

$$E_{\text{int}} = -\frac{e\hbar}{2m}\sigma^{(3)} \cdot \mathbf{B}^{(3)} \tag{87}$$

A complete derivation involves a complete expansion of the Pauli Hamiltonian and the recognition that for the two complexified vector potentials $\mathbf{A}^{(1)}$ and $\mathbf{A}^{(2)}$ that one has the term

$$(\sigma \cdot \mathbf{A})^2 = \mathbf{A} \cdot \mathbf{A}^* + i\sigma \cdot \mathbf{A} \times \mathbf{A}^* \tag{88}$$

This ansatz tends to conform to various data, and, as will be later pointed out, gives predictions of various nonlinear optical effects as well as vortex effects and photon bunching.

This 3-magnetic field has some striking effects. It is easy to see that there are the complex valued electric fields $\mathbf{E}^{(1,2)} = \frac{\partial \mathbf{A}^{(1,2)}}{\partial t} = \omega \mathbf{A}^{(1,2)}$ So we then see that the magnitude of the optical conjugate product is then I/ω^2 for $I = |\mathbf{A} \times \mathbf{A}^*|$ defined as the intensity of electromagnetic radiation or optical beam. An exact expression for this magnetic field is then seen to be

$$\mathbf{B}^{(3)} = \frac{e\mu_0 c}{\hbar}\frac{I}{\omega^2}\mathbf{e}^{(3)} = 5.723 \times 10^{17}\frac{I}{\omega^2}\mathbf{e}^{(3)} \tag{89}$$

where the constants are evaluated with SI unites. This has some rather aggregious consequences. For visible light this effect is quite small. For a beam of 10 W/cm^2 at the visible wavelength $\lambda = 500$ nm the magnetic field is on the order of a nanotesla. However, for a 10-MHz radiofrequency wave this magnetic field is 14.5 MT (megatesla). This apparently is a way of generating rather large magnetic fields without the need of massive electromagnets.

The occurrence of the nonlinear Schrödinger equation is then a fairly generic result. For the $\mathbf{A}^{(1)}$ potential we have the magnetic field that is easily seen to be

$$\mathbf{B}^{(1)} = \nabla \times \mathbf{A}^{(1)} + \frac{ie}{\hbar}(\mathbf{A}^{(2)} + \mathbf{A}^{(3)}) \times \mathbf{A}^{(1)} \tag{90}$$

The last term vanishes since the $\mathbf{A}^{(3)}$ photon is found to be very massive in an examination of this approach to electromagnetism embedded in an extended standard model. These issues will be discussed later. This photon decays away and so the $\mathbf{A}^{(3)}$ potential is very short ranged $\simeq 10^{-17}$ cm and is of no consequence to quantum optics. Let $\nabla \times \mathbf{A}^{(1)} = \mathbf{B}_0^{(1)}$. Now compute Maxwell's equation, where $\mathscr{D} = \nabla + (ie/\hbar)(\mathbf{A}^{(1)} + \mathbf{A}^{(2)})$ is a covariant form of ∇

$$\mathscr{D} \times \mathbf{B}^1 = \nabla \times \mathbf{B}_0^{(1)} + \frac{ie}{\hbar}(\mathbf{A}^{(1)} + \mathbf{A}^{(2)}) \times \mathbf{B}_0^{(1)} \tag{91}$$

where $\nabla \times \mathbf{B}^{(3)} = 0$. Now compute $\mathscr{D} \times \mathscr{D} \times \mathbf{B}^{(1)}$ to find the covariant wave equation:

$$\mathscr{D} \times \mathscr{D} \times \mathbf{B}^{(1)} = \nabla^2 \mathbf{B}_0^{(1)} + \left(\frac{e}{\hbar}\right)^2 (|\mathbf{A}^{(1)}|^2 + |\mathbf{A}^{(2)}|^2) \mathbf{B}_0^{(1)} \tag{92}$$

Now use $|\mathbf{A}^{(1,2)}| = (1/k)|\mathbf{B}^{(1,2)}|$ to find

$$\mathscr{D} \times \mathscr{D} \times \mathbf{B}^{(1)} = \nabla^2 \mathbf{B}_0^{(1)} + 2\left(\frac{e}{k\hbar}\right)^2 |\mathbf{B}_0^{(1)}|^2 \mathbf{B}_0^{(1)} \tag{93}$$

Now $\mathscr{D} \times \mathscr{D} \times \mathbf{B}^{(1)} = (1/c^2)\mathscr{D}^2 \mathbf{E}_0^{(1)}/\partial t^2$, which means that we arrive at the nonlinear equation

$$\nabla^2 \mathbf{B}^{(1)} + 2\left(\frac{e}{k\hbar}\right)^2 |\mathbf{B}^{(1)}|^2 \mathbf{B}^{(1)} = \epsilon\mu \frac{\mathscr{D}^2 \mathbf{E}^{(1)}}{\partial t^2} \tag{94}$$

Now we write the same Fourier of expansion for the electric field and write everything according to the magnetic field intensity $\mathbf{H} = \frac{1}{\mu}\mathbf{B}$, and we find with the case that $(e/\hbar)A_0 \simeq \omega$ the amplitude fixed to the wavelength as is the case for some solitons, for Gaussian packets, we arrive at the same cubic Schrödinger equation:

$$\frac{c}{k}\partial_x^2 \mathscr{H}^{(1)} + 2\left(\frac{e\mu}{k\hbar}\right)^2 \omega |\mathscr{H}^{(1)}|^2 \mathscr{H}^{(1)} = -\frac{i}{c}\frac{\partial \mathscr{H}^{(1)}}{\partial t} \tag{95}$$

The solution to this equation is $A \operatorname{sech}(kx)e^{i\omega t}$, which is a soliton solution. In the case where we have nonlinear optics and the occurrence of the cyclic electromagnetic fields, the Maxwell equations for the propagation of an electromagnetic wave are covariant and then give rise to soliton wave equations.

The difference this derivation has in comparison to the previous derivation of the nonlinear Schrödinger equation is that the nonlinearity is more fundamentally due to the non-Abelian wavefunction rather than from material coefficients. In effect these material coefficients and phenomenology behave as they do because the variable index of refraction is associated with non-Abelian electrodynamics. Ultimately these two views will merge, for the mechanisms on how photons interact with atoms and molecules will give a more complete picture on how non-Abelian electrodynamics participates in these processes. However, at this stage we can see that we obtain nonlinear terms from a non-Abelian electrodynamics that is fundamentally nonlinear. This is in contrast to the phenomenological approach that imposes these nonlinearities onto a fundamentally linear theory of electrodynamics.

B.   The Quantized $U(1)$–$O(3)_b$ Electromagnetic Field

The electromagnetic field is quantized as a set of harmonic oscillators. Maxwell's equations, and the resulting wave equations, are described by partial differential equations that formally have an infinite number of degrees of freedom. Physically this means that the electromagnetic field is described by an infinite number of harmonic oscillators, where one sits at every point in space. The modes of the electromagnetic field are then completely described by this ensemble of harmonic oscillators.

The harmonic oscillator has a long history in physics. Galileo noticed, starting as a youth who watched a chandelier swing in the cathedral at Pisa, that a mass attached to a lightweight string executed swings through a small angle with a period that was independent of the mass. This oscillation was completely understood with Newton's laws by Robert Hooke. The Hamiltonian for this one dimensional system is

$$H = \frac{1}{2}(p^2 + \omega^2 q^2) \tag{96}$$

where p and q are the momentum and position variables of the system. Quantum-mechanically, these variables are replaced by quantum operators $p \rightarrow \hat{p}$ and $q \rightarrow \hat{q}$. These variables are combined to form ladder operators known as the "lowering" and "raising operators," more often called *absorption* or *annihilation* and *emission* or *creation operators*:

$$a = \frac{1}{\sqrt{2\hbar\omega}}(\omega\hat{q} + i\hat{p}), \qquad a^\dagger = \frac{1}{\sqrt{2\hbar\omega}}(\omega\hat{q} - i\hat{p}) \tag{97}$$

These operators allow for the description of the quantum harmonic oscillator that is very parsimonious. The quantum harmonic oscillator has evenly spaced eigenstates, and the state of the system may be changed according to

$$a|n\rangle = \sqrt{n}|n-1\rangle, \qquad a^\dagger|n\rangle = \sqrt{n+1}|n+1\rangle \tag{98}$$

It is easy to see that the number operator is written as $N = a^\dagger a$ that are diagonal with respect to the eigenvalues $N|n\rangle = n|n\rangle$ and also define the energy levels for the system since the Hamiltonian is

$$H = \hbar\omega\left(N + \frac{1}{2}\right) = \hbar\omega\left(a^\dagger a + \frac{1}{2}\right) = \hbar\omega\left(n + \frac{1}{2}\right) \tag{99}$$

A curious aspect of this is that the $n = 0$ state is one that has a nonzero energy $\frac{1}{2}\hbar\omega$.

Now consider an ensemble of harmonic oscillators in three dimensions. Each of these harmonic oscillators has a different frequency $\omega = |\mathbf{k}|c$, their own Hamiltonian and raising and lowering operators

$$a_{\mathbf{k}} = \frac{1}{\sqrt{2\hbar\omega}}(\omega\hat{q}_{\mathbf{k}} + i\hat{p}_{\mathbf{k}}), \qquad a_{\mathbf{k}}^{\dagger} = \frac{1}{\sqrt{2\hbar\omega}}(\omega\hat{q}_{\mathbf{k}} - i\hat{p}_{\mathbf{k}}) \qquad (100)$$

We then have a description of an infinite number of harmonic oscillators with every possible mode at every point in space. The electromagnetic field is quantized in a cavity with a volume V by defining annihilation and creation operators by redefining these raising and lower operators as

$$a_{\mathbf{k}} \rightarrow \sqrt{\frac{\hbar}{4\epsilon_0 V}}a_{\mathbf{k}}, \qquad a_{\mathbf{k}}^{\dagger} \rightarrow \sqrt{\frac{\hbar}{4\epsilon_0 V}}a_{\mathbf{k}}^{\dagger} \qquad (101)$$

This allows for the expansion of the vector potential into spacial eigenmodes

$$\mathbf{A} = i\sum_k \sqrt{\frac{\hbar}{2\omega\epsilon V}}\mathbf{e}(a_k^{\dagger}e^{-i\mathbf{k}\cdot\mathbf{r}} - a_k e^{i\mathbf{k}\cdot\mathbf{r}}) \qquad (102)$$

Here ϵ is the electric permittivity and ω is the frequency of the eigenmodes. The Abelian magnetic field is then defined by

$$\mathbf{B} = \sum_k \sqrt{\frac{\hbar}{2\omega\epsilon V}}\mathbf{k} \times \mathbf{e}(a_k^{\dagger}e^{-i\mathbf{k}\cdot\mathbf{r}} + a_k e^{i\mathbf{k}\cdot\mathbf{r}}) \qquad (103)$$

and the electric field is defined by

$$\mathbf{E} = \sum_k \sqrt{\frac{\hbar}{2\omega\epsilon V}}\mathbf{e}\omega(a_k^{\dagger}e^{-i\mathbf{k}\cdot\mathbf{r}} + a_k e^{i\mathbf{k}\cdot\mathbf{r}}) \qquad (104)$$

This is the Abelian theory of quantum electrodynamics as a free field uncoupled to charged particles and fermions.

Since there is a non-Abelian nature to this theory, we return to the nonrelativistic equation that describes the interaction of a fermion with the electromagnetic field. The Pauli Hamiltonian is modified with the addition of a $\mathbf{B}^{(3)}$ interaction term [9]

$$H_{B^{(3)}} = H + \frac{e^2}{2m}(\sigma \cdot \mathbf{A})(\sigma \cdot \mathbf{A}^*) \qquad (105)$$

which may be rewritten according to the algebra of Pauli matrics

$$H_{B^{(3)}} = \frac{e^2}{2m}(\mathbf{A} \cdot \mathbf{A} + i\boldsymbol{\sigma} \cdot \mathbf{A} \times \mathbf{A}^*) \tag{106}$$

If we write this interaction Hamiltonian according to creation and annihilation operators, we find that this term can be written as

$$H_{B^3_{\text{int}}} = \frac{e^2}{4m\hbar\epsilon_0 V}\sum_k\left(\omega_k^{-1}Ia_k^\dagger a_k + \sum_q \omega_q^{-1}\sigma^{(3)}(a_q^\dagger a_{k-q} + a_q a_{k-q}^\dagger)\right) \tag{107}$$

This interaction Hamiltonian describes the exchange of a photon that results in the change of the spin of the fermion. This process is equivalent to the absorption of a photon in the atomic state transition $i \rightarrow j$ and the absorption of a photon in the atomic transition $j \rightarrow i$.

Normally one does not worry about the free Hamiltonian term $\frac{1}{2}B^2$, but in the case of the $\mathbf{B}^{(3)}$ field, we cannot afford this luxury. This term is written according to the field operators as

$$H_{B^{(3)}} = \frac{e}{2\omega_q\epsilon_0 V}\sum_{k,k',q}(a_{k+q}^\dagger a_k a_{k'-q}^\dagger a_{k'}) \tag{108}$$

This term is crucial to the concept of non-Abelian electrodynamics. Essentially, it describes the interaction between four photons. It describes the absorption of photons with the modes $k + q$ and $k' - q$ and the emission of photons with the modes k and k'. Physically this is a process where two photons mutually interact and exchange momenta. A classical analog of this process is to think of two photons as possessing $\mathbf{B}^{(3)}$ fields that are mutually coupled. This is one aspect of non-Abelian electrodynamics that is different from standard electrodynamics. An analogous situation occurs with gluons in quantum chromodynamics. Here gluballs can exist that are self-bound states of gluons that are mutually interacting. The non-Abelian electrodynamic effect is far simpler since there is no issue of confinement, but the situation is one where photons can interact. This effect is what is a part of the $|\mathscr{H}|^2\mathscr{H}$ term that counters the dispersive effects of an electromagnetic wave as governed by the nonlinear or cubic Schrödinger equation. This is a form of self-focusing or photon bunching that results from this form of mutual interaction between photons.

This is the nature of non-Abelian electrodynamics in a nonrelativistic regime. It leads to various predictions that appear to obtain for electromagnetic fields in media. As yet there have not been the appearance of these types of effects for fields in a vacuum. Just why it is that nonlinear optics appears to be associated with the $B^{(3)}$ field is unclear. A medium acts as a renormalized vacuum, and it is possible that the appearance of atoms with charge separations acts to cause the appearance of the $B^{(3)}$ field. It appears that the $B^{(3)}$ field vanishes in vacua, but it may manifest itself in various media. On the other hand this could just be an accident of nature. In this case the $B^{(3)}$ field is simply a mathematical manifestation that permits the calculation of various nonlinear effects.

C. Relativistic $O(3)_b$ QED

Non-Abelian electrodynamics has been presented in considerable detail in a nonrelativistic setting. However, all gauge fields exist in spacetime and thus exhibits Poincaré transformation. In flat spacetime these transformations are global symmetries that act to transform the electric and magnetic components of a gauge field into each other. The same is the case for non-Abelian electrodynamics. Further, the electromagnetic vector potential is written according to absorption and emission operators that act on element of a Fock space of states. It is then reasonable to require that the theory be treated in a manifestly Lorentz covariant manner.

The theory is defined by the Lagrangian density

$$\mathscr{L} = -\frac{1}{4}F^a{}_{\mu\nu}F^{a\mu\nu} \tag{109}$$

with the stress–energy tensor components defined according to the gauge-covariant derivative

$$F^a{}_{\mu\nu} = \partial_\nu A^a{}_\mu - \partial_\mu A^a{}_\nu + ig\epsilon^{abc}[A^b{}_\nu, A^c{}_\mu] \tag{110}$$

where the spacial components of the 4-vector potential are Hermition $A^{a\dagger}{}_i = A^a{}_i$, $i \in \{1, 2, 3\}$, and the temporal parts are anti-Hermitian $A^{a\dagger}{}_0 = -A^a{}_0$. Here g is the coupling constant for the gauge theory. The upper Latin index refers to the internal degrees associated with the gauge theory. The variational calculus with this Lagrangian density leads to the field equation

$$\partial_\mu F^{a\mu\nu} + ig\epsilon^{abc}A^b{}_\mu F^{c\mu\nu} = 0 \tag{111}$$

From the field stress tensor components, we may write the electric and magnetic field components as

$$E^a{}_i = F^a{}_{i0} = -\dot{A}^a_i - \nabla_i A^a{}_0 + ig\epsilon^{abc}A^b{}_0 A^c{}_i \tag{112a}$$

$$\epsilon^k_{ij}B^a{}_k = \nabla_i A^a{}_j - \nabla_j A^a{}_i + ig\epsilon^{abc}A^b{}_i A^c{}_j \tag{112b}$$

The components of the vector potential are then expanded in a Fourier series of modes with creation and annihilation operators that act on the Fock space of states. If this is done according to a box normalization, in a volume V, with periodic boundary conditions, we have

$$A^a{}_i(\mathbf{r}, t) = \sum_k \frac{1}{\sqrt{2\omega V}} \left(e_i a^a(k) e^{i\mathbf{k}\cdot\mathbf{r}} + e_i a^{a\dagger}(k) e^{-i\mathbf{k}\cdot\mathbf{r}} \right) \qquad (113)$$

Here we are considering only the transverse components of the vector potential. With these vector potential components written according to these operators, the electric and magnetic fields within $O(3)_b$ electrodynamics are then

$$E^a{}_i = \sum_k \frac{1}{\sqrt{2\omega V}} \left(\frac{|k|}{c} e_i a^a(k) e^{i\mathbf{k}\cdot\mathbf{r}} + \frac{|k|}{c} e_i a^{a\dagger(k)} e^{-i\mathbf{k}\cdot\mathbf{r}} \right)$$

$$\epsilon^k_{ij} B^a{}_k = \sum_k \frac{1}{\sqrt{2\omega V}} (k_{[j} e_{i]} a^a(k) e^{i\mathbf{k}\cdot\mathbf{r}} + k_{[j} e_{i]} a^{a\dagger}(k) e^{-i\mathbf{k}\cdot\mathbf{r}}) \qquad (114)$$

$$+ ig\epsilon^{abc} \sum_{kk'} e_{[j} e_{i]} (a^b(k) e^{i\mathbf{k}\cdot\mathbf{r}} + a^{b\dagger}(k) e^{-i\mathbf{k}\cdot\mathbf{r}})(a^c(k') e^{i\mathbf{k}'\cdot\mathbf{r}} + a^{c\dagger}(k)' e^{-i\mathbf{k}'\cdot\mathbf{r}})$$

It is then apparent that the Hamiltonian for this non-Abelian field theory is going to contain quartic terms in addition to the quadratic terms seen in abelian field theory, such as $U(1)$ electromagnetism.

If we consider non-Abelian electromagnetism, we have a situation where the vector potential component $A^3{}_i$ vanish and where $A^{(1)}{}_i = A^{(2)*}{}_i$. The annulment of the components $A^3{}_i$ has been studied in the context of the unification of non-Abelian electromagnetism and weak interactions, where on the physical vacuum of the broken symmetry $SU(2) \times SU(2)$ the vector boson corresponding to $A^3{}_i$ is very massive and vanishes on low-energy scales. This means that the 3-component of the magnetic field is then

$$\mathbf{B}^3 = i\frac{e}{\hbar} \mathbf{A}^1 \times \mathbf{A}^2 \qquad (115)$$

It is apparent that for $A^3{}_i = 0$, the electric field component does not contain a product of potential terms. In general the vanishing of this term occurs if there are no longitudinal electric field components. Within the framework of most quantum electrodynamic, or quantum optical, calculations this is often the case. The $\mathbf{B}^{(3)}$ field then is a Fourier sum over modes with operators $a^\dagger_{k-q} a_q$. The $\mathbf{B}^{(3)}$ field is then directed orthogonal to the plane defined by $\mathbf{A}^1$ and $\mathbf{A}^2$. The four-dimensional dual to this term is defined on a time-like surface that has the interpretation, under dyad–vector duality in three dimensions as, as an electric field or $\mathbf{E}^{(3)}$. The vanishing of the $\mathbf{E}^{(3)}$ can then be seen by the nonexistence of the raising and lowering operators $a^3, a^{3\dagger}$, where the $\mathbf{B}^{(3)}$ exists solely due to the occurrence of raising and lowering operators that $\mathbf{A}^{(1)}$ and $\mathbf{A}^{(2)}$ are expanded according to. This represents a breakdown of duality in four dimensions and the requirement that $\mathbf{B}$ be a longitudinal field.

This non-Abelian gauge theory satisfies the usual transformation properties. If $\mathcal{M}$ is the base manifold in four dimensions, then the gauge theory is determined by an internal set of symmetries described by a principal bundle. Let U_α, where $\alpha = 1, 2, \ldots, n$, be an atlas of charts on the $\mathcal{M}$. The transitions from one chart to another is given by $g_{\alpha\beta} : U_\beta \rightarrow U_\alpha$, where these determine the transition functions between sections on the principal bundle. The transform between one section to another is given by

$$s_\alpha = g_{\alpha\beta}s_\beta = e^{iX_{\alpha\beta}}s_\beta \tag{116}$$

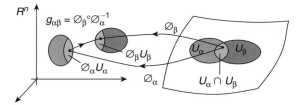

Figure 3.   Transition functions between two charts on a manifold.

From this point we will suppress the chart indices to indicate sections and use the notation s, s' for the two charts with $gs = s'$. Now let the differential operator d act on s'

$$ds' = (gds + sdg) \tag{117}$$

Now define $g^{-1}dg$ as a connection coefficient A on the section s:

$$ds' = g(ds + ig^{-1}dg)s \tag{118}$$

Now consider the action of g on $(d + A)s$ which equals $(d + A')s'$:

$$\begin{aligned}(d + A')s' &= g(d + A)s \\ &= g(d + A)g^{-1}gs = (d + gAg^{-1} + gdg^{-1})s' \end{aligned} \tag{119a}$$

This is a fundamental definition for how a gauge connection transforms:

$$A' = gAg^{-1} + gdg^{-1} \tag{119b}$$

Now we consider the group element g to be defined by algebraic generators so that $g = e^{iX}$. Further consider the transformation to be sufficiently small so that $e^{iX} \simeq 1 + iX$:

$$A's' = ((1 + iX)A(1 - iX) - idX)s' = (A + i[X, A] - idX)s' \quad (120)$$

If we are working with local gauge transformations where A is flat, we can work with the pure gauge term $(dg)d^{-1} = idX$ as the gauge connection.

Now to get the fields from this definition, we are given the fact that the fields are defined to be under a gauge transformation

$$dA' = d(gAg^{-1} + (dg)g^{-1}) \quad (121)$$

From this we find that

$$dA' = g(dA + A \wedge A)g^{-1} \quad (122)$$

which means that the fields transform homogenously under local gauge transformations. Just as the chart indices have been suppressed, so have the indices for the internal symmetry space.

Now for non-Abelian electromagnetic field theory, we have the 3-Lie index component of the field, and for the magnetic field $\mathbf{B}^{(3)}$, it equals

$$\mathbf{B}^{[3]} = i\mathbf{A} \times \mathbf{A}^* \quad (123)$$

where this is a component that emerges from the $A \wedge A$ term. We are working here with $\hbar = c = 1$. Then under local gauge transformations we will have

$$\mathbf{B}^{(3)\prime} = ig(\mathbf{A} \times \mathbf{A}^*)g^{-1} \quad (124)$$

where g is the group element for the $O(3)_b$ theory. Then one can go on and write $g \simeq 1 + iX$ and find that

$$\mathbf{B}^{(3)\prime} = i(1 + iX)(\mathbf{A} \times \mathbf{A}^*)(1 - iX) = i\mathbf{A} \times \mathbf{A}^* - [X, [\mathbf{A}, \mathbf{A}^*]] + O(X^2) \quad (125)$$

This can be written according to Lie derivative, and if X is a generator for a global gauge transformation, then this double commutator vanishes. We are then left with

$$\mathbf{B}^{(3)\prime} = \mathbf{B}^{(3)} - iL_X\mathbf{B}^{(3)} \quad (126)$$

where the last term is the Lie derivative of $\mathbf{B}^{(3)}$ with respect to the variable X, here parameterized along a path.

In the case of quantum field theory the section determines the Hilbert space of states under a certain gauge. This choice of gauge then determines the unitary representation of the Hilbert space. We may then replace the section with the fermion field ψ, which acts on the Fock space of states. It is then apparent that a gauge transformation $A^a{}_\mu \to A^a{}_\mu + \delta A^a{}_\mu$ is associated with a unitary transform of the fermion field $\psi \to \psi + \delta\psi$. The unitary transformation of the fermion may be written according to $\psi' = U\psi$ where the unitary matrix is represented as the line integral along a path

$$U = \mathscr{T} e^{-ig \int^\Gamma A_\mu dx^\mu} \tag{127}$$

where $\mathscr{T}$ is the time-ordering operator that arranges fields in a product in a time-ordered sequence. The application of the differential operator $\mathbf{d}$ on the unitary matrix gives

$$\mathbf{d}U = -ig(A_\mu - A'_\mu)\mathbf{d}x^\mu U \tag{128}$$

which leads to the result

$$\frac{i}{g} U^\dagger \mathbf{d}U - U^\dagger(A_\mu - A'_\mu)\mathbf{d}x^\mu U = 0 \tag{129}$$

This demonstrates the association between the unitary transformation of the fermion field and the gauge theory.

More work is required to couple the gauge theory to the fermion. We have the gauge field determined by its Lagrangian density, and the fermion field determined by the Dirac Lagrangian density

$$\mathscr{L}_D = -\bar{\psi}(\gamma^\mu \partial_\mu + m)\psi \tag{130}$$

However, these two Lagrangian densities do not couple the two fields together. This requires that the free-field equation for the gauge field becomes

$$\partial_\mu F^{a\mu\nu} + ig\epsilon^{abc} A^b{}_\mu F^{c\mu\nu} = j^\nu \tag{131}$$

Since this field equation is obtained by the Euler–Lagrange equation the inhomogenous term is the result of

$$j^\nu = \frac{\partial\mathscr{L}}{\partial A_\nu} \tag{132}$$

this implies the addition of an interaction Lagrangian density $\mathscr{L}_i = j^\nu A_\nu$. The current term is then determined by the Dirac field and is $j^\nu = \bar{\psi}\gamma^\nu\psi$. The subject of mass remormalization also requires that an additional interaction term be included: $\bar{\psi}\gamma^{\nu}\psi\delta m$, where δm is the difference between the physical mass and the bare mass [3].

The total Lagrangian $\mathcal{L} = \mathcal{L}_G + \mathcal{L}_D + \mathcal{L}_i$ then involves the interaction between fermions and the gauge field. The Dirac field will be generically considered to be the electron and the gauge theory will be considered to be the non-Abelian electromagnetic field. The theory then describes the interaction between electrons and photons. A gauge theory involves the conveyance of momentum form one particle (electron) to another by the virtual creation and destruction of a vector boson (photon) that couples to the two electrons. The process can be diagrammatically represented as

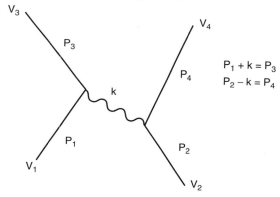

Figure 4. Feynman tree diagram for electron-electron scattering.

The process $p_1 + p_2 \rightarrow p_3 + p_4$ then involves the conservation of momentum, for there is no creation of any averaged momentum from the virtual quantum fluctuation. This process can be examined within the Coulomb gauge $\nabla \cdot \mathbf{A} = 0$. The field equation is then

$$\nabla \cdot \mathbf{E} = -\nabla^2 A_0 = ie\bar{\psi}\gamma^0\psi = e\rho \tag{133}$$

which has the solution

$$A_0(\mathbf{r}, t) = e \int d^3 r \frac{\rho(\mathbf{r}', t)}{4\pi|\mathbf{r} - \mathbf{r}'|} \tag{134}$$

The amplitude for this simple scattering process consists of the electromagnetic Hamiltonian and the interaction process. These two terms produce the amplitude

$$\frac{(-i)^2 e^2}{2} \int d^4x \, d^4x' \, \mathcal{T} j_\mu A^\mu j_\nu A^\nu = \frac{(-i)^2 e^2}{2} \int d^4x \, d^4x' j_\mu G^{\mu\nu}(x - x') j_\nu \quad (135)$$

where $G^{\mu\nu}(x - x')$ is the propagator of the field that satisfies

$$\mathcal{T} A^\mu A^\nu G^{\mu\nu}(x - x') = \frac{-i\delta(\mathbf{r} - \mathbf{r}')}{4\pi|\mathbf{r} - \mathbf{r}'|} \quad (136)$$

For the purely transverse field, the spacial components of the propagator are

$$G^{ij}(\mathbf{r} - \mathbf{r}') = \frac{-i}{4\pi^2} \int \frac{d^4k}{k^2 - i\epsilon} \left(\delta^{ij} - \frac{k^i k^j}{\mathbf{k}^2} \right) e^{i\mathbf{k}(\mathbf{r} - \mathbf{r}')} \quad (137)$$

where $k^2 = \mathbf{k}^2 - k_0^2$. This is seen to be the Fourier transform of the propagator in momentum space. The temporal components are then seen to be

$$\frac{1}{8\pi^3} \int \frac{e^{i\mathbf{k} \cdot \mathbf{r}}}{\mathbf{k}^2} = \frac{1}{r} \quad (138)$$

The amplitudes for the process are then evaluated on the initial and final states of the electrons. This then results in the matrix elements

$$\frac{(-i)^2 e^2}{2} \int d^4x \, d^4x' \langle p_2 | j_\mu | p_1 \rangle G^{\mu\nu}(x - x') \langle p_4 | j_\nu | p_3 \rangle \quad (139)$$

for the amplitudes. The amplitudes $\langle p_2 | j_\mu | p_1 \rangle$ and $\langle p_4 | j_\nu | p_3 \rangle$ are then represented as plane waves

$$\langle p_2 | j_\mu | p_1 \rangle = e^{i(p_1 - p_2)r} X_\mu \quad (140a)$$

$$\langle p_4 | j_\mu | p_3 \rangle = e^{i(p_3 - p_4)r} Y_\mu \quad (140b)$$

where X_μ and Y_μ are independent of the position coordinates. By momentum conservation we demand that $k_\mu = p_{1\mu} - p_{2\mu} = p_{4\mu} - p_{3\mu}$. The propagator acts on these matrix elements to give the amplitude:

$$\frac{(-i)^2 e^2}{2} \int d^4x \, d^4x' \frac{-i}{k^2} X^\mu Y_\mu \quad (141)$$

Finally, this expression can be evaluated for many possible gauges according to

$$\frac{(-i)^2 e^2}{2} \int d^4x \, d^4x' \frac{-i}{k^2} \left(X_i Y_i - \beta X_0 Y_0 \right) e^{i\mathbf{k} \cdot r} \quad (142)$$

where for $\beta = 0$ this is evaluated in the Feynman gauge, and for $\beta = -1$ this is evaluated in the Landau gauge.

This example, within $U(1)$ electrodynamics can then be seen in the light of non-Abelian electrodynamics. This may simply be seen by the replacement $A_\mu \rightarrow t^a A^a{}_\mu$, where t^a is a structure constant that obeys $[t^a, t^b] = 2\epsilon^{abc} t^c$. The time ordered product is written as

$$\mathscr{T} j_\mu A^\mu j_\nu A^\nu = \frac{1}{2} \mathscr{T}(\{t^a, t^b\} + [t^a, t^b]) j_\mu A^{a\mu} j_\nu A^{b\nu} \tag{143}$$

where the product of the structure constants is written according to its symmetric and antisymmetric parts. The physical requirement that $A^3{}_\mu = 0$ is then imposed. From this the symmetric part of the time-ordered product yields the same result as found in the $U(1)$ case. The antisymmetric part is then easily seen to be

$$\mathscr{T} j_\mu A^\mu j_\nu A^\nu = \frac{1}{2} \mathscr{T}[t^a, t^b] j_\mu A^{a\mu} j_\nu A^{b\nu} = 0 \tag{144}$$

This means that on the tree level there are no contributions from the $\mathbf{B}^{(3)}$ field.

In order to compute an amplitude contribution from the $\mathbf{B}^{(3)}$ field, a process that is second-order must be considered. This involves a loop diagram of the form

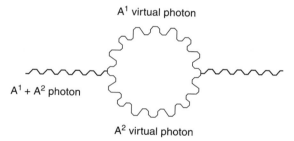

A^1 virtual photon

A^1 + A^2 photon

A^2 virtual photon

Figure 5.   Virtual photon loop correlated with $\mathbf{B}^{(3)}$ field fluctuation.

The propagator then assumes the form

$$G_{\mu\nu}(\mathbf{r} - \mathbf{r}') = \alpha^2 \mathscr{T} t^a t^b t^c t^d t^e t^f A^a{}_\mu(\mathbf{r})$$

$$\times \int \frac{d^4 k'}{(2\pi)^2} (A^{b\rho}(k') A^c{}_\sigma(k')) \int \frac{d^4 k''}{(2\pi)^2} (A^{e\rho}(k'') A^{f\sigma}(k'')) A^f{}_\nu \tag{145}$$

where the integrations exists since the vertices tied to the loop do not constrain momentum conservation. The four fields in the momentum integrals under the action of the antisymmetric portion of the structure constants contribute to

$$\int \frac{d^4 k'}{(2\pi)^2} B^3{}_\rho(k') \int \frac{d^4 k''}{(2\pi)^2} B^{3\rho}(k'') \tag{146}$$

This term is then a sum over all possible fluctuations of the $\mathbf{B}^{(3)}$ that couple to the virtual photon coupled to the electrons. This means that the propagator is of the form

$$G_{\mu\nu}(\mathbf{r} - \mathbf{r}') = \alpha^2 \mathscr{T} A^a{}_\mu(\mathbf{r}) \left(\int \frac{d^4 k'}{(2\pi)^2} B^3{}_\rho(k') \int \frac{d^4 k''}{(2\pi)^2} B^{3\rho}(k'') \right) A^f{}_\nu \tag{147}$$

This then contributes the following amplitude:

$$\langle | \rangle_{B^3} = \frac{i}{k^2} X^\mu Y_\mu \int \frac{d^4 k'}{(2\pi)^2} \int \frac{d^4 k''}{(2\pi)^2} \left(e^{i(\mathbf{k}' - \mathbf{k}'') \cdot \mathbf{r}} \right.$$
$$\left. \times \frac{1}{k^2} \left(\mathbf{X} \cdot \mathbf{Y} - \frac{(\mathbf{k}' \cdot \mathbf{X})(\mathbf{k}'' \cdot \mathbf{Y})}{|\mathbf{k}|^2} \right) - \frac{X_0 Y_0}{|\mathbf{k}|^2} \right) \tag{148}$$

Here $|k|$ is the magnitude of the four vector, and $|\mathbf{k}|$ is the magnitude of the spacial part of the 4-vector k^μ. The integrals in this amplitude suffer from the usual ultraviolet divergence that can be removed through regularization techniques.

This is an introduction to the sort of process that may occur in $O(3)_b$ electrodynamics. In effect, the $\mathbf{B}^{(3)}$ field produces quantum vortices that interact with electrons, as well as other charged particles, where these vortices are quantized states and exist as fluctuations in the QED vacuum. As mentioned earlier the dual of the $\mathbf{B}^{(3)}$ field does not exist as an electric field. These quantum fluctuations are easily seen to be associated with the $\mathbf{E}^{(1)}$ and $\mathbf{E}^{(2)}$ fields:

$$\delta \mathbf{B}^3 = \frac{ie}{\hbar} (\delta \mathbf{A}^1 \times \mathbf{A}^2 + \mathbf{A}^1 \times \delta \mathbf{A}^2)$$
$$= \frac{ie}{\hbar} \frac{1}{\omega^2} (\delta \mathbf{E}^1 \times \mathbf{E}^2 + \mathbf{E}^1 \times \delta \mathbf{E}^2) \tag{149}$$

This indicates a number of things. The first is that the quantum fluctuations of the $\mathbf{B}^{(3)}$ field are accompanied by fluctuations in the standard electric field. Further, the ultraviolet divergence of the above integral is probably unimportant due to the $\frac{1}{\omega^2}$ relationship with the fluctuation. This tends to imply an infrared divergence; however, the analysis with cavity QED indicates that the statistical occurrence of states is such that the divergence is damped. Infrared divergences are known to be of little trouble because of their statistical occurrence.

This approach to QED also suggests that methods for renormalization are applicable. If the quantum fluctuation of the electric field is associated in part with fluctuations in the $\mathbf{B}^{(3)}$ field, then the divergences that occur at the ultraviolet regime can possibly be absorbed into fluctuations with the $\mathbf{B}^{(3)}$ field. Physically these are damped out by the $\frac{1}{\omega^2}$ term in the Feynman path integral. This potentially leads to an additional physical understanding of the disappearance of divergences that occur at the high-frequency domain of QED. This would be possible if the divergences in $U(1)$ electrodynamic processes, which exist as a subset of $O(3)_b$ electrodynamic processes, can be absorbed into integrals that involve photon loop processes associated with fluctuations in the $\mathbf{B}^{(3)}$ field. These fluctuations appear to quench ultraviolet divergences by its $\frac{1}{\omega^2}$ behavior, and it may quench divergences for all processes if these divergences can be absorbed into $\mathbf{B}^{(3)}$ fluctuations.

At high energies it is reasonable to think that the electroweak theory is $SU(2) \times SU(2)$. The current $SU(2) \times U(1)$ theory is renormalizable since the vector boson propagator is "mixed" with the $U(1)$ field that is renormalizable. With an electroweak theory extended to include non-Abelain electrodynamics essentially the same will occur where the unphysical term $m\sigma^a A^{a\mu}\partial_\mu \zeta$, for ζ an unphysical field that oscillates around the Higgs minimum, is canceled by the following gauge fixing Lagrangian density:

$$\mathscr{L}_{gf} = \frac{1}{2\zeta}(T^a \partial_\mu A^{a\mu} + \xi m \zeta)^2 \tag{150}$$

Here for $\xi = 1$ we have the Feynman gauge, and $\xi = 0$ is the Landau gauge. This gauge fixing term will enter into the massive boson propagators for the $\mathbf{A}^{(3)}$ field. The propagator will be of the form

$$-\frac{i}{p^2 - m^2 + i\epsilon}\left(\delta_{\mu\nu} - \frac{\xi p_\mu p_\nu}{p^2}\right) \tag{151}$$

The existence of this propagator will be the largest addition to the physics of electroweak interactions when electromagnetism is nonAbelian. Further discussion on the subject of $SU(2) \times SU(2)$ electroweak theory is given by the authors in [4]. Estimates on the mass of this boson are around four times the mass of the Z_0 boson and should be observable with the CERN Large Hadron Collider.

D.   Renormalization of $O(3)_b$ QED

Quantum electrodynamics involves the interaction of electrons, or other charged particles, and photons, where the interaction between two electrons involves the exchange of a virtual photon. Based upon the equation for a propagator that the interaction between electrons and photons means that the potential function may be written as [5]

$$A = \frac{\bar{v}_3 \gamma^\mu v_1 \bar{v}_4 \gamma_\mu v_2}{q^2} \tag{152}$$

Here v_i are the Dirac spinors for the electrons. This leads to the expectation that the potential in the Coulomb case is $\Phi \simeq e^2/r$. The issue of renormalization is apparent in that the potential and propagator is divergent in the limit that the distance between the electrons approaches zero. Further, as this distance decreases, then, by the Heisenberg uncertainty principle, $\Delta p \, \Delta x \geq \hbar$ means that the momentum exchanged by the electrons due to fluctuations on that small scale becomes divergent. Carried further, this means that the vacuum is filled with virtual quanta that have enormously high momentum fluctuations. When these virtual quanta couple to systems they contribute divergences in the limit their wavelengths approach zero: the ultraviolet divergence.

Three types of processes are divergent as a result of this coupling to virtual quanta: the self-energy of the electron, vacuum polarizations, and vertex functions.

With $O(3)_b$ QED the major difference emerges from the effective photon bunching or interactions that can result in a photon loop, composed of an $A^{(1)}$ photon and an $A^{(2)}$ photon. This loop will be associated with a quanta of $B^{(3)}$ field. Equation (149) illustrates how this fluctuation in the $A^{(1)}$ and $A(2)$ potentials are associated with this magnetic fluctuation. The other renormalization techniques in $U(1)$ QED still apply, and are demonstrated below, and the renormalization of divergences associated with the $B^{(3)}$ magnetic fluctuation is also illustrated.

We will discuss at some length the interaction of a free electron with the vacuum, for this is similar to the renormalization problem presented by $O(3)_b$ electrodynamics. An electron interacts with the vacuum according to the Dirac equation

$$(\gamma^\mu(\partial_\mu - ieA_\mu) - m)\psi = 0 \tag{153}$$

Even if there is no electromagnetic field present, the vector potential exhibits fluctuations $A_\mu = \langle A_\mu \rangle + \delta A_\mu$, so that even if there is only the vacuum, physics still involves this fluctuation. This is also seen in the zero-point energy of the harmonic oscillator expansion of the fields. So an electron will interact with virtual photons. If we represent all of these interactions as a blob coupled to the path of an electron, this blob may be expanded into a sum of diagrams where the electron interacts with photons. Each term is an order expansion and contributes a term on the order of $\alpha = e^2/\hbar c$. A single loop contributes the integral

$$\simeq \int_{-\infty}^{\infty} \frac{\gamma \cdot p - m}{p + k} \tag{154}$$

which has an ultraviolet divergence as $k \to \infty$ for $\int_0^{\infty} dk\, k$. The standard approach amounts to imposing a cut off in the integral Λ so the integration is

$$\int_0^{\infty} dk\, k \to \int_0^{\Lambda} dk\, k \tag{155}$$

so that for an electron of mass m this defines a mass counter term

$$\delta m = \frac{3\alpha m}{2\pi} \log\left(\frac{\Lambda}{m}\right) \tag{156}$$

Then, given the bare mass of the electron as m_0, we have the mass of the electron as $m = m_0 - \delta m$. By the Dirac equation this also contributes a counter term into the Lagrangian $\delta m \bar{\psi} \psi$.

The counterterm is computed by performing a perturbation expansion of Green's function, or propagator for the free electron. The entire process is represented by $\Gamma^{(n)}$, which in general is determined by a time ordered product of fields

$$\Gamma^{(n)}(p_1, p_2, \ldots, p_{n-1}) = \int \prod_i dx_i e^{ip_i x^i} \langle 0 | \mathcal{T} \phi_1 \phi_2 \cdots \phi_{n-1} | 0 \rangle \tag{157}$$

and generally describes processes of the type illustrated in Fig. 6.

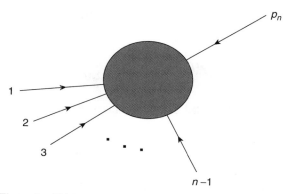

Figure 6. "Blob" diagram for a time ordered product of fields.

This process may also be computed from the path integral

$$Z[J(x)] = \int \mathcal{D}\{\phi\} e^{iS + J\phi} \tag{158}$$

by the functional derivative the path integral according to the test source function $J(x)$:

$$\Gamma^{(n)} = Z^{-1}[J(x)] \frac{\delta^n}{\delta J(x)^n} Z[J(x)]|_{J=0} \tag{159}$$

The propagator $\Gamma^{(2)}$ for the free electron is approximated by loops that are given by the function $-i\Sigma(k^2)$ connected by electron propagators of the form $i/(k^2 - m^2)$. So the propagator that computes the free electron with the mass counter term is then given by

$$\frac{i}{k^2 - m^2 - i\epsilon} = \frac{i}{k^2 - m^2} + \frac{i}{k^2 - m^2}(-i\Sigma(k^2)) \frac{i}{k^2 - m^2}$$
$$+ \frac{i}{k^2 - m^2}(-i\Sigma(k^2)) \frac{i}{k^2 - m^2}(-i\Sigma(k^2)) \frac{i}{k^2 - m^2} + \cdots \tag{160}$$

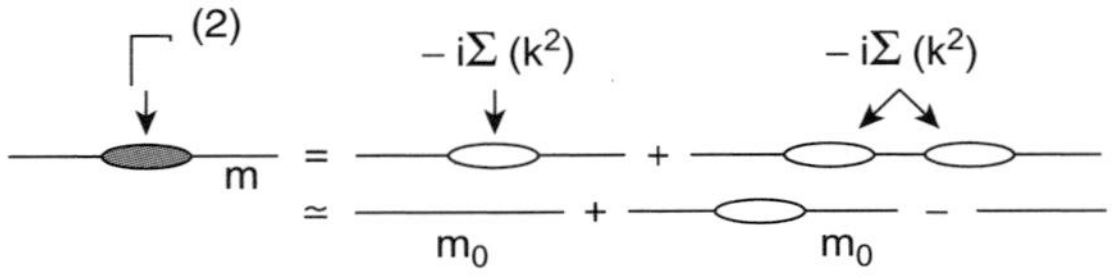

Figure 7. Self energy of the free electron.

This is illustrated in Fig. 7. This series may then be written in a more compact form with

$$\Gamma^{(2)}(k^2) = \frac{i}{k^2 - m_0^2 - \Sigma(k^2)} \tag{161}$$

where $\Sigma(k^2)$ is Taylor-expanded around the mass m_0 with the result that

$$\Gamma^{(2)}(k^2) = \frac{i}{k^2 - m_0^2 - \delta m^2} \tag{162}$$

This is a matter of replacing all of the correction terms with a finite number, in this case one, counter terms that may be evaluated.

We now add a term to the harmonic oscillator Lagrangian of the form $\lambda\phi^4$, where ϕ is a field that represents the field coupled to the electron. To evaluate the amplitude, we then have the integral of the form

$$I^2 = \lambda^2 \int \frac{d^4k}{(2\pi)^4} \frac{1}{k^2 - \mu^2} \tag{163}$$

To perform a dimensional regularization of this integral we replace the integral with

$$I^2(\omega) = \frac{\lambda}{2} \int \frac{d^{2\omega}k_e}{(2\pi)^4} \frac{1}{k_e^2 - \mu^2} \tag{164}$$

where the dimension of the system has been replaced by 2ω. We then use

$$d^{2\omega}k_e = \frac{2\pi^\omega}{\Gamma(\omega)} k_e^{2\omega-1} dk_e \tag{165}$$

and $\lambda = (M^2)^{2-\omega}\lambda_\omega$ to obtain the solution to the integral as

$$I^2(\omega) = -\frac{\lambda_\omega}{2} \left(\frac{\pi}{2\pi}\right)^\omega (M^2)^{2-\omega}\lambda_\omega (m^2)^{\omega-1} \frac{\Gamma(1-\omega)}{\Gamma(1)} \tag{166}$$

Now the "trick" used is to identify $\omega = 2 - \epsilon$ to obtain

$$I^2(\omega) = \frac{\lambda m^2}{32\pi^2} \left(1 - \gamma - \log\left(\frac{m^2}{4\pi M^2}\right)\right) \tag{167}$$

from which the mass counterterm is defined. Here γ is the Euler-Mascheroni constant $\gamma = .5772\ldots$.

A similar divergent process exists with $O(3)_b$ electrodynamics with fluctuations associated with the $\mathbf{B}^{(3)}$ field. It is associated with computing the propagator for a photon loop, as illustrated in Fig. 5. The integral involved is of the form

$$I^4 = I^2 \int \frac{d^4k}{(2\pi)^4} \frac{1}{(k^2 + ie)((k+q)^2 + ie)} \tag{168}$$

By letting $4 \to 2\omega$ we arrive at an integral of the form

$$I^4 = \lambda^2 \int \frac{d^{2\omega}k}{(2\pi)^4} \frac{1}{(k^2 + ie)((k+q)^2 + ie)} \tag{169}$$

then if the self-interaction term is written as

$$\lambda = 4 - 2\omega = (Q^2)^{2-\omega}\lambda_\omega \tag{170}$$

and the integration measure is redefined as

$$d^{2\omega}k = k^{2\omega-1}dk_\omega \tag{171}$$

then

$$I^4 = l_\omega(Q^2)^{2-\omega} \times \int \frac{dk_\omega}{(2\pi)^4} \frac{k^{2\omega-1}}{(k^2 + i\epsilon)((k+q)^2 + i\epsilon)} \tag{172}$$

This integral may then be evaluated as

$$I^4 = const * q^{\omega-2} \frac{\Gamma(\omega)\Gamma(2-\omega)}{\Gamma(2)} \tag{173}$$

where const is the constant

$$\kappa = \frac{l_\omega(Q^2)^{2-\omega}}{(2\pi)^{2\omega}} \tag{174}$$

So in a more compact form, this leads to the result

$$I^4 = \frac{\lambda_\omega}{(2\pi)^{2\omega}} \left(\frac{Q}{q}\right)^{4-\omega} \times \frac{\Gamma(\omega)\Gamma(2-\omega)}{\Gamma(2)} \tag{175}$$

or

$$I^4 = \frac{\lambda_\omega}{(2\pi)^{2\omega}} \left(\frac{Q}{q}\right)^{4-\omega} \times \left(\frac{1}{3+\omega} + 1 - \gamma\right)\left(\frac{1}{1-\omega} + 1 - \gamma\right) \tag{176}$$

Then, since $\omega = 2 - \epsilon$, the trick of dimensional regularization, we see that

$$\Gamma(2 - \omega) = \Gamma(\epsilon) \\ \Gamma(\omega) = \Gamma(2 - \epsilon) \tag{177}$$

Now this integral can be expressed in the form

$$I^4 = \frac{1}{5}\frac{q}{32\pi}^4 \times \left(1 - 6\left(\left(\gamma + 4\log\left(\frac{q}{Q}\right)\right)\right)\right) \tag{178}$$

which is a finite quantity. Here γ is the Euler–Mascheroni constant.

This calculation demonstrates that the loop fluctuation of a photons, which correlated to a virtual quanta of $\mathbf{B}^{(3)}$ field, can be calculated to be finite with out divergence. So the virtual fluctuation of a $\mathbf{B}^{(3)}$ field does not lead to an ultraviolet divergence, and thus $O(3)_b$ QED is renormalizable by dimensional regularization.

The issues of vacuum polarization and vertices may be computed in the same manner as seen with $U(1)$ electrodynamics. Effectively $O(3)_b$ quantum electrodynamics appears to be, based on these initial regularization exercises, to be free of intractable ultraviolet divergences. The calculation of the Lamb shift also indicates that $O(3)_b$ QED is also free of such divergences in the infrared region. This is a good sign that the theory at least is not frought with computational intractabilities that cast sever doubts on a theoretical level.

E. $B^{(3)}$ Field as a Vacuum Symmetry

It is standard practice to adopt the rule in $U(1)$ electrodynamics that with the Hamiltonian of the form

$$H = \frac{1}{2m}|\mathbf{p} - e\mathbf{A}|^2 \tag{179}$$

that the gauge potential acts only once. The quadratic term is basically eliminated and ignored. This rule is essentially what $O(3)_b$ electrodynamics challenges. We then consider the role of $\mathbf{A} \cdot \mathbf{A}^* = \mathbf{A}^{(1)} \cdot \mathbf{A}^{(2)}$ within this Hamiltonian. This Hamiltonian leads to the evolution operator $U = e^{-iHt}$, which has the form

$$U = e^{-iH_0 t} e^{\mathbf{A}^{(1)} \cdot \mathbf{A}^{(2)}} \tag{180}$$

where H_0 is the Hamiltonian without the term quadratic in the potentials. The vector potentials may then be written as

$$\mathbf{A}^{(1)} = \frac{A^{(0)}}{\sqrt{2}}(e_x + ie_y)(a_k e^{i\mathbf{k}\cdot\mathbf{r}-i\omega t} - a_k^\dagger e^{-i\mathbf{k}\cdot\mathbf{r}+i\omega t}) \tag{181}$$

so that the modulus square of this operator is

$$\mathbf{A}^{(1)} \cdot \mathbf{A}^{(2)} = A^{(0)2}\left(a^\dagger a + \frac{1}{2} - \frac{1}{2}(a^{\dagger 2} e^{-2i(\mathbf{k}\cdot\mathbf{r}-\omega t)} + a^2 e^{2i(\mathbf{k}\cdot\mathbf{r}-\omega t)})\right) \tag{182}$$

This result is very interesting, for the first two terms on the right-hand side are just the standard harmonic oscillator Hamiltonian for the electromagnetic field H_{em}, and the latter are terms easily seen to be incommensurate with that Hamiltonian under commutation. As a result the evolution operator is then

$$U = e^{-i(H_0 + H_{em})t} e^{(za^{\dagger 2} + z^* a^2)} \tag{183}$$

for $z = te^{-2i(\mathbf{k}\cdot\mathbf{r}-\omega t)t}$. The operator $S(z) = e^{(za^{\dagger 2} + z^* a^2)}$ is a squeezed-state operator, which involves symmetries that lie outside those defined strictly by the Hamiltonian.

There are some reasons for supposing that the $\mathbf{B}^{(3)}$ may correspond to such symmetries. In the next two sections we discuss how non-Abelian electrodynamics is unified with the weak interactions. Below it will be concluded that there is a duality between the $\mathbf{B}^{(3)}$ and $\mathbf{E}^{(3)}$ fields. From this it can be easily seen that the Lagrangian for these two fields vanishes. This is evidently a curious situation where there should exist a field, but where it has no Lagrangian. This would imply that there is no dynamics associated with this field. The argument is made that the existence of a massive $A^{(3)}$ field breaks this duality. This is then invoked to justify that $\mathbf{E}^{(3)} = 0$. However, this then creates a further difficulty. Electric and magnetic fields transform by the Lorentz group as $E_z' = \gamma(E_z - \beta B_y)$ and $B_x' = \gamma(B_x - \beta E_y)$. This leads to the unsettling prospect that if $\mathbf{B}^{(2)} > 0$ and $\mathbf{E}^{(2)} = 0$, there is then a breaking of the Lorentz symmetry to spacetime. This means that unless there is an associated 3-electric field, we may have to conclude that $\langle \mathbf{B}^{(3)2} \rangle = \langle \mathbf{E}^{(3)2} \rangle = 0$. This then gives weight to the prospect that non-Abelian electrodynamics corresponds to non-Lagrangian symmetries or operators in electrodynamics.

It is still possible to have the Sagnac effect. If we consider the counter-rotating portion of the non-Abelian contribution to

$$\theta = i \oint \mathscr{D}_\mu \, dx^\mu = i \int \int [\mathscr{D}_\mu, \mathscr{D}_\nu] \, d\sigma^{\mu\nu} \tag{184}$$

we obtain

$$\theta = i \int (a^{\dagger 2} e^{-2i(\mathbf{k}\cdot\mathbf{r}-\omega t)} + a^2 e^{2i(\mathbf{k}\cdot\mathbf{r}-\omega t)}) J^3 \tag{185}$$

On evaluation of the integral and using the fact that $J^3 \propto \Omega$, the rotation of the platform, it is then apparent that

$$\Delta\theta = \langle \ln S(z) \rangle \tag{186}$$

where $z \simeq e^{4\omega\Omega\mathscr{A}/c^2}$, where the doubling of the frequency argument occurs from the existence of two paths, A and C. This apparently gives the Sagnac effect according to the squeezing of light.

Would this mean that almost everything presented in this chapter is wrong? No, but it does mean that the classical results are purely pedagogical tools. The quantum results may still hold. For instance, with the cavity QED work the term $H_{B^3\text{int}}$, Eq. (107) corresponds to the absorption of one photon and the emission of another as the atom changes its internal state. Similarly, the term $H_{B^{(3)}}$ in Eq. (108) corresponds to the absorption of two k, k' mode photons and the emission of $k + q$ and $k' - q$ by the atom during a quantum fluctuation between its two atomic states.

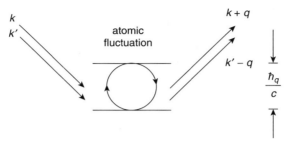

Figure 8. $B^{(3)}$ Hamiltonian described according to photon-atom interaction.

This indicates that the quantum mechanical aspects of this theory is valid, and the Hamiltonian, $H_{B^{(3)}}$, then involves quantum fluctuations in the atomic states. It must also be noticed that this interaction Hamiltonian is real only since it involves the introduction of a quantum system. In the absence of this atom, we would no longer obtain this photon–photon coupling. In the case of photon loops, this process must then be considered as attached to a fermion line, where the fermion has a fluctuation in its momentum to give rise to this photon graph.

This means that states are not completely described as eigenstates of the Hamiltonian. While they posses kinematical properties of states, but the squeezing of these states are not determined by diagonal operators and are thus not Hamiltonian or Lagrangian symmetries. This is an interesting result, for this implies that squeezed states in QED are connected to an underlying non-Abelian symmetry. This would continue to be the case even if there were no classical $\mathbf{B}^{(3)}$ field, or if $H_{\mathbf{B}^{(3)}} = 1/2|B^{(3)}|^2 = 0$. Under this condition the non-Abelian symmetry of QED would be manifested as vacuum squeezed states. In this case electrodynamics is then a $U(1)$ gauge theory, as described by a Lagrangian, plus additional non-Lagrangian symmetries.

In face of the relative paucity of measurement data for the occurrence of the $\mathbf{B}^{(3)}$ field this appears to be the most reasonable conclusion that can be drawn, and yet still uphold the basic premise that electrodynamics has an extended gauge group structure. This would mean that research along these lines should be directed towards a more complete understanding of the interaction of electromagnetic fields and nonlinear media. This may imply that while $\mathbf{B}^{(3)}$ vanishes, while its underlying symmetries still exist, this field may become present as the QED vacuum is charge polarized by the presence of atoms in certain media. As searches for a classical $\mathbf{B}^{(3)}$ field in the vacuum have so far yielded null results, it makes little sense to pursue the subject of non-Abelian electrodynamics towards a search for classical field effects.

F.   A Possible $SO(10)$ Grand Unification that Includes Non-Abelian Electrodynamics

The universe we live in is a low energy world. Much of the fundamental structure or symmetries of the early universe have been frozen out, and our world is one of highly broken symmetry. We have electromagnetism, which is in itself a gauge field of unbroken symmetry. We also observe the weak and strong nuclear interactions. However, in the very early universe these gauge fields were embedded into some form of a unified gauge field. We have a partial unification of electromagnetism and the weak interaction in the standard model. However, this theory is a twisting of the two principal bundles and suffers from some problems with the adjustment of coupling constants required in its Lagrangians to account for physics on the physical vacuum. This is the case with both the $U(1) \times SU(2)$ model and the $SU(2) \times SU(2)$ model presented here. For these reasons the standard model, along with any extended variant, can only be regarded as an approximation. Presumably this approximation is recovered from a grand unified theory (GUT) that fully unifies gauge fields at 10^{15} GeV.

How it is that this GUT theory can be tested is a matter of some difficulty. It is still possible that low energy effects of a GUT may be detected. This was the hope for proton decay with the minimal $SU(5)$ model. As there may be issues with chirality, or residual chirality in QCD it may be possible that GUTs can be experimentally tested.

As this concerns the nature of non-Abelian electrodynamics, we will pursue the matter of a GUT that incorporates non-Abelian electrodynamics. This GUT will be an $SO(10)$ theory as outlined above. We have that an extended electroweak theory that encompasses non-Abelian electrodynamics is spin$(4) = SU(2) \times SU(2)$. This in turn can be embedded into a larger $SO(10)$ algebra with spin$(6) = SU(4)$. $SO(10)$ may be decomposed into $SU(2) \times SU(2) \times SU(4)$. This permits the embedding of the extended electroweak theory with $SU(4)$, which may contain the nuclear interactions as $SU(4) \simeq SU(3) \times U(1)$. In the following paragraphs we will discuss the nature of this gauge theory and illustrate some basic results and predictions on how nature should appear. We will also discuss the nature of fermion fields in an $SU(2) \times SU(2) \times SU(4)$ theory.

We start with a discussion of what an $SU(4)$ gauge theory that embeds the nuclear interaction will look like. A Lie algebra consists of operators that are analogous to raising and lowering operators and operators that are analogous to Hamiltonians. We write these operators as X_a and H_i, where the index i ranges within $i = 1, \ldots, m$. The operators H_i act on the basis $\{E_\alpha\}$ as

$$H_i |E_\alpha\rangle = \alpha_i |E_i\rangle \tag{187}$$

where α_i are the roots of the algebra. It is not difficult to show that these operators obey the rules

$$[H_i, E_\alpha] = \alpha_i E_\alpha$$
$$[E_\alpha, E_{-\alpha}] = \alpha_i H_\alpha \tag{188}$$
$$[H_i, H_j] = \delta_{ij} H_i$$

With these operators it may be demonstrated that these roots have reflection symmetries. These reflections are thought of as a reflection within a Weyl chamber. These reflections are determined by the integer that measures the projection of one root α onto another

$$c_{ij} = 2 \frac{\alpha_i \cdot \alpha_j}{\alpha_j^2} \tag{189}$$

or equivalently that the root α is reflected off the Weyl chamber wall be

$$\alpha_i' = \alpha_i - c_{ij}\alpha_j \tag{190}$$

The angle of reflection is defined as

$$\cos\theta = -\frac{1}{2}\sqrt{c_{ij}c'_{ij}}, \qquad c'_{ij} = 2\frac{\alpha_i \cdot \alpha_j}{\alpha_i^2} \tag{191}$$

We first look at the $SU(4)$ part of the gauge field. The weights of the algebra v_i, $i = 1, \ldots, 4$ define the roots by $\alpha_i = v_i - v_{i+1}$

$$\alpha_1 = (1, 0, 0)$$

$$\alpha_2 = \left(\frac{1}{2}, \frac{\sqrt{3}}{2}, 0\right) \tag{192}$$

$$\alpha_3 = \left(0, -\frac{1}{\sqrt{3}}, \sqrt{\frac{2}{3}}\right)$$

These roots define a regular figure in 3D space that is a tetrahedron. The vertices of this tetrahedron correspond to

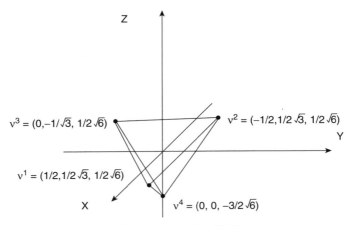

Figure 9.   Roots for $SU(4)$.

Within this theory we might expect that there should be four quark doublets. In this case the weights v^1, v^2, and v^3 correspond to the (u, d), (s, c), and (b, t) quarks. The additional weight does not correspond to an additional quark doublet that occurs at energies higher than currently probed. An alternative explaination is that $SU(4)$ is gauge theory where there are additional degrees of freedom that rotate QCD colors amongst each other at high energy with the introduction of an new massive Z-type particle plus three additional charged massive bosons. These massive bosons are manifested by the breaking of $SU(4) \rightarrow SU(3)$ by Higgs symmetry breaking, as the 4 is an $3 \oplus 1$, and $\bar{4}$ is a $\bar{3} \oplus 1$. This also means that quarks at high energy are complete and do not exist as a new form of particle field. This is in agreement with most conclusions that the number of possible quarks is 6, since any more would mean that the entropy of the early cosmology would be greater than currently thought.

We now consider tensor methods for the representation of $SU(4)$. We start by labeling the basis of states for $SU(4)$ according to its weights

$$|1\rangle = \left| \frac{1}{2}, \frac{1}{2\sqrt{3}}, \frac{1}{2\sqrt{6}} \right\rangle$$

$$|2\rangle = \left| -\frac{1}{2}, \frac{1}{2\sqrt{3}}, \frac{1}{2\sqrt{6}} \right\rangle$$

$$|3\rangle = \left| 0, -\frac{1}{\sqrt{3}}, \frac{1}{2\sqrt{6}} \right\rangle \qquad (193)$$

$$|4\rangle = \left| 0, 0, -\frac{3}{2\sqrt{6}} \right\rangle$$

This basis illustrates that these vectors are eigenstates of the H_1, H_2, and H_3 matrices with diagonal entries $\lambda_a ij/2$. We assign the off diagonal matrices to be

$$\frac{1}{\sqrt{2}}(T_1 \pm iT_2) = E_{\pm 1,0,0}$$

$$\frac{1}{\sqrt{2}}(T_4 \pm iT_5) = E_{1/2,\pm 1/2\sqrt{3},0}$$

$$\frac{1}{\sqrt{2}}(T_6 \pm iT_7) = E_{\mp 1/2,\pm 1/2\sqrt{3},0}$$

$$\frac{1}{\sqrt{2}}(T_9 \pm iT_{10}) = E_{1/2,1/\sqrt{3},\pm 2/\sqrt{6}} \qquad (194)$$

$$\frac{1}{\sqrt{2}}(T_{11} \pm iT_{12}) = E_{\mp 1/2,\sqrt{3}/2,\pm 2/\sqrt{6}}$$

$$\frac{1}{\sqrt{2}}(T_{13} \pm iT_{14}) = E_{0,\mp 1/\sqrt{3},\pm 2/\sqrt{6}}$$

These are then 12 ladder operators that along with the 3 Cartan center operators H_i, we have all of the 15 parameters of the $SU(4)$ algebra. It is up to the reader to put these operators in matrix form. The Cartan center operators are then calculated by the commutator in Eq. (188). We then have a set of weights that form the cuboctahedron in the 3-space spanned by the Cartan center operators.

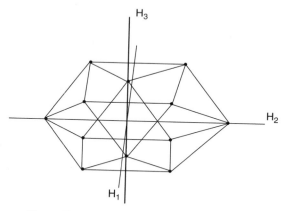

Figure 10.   Cartan center of $SU(4)$ and its weights.

This is the 4 of the $SU(4)$. The $SU(3)$ algebra is seen as the hexagon that lies on the $H_3 = 0$ plane of the space spanned by the Cartan centers. This indicates that there are additional transformation of gluon colors and quark flavors. This is an aspect of the general embedding of color QCD into $SO(10)$. The decomposition of $SO(10)$ into

$$SO(10) \rightarrow SU(4) \times SU(2) \times SU(2) \rightarrow SU(3) \times U(1) \times SU(2)_R \times SU(2)_L \tag{195}$$

There are then, in this decomposition, 16 fermions that transform as

$$16 = (3, 2, 1) + (1, 2, 1) + (\bar{3}, 1, 2) + (1, 1, 2) \tag{196}$$

which correspond to the quark doublets, the lepton doublets, and their complements. There are 45 gauge bosons that transform as

$$45 = (8, 1, 1) + (1, 3, 1) + (1, 1, 2) + (1, 1, 1) + (3, 2, 2)$$
$$+ (\bar{3}, 2, 1) + (3, 2, 2) + (3, 1, 1) + (\bar{3}, 1, 1) \tag{197}$$

which correspond to the $SU(3)$ gauge transformations, the $SU(2)_{L,R}$ transformations by $W^{0,\pm}$ gauge bosons, the auxilliary B gauge bosons, and the last four terms are pairs of the flavor–flavor and the lepto-quark transformation bosons at the GUT energy. The neutral gauge bosons $W_R^{\pm}$ combine to form photons and there are two neutral gauge bosons that form the Z^0 and Z_γ. These gauge bosons enter into a Lagrangian seen in equation 46. The quark flavor ard quark to lepton transforming gauge bosons, represented as X^α, Y^α, $\bar{X}^\alpha$, $\bar{Y}^\alpha$ and V^α and $\bar{V}^\alpha$. The Lagrangian for the flavor transformations are

$$\mathscr{L} = g\epsilon_{\alpha\beta\gamma}(X_\mu{}^\alpha \bar{d}_L^{*\beta} \gamma^\mu d_L{}^\gamma - Y_\mu{}^\alpha \bar{d}_L^{*\beta} \gamma^\mu d_L{}^\gamma) \tag{198}$$

and the Lagrangian that executes gauge rotations between quarks into leptons is

$$\mathscr{L} = g\epsilon_{\alpha\beta\gamma}(X_\mu{}^\alpha(\bar{u}_L^{*\beta}\gamma^\mu \nu_R{}^\gamma + \bar{u}_L^{*\beta}\gamma^\mu \nu_L{}^\gamma) - Y_\mu{}^\alpha(\bar{u}_R^{*\beta}\gamma^\mu \nu_R{}^\gamma + \bar{d}_L^{*\beta}\gamma^\mu \nu_L{}^\gamma))$$
$$+ g\epsilon_{\alpha\beta\gamma}(V_\mu{}^\alpha(\bar{u}_R^{*\beta}\gamma^\mu e_R{}^\gamma + \bar{d}_L^{*\beta}\gamma^\mu e_L{}^\gamma) - Y_\mu{}^\alpha(\bar{u}_R^{*\beta}\gamma^\mu e_R{}^\gamma + \bar{d}_L^{*\beta}\gamma^\mu e_L{}^\gamma)$$
$$+ H.C.) \tag{199}$$

This theory has the advantage over the $SU(5)$ minimal model that the various right- and left-handed quarks and leptons are treated equivalently in the 16 representation. This leaves the $B - L$ boson is a $U(1)$ transformation that acts on the quarks, but not the leptons. This term is then a $\sigma^{(3)}$ that acts on each of the (u, d) doublets. This boson is referred to in the $B - L$ boson in the Pati–Salam model.

The issue of proton decay rests upon the mass of the X, Y, V bosons, and the flavor mixing Cabibbo angles. The *Cabibbo angle* is defined as

$$d_{\alpha\theta} = d_\alpha \cos\theta_c + u_{\alpha+1}\cos\theta_c, \quad \cos\theta_c \simeq .974 \tag{200}$$

This angle defines the flavor mixing in processes. For small mixing angles the terms in the Lagrangian [Eq. (198)] assume the form

$$\simeq V_\mu{}^\alpha((\bar{u}_R^{*\beta}\cos\theta_c + \bar{d}^\beta\sin\theta_c \gamma^\mu e_R{}^\gamma \tag{201}$$

where the first term is illustrated for brevity. This leads to an estimte of the lifetime of a proton that is on the order of

$$T \simeq 5. \times 10^{30}\left(\frac{M_X + M_Y + M_V}{5 \times 10^{14}}\right) \tag{202a}$$

or that the lifetime is within

$$10^{28} < T < 10^{32} \tag{202b}$$

However, if the mass of these bosons is given according to Higgs fields that determine mass spectra–spectra for gauge bosons, similar to what occurs with the Fermi masses given in Eq. (45), this estimate may be adjusted upward.

For these reasons there are reasons to consider this model, or a similar variant, as a reasonable model for the unification of gauge fields outside of gravitation. The extension of the gauge symmetries for electromagnetism at high energy, even if the field is $U(1)$ on the physical vacuum, leads to a standard model with a nice symmetry between chiral fields, and this symmetry is further contained in GUT.

XI. COSMOLOGICAL CONSIDERATIONS

The nature of a grandly unified gauge theory has implications for cosmology. The initial event that brought forth the spacetime manifold of the universe, a Robertson–Walker solution, there existed some form of a gauge field that was a unification of the gauge fields we are now familiar with. This observable universe was smaller than the nucleus of an atom, and its contents were exceedingly hot. This heat is liberated from the phase transition in the generation of the universe. This heat is then energy that is distributed amongst various degrees of freedom. If there are many degrees of freedom this energy is associated with a lower temperature. This means that the number and mass of elementary particles in the early universe is a determinant of the pressure of the early material in the universe. This means that the sort of universe that we live in and understand does place certain boundaries on the nature of grand unification theories.

The Roberston–Walker cosmology considers a spacial 3-sphere that evolves through space. It is the simplest of possible models that assumes the universe is isotropic and homogeneous. The metric for this universe is written as

$$ds^2 = dt^2 - R^2(t)\left(\frac{dr^2}{1 - kr^2} + r^2 d\Omega^2\right) \tag{203}$$

The term k in this metric is a constant that determines the spacial curvature of the cosmology. For $k = 1$ the cosmology is a closed spherical universe, for $k = 0$ the cosmology is flat, and for $k = -1$ the cosmology is open. The Einstein field equations give a constraint equation and a dynamical equation for the rate the radius changes with time. If we define a velocity as $\mathbf{v} = (\dot{R}/R)H(t)\mathbf{r}$, where $H(t)$ is the Hubble parameter, a constant locally, the constraint equations is

$$\left(\frac{\dot{R}}{R}\right) = \frac{8\pi}{3}G\rho - \frac{k}{R^2} \tag{204a}$$

and the dynamical equation is

$$\frac{\ddot{R}}{R} = -4\pi G\left(p + \frac{p}{3}\right) \tag{204b}$$

It is apparent that the evolution of this cosmology is then reduced to a problem common in classical mechanics. An integration of the constraint equation, a statement of energy conservation within a sphere of radius R, results in the dynamical equation. It may then be seen that k is a constant of integration.

For simplicity, we consider the universe to be radiation dominated. This is because the particles, even the highly massive X, Y, and V particles have such large kinetic energies that they behave similar to photons or massless bosons. This state of affairs in the early universe was know to exist up until the universe dropped to a temperature below 10^3 K 100,000 years into its evolution. For the radiation dominated period in the evolution of the universe the pressure and density were related by

$$p = \frac{1}{3}\rho \tag{205}$$

The density is related to the total number of bosonic and fermionic helicities

$$\rho = \frac{1}{2}g(T)\rho_{bb} = 3TN(T), \qquad g(T) = H_{\text{Boson}} + \frac{7}{8}H_{\text{Fermi}} \tag{206}$$

where H are the numbers of helicity states, T is the temperature, and $N(T)$ is the number density of particles. If we have a partition function for the system of particles in the early universe, we have the pressure determined by

$$p = -\left(\frac{\partial A}{\partial V}\right)_{N,T} = kT\left(\frac{\partial}{\partial V}\ln Z\right)_{N,T} \tag{207}$$

If we write the partition function in the standard Boltzmann form $Z = \sum_n \exp\left(-\beta E_n\right)$, the pressure is then

$$p = -Z^{-1}\sum_n \frac{\partial E_n}{\partial V}\exp\left(-\beta E_n\right) \tag{208}$$

If we then consider the constraint equation 203a as the conservation of energy density in a sphere

$$\frac{d}{dt}\left(\frac{4\pi}{3}\rho R^3\right) = -p\frac{d}{dt}\left(\frac{4\pi}{3}R^3\right) \tag{209}$$

the result in Eq. (207) is then constrained by the dynamics of the cosmology.

If we now include the issue of inflation, the situation becomes somewhat more complex. During the inflationary period of expansion the radius of the cosmology expands exponentially

$$R(t) = R_0 \exp\left(\gamma t\right) \tag{210}$$

where

$$\gamma = \sqrt{\frac{8\pi\rho_0}{3M}} \simeq \frac{T_c^2}{M} \tag{211}$$

Here M is the mass of the particles in the universe. We then see that the universe exhibits a scale change $\sigma = R(t + \Delta t)/R(t)$, where Δt is the duration of the inflationary period. The parameter for the geometry of the spacial universe is $\Omega = \rho/\rho_c$, for $\rho_c = (\dot{R}/R)3/8\pi G$. For the condition for flatness $\Omega = 1$ in the early universe, we require that $\sigma > 10^{27}$. The expansion parameter will reduce $\omega - 1$ by a factor of σ, and thus guarantee that there is cosmic flatness, or close to flatness.

At this stage in astronomy the value of Ω is not completely known. There is the problem of dark matter, which is a putative form of matter that increases $\Omega = .1$, based on the observation of luminous matter to unity to account for flatness. However, there is some debate as to whether the universe is one with acceleration. This will change the nature of the flatness issue if these data turn out to be well established. However, the value of the mass of the elementary particles in the early universe is dominated by massive particles such as the flavor and quark–lepton transforming bosons. This means that cosmology, if we garner a complete observational understanding of the universe, will impose bounds on the value of σ if the value of ω is known. This will provide a constraint on the types of theories that can be proposed for the grand unification of gauge theories.

It is apparent that the numbers and masses of the flavor and quark–lepton transforming gauge bosons are larger than those of the $SU(5)$ minimal model. This means that the value of σ is lower, and assuming that the duration of the inflationary period is fixed, the scale for the expansion of the universe is reduced. This means that there is the enhanced prospect for deviations from flatness. So one may presume that the universe started as a small 3-sphere with a large curvature, where the inflationary period flattened out the universe, but maybe not completely. This leaves open the prospect that if before inflation that if the universe were open or closed, $k = \pm 1$, that the universe today still contains this structure on a sufficiently large scale. The closer to flatness the universe is, the tighter are the constraints on the masses of particles in the early universe.

DISCUSSION

This section within a book devoted to non-Abelian electrodynamics may appear to be comparatively pessimistic for those who are looking for a classical effect. However, out of the conclusion that there are no photon states that correspond to the $\mathbf{B}^{(3)}$ field comes good news. If nature is such that at high energy the symmetry of nature is extended there will be implications for high energy physics and that at low energy these additional symmetries manifest themselves as non-Hamiltonian effects in quantum optics. These types of physics are of far greater impact than attempts to uphold the hypothesis that the $\mathbf{B}^{(3)}$ field is a real field that can be measured classically. Further, as the data sighted to uphold the existence of a classical $\mathbf{B}^{(3)}$ field are quite dated, have not been duplicated, and that more recent attempts at a classical measurement of $\mathbf{B}^{(3)}$ field have given null result, it is most likely that there is no such classical field effect.

The major thrust is that vacuum symmetries, those that do not affect the eigenstates of a system involving atoms and photons, are residual aspects of high energy physics that is more symmetric than the canonical standard model of electroweak interactions and the minimal $SU(5)$ grand unification scheme. This approach to the fundamental nature of gauge theories suggests that gauge duality is a reasonable conjecture for the foundations of physics. This also has implications for the structure of gauge theory associated with strings and membranes. An understanding of the nature of gauge fields in the TeV range, along with potential hints of physics at the GUT level, will give better directions on where to pursue theories that involve Planck-scale physics.

References

1. R. P. Feynman, *Phys. Rev.* **76**, 769 (1949).
2. F. J. Dyson, *Phys. Rev.* **75**, 486 (1949).
3. J. D. Bjorken and S. D. Drell, *Relativistic Quantum Fields*, McGraw-Hill, New York, 1965.
4. L. B. Crowell and M. Evans, *Found. Phys. Lett.* **12**, 4 (1999).
5. A. O. Barut in notes from his lectures on QED (1988).
6. M. W. Evans and J. P. Vigier, *The Enigmatic Photon*, Vol. 2. *Non-Abelian Electrodynamics*, (Kluwer Academic, Dordrecht, 1995).
7. M. W. Evans, J. P. Vigier, S. Roy, and S. Jeffers, *The Enigmatic Photon*, Vol. 3. *Theory and Practice of the* **B**$^{(3)}$ *Theory*, Kluwer, Dordrecht, 1996.
8. P. Ramond, *Field Theory*, Princeton Series in Physics, Princeton Univ. Press, Princeton, NJ, 1982.
9. S. Weinberg, *Phys. Rev. Lett.* **19**, 1264 (1967).
10. C. Montonen and D. Olive, *Nucl. Phys.* **B110**, 237 (1976).
11. N. Seiberg, *Phys. Lett.* **206B**, 75 (1988).
12. N. Seiberg and E. Witten, *Nucl. Phys.* **B426**, 19 (1994).
13. N. Seiberg and E. Witten, *Nucl. Phys.* **B431**, 484 (1994).
14. P. S. Pershan, J. van der Ziel, and L. D. Malmstrom, *Phys. Rev.* **143**, 574 (1966).
15. L. Crowell, M. Evans, *Found. Phys. Lett.* **12**, 5 (1999).
16. M. Gell-Mann, *Phys. Rev.* **125**, 1067 (1962).
17. M. Gell-Mann and M. Levy, *Nuovo Cimento* **16**, 705 (1960).
18. K. Cahill, L. Crowell, and D. Khetselius *Proc. 8th Div. Part. Fields*, 1994, p. 1213.
19. J. Erler, Paul Langacker, *Phys. Rev. Lett* **84**, 212–215 (2000). http://xxx.lanl.gov/abs/hep-ph/9910315.
20. LEP and SLD Collaborations, CERN report CERN-EP/9915, Feb. 8, 1999.

FLUCTUATIONAL ESCAPE AND RELATED PHENOMENA IN NONLINEAR OPTICAL SYSTEMS

I. A. KHOVANOV

Department of Physics, Saratov State University, Saratov, Russia

D. G. LUCHINSKY

Department of Physics, Lancaster University, Lancaster, United Kingdom and Russian Research Institute for Metrological Service, Moscow, Russia

R. MANNELLA

Dipartimento di Fisica, Università di Pisa and Istituto Nazionale Fisica della Materia UdR Pisa, Pisa, Italy and Department of Physics, Lancaster University, Lancaster, United Kingdom

P. V. E. McCLINTOCK

Department of Physics, Lancaster University, Lancaster, United Kingdom

CONTENTS

Modern Nonlinear Optics, Part 3, Second Edition, Advances in Chemical Physics, Volume 119, Edited by Myron W. Evans. Series Editors I. Prigogine and Stuart A. Rice.
ISBN 0-471-38932-3 © 2001 John Wiley & Sons, Inc.

I. INTRODUCTION

A characteristic feature of nonlinear science generally, and of nonlinear optics in particular, is the common necessity of having to make simplifications, and then approximations in order to solve the equations of even the simplified models. These considerations apply a fortiori to the study of fluctuation phenomena in nonlinear systems, and thus account for the increasing role being played by analog and digital simulations, which enable the behaviour of the model systems to be investigated in considerable detail.

Several years have now passed since a contribution [1] to an earlier volume in this series illustrated some of these ideas. It was shown in particular that detailed analyses of fluctuations in model systems not only provide a deeper understanding of complex phenomena but often also pave the way to the development of new experimental techniques and new ideas of technological significance.

In this chapter, we discuss the application of simulation techniques to the study of *fluctuational escape and related phenomena* in nonlinear optical systems: that is, situations where a large deviation of the system from an equilibrium state occurs under the influence of relatively weak noise present in the system. We will be interested primarily in the analysis of situations where large deviations lead to new nontrivial behaviour or to a transition to a different state. The topics to be discussed have been selected mainly for their own intrinsic scientific interest, but also in order to provide an indication of the power and utility of the simulation approach as a means of focusing on, and reaching an understanding of, the essential physics underlying the phenomena under investigation; they also provide examples of different theoretical approaches and situations where numerical and analogue simulations have led to the development of new experimental techniques and new ideas with potential technological significance. Although the different sections all share the same general theme—of fluctuational escape phenomena in model nonlinear optical systems—they deal with quite different aspects of the subject; each of them is therefore to a considerable extent self-contained (Section IV is an exception, because it should be read after Section III) and thus can be read almost independently of the others. Before considering particular systems, we review briefly the scientific context of the work and discuss in a general way the significance of escape phenomena in nonlinear optics.

The investigation of fluctuations by means of analog or digital simulation is usually found most useful for those systems where the fluctuations of the quantities of immediate physical interest can be assumed to be due to noise. The latter perception of fluctuations goes back to Einstein, Smoluchowski, and Langevin [2–4] and has often been used in optics [5–8]. In nonlinear optics, the noise can be regarded as arising from two main sources. First there are internal fluctuations in the macroscopic system itself. These arise because spontaneous emission of light by individual atoms occurs at random, and because of fluctuations in the populations of atomic energy levels. The physical characteristics of such noise are usually closely related to the physical characteristics of the model that describes the "regular" dynamics of the system, namely, in the absence of noise. In particular, the power spectrum of thermal noise and its intensity can be expressed in terms of the dissipation characteristics via the fluctuation–dissipation relations [9] and, if the dissipation is nonretarded so that the corresponding dissipative forces (e.g. the friction force) depend only on the instantaneous values of dynamical variables, the noise power spectrum is independent of frequency, thus the noise is white. The model in which noise is white and Gaussian is one of the most commonly used in optics because the quantities of physical interest often vary slowly compared with the fast random processes that give rise to the noise, such as emission or absorption of a photon [5–8]. The second very important source of noise is external: for example, fluctuations of the pump power in a laser. The physical characteristics of such noise naturally vary from one particular system to another; its correlation time is often much longer than that of the internal noise, and its effects can be large and sometimes quite unexpected [10].

In general, the fluctuations observed in nonlinear optics are both spatial *and* temporal; the variations of the quantities of interest occur to a large extent independently in time and in space. However, in many cases the spatial modes in a system are well separated; the dynamics of interest is then just that of a few dominant modes. The appearance of such modes is typical for high-Q active and passive optical cavities. In view of progress in microelectronics (quantum-dot technology, semiconductor–laser arrays, etc.), the investigation of systems with a discrete set of spatial structures (modes) is particularly interesting and important [11]. The amplitudes and phases of the actual modes (or other appropriate characteristics of a system that do not depend on coordinates) make a set of purely *dynamical* variables, and the analysis of fluctuations in a system reduces to the investigation of the kinetics of a *dynamical system*.

One of the most remarkable phenomena where fluctuational transitions play a key role is *stochastic resonance* (SR), a phenomenon in which a weak periodic signal in a nonlinear system is enhanced by an increase of the ambient noise intensity; a stronger definition requires that the signal/noise *ratio* (SNR) should also increase. The SR phenomenon appears to be widespread. After being introduced as a possible explanation of the earth's ice-age cycle [12,13], SR has subsequently been observed or invoked in a large variety of contexts (see, e.g., Refs. 14–18 for reviews). SR has also been extensively investigated in nonlinear optical systems including lasers [19–22], passive optical systems [23–26], and a Brownian particle in an optical trap [27]. In this chapter, following a brief introduction to the SR phenomenon in an optical bistable system, a new form of optical heterodyning related to stochastic resonance is described, in which two high-frequency signals (input and reference signals) are applied to a bistable system. We note that the effect of noise-enhanced heterodyning was first predicted theoretically and investigated in analog models in a broad range of parameters [28]. These investigations in turn made it possible to observe a noise-induced enhancement of heterodyning in an optical bistable device [29]. At the same time, noise-induced increase of the SNR (rather than of the signal only) can occur only in certain classes of nonlinear systems [31].

When it was first discovered, and for some time afterward, SR seemed a rather mysterious phenomenon and a number of highly sophisticated theoretical approaches were proposed (see citations in, e.g., the reviews Refs. 16,30,32, and 33). All these theories assumed that bistability is an essential prerequisite for the SR phenomenon to occur. Only some years later was it appreciated [34,35] that a much simpler formalism—*linear response theory* (LRT)—would suffice to describe what was often the most interesting limit in practice, where the signal was relatively small and the noise was relatively strong. An analytic theory of the more complicated effects that occur for stronger signal strengths [34,36–38] has also been developed, and has been confirmed in considerable detail through analogue electronic experiments.

The perception of SR as a linear response phenomenon led naturally, however, to the realisation that SR can also occur without bistability [31,39,40] and to observation of the phenomenon in an underdamped, monostable, non-linear oscillator [39]. In fact, it is well known that the response of a monostable system to signals in certain frequency ranges can be strongly increased by noise, such as just by raising the temperature. Examples range from currents in electron tubes to optical absorption near absorption edges in semiconductors. For underdamped oscillators, a temperature-induced shift and broadening of the absorption peaks, or "tuning" by external driving due to the oscillator nonlinearity, was first discussed in [41]; complete classical and quantum theories of these effects have been given [42]. Underdamped systems have also been considered [43].

Describing SR in terms of a susceptibility is particularly advantageous for systems that are in thermal equilibrium, or in quasiequilibrium. In such cases the fluctuation-dissipation relations [9] can be used to express the susceptibility in terms of the spectral density of fluctuations in the absence of the periodic driving. This was used explicitly in the case of noise-protected heterodyning. It is true in general that the analysis of fluctuations is greatly facilitated by the presence of thermal equilibrium when the conditions of detailed balance and of the time reversal symmetry are satisfied [44].

However, in many cases the fluctuating systems of interest are far from thermal equilibrium. Examples include optical bistable devices [45], lasers [23,46], pattern forming systems [47], trapped electrons that display bistability and switching in a strong periodic field [48–50], and spatially periodic systems (ratchets) that display a unidirectional current when driven away from thermal equilibrium [51–56].

A powerful tool for analyzing fluctuations in a nonequilibrium systems is based on the Hamiltonian [57] theory of fluctuations or alternatively on a path-integral approach to the problem [44,58–62]. The analysis requires the solution of two closely interrelated problems. The first is the evaluation of the probability density for a system to occupy a state far from the stable state in the phase space. In the stationary regime, the tails of this probability are determined by the probabilities of large fluctuations.

The other problem is that of the *fluctuational paths* along which the system moves when a large fluctuation occurs. The distribution of fluctuational paths is a fundamental characteristic of the fluctuation *dynamics*, and its understanding paves the way to developing techniques for controlling fluctuations. Its importance for gaining insight into the physics of fluctuations from a dynamical perspective was recognized back in 1953 by Onsager and Machlup [44]. A theoretical understanding, and basic techniques for treating the problem, have been developed since that time; but it was not until the early 1990s [60] that the distribution of fluctuational paths for large fluctuations was observed in an actual experiment, through an analog simulation.

A simple qualitative idea behind the theory of large fluctuations in noise-driven systems is that such fluctuations result from large outbursts of noise that push the system far from the attractor. The probabilities of large outbursts are small, and will actually be determined by the probability of the *most probable* outburst of noise capable of bringing the system to a given state. This particular realisation of noise is just the optimal fluctuational force. Because a realisation (a path) of noise results in a corresponding realisation of the dynamical variable [63], there also exists an optimal path along which the system arrives to a give state with overwhelming probability. From a different perspective, optimal paths were first described for nonlinear non-equilibrium Markov systems in [57]. Using another approach, the analysis of the tails of the distribution was also done in [64], whereas the approach described above was discussed in [65] in the context of escape from a metastable state. This approach is not limited to Markov systems [42]. For systems driven by Ornstein–Uhlenbeck noise, the problem of optimal paths was discussed in [58,66–73]; an equivalent eikonal formulation has been developed [74–77]. The general case of Gaussian noise has been discussed [78–80]; see also Section III. For reviews of related work on fluctuations in colored noise driven systems, see Refs. 81 and 82.

A brief introduction to the theory of large fluctuations is given in Section III together with the results of some direct observations of the optimal paths in model systems. It is very important to note that, following the first observations of optimal paths in analog electronic models, fluctuational paths have been investigated in optical systems, including measurements of the *prehistory probability distribution* (PPD) of the radiation intensity I for dropout events in a semiconductor laser [83], and the time-resolved measurement of polarization fluctuations in a semiconductor vertical-cavity surface-emitting laser [84].

The preliminary analogue and numerical simulations made it possible to test fundamental tenets of the theory of large fluctuations, and thus to provide an experimental basis on which the theory could advance. At the end of Section III we present two examples of advances in the theory of large fluctuations. In the first example, the time evolution of the escape flux over a barrier on a short timescale is considered. It is a problem of fundamental importance [85] and, furthermore, of immediate practical interest given that new methods of spectroscopy with femtosecond resolution have now become available [86]. The technique of nonstationary optimal paths can be employed to solve the problem and numerical simulations verify the theoretical predictions. The striking feature predicted theoretically and demonstrated in simulations is that, for a system initially at the bottom of the well, the escape flux over the barrier on times of the order of a period of an eigenoscillation grows in a stepwise manner, provided that friction is small or moderate. If the initial state is not at the bottom of the well, the steps at large enough times transform into oscillations. The stepwise/oscillatory evolution at short times appears to be a generic feature of a noise-induced flux.

The second example is related to an analytical solution [56,87] of the longstanding problem of escape from a potential well in the presence of nonadiabatic periodic driving. It was shown [56,87] that, over a broad range of driving field magnitudes, the logarithm of the fluctuation probability is linear in the field, and the response can therefore be characterized by a *logarithmic susceptibility* (LS). We evaluate the activation energies for escape, with account taken of the field-induced lifting of the time and spatial degeneracy of instantonlike nucleation trajectories. The immediate advantage of the theory is that it provides the solution of a complicated theoretical problem in a simple analytical form that describes the dependence of the "activation energy" on both the amplitude and frequency of the driving field and can be extended immediately to a periodic driving field of arbitrary form.

Analogue experiments and digital simulations confirmed that the variation of the activation energy for escape with driving force parameters is accurately described by the LS. Experimental data on the dispersion are in quantitative agreement with the theory. And, again, it is interesting to note that, after the LS was investigated in analogue and numerical simulations, it was then also measured in optical experiments on a submicrometer Brownian particle in a bistable three-dimensional optical trap [88,89]. This research emphasizes the fundamental importance of the logarithmic susceptibility, a new physical quantity that relates the response of the system in the absence of detailed balance to its characteristics in thermal equilibrium. It yields quantitative agreement with experiment and expresses the corrections to the "activation energy" in a simple integral form analogous to that wellknown from linear response theory.

In the preceding example, analog and numerical simulations were used to verify existing theoretical predictions. However, in reality the significance of analog and digital simulations goes far beyond this modest role. The analog circuit combines features of a real physical system and of the computer model and an attentive researcher can very often make important discoveries by analyzing its behavior. Perhaps the most striking example is given in the review by Kautz [90]: "In discussing analogue simulations of a RF-biased Josephson junction, performed by Levinsen and others at Berkley, Levinsen and Sullivan conceived a new type of voltage standard...." From our own experience, examples of theory being led by the analog simulation include the discoveries of noise-induced spectral narrowing [91] and of SR in monostable systems [31,39], leading to extensive research by many groups and correspondingly to substantial theoretical progress.

We then report and discuss the results of recent investigations of fluctuational escape from the basins of attraction of chaotic attractors (CAs). The question of noise-induced escape from a basin of attraction of a CA has remained a major scientific challenge ever since the first attempts to generalize the classical escape problem to cover this case [92–94]. The difficulty in solving these problems stems from the complexity of the system's dynamics near a CA and is related, in particular, to the delicate problem of the uniqueness of the solution and the boundary conditions at a CA. The approach proposed here is based on the analysis of the prehistory probability distribution. It is shown in particular that both the existence and uniqueness of a solution can be verified experimentally using measurements of a PPD. Moreover, using this technique and its extension [95] to measure both the optimal paths and the corresponding optimal fluctuational force, one can identify the initial conditions on a chaotic attractor and find an approximation to the energy-optimal control function of escape from a CA, thus paving the way to exciting new applications in the field of nonlinear control. One such application to the energy-optimal control of escape from the basin of attraction of the CA of a periodically driven nonlinear oscillator will be described. Finally, fluctuational escape from the Lorenz attractor, a well-known system that is of importance in modelling the dynamics of real optical systems, will be discussed.

The chapter is organized as follows: Section II describes an investigation of the SR phenomenon and of noise-protected heterodyning in an electronic circuit and in an optical bistable device. Section III discusses the results of investigation of optimal paths for large fluctuations and their relationships to the analysis of fluctuations in real optical systems. It then presents two examples of advances in the theory of large fluctuations related to the time evolution of the escape flux over a barrier in a potential system on a short timescale, and to the nonadiabatic escape problem. The results of numerical and analogue simulations are compared with theory. Section VI describes investigations of the escape from a CA and the applications of these results to the solution of the nonlinear optimal control problem. Finally, in Section V, we summarise the results and consider future perspectives.

II.   STOCHASTIC RESONANCE AND NOISE-PROTECTED HETERODYNING

A.   Introduction

The idea of stochastic resonance (SR) was introduced by Benzi et al. [12], and Nicolis [13], who showed that a weak periodic signal in a nonlinear system can be enhanced by the addition of external noise of appropriate intensity; it was demonstrated subsequently that the same is often true of the signal-to-noise ratio (SNR) as well [19,96]. The quest for practical applications of SR has become a subject of intensive investigation [97]. An important restriction in this respect is [28,97] that the frequency of the input signal should be low compared to the characteristic frequencies of the system under study. Indeed, most investigations of SR to date [20,23,34,35,98–101] (see also Ref. 102 and references cited therein) have related to low-frequency signals driven bistable systems. The origin of the SR in such cases lies in the fact that the low-frequency driving force modulates the probabilities of fluctuational transitions W_{nm} between the coexisting stable states, and hence the populations of the states, which gives rise to a comparatively strong modulation of a coordinate of the system with an amplitude proportional to the distance between the stable positions. This mechanism of strong response of a symmetric bistable system to an external forcing was first suggested by Debye [103] in the context of molecules that have several different equivalent orientations in solids and may reorient between them. Since the transition probabilities increase sharply (exponentially, for Gaussian noise) with noise intensity D, the efficiency of modulation and the SNR are also sharply increased. The mechanism is operative provided: (1) the stationary populations of the states are nearly equal to each other; and (2) the frequency of the force is much smaller than the reciprocal relaxation time t_r^{-1} of the system [15]. It was suggested [28], however, and demonstrated in analog simulations, that a related phenomenon can occur when a nonlinear system is driven by two *high-frequency* signals: if the resultant heterodyne signal is of sufficiently low frequency, both it and its SNR can be enhanced by the addition of noise.

Here, we use the ideas of SR and heterodyning to demonstrate the new phenomenon of *noise-enhanced optical heterodyning* in an optically bistable (OB) device driven by two modulated laser beams at different wavelengths. An optical system was chosen for the investigations for two main reasons. First, because of progress in optical data processing and communication [104,105] and of possible applications of optical bistability in this context [106], the trend to miniaturize OB devices and to reduce their threshold power [107] has highlighted the problem of controlling the signal and the SNR in optical systems. Second, OB systems provide an opportunity to investigate a wide range of quite general fluctuation phenomena associated with coexisting stable states far from thermal equilibrium. Thus the investigation of fluctuations in these systems is of fundamental interest and significance.

In Section II.B the fluctuations and fluctuational transitions in an OB system subject to white noise are analyzed. In Section II.C the phenomenon of stochastic resonance in the OB system is discussed in terms of linear response theory and the corresponding experimental results are presented. In Section II.D we discuss theory and experimental results for the new form of optical heterodyning noise-protected with stochastic resonance. Finally, Section II.E contains concluding remarks.

B. Fluctuations and Fluctuational Transitions in an OB (Optically Bistable) System

1. Theory

A simple model that makes it possible to describe optical bistability in a variety of systems is a plane nonlinear Fabry–Perot interferometer, filled with a medium whose refractive index is intensity dependent [106]. The "slow" kinetics of a

nonlinear interferometer may be often described by a Debye relaxation equation
for the phase gain ϕ, of form

$$\dot{\phi} + \frac{1}{\tau}(\phi - \phi_0) = I_{in}(t)M(\phi) + I_{ref}(t)$$
$$I_T(t) = N(\phi)I_{in}(t), \quad N(\phi) = N(\phi + 2\pi), \quad M(\phi) = M(\phi + 2\pi)$$
$$(1)$$

Here $I_{in}(t)$ is the intensity of the incident radiation and ϕ_0 is the phase of the
interferometer in the dark. The functions $N(\phi)$ and $M(\phi)$ relate the intensities of
the transmitted and intracavity fields to that of the incident light. The function
$I_{ref}(t)$ corresponds to the intensity of radiation from an additional source, which
is very likely to be present in a real device to control the operating point. This
description is valid in a plane-wave approximation, provided that we neglect
transverse effects and the intracavity buildup time in comparison with the
characteristic relaxation time of nonlinear response in the system. It has been
shown that the Debye approximation holds for many OB systems with different
mechanisms of nonlinearity.

Let us now consider stochastic motion in an OB system. In general, noise in
an OB system may result from fluctuations of the incident field, or from thermal
and quantum fluctuations in the system itself. We shall consider the former. The
fluctuations of the intensities of the input or reference signals give rise
respectively to either multiplicative or additive noise driving the phase. Both
types of fluctuations can be considered within the same approach [108]. Here we
discuss only the effects of zero-mean white Gaussian noise in the reference
signal:

$$I_{ref}(t) = \bar{I}_{ref} + \Delta I(t), \quad \langle \Delta I(t) \rangle = 0, \quad \langle \Delta I(t)\Delta I(0) \rangle = 2D\delta(t)$$

In this case, for a constant intensity of the input signal, $I_{in}(t) = \bar{I}_{in} = $ constant,
Eq. (1) describes the Brownian motion of the phase ϕ in a bistable potential

$$U(\phi) = \frac{1}{\tau}\left(\frac{1}{2}\phi^2 - \phi\phi_0\right) - \bar{I}_{ref}\phi - \bar{I}_{in}\int_0^\phi d\phi' M(\phi') \qquad (2)$$

Stable states can be found, for example, by graphical solution of the equation
$1/\tau(\phi - \phi_0) = M(\phi)\bar{I}_{in} + \bar{I}_{ref}$ for the potential minima [42,65], and it can be
shown immediately that OB arises only if the system is biased by a sufficiently
strong external field, that is, when it is far away from thermal equilibrium. If the
noise intensity is weak, the system, when placed initially in an arbitrary state,
will, with an overwhelming probability, approach the nearest potential minimum and will fluctuate near this minimum. Both the fluctuations and relaxation

will be characterized by the relaxation time of the system τ_r. So within a time $\sim \tau_r$ the system forgets about its initial state and a quasistationary distribution is formed near the stable position. It is of Gaussian shape near its maximum and of width $\propto (D\tau_r)^{1/2}$. If the noise intensity is small compared to the potential barrier height, fluctuational transitions between the stable states occur rarely and the probabilities W_{nm} of transitions are given by Kramers' [109] relation

$$W_{nm} \propto \exp\left(-\frac{\Delta U_n}{D}\right) \qquad (3)$$

The stationary distribution over the wells is formed over a time $\sim \max\{W_{nm}^{-1}\}$. For the case of white Gaussian noise this distribution has the well-known form of the Gibbs distribution:

$$p(\phi) = Z^{-1}\exp\left(-\frac{U(\phi)}{D}\right), \qquad Z = \int d\phi \exp\left[-\frac{U(\phi)}{D}\right] \qquad (4)$$

For small noise intensities the distribution has sharp maxima near the stable states and their populations $w_{1,2}$ are described by the balance equations

$$\dot{w}_1 = -W_{12}w_1 + W_{21}w_2, \qquad w_2 = 1 - w_1 \qquad (5)$$

For arbitrary parameters of the system, w_1 and w_2 differ dramatically from each other; one of them is ~ 1, and the other is close to zero. Within a narrow range of parameters, however, they have the same order of magnitude and one can refer to a *kinetic phase transition* between the two stable states; it is analogous to the first-order phase transition in an equilibrium system with a potential (in the absence of quantum fluctuations) playing the role of the generalized free energy of the system [42,65,110]. This is the range of parameters that is of particular interest in the present chapter.

The model (1)–(5) describes stochastic motion in a general OB system for white Gaussian noise in the low noise intensity limit. We now apply this model to the description of some experimental results on fluctuations and fluctuational transitions in some particular OB devices.

2. Experiment

In the experiments we have used two approaches. First, we have simulated the kinetics of a bistable optical system in the Debye relaxation appoximation for different forms of potential by means of electronic analog simulation. Secondly, we have investigated the kinetics of a double-cavity membrane system (DCMS) driven by two modulated laser beams at different wavelengths. This system is known to display optical bistability [111].

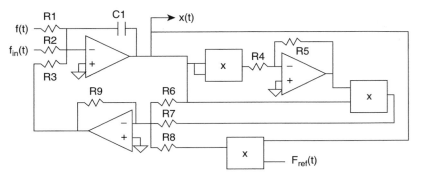

Figure 1. Analog circuit used in the heterodyning simulations of Ref. 28.

An example of an electronic circuit is shown in Fig. 1. It is similar to the circuit used to model an OB system with a dispersive mechanism of nonlinearity [112]. The circuits were driven by noise from a feedback-shift-register noise generator and in addition, if necessary, by sinusoidal periodic forces from a pair of Hewlett-Packard Model 3325B frequency synthesizers. In the DCMS used for the optical experiments (see Fig. 2), the first resonator is formed by a membrane consisting of a thin film ($\approx 1\,\mu m$ thick) of semiconducting GaSe single crystal, separated from a plane dielectric mirror by a metal diaphragm $\approx 500\,\mu m$ in diameter. The air-filled gap between the mirror and the membrane is $\approx 10\,\mu m$ wide and forms a second resonator. The incident beam from an argon

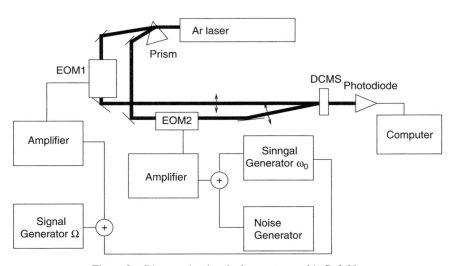

Figure 2. Diagram showing the laser setup used in Ref. 29.

laser, of wavelength 514.5 nm, propagating along the normal to the mirror, provides an input signal. An additional beam of wavelength 488 nm from an argon laser is inclined with respect to the DCMS axis and provides a reference signal. The intensities of the laser beams are modulated by two electro-optic shutters, to which periodic signals and noise are applied. The optical bistability arises because of thermoelastic bending of the membrane caused by the 514.5 nm laser beam; this particular mechanism has been found to be very effective for the investigation of a variety of OB effects [111,113,114]. The form of the periodic function $M(\phi)$ in (1) depends on the mechanism of thermal relaxation and the boundary conditions at the edge of the film; an approximate expression was found in [111] on the basis of variational analytical approaches developed for describing the thermoelasticity of shells. The phase gain ϕ is linear in bending and follows adiabatically the thermal relaxation of the film, thus ensuring the validity of the Debye relaxation approximation. Heating of the DCMS by the 488-nm reference signal is directly proportional to its intensity.

It follows from the above discussion that an indicator of applicability of the description of stochastic motion in an OB system is an activation dependence of the transition probabilities W_{nm} on the noise intensity. Using level-crossing measurements (shown to be independent on the level positions), we found in our previous experiments [108] that the activation law applies over the whole range of noise intensities that we are using.

For weak noise the spectral density of fluctuations (SDF) at the output of the OB system is defined as

$$Q(\omega) = \frac{1}{4\pi T} \left| \int_{-T}^{+T} dt\, e^{i\omega t} I_T(t) \right|^2 \qquad T \to \infty \tag{6}$$

For small noise intensities the system spends most of the time fluctuating near the stable positions, and interwell transitions occur only occasionally. $Q(\omega)$ can then be represented as the sum of partial contributions from vibrations about the equilibrium positions x_n weighted with the populations of the corresponding stable states w_n, and from interwell transitions. The intrawell contribution takes the form

$$Q_n^{(0)}(\omega) = N'^2(\phi_n)\bar{I}_{in}^2 \frac{D}{\pi} \frac{1}{U_n'' + \omega^2} \tag{7}$$

where ϕ_n is the value of the phase ϕ in the ith stable state, $U'(\phi_n) = 0$, $U''(\phi_n) > 0$.

One of the most important general features of fluctuations in a bistable system is the onset of a narrow zero-frequency spectral peak for parameter values lying in the range of the kinetic phase transition. This peak arises from the fluctuation-induced transitions between the stable states of the system and is of Lorentzian shape

$$Q_{\mathrm{tr}}^{(0)}(\omega) = \frac{w_1 w_2}{\pi} (\bar{I}_{T1} - \bar{I}_{T2})^2 \frac{W_{12} + W_{21}}{(W_{12} + W_{21})^2 + \omega^2}, \qquad \bar{I}_{Tn} \equiv N(\phi_n)\bar{I}_{\mathrm{in}} \qquad (8)$$

The onset of this peak is closely related to stochastic resonance, which can occur if a weak periodic signal is added to the input.

C.   Stochastic Resonance in an OB System

For an OB system driven by a combination of the stochastic reference beam and the periodically modulated input beam ($I_{\mathrm{in}}(t) = \bar{I}_{\mathrm{in}} + A\cos\Omega t$), the equation for the phase takes on the form

$$\dot{\phi} + U'(\phi) = M(\phi)A\cos(\Omega t) + \Delta I(t) \qquad (9)$$

To first order in A the intensity of the transmitted radiation is given by

$$\langle I_T(t)\rangle = \bar{I}_T + A\operatorname{Re}\left[\chi(\Omega)\exp(-i\Omega t)\right]$$

where $\chi(\Omega)$ is the susceptibility. As a result of the periodic term in the intensity of the outgoing radiation there arises a δ-spike in the SDF (6), of area equal to $(1/4)A^2|\chi(\Omega)|^2$. For low noise intensity D, when the system spends most of its time fluctuating about the stable states $n = 1, 2$, the susceptibility (like the SDF) is given by the sum of contributions from the vibrations about these states $\chi_n(\Omega)$ and the term $\chi_{\mathrm{tr}}(\Omega)$ that results from the periodic modulation of the populations by the force $A\exp(-i\Omega t)$.

For $\Omega \ll \tau^{-1}$ the intrawell susceptibilities correspond to quasistatic forcing, and can easily be obtained by linearising the equation of motion near the stable states. To calculate the interwell contribution to lowest order in A/D, one has to find corrections to the escape probabilities W_{nm}, which can be easily done using a path-integral formulation, solving the corresponding variational problem [82], and calculating the periodic redistribution over the wells, using balance equations. The resulting expression in the case of additive noise in the OB system takes the form

$$\chi(\Omega) = \sum_{n=1,2} M(\phi_n)\frac{\partial \bar{I}_{Tn}}{\partial \bar{I}_{\mathrm{ref}}} w_n$$

$$+ \frac{w_1 w_2}{D}(\bar{I}_{T1} - \bar{I}_{T2})\frac{W_{12} + W_{21}}{(W_{12} + W_{21}) - i\Omega}\int_{\phi_1}^{\phi_2} d\phi' M(\phi') \qquad (10)$$

In general, the SDF at low frequency is a superposition of a δ-function peak at the frequency Ω, the zero-frequency peak, and a broad, smoothly varying background (at $\omega\tau \ll 1$), which is proportional to D and is small if D is small.

According to several studies [12,13,19,96], the two principal features of stochastic resonance phenomena are that the signal and/or the signal-to-noise ratio R

$$R = \frac{1}{4}\frac{A^2|\chi(\Omega)|^2}{Q^{(0)}(\Omega)} \qquad (A \to 0) \qquad (11)$$

can be enhanced by adding noise to the system, and display resonance-like behavior in a certain range of noise intensities. It follows from Eqs. (7), (8), (10), and (11) that the signal and R in OB system indeed increase sharply with D if the heights of the "potential barriers" satisfy $\Delta U_{1,2} \gg D$, because the probabilities of fluctuational transitions (3) sharply increase with noise intensity.

These particular effects have been observed experimentally [115]. A sinusoidal signal at a frequency of 3.9 Hz was applied to an electrooptic modulator to modulate the input signal at wavelength 514.5 nm, while the intensity of the 488 nm radiation was modulated with noise. It is clearly seen from Fig. 3 that the signal and R (for the transmitted light intensity at wavelength 514.5 nm) increase sharply in certain range of the noise amplitude D. Outside this range R decreases with increase of D.

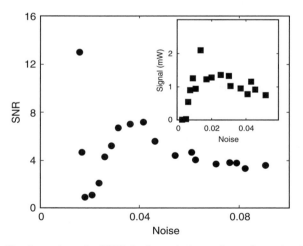

Figure 3. Signal-to-noise-ratio (SNR) in the optical experiment for a signal at frequency $\Omega = 3.9$ Hz as a function of the internal noise intensity [115]. Inset: the corresponding signal amplification.

But, as mentioned above in Sec. I, the mechanism of bistable stochastic resonance requires that the frequency of the input signal is much less than reciprocal relaxation time of the system.

D.   Noise-Enhanced Optical Heterodyning

We now consider the case where two high-frequency fields are mixing non-linearly in the OB system to generate a heterodyne signal. The equation of motion for the OB system takes the form

$$\dot{\phi} + U'(\phi) = M(\phi)A_{in}(t)\cos(\omega_0 t + \psi(t)) + A_{ref}\cos(\omega_0 t) + \Delta I(t) \quad (12)$$

where ω_0 is a high frequency ($\gg \tau^{-1}$) and $A_{in}(t), \psi(t)$ are the slowly-varying amplitude and phase of the modulated input signal, respectively.

In the most interesting and practically important case, when the frequency ω_0 is much higher than the reciprocal relaxation time of the system, simple analytical results can be obtained in the spirit of Ref. 116. If the characteristic frequency of the modulation $\Omega \equiv \dot{\psi}$, and $\dot{A}_{in}/A_{in} \ll \tau^{-1} \ll \omega_0$, the response consists of a comparatively slow motion $\phi^{(sl)}$ with fast oscillations at frequency ω_0 superimposed on it. We therefore seek a solution in the form

$$\phi(t) = \phi^{(sl)}(t) + \phi^{(1)}(t)$$
$$\phi^{(1)}(t) = \omega_0^{-1}[A_{ref}\sin\omega_0 t + M(\phi^{(sl)})A_{in}\sin(\omega_0 t + \Omega t))] \quad (\dot{\psi} \equiv \Omega) \quad (13)$$

Because of the nonlinearity of $M(\phi)$, the oscillations induced by the two beams produce a slowly varying heterodyne force driving the slow motion

$$\dot{\phi}^{(sl)} + U'(\phi^{(sl)}) = -A_{eff}M'(\phi^{(sl)})\sin\Omega t + \Delta I(t)$$
$$A_{eff} = \frac{A_{ref}A_{in}(t)}{2\omega_0} \quad (14)$$

Thus we have reduced this problem to the form of conventional SR (see Section II.C) with only a renormalized effective amplitude for the input signal A_{eff} [cf Eq. 9)] and the function $M(\phi)$ replaced by its derivative $M'(\phi)$ in the first term on the right hand side. By analogy with standard SR, the SNR for heterodyning can be characterized by the ratio R of the low-frequency signal in the intensity of the transmitted radiation, given by $\frac{1}{4}A_{eff}^2|\chi(\Omega)|^2$, to the value of the power spectrum $Q^{(0)}(\Omega)$ [with $Q^{(0)}(\Omega)$ given by (7)–(8)]. The susceptibility of the system can be easily calculated and takes the form

$$\chi(\Omega) = \sum_{n=1,2} w_n\left(N'(\phi_n) + M'(\phi_n)\frac{\partial\bar{I}_{Tn}}{\partial\bar{I}_{ref}}\right) + \frac{w_1 w_2}{D}(\bar{I}_{T1} - \bar{I}_{T2})(M(\phi_1)$$
$$- M(\phi_2))\frac{W_{12} + W_{21}}{(W_{12} + W_{21}) - i\Omega} \quad (15)$$

Similar to what happens in conventional SR, the heterodyne signal and its SNR can be amplified by adding noise to the system, thus manifesting the new phenomenon of *noise-enhanced optical heterodyning*.

These theoretical predictions were first tested in analog simulations for Brownian motion in the symmetric Duffing potential with $M(\phi) \equiv \phi$ [28]. It was found that the heterodyne signal amplitude and corresponding R could be enhanced by adding noise to the system for the cases both of white noise and of broadband high-frequency noise (i.e., noise with a power spectrum centered near the high-frequency ω_0 with half width $\Delta\omega$: $\tau^{-1} \ll \Delta\omega \ll \omega_0$). The specific dependences of the renormalized amplitude of the heterodyne signal A_{eff} on the amplitudes and frequency of the input and reference signals were found to be in good agreement with the theory as shown in Fig. 4. To investigate noise-enhanced optical heterodyning in the DCMS, the 488 nm reference signal was modulated periodically at frequency $\omega_0 = 2.1$ kHz and in addition by noise with a cutoff frequency of 5 kHz. The 514.5 nm input signal was modulated at frequencies $\omega_0 \pm \Omega = 2.1 \pm 0.0039$ kHz. A heterodyne signal at frequency $\Omega = 3.9$ Hz was detected in the transmitted light intensity I_T at wavelength 514.5 nm. The characteristic relaxation time τ_r of the DCMS measured in experiment was order of 2 ms, thus meeting the assumption that $\Omega \ll \tau_r^{-1} \ll \omega_0$.

We have observed strong noise-induced enhancements of both the heterodyne signal (by a factor of 1000) and the signal-to-noise ratio, in Fig. 5. The

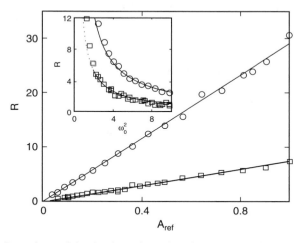

Figure 4. Dependence of the signal-to-noise-ratio R dependence on the squared amplitude of the reference signal, measured [28] in an analog electronic experiment for noise intensities $D = 0.015$ (circles) and $D = 0.14$ (squares). The inset shows the dependence of R on the squared frequency ω_0 for the same two noise intensities.

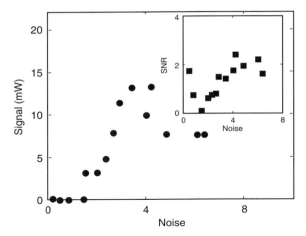

Figure 5. Signal amplification in the optical heterodyning experiment, with $\omega_0 = 2.1$ kHz and $\Omega = 3.9$ Hz, as a function of the internal noise intensity [29]. Inset: the corresponding signal-to-noise-ratio (SNR).

dependence of R on the noise intensity is of the characteristic reversed-N shape familiar from SR in bistable systems and consistent with the theory given above. The enhancement of the SNR occurs within a restricted range of noise intensity, as expected, and the ratio between the value of R at the minimum to that at the local maximum (i.e., the maximum noise-induced "amplification" of the SNR) is ~10.

E. Conclusions

It will be apparent from the above discussion that the double-cavity membrane system is ideally suited to investigations of fluctuations and fluctuational transition phenomena. Stochastic resonance and huge noise-induced amplification of a heterodyne signal have been observed. We would emphasize that noise-protected heterodyning is a general phenomenon that may occur in bistable systems of various sorts, and that it may therefore be of interest for applications in engineering.

III. OPTIMAL PATHS, LARGE FLUCTUATIONS, AND IRREVERSIBILITY

A. Introduction

A fluctuating system typically spends most of its time in the close vicinity of a stable state. Just occasionally, however, it will undergo a much larger departure before coming back or perhaps, in some cases, making a transition to the vicinity of a different stable state. Despite their rarity, these large fluctuations are of great importance in diverse contexts including, for example, nucleation at phase transitions, chemical reactions, mutations in DNA sequences, protein transport in biological cells, and failures of electronic devices. As already mentioned above, there are many cases of practical interest where the fluctuating system is far from thermal equilibrium. Examples include lasers [46], pattern-forming systems [47], trapped electrons that display bistability and switching in a strong periodic field [48,50], and Brownian ratchets [117] which can support a unidirectional current under nonequilibrium conditions. In general, the analysis of the behavior of nonequilibrium systems is difficult, as there is no general relations from which the stationary distribution or the probability of fluctuations can be obtained.

The most promising approach to the analysis of large fluctuations is through the concept of the *optimal path* [42,57,61,65,118–122]. This is the path that the system is predicted to follow with overwhelming probability during the course of the fluctuation. For many years it remained unclear how the optimal path— calculated as a trajectory of an auxiliary Hamiltonian system (see below)—is related to the behaviour of real fluctuating systems. However, through the introduction and use of the prehistory probability distribution [60] (see also Ref. 123), it has been demonstrated that optimal paths are physical observables that can be measured experimentally for both equilibrium [60] and nonequilibrium [124] systems. In what follows we review briefly what has been achieved and point out the opportunities that have now appeared for making rapid scientific progress in this burgeoning research field.

B.   Theory

Consider an overdamped system driven by a periodic force $K(q; \phi)$ and white noise $\xi(t)$, with equation of motion

$$\dot{q} = K(q; \phi) + \xi(t), \qquad K(q; \phi) = K(q; \phi + 2\pi)$$
$$\phi \equiv \phi(t) = \omega t + \phi_0, \qquad \langle \xi(t)\xi(t') \rangle = D\delta(t - t') \tag{16}$$

The familiar *overdamped* bistable oscillator driven by a periodic force provides a simple example of the kind of system we have in mind:

$$\dot{q} = -U'(q) + A \cos \omega t + \xi(t)$$
$$U(q) = -\frac{1}{2}q^2 + \frac{1}{4}q^4 \tag{17}$$

We consider a situation that is both *nonadiabatic* and *nonlinear* in which neither ω nor A need be small; only the intensity D of the Gaussian noise will be assumed small. We investigate rare fluctuations to a remote point (q_f, ϕ_f), coming from the metastable state within whose domain of attraction (q_f, ϕ_f) is located. The position of the stable state $q^{(0)}(t)$ is itself a periodic function of time;

$$\dot{q}^{(0)} = K(q^{(0)}; \phi), \qquad q^{(0)}(t + 2\pi\omega^{-1}) = q^{(0)}(t) \tag{18}$$

The equations for optimal paths can be found using the eikonal approximation to solve the corresponding Fokker–Planck equation, or by using a path-integral formulation and evaluating the path integral over the fluctuational paths in the steepest-descent approximation (for details and discussion, see Refs. 42,57,64, 65,71–73, and 118). The optimal path corresponds to the locus traced out by the maximum in the prehistory probability density, $p_h(q, \phi | q_f, \phi_f)$ [60,124]. This is the probability density that a system arriving at the point (q_f, ϕ_f) at the instant t_f $(\phi(t_f) = \phi_f)$ had passed through the point q, ϕ at the instant t $(t < t_f)$. A particular advantage of this formulation is that p_h is a physical quantity that can be measured experimentally. The approach can be extended to include the analysis of singular points in the pattern of optimal paths.

Using the path-integral expression for the transition probability density [64], one can write p_h in the form [60]

$$p_h(q, \phi | q_f, \phi_f) = C \int_{q(t_i) \approx q^{(0)}(t_i)}^{q(t_f) = q_f} \mathscr{D}q(t') \, \delta(q(t) - q)$$

$$\times \exp\left[-\frac{S[q(t)]}{D} - \frac{1}{2}\int_{t_i}^{t_f} dt' \frac{\partial K}{\partial q}\right] \qquad t_i \to -\infty \tag{19}$$

$$\phi \equiv \phi(t), \qquad \phi_f \equiv \phi(t_f)$$

Here, C is a normalization constant determined by the condition

$$\int dq \, p_h(q, \phi | q_f, \phi_f) = 1$$

$S[q(t)]$ has the form of an action functional for an auxiliary dynamical system with time-dependent Lagrangian $L(\dot{q}, q; \phi)$:

$$S[q(t)] = \int_{t_i}^{t_f} dt \, L(\dot{q}, q; \phi), \qquad L(\dot{q}, q; \phi) = \frac{1}{2}[\dot{q} - K(q; \phi)]^2 \tag{20}$$

In the range of small noise intensities D, the optimal path $q_{\text{opt}}(t | q_f, \phi_f)$ to the point (q_f, ϕ_f) is given by the condition that the action S be minimal. The variational problem for S to be extremal gives Hamiltonian equations of motion for the coordinate q and momentum p of the auxiliary system

$$\frac{dq}{dt} = \frac{\partial H}{\partial p}, \qquad \frac{dp}{dt} = -\frac{\partial H}{\partial q}, \qquad \frac{dS}{dt} = \frac{1}{2}p^2$$

$$H \equiv H(q,p;\phi) = \frac{1}{2}p^2 + pK(q;\phi) \qquad (21)$$

$$H(q,p;\phi) = H(q,p;\phi + 2\pi)$$

The boundary conditions for the extreme paths (21) follow from (19) and (20)

$$q(t_f) = q_f$$
$$q(t_i) \to q^{(0)}(t_i), \quad p(t_i) \to 0, \quad S(t_i) \to 0 \qquad \text{for} \quad t_i \to -\infty \qquad (22)$$

Since the Hamiltonian $H(q,p;\phi)$ is periodic in ϕ, the set of paths $\{q(t), p(t)\}$ is also periodic: the paths that arrive at a point $(q_f, \phi_f + 2\pi)$ are the same as the paths that arrive at the point (q_f, ϕ_f), but shifted in time by the period $2\pi/\omega$. The action $S(q_f, \phi_f)$ evaluated along the extreme paths is also periodic as a function of the phase ϕ_f of the final point (q_f, ϕ_f). The function $S(q, \phi)$ satisfies the Hamilton–Jacobi equation

$$\omega \frac{\partial S}{\partial \phi} = -H\left(q, \frac{\partial S}{\partial q}; \phi\right), \qquad p \equiv \frac{\partial S}{\partial q} \qquad (23)$$

$$S(q, \phi) = S(q, \phi + 2\pi)$$

It is straightforward to see that the extreme paths obtained by solving (21) form a one-parameter set. It is known from the theory of dynamical systems [125] that trajectories emanating from a stationary state lie on a Lagrangian manifold (LM) in phase space $(q, \phi, p = \partial S/\partial q)$ (the unstable manifold of the corresponding state) and form a one-parameter set. The action $S(q, t)$ is a smooth single-valued function of position on the LM. It is a Lyapunov function: it is nondecreasing along the optimal trajectories. Therefore $S(q, t)$ may be viewed as a generalised nonequilibrium thermodynamic potential for a fluctuating dynamical system [64]. The projections of trajectories in phase space onto configuration space form the extreme paths. Optimal paths are the extreme paths that give the minimal action to a given point in the configuration space. These are the optimal paths that can be visualised in an experiment via measurements of the prehistory probability distribution.

The pattern of extreme paths, LM, and action surfaces for an overdamped periodically driven oscillator (17) are shown in Fig. 6. The figure illustrates generic topological features of the pattern in question. It can be seen from Fig. 6 that, although there is only one path to a point (q, ϕ, p) in phase space, several

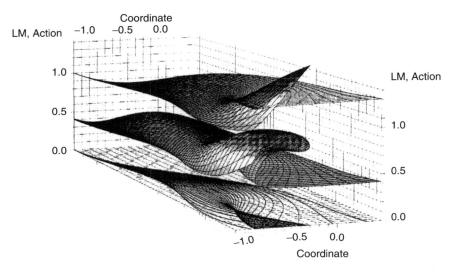

Figure 6. From top to bottom: action surface; Lagrangian manifold (LM); and extreme paths calculated [80] for the system (17) using equations (21). The parameters for the system were $A = 0.264$ and $\omega = 1.2$. To clarify interrelations between singularities in the pattern of optimal paths, action surface, and LM surface, they are shown in a single figure, as follows, the action surface has been shifted up by one unit; and the LM has been scaled by a factor $\frac{1}{2}$ and shifted up by 0.4.

different extreme paths may come from the stationary periodic state to the corresponding point (q, ϕ) in configuration space. These paths cross each other. This is a consequence of the folding of the Lagrangian manifold.

A generic feature related to folding of LMs is the occurrence of *caustics* in the pattern of extreme paths. Caustics are projections of the folds of an LM. They start at cusp points. It is clear from Fig. 6 that an LM structure with two folds merging at the cusp must give rise to a local swallowtail singularity in the action surface. The spinode edges of the action surface correspond to the caustics. A *switching line* emanates from the cusp point at which two caustics meet. This is the projection of the line in phase space along which the two lowest sheets of the action surface intersect. The switching line separates regions that are reached along different *optimal* paths, and the optimal paths intersect on the switching line. The intersection occurs *before* a caustic is encountered by the optimal path. The formation of the singularities, avoidance of caustics, and formation of switching lines have been analyzed numerically [119], and a complete theory has been given [120]. Until 1996, the generic topological features of the pattern of optimal paths had not been observed in any experiment. We now describe briefly the experimental technique [124] that enables the pattern of optimal paths and its singularities to be observed, and we present and discuss some of our initial results.

C. Experiments

The experiments are based on analog electronic circuits designed in the usual way [112,126] to model the system of interest, and then driven by appropriate external forces. Their response is measured and analysed digitally to create the statistical quantity of interest which, in the present case, was usually a prehistory probability distribution [60,124]. We again emphasize that such experiments provide a valid test of the theory, and that the theory should in this case be universally applicable to *any* system described by (16), including natural systems, technological ones, or the electronic models studied here. Some experiments on a model of (17) are now described and discussed as an illustrative example of what can already be achieved.

The model was driven continuously by external quasi–white noise from a noise generator and by a periodic force from a frequency synthesiser. The fluctuating voltage representing $q(t)$ was digitized and analysed in discrete blocks of 32,768 samples using a Nicolet NIC-1180 data processor. The input sweeps were triggered by the frequency synthesizer so that information about the phase of the periodic force could be retained. Whenever $q(t)$ entered a designated square centered on a particular (q_f, ϕ_f) value, the immediately preceding part of the trajectory was collected and stored; in cases where relaxation trajectories were also of interest, the immediately following part of the trajectory was preserved, too. The trajectories that had arrived in any chosen square could subsequently be ensemble-averaged together to create the prehistory probability distribution $p_h(q, \phi | q_f, \phi_f)$ corresponding to the chosen (q_f, ϕ_f), with or without the relaxational tail back toward the stable state.

Because the fluctuations of interest were—by definition—rare, it was usually necessary to continue the data acquisition process for several weeks in order to build up acceptably smooth distributions. For this reason, the analysis algorithm was designed to enable trajectories to several termination squares (not just one) to be sought in parallel: an 8×8 matrix of 64 adjacent termination squares, each centered on a different (q_f, ϕ_f) was scanned.

Experimentally measured p_h for the system (17) for two qualitatively different situations are shown in Figs. 7 and 8. It is immediately evident: (1) that the prehistory distributions are sharp and have well-defined ridges; (2) that the ridges follow very closely the theoretical trajectories obtained by solving numerically the equations of motion for the optimal paths, shown by the full curves on the top planes. It is important to compare the fluctuational path bringing the system to (q_f, ϕ_f) with the relaxational path back towards the stable state in thermal equilibrium, Fig. 7, and away from it, Fig. 8. Figure 7 plots the distribution for the system (17) in thermal equilibrium, namely $A = 0$. The

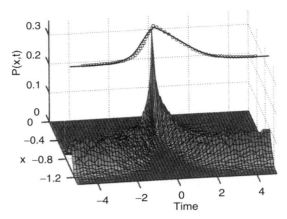

Figure 7. The prehistory probability density $p_h(x, t; x_f, 0)$ [60] for (17), measured [62] for $A = 0$ in the analog electronic experiment for a final position $x_f = -0.30$ with $D = 0.0701$.

ridges of a distribution are compared with the calculated fluctuational and relaxational paths at the top of the figure. The time reversal symmetry [44] between these paths can be clearly seen. Figure 8 plots the p_h and the ridges of a distribution recorded for the special nonequilibrium situation that arises when the termination point lies on the switching line [124]. In Fig. 8, the time-dependent

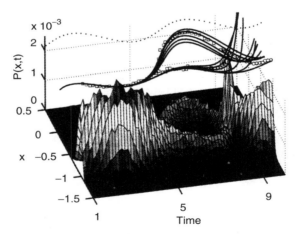

Figure 8. Fluctuational behavior measured and calculated for an electronic model of the non-equilibrium system (17) with $A = 0.264$, $D = 0.012$. The man figure plots the prehistory probability density $(p_h(x, t; x_f, 0)$ and posthistory distribution to/from the remote state $x_f = -0.63, t = 0.83$, which lies on the switching line. In the top plane, the fluctuational (squares) and relaxational (circles) optimal paths to/from this remote state were determined by tracing the ridges of the distribution [62].

stable and unstable states bear $x = -1$ and $x = 0$ are shown by dashed lines on the top. The data are compared to the (theoretical) fluctuational paths, calculated from (21), shown as full lines. It can be seen: that there are *two* distinct paths via which the system can arrive at (q_f, ϕ_f) but only *one* relaxational path taking it back to the stable state. Unlike the behaviour expected and seen [62] in equilibrium systems, neither of the fluctuational paths is a time-reversed image of the relaxational one. Note that there are *two* equally probable fluctuational paths to arrive on a switching line, they form a so called *corral* [124].

Although the system (17) is relatively simple, it describes very well the fluctuational dynamics of many real physical systems. In particular, a behavior qualitatively similar to the one shown in Fig. 7 was observed recently in the experiments with semiconductor lasers [83,84].

In the work by Hales and co-authors [83] the prehistory distribution was observed experimentally using a semiconductor laser with optical feedback. Near the solitary threshold, the system was unstable: after a period of nearly steady operation, the radiation intensity decreased; then it recovered comparatively quickly, growing to regain its original value; decreased again; and the cycle repeated. In the experiment, the output intensity was digitized with 1 ns resolution. The p_h obtained in [83] from 1512 events is shown in Fig. 9. The results were compared with the results of numerical simulation for the system (17).

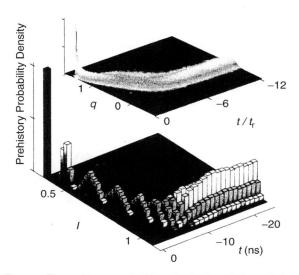

Figure 9. Bottom: The prehistory probability distribution of the radiation intensity I (in arbitrary units) for dropout events in a semiconductor laser. Top: The PPD for a Brownian particle, obtained from simulations [83].

In the work by Willemsen and co-authors [84] the three Stokes polarization parameters were studied during polarization switches in a vertical-cavity semiconductor laser. It was demonstrated that when the linear part of the absorptive anisotropy is close to zero [127], the laser is bistable and switches stochastically between two polarisations [128]. The analysis of large fluctuations of polarizations in this system [84] reveals what authors have called a "stochastic inversion symmetry" (see Fig. 10), which is analogous to the time-reversal symmetry observed for the model (17) and shown in Fig. 7.

D. Optimal Paths on a Finite Time Range, and Conclusions

The previous discussion, and the results of Refs. [62,124, and 129–132] among others, show that our analog electronic technique makes it possible to test fundamental tenets of fluctuation theory, and thus provide an experimental basis on which the theory can advance. We can investigate the pattern of optimal paths for thermally nonequilibrium systems and reveal its singularities including, in particular, switching lines and strong (nonanalytic in the noise intensity) smearing of the prehistory probability distribution near cusp points. The particular system we have investigated has the least number of degrees of

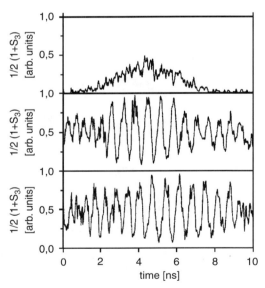

Figure 10. Time-resolved measurements of a very large polarization fluctuation, where the size of the fluctuation is about half (in fact, 45%) of that of a complete polarization switch [84]. s_1, s_2 and s_3 are the normalized Stokes parameters representing the polarization state on the Poincaré sphere [84].

freedom necessary to observe these singularities, and therefore it is most appropriate for analysis in these initial investigations. The approach that we have described is in principle applicable to any nonequilibrium system, and we believe it will be found useful in a wide range of applications.

It should also be clear that the structures predicted by the theory are indeed observed in real systems. Reasoning along these lines, researchers have started predicting peculiar features that should be observable in real systems, on the basis of the topology of corresponding Lagrangian manifold. These include features predicted on the assumption that the optimal path (and the corresponding fluctuation in the real system) takes place over a finite time range [133–137].

One of the most striking effects predicted on this basis is perhaps what occurs in noise induced escape from a metastable well on a timescale preceding the formation of a quasiequilibrium distribution within the metastable part of the potential (see Refs. 136 and 137 for more details), which we now review briefly.

In his seminal work [109], Kramers considered the noise-induced flux from a single metastable potential well i.e. he considered a Brownian particle

$$\ddot{q} + \Gamma\dot{q} + \frac{dU}{dq} = f(t)$$
$$\langle f(t) \rangle = 0 \qquad\qquad (24)$$
$$\langle f(t)f(t') \rangle = 2\Gamma T\delta(t - t')$$

which was put initially at the bottom of a metastable potential well $U(q)$ and then he calculated the quasistationary probability flux beyond an absorbing barrier. There have been many developments and generalizations of the Kramers problem (see Refs. 138 and 139 for reviews), but both he and most of those who followed him considered just the *quasistationary* flux, i.e. the flux established after the formation of a quasistationary distribution within the well (up to the barrier). The quasistationary flux is characterized by a slow exponential decay, an Arrhenius dependence on temperature T, and a relatively weak dependence on friction Γ:

$$J_{qs}(t) = \alpha_{escape}e^{-\alpha_{escape}t}, \qquad \alpha_{escape} = Pe^{-(\Delta U/T)} \qquad (25)$$

where P depends on Γ and T in a nonactivation manner.

But how does the flux evolve from its zero value at the initial moment to its quasistationary value at timescales exceeding the time t_f for the formation of quasiequilibrium? It is obvious that the answer may depend on initial conditions. The most natural are those corresponding to the stable stationary state of the noise-free system i.e. $(q = q_{bottom}, \dot{q} = 0)$ where q_{bottom} is the coordinate of the bottom of the potential well. We assume such an initial state here. If the noise is switched on suddenly (e.g., if the thermal isolation of a frozen system is broken), then the time evolution of the escape flux from the noise-free metastable initial state is highly relevant. It might seem natural that the evolution from zero to the quasistationary value should be *smooth*. Such an assumption might also seem to have been confirmed recently by Schneidman [140] who found that, for both the strongly underdamped and overdamped cases, the escape flux from a single metastable well grows with time t smoothly, at $t \sim t_f$. But does this exhaust the problem? We can prove theoretically, and demonstrate experimentally, that there are some generic situations when the escape flux behaves in a quite different manner.

Our prediction are based, as mentioned, on an extensive use of the method of *optimal fluctuation* within which an escape rate is sought in the form

$$\alpha_{escape} = Pe^{-(S/T)} \tag{26}$$

where the action S does not depend on T; the prefactor P does depend on T, but relatively weakly. The action S is related to a certain optimal fluctuation that, in turn, corresponds to the *most probable escape path* (MPEP).

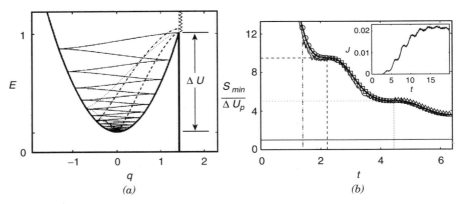

Figure 11. (a) Examples of MPEPs (plotted in the energy-coordinate plane $E - q$ where $E = \dot{q}^2/2 + U(q)$) to escape from the bottom of the metastable well $U(q) = q^2/2$ with $q < \sqrt{2}$ (thick solid line) to beyond the barrier at $q = \sqrt{2}(U(q) = -\infty$ at $q > \sqrt{2}$, which is equivalent to the absorbing wall indicated by triangles), for $\Gamma = 0.05$; (b) the corresponding theoretical (thick solid line) and experimental (thin jagged line) dependences of the action S on the escape time t. Circles, squares, and triangles indicate bits corresponding to respectively 0, 1, and 2 turning points in the MPEP. The dashed and dotted lines indicate: in (b) the first and second inflection points with $dS/dt = 0$; and in (a) the corresponding MPEPs. The thin solid line shows: in (b) the large-time asymptotic levels $S = \Delta U$; and in (a) the corresponding MPEP (which is the time reversal of the noise-free trajectory from the top of the barrier into the bottom of the well). The dash–dotted line shows in (a) hte MPEP corresponding to some arbitrarily chosen time $t = 4.51$ [see (b)] and demonstrates, in particular, that the escape velocity is generally nonzero. The inset shows the experimental dependence of the flux on time, for $T = \Delta U$ [136,137].

The quasistationary flux is formed by optimal fluctuations which bring the system from the bottom of the well to the saddle during an *optimal time*

$$t_{opt} \sim \frac{1}{\min(\Gamma, \omega_0)} \ln\left(\frac{\Delta U}{T}\right) \qquad (27)$$

where ω_0 is the frequency of eigenoscillation in the bottom of the well.

At much shorter timescales, $t \ll t_{opt}$, the flux is necessarily formed by optimal fluctuations strongly differing from those of duration t_{opt}, and the smaller t the more marked this difference becomes. Thus, in the range (27), S depends on t. Moreover, it can be shown rigorously that if $\Gamma < \Gamma_c$ where Γ_c is typically equal to $2\omega_0$, then $S(t)$ is a *stepwise* function: see the example in Fig. 11. The vertical and horizontal positions of the center of the step $S(t)$ number n (counted from the left) equal respectively $\Delta U \omega_0 / (n\pi\Gamma)$ and $n\pi/\omega_0$, provided $n\pi\Gamma \ll \omega_0$. Generally, when the shape of the potential well departs from parabolicity, the equalities turn into approximations. Thus, in the range (27), J depends exponentially strongly on both Γ and t (cf. the inset of Fig. 11).

E. Logarithmic Susceptibility

A very good example of the usefulness of the concept of the optimal path is the idea of the logarithm susceptibility (LS) [56,87,141].

Underlying the theory of the LS [56,87] is the realization that, although the motion of the fluctuating system is random, large rare fluctuations from a metastable state to a remote state, or during escape, take place in an almost deterministic manner: the system is overwhelmingly most likely to move along a particular trajectory known as the optimal path (see Refs. 42,57,64,69, and 121 and references cited therein). The effect of a comparatively weak field on the escape probability can therefore be understood in terms of the work that the field does on the system as it moves along the optimal path. One may expect this work to be related to the field-induced change in the activation energy R for the corresponding large fluctuation. This change is linear in the field, provided that the field-induced change of the optimal path itself is negligible. It follows from these arguments that in the case of periodic driving $F(t) = \sum_k F_k \exp(ik\Omega t)$, the leading-order correction δR to the activation energy of escape is

$$\delta R = \min_{t_c} \delta R(t_c), \qquad \delta R(t_c) = \sum_k F_k \tilde{\chi}(k\Omega) e^{ik\Omega t_c}$$

$$\tilde{\chi}(\Omega) = -\int_{-\infty}^{\infty} dt \, \dot{q}^{(0)}(t) e^{i\Omega t}, \qquad \dot{q}^{(0)} = U'(q^{(0)}) \qquad (28)$$

Here, $\tilde{\chi}(\Omega)$ is the LS for escape. It is given [56,87] by the Fourier transform of the velocity along the most probable escape path $q^{(0)}(t)$ in the absence of driving

$(F(t) = 0)$. The path $q^{(0)}(t)$ is an instanton [123]: it starts from $t \to -\infty$ at the metastable minimum q_s of the potential $U(q)$ and for $t \to \infty$ arrives at the top q_u of the potential barrier over which the particle escapes. The minimization over t_c corresponds to choosing the position of the center of the instanton so as to maximize the work the field $F(t)$ does on the system along the escape path $q^{(0)}(t - t_c)$. We have already noted that, for Markov systems in thermal equilibrium, optimal fluctuational paths are the time-reversed relaxational paths in the absence of noise [44,131,142]. Unlike the standard linear susceptibility [9], which, on causality arguments, is given by a Fourier integral over time from 0 to ∞, the LS $\tilde{\chi}(\Omega)$ is given by an integral from $-\infty$ to ∞. The analytical properties of $\tilde{\chi}(\Omega)$ therefore differ from those of the standard susceptibility, and in particular their high-frequency asymptotics are *qualitatively* different. The standard susceptibility for damped dynamical systems decays as a power law for large Ω (e.g., as $1/[U''(q_s) - i\Omega]$, for the model of damped Duffing oscillator). In contrast, from (28) the LS decreases *exponentially* rapidly:

$$\tilde{\chi}(\Omega) = M e^{-|\Omega|\tau_p}, \qquad \tau_p = \min \left| \mathrm{Im} \int dq/U'(q) \right| \qquad (29)$$

Here, the integral is taken from any point in the interval (q_s, q_u) to the (complex) position q_p of the appropriate singularity of $U'(q)$. Note that $\dot{q}^{(0)}(t - t_c)$ for given real t_c has a pole or a branching point at $\mathrm{Im}\, t = \tau_p$. The prefactor M depends on the form of $U(q)$ near q_p and can be obtained in a standard way. In particular, for a polynomial potential $(|q_p| \to \infty)$ with $U(q) = Cq^n/n$ for $|q| \to \infty$, we have

$$|M| = 2\pi \left| \frac{\Omega}{C} \right|^\nu \frac{|\nu|^{\nu+1}}{\nu!}, \qquad \nu = \frac{1}{n - 2} \qquad (30)$$

This expression applies also for finite $|q_p|$, with $U(q) \approx C/\mu(q - q_p)^\mu$ for $q \to q_p$, if n in (30) is replaced by $-\mu$: note that $|M|$ then decreases with increasing Ω.

To test these predictions, we used an analog electronic model [112] of the overdamped motion of a Brownian particle in the double-well Duffing potential. We drove it with zero-mean quasi–white Gaussian noise from a shift-register noise generator, digitized the response $q(t)$, and analyzed it with a digital data processor. We also carried out a complementary digital simulation [143]. Numerical simulations in the case of small damping are currently in progress; preliminary results indicate a resonant behavior of the LS. The analog and digital measurements of R involved noise intensities in the ranges $D = 0.028-0.036$ and $D = 0.020 - 0.028$, respectively; the lowest (real time [112]) driving frequency used was 460 Hz. The results are plotted in Fig. 12. The major observation is that, as expected, R is indeed *linear* in the force amplitude ($R = \frac{1}{4}$

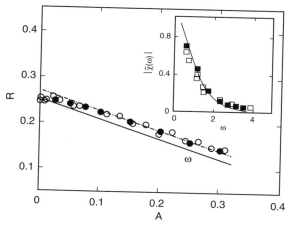

Figure 12. The dependence of the activation energy R on the amplitude A of the harmonic driving force $F(t) = A \cos(1.2t)$ as determined [141] by electronic experiment (filled circles), numerical sumulations (open circles) and analytical calculation (solid line), based on (28) for an overdamped duffing oscillator $U(q) = -q^2/2 + q^4/4$; the dashed–dotted line, drawn parallel to the full curve, is a guide to the eye. The inset shows the absolute value of the LS of the system $|\tilde{\chi}(\omega)|$ (28) measured (filled and open squares for experiment and numerical simulation, respectively) and calculated (full curve) as a function of frequency Ω using (29) with $\tilde{\chi}(0) = -1$ and $\tau_p = \pi/2, M = -(1+i)(\pi\omega)^{1/2}$ in (29).

for $A = 0$). The slope yields the absolute value of the LS. Its frequency dependence, a fundamental characteristic of the original equilibrium system, is compared with the theoretical predictions (29) in the inset of Fig. 12.

The LS theory was applied to the localization of a Brownian particle in a three-dimensional optical trap [89]: a transparent dielectric spherical silica particle of diameter 0.6 μm suspended in a liquid [88]. The particle moves at random within the potential well created with a gradient three-dimensional optical trap—a technique widely used in biophysical studies. The potential was modulated by a biharmonic force. By changing the phase shift between the two harmonics it was possible to localize the particle in one of the wells in very good quantitative agreement with the predictions based on the LS.

F. Conclusions

It is evident from the preceding discussion that the theory of the optimal paths provides a deep physical insight into the dynamics of fluctuations and is in good agreement both with the results of analog and numerical simulations and with the results of the experiments in optical systems. It has now become possible to use the prehistory formulation [60] as a basis for experiments on fluctuational dynamics. The work on Markov systems presented in this section has already verified several longstanding theoretical predictions, including symmetry between the growth and decay of classical fluctuations [44], the breaking of this symmetry under nonequilibrium conditions [57,59,65,119], the relationship between lack of detailed balance and onset of singularities in the pattern of optimal paths, as well as the character of these singularities [120–122,124,144], including occurrence of switching between optimal paths and critical broadening of the paths distribution. It has now become possible to apply this theory and the corresponding experimental methods to the analysis of the fluctuational dynamics in optical systems and to develop new methods of controlling them.

V.   CHAOTIC ESCAPE AND THE PROBLEM OF OPTIMAL CONTROL

One of the main problem in the dynamics of optical systems is that of controlling the system dynamics [145]. The difficulties in solving such a problem depend on many factors. A typical optical system is characterized by the phenomenon of multistability [146–149], specifically, the coexistence of a relatively small number of distinct dynamical regimes that are defined by the initial conditions. Because real optical systems are always subject to random fluctuations [46,147], spontaneous transitions of the system take place from one regime to another. It is obviously desirable to be able to control these transitions. Moreover, in optical systems nonregular oscillations are often observed that can be described by the theory of deterministic chaos [145,147, 149]; such nonregular oscillations in the phase space of the system can be characterized by a chaotic attractor. The transformation of the system dynamics from a chaotic regime to a regular regime is also an interesting problem in dynamical control. In solving it for real systems, it is essential to take account of fluctuations.

The need to be able to control chaos has attracted considerable attention. Methods already available include a variety of minimal forms of interaction [150–155] and methods of strong control [156,157] that necessarily require a large modification of the system's dynamics, for at least a limited period of time. For example, in Refs. 158 and 159, the procedure of controlling chaos by means of minimal forms of interaction (saddle cycle stabilization) is realized for different laser systems.

At the same time the energy-optimal directing of the motion away from a chaotic attractor (CA) to another coexisting attractor has remained an important unsolved problem of long standing. Its solution would be an important extension of the range of model-exploration objectives [156,154] achievable through minimal control techniques and has a variety of applications for controlling the dynamics of multistable optical systems [147].

In this section the application of the optimal path approach to the problem of escape from a nonhyperbolic and from a quasihyperbolic attractor is examined. We discuss these two different types of chaotic attractor because it is known [160] that noise does not change very much the structure and properties of quasi-hyperbolic attractors, but that the structure of non-hyperbolic attractors is abruptly changed in the presence of noise, with a strong dependence on noise intensity. Note that for optical systems both types of chaotic attractor [161–163] (nonhyperbolic and quasihyperbolic) are observed, but a nonhyperbolic attractor is much more typical.

A. Escape from a Nonhyperbolic Attractor

1. Introduction to the Optimal Control Problem

Consider a system of the form

$$\dot{x} = f(x, u, t) \tag{31}$$

with the state variable $x \in R^n$, and an admissible control function $u \in R^m$ in the control set U. Assume that it is desired to transfer the system from the state $X_0 = x(t_0)$ to the terminal state $X_1 = x(t_1)$ in such a way that the ("cost") functional

$$J = \min \int_{t_0}^{t_1} f_0(x, u, t) \, dt \tag{32}$$

is minimized, with t_1 unspecified. Let $(u(t), x(t))$ be a solution of this problem. Then there exist continuous piecewise differentiable functions $y_0(t), \ldots, y_n(t)$ that are not simultaneously zero and that satisfy together with the functions $x_i(t)$ the differential equations [164]

$$\dot{x}_i = \frac{\partial H}{\partial y_i}$$
$$\dot{y}_i = -\frac{\partial H}{\partial x_i} \tag{33}$$

with the Hamiltonian

$$H(x_1, \ldots, x_n; y_0, \ldots, y_n; u(t), t) = \sum_{i=0}^{n} y_i f_i(x_1, \ldots, x_n; u(t), t) \tag{34}$$

An optimal control function $u(t)$ maximizes H at each instant. H is a continuous function of the time and one has $H(t_1) = 0$. If the functions $f_i, i = 0, \ldots, n$ do not depend on time explicitly, then H is a constant and equal to zero.

It can be seen that the solution of the problem of the energy-optimal guiding of the system from a chaotic attractor to another coexisting attractor requires the solution of the boundary-value problem (33)–(34) for the Hamiltonian dynamics. The difficulty in solving these problems stems from the complexity of the system dynamics near a CA and is related, in particular, to the delicate problems of the uniqueness of the solution, its behaviour near a CA, and the boundary conditions at a CA.

Below we show how the energy-optimal control of chaos can be solved via a statistical analysis of fluctuational trajectories of a chaotic system in the presence of small random perturbations. This approach is based on an analogy between the variational formulations of both problems [165]: the problem of the energy-optimal control of chaos and the problem of stability of a weakly randomly perturbed chaotic attractor. One of the key points of the approach is the identification of the optimal control function as an optimal fluctuational force [165].

We emphasize that the question of stability of a CA under small random perturbations is in itself an important unsolved problem in the theory of fluctuations [92–94] and the difficulties in solving it are similar to those mentioned above. Thus it is unclear at first glance how an analogy between these two unsolved problems could be of any help. However, as already noted above, the new method for statistical analysis of fluctuational trajectories [60,62,95,112] based on the prehistory probability distribution allows direct experimental insight into the almost deterministic dynamics of fluctuations in the limit of small noise intensity. Using this techique, it turns out to be possible to verify experimentally the existence of a unique solution, to identify the boundary condition on a CA, and to find an accurate approximation of the optimal control function.

Let us now formulate the problem of the energy-optimal steering of the motion from a chaotic attractor to the coexisting stable limit cycle for a simple model, a noncentrosymmetric Duffing oscillator. This is the model that, in the absence of fluctuations, has traditionally been considered in connection with a variety of problems in nonlinear optics [166]. Consider the motion of a periodically driven nonlinear oscillator under control

$$\dot{q}_1 = K_1(\mathbf{q}(t)) = q_2$$
$$\dot{q}_2 = K_2(\mathbf{q}(t)) + u(t) \tag{35}$$
$$= -2\Gamma q_2 - \omega_0^2 q_1 - \beta q_1^2 - \gamma q_1^3 + h\sin(\omega t) + u(t)$$

Here $u(t)$ is the control function. It is a system where chaos can be observed at relatively small values $h \approx 0.1$ of the driving force amplitude and the chaotic attractor is a nonhyperbolic attractor or a quasiattractor [167].

We have considered the following energy-optimal control problem. The system (35) with unconstrained control function $u(t)$ is to be steered from a CA to a coexisting stable limit cycle (SC) in such a way that the "cost" functional J is minimized, with t_1 unspecified

$$R = \inf_{u \in U} \frac{1}{2} \int_{t_0}^{t_1} u^2(t)\, dt \qquad (36)$$

Here the control set U consists of functions (control signals) able to move the system from the CA to the SC. The Pontryagin Hamiltonian (34) and the corresponding equations of motion take the form

$$\dot{q}_i = \frac{\partial H_c}{\partial p_i}, \quad \dot{p}_i = -\frac{\partial H_c}{\partial q_i}, \qquad i = \{1, 2\}$$

$$H_c = \frac{1}{2} p_2^2 + p_1 K_1 + p_2 K_2 \qquad (37)$$

Here it is assumed that the optimal control function $u(t)$ at each instant takes those values $u(t) = p_2$ that maximize H_c over U.

We note that for $p_1 = 0$ and $p_2 = 0$ the dynamics of (37) reduces to the deterministic dynamics of the original system (35) in the absence of control ($u(t) = 0$). So we begin our analysis by considering some relevant properties of the deterministic dynamics of a periodically driven nonlinear oscillator.

The parameters of the system (35) were chosen such that the potential is monostable ($\beta^2 < 4\gamma\omega_0^2$), the dependence of the energy of oscillations on their frequency is nonmonotonic ($\frac{\beta^2}{\gamma\omega_0^2} > \frac{9}{10}$), and the motion is underdamped $\Gamma \ll \omega \approx 2\omega_0$.

A simplified parameter space diagram obtained numerically [168] is shown in Fig. 13. The dashed lines bound the region in which both the linear and nonlinear responses of period 1 coexist. The upper line marks the boundary of the linear response, and the lower line marks that for the nonlinear responses. The boundaries of hysteresis for the period 1 resonance are shown by solid lines. The region in which linear response coexists with one or two nonlinear responses of period 2 is bounded by dotted lines. This region is similar to the one bounded by dashed lines. The region of coexistence of the two resonances of period 2 is bounded by the dashed–dotted line. Chaotic states are indicated by small dots. The chaotic state appears as the result of period-doubling bifurcations, and thus corresponds to a nonhyperbolic attractor [167]. Its boundary of attraction $\partial\Omega$ is nonfractal and is formed by the unstable manifold of the saddle cycle of period 1 (S1).

For a given damping ($\Gamma = 0.025$) the amplitude and the frequency of the driving force were chosen so that the chaotic attractor coexists with the stable limit cycle (SC): $h = 0.13, \omega_f = 0.95$ (see Fig. 13).

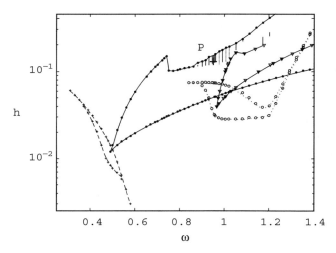

Figure 13. Phase diagram of the system (35) on the (ω, h) plane obtained numerically for the parameter values $\Gamma = 0.025, \omega_0 = 0.597, \beta = 1, \gamma = 1$. See text for a description of the symbols; the various lines are guide to the eye. The working point P, with $\omega_f = 0.95, h = 0.13$, shown by a thick plus, was chosen to lie in the region of coexistence of the period 1 stable limit cycle and of the strange attractor [168].

The basins of attraction of the coexisting CA (strange attractor) and SC are shown in the Fig. 14 for the Poincaré crosssection $\omega_f t = 0.6\pi (\text{mod} 2\pi)$ in the absence of noise [169]. The value of the maximal Lyapunov exponent for the CA is 0.0449. The presence of the control function effectively doubles the dimension of the phase space (compare (35) and (37)) and changes its geometry. In the extended phase space the attractor is connected to the basin of attraction of the stable limit cycle via an unstable invariant manifold. It is precisely the complexity of the structure of the phase space of the auxiliary Hamiltonian system (37) near the nonhyperbolic attractor that makes it difficult to solve the energy-optimal control problem.

However, using a method proposed [60,62,95,112] for experimental analysis of the Hamiltonian flow in an extended phase space of the fluctuating system, we can exploit the analogy between the Wentzel–Freidlin and Pontryagin Hamiltonians arising in the analysis of fluctuations, and the energy-optimal control problem in a nonlinear oscillator. To see how this can be done, let us consider the fluctuational dynamics of the nonlinear oscillator (35).

Let us analyze the motion of an oscillator interacting with a thermal bath:

$$\dot{q}_1 = q_2$$
$$\dot{q}_2 = -2\Gamma q_2 - \omega_0^2 q_1 - \beta q_1^2 - \gamma q_1^3 + h\sin(\omega t) + \xi(t) \qquad (38)$$
$$\langle \xi(t) \rangle = 0, \qquad \langle \xi(t)\xi(0) \rangle = D\delta(t) = 4\Gamma kT\delta(t)$$

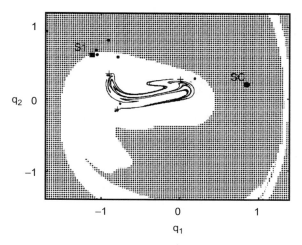

Figure 14. The basins of attraction of the SC (shaded) and CA (white) for a Poincaré cross section with $\omega_f t = 0.6\pi(\mathrm{mod}\,2\pi)$, $\omega_f = 0.95$ in terms of q_1 at q_2. The boundary of the CA's basin of attraction, the saddle cycle of period 1, S1, is shown by the filled square. The saddle cycle of period 3, S3, is shown by pluses. The intersections of the actual escape trajectory with the Poincaré cross section are indicated by the filled circles [169].

In the zero-noise-intensity limit, a consistent theoretical development [42,170] from the microscopic to the macroscopic equations of motion leads to descriptions of both its deterministic (dissipative) and fluctuational dynamics within the framework of Hamiltonian formalism [57]. The comparison of the Hamiltonian approach to large fluctuations, described in Section III, and the approach to the optimal control problem show that, both on physical grounds and rigorously, the Wentzel–Freidlin Hamiltonian [57] (37) is equivalent to the Pontryagin Hamiltonian (34) [164] and the corresponding optimal control function is equivalent to the optimal fluctuational force. The analogy between the two problems opens up the possibility of direct experimental insight into the geometry of the phase space of system (34) using a statistical analysis of the fluctuational trajectories in this system when a control function $u(t)$ is substituted for the random function $\xi(t)$. In particular, the optimal control signal $\bar{u}(t)$ can be identified with the optimal fluctuational force that drives the system from the chaotic attractor to the stable limit cycle [165]. We note that both $\bar{u}(t)$ and the optimal force are related to p_2 in (37) [144].

We therefore suggested that the optimal control function $\bar{u}(t)$ can be found experimentally by measurement of the optimal fluctuational force [95,112].

This interrelationship is intuitively clear because, in thermal equilibrium $(D = 4\Gamma k_B T)$, the probability of fluctuations is determined by the minimum work of the external source needed to produce the given change in the thermodynamic quantities $\rho \propto \exp(-R_{\min}/k_B T)$ [9]. We emphasize that the analysis presented above draws an analogy between two quite distinct and separate problems: the deterministic energy-optimal control problem and the problem of the stability of the system in the presence of small random perturbations. Very similar conclusions can be drawn using a more general formulation of the stochastic optimal control problem [165].

2. Statistical Analysis of Fluctuational Trajectories

A statistical analysis of the fluctuational trajectories is based on the measurements of the *prehistory probability distribution* [60] $p_h(q, t; q_f, t_f)$ (see Section IIIC). By investigating the prehistory probability distribution experimentally, one can establish the area of phase space within which optimal paths are well defined, specifically, where the tube of fluctuational paths around an optimal path is narrow. The prehistory distribution thus provides information about both the optimal path and the probability that it will be followed. In practice the method essentially reduces to continuously following the dynamics of the system and constructing the distribution of all realizations of the fluctuational trajectories that transfer it from a state of equilibrium to a prescribed remote state.

To find the optimal control function $\bar{u}(t)$, we performed digital simulations of (38) using the Heun algorithm, with particular care given to the random-number generator [143,171], because simulation times necessarily grow exponentially as $D \to 0$. We have also carried out a complementary analog electronic modeling [112] of (38). We drive the model with zero-mean quasi–white Gaussian noise from a noise generator, digitize the response $q_1(t)$, $q_2(t)$, $\xi(t)$; and analyze it with a digital data processor. In both analog and digital simulations, trajectories moving the system from the chaotic attractor to the stable limit cycle were collected, and the corresponding distributions of the escape trajectories were built and analyzed. Qualitatively similar results were obtained but, because precision is of particular importance here, most of the data reported below are those from the digital simulations.

For the technique to be applicable, a solution of (37) moving the system from the CA to $\partial\Omega$ must exist, and one has to be able to identify the boundary conditions for this solution on the CA.

In the presence of weak noise there is a finite probability of noise-induced transitions between the chaotic attractor and the stable limit cycle. In Fig. 14 the filled circles show the intersections of one of the real escape trajectories with the given Poincaré section. The following intuitive escape scenario can be expected in the Hamiltonian formalism. Let us consider first the escape of the system from the basin of attraction of a stable limit cycle that is bounded by an saddle cycle. In general, escape occurs along a single optimal trajectory $q_{\mathrm{opt}}(t)$ connecting the two limit cycles.

The trajectory $q_{opt}(t)$ is determined by minimizing S in (20) on the set of all classical deterministic trajectories determined by the Hamiltonian H (37), that start on a stable limit cycle as $t \to -\infty$ and terminate on a saddle cycle as $t \to \infty$. That is, $q_{opt}(t)$ is a heteroclinic trajectory of the system (37) with minimum action, where the minimum is understood in the sense indicated, and the escape probability assumes the form $P \approx \exp(-S/D)$. We note that the existence of optimal escape trajectories and the validity of the Hamiltonian formalism have been confirmed experimentally for a number of nonchaotic systems (see Refs. 62, 95, 112, 132, and 172 and references cited therein).

If the noise is weak, then the probability $P \approx \exp(-S/D)$ to escape along the optimal trajectory is exponentially small, but it is exponentially greater than the escape probability along any other trajectory, including along other heteroclinic trajectories of the system (37).

Since the basin of attraction of the CA is bounded by the saddle cycle S1, the situation near S1 remains qualitatively the same and the escape trajectory remains unique in this region. However, the situation is different near the chaotic attractor. In this region it is virtually impossible to analyze the Hamiltonian flux of the auxiliary system (37), and no predictions have been made about the character of the distribution of the optimal trajectories near the CA. The simplest scenario is that an optimal trajectory approaching (in reversed time) the boundary of a chaotic attractor is smeared into a "cometary tail" and is lost, merging with the boundary of the attractor.

However, statistical analysis of real fluctuation-induced escape trajectories gives a more detailed picture of the noise-induced escape from a chaotic attractor. Several thousand real escape trajectories of the system (38) from the basin of attraction of a CA in various operating regimes were investigated [173]. The typical situation as measured in analog simulations is displayed in Fig. 15 for system parameters close to the point P in Fig. 13 and a noise intensity $D \approx 0.0005$. The figure shows 65 measured fluctuational escape trajectories. All the trajectories have been shifted in time so that the characteristic regions of the trajectories corresponding to the transition from chaotic to regular motion coincide with each other.

It is evident that all real trajectories pass through the close neighborhood of some optimal trajectory in a tube with a radius $\propto \sqrt{D}$. Therefore it is possible to determine the optimal escape paths by simple averaging performed separately for each group of trajectories. The number of different optimal escape paths obtained for the transition CA $\to$ S3 depends on the choice of the working point. From one to three distinct optimal escape paths for operation in various regimes were observed experimentally. The escape probabilities along different paths are different, and, as the noise intensity is reduced, one of the escape paths becomes exponentially more probable then the others. In what follows we concentrate on the properties of this most probable escape path.

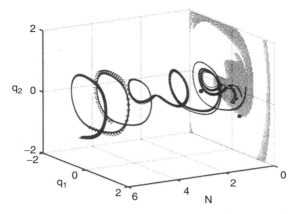

Figure 15. Escape trajectories found [173] in the analog simulations for the parameters $h = 0.19, \omega_f \approx 1.045, \omega_0 \approx 0.597, D \approx 0.0005$ are shown in comparison with the Poincaré cross section of a quasiattractor and its basins of attraction for $\omega_f t = 0$.

To find the boundary conditions on the CA, we analyze the prehistory probability distribution $p_h(q, t; q_f, t_f)$ of the escape trajectories. The corresponding distribution is shown in the Fig. 16. It can be inferred by the inspection of how the ridge of the most probable escape path merges the CA that most of the escape trajectories pass close to the saddle cycle of the period 5 embedded into the CA.

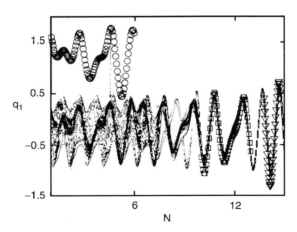

Figure 16. The prehistory probability distribution of the escape trajectories for the parameters as in Fig. 13. The circles, squares, and triangles show single periods of the saddle cycles of periods 5 (S5), 3 (S3), and 1 (S1), respectively [173].

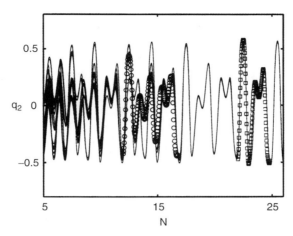

Figure 17.   Escape trajectories for the parameters as in Fig. 16. The squares and circles show one period of the saddle cycle S3 and one period of S5, respectively.

This hypothesis can be elaborated further using a statistical analysis of the trajectories arriving a small tube around S3 with the noise intensity reduced by a few orders of magnitude up to $D = 1.5 \times 10^{-6}$, see Fig. 17 [173]. The analysis reveals that the energetically favorable way to move the system from the CA to the stable limit cycle starts at the saddle cycle of period 5 (S5) embedded in the CA, passes through saddle cycle S3 and finishes at the saddle cycle S1 at the boundary of the basin of attraction of the CA. Subsequent motion of the system towards the stable limit cycle does not require external action.

To find an approximation to the optimal control function we collect all successful realizations $(q_1^{\mathrm{esc}}(t), q_2^{\mathrm{esc}}(t), \xi^{\mathrm{esc}}(t))$ that move it from the CA to $\partial\Omega$. An approximate solution $\tilde{u}(t)$ is then found as an ensemble average over the corresponding realizations of the random force $\langle \xi^{\mathrm{esc}}(t) \rangle$ (the exact solution is $\bar{u}(t) = \lim_{D \to 0} \tilde{u}(t)$). The results of this procedure are shown in the upper trace of Fig. 18. To remove the irrelevant high-frequency component left after averaging, we filtered through a zero-phase low-pass filter with frequency cutoff $\omega_c = 1.9$.

It can be seen from the figure that the optimal force switches on at the moment when the system leaves S5 along its unstable manifold. The optimal force returns to zero when the system reaches the saddle cycle S1.

Thus we conclude that the solution $\tilde{u}(t)$ and the corresponding boundary conditions can be found using our new experimental method. Moreover the problem of escape from the CA of a periodically driven nonlinear oscillator can essentially be reduced to the analysis of a transition between three saddle cycles

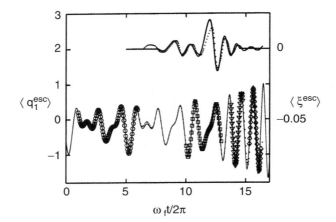

Figure 18. The most probable escape path (bottom solid curve) from S5 to the S1, found in the numerical simulations. The stable limit cycle is shown by rombs; see Fig. 16 for other symbols. Parameters were $h = 0.13, \omega_f = 0.95, \omega_0 \approx 0.597, D = 0.0005$. Top: optimal force (solid line) corresponding to the optimal path after filtration [169]. The optimal path and optimal force from numerical solution of the boundary-value problem are shown by dots.

S5 $\rightarrow$ S3 $\rightarrow$ S1. We note that the latter result is in qualitative agreement with the well known statement that unstable cycles provide detailed invariant characterizations for dynamical systems of low intrinsic dimension [174–176].

This result opens up the possibility of the numerical solution of the corresponding boundary value problem for energy-optimal control formulated above.

It can be shown [173] that the average time for the system to approach S5 is much smaller then the average escape time and thus the optimal escape paths found from the statistical analysis of the escape trajectories is independent of the initial conditions on the attractor and provides an approximation to the global minimum of the corresponding deterministic control problem.

3. Numerical Solution of the Boundary Value Problem

In principle, it is possible to find the optimal path by direct solution of the Pontryagin Hamiltonian (37), with appropriate boundary conditions. We must stress that even for this relatively simple system, the solution is a formidable, and almost impossible, task. First of all, in general one has no insight into the appropriate boundary conditions, in particular into those at the starting time (which belong to the strange attractor). But even if the boundaries were known, in practice the determination of the optimal path is impossible: the functional R of Eq. (36) has so many local minima, that it proved impractical to attempt a (general) search for the optimal path.

However, once the fluctuational trajectories were available, we did indeed manage to find the optimal path by direct solution of the Pontryagin Hamiltonian. The idea is to study the escape scenario that emerges from the fluctuational trajectories: as we mentioned, the escape takes place through S3, S5, and S1. We then built an initial trial function, taking a linear combination of the structures involved in the escape. The combination was such that at short times the trial function coincided with S5, while at large times it coincided with S1. At intermediate times, we had a mixture of S5 and S1 in the initial trial function, but no S3. Using a well-known algorithm for two-point boundary conditions (TWPBVP, obtained via netlib [177]; see Ref. 178 for details), we then relaxed this trial function to find the optimal path, defined as the path that minimizes the functional (36). It is striking that the relaxational optimal path that we found does go through S3, in good agreement with what was observed for the optimal path obtained via the fluctuational trajectories [173].

4. The Energy-Optimal Migration Control of a Chaotic Oscillator

Here we examine the control of migration in a periodically driven nonlinear oscillator. Our aim is to demonstrate that application of the approximate solution found from the statistical analysis of fluctuational trajectories optimizes (minimizes) the energy of the control function. We compare the performance of some known adaptive control algorithms to that of the control function found through our analysis.

To verify that the optimal force $\tilde{u}(t)$ found in the experiment does minimize the energy of the control function steering the system (35) from the CA to the S1, we set it to arbitrary initial conditions in the basin of attraction of CA and let it evolve deterministically until it passed through the initial part of the unstable manifold of S5. At this moment the deterministic control function was switched on. For small variations in the shape of the control function and/or initial conditions, the amplitude of the control function was set to the threshold of the switching for the system from chaotic motion to regular motion on the stable limit cycle. It was found that the system is very sensitive to variation of both the shape of the control function and the initial conditions. It was also demonstrated that any deviation from the shape of $\tilde{u}(t)$ or from the initial conditions found in the experiment leads to a substantial increase in the energy of the control function required to steer system from a CA to S1. Some experimental results are shown in Fig. 19. Thus it can be seen that the energy of the control function is approximately twice larger if the optimal force is approximated by the sin function modulated by the Gaussian $u(t) = a_1 \sin(a_2 t) \exp(-(t - a_3)^2 a_4)$ and it is ~ 4 and 20 times larger if the optimal force is approximated by rectangular pulses or perturbed with arbitrary low-frequency perturbations, respectively.

We have also performed experiments using an open-plus-closed-loop control technique [156] and adaptive control algorithm [157] to steer the system from

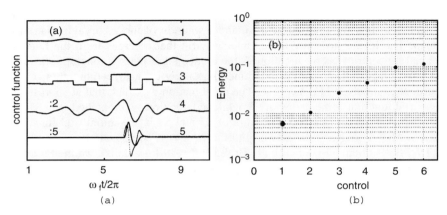

Figure 19. (a) The shapes of the control functions (not drawn to scale) used in the numerical experiment: 1—optimal force found from the statistical analysis of the fluctuational escape trajectories; 2—approximation of the optimal force by the $u(t) = a_1 \sin(a_2 t) \exp\left(-(t - a_3)^2 a_4\right)$ where a_i are constants; 3—approximation of the optimal force by the rectangular pulses; 4—arbitrary perturbation of the optimal force with a low-frequency perturbation; 5—control functions produced by the OPCL alogrithm; 6—control function for the adaptive control. (b) Energies of the control functions shown in (a) [169].

the CA to the S1. The equations of motion are taken in the form

$$\dot{q}_1 = q_2 + F_1(q, g, t)$$
$$\dot{q}_2 = -2\Gamma q_2 - \omega_0^2 q_1 - \beta q_1^2 - \gamma q_1^3 - f \cos(\omega t) \qquad (39)$$
$$+ F_2(q, g, t), \qquad q = q_1, q_2, \quad g = g_1, g_2$$

Here $F(q, g, t)$ is the control function in the form

$$F(q, g, t) = (\dot{g} - K(g)) + S(t)(K'(g) - A)(g(t) - q(t)) \qquad (40)$$

We will be interested in the situation when the "goal dynamics" $g(t)$ is a solution of (35) with $u(t) = 0$, namely, $\dot{g} = K(g)$. Specifically, $g(t)$ describes the stable limit cycle SC of period 1 coexisting with the CA. Thus the first term in (40) vanishes. And $F(q, g, t)$ takes the following explicit form

$$F_i(q, g, t) = S(t) \sum_{j=1,2} (K_{ij} - a_{ij})(g_j - q_j) \qquad (41)$$

Here $i = 1, 2$ and $K_{ij} = \partial K_i / \partial q_j$. We have considered only the case $a_{ij} = -|a_{ij}|\delta_{ij}$ and $S(t) = 1 - \exp(-\lambda t)$ as has been suggested [157]. Parameters $|a|$ and λ were varied to optimize the energy of the control function.

The energy of the control functions obtained by these methods varies from 0.14 to 0.6 and thus it is more then one order of magnitude larger then the energy of the optimal control function $\tilde{u}(t)$ found by our new technique, (see Fig. 19). Similar results were obtained using the algorithm for adaptive chaos control [151] for the migration of the nonlinear oscillator from the CA to SC (see Fig. 19).

We note that neither the OPCL nor the adaptive control algorithms were devised to optimize the energy of the control, but rather the recovery time. It is clear that these methods are insensitive to the initial conditions at the CA. The shapes of the control functions are, to a large extent, also prescribed by the algorithms and are not optimized. In this sense the high energy of the control functions is not a surprise: the results presented serve the purpose of illustrating the main point: the sensitivity of the optimal control to the shape of the control function and to the initial conditions, discussed above.

B. Fluctuational Escape from a Quasihyperbolic Attractor

We now consider, for comparison, fluctuational escape from the Lorenz attractor, which, for a certain range of parameters, is a quasihyperbolic attractor consisting of unstable sets only [161]:

$$\dot{q}_1 = \sigma(q_2 - q_1)$$
$$\dot{q}_2 = rq_1 - q_2 - q_1q_3 \qquad (42)$$
$$\dot{q}_3 = q_1q_2 - bq_3 + \xi(t)$$
$$\langle \xi(t) \rangle = 0, \qquad \langle \xi(t)\xi(0) \rangle = D\delta(t) \qquad (43)$$

In the absence of noise, the system [179] describes the generation of a single-mode laser field interacting with a homogeneously broadened two-level medium [180]. The variables and parameters of the Lorenz system can be interpreted in terms of a laser system as: q_1 is the normalized electric field amplitude, q_2 the normalized polarization, q_3 the normalized inversion, $\sigma = k/\gamma_1$, $r = \Lambda + 1$, $b = \gamma_2/\gamma_1$, with k the decay rate of the field in the cavity, γ_1 and γ_2 the relaxation constants of the inversion and polarization, and Λ the pump parameter. Far-infrared lasers have been proposed as an example of a realization of the Lorenz system [162]. A detailed comparison of the dynamics of the system (42) and a far-infrared laser, plus a discussing the validity of the Lorenz system as laser model, can be found in Ref. 163.

The Lorenz equations have a simple structure and contain two nonlinear terms only. Let us briefly consider the main bifurcations in the system (42) (a more detailed analysis can be found in Ref. 181). We fix the parameters $\sigma = 10$, $b = \frac{8}{3}$ and vary the parameter r: in this case two global bifurcations take place (see the bifurcation diagram in Fig. 20). For $r = 1$, a supercritical

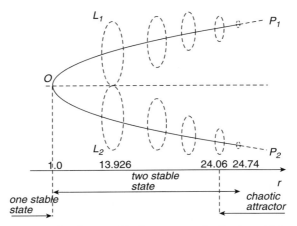

Figure 20. The bifurcation diagram of the Lorenz system for fixed $\sigma = 10, b = \frac{8}{3}$. The unstable and stable sets are shown by dashed and solid lines, respectively.

pitchfork bifurcation happens: the stationary state at the origin $O \equiv (0,0,0)$ becomes a saddle state and two new stationary states appear: $P_1 \equiv (\sqrt{b(r-1)}, \sqrt{b(r-1)}, r-1)$ and $P_2 \equiv (-\sqrt{b(r-1)}, -\sqrt{b(r-1)}, r-1)$. In the system phase space there are two stable points P_1 and P_2, a saddle point O at the origin, and their one-dimensional (separatrixes) and two-dimensional manifolds.

The second bifurcation occurs at

$$r = \frac{\sigma(\sigma + b + 3)}{(\sigma - b - 1)} \approx 24.74, \ldots \tag{44}$$

and it is a subcritical Hopf bifurcation, when states P_1 and P_2 loss their stability and in the phase space there is the unique chaotic quasihyperbolic attractor.

There are also two local bifurcations. The first one takes place for $r \approx 13.926 \ldots$, when a homoclinic tangency of separatrixes of the origin O occurs (it is not shown in Fig. 20) and a hyperbolic set appears, which consists of a infinite number of saddle cycles. Beside the hyperbolic set, there are two saddle cycles, L_1 and L_2, around the stable states, P_1 and P_2. The separatrices of the origin O reach the saddle cycles L_1 and L_2, and the attractors of the system are the states P_1 and P_2. The second local bifurcation is observed for $r \approx 24.06$. The separatrices do not any longer reach to the saddle cycles L_1 and L_2. As a result, in the phase space of the system a stable quasihyperbolic state appears— the Lorenz attractor. The chaotic Lorenz attractor includes separatrices, the saddle point O and a hyperbolic set, which appears as a result of homoclinic tangency of the separatrices. The presence of the saddle point in the chaotic attractor defines the prefix "quasi" in the definition of the chaotic attractor as *quasihyperbolic* [161]. The states P_1, P_2 remain stable. Thus, in the range $r \in [24.06 : 24.74]$ the coexistence of the chaotic attractor and two stable point attractors is observed in the phase space of the Lorenz system. Let us fix the parameter $r = 24.08$ in this range and consider the noise-induced escape for the chaotic attractor to the basins of attraction of the stable points. Note that in Ref. 64 the invariant measure of the noisy Lorenz attractor was found within a Hamiltonian formalism, but large deviations from a chaotic attractor were not considered.

First, we examine [182] the structure of the system phase space for chosen parameters $\sigma = 10$, $b = \frac{8}{3}$, $r = 24.08$ (Fig. 21).

The saddle cycles L_1 and L_2 surround the stable states P_1 and P_2 and are located at the intersection of the unstable W^u and stable W^s manifolds. The unstable manifold goes to the stable state P_1 from one side and to the chaotic attractor from the other side. The stable manifold W^s forms a tube in the vicinity of the stable state [183]. The saddle cycles L_1 and L_2 have the multipliers $(1.0000, 1.0280, 0.0001)$, and therefore trajectories will go slowly away along the unstable manifold, and they will approach quickly along the stable manifold.

For simplicity, we add noise in the form of a white noise $\xi(t)$ to the third equation of system (42), preserving the original system symmetry.

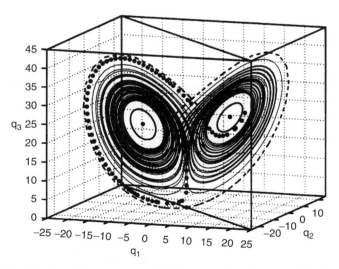

Figure 21. Structure of the phase space of the Lorenz system. An escape trajectory measured by numerical simulation is indicated by the filled circles. The trajectory of the Lorenz attractor is shown by a thin line; the separatrixes Γ_1 and Γ_2 by dashed lines [182].

As in the case of escape from a nonhyperbolic attractor, there is no theoretical prediction about the process of fluctuational escape from the Lorenz attractor. But the process is readily studied via numerical simulation and via analysis of the prehistory probability distribution built using the fluctuational escape trajectories. For definiteness, we examine escape to the stable point P_1. The averaged escape trajectory and the corresponding averaged fluctuational force obtained in this way are shown in Fig. 22. We have found that the escape occurs via the following scenario. The escape trajectory starts from the stable manifold of the saddle point O. Under the action of a fluctuation, an escape trajectory tends to point O along the two-dimensional stable manifold. Then, without reaching the saddle point O, the trajectory departs from it again, following a path close to the separatrix Γ_2, and falling into the neighborhood of the saddle cycle L_1. In the absence of an external force, the trajectory goes away from the cycle L_1, slowly untwisting. The fluctuational force induces a crossing through the saddle cycle, and the trajectory then relaxes to the stable point P_1. We can thus split the escape process into two parts: fluctuational and relaxational. In practice all the fluctuational part belongs to the Lorenz attractor, and itself consists of two stages: at first, the fluctuational force throws the trajectory as close as possible to the cycle L_1; then, the trajectory crosses this cycle under the action of fluctuations. The first stage is defined by the stable and unstable manifolds of the saddle point O, and the time dependence of the fluctuational force is similar to that of the coordinate q_3 (Fig. 22). During the second stage, the fluctuations have a component that oscillates in antiphase to

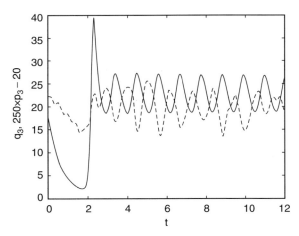

Figure 22. The averaged escape trajectory (solid line) and the averaged fluctuational force (dashed line) during escape from the Lorenz attractor [182].

the coordinate q_3. Because the trajectory of the noise-free system departs from the cycle L_1 very slowly, the fluctuational force inducing the crossing through the cycle may start to act at any time during a long interval. For this reason the averaged fluctuational force itself consists of a long oscillating function.

It is clear that all of the escape trajectory from the Lorenz attractor lies on the attractor itself. The role of the fluctuations is, first, to bring the trajectory to a seldom-visited area in the neighborhood of the saddle cycle L_1, and then to induce a crossing of the cycle L_1. So we may conclude that the role of the fluctuations is different in this case, and the possibility of applying the Hamiltonian formalism will require a more detailed analysis of the crossing process.

Thus, we have found that the mechanisms of escape from a nonhyperbolic attractor and a quasihyperbolic (Lorenz) attractor are quite different, and that the prehistory of the escape trajectories reflects the different structure of their chaotic attractors. The escape process for the nonhyperbolic attractor is realized via several steps, which include transitions between low-period saddle-cycles coexisting in the system phase space. The escape from the Lorenz attractor consist of two qualitatively different stages: the first is defined by the stable and unstable manifolds of the saddle center point, and lies on the attractor; the second is the escape itself, crossing the saddle boundary cycle surrounding the stable point attractor. Finally, we should like to point out that our main results were obtained via an *experimental* definition of optimal paths, confirming our experimental approach as a powerful instrument for investigating noise-induced escape from complex attractors.

VI. CONCLUSIONS

The rapid advances in the understanding of fluctuating nonlinear systems, including optical systems, have come about in large part through the mutually supportive relationship between analytic calculation and analog and digital simulations. This has been especially true of problems involving large fluctuations, where use of simulations, coupled with the introduction of the prehistory probability distribution, have set the area on an experimental basis for the first time, and helped to stimulate new advances in the theory. These have included the logarithmic susceptibility, described above, which promises to do for optimal paths what the conventional linear susceptibility has done for linear response theory. The theory of the logarithmic susceptibility, in turn, has been tested, and its limits of applicability explored, through simulations. And the same is true of developments in understanding Kramers' problem on short timescales. Studies of the fluctuational escape from chaotic attractors, of which two examples are described above, are entirely simulation-led at present. But the results of the analog and digital experiments have already provided strong guidance for future developments in the theory. It seems certain that the close symbiotic relationship between simulation and theory in fluctuational dynamics will continue, and that the emergence of many new results and phenomena may be anticipated in the near future.

Acknowledgments

We warmly acknowledge the assistance and support of our many collaborators in the research reviewed in this chapter including, especially, I. Kh. Kaufman, V. N. Smelyanskiy, S. M. Soskin, N. D. Stein and N. G. Stocks. We are grateful to S. Beri for his help and advice. But, most of all, we would like to acknowledge our continuing indebtedness to M. I. Dykman, whose insights and ideas now permeate the entire field of fluctuational dynamics, and with whom it has been such a privilege and pleasure to collaborate over many years. The research has been supported in part by the Engineering and Physical Sciences Research Council (UK), INTAS, the EC, the Royal Society of London, and by Award No. REC-006 of the U.S. Civilian Research & Development Foundation for the Independent States of the Former Soviet Union (CRDF).

References

1. M. I. Dykman, D. G. Luchinsky, R. Mannella, P. V. E. McClintock, N. D. Stein, and N. G. Stocks, *Adv. Chem. Phys.* **LXXXV**, 265 (1994).

2. A. Einstein, *Ann. Phys.* **17**, 549 (1905).

3. M. von Smoluchowski, *Bull. Intern. Acad. D. Sci. Cracovie (A)*, 418 (1913).

4. P. Langevin, *Compt. Rend.* **146**, 530 (1908).

5. W. H. Louisell, *Radiation and Noise in Quantum Electronics*, McGraw-Hill, New York, 1964.

6. M. O. Scully and W. E. Lamb Jr., *Phys. Rev.* **159**, 208 (1967).

7. M. Lax, *Fluctuation and Coherence Phenomena in Classical and Quantum Physics*, Gordon & Breach, New York, 1968.

8. H. Haken, in S. Flugge (Ed.), *Encyclopaedia of Physics*, Springer, Berlin, 1970, Vol. XXV/2c.

9. L. D. Landau and E. M. Lifshitz, *Statistical Physics*, 3rd ed., Part 1, Pergamon, New York, 1980.

10. R. Short, L. Mandel, and R. Roy, *Phys. Rev. Lett.* **49**, 647 (1982).

11. See, for example, *OSA Proceedings on Nonlinear Dynamics in Optical Systems*, N. B. Abraham, E. Garmire, and P. Mandel (Eds.), Optical Society of America, Washington, DC, 1991; and subsequent volumes in this proceedings series.

12. R. Benzi, A. Sutera, and A. Vulpiani, *J. Phys. A* **14**, L453 (1981); R. Benzi, G. Parisi, A. Sutera, and A. Vulpiani, *Tellus* **34**, 10 (1982).

13. C. Nicolis, *Tellus* **34**, 1 (1982).

14. M. I. Dykman, D. G. Luchinsky, R. Mannella, P. V. E. McClintock, N. D. Stein, and N. G. Stocks, *J. Stat. Phys.* **70**, 479 (1993).

15. M. I. Dykman, D. G. Luchinsky, R. Mannella, P. V. E. McClintock, N. D. Stein, and N. G. Stocks, *Nuovo Cimento D* **17**, 661 (1995).

16. L. Gammaitoni, P. Hänggi, P. Jung, and F. Marchesoni, *Rev. Mod. Phys.* **70**, 223 (1998).

17. D. G. Luchinsky, R. Mannella, P. V. E. McClintock, and N. G. Stocks, *IEEE Trans. Circuits and Systems II: Analog and Digital Signal Processing* **46**, 1205 (1999).

18. D. G. Luchinsky, R. Mannella, P. V. E. McClintock, and N. G. Stocks, *IEEE Trans. Circuits and Systems II: Analog and Digital Signal Processing* **46**, 1215 (1999).

19. B. McNamara, K. Wiensenfeld, and R. Roy, *Phys. Rev. Lett.* **60**, 2626 (1988).

20. G. Vemuri and R. Roy, *Phys. Rev. A* **39**, 4668 (1989).

21. A. Fioretti, L. Guidoni, R. Mannella, and E. Arimondo, *J. Stat. Phys.* **70**, 403 (1993).

22. J. M. Iannelli, A. Yariv, T. Chen, and Y. Zhuang, *Appl. Phys. Lett.* **65**, 1983 (1994).

23. M. I. Dykman, A. L. Velikovich, G. P. Golubev, D. G. Luchinskii, and S. V. Tsuprikov, *Sov. Phys. JETP Lett.* **53**, 193 (1991).

24. J. Grohs, S. Apanasevich, P. Jung, H. Issler, D. Burak, and C. Klingshirn, *Phys. Rev. A* **49**, 2199 (1994).

25. B. Jost and B. Saleh, *Opt. Lett.* **21**, 287 (1996).

26. F. Vaudelle, J. Gazengel, G. Rivoire, X. Godivier, and F. Chapeau-Blondeau, *J. Opt. Soc. Am. B* **15**, 2674 (1998).

27. A. Simon and A. Libchaber, *Phys. Rev. Lett.* **68**, 3375 (1992).

28. M. I. Dykman, G. P. Golubev, D. G. Luchinsky, P. V. E. McClintock, N. D. Stein, and N. G. Stocks, *Phys. Rev. E* **49**, 1935 (1994).

29. M. I. Dykman, G. P. Golubev, I. K. Kaufman, D. G. Luchinsky, P. V. E. McClintock, and E. A. Zhukov, *Appl. Phys. Lett.* **67**, 308 (1995).

30. P. Jung, *Phys. Rep.* **234**, 175 (1993).

31. N. G. Stocks, N. D. Stein, S. M. Soskin, and P. V. E. McClintock, *J. Phys. A* **25**, L1119 (1992).

32. F. Moss, in G. H. Weiss (Ed.), *Contemporary Problems in Statistical Physics*, SIAM, Philadelphia, 1994, pp. 205–253.

33. K. Wiesenfeld and F. Moss, *Nature* **373**, 33 (1995).

34. M. I. Dykman, R. Mannella, P. V. E. McClintock, and N. G. Stocks, *Phys. Rev. Lett.* **65**, 2606 (1990).

35. M. I. Dykman, R. Mannella, P. V. E. McClintock, and N. G. Stocks, *JETP Lett.* **52**, 141 (1990).

36. T. Zhou, F. Moss, and P. Jung, *Phys. Rev. A* **42**, 3161 (1990).

37. M. I. Dykman, R. Mannella, P. V. E. McClintock, N. D. Stein, and N. G. Stocks, *Phys. Rev. E* **47**, 1629 (1993).

38. N. G. Stocks, *Nuovo Cimento D* **17**, 925 (1995).

39. N. G. Stocks, N. D. Stein, and P. V. E. McClintock, *J. Phys. A* **26**, L385 (1993).

40. I. Kh. Kaufman, D. G. Luchinsky, P. V. E. McClintock, S. M. Soskin, and N. D. Stein, *Phys. Rev. E* **57**, 78–87 (1998); I. Kh. Kaufman, D. G. Luchinsky, P. V. E. McClintock, S. M. Soskin, and N. D. Stein, *Phys. Lett. A* **220**, 219–223 (1996).

41. M. A. Ivanov, L. B. Kvashnina, and M. A. Krivoglaz, *Sov. Phys. Solid State* **7**, 1652 (1966).

42. M. A. Dykman and M. I. Krivoglaz, in M. Khalatnikov (Ed.), *Soviet Physics Reviews*, Harwood, New York, 1984, Vol. 5, pp. 265–442.

43. L. Alfonsi, L. Gammaitoni, S. Santucci, and A. R. Bulsara, *Phys. Rev. E* **62**, 299 (2000).

44. L. Onsager and S. Machlup, *Phys. Rev.* **91**, 1505 (1953).

45. E. Arimondo, D. Hennequin, and P. Gloriuex, in F. Moss and P. V. E. McClintock (Eds.), *Noise in Nonlinear Dynamical Systems*, Cambridge Univ. Press, Cambridge, UK, 1989, Vol. 3, pp. 119–158.

46. R. Roy, A. W. Yu, and S. Zhu, in F. Moss and P. V. E. McClintock (Eds.), *Noise in Nonlinear Dynamical Systems*, Cambridge Univ. Press, Cambridge, UK, 1989, Vol. 3, pp. 90–118.

47. M. C. Cross and P. C. Hohenberg, *Rev. Mod. Phys.* **65**, 851 (1993).

48. G. Gabrielse, H. Dehmelt, and W. Kells, *Phys. Rev. Lett.* **54**, 537 (1985).

49. H. Dehmelt, *Rev. Mod. Phys.* **62**, 525 (1990).

50. J. Tan and G. Gabrielse, *Phys. Rev. Lett.* **67**, 3090 (1991); J. Tan and G. Gabrielse, *Phys. Rev. A* **48**, 3105 (1993).

51. M. O. Magnasco, *Phys. Rev. Lett.* **71**, 1477 (1993).

52. R. D. Astumian and M. Bier, *Phys. Rev. Lett.* **72**, 1766 (1994).

53. M. M. Millonas and M. I. Dykman, *Phys. Lett. A* **185**, 65 (1994).

54. J. Prost, J. F. Chauwin, L. Peliti, and A. Ajdari, *Phys. Rev. Lett.* **72**, 2652 (1994).

55. C. R. Doering, W. Horsthemke, and J. Riordan, *Phys. Rev. Lett.* **72**, 2984 (1994).

56. M. I. Dykman, H. Rabitz, V. N. Smelyanskiy, and B. E. Vugmeister, *Phys. Rev. Lett.* **79**, 1178 (1997).

57. A. D. Wentzell and M. I. Freidlin, *Russ. Math. Surveys* **25**, 1 (1970); M. I. Freidlin and A. D. Wentzell, *Random Perturbations in Dynamical Systems*, Springer, New York, 1984.

58. A. J. McKane, *Phys. Rev. A* **40**, 4050 (1989).

59. R. S. Maier and D. L. Stein, *Phys. Rev. E* **48**, 931 (1993).

60. M. I. Dykman, P. V. E. McClintock, V. N. Smelyanski, N. D. Stein, and N. G. Stocks, *Phys. Rev. Lett.* **68**, 2718 (1992).

61. M. I. Dykman, E. Mori, J. Ross, and P. M. Hunt, *J Chem. Phys.* **100**, 5735 (1994).

62. D. G. Luchinsky and P. V. E. McClintock, *Nature* **389**, 463 (1997).

63. R. P. Feynman and A. R. Hibbs, *Quantum Mechanics and Path Integrals*, McGraw-Hill, New York, 1965.

64. R. Graham, in G. Höbler (Ed.), *Quantum Statistics in Optics and Solid-State Physics*, Springer Tracts in Modern Physics, Springer, Berlin, 1973, Vol. 66, pp. 1–97; R. Graham and T. Tél, *Phys. Rev. Lett.* **52**, 9 (1984); R. Graham and T. Tél, *Phys. Rev. A* **31**, 1109 (1985); R. Graham, in F. Moss and P. V. E. McClintock (Eds.), *Noise in Nonlinear Dynamical Systems*, Cambridge Univ. Press, Cambridge, UK, 1989, Vol. 1, pp. 225–278.

65. M. I. Dykman and M. A. Krivoglaz, *Sov. Phys. JETP* **50**, 30 (1979).

66. J. F. Luciani and A. D. Verga, *Europhys. Lett.* **4**, 255 (1987).

67. J. F. Luciani and A. D. Verga, *J. Stat. Phys.* **50**, 567 (1988).

68. G. Tsironis and P. Grigolini, *Phys. Rev. Lett.* **61**, 7 (1988).

69. A. J. Bray and A. J. McKane, *Phys. Rev. Lett.* **62**, 493 (1989).

70. H. S. Wio, P. Colet, M. San Miguel, L. Pesquera, and M. A. Rodríguez, *Phys. Rev. A* **40**, 7312 (1989).

71. A. J. McKane, H. C. Luckock, and A. J. Bray, *Phys. Rev. A* **41**, 644 (1990).

72. A. J. Bray, A. J. McKane, and T. J. Newman, *Phys. Rev. A* **41**, 657 (1990).

73. H. C. Luckock and A. J. McKane, *Phys. Rev A* **42**, 1982 (1990).

74. J. K. Cohen and R. M. Lewis, *J. Inst. Math. Appl.* **3**, 266–290 (1967).

75. M. M. Kłosek-Dygas, B. J. Matkowsky, and Z. Schuss, *SIAM J. Appl. Math.* **48**, 425 (1988).

76. M. M. Kłosek-Dygas, B. J. Matkowsky, and Z. Schuss, *Phys. Rev. A* **38**, 2605 (1988).

77. M. M. Kłosek-Dygas, B. J. Matkowsky, and Z. Schuss, *J. Stat. Phys.* **54**, 1309–1320 (1989).

78. M. I. Dykman, *Phys. Rev. A* **42**, 2020 (1990).

79. S. J. B. Einchcomb and A. J. McKane, *Phys. Rev. E* **51**, 2974 (1995).

80. M. I. Dykman, V. N. Smelyanskiy, D. G. Luchinsky, R. Mannella, P. V. E. McClintock, and N. D. Stein, *Int. J. Bifurc. Chaos* **8**, 747 (1998).

81. K. Lindenberg, B. J. West, and J. Masoliver, in F. Moss and P. V. E. McClintock (Eds.), *Noise in Nonlinear Dynamical Systems*, Cambridge Univ. Press, Cambridge, UK, 1989, Vol. 1, pp. 110–160.

82. M. I. Dykman and K. Lindenberg, in G. H. Weiss (Ed.), *Contemporary Problems in Statistical Physics*, SIAM, Philadelphia, 1994, pp. 41–101.

83. J. Hales, A. Zhukov, R. Roy, and M. I. Dykman, *Phys. Rev. Lett.* **85**, 78 (2000).

84. M. B. Willemsen, M. P. van Exter, and J. P. Woerdman, *Phys. Rev. Lett.* **84**, 4337 (2000).

85. M. Topaler and N. Makri, *J. Chem. Phys.* **101**, 7500 (1994).

86. E. W. G. Diau, J. L. Herek, Z. H. Kim, and A. H. Zewail, *Science* **279**, 847 (1998); D. E. Folmer, L. Poth, E. S. Wisniewski, and A. W. Castleman Jr., *Chem. Phys. Lett.* **287**, 1 (1998); A. H. Zewail, *Femtochemistry: Ultrafast Dynamics of the Chemical Bond*, World Scientific, Singapore, 1994.

87. V. N. Smelyanskiy, M. I. Dykman, H. Rabitz, and B. E. Vugmeister, *Phys. Rev. Lett.* **79**, 3113 (1997).

88. L. I. McCann, M. I. Dykman, and B. Golding, *Nature* **402**, 785 (1999).

89. B. Golding, L. I. McCann, and M. I. Dykman, in D. S. Broomhead, E. A. Luchinskaya, P. V. E. McClintock, and T. Mullin (Eds.), *Stochastic and Chaotic Dynamics in the Lakes*, American Institute of Physics, Melville, 2000, pp. 34–41.

90. R. L. Kautz, *Rep. Prog. Phys.* **59**, 935 (1996).

91. M. I. Dykman, R. Mannella, P. V. E. McClintock, S. M. Soskin, and N. G. Stocks, *Phys. Rev. A* **42**, 7041 (1990).

92. R. L. Kautz, *Phys. Lett. A* **125**, 315 (1987).

93. P. Grassberger, *J. Phys. A* **22**, 3283 (1989).

94. R. Graham, A. Hamm, and T. Tél, *Phys. Rev. Lett.* **66**, 3089 (1991).

95. D. G. Luchinsky, *J. Phys. A* **30**, L577 (1997).

96. S. Fauve and F. Heslot, *Phys. Lett. A* **97**, 5 (1983); B. McNamara and K. Wiesenfeld, *Phys. Rev. A* **39**, 4854 (1989).

97. (a) R. N. Mantegna and B. Spagnolo, *Phys. Rev. E* **49**, R1792 (1994); (b) A. D. Hibbs, A. L. Singsaas, E. W. Jacobs, A. R. Bulsara, and J. J. Bekkedahl, *J. Appl. Phys.* **77**, 2582 (1995); A. D. Hibbs, E. W. Jacobs, A. R. Bulsara, J. J. Bekkedahl, and F. Moss, *Nuovo Cimento D* **17**, 811 (1995).

98. L. Gammaitoni, M. Martinelli, L. Pardi, and S. Santucci, *Phys. Rev. Lett.* **67**, 1799 (1991); L. Gammaitoni, F. Marchesoni, M. Martinelli, L. Pardi, and S. Santucci, *Phys. Lett. A* **158**, 449 (1991).

99. E. Ippen, J. Lindner, and W. L. Ditto, *J. Stat. Phys.* **70**, 437 (1993).

100. G. Debnath, T. Zhou, and F. Moss, *Phys. Rev. A* **39**, 4323 (1989); L. Gammaitoni, F. Marchesoni, E. Menichella-Saetta, and S. Santucci, *Phys. Rev. Lett.* **62**, 349 (1989).

101. J. K. Douglass, L. Wilkens, E. Pantazelou, and F. Moss, *Nature* **365**, 337 (1993).

102. Special Issue of the *J. Stat. Phys.* **70**(1/2) (1993); Special Issue of *Nuovo Cimento D* **17**(7–8) (1995).

103. P. J. W. Debye, *Polar Molecules*, Dover, New York, 1929.

104. G. P. Agrawal, *Fiber-Optic Communication Systems*, Wiley, New York, 1992.

105. *Image and Video Processing II*, Proceedings of SPIE **2182**, Bellingham, Washington, 1994; J. W. Bilbo and E. Parks (Eds.), *Quality and Reliability for Optical Systems*, Bellingham, Washington, DC, 1993.

106. H. Gibbs, *Optical Bistability: Controlling Light with Light*, Academic, New York, 1985.

107. T. Rivera, F. R. Ladan, A. Israël, R. Azoulay, R. Kuszelevicz, and J. L. Oudar, *Appl. Phys. Lett.* **64**, 869 (1994).

108. M. I. Dykman, G. P. Golubev, D. G. Luchinsky, A. L. Velikovich, and S. V. Tsuprikov, *Phys. Rev. A* **44**, 2439 (1991).

109. H. Kramers, *Physica* **7**, 284 (1940).

110. L. A. Lugiato, and R. Bonifacio, in F. T. Arecchi, R. Bonifacio, and M. O. Scully (Eds.), *Coherence in Spectroscopy and Modern Physics*, Plenum, New York, 1978, p. 85; L. A. Lugiato, *Contemp. Phys.* **24**, 333 (1983).

111. G. P. Golubev, D. G. Luchinsky, A. L. Velikovich, and M. A. Liberman, *Opt. Commun.* **64**, 181 (1987); A. L. Velikovich, G. P. Golubev, V. P. Golubchenko, and D. G. Luchinsky, *Opt. Commun.* **80**, 444 (1991).

112. D. G. Luchinsky, P. V. E. McClintock, and M. I. Dykman, *Rep. Prog. Phys.* **61**, 889 (1998).

113. A. D. Lloyd, I. Janossy, H. A. Mackenzie, B. S. Wherret, *Opt. Commun.* **61**, 339 (1987); D. C. Hutchings, A. D. Lloyd, I. Janossy, B. S. Wherret, ibid. p. 345; P. Pirani, V. Briguet, and W. Lucosz, *Helvetica Phys. Acta* **60**, 836 (1987).

114. M. I. Dykman, D. G. Luchinsky, P. V. E. McClintock, N. D. Stein, and N. G. Stocks, *Phys. Rev. A* **45**, R7678 (1992).

115. M. I. Dykman, G. P. Golubev, I. Kh. Kaufman, D. G. Luchinsky, P. V. E. McClintock, and E. A. Zhukov, *Nuovo Cimento D* **17**, 743 (1995).

116. L. D. Landau and E. M. Lifshitz, *Mechanics*, 3rd ed., Pergamon, Oxford, 1976.

117. Beside Refs. 53–55, see also M. Magnasco, *Phys. Rev. Lett.* **71**, 1477 (1993); R. D. Astumian and M. Bier, *Phys. Rev. Lett.* **72**, 1766 (1994); F. Marchesoni, *Phys. Rev. Lett.* **77**, 2364 (1996); and references cited therein.

118. D. Ludwig, *SIAM Rev.* **17**, 605 (1975).

119. H. R. Jauslin, *J. Stat. Phys.* **42**, 573 (1986); *Physica* **144A**, 179 (1987).

120. M. I. Dykman, M. M. Millonas, and V. N. Smelyanskiy, *Phys. Lett. A* **195**, 53 (1994).

121. R. S. Maier and D. L. Stein, *J. Stat. Phys.* **83**, 291 (1996).

122. R. S. Maier and D. L. Stein, *SIAM J. Appl. Math.* **57**, 752 (1997).

123. L. S. Schulman, *Techniques and Applications of Path Integration*, Wiley, New York, 1981.

124. M. I. Dykman, D. G. Luchinsky, P. V. E. McClintock, and V. N. Smelyanskiy, *Phys. Rev. Lett.* **77**, 5229 (1996).

125. S. Wiggins, *Global Bifurcations and Chaos: Analytical Methods*, Springer-Verlag, New York, 1988.

126. L. Fronzoni, in F. Moss and P. V. E. McClintock (Eds.), *Noise in Nonlinear Dynamical Systems*, Cambridge Univ. Press, Cambridge, UK, 1989, Vol. 3, pp. 222–242; P. V. E. McClintock, and F. Moss, ibid. pp. 243–274.

127. M. B. Willemsen, M. P. van Exter, and J. P. Woerdman, *Phys. Rev. A* **60**, 4105 (1999).

128. M. B. Willemsen, M. U. F. Khalid, M. P. van Exter, and J. P. Woerdman, *Phys. Rev. Lett.* **82**, 4815 (1999).

129. M. I. Dykman, V. N. Smelyanskiy, D. G. Luchinsky, R. Mannella, P. V. E. McClintock, and N. D. Stein, *Int. J. Bifurc. Chaos* **4**, 747 (1998).

130. M. I. Dykman, V. N. Smelyanskiy, D. G. Luchinsky, R. Mannella, P. V. E. McClintock, and N. D. Stein, *Nonlinear Phenomena Complex Syst.* **2**, 1 (1999).

131. D. G. Luchinsky, R. S. Maier, R. Mannella, P. V. E. McClintock, and D. L. Stein, *Phys. Rev. Lett.* **79**, 3109 (1997).

132. D. G. Luchinsky, R. S. Maier, R. Mannella, P. V. E. McClintock, and D. L. Stein, *Phys. Rev. Lett.* **82**, 1806 (1999).

133. B. E. Vugmeister, J. Botina, and H. Rabitz, *Phys. Rev. E* **55**, 5338 (1997).

134. R. Mannella, *Phys. Rev. E* **59**, 2479 (1999); B. E. Vugmeister, J. Botina, and H. Rabitz, *Phys. Rev. E* **59**, 2481 (1999).

135. S. M. Soskin, *J. Stat. Phys.* **97**, 609 (1999).

136. S. M. Soskin, V. I. Sheka, T. L. Linnik, M. Arrayás, I. Kh. Kaufman, D. G. Luchinsky, P. V. E. McClintock, and R. Mannella, in D. S. Broomhead, E. A. Luchinskaya, P. V. E. McClintock and T. Mullin (Eds.), *Stochaos: Stochastic and Chaotic Dynamics in the Lakes*, Vol. 502, American Institute of Physics, Woodbury, NY, 2000, pp. 60–65.

137. S. M. Soskin, V. I. Sheka, T. L. Linnik, and R. Mannella, *Phys. Rev. Lett.* **86**, 1665 (2001); M. Arrayás, I. Kh. Kaufman, D. G. Luchinsky, P. V. E. McClintock, and S. M. Soskin, *Phys. Rev. Lett.* **84**, 2556 (2000).

138. P. Hanggi, P. Talkner, and M. Borkovec, *Rev. Mod. Phys.* **62**, 251 (1990).

139. V. I. Mel'nikov, *Phys. Rep.* **209**, 1 (1991).

140. V. A. Shneidman, *Phys. Rev. E* **56**, 5257 (1997).

141. D. G. Luchinsky, R. Mannella, P. V. E. McClintock, M. I. Dykman, and V. N. Smelyanskiy, *J. Phys. A* **32**, L321 (1999).

142. M. Marder, *Phys. Rev. Lett.* **74**, 4547 (1995).

143. R. Mannella, in L. Vázquez, F. Tirando, and I. Martin (Eds.), *Supercomputation in Nonlinear and Disordered Systems*, World Scientific, Singapore, 1997, pp. 100–130.

144. V. N. Smelyanskiy, M. I. Dykman, and R. S. Maier, *Phys. Rev. E* **55**, 2369 (1997).

145. R. G. Harrison, J. S. Uppal, and P. Osborne, (Eds.), *Nonlinear Dynamics and Spatial Complexity in Optical Systems*, SUSSP/Institute of Physics, Bristol, 1993.

146. L. A. Lugiato, in E. Wolf (Ed.), *Progress in Optics*, North-Holland, Amsterdam, 1984, Vol. 21, pp. 71–211.

147. G. Carpintero and H. Lamela, *Opt. Lett.* **24**, 1711 (1999).

148. I. B. Schwartz and T. W. Carr, *Phys. Rev. E* **59**, 6658 (1999).

149. Y. Liu and P. Davis, *Opt. Lett.* **25**, 475 (2000).

150. T. Shinbrot, C. Ott, C. Grebogi, and A. J. Yorke, *Phys. Rev. Lett.* **65**, 3215 (1990).

151. S. Boccaletti, C. Grebogy, Y. C. Lai, H. Mancini, and D. Maza, *Phys. Rep.* **329**, 103 (2000).

152. T. Shinbrot, *Adv. Phys.* **44**, 73 (1995).

153. T. Shinbrot, C. Grebogi, J. Ott, and E. Yorke, *Nature* **363**, 411 (1993).

154. B. Hubinger, R. Doerner, and W. Martienssen, *Phys. Rev. E* **50**, 932 (1994).

155. E. Barreto, E. J. Kostelich, C. Grebogy, J. Ott, and E. Yorke, *Phys. Rev. E* **51**, 4169–4172 (1995).

156. E. A. Jackson, *Chaos* **7**, 550–559 (1997).

157. S. P. Raj and S. Rajasekar, *Phys. Rev. E* **55**, 6237–6240 (1997).

158. C. Reyl, L. Flepp, R. Badii, and E. Brun, *Phys. Rev. Lett.* **67**, 267 (1991).

159. R. Roy, T. W. Murphy, T. D. Maier, Z. Gillis, and E. R. Hunt, *Phys. Rev. Lett.* **68**, 1259 (1992).

160. V. S. Anishchenko, A. S. Kopeikin, T. E. Vadivasova, G. I. Strelkova, and J. Kurths, *Phys. Rev. E* **62** (2000) (in press).

161. A. L. Shilnikov, *Comput. Math. Appl.* **34**, 245 (1997).

162. C. O. Weiss, N. B. Abraham, and U. Hubner, *Phys. Rev. Lett.* **61**, 1587 (1988).

163. C. O. Weiss, U. Hubner, N. B. Abraham, and D. Tang, *Infrared. Phys. Technol.* **36**, 489 (1995).

164. L. M. Hocking, *Optimal Control*, Oxford Applied Math and Computing Science, Clarendon Press, Oxford, 1997.

165. V. N. Smelyanskiy and M. I. Dykman, *Phys. Rev. E* **55**, 2516 (1997).

166. Y. R. Shen, *The Principles of Nonlinear Optics*, Wiley, New York, 1984.

167. S. V. Gonchenko, L. P. Shilnikov, and D. V. Turaev, *Comput. Math. Applic.* **34**, 195 (1997).

168. D. G. Luchinsky, I. A. Khovanov, and P. V. E. McClintock, *Prog. Theor. Phys. Suppl.* **139**, 152 (2000).

169. D. G. Luchinsky, I. A. Khovanov, R. Mannella, and P. V. E. McClintock, *Phys. Rev. Lett.* **85**, 2100 (2000).

170. R. Zwanzig, *J. Stat. Phys.* **9**, 215 (1973).

171. G. Marsaglia and W.-W. Tsang, *SIAM J. Sci. Stat. Comput.* **5**, 349 (1984).

172. D. G. Luchinsky, P. V. E. McClintock, S. M. Soskin, and R. Mannella, *Phys. Rev. Lett.* **76**, 4453 (1996).

173. I. A. Khovanov, D. G. Luchinsky, R. Mannella, and P. V. E. McClintock, "Optimal fluctuations and the control of chaos," *Int. J. Bifurc. Chaos.* (in press).

174. D. Auerbach, P. Cvitanović, J-P. Eckmann, G. Gunaratne, and I. Procaccia, *Phys. Rev. Lett.* **58**, 2387 (1987).

175. C. Grebogi, E. Ott, and J. A. Yorke, *Phys. Rev. A* **37**, 1711 (1988).

176. P. Schmelcher, and F. K. Diakonos, *Phys. Rev. Lett.* **78**, 4733 (1997).

177. Codes can be obtained from http://www.netlib.no/.

178. J. R. Cash, *Comput. Math. Appl.* **12A**, 1029 (1986); J. R. Cash, *SIAM J. Numer. Anal.* **25**, 862 (1988); J. R. Cash and M. H. Wright, *Computing* **45**, 17 (1990); J. R. Cash and M. H. Wright, *SIAM J. Sci. Stat. Comput.* **12**, 971 (1991).

179. E. N. Lorenz, *J. Atmos. Sci.* **20**, 130 (1963).

180. H. Haken, *Phys. Lett.* **53A**, 77 (1975).

181. C. Sparrow, *The Lorenz Equations: Bifurcations, Chaos and Strange Attractors*, Springer, New-York, 1982.

182. I. A. Khovanov, D. G. Luchinsky, R. Mannella, and P. V. E. McClintock, in J. A. Freund and T. Pöschel (Eds.), *Stochastic Processes in Physics, Chemistry and Biology*, Springer, Berlin, 2000, pp. 378–389.

183. M. E. Johnson, M. S. Jolly, and I. G. Kevrekidis, *Numerical Algorithms* **14**, 125 (1997).

BELTRAMI VECTOR FIELDS IN ELECTRODYNAMICS—A REASON FOR REEXAMINING THE STRUCTURAL FOUNDATIONS OF CLASSICAL FIELD PHYSICS?

DONALD REED

CONTENTS

Modern Nonlinear Optics, Part 3, Second Edition, Advances in Chemical Physics, Volume 119, Edited by Myron W. Evans. Series Editors I. Prigogine and Stuart A. Rice.
ISBN 0-471-38932-3   © 2001 John Wiley & Sons, Inc.

I. INTRODUCTION

When we come to examine the annals of classical hydrodynamics and electrodynamics, we find that the foundations of vector field theory have provided some key field structures whose role has repeatedly been acknowledged as instrumental in not only underpinning the structural edifice of classical continuum field physics, but in accounting for its empirical exhibits as well.

However, there is one equally important vector field configuration that, despite its ubiquitous exhibit throughout a panoply of applications in classical field physics, has strangely enough remained less understood, underappreciated and currently consigned to linger in relative obscurity, known only to specialists in certain fields.

The following exposition seeks to remedy this situation by calling attention to the extensive but little recognized applications of this field structure in various areas of hydrodynamics and electrodynamics. Also, many of these empirical exhibits, particularly in classical electrodynamics, have continued to be accompanied by recorded energy anomalies or other related phenomena currently unexplainable by accepted scientific paradigms. This is one indication which strongly implies a disturbing lack of completeness in the foundations of electromagnetic theory.

An aim of this chapter is to show that these deficiencies could possibly be better understood once it is acknowledged that certain exhibits of classical EM—much like those associated with the Aharonov–Bohm effect—may in fact be a function of topological field symmetries higher than those of the standard U(1) variety. Consequently, our inability to quantify some of the EM phenomena associated with the field structure in question, could be due to a heretofore lack of understanding of its possible significance as a fundamental topological field archetype, universal throughout physics in general. Accordingly, in support of this contention, a smaller focus of this chapter explores the speculation that our failures up to present in constructing a viable deterministic model or theory of turbulence in hydrodynamics, might also rest on the heretofore unsuspected role this field structure might play at a more archetypal level of nature.

II. PROPERTIES OF BELTRAMI FLOW FIELDS
IN HYDRODYNAMICS

This vector field condition is sometimes referred to as Beltrami fluid flow, and was previously treated in a similar exposition by the author in 1995 [1]. There it was indicated that Beltrami vector field flow is representative of a certain class of vector fields that are termed *force-free*. This type of field topology was first brought to prominence by Eugenio Beltrami in his 1889 paper "Considerations on Hydrodynamics." [2]. This type of morphology describes a regime of fluid flow in which both the velocity field **v** (flux of streamlines) and the vorticity (**w** = curl **v**) are either parallel or antiparallel.

This also describes a Magnus force-free flow, which is expressed by the following relationship:

$$\mathbf{v} \times (\text{curl } \mathbf{v}) = 0 \tag{1}$$

As pointed out by Bjorgum and Godal [3], this relationship represents one of the eight types of possible vector fields. These are further derived from three basic field types:

1. Solenoidal vector fields for which

$$\text{div } \mathbf{v} = 0 \tag{2}$$

2. Complex lamellar vector fields for which

$$\mathbf{v} \cdot (\text{curl } \mathbf{v}) = 0 \tag{3}$$

3. Beltrami vector field condition in (1) above

An alternative formulation of the Beltrami condition is given as follows. For any vector field there exists the identity

$$(\mathbf{v} \cdot \text{grad})\mathbf{v} = \text{grad} \frac{(v^2)}{2} - \mathbf{v} \times (\text{curl } \mathbf{v})$$

where v denotes the magnitude of **v**. Now, the Beltrami condition is satisfied if

$$(\mathbf{v} \cdot \text{grad})\mathbf{v} = \text{grad} \frac{(v^2)}{2} \tag{4}$$

Consequently, (4) represents a necessary and sufficient Beltrami condition. Since the Beltrami flow (1) describes parallel or antiparallel vorticity and velocity vectors, another useful formulation of the Beltrami condition is represented by the relation

$$\text{curl } \mathbf{v} = c\mathbf{v} \tag{5}$$

where c denotes a scalar point function of position. This factor c assumes a certain degree of importance in association with Beltrami fields. Certain formulas can be derived from (5). First, $|c| = w/v$, where w denotes the magnitude of the vorticity **w**. By taking the scalar (dot) product of **v** with (5), we obtain

$$c = \frac{\mathbf{v} \cdot (\text{curl } \mathbf{v})}{v^2} \tag{6}$$

Beltrami derived a formula similar to (6) in which c is expressed by the vorticity $\mathbf{w}$ only. By introducing $\mathbf{v} = \mathbf{w}/c$, we obtain

$$c = \frac{(\mathbf{w}/c) \cdot (\operatorname{curl} \mathbf{w}/c)}{(w/c)^2} = \frac{\mathbf{w} \cdot (\operatorname{curl} \mathbf{w})}{w^2} \tag{7}$$

III. SPECIALIZED BELTRAMI FIELDS

Now, if a Beltrami field is simultaneously complex lamellar, (1) combined with (3), then curl $\mathbf{v}$ is both perpendicular and parallel to $\mathbf{v}$. This can happen only if curl $\mathbf{v}$ is zero (that is, the field $\mathbf{v}$ is curl-less, or lamellar). Hence a vector field that is simultaneously a complex lamellar and a Beltrami field is necessarily lamellar.

If the divergence of (5) is taken, we obtain

$$c(\operatorname{div} \mathbf{v}) + \mathbf{v} \cdot (\operatorname{grad} c) = 0 \tag{8}$$

If a Beltrami field (1) is simultaneously solenoidal (2), then (8) reduces to

$$\mathbf{v} \cdot (\operatorname{grad} c) = 0 \tag{9}$$

In other words, in a solenoidal Beltrami field the vector lines are situated in the surfaces $c =$ constant. This theorem was originally derived by Ballabh [4] for a Beltrami flow proper of an incompressible medium. For the sake of completeness, we mention that the combination of the three conditions (1), (2), and (3) only leads to a Laplacian field, that is better defined by a vector field that is both solenoidal (divergence-less) and lamellar (curlless).

Now, we consider a Beltrami field whose curl is also a Beltrami field. For this case, in addition to (1), the field must satisfy the following condition:

$$(\operatorname{cur} \mathbf{v}) \times (\operatorname{curl} \operatorname{curl} \mathbf{v}) = 0 \qquad \text{or} \qquad \mathbf{v} \times (\operatorname{curl} \operatorname{curl} \mathbf{v}) = 0 \tag{10}$$

By taking the curl of (5), we obtain

$$\operatorname{curl} \operatorname{curl} \mathbf{v} = (\operatorname{grad} c) \times \mathbf{v} + c(\operatorname{curl} \mathbf{v}) \tag{11}$$

Now, if c is uniform [i.e., $c(x, y, z) = k$(a constant)], then (11) becomes

$$\operatorname{curl} \operatorname{curl} \mathbf{v} = c(\operatorname{curl} \mathbf{v}) \tag{12}$$

In the case of uniform $c(\operatorname{grad} c = 0)$, curl $\mathbf{v} = \mathbf{w}$ (vorticity) is also a Beltrami field, possessing the same coefficient c. This type of vector field is called a *Trkalian field*, after Trkal, a Russian researcher who studied Beltrami flows proper where $c = $ constant. As pointed out by Nemenyi and Prim [5], all successive curls of Trkalian fields will also represent Trkalian fields, with the same coefficient c as the original field. Also, under these conditions, taking the divergence of (5), we get div(curl $\mathbf{v} = 0 = $ div($c\mathbf{v}$) $= $ (grad c) $\cdot \mathbf{v} + c$(div $\mathbf{v}$), or $0 = c$(div $\mathbf{v}$), showing div $\mathbf{v} = 0$ ($\mathbf{v}$ is solenoidal). Thus, we conclude that the Trkalian field is also a special solenoidal Beltrami field.

IV. PROPERTIES OF THE COEFFICIENT c

Because of (5), the equation for c (6) becomes

$$c = \frac{\mathbf{v} \cdot \text{curl} \frac{(\text{curl } \mathbf{v}))}{c}}{v^2} = \frac{c\mathbf{v} \cdot \text{curl} \frac{(\text{curl } \mathbf{v}))}{c}}{cv^2} = \frac{c\,\mathbf{v} \cdot (\text{curl}(\text{curl } \mathbf{v}))}{(c\,\mathbf{v})^2} \tag{13}$$

Finally, (13) becomes

$$1 = \frac{\mathbf{v} \cdot \text{curl}(\text{curl } \mathbf{v})}{(\text{curl } \mathbf{v})^2} \qquad \text{or} \qquad (\text{curl } \mathbf{v})^2 = \mathbf{v} \cdot (\text{curl}(\text{curl } \mathbf{v})) \tag{14}$$

Since the left side of (14) is necessarily positive, we conclude that a Beltrami field $\mathbf{v}$ and the curl of its curl [right side of (14)] always meet at an acute angle. If the angle is zero, then curl $\mathbf{v}$ is also a Beltrami field. In this last case, because of considerations above, $\mathbf{v}$ is a Trkalian field.

For any Beltrami field, c can also be given a geometric interpretation. Calling $\mathbf{t}$ the unit vector along $\mathbf{v}$, we apply Stokes' theorem to a curve (ds) determined by an orthogonal cross section (da) of an infinitesimal circular vector tube. If r denotes the radius, we find (see Fig. 1)

$$\int_{(c)} \mathbf{t} \cdot d\mathbf{s} = \int_{(s)} \text{curl } \mathbf{t} \cdot d\mathbf{a} = \pi r^2 \mathbf{t} \cdot \text{curl } \mathbf{t} \tag{15}$$

Since $\mathbf{v} = v\mathbf{t}$, we have for (6)

$$\frac{\mathbf{v} \cdot \text{curl } \mathbf{v}}{v^2} = \mathbf{t} \cdot \text{curl } \mathbf{t} = c \tag{16}$$

Applying (16) to (15), we obtain

$$c = \frac{1}{\pi r^2} \int_{(s)}^{\mathbf{t} \cdot d\mathbf{s}} = \frac{1}{\pi r^2} \int_{(c)} \left(\mathbf{t}_0 + \frac{\delta \mathbf{t} r}{\delta r} \right) \cdot d\mathbf{s}$$

$$= \frac{1}{\pi r^2} \int_{(c)} \frac{\delta \mathbf{t} \, r}{\delta r} \cdot \frac{ds\,ds}{ds} = \frac{1}{\pi r^2} \int \theta \, ds = \frac{2\bar{\theta}}{r}$$

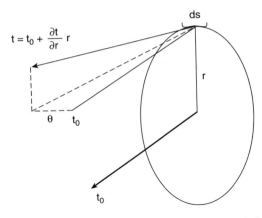

Figure 1. Diagram showing that $\theta = (\partial t / \partial r) r \cdot (ds/ds)$.

Here $\bar{\theta}$ denotes the average angle θ between $\mathbf{t}_0$ and the projection of $\mathbf{t}_0 + \delta \mathbf{t}/\delta r r$ on the plane determined by ds and $\mathbf{t}$ (see Fig. 1); $\theta = \delta \mathbf{t}/\delta r r \cdot d\mathbf{s}/ds$. From this analysis, we see that the factor c can be considered geometrically as the torsion of neighboring vector lines $\mathbf{v}$ in any Beltrami field. Truesdell [6] calls c the "abnormality" of the vector line system of a Beltrami field.

V. FIELD MORPHOLOGY IN A BELTRAMI CONFIGURATION

In a general hydrodynamic system, the vorticity $\mathbf{w}$ is perpendicular to the velocity field $\mathbf{v}$, creating a so-called Magnus pressure force. This force is directed along the axis of a right-hand screw as it would advance if the velocity vector rotated around the axis toward the vorticity vector. The conditions surrounding a wing that produce aerodynamic lift describe this effect precisely (see Fig. 2).

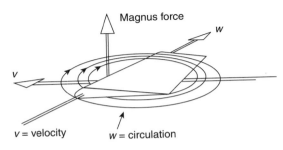

Figure 2. Indicating the Magnus force and vortices surrounding a wing.

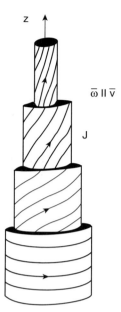

$\overline{\omega} \parallel \overline{v}$

J

Figure 3. Classical sheared vortex configuration in a Beltrami field (axisymmetric mode).

However, in a Beltrami field, the vorticity and velocity vectors are parallel or antiparallel, resulting in a zero Magnus force. The Beltrami condition (1) is therefore an equivalent way of characterizing a force-free flow situation, and vice versa.

In order to model this type of flow field geometrically, Beltrami found that it was necessary to consider a three-dimensional circular axisymmetric flow in which the velocity and vorticity field lines described a helical pattern. This helicoidal flow field was unique in that the pitch of the circular helices decreased as the radius from the central axis increased. This produces a specialized shear effect between the field lines of successively larger cylindrical tubes constituting the respective helices. In the limit of such a field, the central axis of the flow also serves as a field line (see Fig. 3).

VI. A NEW LOOK AT VECTOR FIELD THEORY

Although Beltrami fields have featured prominently in hydrodynamics for over a century, only until relatively recently have they received much attention in experimental and theoretical classical electrodynamics. The reason for the omission of this link in the standard development of electric/magnetic field theory can possibly be traced to a key related deficiency in the structure of vector field theory—which, in turn, constitutes the very foundation on which all edifices of classical continuum field physics are constructed. Here we refer to the self-imposed limitations of standard Gibbsian vector analysis for modeling the evolution of vector fields with higher-topology structural features. This deficiency can partially be attributed to the decision by the architects of vector analysis at the beginning of the twentieth century (notably Gibbs and Heaviside) to remove its *quaternion-based* foundation for the purpose of modeling vector fields in simply-connected three-dimensional (3D) Euclidean space. However, with vector fields limited to simple topologies, key topological features such as *helicity,* which feature prominently in the dynamics of the Beltrami flow (in both fluid turbulence and plasma dynamic contexts), remain untreated by these methods, and are better understood topologically through treatment by Cartan's calculus of differential forms. Specifically, the inherent spiral geometry exhibited by Beltrami flow cannot be clearly ascertained from use of the standard traditional methods for the construction of the general vector field.

There are two known standard methods for decomposition of any smooth (differentiable) vector field. One is that attributed to Helmholtz, which splits any vector field into a lamellar (curl-free) component, and a solenoidal (divergence-less) component. The second, which divides a general vector field into lamellar and complex lamellar parts, is that popularized by Monge. However, the relatively recent discovery by Moses [7] shows that any smooth vector field—general or with restraints to be determined—may also be separable into *circularly polarized* vectors. Furthermore, this third method simplifies the otherwise difficult analysis of three-dimensional classical flow fields. The Beltrami flow field, which has a natural chiral structure, is particularly amenable to this type analysis.

Beltrami flow is a particular case whose field generalization is popularly referred to as *eigenfunctions of the curl operator.* In his investigations using curleigenfunctions, Moses shows that the expansion of vector fields in terms of these operators, leads to a decomposition into *three* modes, as opposed to the customary two. One is the lamellar vector field (implying the existence of a scalar potential) with eigenvalue zero. However, the solenoidal vector field divides into two chiral circularly polarized vector potential fields of opposite signs of polarization and eigenvalues $+k$ and $-k$, respectively. Not only does this new decomposition considerably simplify problems dealing with vector fields that are defined over the whole coordinate space, but Moses describes how this decomposition is rotationally invariant. Thus, under a rotation of the coordinates, the vector modes that are introduced in this manner, remain individually invariant under this transformation. This allows for substantial improvement on the traditional Helmholtz or Monge decompositions. In several instructive examples, Moses [7] displays the versatile utility of this method. For instance, in the arena of incompressible and viscuous fluid dynamics, the curl-eigenfunction method shows how fluid motion with vorticity can be described as a super-position of circularly polarized modes only. A consequence of this approach is that the normally nonlinear convection term in the Navier-Stokes equation drops out, allowing for exact solutions of this relation. These are impossible under the conditions of turbulence, which allows for only approximate solutions to be derived.

VII. FLUID TURBULENCE AND BELTRAMI FIELD MODES

In the specialized area of turbulent fluid flow research, it is therefore not surprising that the Beltrami vortex flow itself, may eventually come to play a significant role. Accordingly, on the forefront of frontiers leading to a heretofore elusive deterministic theory of turbulence, it has been suggested [8], as a result of empirical evidence, that in regions of space, turbulent flows spontaneously organize into a coherent hierarchy of weakly interacting superimposed approximate Beltrami flows. Detailed numerical experiments for channel flows and decaying homogeneous turbulence using spectral methods, have provided evidence for such behavior [9]. The implications of this for fluid dynamics is that every incompressible fluid flow (divergenceless velocity field), as well as every solution to the Navier–Stokes equation, is a superposition of interacting Beltrami flows. It has been further postulated that full 3D dimensional turbulent flows may exhibit regimes of both weak local interaction and strong local interaction (due, e.g., to vortex stretching) between their Beltrami components [8]. These facts are suggestive of a nonlinear *uncertainty principle* for macroscopic turbulent fluid flows that mediates between regions of effective description of the flow field in physical (coordinate) space and that in wavenumber (momentum) space [8]. The discovery and certification of such an uncertainty principle in a macroscopic context, analogous to that canonized in quantum physics, could have astounding cross-disciplinary implications. Not only might it provide for the long sought-for key to establishing a deterministic model for turbulent fluidflow, but might indirectly provide a major breakthrough toward a deeper understanding of both the classical and quantum descriptions of nature. At any rate, the novel *circularly polarized* decomposition of vector fields in this context might possibly pave the way to this goal.

VIII. COMPLEX HELICAL WAVE DECOMPOSITION AND VORTEX STRUCTURES

In this regard, Melander and Hussain [10] have made significant advances in casting fresh light on the perplexing problems of vortex core dynamics and the coupling between large scales and fine scales in the vicinity of a coherent structure. Essentially, they postulated that the dynamics of coherent structures is better understood within the framework of vortex–vorticity dynamics rather than in terms of primitive variables such as pressure and velocities. On this basis this detailed analysis of helical fluid structures was acccomplished by decomposing the flow field into *complex curl eigenfunction modes,* originally inaugurated by Lesieur [11] and called *complex helical wave decomposition.* Lesieur initially utilized this method for easier access to a spectral (Fourier) analysis of turbulent flow modes. The following brief interlude on the particulars of the helical wave decomposition for description of aspects of fluid turbulence will be time well spent as this model will resurface later when considering novel electrodynamic applications of the Beltrami vector field.

We start by constructing an orthonormal basis **a**, **b**, **c** (where **c** is a unit vector along the wavevector **k**, with **a** and **b** in the two-dimensional vector space orthogonal to **k**). The significance of this ansatz is that any vector function $F(x,y,z)$ is divergence free if and only if its Fourier coefficients **F(k)** are orthogonal to **k**, that is if $\mathbf{k} \cdot \mathbf{F(k)} = 0$. Thus, **F(k)** is a linear combination of **a(k)** and **b(k)**. Lesieur defines the complex helical waves as

$$\mathbf{V}^+(\mathbf{k}, \mathbf{x}) = [\mathbf{b(k)} - i\mathbf{a(k)}] \exp{(i\mathbf{k}, \mathbf{x})} \tag{17}$$

$$\mathbf{V}^-(\mathbf{k}, \mathbf{x}) = [\mathbf{b(k)} + i\mathbf{a(k)}] \exp{(i\mathbf{k}, \mathbf{x})} \tag{18}$$

which are eigenfunctions of the curl operator corresponding to the eigenvalues $|\mathbf{k}| \lessgtr 0$, namely, curl $\mathbf{V}^+ = |\mathbf{k}|\mathbf{V}^+$, and curl $\mathbf{v}^- = -|\mathbf{k}|\mathbf{V}^-$. These eigenfunctions are orthogonal with respect to the inner product:

$$\langle \mathbf{f}, \mathbf{g} \rangle = \int \mathbf{f} \cdot \mathbf{g}^* \, d\mathbf{x} \tag{19}$$

where $*$ denotes complex conjugation and the integration extends over all space (or a periodic box). In fact, all eigenfunctions of the curl operator, including those corresponding to the eigenvalue zero, are orthogonal, with respect to the inner product (19). Moreover, the set of all linearly independent eigenfunctions of the curl operator form a complete set of vector functions in R3 and all eigenvalues are real. The complex helical waves span the subspace of solenoidal vector functions, for the only vector field that cannot be expanded in terms of complex helical waves is the gradient of a potential. Thus, the divergence-free velocity field $u(x,t)$ can be expressed in terms of complex helical waves and the gradient of a harmonic potential:

$$\mathbf{u}(\mathbf{x},t) = \int_{k \neq 0} u^+(\mathbf{k}, t) \mathbf{v}^+(\mathbf{k}, \mathbf{x}) \, d\mathbf{k} + \int_{k \neq 0} u^-(\mathbf{k},t) \ \mathbf{v}^-(\mathbf{k}, \mathbf{x}) \, d\mathbf{k} + \ \text{grad} \ \emptyset$$

$$= \mathbf{u}_R + \mathbf{u}_L + \text{grad} \ \emptyset \tag{20}$$

Here, the first integral ($\mathbf{u}_R$) is the projection of $\mathbf{u}$ onto the vector space spanned by all eigenfunctions corresponding to positive eigenvalues ($+|\mathbf{k}|$) of the curl operator *(right-handed component* of $\mathbf{u}$). Likewise, the second integral ($\mathbf{u}_L$) is a linear combination of eigenfunctions corresponding to negative eigenvalues *(left-handed component* of $\mathbf{u}$). Also, grad $\emptyset$ is the projection of $\mathbf{u}$ onto the null space of the curl operator (zero eigenvalue of the curl operator). Since $\mathbf{u}$ is divergence free we have div(grad $\emptyset$)$=0$. Assuming that the potential part of the flow is constant at infinity, we have grad $\emptyset$ as a constant vector field that can be removed by choosing an appropriate inertial frame. For the vorticity field $\mathbf{w}(\mathbf{x},\,t) = $ curl $\mathbf{u}(\mathbf{x},t)$, we have a similar decomposition.

In their milestone work, Melander and Hussain found that the method of complex helical wave decomposition was instrumental in modeling both laminar as well as turbulent shear flows associated with coherent vortical structures, and revealed much new important data about this phenomenon than had ever been known before through standard statistical procedures. In particular, this approach plays a crucial role in the description of the resulting intermittent fine-scale structures that accompany the core vortex. Specifically, the large-scale coherent central structure is responsible for organizing nearby fine-scale turbulence into a family of highly polarized vortex threads spun azimuthally around the coherent structure.

In addition, this method shows that the polarization alternates between adjacent threads along the column. It is found that for a localized central axisymmetric vortex, the polarized structures constituting the central column are also localized and deform slowly (compared to unpolarized structures) and behave almost like solitary waves when isolated. This is because the non-linearity in the Navier–Stokes equation is largely suppressed between eigen-modes of the same polarity. Thus, it is found that the rapid changes in the total vorticity field result from the superposition of two slowly deforming wavetrains moving in opposite directions with different propagation velocities. The ability of the helical wave decomposition to extract the wavepackets propagating in opposite directions on a vortex allows us to view the vortex evolution in terms of the motion of polarized wavepackets and their non-linear interaction.

IX.   BELTRAMI FLOW AS ARCHETYPAL FIELD STRUCTURE: A SCHAUBERGER–BELTRAMI CONNECTION?

One of the underlying themes of this exposition is the suggestion that the Beltrami flow field could play an important but yet dimly suspected archetypal role in organizing matter and energy at a deeper level of nature. One indication of this might be the possible non-linear uncertainty principle cited earlier, which may be a feature exhibited by Beltrami topology in turbulent fluid media. Another could be the ability of a central fluid vortex core to organize (polarize) surrounding fine-scale thread-like turbulence, as revealed by the use of curl eigenfunction methods. However, as we will shortly see, this possible unexplored fundamental aspect of Beltrami flow geometry turns out to be indicated in other field contexts in which it manifests, such as electrodynamics. In particular, in many empirical exhibits where Beltrami flow plays a role it will be found that phenomena have been recorded that are unexplainable by orthodox field methods. Just such a result comes from an unexpected source: V. Schauberger's work, which uncovered the energetic properties of water flow. In the early part of the twentieth century, Schauberger demonstrated that particular vortical patterns formed by the streamlines of natural water flow, resulted in a water quality that was pure and health promoting [12]. In water pipes which were constructed with a vortical cross section, not only was excess energy delivered to the system (process of "implosion" [12]), but the water flowed in the pipe with negative resistance [12]. Figures 4 and 5 depict one of these spiral configurations: Schauberger's so-called *longitudinal vortex*. A key observation here is that this streamline geometry is virtually identical to the sheared-helical cylindrical vector field flow pattern exhibited by the Beltrami vortex. In both of these figures, the water is colder and more dense the closer the spiral approaches the central axis. This fact cannot be accounted for by standard classical fluid dynamics, which postulates that fluid temperature is independent of density of the medium. Certainly, most laws of mechanics forbid the occurrence of negative resistance anyway. It becomes apparent that Schauberger's work shows possible support for considering the sheared-helical (Beltrami) vortex morphology in an archetypal context.

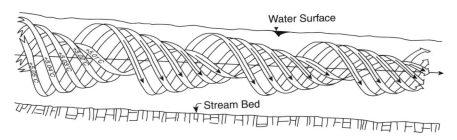

Figure 4. The longitudinal vortex. A longitudinal vortex showing laminar flow about the central axis. The coldest water filaments are always closest to the central axis of flow. Thermal stratification occurs even with minimal differences in water temperature. The central core water is subjected to the least turbulence and acclerates ahead, drawing the rest of the water body in its wake.

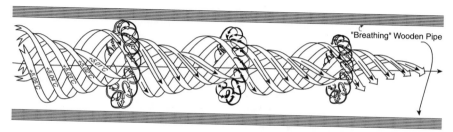

"Breathing" Wooden Pipe

Figure 5. The double-spiral longitudinal vortex. A longitudinal vortex showing the development of toroidal countervortices. These occur on interaction with the pipe walls and have an effect similar to ball bearings, enhancing the forward movement. Their interior rotation follows the direction of rotation and forward motion of the central vortex, whereas the direction of their exterior rotation and translatory motion are reversed. These toroidal vortices act to transfer oxygen, bacteria, and other impurities to the periphery of the pipe, where, because of the accumulation of excessive oxygen, the inferior, pathogenic bacteria are destroyed and the water rendered bacteria-free.

X. HYPOTHESIS OF FORCE-FREE MAGNETIC FIELDS (FFMFs)

Up until the twentieth century, helicoidal fields of the Beltrami variety had only been recognized to exist in hydrodynamic phenomena. For instance, wingtip vortices as well as the flow engendered by atmospheric vortices such as tornadoes, are two examples in which fluid flow approximates a force-free configuration [13]. Then in 1901, the Norweigian physicist Birkeland observed filaments in the earth's aurora and modeled them with his terella experiments. He postulated that the filaments were formed by the helicoidal flow of electrons around bundles of the earth's magnetic field lines in a force-free arrangement.

Early interest in force-free magnetic fields also arose in the study of astronomy. Stars that exhibited large magnetic fields have been known to exist. In the gaseous envelopes surrounding these magnetic Stars, strong magnetic fields and currents are known to exist simultaneously. Early investigations indicated that pressure gradients or gravitational fields appear to be of insufficient magnitude to counteract the reaction forces between the magnetic fields and the currents present.

In 1954 Lust and Schluter [14] introduced force-free magnetic fields (FFMFs) into a theoretical model for stellar media in order to allow intense magnetic fields to coexist with large currents in stellar matter with vanishing Lorentz force. Notice should be taken that the Lorentz force is the electrodynamic analogue of the Magnus force alluded to above (see Fig. 6 and compare with Fig. 2).

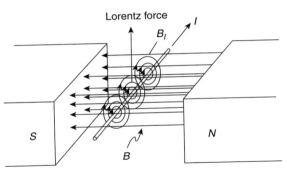

I = electrical current
B = magnetic field due to N-S poles
B_I = magnetic field due to current I

Figure 6. Illustrating electromagnetic action of the Lorentz force in a current-carrying conductor.

Taking this model for astrophysics further by postulating a Beltrami field morphology to electromagnetics, Chandrasekhar and Woltjer [15] posited that similar Lorentz force-free fields might exist which are quantified by the relation

$$\operatorname{curl} \mathbf{B} = k\mathbf{B} \tag{21}$$

(where k is a constant or function of position; $\mathbf{B}$ = real) in a magnetostatic field where the current density $\mathbf{j} = \operatorname{curl} \mathbf{B}$, and the magnetic field induction vector $\mathbf{B}$ are everywhere parallel to each other. Similar to the hydrodynamic case (5), currents do not have do any work against a force-free magnetic field. For these fields to occur and to continue to exist in the envelopes of magnetic stars, it is necessary that FFMF solutions be stable. Chandrasekhar and Woltjer also showed that for a given mean-square current density, exclusive of surface currents, the maximum magnetic energy per unit volume can exist in a stationary state only if the magnetic field is force-free with constant value of k. Second, it was shown that for a given amount of magnetic energy, the minimum dissipation occurs for force-free fields with constant k. Woltjer [16] further showed that for infinite conductivity, the force-free fields with constant k represent the lowest state of magnetic energy that a closed system may attain. Woltjer also demonstrated that for spherically symmetric perturbations, which are zero on the boundary, the axisymmetric fields with constant k are stable.

Additional models of FFMF for interstellar physics also postulated that the spiral arms of galaxies, as well as solar flares and prominences, could be constructed of such force-free fields [17]. Similar Beltrami field structures have

Elphic and Rusell: Flux Rope Observations and Models

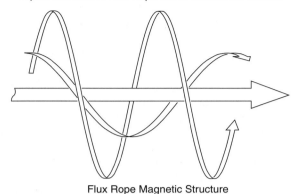

Flux Rope Magnetic Structure

Figure 7. Flux rope observations and models by Elphic and Russell [21]. Schematic representation of flux rope magnetic structure. The breadth of each arrow denotes the field strength; the central axial field is strong, while the outer, more nearly azimuthal field is weaker.

been formulated and developed as models for many other astrophysical phenomena in recent years. For instance, the magnetic clouds ejected from the Sun which have produced major perturbations in the Earth's radiation belts during the satellite era seem to possess FFMFs that have budded from the solar magnetic field [18,19]. The topology of filaments and chromospheric fibrils near sunspots have also been interpreted in terms of configuration lines of force of an axisymmetric force-free chromospheric magnetic field [20]. Magnetic flux ropes in the ionosphere of Venus also appear to possess force-free topology [21]. These theoretical findings have been corroborated by satellite data on magnetic field strength in various parts of these magnetic field structures. Figure 7 depicts the topology postulated for the FFMF believed common to magnetic clouds, chromospheric fibrils, and Venusian magnetic flux ropes. Immediately apparent is the signature sheared helical Beltrami morphology for the lines of force. Notice that the magnetic field strength (thickness of field lines) increases as the central axis is approached while the outer, more nearly azimuthal field is weaker. Compare this with the Schauberger water flow geometry which is quite similar. Again, here is possibly another hint relating to an archetypal role for Beltrami morphology in nature.

XI. EMPIRICAL CONFIRMATION OF BELTRAMI FFMF IN PLASMA RESEARCH

Prior to 1966, the properties of FFMF had only been theoretically predicted [22]. However, in the late 1960s, the morphology of electromagnetic Beltrami vortices was observed during the plasma focus experiments originally designed to investigate the possibilities for plasma confinement in fusion technology. A plasma focus group headed by W. Bostick demonstrated that the current sheath of the plasma focus is carried by contra-rotating pairs of plasma vortex filaments that exhibit a force-free morphology [23]. Under the intense discharge created in the plasma focus, it was found that these non-linear field structures cause interesting energy anomalies to occur, unaccountable by standard theory. For instance, due to the helical path followed by the current around the cylinder axis, as opposed to a linear path, these plasma vortices routinely permit the standard Alfven current limit to be surpassed, as well as transform a significant fraction of the input electrical energy into intense magnetic fields [24], In her thought-provoking book, White [25] discusses these plasma filaments as examples which violate the second law of thermodynamics by increasing the energy gradient within the field, while orthodox continuum classical physics predicts the entropic dissipation of energy. Moreover, these structures violate Heaviside–Lorentz relationships by creating ion currents that capture and concentrate their own and surrounding magnetic fields. While the electron and ion currents are in the process of forming into vortical filaments, a nested series of spiral paths is set up such that the generated magnetic fields cancel out in terms of their effects on the motion of the current flow. In fact, as shown in Fig. 8, all the force fields of the plasma—the local magnetic field, the current density, the fluid flow velocity, and the fluid vorticity—are locally collinear in each filament. In Fig. 9 it should be noted that the Poynting vector ($\sim \mathbf{E} \times \mathbf{B}$), which is taken to represent the power flow in an electromagnetic field, is directed parallel to the filament axis in a

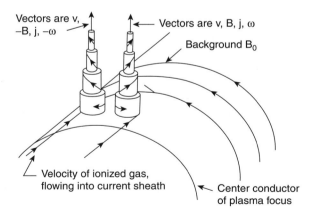

Figure 8. Dissected diagram of the vector configuration of a pair of Beltrami vortex filaments formed in the current sheath of the plasma focus ($\mathbf{v}$ = flow velocity, $\mathbf{B}$ = local magnetic field, $\mathbf{j}$ = current density, ω = vorticity, $\mathbf{B}_0$ = background magnetic field).

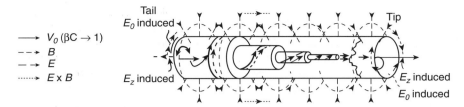

Figure 9. Morphologies of $\mathbf{V}, \mathbf{B}, \mathbf{E}$, and $\mathbf{E} \times \mathbf{B}$ as hypothesized by Bostick [23] for a propagating vortex-shaped electron vortex structure.

single vortex. Now, since this figure represents one-half of such a vortex pair, the total Poynting vector sums to zero due to the anti-parallel magnetic fields existing in the complete structure. Figures 8 or 9 should be consulted to see this relationship.

XII.   BELTRAMI FFMF TOPOLOGY AS A MODEL FOR EMPIRICAL ELECTROMAGNETIC PHENOMENA

Superconductivity is one of the best- known empirically quantified macroscopic electromagnetic phenomena whose basis is currently recognized to be quantum-mechanical. The behavior of the electric and magnetic fields under super-conductivity is governed by the London equations. However, there have been a series of papers questioning whether the originally phenomenologically theorized but now quantum-mechanically canonized London equations can be given a purely classical derivation [26]. Bostick [27,28], for instance, has claimed to show that the London equations do indeed have a classical origin that applies to superconductors and to some collisionless plasmas as well. In particular, it has been asserted that the Beltrami vortices in the plasma focus display the same paired flux-tube morphology as type II superconductors [29,30]. Others have also pointed out this little explored connection. Frohlich [31] has shown that the hydrodynamic equations of compressible fluids, together with the London equations, lead to the macroscopic Ginzburg–Landau equation, and in the presence of many fuxoids (quantum units of flux), all relevant equations can be expressed with the aid of the velocity potential (v) and the macroscopic parameter (μ = electric charge/mass density), without involving either quantum phase factors or Planck's constant.

In essence, it has been asserted that Beltrami plasma vortex filaments are able to at least simulate the morphology of types I and II superconductors. This occurs because the organized energy of the vortex configurations comprising the ions and electrons far exceeds the disorganized or thermal energy, and that the transition from disorganized turbulence to organized helical structures is a phase transition involving condensation without the rise of temperature. As we have pointed out, due to the feature of zero Lorentz force, the Beltrami vortex structures have less resistance than any other morphology. For ideal conditions in superconductivity, the elimination or reduction of the Lorentz force without reducing the field or the current is definitely desirable. These facts would tend to lend credence to the hypothesis of the concept that superconductivity in macroscopic classical physics can essentially exist, even though quantum effects have not been invoked to canonize the process.

XIII. HELICITY OF BELTRAMI TOPOLOGY IN FFMF

In all applications of the Beltrami vector field, whether in fluid or electro-magnetic contexts, it is found that the existence of *helicity* is the common key. In fluid dynamics, helicity is present because of the nonzero value of the "abnormality," c. In a general context, helicity measures the topological knotted-ness of a given vector field: vorticity ($\mathbf{w}$) in fluid dynamics, magnetic field induction ($\mathbf{B}$) in plasma dynamics, and has been shown to be related to the mathematical Hopf invariant [32,33]. The significance of topological parameters such as helicity in engendering a key role for FFMF in nature, comes to the fore when considering an ideal plasma possessing small viscosity and infinite conductivity on the boundary. Here, the minimization of magnetic energy subject to the constraint of magnetic helicity conservation (invariance in the linkage of field lines), produces through plasma self-organization a magnetic relaxation of the system into an equilibrium satisfying a Beltrami equation [34]. The following derivation showing this relationship, bears further witness to an archetypal role for the Beltrami field condition. Indeed, given the opportunity in certain environments, nature seems to have the pronounced inclination to organize itself according to force-free least action systems.

The helicity, defined by the following relation, is an invariant for every infinitesimal flux tube surrounding a closed line of force:

$$H = \int \mathbf{A} \cdot \mathbf{B} \, dv \qquad (22)$$

where $\mathbf{A}$ represents the vector potential and $\mathbf{B} = \text{curl } \mathbf{A}$, and dv is an infinitesimal volume element If we minimize the magnetic energy:

$$W = \frac{1}{2\mu_0 \int B^2 \, dv} \qquad (23)$$

subject to the constraint of magnetic helicity conservation described above, then for a plasma confined by a perfectly conducting toroidal shell the equilibrium satisfies curl $\mathbf{B} = k\mathbf{B}$. From the first variation of $W - kH/2$:

$$\delta\left(W - \frac{kH}{2}\right) = \int \frac{\mathbf{B} \cdot \delta\mathbf{B}}{\mu_0}\, dv - \frac{k}{2}\int (\delta\mathbf{A} \cdot \mathbf{B} + \mathbf{A} \cdot \delta\mathbf{B})\, dv \tag{24}$$

where k is a Lagrange multiplier. The first term is written as

$$\frac{1}{\mu_0}\int \mathbf{B} \cdot \delta\mathbf{B}\, dv = \frac{1}{\mu_0}\int \mathbf{B} \cdot \text{curl } \delta\mathbf{A}\, dv$$

$$= \frac{-1}{\mu_0}\left(\int \mathbf{B} \times \delta\mathbf{A} \cdot \mathbf{B}\, dS - \int \delta\mathbf{A} \cdot \text{curl } \mathbf{B}\, dv\right) \tag{25}$$

and the second term is

$$\int (\delta\mathbf{A} \cdot \mathbf{B} + \mathbf{A} \cdot \delta\mathbf{B})\, dv = \int (\delta\mathbf{A} \cdot \mathbf{B} + \mathbf{A} \cdot \text{curl } \delta\mathbf{A})\, dv$$

$$= 2\int (\delta\mathbf{A} \cdot \mathbf{B})\, dv - \int \mathbf{A} \times \delta\mathbf{A}) \cdot \mathbf{n}\, dS \tag{26}$$

Here we assume the wall is perfectly conducting. Then $\mathbf{n} \times \delta\mathbf{A} = 0$ is imposed on the wall, which corresponds to $\mathbf{n} \times \mathbf{E} = 0$, and the surface integrals of (25) and (26) vanish. Therefore (24) becomes

$$\delta\left(W - \frac{kH}{2}\right) = \int \delta\mathbf{A} \cdot [(\text{curl } \mathbf{B}) - k\mathbf{B}]\, dv \tag{27}$$

For an arbitrary choice of $\delta\mathbf{A}$, (27) becomes zero when (21) is satisfied. When (27) is zero, the magnetic energy is minimized and the helicity is held constant. Thus the state of minimum magnetic energy is a force-free equilibrium.

However, the discussion above cannot be the appropriate description of the quiescent state. In order to determine k, the invariant H for each closed field line would have to be calculated and related to its initial state. Hence, far from being universal and independent of initial conditions, the state defined by (22) depends on every detail of the initial state. To resolve this difficulty, it must be recognized that real turbulent plasmas are never perfectly conducting as in the ideal example above, but possess a certain amount of resistivity. Consequently, all topological invariants H cease to be relevant. Nevertheless, as long as the resistivity is small, the *sum* of all the invariants, that is, the integral of $\mathbf{A} \cdot \mathbf{B}$ over the total plasma volume is independent of any topological considerations and the need to identify particular field lines. The resulting configuration of a slightly resistive plasma after minimizing the energy subject to the constraint that total magnetic helicity be invariant is also the force-free state (21).

XIV. TRANSFORMATION PROPERTIES OF FFMF

As mentioned previously, misplaced focus throughout the years on the limitations inherent in the Gibbs–Heaviside vector analysis for modeling the real world, has in many ways done a disservice to formulating a proper understanding of classical vector field theory, especially when vector fields are to be associated with non-trivial topologies. In particular, many false conclusions have been repeatedly drawn, pertaining to the Beltrami FFMF condition by employing standard vector analysis. Such incorrect findings have, in turn, been partially instrumental in helping to promulgate confusion and in prejudicing physicists from considering the Beltrami vector field in any but the most superficial field contexts. One prime example clearly illustrating the confusion that has ensued, is the matter of the transformation properties of the FFMF. We shall consider specifically gauge invariance, parity, Lorentz invariance and time reversal.

Originally it was erroneously concluded that the FFMF condition was not invariant under parity [35]. This mistake was due to a misinterpretation caused by the ambiguity in standard vector analysis which does not clearly distinguish between a polar vector and a so-called pseudo- (or axial) vector, which is obtained as the vector product of two polar vectors (or curl of a polar vector). Later, correction of this error demonstrated the parity invariance of the FFMF equation through the use of the calculus of differential forms [36]. This formalism is free from the inconsistencies noted above and is also inherently topological, independent of coordinate system or metric. In particular, it was shown [37,38] that the force-free condition (curl $\mathbf{B} = k\mathbf{B}$) is parity-invariant, where k must be a pseudoscalar since it represents the ratio between a polar vector (curl $\mathbf{B}$), and an axial vector ($\mathbf{B}$). The force-free relation is also gauge-invariant [39], and invariant under time-reversal, where k is a proper pseudoscalar [40]. However, it is not necessarily Lorentz-invariant, as was shown in Ref. 41. All correct invariance properties of the FFMF relation are summarized in [41,42].

XV. MAGNETIC FIELD SOLUTIONS TO THE FFMF EQUATION

A divergenceless magnetic field (div $\mathbf{B} = 0$) is assumed in most classical electromagnetic field applications. Thus, this constraint is usually applied in the Beltrami FFMF condition (21) when determining solutions to this equation. From previous considerations, the solenoidal FFMF describes a Trkalian field with k = constant. Taking the curl of (21) under the constraint of solenoidal magnetic field, results in the vector Helmholtz equation:

$$\nabla^2 \mathbf{B} + k^2 \mathbf{B} = 0 \tag{28}$$

To solve this we consider a scalar function and its related scalar Helmholtz equation:

$$\nabla^2 \psi + k^2 \psi = 0 \tag{29}$$

Now, from any function that satisfies (29), can be formed three independent vectors that satisfy the vector wave equation (28) [42]. They are traditionally signified by $\mathbf{L} = \mathrm{grad}\,(\psi)$, $\mathbf{P} = \mathrm{curl}\,(\psi\mathbf{a})$, and $\mathbf{T} = \mathrm{curl}\,\mathrm{curl}\,(\psi\mathbf{a})$, where $\mathbf{a}$ is an arbitrary constant vector. Thus, to find solutions to (21), we express the vector $\mathbf{B}$ as a linear superposition of the three vectors

$$\mathbf{B} = x\,\mathrm{grad}(\psi) + y\,\mathrm{curl}(\psi\mathbf{a}) + z\,\mathrm{curl}\,\mathrm{curl}(\psi\mathbf{a}) \tag{30}$$

where x, y and z are some constants to be determined. To find the relationship between the constants, we first apply div $\mathbf{B} = 0$ to (30). This implies $x = 0$. Then applying curl $\mathbf{B} = k\mathbf{B}$ to (30) gives

$$\mathrm{curl}\,\mathbf{B} = y\,\mathrm{curl}\,\mathrm{curl}(\psi\mathbf{a}) + z\,\mathrm{curl}\,\mathrm{curl}\,\mathrm{curl}\,(\psi\mathbf{a})$$

$$= y\,\mathrm{curl}\,\mathrm{curl}\,(\psi\mathbf{a}) - z\nabla^2\,\mathrm{curl}\,(\psi\mathbf{a}) \quad [\text{since } (\mathrm{curl}\,(\mathrm{grad}\,\psi)) = 0)]$$

$$= y\,\mathrm{curl}\,\mathrm{curl}\,(\psi\mathbf{a}) + zk^2\,\mathrm{curl}(\psi\mathbf{a}) \quad [\text{since } \nabla^2 = -k^2] \tag{31}$$

Now, applying (30) again

$$k\mathbf{B} = ky\,\mathrm{curl}(\psi\mathbf{a}) + kz\,\mathrm{curl}\,\mathrm{curl}\,(\psi\mathbf{a}) \tag{32}$$

Equating coefficients in (31)–(32) gives $y = kz$. Consequently, we seek solutions of the form:

$$\mathbf{B} = kz\,\mathrm{curl}(\psi\mathbf{a}) + z\,\mathrm{curl}\,\mathrm{curl}\,(\psi\mathbf{a})$$

$$\mathbf{B} = B[k\,\mathrm{curl}(\psi\mathbf{a}) + \mathrm{curl}\,\mathrm{curl}\,(\psi\mathbf{a})] \tag{33}$$

With this expression for $\mathbf{B}$, we find that the field lines for the solution assume a key geometric relationship in addition to the previously considered axisymmetric helicoidal solutions exemplified in the vortex filaments of the plasma focus. Several researchers [43,44] have termed $\mathbf{P}$ the poloidal solution and $\mathbf{T}$ the toroidal solution. Accordingly, if the equation for $\mathbf{B}$ is expressed in terms of cylindrical polar coordinates, we find that the FFMF field solutions to (21) with k constant, are spiral field lines that lie on axisymmetric tours surfaces, with the pitch of the spirals changing gradually from totally poloidal (surface C) to completely toroidal (surface A). Consult Fig. 10 for a clearer depiction of this field morphology.

The toroidal solution to the Beltrami equation is more than a theoretical possibility. Indeed, in addition to the contrarotating Beltrami vortex filaments discovered by Bostick, plasma vortex rings called *plasmoids* were also recognized experimentally. Wells [45,46] in particular, studied the plasma topology generated by a conical theta-pinch device and verified that the field geometry of these plasma vortex rings approximated the quasi-force-free topology described above and illustrated in Fig. 10. Moreover, Chandrasekhar [10] showed that within a spherical boundary there is equipartition of energy between the toroidal and poloidal components of any force-free magnetic field.

The general solution of equation (33) in cylindrical coordinates can be written as a series of modes of the form [47]

$$\mathbf{B} = \Sigma_{m,n} \, B_{mn} \, \mathbf{b}_{mn} \, (r, \theta, z) \tag{34}$$

where m is a nonnegative integer, and where individual modes $\mathbf{b}_{mn}$ depend upon θ and z through the phase function $\emptyset = (m\theta + nz)$. The explicit expressions for the modes generally involve a linear combination of the Bessel functions J and Neumann functions N. However, when the domain of the solution involves the axis $r = 0$, and we restrict ourselves to an axisymmetric wave equation:

$$\frac{1}{r}\frac{d}{dr}\left(r\frac{d\psi}{dr}\right) + k^2\psi = 0 \tag{35}$$

the solution to the scalar function is $\psi = C J_0 \, (kr)$, where C is any constant. Using this and substituting in (34) with mode $m = 0, n = 0$, and $\mathbf{a} = (0,0,1)$, we

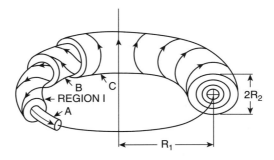

Figure 10.   Illustrating the toroidal mode solution to soleniodal FFMF equation.

get for the field components of the magnetic induction:

$$\mathbf{B} = B_0(0, J_1 \ (kr), J_0(kr)) \tag{36}$$

This is the solution originally demonstrated by Chandrasekhar and Kendall [48]. A pictorial representation of of this axisymmetric FFMF is shown in Fig. 3.

The solution to (34) has also been found in rectangular and spherical coordinates, but due to these complicated expressions, will not be further considered here. For details, readers are referred to the excellent review work by Zaghloul and Barajas [49], as well as the comprehensive recent book by Marsh [50]. These references also include information on FFMFs when k is not constant, depending upon space and time.

Nevertheless, the solution to the FFMF equation in cylindrical coordinates with certain constraints on the wave equation, most clearly illustrates the geometrical relation between the field lines. Also, the solution to the FFMF equation given in Beltrami's 1889 paper and in Bjorgum and Godal, which has been experimentally verified by Wells [51], Taylor [52] and others, illustrates that this is an eigenvalue equation. It possesses a whole spectrum of solutions corresponding to varying energy states with correspondingly different values of k. However, the geometry of the lowest state is that of a sheared helical structure across the whole cross section of the plasma column. This is illustrated in Fig. 3 for an axisymmetric field in cylindrical coordinates, or as in Fig. 10, for the general solution in toroidal coordinates or cylindrical polar coordinates. There are two separate field configurations in the next highest energy state. At the center of the cross section is a modified form of the first eigenvalue field, whereas the second is a helical field of opposite handed-ness, that is, a "reversed field."

Neverthesess, in the case of the solenoidal FFMF with k constant (Trkalian field), the lowest state of magnetic energy that a closed system may attain always assumes the topological feature of the sheared vortical structure, either in cylindrical or vortex ring geometry. This fact implies the stability of the Trkalian field flow. Moreover, for a plasma system in which magnetic forces are dominant, and in which there is some mechanism to dissipate the fluid motions, the FFMF with Trkalian field flow is the only form of stable magnetic field that can decay without giving rise to material fluid motions. The solenoidal FFMF solution with k constant appears to be the natural end configuration.

XVI.   TRKALIAN FIELD SOLUTIONS
TO MAXWELL'S EQUATIONS

A complete set of standard time-harmonic solutions to Maxwell's equations usually involve the plane wave decomposition of the field into transverse electric

(TE) and transverse magnetic (TM) parts. However, Rumsey [53] detailed a secondary method of solving the same equations that effected a decomposition of the field into left-handed and right-handed *circularly polarized* parts. For such unique field solutions to the time-harmonic Maxwell equations (ϵ = electric permittivity, μ = magnetic permeability);

$$\text{curl } \mathbf{H} = iw\epsilon\mathbf{E} \tag{37a}$$

$$\text{curl } \mathbf{E} = -iw\mu\mathbf{H} \tag{37b}$$

instead of expressing the electric and magnetic field intensities ($\mathbf{E}$ and $\mathbf{H}$) of Eqs. (37a,b) customarily as a superposition of plane waves, Rumsey showed how the field solutions could be expressed as circularly-polarized waves in terms of only one scalar potential function U where, for instance

$$\mathbf{E} = \text{curl curl } (\mathbf{z}U) + k \text{ curl } (\mathbf{z}U) \tag{38}$$

where $\pm$ indicates chiral circularly-polarized vectors. This is highly similar to the previously given solution to the FFMF Eq. (33), and at the same time also reminiscent of the chiral complex vector solutions found by Moses [7]. Since the wavevector is along the z-axis (k is some constant to be determined) (38) reduces to

$$\mathbf{E} = \text{grad } \frac{(dU)}{dz} \pm k \text{ curl } (\mathbf{z}U) + k^2(\mathbf{z}U) \tag{39}$$

In the quite different physical context of fields in ideal fluid media (incompressible, inviscid, homogeneous, with external force (β) conservative), Bjorgum and Godal [2] derived a solution for the velocity field ($\mathbf{v}$) for a general Trkalian flow which bears a striking resemblance to (39):

$$\mathbf{v} = \frac{\text{grad } (dH)}{dl} \pm \beta \text{ curl } (\mathbf{I}H) + \beta^2(\mathbf{I}H) \tag{40}$$

where $\mathbf{I}$ is the vector of propagation and H is a scalar potential. We can see that (39) and (40) are structurally equivalent if we set $\mathbf{1} = \mathbf{z}, H = U$, and $k = \beta$. In addition to the conclusions drawn by Rumsey in regard to the circularly polarized solutions to (37a,b), we observe that these solutions also possess the standard sheared vortex field topology we have alluded to in connection with any general Trkalian field solution. Whether Rumsey was aware of this connection between fluid and electrodynamics is unknown.

These facts clearly demonstrate a possible significant, albeit little explored correlation between foundational classical electrodynamics and fluid dynamics via the general Trkalian (sheared vortex) topology. At any rate, a potential fruitful avenue has been opened for future exploration of this link.

XVII. BELTRAMI FIELD RELATIONS FROM TIME-HARMONIC EM IN CHIRAL MEDIA

When we consider time-harmonic electrodynamics in more general media (chiral-biisotropic), A. Lakhtakia also underscored the importance of the Beltrami field condition [54]. In particular, he found that time-harmonic EM fields in a homogeneous reciprocal biisotropic medium are circularly polarized, and must be described by Beltrami vector fields.

The motivation for this work is the potential use of chiral (and maybe biisotropic) cylinders as rod antennas and scatterers. Accordingly, Lakhtakia investigated the boundary value problem relevant to the scattering of an incident oblique plane EM wave by an infinitely long homogeneous biisotropic cylinder. This medium is described by the so-called Fedorov representation through the monochromatic frequency-domain constitutive relations:

$$\mathbf{D} = \epsilon[\mathbf{E} + \alpha\,\mathrm{curl}\,\mathbf{E}], \qquad \mathbf{B} = \mu[\mathbf{H} + \beta\,\mathrm{curl}\,\mathbf{H}] \tag{41}$$

where μ is the magnetic permeability scalar, ϵ is the electric permittivity scalar, and α and β are the biisotropy pseudoscalars. By applying the specialized Bohren diagonalization transformation [55] to this synthetic field composition, when $\mathbf{E}$ and $\mathbf{H}$ are substituted into the time-harmonic source free Maxwell equations, with a time dependence assumed to be exp ($i\,wt$):

$$\mathrm{curl}\,\mathbf{H} = \frac{\delta\mathbf{D}}{\delta t,} \qquad \mathrm{curl}\,\mathbf{E} = \frac{-\delta\mathbf{B}}{\delta t} \tag{42}$$

we get

$$\mathbf{E} = (\mathbf{Q}_1 + \mathbf{Q}_2), \qquad \mathbf{H} = i\left(\frac{\mathbf{Q}_1}{\eta_1} + \frac{\mathbf{Q}_2}{\eta_2}\right) \tag{43}$$

where η_1 and η_2 are impedances, and $\mathbf{Q}_1$ and $\mathbf{Q}_2$ are Beltrami vector fields satisfying the relations

$$\mathrm{curl}\,\mathbf{Q}_1 = \gamma_1\mathbf{Q}_1, \qquad \mathrm{curl}\,\mathbf{Q}_2 = -\gamma_2\mathbf{Q}_2 \tag{44}$$

where γ_1 and γ_2 are wavenumbers that are functions of the frequency and constitutive scalar and pseudoscalar parameters.

It was found that in order to derive an accurate description of both incident and scattered EM radiation in a chiral medium, the use of Beltrami field relations is essential. Further details may be found in Refs. 56–58.

XVIII.   BELTRAMI VECTOR POTENTIAL ASSOCIATED
WITH TEM (TRANSVERSE ELECTROMAGNETIC)
STANDING WAVES WITH E//B

Besides its appearance in the FFMF equation in plasma physics, as well as associated with time-harmonic fields in chiral media, the chiral Beltrami vector field reveals itself in theoretical models for classical transverse electromagnetic (TEM) waves. Specifically, the existence of a general class of TEM waves has been advanced in which the electric and magnetic field vectors are parallel [59]. Interestingly, it was found that for one representation of this wave type, the magnetic vector potential ($\mathbf{A}$) satisfies a Beltrami equation:

$$\text{cur } \mathbf{A} = k\mathbf{A} \tag{45}$$

A solution of (45) is

$$\mathbf{A} = a[\mathbf{i}\sin(kz) + \mathbf{j}\cos(kz)]\cos(wt) \tag{46}$$

where a is a constant. The associated electric and magnetic fields are

$$\mathbf{E} = \frac{-1}{c}\frac{\partial \mathbf{A}}{\partial t} = wa/c[[\mathbf{i}\sin(kz) + \mathbf{j}\cos(kz)\sin(wt) \tag{47}$$

$$\mathbf{B} = \text{curl } \mathbf{A} = ka[\mathbf{i}\sin(kz) + \mathbf{j}\cos(kz)]\cos(wt) \tag{48}$$

One immediately sees that $\mathbf{E}$ and $\mathbf{B}$ (and also $\mathbf{A}$) are everywhere parallel, and all are perpendicular to the propagation vector ($k\ \mathbf{z}$). Consequently every plane wave solution to (45) corresponds to two circularly polarized waves propagating oppositely to each other and combining to form a standing wave. This standing wave does not possess the standard power flow feature of linear- or circularly polarized electromagnetic waves with $\mathbf{E} \perp \mathbf{B}$, since the combining Poynting vectors of the circularly polarized waves cancel each other similar to the situation we earlier met in connection with Beltrami vortex filaments. Essentially, the combination of these two waves produces a standing wave propagating non-zero magnetic helicity. In the book by Marsh [50], the relationship is shown between the helicity and energy densities for this wave as well as the very interesting fact that any magnetostatic solution to the FFMF equations can be used to construct a solution to Maxwell's equations with $\mathbf{E}//\mathbf{B}$. The current author also shows the geometric relationship of these waves with respect to space and time [60]. It is noted from this analysis that, in contrast to standard linearly-polarized waves, these unique standing waves have no nodes, constant amplitude, and describe a surface of minimum area called the helicoid (see Fig. 11. Moreover, such waves with these counter-intuitive properties have not only been theoretically predicted

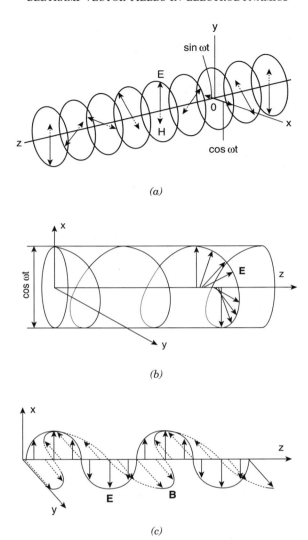

Figure 11. (a) Helical distribution of linearly oscillating parallel electric and magnetic fields shown by solid and broken arrows, respectively; (b) electric field distribution of a helicoidally polarized standing wave [the magnetic field **B** (not shown) is parallel to **E** and oscillates in time in quadrature with **E**]; (c) electric (solid line) and magnetic (dashed line) field distributions of a linearly polarized standing wave.

but have been experimentally realized in the so-called "twisted-mode technique" for obtaining uniform energy density in a laser cavity [61,62]. Experimental protocols to produce **E//B** waves are also illustrated in the paper [60].

XIX.   CONNECTION BETWEEN SPINORS, HERTZ POTENTIAL, AND BELTRAMI FIELDS

A.   Hillion–Quinnez Model

In utilizing a complex three-vector (self-dual tensor) rather than a real antisymmetric tensor to describe the electromagnetic field, Hillion and Quinnez discussed the equivalence between the 2-spinor field and the complex electromagnetic field [63]. Using a Hertz potential [64] instead of the standard 4-vector potential in this model, they derived an energy momentum tensor out of which Beltrami-type field relations emerged. This development proceeded from the Maxwell equations in free homogeneous isotropic space

$$\text{curl } \mathbf{E} = \frac{-\mu}{c\partial t \mathbf{H}} \tag{49a}$$

$$\text{curl } \mathbf{H} = \frac{\epsilon}{c\partial t \mathbf{E}} \tag{49b}$$

where $\mathbf{E}(x, t)$ and $\mathbf{H}(x, t)$ are the components of the electric and magnetic fields; ϵ and μ are the permittivity and permeability, respectively; c is the velocity of light, ∂_t and ∂_j are the derivatives with respect to time and x_j, and x is an arbitrary point in R3. If we introduce the complex vector

$$\Lambda = -(\sqrt{\epsilon}\mathbf{E} - i\sqrt{\mu}\mathbf{H}) \tag{50}$$

then Eqs. (49a,b) become

$$i\,\text{curl}\,\Lambda = \frac{n}{c\partial t \Lambda} \tag{51}$$

[where $n = (\epsilon\mu)^{1/2}$, the refractive index]. From taking the x component of (49b) and adding i times the y component of (49b), we get

$$\frac{i\epsilon}{c\partial t(E_x \pm iE_y)} + \partial z(H_x \pm iH_y) = (\partial x \pm i\partial y)H_z \tag{52a}$$

Doing the same with (49a), we get

$$\partial z(E_x \pm iE_y) - \frac{i\mu}{c\partial t}(H_x \pm iH_y) = (\partial x \pm i\partial y)E_z \tag{52b}$$

which becomes, in terms of (50)

$$(\partial x + i\partial y)\Lambda z = \left(\partial z + \frac{n}{c\partial t}\right)(\Lambda x + i\Lambda y) \tag{53a}$$

$$(\partial x - i\partial y)\Lambda = \left(\partial z - \frac{n}{c\partial t}\right)(\Lambda x - i\Lambda y) \tag{53b}$$

If we now consider a set of 2-spinors $\psi^a(x,t), a = 1, 2$, with complex components $\psi_\alpha^a(x,t), \alpha = 1, 2$, satisfying the Pauli equation

$$\left(\sigma^j \partial_j - \frac{n}{c\partial t}\right)\psi_\alpha^a(x,t) = 0 \qquad a = (1,2) \tag{54}$$

where σ_j are the Pauli matrices and we use the summation convention, $\sigma^j \partial_j = \sigma_1 \partial_x + \partial_2 \partial_y + \sigma_2 \partial_z$.

Equation (54) takes the form

$$(\partial x + i\partial y)\psi^a 1 - \left(\partial z + \frac{n}{c\partial t}\right)\psi^a 2 = 0 \tag{55a}$$

$$(\partial x - i\partial y)\psi_2^a + \left(\partial z - \frac{n}{c\partial t}\right)\psi^a 1 = 0 \tag{55b}$$

Now, by comparing (53a,b) with (55a,b), we get the following identifications:

$$\psi_2^1 = \Lambda_z, \qquad \psi_2^1 = \Lambda_x + i\Lambda_y; \qquad \psi_2^2 = -\Lambda_z \quad \psi_1^2 = \Lambda_x - i\Lambda_y$$

or

$$\Lambda_x = \frac{1}{2}(\psi_2^1 + \psi_1^2); \qquad \Lambda_y = \frac{1}{2}i(\psi_2^1 - \psi_1^2); \qquad \Lambda_z = \frac{1}{2}(\psi_1^1 - \psi_2^2) \tag{56}$$

with the constraint on the spinor field:

$$\psi_1^1 + \psi_2^2 = 0 \tag{57}$$

We now introduce a complex scalar φ such that

$$\Lambda_z = \psi_1^1 = \left(\frac{\partial^2 z - n^2}{c^2 \partial^2 t}\right)\varphi \tag{58}$$

Then using (55) and (56), we get

$$\Lambda_x = \frac{1}{2}\left[(\partial x + i\partial y)\left(\frac{\partial z - n}{c\partial t}\right)\varphi + (\partial x - i\partial y)\left(\frac{\partial z + n}{c\partial t}\right)\varphi\right] \tag{59a}$$

$$\Lambda_y = \frac{1}{2}i\left[(\partial x + i\partial y)\left(\frac{\partial z - n}{c\partial t}\right)\varphi - (\partial x - i\partial y)\left(\frac{\partial z + n}{c\partial t}\right)\varphi\right] \tag{59b}$$

Expanding these, we get

$$\Lambda_x = \left(\frac{\partial x \partial z - in}{c \partial y \partial t}\right)\varphi \tag{60a}$$

$$\Lambda_y = \left(\frac{i\partial y \partial z - n}{c \partial x \partial t}\right)\varphi \tag{60b}$$

which are recognized as the x and y components of a Hertz potential $\mathbf{\Pi}$ (complex 3-vector) with the following relationship to the complex field vector:

$$\Lambda = \operatorname{curl}\operatorname{curl}\mathbf{\Pi} + \frac{in}{c\partial}t\operatorname{curl}\mathbf{\Pi} \tag{61}$$

Thus, defining $\mathbf{\Pi}$ by $\varphi\mathbf{k}$, where $\mathbf{k}$ is a unit vector along the z axis, from (58) and (59) we get

$$\mathbf{\Pi} = \mathbf{M} + i\mathbf{N} \tag{62}$$

where $\mathbf{M}$ and $\mathbf{N}$ are, respectively, the electric and magnetic Hertz vectors [65]. Thus, it follows that any electromagnetic field in a homogeneous isotropic medium in free space, away from charges and currents, can be expressed either in terms of the spinors, ψ^a, or in terms of the complex Hertz vector $\mathbf{\Pi}$.

Now, in order to derive a scalar Lagrangian density, we introduce the matrix Ω and the spinors ϕ_0 and $\hat{\phi}_0$:

$$\Omega = \begin{vmatrix} \psi_1^1 & \psi_1^2 \\ \psi_2^1 & \psi_2^2 \end{vmatrix}; \quad \phi_0 = \begin{pmatrix} 1 \\ 0 \end{pmatrix} \hat{\phi}_0 = \begin{pmatrix} 0 \\ 1 \end{pmatrix} \tag{63}$$

and we define the Proca–Pauli field $\psi^a, a = 1, 2$ by the relations $\psi^1 = \Omega\phi_0$ and $\psi^2 = \Omega\hat{\phi}_0$, and we get the Proca–Pauli equation $\sigma^u\partial u\psi^a = 0$, $a = 1, 2$, which is Eq. (54) written in a manifestly covariant form.

Then the Lagrangian density:

$$\mathscr{L} = \frac{ic}{2}\sum_{a=1}^{2}(\psi^{a+}\sigma^u\partial u\psi^a - \partial u\psi^{a+}\sigma^\mu\psi^a) \tag{64}$$

where ψ^{a+} represents the Hermitian conjugate of ψ^a, is a real scalar invariant under the proper orthochronous Lorentz group $L_+^\uparrow$. Let ju be the energy flow vector. Then, from (64) we get

$$\sum_{ju=a=1}^{2}\psi^{a+}\sigma u\psi^a, \qquad u = 0\text{--}3 \tag{65}$$

The energy-momentum tensor derived from (65) is

$$T_{u,v} = \frac{ic}{2} \sum_{a=1}^{2} (\psi^{a+} \sigma u \, \partial v \psi^a - \partial v \, \psi^{a+} \sigma u \, \psi^a) \tag{66}$$

In particular,

$$T_{ov} = \frac{ic}{2} \sum_{a=1} [\psi^{a+} \partial v \, \psi^a - (\partial v \, \psi^{a+}) \psi^a]$$

$$= -2nc \, (H_k \partial v E^k - E_k \partial v H^k) \tag{67}$$

Now, T_{oo}, the energy density is:

$$T_{oo} = -2nc \, (H_k \partial_0 E_k - E_k \partial_0 H^k) \tag{68}$$

The surprising connection with the Beltrami relation emerges when we use (49a,b) to transform (68) to

$$T_{oo} = -c(\mu\mathbf{H} \cdot \text{curl } \mathbf{H} + \varepsilon \mathbf{E} \cdot \text{curl } \mathbf{E}) \tag{69}$$

Examining this equation and returning to previous considerations, both of these terms have the form for the "abnormality" of a twice differentiable vector field. See Eq. (6), where we saw that if the abnormality factor for a specific vector field is nonzero, it represents the departure of that vector field from the property of having a normal congruence of surfaces. So, according to (69), the energy-momentum tensor has to do with the *vorticity* or a type of *helicity* displayed by the electromagnetic field itself. Assuming a constant value for the abnormality (k), then both fields $\mathbf{E}$ and $\mathbf{H}$ must then conform to the Beltrami equations:

$$\mathbf{H} \times \text{curl } \mathbf{H} = 0, \qquad \mathbf{E} \times \text{curl } \mathbf{E} = 0 \tag{70}$$

This then implies:

$$\text{curl } \mathbf{H} = k\mathbf{H}, \qquad \text{curl } \mathbf{E} = k\mathbf{E} \tag{71}$$

In this application we consider EM fields in free space; consequently both $\mathbf{E}$ and $\mathbf{H}$ are solenoidal and satisfy Trkalian field relations. Thus, taking the curl of (71), both vector fields satisfy Helmholtz vector wave equations:

$$\nabla^2 \mathbf{H} + k^2 \mathbf{H} = 0, \qquad \nabla^2 \mathbf{E} + k^2 \mathbf{E} = 0 \tag{72}$$

Now, from (49a,b) and (70), we get

$$\mathbf{H} \times \frac{d\mathbf{E}}{dt} = 0 = \mathbf{E} \times \frac{d\mathbf{H}}{dt} \tag{73a}$$

$$k\mathbf{E} = -\frac{\mu}{c} \frac{d\mathbf{H}}{dt} \qquad k\mathbf{H} = \frac{\epsilon}{c} \frac{d\mathbf{E}}{dt} \tag{73b}$$

and these relations imply

$$\frac{d^2\mathbf{H}}{dt^2} + \frac{k^2 c^2}{n^2 \mathbf{H}} = 0, \qquad \frac{d^2\mathbf{E}}{dt^2} + \frac{k^2 c^2}{n^2 \mathbf{E}} = 0 \tag{74}$$

Now, a solution of (73) and (74) takes the form

$$\mathbf{E} = \frac{\mathbf{A}_1}{\sqrt{\epsilon}} \cos \frac{(knt)}{c} + \frac{\mathbf{A}_2}{\sqrt{\epsilon}} \sin \frac{(knt)}{c} \tag{75a}$$

$$\mathbf{H} = \frac{\mathbf{A}_1}{\sqrt{\mu}} \cos \frac{(knt)}{c} + \frac{\mathbf{A}_2}{\sqrt{\mu}} \sin \frac{(knt)}{c} \tag{75b}$$

As is well known, this transformation has the form of a duality transformation [66–68], where, in the above special case, the vectors $\mathbf{A}_1$ and $\mathbf{A}_2$, similar to $\mathbf{E}$ and $\mathbf{H}$, are compelled to satisfy Trkalian field relations:

$$\text{curl}\,\mathbf{A}_1 = k\mathbf{A}_1, \qquad \text{curl}\,\mathbf{A}_2 = k\mathbf{A}_2, \qquad \text{div}\,\mathbf{A}_1 = 0 = \text{div}\,\mathbf{A}_2 \tag{76}$$

Let us carefully note that the emergence of Beltrami field relations from the Hillion–Quinnez model, is not due to an ad hoc rendering of the EM field relations, but is a logically consistent result that follows from recognizing that any electromagnetic field in a homogeneous, isotropic medium in free space, away from charges and currents, can be expressed either in terms of the spinors ψ^a, or in terms of the complex Hertz-vector $\mathbf{\Pi}$.

The possible ultimate significance these facts might have for reshaping the edifice of classical electrodynamics through the compatible incorporation of non-Abelian SU(2) symmetries such as those represented by the above spinor–Hertz potential rendering of free space EM, can at this point only be speculated on. Moreover, even more amazing, as we shall see next, the seemingly surprising connection revealed between Beltrami vector fields and SU(2) transformation groups is not an isolated phenomenon, relevant to this one special model developed by Hillion and Quinnez. In fact, there appears to be a recurring theme of Beltrami (specifically Trkalian) field relations that emerge when EM fields are derived from a Hertz potential in the context of the general Clifford algebra formalism, which subsumes the complex unimodular group transformations that spinors encompass. Also, unlike many of the previous EM models we have considered that are associated with the Beltrami relation, Clifford algebra rendering of electromagnetics, such as is next examined in the Rodrigues–Vaz model, can produce fields possessing many counterintuitive properties, similar to those of the **E//B** wave, and others that even appear to violate established physical laws.

B. Rodrigues–Vaz Model

Up to now, we have examined how the Beltrami vector field relation surfaces in many electromagnetic contexts, featuring predominantly plane-wave solutions (PWSs) to the free-space Maxwell equations: in conjunction with biisotropic media (Lakhtakia–Bohren), in homogeneous isotropic vacua (Hillion/Quinnez), or in the magnetostatic context exemplified by FFMFs associated with plasmas (Bostick, etc.).

A feature common to all the above PWS to the Maxwell equations is that they all share the attribute of exhibiting *null-field* behavior [69]. That is, for this type of field, **E** and **H** vectors are orthogonal and transverse to the propagation vector at all times and locations, as well as proportional in magnitude. The factor of proportionality c, the speed of light, which is also the group velocity of the wave, is the rate at which energy is propagated through space. These relations are described mathematically by zero value manifested by the Lorentz–Poincaré field invariant scalars. Using the Hillion–Quinnez model, The two scalars are defined in the following manner, in tejrms of the field vectors **E** and **H**:

$$I_1 = \epsilon|E|^2 - \mu|H|^2; \qquad I_2 = \mathbf{E} \cdot \mathbf{H} \tag{77}$$

where each of these values is invariant under any proper (orthochronous) Lorentz transformation [70]. Now, the product of the complex field vector (50) with itself is a combination of I_1 and I_2:

$$(\sqrt{\epsilon}\,\mathbf{E} - i\sqrt{\mu}\,\mathbf{H})^2 = I_1 - i(\epsilon\mu)^{1/2}I_2 \tag{78}$$

As we can see, it is the constraint on the spinor field in this case (57) that leads to zero value of I_1 and I_2 if it is also postulated that $\psi_1^1 = f_1 g_1, \psi_1^2 = f_1 g_2,$ $\psi_2^1 = f_2 g_1, \psi_2^2 = f_2 g_2$, where (f_1, f_2) and (g_1, g_2) are the components of spinors satisfying the Pauli equation. Then (57) becomes

$$f_1 g_1 + f_2 g_2 = 0 \tag{79}$$

By substituting the previous definition for these spinors into (56) and using (79), it follows that Λ_j is a null vector, (e.g., $\Lambda_j \Lambda^j = 0$), which, according to (78), implies $I_1 = I_2 = 0$.

Similar to the Hillion–Quinnez model, Rodrigues and Vaz defined an EM field that is a function of a specific Hertz potential:

$$\mathbf{\Pi} = \emptyset(\mathbf{x}) \, \exp \, (\gamma_5 \Omega t) \gamma^1 \gamma^2 \tag{80}$$

where, γ_1, γ_2 are two of the four basis vectors $\gamma_0, \gamma_1, \gamma_2, \gamma_3$, in the so-called spacetime algebra $Cl_{1,3}$, which satisfy the commutation relations $\gamma^u \gamma^v + \gamma^v \gamma^u = 2n_{nv}$, and $n_{nv} = \text{diag}(1, -1, -1, -1), u, v = 3\text{–}0$ and $\gamma_5 \gamma_0 \gamma_1 \gamma_2 \gamma_3$. In this model the electromagnetic field tensor F^{uv} is represented by a 2-form F, where

$$F = \frac{1}{2} F^{uv} \gamma_u \gamma_v \tag{81}$$

In this model c, the velocity of light is equal to 1, and the invariants of the EM field are obtained from

$$F^2 = F \cdot F + F \wedge F$$

where

$$I_1 = F \cdot F = -\frac{1}{2} F^{uv} \, F_{uv}, \qquad I_2 = F \wedge F = -\gamma_5 F^{uv} F^{\alpha\beta} e_{\mu\nu\alpha\beta} \tag{82}$$

where $e_{\mu\nu\alpha\beta}$ is the antisymmetric Levi–Cevita symbol. These assume the form of the familiar invariant expressions once we recognize the even subalgebra of the Clifford algebra $Cl^+ 1, 3, = Cl \, 3,0$, the Pauli algebra, in which the pseudoscalar $\mathbf{i} = \sigma_1 \sigma_2 \sigma_3 = \gamma_0 \gamma_1 \gamma_2 \gamma_3$. In the Pauli algebra we have $F = \mathbf{E} + \mathbf{i}\mathbf{B}$, so that:

$$F^2 = (|\mathbf{E}|^2 - |\mathbf{B}|^2) + 2i\mathbf{E} \cdot \mathbf{B} = F \cdot F + F \wedge F \tag{83}$$

For their Hertz potential, Rodrigues and Vaz chose the factor $\phi(t, \mathbf{x}) = \phi(\mathbf{x}) \exp (\gamma_5 \Omega \, t)$. Now, since $\mathbf{\Pi}$ satisfies the wave equation, we conclude that the factor $(\phi \mathbf{x})$ in turn satisfies the Helmholtz equation:

$$\nabla^2 \phi(\mathbf{x}) + \Omega^2 \phi(\mathbf{x}) = 0 \tag{84}$$

In this case we consider the simplest solutions of (84) in spherical coordinates:

$$\phi(\mathbf{x}) = \frac{C \sin (\Omega r)}{r}, \qquad r^2 = x^2 + y^2 + z^2 \tag{85}$$

Once again, using the Pauli algebra, we express the Hertz potential as a sum of its electric and magnetic parts:

$$\mathbf{\Pi} = \mathbf{M} + i\mathbf{N} \tag{86}$$

In terms of these vectors the electric and magnetic field vectors are expressed as

$$\mathbf{E} = -\partial_0 (\text{curl } \mathbf{N}) + \text{curl curl } \mathbf{M}; \qquad \mathbf{B} = -\partial_0 (\text{curl } \mathbf{M}) - \text{curl curl } \mathbf{N} \quad (87)$$

Using (85), (80), and (86), and substituting into (87), we get for the resulting EM 2-form F_0:

$$
\begin{aligned}
F_0 = \frac{C}{r^3} [& \sin(\Omega t)(\alpha \Omega r \sin\theta \sin\varphi - \beta \sin\theta \cos\theta \cos\varphi)\gamma_0\gamma_1 \\
& - \sin(\Omega t)(\alpha \Omega r \sin\theta \cos\varphi + \beta \sin\theta \cos\theta \sin\varphi)\gamma_0\gamma_2 \\
& + \sin(\Omega t)(\beta \sin^2\theta - 2\alpha)\gamma_0\gamma_3 + \cos(\Omega t)(\beta \sin^2\theta - 2\alpha)\gamma_1\gamma_2 \\
& + \cos(\Omega t)(\beta \sin\theta \cos\theta \sin\varphi + \alpha \Omega r \sin\theta \cos\varphi)\gamma_1\gamma_3 \\
& + \cos(\Omega t)(-\beta \sin\theta \cos\theta \cos\varphi + \alpha \Omega r \sin\theta \sin\varphi)\gamma_2\gamma_3]
\end{aligned}
\quad (88)
$$

where $\alpha = \Omega r \cos(\Omega r) - \sin(\Omega r)$ and $\beta = 3\alpha + \Omega^2 r^2 \sin(\Omega r)$. Observe that F_0 is regular at the origin and vanishes at infinity. Rewriting the solution using the Pauli algebra

$$F_0 = \mathbf{E}_0 + i\mathbf{B}_0 \quad (89)$$

we get the result

$$\mathbf{E}_0 = \mathbf{W} \sin(\Omega t), \quad \mathbf{B}_0 = \mathbf{W} \cos(\Omega t) \quad (90)$$

with

$$\mathbf{W} = -C\left(\frac{\alpha\Omega y}{r^3} - \beta\frac{xz}{r^5}, \quad -\alpha\frac{\Omega x}{r^3} - \beta\frac{yz}{r^5}, \quad \beta\frac{(x^2 + y^2)}{r^5} - \frac{2\alpha}{r^3}\right)$$

We verify that $\text{div } \mathbf{W} = 0$, $\text{div } \mathbf{E}_0 = 0$, $\text{div } \mathbf{B}_0 = 0$, $\text{curl } \mathbf{E}_0 + \partial\mathbf{B}_0/\partial t = 0$, $\text{curl } \mathbf{B}_0 - \partial \mathbf{E}_0/\partial t = 0$, and the key relation

$$\text{curl } \mathbf{W} = \Omega\mathbf{W} \quad (91)$$

This is clearly a Beltrami equation, but what is more amazing is that the field result (88) describes a solution to the free-space Maxwell equations that, in contrast to standard PWS, the electric ($\mathbf{E}_0$) and magnetic ($\mathbf{B}_0$) vectors are parallel [e.g., $\mathbf{E}_0 \times \mathbf{B}_0 = 0$, where $\mathbf{E}_0 \times \mathbf{B}_0 = -i(\mathbf{E}_0 \wedge \mathbf{B}_0)$], the signal (group) velocity of the wave is *subluminal* ($v < c$), the field invariants are non-null, and as (91) clearly shows, this wave is not transverse but possesses *longitudinal* components. Moreover, Rodrigues and Vaz found similar solutions to the free-space Maxwell equations that describe a *superluminal* ($v > c$) situation [71].

Consequently, much like the Beltrami vortex filaments discussed earlier in conjunction with the magnetostatic FFMF, the Beltrami vector relations associated with nonluminal solutions to the free space Maxwell equations, are directly related to physical classical field phenomena currently unexplainable by accepted scientific paradigms. For instance, such non-PWS of the free-space Maxwell equations are direct violations of the sacrosanct principle of special relativity [72], as well as exhibit other counterintuitive properties. Yet, even more extraordinary, these non-PWS are not only theoretical possibilities, but have been demonstrated to exist empirically in the form of the so-called *evanescent mode* propagation of electromagnetic energy [72–76].

Although SU(2) groups associated with spinors are not directly incorporated into the Rodrigues–Vaz model to describe the EM field, they surprisingly turn out to be implied as an integral part of this edifice, in order to describe super- and subluminal waves with nonnull behavior. In this regard, the EM field two-form is shown to be a function of the so-called Dirac–Hestenes spinors [77] through the relation:

$$F = \psi \gamma_1 \gamma_2 \tilde{\psi} \tag{92}$$

where $\tilde{\psi}$ represents the spinor obtained by reversion of the original spinor ψ [78]. To see how (92) comes about, we need to understand the meaning of an *extremal* EM field. The latter is a field for which the electric [magnetic] field vanishes and the magnetic [electric] field is parallel to a given spatial direction. Now, there is a well-known proven theorem discovered by Rainich [79] and reconsidered by Misner and Wheeler [80,81], that at any point of spacetime, any nonnull ($F^2 <> 0$) electromagnetic field can be transformed into an extremal field by performing a Lorentz transformation combined with a *duality* transformation. A duality-transformed electromagnetic field can be most simply described as the product of the original field F and the *quasiscalar* factor (sum of scalar plus pseudoscalar): $F' = \exp(\gamma_5 \lambda) F$, where λ is the "angle of rotation" in "duality-space." It is known that the two scalar Lorentz–Poincaré field invariants, I_1 and I_2, do not remain individually constant under a duality rotation [82], although the combination $[(I_1)^2 + (I_2)^2]$ will stay fixed.

The relative significance of the role played by the Beltrami relation (91) in formulating this SU(2) structure of *nonluminal* electrodynamics is at present uncertain. Nevertheless, there is enough evidence to suggest that the Beltrami (specifically, Trkalian) vector fields might possibly be intimately associated with higher symmetry field physics, and related to multiply connected field topologies because of their intrinsic nonzero helicity. One final additional study, in which longitudinal field components also play a key role, indicates that this might be the case is described below.

XX. EVANS–VIGIER LONGITUDINAL $\mathbf{B}^{(3)}$ FIELD AND TRKALIAN VECTOR FIELDS

Developed since the early 1990s, concurrently with both Hillion–Quinnez and Rodrigues–Vaz SU(2) EM field models, but based on a different non-Abelian gauge group, is the so-called Evans–Vigier longitudinal $\mathbf{B}^{(3)}$ field representation [83–93]. In this model, a Yang–Mills gauge field theory [94] with an internal O(3) gauge field symmetry [95] is invoked to account for various magnetooptical effects that are claimed to be a function of a third magnetic field vector component that has been termed $\mathbf{B}^{(3)}$. One of the central theorems of O(3) electrodynamics is the B cyclic theorem:

$$\mathbf{B}^{(1)} \times \mathbf{B}^{(2)} = i\mathbf{B}^{(0)} \, \mathbf{B}^{(3)*} \tag{93}$$

a conjugate product which relates three basic magnetic field components in vacuo defined as

$$\mathbf{B}^{(1)} = \frac{\mathbf{B}^{(0)}}{\sqrt{2}} (i\mathbf{i} + \mathbf{j}) \exp(i\phi) \tag{93a}$$

$$\mathbf{B}^{(2)} = \frac{\mathbf{B}^{(0)}}{\sqrt{2}} (-i\mathbf{i} + \mathbf{j}) \exp(-i\phi) \tag{93b}$$

$$\mathbf{B}^{(3)} = \mathbf{B}^{(0)} \mathbf{k} \tag{93c}$$

where $\phi = wt - kz$, a phase factor, and $\mathbf{i}, \mathbf{j}, \mathbf{k}$, are the three unit vectors in the directions of the axes x, y, and z, respectively. Although the existence of the $\mathbf{B}^{(3)}$ field has been a subject of controversy, both pro and con, Evans has claimed [96] that these three magnetic field components encompassed by the relations (93a–c), along with the electric field components as well as the components of the magnetic vector potential ($\mathbf{A}$), are themselves components of a Beltrami–Trkalian vector field relation. This is readily verified in the case of (93a,b), since they present the form of the circularly polarized solution to the Moses eigenfunctions of the curl operator we have discussed formerly in connection with turbulence in fluid dynamics.

Implied by these developments is the increasing importance given to hypercomplex formalisms for modeling the symmetries in elementary particle physics and quantum vacuum morphology. As discussed in former earlier papers [97,98], the author believes that the most appropriate algebra for describing a hypothesized vortical structure for quantum-level singularities, as well as their macroscopic counterparts (Beltrami-type fields), is the *biquaternion* algebra (hypercomplex numbers of order 8)—the Clifford algebra of order 3, represented by the Pauli algebra *Cl*3,0 such as previously examined in the

Rodrigues–Vaz model. For instance, it is known that in a macroscopic Euclidean context, biquaternions are required to describe the kinematics and dynamics for the most general twisting movement of a rigid body in space [99,100]. It is therefore suggested that the most suitable mathematical formalism for *screw-type* EM fields of the Beltrami variety should transcend a traditional vectorial treatment, encompassing a *paravectorial* hypercomplex formalism akin to the Clifford (Dirac) algebra used effectively to describe the electron spin in a relativistic context [101].

It is a conclusion in this regard that the founders of vector field analysis were remiss in failing to take full account of the significance of Beltrami field topology in addition to the traditional solenoidal, lamellar, and complex lamellar fields. An inclusion of the examination of the Beltrami condition in the development of the vector calculus would possibly have brought attention to the important intimate association of this field configuration with non-Abelian mathematical structure. If the history of vector analysis had taken this path, it is possible that the architects of vector field theory and classical electrodynamics, would not have been so quick to indiscriminately sever its connection from the natural quaternion-based foundation. Perhaps the work by Hillion–Quinnez, Rodrigues–Vaz, Evans, and others [102,103] showing the need to consider non-Abelian models in electromagnetism, will be instrumental in helping to set the future of classical EM theory and vector field theory in general, on a firmer foundation.

XXI.   CONCLUSIONS AND PROSPECTS

Both theoretical and empirical investigations are necessary in any study of the Beltrami field relation, since at present much more is unknown than is known about the phenomenon of Beltrami electromagnetic fields. Consequently, it is hoped that this present exploration will spur on specialists in the technical arena to focus their attention on the mysteries surrounding FFMFs and vortices in general—especially the properties attributed to plasma vortex filaments that demonstrate a violation of the various established laws of thermodynamics. To this end, we have reviewed much past literature on the FFMF, and Beltrami flow fields in general, underscoring key established characteristics of these special vector fields. Beltrami field anomalies associated with plasma configurations are, in this author's opinion, not to be viewed as merely a fluke or trivial footnote in the annals of empirical science, but represent a profound enigma whose resolution may be intimately related to the very structural foundations of classical electrodynamics itself. Our second focus, therefore, has been of a more conjectural nature, examining Beltrami vortex fields and their specialized Trkalian modes, as a possible foundation for a more expansive view of electromagnetism that incorporates a non-Abelian structural edifice.

In accordance with these notions, the current chapter has focused on a visual approach with a primarily pictorial representation of Beltrami fields and their relatives. It is hoped that this emphasis on intuitive pictures will perhaps inspire researchers of fundamental field physics to consider the benefits of a return to more concrete theoretical models in the future. Certainly, in the macroscopic context, the anomalies found associated with classical vortex study applied to empirical exhibits in plasma focus entities, tornadoes and astrophysical phenomena, need reexamination.

The limited scope and space allotted to this current study precludes a detailed investigation into all aspects of the Beltrami vector field and its applications in science. Much more needs to be done as we have barely scratched the surface of this subject. For readers who would like to continue the search for the possible unplumbed secrets underlying Beltrami vector fields and their value to science, the following extra reference studies are recommended.

Z. Yoshida has been instrumental in further uncovering the mathematical properties of the curl eigenfunctions, the solution to the magnetic force-free equations, and the importance of topology in dealing with vector fields possessing nonzero helicity [104–106]. M. MacLeod used Moses curl eigenfunctions to describe FFMFs, showing such fields can be defined entirely by the value of their curl transform on the unit hemisphere in transform space [107,108]. This change of viewpoint suggests an orderly approach to the classification of the properties of such fields. P. Baldwin has done some significant interesting work exploring the properties of the solution to the Trkalian field equation for complex–dual vector fields, in terms of Monge-Clebsch potentials [109,110]. R. Kiehn has underscored the importance of topology in modeling the evolution of vector field 1-form potentials of rank ≥3. Using Cartan's calculus of differential forms and the fact that vector fields of positive helicity (such as Beltrami fields) correspond to one-form potentials of rank at least 3 (Pfaff dimension of 3 or 4), Kiehn has explored both turbulence in fluid dynamics and thermodynamic irreversibility for 1-form potential fields of rank 4 [111], as well as helicity–torsion properties of unique EM waves with non-zero value for Lorentz-Poincare field invariants [112]. D. R. Wells [113] and T. Waite have used the Beltrami–Trkalian field relations to describe extended elementary particles (Waite), and astrophysical plasmas (Wells), whose quantum dynamics are a function of continuous vector fields. Waite, in particular [114], and in association with Barut and Zeni [115], has shown how solitonic behavior of pure electromagnetic particles, such as electrons, can arise from the assumption of a continuous fluid-like primordial aether whose dynamics conform to a Beltrami–Trkalian field relation. The toroidal topology depicted in Fig. 10 describes this ether-like fluid precisely. Considering the possible further connection of EM Hertz–Debye potentials to Beltrami fields, Benn and Kress [116] have applied the generalized Hertz–Debye potential scheme to the force-free field problem, by showing that a nonconstant eigenvalue of the curl operator produces an equivalent curved space source-free Maxwell equation set. Martinez [117] used differential forms to show a previously unnoticed important connection between force-free fields and minimal surfaces.

Beltrami field fluid dynamics research into turbulence has been underscored in the so-called chaotic ABC flows [118,119], and Viktor Trkal's original seminal paper on incompressible fluid dynamics and curl eigenfunctions with constant "abnormality," has been translated and reprinted in [120]. Additional references concerning research into FFMF are those by Freire [121] and Vainshtein [122]. Key properties of Beltrami fields in hydrodynamics and magneto-hydrodynamics are indicated by Dritschel [123], McLaughlin & Pironneau [124], Montgomery et al. [125], and Marris [126–128]. Finally, Bjorgum has written a seminal paper on 3-dimensional Trkalian flows as solutions to the non-linear hydrodyanmical equations [129].

With all these exhibits of the Beltrami vector field in nature previously expounded, some of which appear to emerge as a total surprise out of disparate empirical applications, we now return to examine in more depth, the original thesis of this chapter. This is the suggestion that the Beltrami–Trkalian vector field might possibly be an archetypal structure or field morphology that is universal and therefore ubiquitous throughout nature. One feature that would tend to support this tenet is the fact that this field structure, inwhatever topological contesxt, exemplifies the characteristic of perfect *balance*, not only between kinematical parameters exhibited through the association of geometric relationships, but also in regard to the dynamic relationship between the two modes (poloidal) and (toroidal) in the toroidal representation. First, a balance is noted between the distance from the axis of the velocity vector tangent to the helical field lines, and the inclination angle of this vector's normal plane from the vortex asis. Referring to Fig. 12, we observe that as the radius from the central asix increases from a to b, the corresponding angle of the normal plane with respect to the vertical axis decreases from α to β. Thus, with respect to the equally pitched helices, $p = a \tan \alpha = b \tan \beta$. Given the pitch of the helices as p, and inclination angle λ, we have $p = r \tan \lambda$. Thus, every axisymmetric (or topologically equivalent toroidal) Beltrami vector field system, determines a helicoidal velocity field describing motion on an infinite set of coaxial helices, with the limiting motions of translation [toroidal rotation] where $r = 0, p = 0$ along, and pure rotation [poloidal motion] where $r = \infty, p = \infty$ around, the central [toroidal] axis. In fact, as the author has shown in [1], that the helical axisymmetric solution attributed to the lowest energy mode of the Beltrami vortex filaments, is similar to the helicoidal velocity/forece field associated with the structure known from antiquarian geometric mechanics as the *screwfield*, generated from projective line geometry—also known as the *linear line*

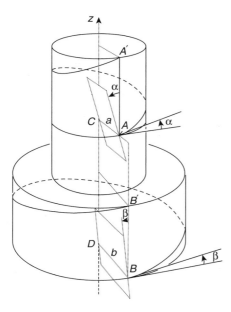

Figure 12. Two equal-pitched helices (radii a and b) of the same coaxial system. The planes at A and B are normal to them.

complex. In the book previously noted [25], White also noted this possible correspondence.

In this regard the unique property of Trkalian fields in which the curl of such a field is also Trkalian, suggests a unique permanece or resiliency of field structure that is possessed by no other vector field, making it rich with promise for application to fundamental vacuum EM field structures. The related property that poloidal and toroidal modes feedback on each other with equipartition of energy [10] in an infinite loop through the curl operation, also suggests this possibility. The Trkalian feature of chiral modes of inherent non-zero helicity, might have a direct application to explicating the associated invariances already established in fundamental field and particle physics, in which helicity/spin play a predominant role. However, such an investigation, as well as further insight into possible connection between Beltrami fields and multiply connected topologies other than toroidal, as well as non-Abelian field symmetries, is beyond the scope of the present chapter and await future research.

References

1. D. Reed, in: D. Grimes and T. W. Barrett (Eds.), *Advanced Electromagnetism: Foundations, Theory, Applications,* World Scientific, Singapore, 1995.

2. E. Beltrami, *Int. J. Fusion Energy* **3**(3), 51–57 (1985).

3. O. Bjorgum and T. Godal, *On Beltrami Vector Fields and Flows,* Part II, Universitet I Bergen Arbok, 1952.

4. R. Ballabh, *Ganita*, **I**(2), 1–4, (1948).

5. P. F. Nemenyi and R. C. Prim, "On the steady Beltrami flow of a perfect gas," *Proc VII Int. Cong. Appl. Mech. Lond.*, Vol., 2, Part I, 1948, pp. 300–314.

6. C. A. Truesdell,*The Kinematics of Vorticity*, Oxford Univ. Press, London, 1957.

7. H. E. Moses, *SIAM J. Appl. Mech.* **21**(1), 114–144 (1971).

8. P. Constantin and A. Majda, *Commun. Math. Phys.* **115**, 435–456 (1988).

9. R. Pelez, V. Yakhot, S. A. Orszag, L. Shtilman, and E. Levitch, *Phys. Rev. Lett.* **54**, 2505–2509 (1985).

10. M. V. Melander and F. Hussain, *Phys. Fluids A* **5**(8), 1992–2003 (1993).

11. M. Lesieur, *Turbulence in Fluids*, 2nd ed., Kluwer, Dordrecht, 1990.

12. C. Coats, *Living Energies: Viktor Schauberger's Brilliant Work with Natural Energy Explained*, Gateway Books, Bath, UK, 1996.

13. P. S. Ray, *Mesoscale Meteorology and Forecasting*, American Meteorological Society, Boston, 1986.

14. K. Lust and A. Schluter, *Astrophys.* **34**, 263–282 (1954).

15. S. Chandrasekhar and L. Woltjer, *Proc. Natl. Acad. Sci.* (USA) **44**(4), 285–289 (1958).

16. L. Woltjer, *Proc. Natl. Acad. Sci.* (USA) **44**(6), 489–491 (1958).

17. H. Alfven, *Cosmical Electrodynamics*, 2nd ed., Clarendon Press, Oxford, 1963.

18. L. F. Burlaga and R. P. Lepping, *Global Configuration of a Magnetic Cloud,* Geophysical Monograph 58, AGU, 1990, pp. 373–377.

19. R. P. Lepping and L. F. Burlaga, *J. Geophys. Res.* **95**(11), 957 (1990); L. F. Burlaga, *J. Geophys. Res.* **93**(A7), 7217–7224 (1988).

20. Y. Nakagawa et al., *Solar Phys.* (1971).

21. R. C. Elphic and C. T. Russell, *J. Geophys. Res.* **8**(A1), 58–72 (1974).

22. S. Chandrasekhar, *Proc. Natl. Acad. Sci.* (USA) **42**(1), 1–5 (1956).

23. W. Bostick, *Phys. Fluids* 2078–2079 (1966); *IEEE Trans. Plasma Sci.* **PS-41**(6), 703–717 (1986).

24. C. B. Stevens, *21st Cent. Sci. and Technol.* **13**(4), 38–45 (Sept.–Oct. 1988).

25. C. White, *Energy Potential—Toward a New Electromagnetic Field Theory*, Campaigner Publishers, New York, 1977.

26. "Classical physics makes a comeback," *The Times*, London (Jan. 19, 1982).

27. W. Bostick, *Int. J. Fusion Energy* **3**(2), 47–51 (1985).

28. W. Bostick, *Int. J. Fusion Energy* **3**(1), 9–52 (1985).

29. W. Bostick, *Int. J. Fusion Energy* **3**(1), 68 (1985).

30. W. Edwards, *Phys. Rev. Lett.* **47**(26), 1863, (1981).

31. H. Frohlich, *Proc. Phys. Soc.* **87**, 330 (1966).

32. H. Pfister and W. Gekelman, *Am. J. Phys.* **59**(6), 497–502 (1991).

33. G. E. Marsh, in G. Grimes and T. W. Barrett (Eds.), *Advanced Electromagnetsm: Foundation, Theory, Applications*, World Scientific, Singapore, 1995, pp. 52–76.

34. K. Nishikawa and M. Wakatani, *Plasma Physics: Basic Theory with Fusion Applications*, Springer-Verlag, Berlin, 1990.

35. N. A. Salingaros, *J. Phys. A* **19**, L101–L104 (1986).

36. H. Flanders, *Differential Forms with Applications to the Physical Sciences*, Dover, New York, 1989.

37. H. Zaghloul and O. Barajas, *Am. J. Phys.* **58**(8), 783–788 (1990).

38. C. D. Collinson, *J. Phys. A* **20**, L193–194 (1987).

39. M. Maheswaran, *J. Phys. A* **19**, L761–L762 (1986).

40. H. Zaghloul and O. Barajas, *Am. J. Phys.* **58**(8), 783–788 (1990).

41. K. R. Brownstein, *Phys. Rev. A* **35**(11), 4856–4858 (1989).

42. W. W. Hansen, *Phys. Rev.* **47**, 139–143 (1935).

43. V. Ferraro and C. Plumpton, *An Introduction to Magneto-Fluid Mechanics*, Oxford Univ. Press, London, 1961.

44. W. M Elsasser, *Phys. Rev.* **69**, pp. 106–166, 1946; G. Buck, *Force-Free Magnetic Fields Solution in Toroidal Coordinates*, Ph.D. dissertation, Univ. Michigan, Univ. Microfilms, 1964.

45. D. R. Wells, *Int. J. Fusion Energy* **3**(4), 17–24 (1985).

46. D. R. Wells, *Phys. Fluids* **7**(6), 826–832 (1964).

47. A. Konigl and A. R. Choudhari, *Astrophys. J.* **289**, 173–187 (1985).

48. S. Chandrasekhar and P. C. Kendall, *Astrophys. J.* **126**, 457–460 (1957).

49. H. Zaghloul and O. Barajas, *Am. J. Phys.* **58**(8), 783–788 (1990).

50. G. E. Marsh, *Force-Free Magnetic Fields*, 1994.

51. D. R. Wells, *Phys. Fluids* **9**(5), 1010–1021 (1966).

52. J. B. Taylor, *Phys. Rev Lett.* **33**, 1139 (1974).

53. V. H. Rumsey, *IRE Trans Ant. Propagation*, 461–465 (1961).

54. A. Lakhtakia, *Specul. Sci. Technol.* **14**(1), 2–17 (1991).

55. C. F. Bohren, *Chem. Phys. Lett.* **29**, 458–462 (1974).

56. A. Lakhtakia et al., *Time-Harmonic Electromagnetic Fields in Chiral Media*, Springer-Verlag, Berlin, 1989.

57. A. Lakhtakia, in H. Istvan and C. A. Pickover (Eds.), *Spiral Symmetry*, World Scientific, Singapore, 1992.

58. A. Lakhtakia, *Specul. Sci. Technol.* **16**(2), 145–153 (1993).

59. C. Chu and T. Ohkawa, *Phys. Rev. Lett.* **48**(3), 837–838 (1982).

60. D. Reed, "The Beltrami field as archetypal vortex topology," *Proc. 1st Int. Symp. New Energy*, Rocky Mountain Research Inst., Fort Collins, CO, 1994.

61. V. Evtuhov and A. E. Seigman, *Appl. Optics* **4**(1), 142–143 (1965).

62. H. Zaghloul and H. A. Buckmaster, in A. Lakhtakia (Ed.), *Essays on the Formal Aspects of Electromagnetic Theory*, World Scientific, Singapore, 1993, pp. 183–206.

63. P. Hillion and S. Quinnez, *Int. J. Theor. Phys.* **25**(7), 727–736 (1986).

64. J. A. Stratton, *Electromagnetic Theory*, McGraw-Hill, New York, 1941.

65. H. Nisbet, *Proc. Roy. Soc. Lond. A* **231**, 250–263 (1955).

66. C. W. Misner and J. A. Wheeler, *Ann. Phys.* **2**, 525 (1957).

67. N. Salingaros and I. Y. Ilamed, *Found. Phys.* **14**(8), 777–797 (1984).

68. N. Salingaros, *Am. J. Phys.* **53**(4), 361–363 (1985).

69. J. L. Synge, *Relativity: The Special Theory*, 1965.

70. N. Salingaros, *Am. J. Phys.* **55**(4), 352–356 (1987).

71. W. A. Rodrigues and J. Vaz Jr., *Adv. Appl. Clifford Algebras* **7**(S), 457–466 (1997).

72. W. A. Rodrigues and J.-Y. Lu, *Found. Phys.* **27**(3), 435–508 (1997).

73. G. Barton, *Phys. Lett. B* **237**(34), 559 (1990).

74. M. V. Cuogo-Pinto et al., *Phys. Lett. B* **446**, 170 (1999).

75. G. Nimtz, *Gen. Rel. Grav.* **31**(5), 737 (1999).

76. T. Bearden, *J. New Energy* (1998).

77. J. Vaz and W. A. Rodrigues, *Adv. Appl. Clifford Algebras* **7**(**S**), 369–386 (1997).

78. J. Vaz and W. A. Rodrigues, *Int. J. Theor. Phys.* **32**, 945 (1993).

79. G. Y. Rainich, *Trans Am. Math Soc.* **27**, 106 (1925).

80. C. W. Misner and J. A. Wheeler, *Ann, Phys.* **2**, 525 (1957).

81. J. A. Wheeler, *Geometrodynamics*, Academic, New York, 1962.

82. N. Salingaros, *Adv. Appl. Math.* **4**, 1–30 (1983).

83. M. W. Evans, *Physica B* **182**, 227,237 (1992).

84. M. W. Evans, *The Photon's Magnetic Field*, World Scientific, Singapore, 1992.

85. M. W. Evans and S. Keilich (Eds.), *Modern Non-Linear Optics*, Wiley, New York, 1997.

86. M. W. Evans and A. A. Hasanein, *The Photomagnetion in Quantum Field Theory*, World Scientific, Singapore, 1994.

87. M. W. Evans and J. P. Vigier, *The Enigmatic Photon*, Vols. 1,2, Kluwer, Dordrecht, 1994.

88. M. W. Evans, J. P. Vigier, S. Roy, and S. Jeffers, *The Enigmatic Photon*, Vols. 3,4, Kluwer, Dordrecht, 1996/1997.

89. M. W. Evans, *The Enigmatic Photon*, Vol. 5, Kluwer, Dordrecht, 1999.

90. M. W. Evans and L. B. Crowell, *Classical and Quantum Electrodynamics and the B(3) Field*, World Scientific, Singapore (in press).

91. M. W. Evans, *Frontier Perspect* (1998).

92. P. K. Anastasovski et al., *Frontier Perspect.* **8**(2), 15–25 (1999).

93. V. V. Dvoeglazov and M. W. Evan *Apeiron* **4**(2–3), 45–47 (1998).

94. P. K. Anastasovski et al., *Found. Phys. Lett.* **12**(3), 257–265 (1999).

95. P. K. Anastasovski et al., *Found. Phys. Lett.* **12**(2), 187–192 (1999).

96. P. K. Anastasovski, M. W. Evan et al., *Found. Phys Lett.* (in press).

97. D. Reed, *Extraordinary Sci.* **4**(2), 28–33 (1992).

98. D. Reed, *Specul. Sci. Technol.* **17**(3), 223 (1994).

99. A. Buchheim, *Am. J. Math* **VII**, 293–326 (1886).

100. F. Dimentburg, *Screw Calculus and Its Application to Mechanics*, Fiziko-Matematischkoy Literatury (Engl. Transl. Natl. Tech. Info. Serv., U.S. Dept. Commerce), 1965.

101. P. A. M. Dirac, *Proc. Roy. Soc. Lond A.* **117A** (1928).

102. T. W. Barrett, in A. Lakhtakia (Ed.), *Essays on the Formal Aspects of Electromagnetic Theory*, World Scientific, Singapore, 1993, pp. 6–86.

103. T. W. Barrett, *Specul. Sci. Technol.* **21**, 291–300 (1999).

104. Z. Yoshida, *J. Math. Phys.* **33**(4), 1252–1256 (1992).

105. Z. Yoshida, *Math. Zeit.* **204**, 235–245 (1990).

106. Z. Yoshida, *J. Plasma Phys.* **45**(3), 481–488 (1991).

107. M. A. MacLeod, *Math. Phys.* **36**(6), 2951 (1995).

108. M. A. MacLeod, *Inverse Problems* **11**, 1087.

109. P. R. Baldwin, *Phys. Rev. E* (1995).

110. P. R. Baldwin and R. M. Kiehn, *Phys. Lett. A* **189**, 161–166 (1994).

111. R. M. Kiehn, in H. K. Moffat and T. S. Tsinober (Eds.), *Topological Fluid Mechanics*, Cambridge Univ. Press, 1990, pp. 449–458.

112. R. M. Kiehn, *Electromagnetic Waves in the Vacuum with Torsion and Spin* www. Cartanscorner.com., 1999.

113. D. R. Wells, *IEEE Trans. Plasma Sci.* **20**(6), 939–943 (1992).

114. T. Waite, *Phys. Essays* **8**, 60–70 (1995).

115. T. Waite, A. O. Barut, and J. R. Zeni, in J. Dowling (Ed.), *Electron Theory and Quantum Electrodynamics*, Plenum, 1995.

116. I. M. Benn and J. Kress, *J. Phys. A: Math. Gen.* **29**, 6295–6304 (1996).

117. J. C. Martinez, *J. Phys A: Math. Gen.* **28**, L317–L322 (1995).

118. T. Dombre et al., *J. Fluid Mech.* **167**, 353–391 (1986).

119. S. Changchun et al., *Chinese Phys. Lett.* **9**(10), 515–518 (****).

120. V. Trkal, *Czech. J. Phys.* **44**(2), 97–105 (1994).

121. G. F. Freire, *Phys. Rev.* 567–570 (Aug. 1965).

122. S. I. Vainshtein, in H. K. Moffatt et al. (Eds.), *Topological Aspects of the Dynamics of Fluids and Plasmas*, Kluwer, Dordrecht, 1992, pp. 177–193.

123. D. G. Dritschel, *J. Fluid Mech.* **222**, 525–541 (1991).

124. D. McLaughlin and O. Pironneau, *J. Math. Phys.* **32**, 297 (1991).

125. D. Montgomery, L. Turner, and G. Vahala, *Phys. Fluids* **21**(5), 757–764 (1978).

126. A. W. Marris, *Arch. Rat. Mech. Anal.* **70**, 47 (1979).

127. A. W. Marris, *Arch. Rat. Mech. Anal.* **80**, 95 (1981).

128. A. W. Marris, *Arch. Rat. Mech. Anal.* **82**, 117 (1983).

129. O. Bjorgum, *Int. Cong. Applied Mech.* **7**, 341 (1948).

CONSTANCY OF VELOCITY OF LIGHT
AND STOCHASTIC BACKGROUND

SISIR ROY

Indian Statistical Institute, Calcutta, India

CONTENTS

I. INTRODUCTION

The constancy of velocity of light attracted the whole scientific community since the publication of the paper on special theory of relativity by Albert

Modern Nonlinear Optics, Part 3, Second Edition, Advances in Chemical Physics, Volume 119,
Edited by Myron W. Evans. Series Editors I. Prigogine and Stuart A. Rice.
ISBN 0-471-38932-3 © 2001 John Wiley & Sons, Inc.

Einstein. It rests on firm experimental basis. The success of quantum field theory and general theory of relativity clearly demonstrate the assertion of constancy of velocity of light. It seems to be futile excersize to reopen the discussion regarding the viability of this assertion and its possible experimental verifiability. Several authors tried to reopen the issue by analyzing some experimental data from gamma-ray bursters (GRBs), active galactic nuclei (AGNs) and pulsars [1]. This quest is motivated mainly by the developments of quantum gravity, quantum foam [2] and several models of spacetime fluctuations [3]. Here the energy-dependent variation in photon velocity is related to the variation in arrival time observed for a photon traveling a distance L from astronomical sources: $\delta t \sim -(\frac{L}{c})(\frac{\delta c}{c})$. Analyzing the different astronomical sources we can envisage the following situation on short timescales $\delta t \leq \Delta t$. This is shown in Table I, where the relevant photon property that might be correlated with the variation of the velocity of light δc and its frequency v or equivalently its energy E are also listed. It is also claimed that any such effect could be expected to increase with energy E, and the simplest possibility , for which there is some theoretical support, is that $\delta c \sim \frac{E}{M}$, where M is some high-energy scale. $\frac{L}{c}\frac{E}{\delta t}$ can be taken as the measure of the experimental sensitivity to such a high-energy scale M. Table I indicates that these astrophysical sources are sensitive to M approaching $M_p \sim 10^{19}$ GeV: the mass scale at which gravity becomes strong. As a consequence, these astrophysical sources may challenge any theory of quantum gravity that predicts such a linear dependence of δc on E. In otherwords it could be that $\frac{\delta c}{c} \sim (\frac{E}{\tilde{M}})$ [2]. Then the astrophysical observations may be sensitive to $\tilde{M} \sim 10^{11}$ TeV.

It is clear from the above mentioned astrophysical data that it raises new opportunity to test the validity of assertion of constancy of velocity of light.

TABLE I
Observational Sensitivities and Limits on $M, \tilde{M}^a$

Source	Distance	E	Δt	Sensitivity of M	Sensitivity to $\tilde{M}$
GRB 920229[b]	3000 Mpc (?)	200 keV	10^{-2} s	0.6×10^{16} GeV(?)	10^6 GeV (?)
GRB 980425[b]	40 Mpc	1.8 MeV	10^{-3} s (?)	0.7×10^{16} GeV(?)	3.6×10^6 GeV (?)
GRB 920925c[b]	40 Mpc (?)	200 TeV (?)	200 s	0.4×10^{19} GeV (?)	8.9×10^{11} GeV (?)
Mrk 421[c]	100 Mpc	2 TeV	280 s	$>7 \times 10^{16}$ GeV	$>1.2 \times 10^{10}$ GeV
Crab pulsar[d]	2.2 kpc	2 GeV	0.35 ms	$>1.3 \times 10^{15}$ GeV	$>5 \times 10^7$ GeV
CRB 990123	5000 Mpc	4 MeV	1 s (?)	2×10^{15} GeV (?)	2.8×10^6 GeV (?)

[a]The linear (quadratic) mass-scale parameters $M, \tilde{M}$ are defined by $\delta c/c = E/M, (E/\tilde{M})^2$, respectively. The question marks in the table indicate uncertain observational inputs. Hard limits are indicated by inequality signs.
[b]Amelino-Camelia et al. [1]; see also Schaefer [16].
[c]Biller et al. [16].
[d]Kaaret [16].

Now we shall indicate the theoretical motivation of questioning the constancy of velocity of light in various contexts. In the context of quantum gravity, any attempt to quantize gravity canonically, must involve a Lorentz noninvariant separation of degrees of freedom and the choice of a preferred reference frame. A local rest frame exists in our approximately Friedman–Robertson–Walker universe, which can be identified with the cosmic microwave background radiation (CMBR). This provides a natural frame where one can consider topological fluctuations in spacetime background that may arise from microscopic black holes or other nonperturbative phenomena in quantum gravity: the so-called spacetime foam [4]. Amelino-Camelia et al. [5] argued that foamy effects might lead the quantum gravitational vacuum to behave as a nontrivial medium, much like a plasma or other environment with nontrivial optical properties. It was also proposed that quantum fluctuations in the lightcone [6] may be expected because of these considerations.

The main idea behind the abovementioned proposals is that quantum-gravitational fluctuations in the vacuum must in general be modified by the passage of an energetic particle and that this recoil will be reflected in the backreaction effects on the propagating particle itself. Three possible optical effects of quantum gravity have been identified:

1. Simple energy-dependent reduction in photon velocity, namely, a frequency-dependent refractive index.

2. The difference between the velocities of photons of different helicities, namely, birefrigence [7]. The possible experimental test of this phenomenon has been discussed by observing the polarized radiation from GRB 990510 [8].

3. The third is a possible energy-dependent diffusive spread in the velocities of different photons of the same energy [9].

Several authors [10] studied the propagation of light in a Maxwell vacuum with small nonzero conductivity and complex refractive index. By solving modified Maxwell equations, one can get the frequency-dependent speed of light, which can be associated with nonzero mass of photon. The nonzero rest mass of photon has a long history, and the astrophysical evidence is discussed in this chapter.

Ng and van Dam [11] explored the quantum foam from a different perspective. They proposed to explore the quantum structure of spacetime using Wigner's clock. It is claimed that classical spacetime breaks down into quantum foam in a manner different from the canonical picture. The main idea is that the uncertainty in spacetime measurement using Wigner's clock can be interpreted as inducing an intrinsic uncertainty in the spacetime metric. Although this fluctuation of spacetime metric is very small, the modern gravitational interferometers [12] may be sensitive enough to detect it. However, fluctuation of the spacetime metric has a long history apart from the more recent interest in quantum gravity. In the mid-twentieth century Karl Menger [13] introduced and developed the idea of statistical geometry by reanalyzing the fundamental concepts of measurement in quantum mechanics. Immediately after that variuos physicists and mathemeticians developed this concept, which can be found in a monograph by the author [3]. The fluctuation of gravitational potential or the existence of random zero-point field might be the cause of the fluctuation of spacetime metric [14,15]. It is to be noted from these observations that even the fluctuation of the velocity of light may cause the fluctuation of the metric, which will be elaborated on in this chapter. The main motivation of the present review is to first analyze the experimental data from the astronomical sources to test the constancy of velocity of light in Section II. Then in Section III we shall discuss the various theoretical models and the possible implications in Section IV.

II. OBSERVATIONAL EVIDENCE

A. Gamma-Ray Bursters

Ellis et al. [4] discussed a model analysis of astrophysical data on GRB pulses that allows us to place a bound motivated by the possible nontrivial medium effects of quantum gravity on the propagation of photon probes. These data were also considered by several groups [1,16] for individual GRB and pulsars. In the analysis of Ellis et al., the data have been analyzed from those GRBs whose redshifts are known after identification of their optical counterparts. This enables them to perform a regression analysis to search for a possible correlation with redshift (distance), to reveal source and medium effects. In the framework of quantum gravity one normally considers short-duration structures in the time profiles of those GRBs whose redshifts and hence distances are known with precision. Then appropriate fits of astrophysical data are done in various energy channels seeking to constrain differences in the timings and widths of peaks for different energy ranges. Simultaneity of peak arrival times at different energies would impose bounds on the induced refractive index of photons. Independence of the widths of peaks from the channel energies would constrain stochastic fluctuations in the velocities of the same energy.

The sample of GRB data considered have been taken from the BATSE catalog [17] and OSSE data [18]. They discussed only on the following GRBs whose redshifts are known:

1. GRB 970508 with BATSE trigger number 6225 and redshift $z = 0.835$
2. GRB 971214 with BATSE trigger number 6533 and redshift $z = 3.14$
3. GRB 980329 with BATSE trigger number 6665 and redshift $z = 5.0$
4. GRB 980703 with BATSE trigger number 6891 and redshift $z = 0.966$
5. GRB 990123 with BATSE trigger number 7343 and redshift $z = 0.966$

The energy ranges in which BATSE generally observes photons are as follows: channel 1, in the energy range 20–50 keV; channel 2, 50–100 keV; channel 3, 100–300 keV; channel 4, 300 keV. In fact, the energies recorded by BATSE are not the exact photon energies, and there is, in particular, some feedthrough from high-energy photons into lower-energy channels. However, this effect can be neglected here. The data for each GRB exhibit nontrivial and nonuniversal structures in time. For each trigger one or two prominent peaks have been fitted in each energy channel with the aim of looking for their difference in time or width between different energy channels. Four different functions have been considered to fit the different peaks:

1. A Gaussian function characterized by the peak location t_p and width parameter σ

2. A Lorentzian function characterized by $\frac{A}{[(t-t_p)^2+(\Gamma/2)^2]}$

3. A tail fit with fitting function

$$N(t) = c_1^*(t - t_0)^m \exp\left[-\frac{(t - t_0)^2}{(2r^2)}\right] \qquad (1)$$

function at $t > t_0$ peaks at $t_p = \tau\sqrt{m} + t_0$, to take into account the tail that tends to appear in the data after the peak.

4. The phenomenological pulse model [18], which has the following functional form:

$$N(t) = c_1 \exp\left[-\frac{(t - t_p)}{\sigma_{r,d}}\right]$$

where t_p is the time at which the photon pulse takes its maximum, σ_r and σ_d are the rise and decay times of the distribution, respectively, and v gives the sharpness or smoothness of the pulse at its peak.

In Fig. 1, the four fits are compared to the data for GRB 970508 in channels 1 and 3. It is evident from the figure that the Gaussian and Lorentzian fits are of lower quality than the tail and pulse fits. It is better to take the latter fits for the remaining GRBs. Figures 2–5 indicate the tail and pulse fits for the remaining GRBs: 971214, 960329, 980703, and 990123, respectively. The values of the tail and pulse fit parameters are compared in Table II. Specifically, we list for both the tail and pulse fitting functions the peak time, t_p and the pulsewidth, σ, defined as half of the width of the pulse at $\exp\left[-\frac{1}{2}\right] \simeq 60$ of its maximum value. If we take this definition, using the tail distribution (1), we get

$$\sigma = \alpha\tau\sqrt{m} \qquad (2)$$

GRB 970508: BATSE data Ch. 1 and Ch. 3

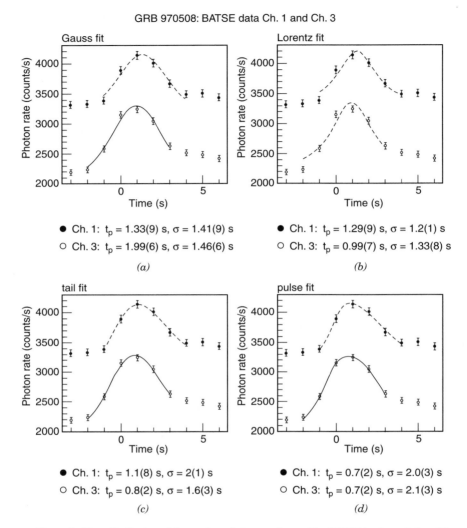

Figure 1. Time distribution of the number of photons observed by BATSE in channels 1 and 3 for GRB 970508, compared with the following fitting functions; (a) Gaussian, (b) Lorentzian, (c) tail function, and (d) pulse function. We list below each panel the positions t_p and widths σ_p (with statistical errors) found for each peak in each fit. We recall that the BATSE data are binned in periods of 1.024 s.

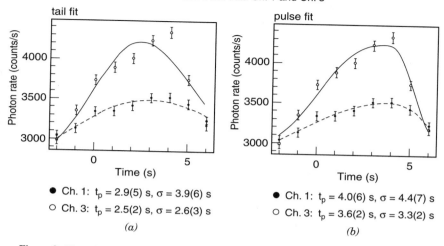

Ch. 1: $t_p = 2.9(5)$ s, $\sigma = 3.9(6)$ s
Ch. 3: $t_p = 2.5(2)$ s, $\sigma = 2.6(3)$ s

(a)

Ch. 1: $t_p = 4.0(6)$ s, $\sigma = 4.4(7)$ s
Ch. 3: $t_p = 3.6(2)$ s, $\sigma = 3.3(2)$ s

(b)

Figure 2. Time distribution of the number of photons observed by BATSE in channels 1 and 3 for GRB 971214, compared with the following fitting functions: (a) tail function, and (b) pluse function. We list below each panel the positions t_p and widths σ_p (with statistical errors) found for each peak in each fit.

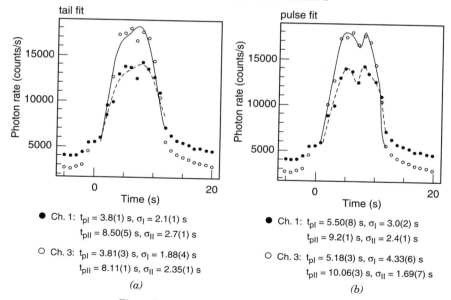

Ch. 1: $t_{pl} = 3.8(1)$ s, $\sigma_l = 2.1(1)$ s
$t_{pll} = 8.50(5)$ s, $\sigma_{ll} = 2.7(1)$ s

Ch. 3: $t_{pl} = 3.81(3)$ s, $\sigma_l = 1.88(4)$ s
$t_{pll} = 8.11(1)$ s, $\sigma_{ll} = 2.35(1)$ s

(a)

Ch. 1: $t_{pl} = 5.50(8)$ s, $\sigma_l = 3.0(2)$ s
$t_{pll} = 9.2(1)$ s, $\sigma_{ll} = 2.4(1)$ s

Ch. 3: $t_{pl} = 5.18(3)$ s, $\sigma_l = 4.33(6)$ s
$t_{pll} = 10.06(3)$ s, $\sigma_{ll} = 1.69(7)$ s

(b)

Figure 3. As in Fig. 2, but for GRB 980329.

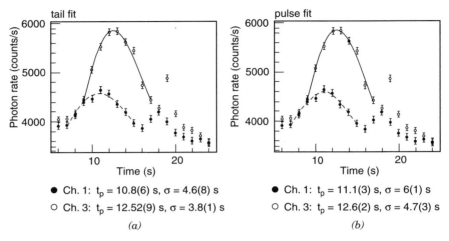

GRB 980703: BATSE data Ch. 1 and Ch. 3

● Ch. 1: t_p = 10.8(6) s, σ = 4.6(8) s

○ Ch. 3: t_p = 12.52(9) s, σ = 3.8(1) s

(a)

● Ch. 1: t_p = 11.1(3) s, σ = 6(1) s

○ Ch. 3: t_p = 12.6(2) s, σ = 4.7(3) s

(b)

Figure 4. As in Fig. 2, but for GRB 980703.

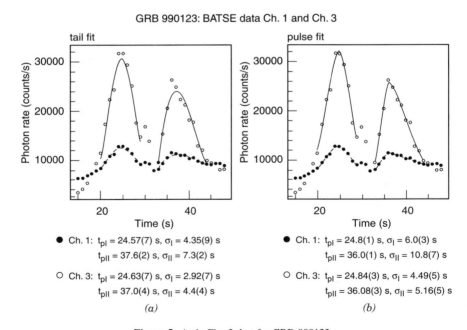

GRB 990123: BATSE data Ch. 1 and Ch. 3

● Ch. 1: t_{pI} = 24.57(7) s, $σ_I$ = 4.35(9) s
    t_{pII} = 37.6(2) s, $σ_{II}$ = 7.3(2) s

○ Ch. 3: t_{pI} = 24.63(7) s, $σ_I$ = 2.92(7) s
    t_{pII} = 37.0(4) s, $σ_{II}$ = 4.4(4) s

(a)

● Ch. 1: t_{pI} = 24.8(1) s, $σ_I$ = 6.0(3) s
    t_{pII} = 36.0(1) s, $σ_{II}$ = 10.8(7) s

○ Ch. 3: t_{pI} = 24.84(3) s, $σ_I$ = 4.49(5) s
    t_{pII} = 36.08(3) s, $σ_{II}$ = 5.16(5) s

(b)

Figure 5. As in Fig. 2, but for GRB 990123.

TABLE II
Results of Fits to GRB Data from BATSE[a]

			GRB 970508	GRB 971214	GRB 980329 (I)	GRB 980329 (II)	GRB 980703	GRB 990123 (I)	GRB 990123 (II)
Ch. 1	Tail	t_p (s)	1.1(8)	2.9(5)	3.8(1)	8.50(5)	10.8(6)	24.57(7)	37.6(2)
		σ (s)	2(1)	3.9(6)	2.1(1)	2.7(1)	4.6(8)	4.35(9)	7.3(2)
	Pulse	t_p (s)	0.7(2)	4.0(6)	5.50(8)	9.2(1)	11.1(3)	24.8(1)	36.0(1)
		σ_r (s)	1.3(3)	5.8(3)	3.65(9)	1.9(1)	7(1)	6.1(2)	4.6(2)
		σ_d (s)	2.7(3)	3(1)	2.3(2)	2.8(1)	6(1)	6.0(3)	17(1)
		σ (s)	2.0(3)	4.47(7)	3.0(0)	2.4(1)	6(1)	6.0(3)	10.8(7)
Ch. 3	Tail	t_p (s)	0.8(2)	2.5(2)	3.81(3)	8.11(1)	12.52(9)	24.63(7)	37.0(4)
		σ (s)	1.6(3)	2.6(3)	1.88(4)	2.35(1)	3.8(1)	2.92(7)	4.4(4)
	Pulse	t_p (s)	0.7(2)	3.6(2)	5.18(6)	10.06(3)	12.6(2)	24.84(3)	36.08(3)
		σ_r (s)	1.8(3)	4.6(2)	3.37(6)	2.3(1)	4.3(3)	4.59(4)	2.85(4)
		σ_d (s)	2.5(3)	2.1(2)	5.28(6)	1.071(7)	5.2(3)	4.39(5)	7.48(5)
		σ (s)	2.1(3)	3.3(2)	4.33(6)	1.69(7)	4.7(3)	4.49(5)	5.16(5)
Δ	Tail	Δt_p(s)	−0.3(8)	−0.4(5)	0.0(1)	−0.39(5)	1.7(6)	0.1(1)	−0.6(4)
		$\Delta\sigma$ (s)	0(1)	−1.3(7)	−0.2(1)	−0.3(1)	−0.8(8)	−1.4(1)	−2.9(4)
	Pulse	Δt_p (s)	0.0(3)	−0.4(6)	−0.3(1)	0.9(1)	1.5(4)	0.04(3)	0.1(1)
		$\Delta\sigma$ (s)	0.1(4)	−1.1(7)	1.3(2)	−0.7(1)	−1(1)	−1.5(3)	−5.6(7)
		$(\Delta t_p)^f$ (s)	0.0(3)(3)	−0.4(6)(0)	−0.3(1)(3)	0.9(1)(13)	1.5(4)(2)	0.04(3)(10)	0.1(1)(7)
		$(\Delta\sigma)^f$ (s)	0.1(4)(1)	−1.1(7)(2)	1.3(2)(15)	−0.7(1)(4)	−1(1)(0.2)	−1.5(3)(1)	−5.6(7)(27)

[a] For $A = t_p$, σ in each case, the quantities ΔA denote the differences $\Delta A = A^{Ch.3} - A^{Ch.1}$, and the quantities $(\Delta A)^f \equiv (\Delta A)^{\text{pulse}}$ ($\sqrt{\delta A_{(1)}^2 + \delta A_{(3)}^2}$)($\delta(\Delta A)$) are our final results for the induced time delays or width differences of the peaks between BATSE channels 1 and 3. The first set of parentheses in the latter expression denotes the statistical error (where $\delta A_{(i)}$ denotes the statistical error in determining A in the ith channel); the second set of parentheses denotes the theoretical "systematic" error, defined as $\delta(\Delta A) \equiv |(\delta A)^{\text{pulse}} - (\Delta A)^{\text{tail}}|$.

579

where $\alpha > 0$ is the solution of the equation and τ and m are defined in (2)

$$\ln(1 + \alpha) - \frac{1}{2}(1 + \alpha)^2 + \frac{1}{2}(1 + m^{-1}) = 0 \tag{3}$$

For the pulse distribution this definition yields $\sigma = (\sigma_\tau + \sigma_d)/2$.

The main interest here is to compare the values of these parameters in the different channels, and use their differences to constrain energy-dependent differences and stochastic fluctuations in photon velocities. It is evident from the Table II that the fitting functions yield constraints on the propagation parameters that are comparable within the statistical errors. Here we use the differences between them as gauges of systematic errors.

The only candidate that one can see for a systematic trend in the data is a tendency for pulses in the higher-energy channels to be narrower than in the lower-energy channels. This effect is seen in Fig. 1 for the case of GRB 970508. However, this narrowing is the opposite of what we would suggest theoretically, which would be a slowing and broadening of the peak at higher energies. The data from channel 3 of the BATSE detector is compared with the data from the OSSE detector, which detects photons in a single channel with energy range $1 < E < 5$–10 MeV. Since the OSSE data are at higher energies, they are more sensitive to the type of energy-dependent effect, in which we are interested. The reason we compare OSSE data with channel 3 of the BATSE data because the latter are free of contamination by the data in lower-energy channels, removing one particular possible source of systematic error. OSSE data are available for the GRBs 980329 and 990123, which we display in Figs. 6 and 7, respectively. The results of numerical analysis of the arrival times and the widths of identified OSSE pulses are given in Table III.

In order to investigate the possible fundamental physical significance of this or any other possible energy-dependent effect, Ellis et al. [4] compiled the data from all GRBs as a function of $\tilde{z}$ where

$$\tilde{z} \equiv 2\left(1 - \frac{1}{(1+z)^{1/2}}\right) \simeq z - \frac{3}{4}z^2 + \cdots$$

which expresses the cosmic-expansion-corrected redshift. Ellis et al. determined limits on the respective quantum gravity scales M_{QG} and M_{stoch} by constraining then possibele magnitudes of the slopes in linear regression analysis of the differences between the arrival times and widths of pulses in different energy ranges from five GRBs with measurable redshifts as functions of $\tilde{z}$. Using the current-value Hubble expansion parameters $H_0 = 100 \times h_0$ km s^{-1} Mpc^{-1}, where $0.6 < h_0 < 0.8$, one can get the following limits

$$M_{QG} \geq 10^{15} \text{ GeV}; \qquad M_{stoch} \geq 2.10^{15} \text{ GeV} \tag{4}$$

on the possible quantum gravity effects.

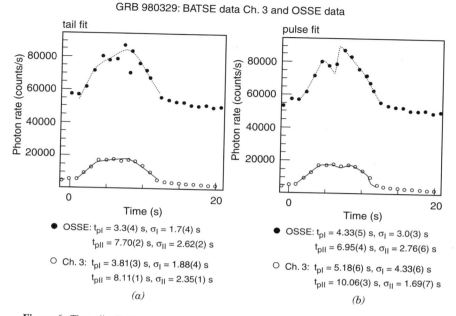

Figure 6. Time distribution of the number of photons observed by OSSE and by BATSE in channel 3 for GRB 980329, compared with the following fitting functions: (a) tail function and (b) pulse function. We list below each panel the positions t_p and widths σ_p (with statistical errors) found for each peak in each fit.

It appears that this kind of analysis yields useful limits on spacetime foam models. However, one should be careful regarding this kind of analysis. First, the regression index should yield reliable information only in case one has a statistically significant population of data with known redshifts, something that at present is not feasible. Again detailed knowledge on the emission mechanisms at the source is essential in order to disentangle effects that may be due to conventional physics, that is, effects not related to foam. For example, nontrivial vacua in nonlinear theories of quantum electrodynamics associated with thermalized fermions or photons (CMBR) lead to nontrivial refractive indices. However, the energy dependence on the probe energy in such cases is different from the foam effect in the sense that it leads to either (1) an energy-independent light velocity, which simply changes value as the universe expands; or (2) energy- and temperature-dependent refractive index, which however, decreases with increasing probe energy and hence leads to the opposite effect. It should be mentioned that there is a another conventional approach where Maxwell's vacua have been associated with small refractive index and the

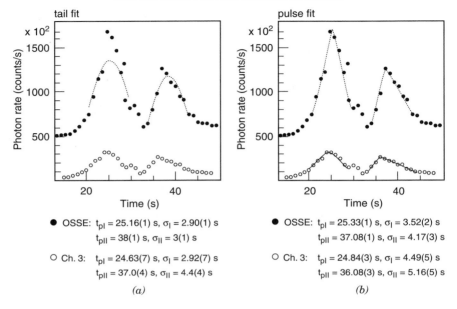

Figure 7. As in Fig. 6, but GRB 990123.

propagation of photon is investigated [10]. This gives rise to small but finite nonzero rest mass of photon. This also plays a significant role at the cosmological scale. So, such conventional effects have to be distinguished from the pure spacetime foam effectes in the relevant analysis. A systematic study of the observed GRB indicates that the pulses of light become narrower and the arrival time shorter as one goes from the low- to the high-energy channels. This is opposite the quantum gravity effect.

B. Ultra-High-Energy Cosmic Rays

Ultra-high energy cosmic rays (UHECRs) with energy [19] higher than 10^{20} eV have drawn much attention as sensitive probes of Lorentz invariance violations and in particular modified dispersion relations [20]. The various models [21] consider the modified dispersion relations so as to explain the violation of GZK cutoff [22]. The possible breakdown of their cutoff due to the quantum gravity effect has also been considered in a series of papers [21]. Within the realm of the quantum gravity framework, it is considered that the quantum gravity effects could provide an explanation for significant increase in the transparency of the universe in such a way so that the sources of UHECRs could be extragalactic,

TABLE III
As in Table II but Comparing Fits to Data from OSSE, in the Range $1 < E < 5$–10 MeV, and
Channel 3 of BATSE

			GRB 980329 (I)	GRB 980329 (II)	GRB 990123 (I)	GRB 990123 (II)
OSSE	Tail	t_p (s)	3.3(4)	7.70(2)	25.16(1)	38(1)
		σ (s)	1.7(4)	2.62(2)	2.90(1)	3(1)
		t_p (s)	4.33(5)	6.95(4)	25.33(1)	37.08(1)
	Pulse	σ_r (s)	2.39(5)	0.74(5)	3.57(2)	2.26(2)
		σ_d (s)	3.6(4)	4.78(7)	3.47(2)	6.07(4)
		σ (s)	3.0(3)	2.76(6)	3.52(2)	4.17(3)
BATSE	Tail	t_p (s)	3.81(3)	8.11(1)	24.63(7)	37.0(4)
(Ch. 3)		σ (s)	1.88(4)	2.35(1)	2.92(7)	4.4(4)
		t_p (s)	5.18(6)	10.06(3)	24.84(3)	36.08(3)
	Pulse	σ_r (s)	3.37(6)	2.3(1)	4.59(4)	2.85(4)
		σ_d (s)	5.28(6)	1.071(7)	4.39(5)	7.48(5)
		σ (s)	4.33(6)	1.69(7)	4.49(5)	5.16(5)
Δ	Tail	Δt_p (s)	$-0.5(4)$	$-0.41(2)$	0.53(7)	1(1)
		$\Delta\sigma$ (s)	$-0.2(4)$	0.27(2)	$-0.02(7)$	$-1(1)$
	Pulse	Δt_p (s)	$-0.85(8)$	$-3.11(5)$	0.49(3)	1.00(3)
		$\Delta\sigma$ (s)	$-1.3(3)$	1.07(9)	$-0.97(5)$	$-0.99(6)$
		$(\Delta t_p)^f$ (s)	$-0.85(8)(35)$	$-3.11(5)(270)$	0.49(3)(4)	1.00(3)(0)
		$(\Delta\sigma)^f$ (s)	$-1.3(3)(11)$	1.07(9)(80)	$-0.97(5)(95)$	$-0.99(6)(1)$

lying much farther away, in contrast with a common belief based on Lorentz-invariant models. This kind of modification of dispersion relation has also been invoked to explain the discrepancies between the observed γ spectrum of Markarian 501 and expectations based on a new estimate of the infrared background [23].

It may be possible that the conventional explanation could account for the spectrum of such ultra-high-enrgy cosmic rays, but the possibility of getting a signature of quantum gravity effects from such data cannot be excluded [24]. For instance, it has been argued that certain models of deformed Lorentz symmetry cannot lead to threshold effects that account for violation of GZK cutoff [25], in contrast to the quantum gravity model, and hence UHECR data can be used to disentangle various models of spacetime foam. In addition, since UHECRs involve charged particles, the possibility of foam-induced transition radiation should be taken into account as a means of excluding models.

It is clear from the preceding analysis of astrophysical sources that it is necessary to analyze the data from other sources such as AGN and/or pulsars to confirm the signature of quantum gravity effects. An alternate possibilty would be to consider more carefully laboratory experiments that might be able to reveal possible variations in the velocity of light.

III. THEORETICAL DEVELOPMENTS

The propagation of electromagnetic waves through vacuum with fluctuations give rise to new effects published in a series of recent papers [26]. This can be broadly classified in the following manner.

A. Quantum Gravity Effects

It is generally believed that quantum gravitational fluctuations are of the typical sizes $\sim l_p \sim 10^{-33}$ cm and timescales $\sim t_p \sim 10^{-43}$ s. It is proposed [26] that particles propagating through the vacuum interact with these fluctuations, inducing nontrivial recoil and associated vacuum polarization effects. As an example they have taken the theoretical model of a recoiling D particle in the quantum gravitational foam. Its recoil due to scattering by a photon has been considered. This leads to a nonzero gravitational field with a metric of the form

$$G_{ij} = \delta_{ij}; \qquad G_{00} = -1; \qquad G_{0i} = \epsilon^2(Y_i + \bar{U}_i t)\Theta_\epsilon(t) \qquad (5)$$

where $0(i)$ denote time (space) components. Here, $\bar{U}_i$ is the recoil velocity of the D particle, which is located at Y_i and ϵ is a small parameter. The metric given above implies that D-brane recoil induces the following perturbation $h_{\mu\nu}$ about flat spacetime:

$$h_{0i} = \epsilon^2 \bar{U}_i t \Theta_i(t) \qquad (6)$$

where only the nonzero components of $h_{\mu\nu}$ are indicated. Let $\epsilon^{-2} \sim t$ at large times, so that the asymptotic form of the gravitational perturbation takes the form

$$h_{0i} = \bar{U}_i \qquad (7)$$

This form corresponds to a breakdown of Lorentz invariance induced by the propagation of the photon. Suppose that the light travels a distance L in time $t \sim \epsilon^{-2}$ in the presence of a metric fluctuation h_{0i}. The effects of such a field in Maxwell's equations may be considered as follows:

$$G_{00} \equiv -h; \qquad \mathcal{G}_i = -\frac{G_{0i}}{G_{00}}; \qquad i = 1, 2, 3 \qquad (8)$$

Maxwell's equations in this background metric in vacua can be written as [27]

$$\text{div}\,\mathbf{B} = 0; \qquad \text{curl}\,\mathbf{H} = \frac{1}{c}\frac{\partial}{\partial t}\mathbf{D} = 0 \qquad \qquad (9)$$

$$\text{div}\,\mathbf{D} = 0; \qquad \mathbf{E} = -\frac{1}{c}\frac{\partial}{\partial t}\mathbf{B} = 0 \qquad (10)$$

where

$$D = \frac{E}{\sqrt{h}} + H \times \mathcal{G}; \qquad B = \frac{H}{\sqrt{h}} + \mathcal{G} \times E \qquad (11)$$

Thus there is a direct analogy with Maxwell's equations in a medium with $\frac{1}{\sqrt{h}}$ playing the role of the electric and magnetic permeability. Considering $h = 1$, one has the same permeability as the classical vacuum. Now for the metric perturbation as considered above, the modified Maxwell equations can be written as

$$\nabla \cdot E + \bar{U} \cdot \frac{1}{c} \partial \partial t E = 0 \qquad (12)$$

$$\nabla \times B - \left(1 - \bar{U}^2\right) \frac{1}{c} \frac{\partial}{\partial t} E + \bar{U} \times \frac{1}{c} \frac{\partial}{\partial t} B + (U \cdot \nabla) E = 0 \qquad (13)$$

$$\nabla \cdot B = 0, \qquad \nabla \times E + \frac{1}{c} \frac{\partial}{\partial t} B = 0 \qquad (14)$$

After simple calculations, one gets the following equations dropping nonleading terms of order $\bar{U}^2$ as

$$\frac{1}{c^2} \frac{\partial^2}{\partial^2 t} B - \nabla^2 B - 2(\bar{U} \cdot \nabla) \frac{1}{c} \frac{\partial}{\partial t} B = 0 \qquad (15)$$

$$\frac{1}{c^2} \frac{\partial^2}{\partial^2 t} E - \nabla^2 E - 2(\bar{U} \cdot \nabla) \frac{1}{c} \frac{\partial}{\partial t} E = 0 \qquad (16)$$

If we consider one-dimensional motion along the x direction, we observe that these equations admit wave solutions of the form

$$E_x = E_z = 0; \quad E_y(x, t) = E_0 e^{ikx - \omega t}; \qquad B_x = B_y = 0; \quad B_z(x, t) = B_0 e^{ikx - \omega t}$$

with a modified dispersion relation

$$k^2 - \omega^2 - 2\bar{U} \cdot k\omega = 0 \qquad (17)$$

As the sign of $\bar{U}$ is that of the momentum vector k along the x direction, the dispersion relation corresponds to *subluminal* propagation with a refractive index:

$$c(E) = c(1 - \bar{U}) + \mathcal{O}(\bar{U}^2) \qquad (18)$$

where we have an estimate of $\bar{U}$ as

$$\bar{U} = \mathcal{O}\left(\frac{E}{M_D c^2}\right) \qquad (19)$$

with M_D as the D-particle mass scale. In string model $M_D = g_s^{-1}M_s$, where g_s is the string coupling constant and M_s is the string scale. Here the refractive index is a mean-field effect that implies a delay in the arrival time of photons to that of an idealized low-energy photon for which quantum gravity effects can be neglected and as of the order of

$$\Delta t \sim \frac{L}{c}|\bar{U}| = \mathcal{O}\left(\frac{EL}{M_D c^2}\right) \tag{20}$$

Ellis et al. [30] discussed the quantum fluctuations about the mean-field solution that would correspond in field theory to quantum fluctuations in the lightcone and could be induced by higher-genus effects in the string approach. Such effects would result in stochastic fluctuations in the velocity of light as of the order of

$$(\delta\Delta t) = \frac{LE}{c\lambda}$$

where $\lambda \sim \frac{M_D c^2}{8g_s}$. It should be noted that in contrast to the variation of the refractive index, which refers to photons of different energy, the fluctuation characterizes the statistical spread in the velocities of photons of the same energy.

B. Propagation of a Pulse of Photons Through Space-Time Foam

GRBs emit photons in pulses containing photons with a combination of different wavelengths, whose sources are believed to be ultrarelativistic shocks with Lorentz factor $\gamma = O(100)$ [28]. Let us consider a wavepacket of photons emitted with a Gaussian distribution in x at the time $t = 0$. One has to find out how such a pulse would be modified at the observation point at a subsequent time t, because of the propagation through the spacetime foam, as a result of the refractive index. This is similar to the motion of a wavepacket in a conventional dispersive medium.

The Gaussian wavepacket may be expressed at $t = 0$ as the real part of

$$f(x) = A e^{-[x^2/(\nabla x_0)^2]} e^{(ik_0 x)}$$

with a modulation envelope, symmetric about the origin and where A is the amplitude. The quantity Δx_0 denotes the root mean square of the spatial spread of the energy distribution in the wavepacket, which is proportional to $|f(x)|^2$. If we assume a dispersion relation $\omega = \omega(k)$, it can be shown that at time t the Gaussian wavepacket will have the form

$$|f(x,t)|^2 = \frac{A^2}{\left(1 + \frac{\alpha^2 t^2}{(\Delta x_0)^4}\right)^{1/2}} \exp\left(-\frac{(x - c_g)^2}{2(\Delta x_0)^2\left[1 + \frac{\alpha^2 t^2}{(\Delta x_0)^4}\right]}\right) \tag{21}$$

where $\alpha \equiv \frac{1}{2}\left(\frac{d^2\omega}{d^2k}\right)$ and $c_g \equiv \frac{d\omega}{dk}$ is the group velocity. This is the velocity with which the peak of the distribution moves in time.

It is obvious from the dispersion relation above that the quadratic term in α does not affect the motion of the peak, but only the spread of the Gaussian wavepacket.

$$|\Delta x| = \Delta x_0 \left(1 + \frac{\alpha^2 t^2}{(\Delta x_0)^4}\right)^{1/2} \tag{22}$$

which increases with time. The quadratic term in α also affects the amplitude of the wave-packet. The amplitude decreases together with an increase in the spread in such a way that the integral of $|f(x,t)|^2$ is cosntant.

In the case of propagation of pulse in quantum gravitational foam [29], the dispersion relation assumes the following form for positive momentum k in units where $c = \hbar = 1$:

$$k = \omega\left(1 + \frac{\omega}{M_D}\right)$$
$$c_g = (1 - \bar{U}) \tag{23}$$
$$\alpha = -\frac{1}{M_D} + \cdots$$

Here, the ellipses $(\cdots)$ denote the higher-order terms in $\frac{1}{M_D}$. Thus the spread of the wavepacket due to the nontrivial refractive index is

$$|\Delta x| = \Delta x_0 \left(1 + \frac{t^2}{M_D^2(\Delta x_0)^4}\right)^{1/2} \tag{24}$$

It should be noted that the spread due to the refractive index $\frac{\delta c}{c} \propto \omega$ is independent of the energy of the photon to leading order in $\frac{1}{M_D}$. Hence, this effect is distinct from the stochastic propagation effect, which depends on the photon energy ω. For astrophysical sources at cosmological distances with redshifts $z \simeq 1$, and with an initial Δx_0 of a few kilometers only if the latter is of order 10^{19} GeV, namely, of order $10^{-30}|\Delta x_0|$, the correction would become of order $|\Delta x_0|$ only if the latter is of the order of 10^{-3} m. Even if one allows M_D to be as low as the sensitivities shown in the Table I , this broadening effect is still negligible for all sources there, at most of order $10^{-22}|\Delta x_0|$. Therefore, in this model, the only broadening effect that needs to be considered is the stochastic quantum gravitational effect on the refractive index.

In the case of a quantum gravitational foam scenario with quadratic refractive index $\frac{\delta c}{c} \sim E^2$, the dispersion relation assumes the following form:

$$k = \omega\left(1 + \frac{\omega}{\tilde{M}}\right)$$

$$c_g = (1 - \bar{U}) \tag{25}$$

$$\alpha = -3\frac{\omega}{\tilde{M}} + \cdots$$

where $\tilde{M} \sim 10^{11}$ TeV. The spread of the wavepacket is

$$|\Delta x| = \Delta x_0 \left(1 + \frac{9\omega^2 t^2}{\tilde{M}^4 (\Delta x_0)^4}\right)^{1/2} \tag{26}$$

Again, if we take the same sensititvities as in Table I, the maximum spreading of the pulse is negligible for $\Delta x_0 \sim 10^{-3}$ m, namely, $10^{-33} \Delta x_0$. One needs only to consider the possible stochastic quantum gravitational effect on the refractive index. However, as the quadratic dependence is not favored theoretically, we will not pursue it further.

C. Cosmological Expansion and Propagation of Photons

Here, we shall discuss the implications of cosmological expansion for the searches of a quantum-gravity-induced refractive index and a stochastic effect. We will consider Friedman–Robertson–Walker (FRW) metrics as an appropriate candidate for standard homogeneous and isotropic cosmology. Let R be the FRW scale factor, and a subscript 0 will denote the value at the present era. H_0 is the present Hubble expansion parameter, and the deceleration parameter q_0 is defined in terms of the curvature k of the FRW metric by $k = (2q_0 - 1)(\frac{H_0^2 R_0^2}{c^2})$, specifically, $\Omega_0 = 2q_0$.

Considering the inflation, we assume a universe with a critical density: $\Omega_0 = 1, k = 0$ and $q_0 = \frac{1}{2}$. The universe is assumed to be matter dominated during all the epoch of interest. Then the scale factor $R(t)$ of the universe expands as:

$$\frac{R(t)}{R_0} = \left(\frac{3H_0}{2}\right)^{2/3} t^{2/3} \tag{27}$$

and the present age of the universe is

$$t_0 = \frac{2}{3H_0} \tag{28}$$

Then no time delay can be larger than this. This relation between redshift and scale factor is

$$\frac{R(t)}{R_0} = \frac{1}{1+z} \tag{29}$$

From these relations we get the age of the universe at any given redshift as

$$t = \frac{2}{3H_0} \frac{1}{(1+z)^{3/2}} = \frac{t_0}{(1+z)^{3/2}} \tag{30}$$

Hence, a photon emitted by an object at redshift z has traveled for a time

$$t_0 - t = \frac{2}{3H_0} \left(1 - \frac{1}{(1+z)^{3/2}}\right) \tag{31}$$

The corresponding differential relation between time and redshift is

$$dt = -\frac{1}{H_0} \frac{1}{(1+z)^{5/2}} dz \tag{32}$$

The total distance L traveled by such a particle after the emission with redshift z is

$$L = \int_t^{t_0} u\, dt = \frac{1}{H_0} \int_0^z \frac{u(z)}{(1+z)^{5/2}} dz \tag{33}$$

Hence, the differences in distances covered by the two particles with velocities differing by Δu is

$$\Delta L = \frac{1}{H_0} \int_0^z \frac{dz}{(1+z)^{5/2}} (\Delta u) \tag{34}$$

In the context of quantum-gravity-induced refractive-index phenomena, we face the same situation. Let the two photons travel with velocities very close to c, whose present energies are E_1 and E_2, respectively. At earlier epochs, their energies would have been blueshifted by a common factor $1 + z$. Let $\Delta E_0 \equiv E_1 - E_2$, then

$$\Delta u = \frac{(\Delta E_0(1+z))}{M}$$

After simple calculations we can get the difference in the arrival time of the two photons as

$$\Delta t = \frac{\Delta L}{c} \sim \frac{2}{H_0} \left[1 - \frac{1}{(1+z)^{1/2}} \right] \frac{\Delta E_0}{M} \tag{35}$$

This expression describes the corrections to the refractive index effect due to the cosmological expansion. For small $z \ll 1$, expression (35) yields

$$\Delta t \sim \frac{(z \cdot \Delta E_0)}{H_0 M}$$

which agrees with the simple expectation

$$\Delta t \sim \frac{r \cdot \Delta E_0}{(c \cdot M)}$$

for a nearby source at distance

$$r = c(t_0 - t) \sim \frac{z}{H_0} + \cdots$$

There would be similar cosmological corrections to the stochastic effect.

D. Lightcone Fluctuations

In modern physics, the unification of the gravity and other interactions poses one of the most challenging problems. Many theories have been proposed since the work of Kaluza and Klein [31]. In case of supergravity and superstring theories, the existence of higher spatial dimensions is postulated. It is assumed that these extra dimensions are not mathematical artifices but represent the physical reality. The question arises as to whether they represent the physical reality and if so, why they are not observed. The usual answer is that they curl into an extremely small compactified manifold, possibly as small as the Planck length scale: $l_p \sim 10^{-33}$ cm. The next question is whether there are any lower bounds on the size of these extra dimensions. It is usually thought that the existence of extra dimensions has no effect on low-energy physics as long as they are extremely small. It has been claimed [32] that this is not the case because of the fluctuations of lightcone arising out of the quantum gravitational vacuum fluctuations due to compactification of spatial dimensions. An explicit calculation was made in a five-dimensional prototypical Kaluza–Klein model that showed that the periodic compactification of the extra spatial dimensions give rise to stochastic fluctuations in the speed of light. This fluctuation grows as the compactification scale decreases and is principle, observable. Indeed, the samller the size of the compactified dimensions, the larger are the fluctuations. This is related to Casimir effect, the vacuum energy arising whenever the boundary conditions are imposed on the quantum field. The gravitational Casimir energy in the five-dimensional case was studied [33] where a nonzero energy density was found that tends to make the extra dimension contract. The lightcone fluctuations due to compactification of spatial dimensions come solely from gravitons (closed-string states).

Yu and Ford [32] studied the effects of compactified extra dimensions on the propagation of speed of light in the uncompactified dimensions. There are non-trivial effects that arise from quantum fluctuations of the gravitational field, in-duced by the compactification. These effects take the form of lightcone fluctuations, variations in the flight times of pulses between a source and a detector.

1. Observability of Lightcone Fluctuations

Let us consider a $(d = 4 + n)$-dimensional spacetime with n extra dimensions. The spacetime metric may be written as

$$ds^2 = (\eta_{\mu\nu} + h_{\mu\nu})dx^\mu dx^\nu = dt^2 - d\mathbf{x}^2 + h_{\mu\nu}dx^\mu dx^\nu$$

where the indices run through $0, 1, 2, 3, \ldots, 3 + n$. Let $\sigma(x, x'$, be one half of the squared geodesic distance between a pair of spacetime points x and x' and $\sigma_0(x, x')$ be the corresponding quantity in the flat background. In the presence of a linearized metric perturbation, $h_{\mu\nu}$, we may expand $\sigma = \sigma_0 + \sigma_1 + O(h_{\mu\nu}^2)$. Here, σ_1 is first-order in $h_{\mu\nu}$. If one quantize $h_{\mu\nu}$, then quantum gravitational vacuum fluctuations will lead to fluctuations in the geodesic separation and therefore induce lightcone fluctuations. In particular, we have $\langle \sigma_1^2 \rangle \neq 0$ since σ_1 becomes a quantum operator when the metric perturbations are quantized. The quantum lightcone fluctuations give rise to stochastic fluctuations in the speed of light, which may produce an observable time delay or advance Δt in the arrival time of pulses.

Now let us consider the propagation of light pulses between a source and a detector separated by a distance r on a flat background with quantized linear perturbations. One can find easily Δt from the following relation:

$$\sigma = \sigma_0 + \sigma_1 + \cdots = \frac{1}{2}[(r + \Delta t)^2 - r^2] \approx r\Delta t \tag{36}$$

Considering the quantum state of gravitons $|\phi\rangle$, that is, the vacuum states associated with the compactification of spatial dimensions, after some calcula-tions we get

$$\Delta t_{obs} \approx \frac{\sqrt{|\langle \sigma_1^2 \rangle R|}}{r} \tag{37}$$

Here, Δt is the ensemble averaged deviation, not necessarily the expected variation in flight time, δt, of two pulses emitted close together in time. The latter is given by Δt only when the correlation time between successive pulses is less than the time separation of the pulses. Physically it means that the gravitational field may not fluctuate significantly in the interval between the two pulses. These stochastic fluctuations in the apparent velocity of light arising from quantum gravitational fluctuations are in principle observable since they may lead to a spread in the arrival times of pulses from distant astrophysical sources or the broadening of the spectral lines.

Yu and Ford [32] found that the dominant contributions to the lightcone fluctuations come from the graviton modes with wavelengths of the order of $\sim L$. In other words, the lightcone fluctuates on a typical time scale of $\sim \frac{1}{L}$. If the travel distance, r, is less than L, successive pulses are uncorrelated only when their time separation is greater than the typical fluctuation timescale. Otherwise they are correlated because the quantum gravitational vacuum fluctuations are not significant enough in the interval between the pulses. If $r > \frac{L^2}{T}$, then successive pulses are in general uncorrelated. Thus the correlation time for large $r \sim \frac{L^2}{r}$, which is much smaller than the compactification scale L. This can be understood as a loss of the correlation as the pulses propagate over an increasing distance. However, it can be shown that in case of one extra dimension, the large quantum lightcone fluctuations due to compactification of the extra dimension require either the size of the extra dimension to be macroscopically large or the rate of change of the extra dimension to be extremely small. In this way the five-dimensional Kaluza–Klein theory may be discarded, or very strong limits on the rate of change of the extra dimension can be placed. One finds that the rate of growth of Δt with r depends crucially on the number of spatial dimensions. In four dimensions, $\Delta t \sim \sqrt{r}$, while in five dimensions $\Delta t \sim r$. One can expect that in case of large number of dimensions, there will be an effect of compactification. It has been demonstrated that the lightcone fluctuation grows logarithmically with distance for one flat compactified extra dimension. But it is not yet proved that this behavior holds for any number of flat extra dimensions.

2.   *Lightcone Fluctuations and Parallel Branes*

Several authors [34] have studied the effect of lightcone fluctuations in the context of large extra dimensions stretching between two parallel brane worlds. The main motivation behind this approach is to resolve the unnatural hierarchy between the weak and Planck scales. The four-dimensional particle theory (e.g., the standard model) is confined to live in one of the branes, but gravity is free to propagate in the higher-dimensional bulk. So the bulk spacetime is dynamical and the 3-branes cannot be rigid but undergo quantum fluctuations in their positions as one assumes distance of separation of the order of Planck length l_p. Suppose that one brane is located at the origin and the other at $(0, 0, 0, z_1,$

$z_2, z_3, \ldots, z_n$). Now one can study the effect of lightcone fluctuations due to gravitons in the bulk by looking at a light ray traveling parallel to one of the boundaries, say, in x axis, but separated from it by a distance $z \sim l_p$. This indicates the effect of quantum fluctuations in the position of the brane.

Cambell-Smith et al. [35] considered one of the branes as the observable world. They have studied this phenomena from both field and string theoretical viewpoints, by analyzing the role of coherent graviton fluctuations on the propagation of photons on one brane.

In the field-theoretic case, they have estimated Δt as $\Delta t \sim \sqrt{L} \ln L$, where L is the distance between the source and the detector. For astrophysical sources such as gamma-ray bursters (GRBs), the order of the effect is about 10^{-12} s, which falls below the sensitivity of observations. Here, Δt is considered as the scale of lightcone broadening. It should be noted that the sensitivity of gravity wave interferometer experiments is much better, and for these experiments one can get lightcone broadening of the order of 10^{-25} s, which may lie within their sensitivity.

In case of string model, they have found the dominant contributions to the phenomenon from the recoil of the D brane due to scattering of closed-string states (gravitons) propagating in the bulk. The recoil distorts the spacetime around the D brane, resulting in a mean-field effect which implies stochastic fluctuations in the arrival time of photons of energy E on the brane of order $\Delta t g_s L (E/M_s)$. Such phenomena place strong restrictions on string-theoretic models of extra dimensions.

E. κ-Deformed Quantum Relativistic Phase Space

The covariant Heisenberg commutation relations in phase space can be written as

$$[x_\mu, p_\nu] = i\hbar g_{\mu\nu}, \qquad g_{\mu\nu} = \text{diag}\,(-1, 1, 1, 1) \tag{38}$$

The spacetime coordinates $x_\mu (\mu = 0,1,2,3)$ can be identified with the translation sector of the Poincaré group, and the four momenta $p_\nu (\nu = 0, 1, 2, 3)$ are given by the translation generators of the Poincaré algebra. In considering quantum deformations of relativistic symmetries as describing the modification of spacetime structure, one is lead to the study of the possible quantum Poincaré groups.

Here, we have considered the genuine 10-generator quantum deformations of $D = 4$ Poincaré symmetries as a particular case. The more general "standard" q deformations have not been considered. These q deformations require adding an 11th (dilatation) generator where one has to deal with the dilatation extended Poincaré algebra [36]. In such cases, the corresponding quantum phase space is much more complicated [37]; also in this case, the deformation parameter is dimensionless, rendering difficult the physical seperation between the ordinary regime of commutative spacetime coordinates and the short-distance regime in which noncommutativity sets in. The classifications of quantum deformations of $D = 4$ Poincaré groups in the framework of Hopf algebras was given by Podlés and Woronowicz [38] and also Ref. 39, which provides the most general class of noncommutative spacetime coordinates $\hat{X}_v$ allowed by the quantum group formalism. If we assume that the quantum deformation does not affect the nonrelativistic kinematics, that is, if we preserve the nonrelativistic $O(3)$ rotations classical and $O(3)$ covariance, the only consistent class of noncommutating spacetime coordinates is described by the relations of the k-deformed Minkowski space with commuting classical space coordinates. In order to describe the relativistic phase space we start with the deformed Hopf algebra of 4-momenta $\hat{p}_\mu$ written as follows

$$
\begin{aligned}
[\hat{p}_0, \hat{p}_k] &= 0 \\
\Delta(\hat{p}_0) &= \hat{p}_0 \otimes 1 + 1 \otimes \hat{p}_0 \\
\Delta(\hat{p}_k) &= \hat{p}_k \otimes e^{\alpha \hat{p}_0} + e^{\beta \hat{p}_0} \otimes \hat{p}_k
\end{aligned}
\tag{39}
$$

with antipode and counit given by

$$
\begin{aligned}
\mathscr{S}(\hat{p}_k) &= -e^{(\alpha+\beta)\hat{p}_0} \hat{p}_k \\
\mathscr{S}(\hat{p}_0) &= -\hat{p}_0 \qquad \epsilon(\hat{p}_\mu) = 0
\end{aligned}
$$

Using the duality relations involving the fundamental constant $\hbar$ (Planck's constant)

$$
\langle \hat{x}_\mu, \hat{p}_k \rangle = -i\hbar g_{\mu v} \qquad g_{\mu v} = (-1, 1, 1, 1)
$$

we obtain the noncommutative deformed configuration space $\mathscr{X}$ as a Hopf algebra with the following algebra and coalgebra structure:

$$
\begin{aligned}
[\hat{x}_0, \hat{x}_k] &= i\hbar(\beta - \alpha)\hat{x}_k, \qquad [\hat{x}_k, \hat{x}_l] = 0 \\
\Delta(\hat{x}_\mu) &= \hat{x}_\mu \otimes 1 + 1 \otimes \hat{x}_\mu \\
\mathscr{S}(\hat{x}_\mu) &= -\hat{x}_\mu, \qquad \epsilon(\hat{x}_\mu) = 0
\end{aligned}
$$

The deformed phase space can be considered as the vector space $\mathscr{X} \otimes \mathscr{P}$ with the product

$$
(x \otimes p)(\tilde{x} \otimes \tilde{p}) = x(p_{(1)} \triangleright \tilde{x}) \otimes p_{(2)} \tilde{p}
\tag{40}
$$

where left action is given by

$$
p \triangleright x = \langle p, x_{(2)} \rangle x_{(1)}
$$

The product of this equation can be written as the commutators between coordinates and momenta by using the obvious isomorphism $x \sim x \otimes 1$, $p \sim 1 \otimes p$. This provides the following commutation relations:

$$[\hat{x}_p, \hat{p}_l] = i\hbar \delta_{kl} e^{\alpha \hat{p}_0}, \qquad [\hat{x}_k, \hat{p}_0] = 0$$

$$[\hat{x}_0, \hat{p}_k] = i\hbar \beta \hat{p}_k, \qquad [\hat{x}_0, \hat{p}_0] = -i\hbar$$

The set of these relations describe the deformed relativistic quantum phase space.

Taking a as the dispersion of the observable in quantum mechanics as

$$\Delta(a) = \sqrt{\langle a^2 \rangle - \langle a \rangle^2} \tag{41}$$

we can write

$$\Delta(a)\Delta(b) \geq \frac{1}{2} |\langle c \rangle| \tag{42}$$

where $c = [a, b]$. The deformed uncertainty relations can be written as

$$\Delta \hat{x}_0 \Delta \hat{x}_k \geq \frac{\hbar}{2} |(\beta - \alpha)| |\langle \hat{x}_k \rangle|$$

$$\Delta \hat{p}_k \Delta \hat{x}_l \geq \frac{\hbar}{2} \delta_{kl} \langle e^{\alpha \hat{p}_0} \rangle$$

$$\Delta \hat{p}_0 \Delta \hat{x}_0 \geq \frac{\hbar}{2}$$

$$\Delta \hat{p}_k \Delta \hat{x}_0 \geq \frac{\hbar}{2} |\langle \beta \hat{p}_k \rangle|$$

One can get the following cases considering the different choices of the parameters α and β:

1. $\alpha = \beta = 0$; standard form of nondeformed covariant phase space
2. $\alpha = \beta$; trivially deformed phase space with commuting configuration space
3. $\alpha = -\beta = \frac{1}{2\kappa c}$; κ-deformed phase space in the standard basis
4. $\alpha = 0, \beta = -\frac{1}{\kappa c}$; κ-deformed phase space in the bi-cross-product basis
5. $\alpha = \frac{1}{\kappa c}, \beta = 0$; κ-deformed phase space in the bi-cross-product basis
6. $\alpha = 0, \beta = \frac{1}{\kappa c}$; κ-deformed phase space in the bi-cross-product basis (case 5) with transposed coproduct

Now the modified covariant Heisenberg uncertainty relations can be written as

$$\Delta \hat{t} \Delta \hat{x}_k \geq \frac{\hbar}{2\kappa c^2} |\langle \hat{x}_k \rangle|$$

$$\Delta \hat{p}_k \Delta \hat{x}_l \geq \frac{\hbar}{2} \delta_{kl}$$

$$\Delta \hat{E} \Delta \hat{t} \geq \frac{\hbar}{2}$$

$$\Delta \hat{p}_k \Delta \hat{t} \geq \frac{\hbar}{2\kappa c^2} |\langle \hat{p}_k \rangle|$$

After algebraic manipulations we get the velocity formula for massless $\kappa-$ deformed quanta as

$$v_i = \frac{\partial E}{\partial \hat{p}_i} \tag{43}$$

or

$$v = |\vec{v}| = c - \frac{|\hat{\mathbf{p}}|}{\kappa} + \mathcal{O}\frac{1}{\kappa^2} \tag{44}$$

It should be noted from these formulas that the 3-momentum-dependent (i.e., energy-dependent) speed of light is a novel phenomenon. It has the same functional form as the energy-dependent speed of light discussed in the context of string theory. However, although this kind of deviation from the ordinary physics arises in both κ Poincaré and string theory contexts, it is rather marginal from the phenomenological viewpoint. Consider the photons of energy of the order of 1 GeV, where it gives rise to a correction of the order of $10^{-19} c$ with respect to the usual speed of light. However, when κ is identified with the Planck scale, Eqs. (43) and (44) are completely consistent with the presently available data.

F. Nonzero Conductivity of Maxwell Vacuum and Energy-Dependent Speed of Light

From a completely different perspective, several authors [10] considered the frequency-dependent speed of light. Here, the main idea is that if one assigns a small but nonzero conductivity to Maxwell vacuum, and considers the propagation of photon in such a vacuum, it gives rise to frequency-dependent speed of light and hence a nonzero but finite photon mass. Here, this small conductivity of the vacuum can be realized to the refractive index of the modified Maxwell vacuum. The modified Maxwell equations can be written as

$$\operatorname{div} \mathbf{E} = 0 \qquad \operatorname{curl} \mathbf{H} = \sigma \mathbf{E} + \epsilon_0 \chi_e \frac{\partial \mathbf{E}}{\partial t} \tag{45}$$

$$\operatorname{div} \mathbf{H} = 0 \qquad \operatorname{curl} \mathbf{E} = -\mu_0 \chi_m \frac{\partial \mathbf{H}}{\partial t} \tag{46}$$

where μ_0 = vacuum permeability constant
χ_e = relative dielectric constant
χ_m = relative permeability constant
σ = conductivity of the vacuum

Here, the 4-current is given by

$$j = (\mathbf{j}, j_0) \quad \text{with} \quad \mathbf{c}j = \sigma \mathbf{E}; \qquad j_0 = 0 \tag{47}$$

Again

$$\nabla \times \nabla \times \mathbf{E} = -\nabla^2 \mathbf{E} \tag{48}$$

which, together with Maxwell's equations, gives

$$\nabla^2 \mathbf{E} = -\frac{\epsilon_0 \chi_e \chi_m}{c^2} \mu_0 \frac{\partial^2 \mathbf{E}}{\partial^2 t} + \sigma \mu_0 \chi_m \frac{\partial \mathbf{E}}{\partial t} \tag{49}$$

This equation is not time-reversal-invariant. The second term on the right-hand side indicates that there will be a dissipation of energy during the propagation of a photon.

In Maxwell's theory, this dispersion of energy is considered to be negligible, and no damping occurs during the propagation of an electromagnetic wave. Let us consider the plane waves propagating in the z direction:

$$E_x = b \exp i\omega \left(t - \frac{z}{v} \right) \qquad H_y = b \left(\frac{\epsilon_0 \chi_e}{\mu_0 \chi_m} \right)^{1/2} \exp i\omega \left(t - \frac{z}{v} \right) \tag{50}$$

Putting $q = 1/v$, we get

$$q^2 = \frac{\epsilon_0 \chi_m \chi_e}{c^2} \left(1 - \frac{i\sigma}{\omega \epsilon_0 \chi_e} \right) \tag{51}$$

The velocity v defined above gives rise to a complex refractive index in vacuum. The real part of q^2 gives rise to phase velocity of propagation of the disturbance through the underlying vacuum. Taking the real and imaginary parts as α and β, respectively, E_x and H_y can be shown to be propotional to

$$\exp(-\omega\beta z)\exp(t - \alpha z) \tag{52}$$

Then the following cases arise:

- Plane waves are progressively damped with the factor $\exp(-kz)$, where $k = \omega\beta$.
- The phase velocity of propagation of the wave is $\frac{1}{\beta}$ and varies with the frequency.

Now, using the de Broglie relation $E = h\nu = mc^2$, one can estimate the mass of the photon as related to the conductivity coefficient σ. If one considers only plane waves in the Z direction, then

$$E_x = be^{i\omega(t-z/v)}$$

Then the complex quantity q can be written in the form

$$q = \alpha - i\beta$$

with

$$\alpha^2 = \frac{\chi_e\chi_m}{2c^2}\left[\left\{1 + \left(\frac{\sigma}{\epsilon_0\chi_e\omega}\right)^2\right\}^{1/2} + 1\right]$$

and

$$\beta^2 = \frac{\chi_e\chi_m}{2c^2}\left[\left\{1 + \left(\frac{\sigma}{\epsilon_0\chi_e\omega}\right)^2\right\}^{1/2} - 1\right]$$

In the limit $\left(\frac{\sigma}{\omega}\right) \to 0$, and so

$$\alpha \simeq 1 + \frac{1}{8}\left(\frac{\sigma^2}{\epsilon_0^2\chi_e^2}\right)\cdot\frac{1}{\omega^2} + O\left(\frac{\sigma^4}{\omega^4}\right); \qquad \beta^2 \simeq \frac{1}{2}\cdot\frac{\sigma^2}{(\epsilon_0\chi_e)^2}\cdot\frac{1}{\omega^2} \tag{53}$$

The velocity $v = \alpha$ is the phase velocity of propagation of disturbance through the underlying vacuum. Henceforth, it will be denoted as v_p. After some calculations the v_p can be written as

$$v_p = \frac{c}{(\chi_e\chi_m)^{1/2}}\left\{1 - \frac{1}{8}\cdot\frac{\sigma^2}{(\epsilon_0\chi_e)^2}\cdot\frac{1}{\omega^2}\right\} \tag{54}$$

and the group velocity v_g is given by

$$v_g = \frac{c}{(\chi_e \chi_m)^{1/2}} \sqrt{1 + \frac{1}{4} \frac{\sigma^2}{(\epsilon_0 \chi_e)^2} \frac{1}{\omega^2}} \tag{55}$$

where the dispersion law gives

$$k^2 = \omega^2 + \frac{\sigma^2}{4} \tag{56}$$

In the limiting case when $\sigma = 0$

$$v_p = v_g = c \tag{57}$$

Now, taking v_g as the velocity of the photon

$$E = \hbar\nu = \frac{m_\gamma c^2}{\sqrt{1 - \frac{v_g^2}{c^2}}} \tag{58}$$

where m_γ is the nonzero rest mass of the photon. From (58) we have

$$m_\gamma^2 c^4 = \hbar^2 \nu^2 \left(1 - \frac{v_g^2}{c^2}\right) \tag{59}$$

or

$$m_\gamma^2 = \frac{\hbar^2 \nu^2}{n^2 c^4} \left[(n^2 - 1) - \frac{\sigma^2}{(\epsilon_0 \chi_e)^2 \frac{1}{\omega^2}}\right]; \qquad n = \sqrt{\chi_e \chi_m} \tag{60}$$

This is unphysical. But if we take the phase velocity in de Broglie relation, we get a physical solution, namely, the real nonzero rest mass of photon. Since the mid 1990s, there has been much interest in the $v_g \neq c$ solutions of Maxwell equations [40]. However, in our framework, taking the phase velocity in de Broglie relation (59), we get

$$m_\gamma^2 c^4 = \frac{\hbar^2 \nu^2}{n^2} \left[(n^2 - 1) + \frac{1}{8} \frac{\sigma^2}{(\epsilon_0 \chi_e)^2} \frac{1}{\omega^2}\right] \tag{61}$$

For $n \sim 1$

$$m_\gamma \simeq \frac{\sigma h}{\sqrt{2(\epsilon_0 \chi_e)}} \cdot \frac{\pi}{c^2} \tag{62}$$

Again, Fuli [41] established a relation between nonzero photon mass and the Hubble constant H. Using that relation, one obtains

$$m_\gamma \simeq 10^{-65}\,\text{g} \tag{63}$$

Combination of Eq. (62) and the relation given by Fuli [41] gives

$$\sigma = \frac{H(\epsilon_0 \chi_e)}{\sqrt{2\pi}} \tag{64}$$

Now measuring the conductivity in vacuo, one can obtain an estimate of Hubble constant. This gives rise to a new possibility to test cosmological models in laboratory experiments.

It is worth mentioning that the conductivity σ has been measured from various perspectives of the redshifts and galactic distances [42] as

$$\sigma = (2.85 \pm 0.15)10^{-29} \quad (\Omega/\text{m})$$

and

$$\left(\frac{R\sigma}{2}\right)^2 = 3 \times 10^{-53}$$

In fact the redshift can be calculated from the following elementary considerations. If $W_0 = K v_0$ be the initial energy of an electromagnetic wave (say, of a single photon) and $W_1 = K v_1$ be residual energy after a path r, we have

$$\frac{W_1}{W_0} = \frac{v_1}{v_0} = e^{(-R\sigma r)}; \qquad v_1 = v_0 e^{(R\sigma r)}$$

So

$$\lambda = \lambda_0 e^{(R\sigma r)}$$

The redshift

$$z = \frac{\Delta\lambda}{\lambda_0} = [e^{(R\sigma r)} - 1]$$

or

$$z + 1 = e^{(R\sigma r)}$$

or

$$r = \frac{1}{R\sigma} \ln(z+1)$$

The galactic redshift could obviously be attributed to the damping of the electromagnetic waves emitted from various galaxies in random motion within a stationary universe. Now, comparison between Hubble relativistic linear law and the logarithmic law that comes out from Maxwell electromagnetic wave equation shows that, in any case, the logarithmic law fits experimental data very well and thus better than linear law.

G. $\sigma \neq 0$ and Space Charges in vacuo

Using plane-wave solutions we get the following dispersion relation in a covariant form

$$(|\mathbf{k}|^2 - n^2 k_0^2) A^\mu(k) = \chi_m \mu_0 \left[g^{\mu\nu} + \left(\frac{n^2-1}{n^2} \right) u^\mu u^\nu \right] J_\nu(k) \qquad (65)$$

where u^μ is the unit time-like vector and $u = (\mathbf{0}, 1)$ for the medium at rest. It is evident from (65) that $\mathbf{A} \neq 0$ but $\phi = 0$ for $\sigma \neq 0$ in Maxwell vacuum. This is nothing more than the usual Coulomb gauge. On the other hand, if we consider Proca equation

$$\Box A_\mu = -\xi^2 A_\mu; \qquad \xi = \frac{m_\gamma c}{\hbar} \qquad (66)$$

this equation can be expressed as

$$\Box A_\mu = J_\mu^{\text{eff}} = (\mathbf{J}^{\text{eff}}, j^{\text{eff}}) = (-\xi^2 c^2 \epsilon_0 \mathbf{A} - \xi^2 \epsilon_0 \phi) \qquad (67)$$

It should be mentioned that in the approach with nonzero electric divergence, the photon mass is also related to the space charges in vacuo. Now, in the approach with $\sigma \neq 0$, we have $\mathbf{j} = \sigma \mathbf{E}$ but $j^{\text{eff}} = 0$. Let us now assume $\mathbf{j} = \sigma \mathbf{E}$ and $j \neq 0$, which means $j^{\text{eff}} \neq 0$. In such a case, j_0 is assumed to be associated with $\bar\rho$, where $\bar\rho$ is the charge density in vacuo. So, in such an approach one can think of the existence of a kind of space charge in vacuo that is to be considered to be associated to nonzero electric field divergence. This will result in a displacement current in vacuum similar to that measured by Bartlett and Corle [43]. The assumption of the existence of space charge in vacuo makes our theory not only fully relativistic but also helps us to understand gauge condition. In the conventional framework of Maxwell's equations

$$\mathbf{j} = \sigma(\mathbf{E} + \mathbf{E}') \qquad (68)$$

where $\mathbf{E}'$ consists of all nonelectrostatic fields and $\mathbf{E}$ is the field derivable from a potential. In our framework, $\mathbf{E} = 0$ and $\mathbf{j} = \sigma\mathbf{E}'$. This indicates that the current distribution and field distribution are entirely defined with respect to the nonconservative field and to the conductivity of the medium. The presence of this nonconservative field is responsible for the loss of energy of the photon when it propagates through this type of vacuum.

In the framework of de Broglie's theory of light, one can get the set of equations in vacuum as

$$\Box F + k_0^2 F = 0 \quad \text{with} \quad F_{ik} = \frac{\partial A_k}{\partial x_i} - \frac{\partial A_i}{\partial x_k}; \qquad k_0 = \frac{m_\gamma c}{\hbar} \qquad (69)$$

Taking $P = \exp i(kct - \mathbf{k} \cdot \mathbf{r})$, we obtain the correspondence between the mechanical quantities (energy W and impulsion P) of the photons. These satisfy the well-known relativistic equation

$$\frac{W^2}{c^2} = P^2 + m_\gamma c^2$$

which is valid for the energy W and impulsion P of the individual photons. De Broglie calculated the solutions explicitly. For a plane wave moving in the Z direction, one gets

$$A_x = \frac{C_1 + C_2}{2} P, \qquad A_y = \frac{C_1 - C_2}{2} iP$$

$$A_z = C_3 P, \qquad V = \frac{C_3 |\vec{k}|}{\vec{k}} P$$

$$E_x = -ik\frac{C_1 + C_2}{2} P, \qquad E_y = k\frac{C_1 - C_2}{2} P$$

$$E_z = -i\frac{k_0^2}{k} C_3 P$$

$$H_x = -|\vec{k}|\frac{C_1 - C_2}{2} P, \qquad H_y = -i|\vec{k}|\frac{C_1 + C_2}{2} P$$

$$H_z = 0$$

For plane transverse waves, that is, for $J_3 = \pm 1$, $E^T = -i\frac{P}{m_\gamma} A^T$ and

$$\frac{|H^T|}{|E^T|} = \frac{k_0}{v}$$

where E^T and A^T are othogonal between them with

$$A^T || E^T \quad \text{and} \quad V^T = 0$$

The density of the wave energy is

$$W^T = |E^T|^2 + |H^T|^2 - (|A^T|^2 + |V^T|^2)$$

These solutions can be divided into four independent monochromatic waves:

1. A Maxwellian transverse right-rotated wave D
2. A Maxwellian transverse left-rotated wave D
3. A Maxwellian longitudinal wave G
4. A non-Maxwellian wave (NM)

One can also normalize such waves in a finite volume V, which gives $k > 0$

$$|C_1| = \left(\frac{hc}{2\pi kV}\right)^{1/2}, \qquad |C_3| = \left(\frac{hck}{4\pi k_0^2 V}\right)^{1/2}$$

$$|C_2| = \left(\frac{hc}{2\pi kV}\right)^{1/2}, \qquad |C_4| = \left(\frac{2k_0}{kV}\right)^{1/2}$$

and define linearly polarized waves (with only E_x and H_y different from zero) by the following relations:

$$A_x = \frac{hc^{1/2}}{4\pi kV}P; \qquad E_x = -ik\frac{hc^{1/2}}{4\pi kV}P; \qquad H_y = -i|k|\frac{hc^{1/2}}{4\pi kV}P$$

$$A_y = A_z = V = E_y = E_z = H_x = H_z = 0$$

Now, instead of $\Box E = k_0^2 E$, we may write

$$\Box E = k_0^2 E + \sigma \mu_0 \chi_m \frac{\partial E}{\partial t} \tag{70}$$

We can safely neglect the k_0^2 term in the usual experiment so that relation (70) is approximated by Eq. (49). This is justified by the fact that m_γ and σ are negligible at the laboratory scale. But the term $\sigma \mu_0 \chi_m \frac{\partial E}{\partial t}$ plays a significant role in Maxwell theory of electromagnetic fields. The term $\frac{\partial E}{\partial t} \simeq \frac{\partial D}{\partial t}$ corresponds to the displacement current in vacuum. The existence of displacement current in vacuum can also be observed in the case of nonzero divergence of the electric field. This is related to the existence of space charge in vacuo. It is worth mentioning that we can get nonzero mass of photon in both approaches. But the group velocity in two cases are different. In vacuum with $\sigma \neq 0$, the group velocity may be greater than the speed of light, whereas in the former approach—with nonzero divergence of electric field—the group velocity is less than the speed of light. In the second approach, the phase velocity is less than the speed of light. We have used phase velocity in calculating nonzero mass of photon. It is to be noted that in many experiments on photon tunneling conducted with different techniques and in different ranges of frequency [44], electromagnetic waves travel a barrier with group velocity $v_g > c$. These results obviously do not violate Einstein causality because according to Sommerfeld and Brillouin [45] it is the front velocity (not group velocity) to be relevant for this and Maxwell theory predicts that electromagnetic waves in vacuum always have a constant front velocity equal to c. The difficulty in the interpretational results lies mainly in the fact that in the barrier traversal no group velocity can be defined, as the wavenumber is imaginary (evanescent waves), so that the time required for the traversal (directly measured) is not univocally definable.

H. Experimental Evidence of Nonzero Photon Mass

In 1992, Vigier [46] surveyed the experimental status of nonzero photon mass. In 1940, de Broglie [46] arrived at an upper limit for the photon mass by estimating the experimental mesaurements of the dispersion of photons and comparing this dispersion with that predicted for a photon with finite mass. Photons with finite rest mass do not all travel at the same velocity, but one, dependent on the ratio of kinetic to rest mass energy.

Starting from the equation with nonzero photon mass m_γ, one gets dispersion relation $k^2 = \omega^2 c^2 - \mu^2$. The group and phase velocities can be found to be

$$|v_g| = \frac{d\omega}{d\kappa} \simeq c\sqrt{1 - \frac{m_\gamma^2 c^2}{\omega^2}} \simeq c\left(1 - \frac{m_0^2 c^2}{2\omega^2}\right)$$

$$|v_p| = \frac{\omega}{\kappa} = c\left[1 - \frac{m_\gamma c^2}{\omega^2}\right]^{-\frac{1}{2}} = c\left(1 + \frac{m_\gamma^2 c^2}{2\omega^2}\right)$$

This dispersion can also be derived from the relation

$$E = \frac{m_\gamma^2 c^2}{\sqrt{1 - \left(\frac{v}{c}\right)^2}}$$

or

$$\frac{v}{c} = \sqrt{1 - \frac{m_\gamma c^4}{\hbar\omega^2}} = \sqrt{1 - \frac{\mu^2 c^2}{\omega^2}}$$

De Broglie was the first to suggest that dispersion of light from stars could be used to set an upper limit. Photons can also be dispersed by other phenomena such as the finite electron intergalactic density. Feinberg [48] sets an upper limit by noting that the dispersion from pulsars is attributed to the finite electron density. His upper limit $m_\gamma \sim 10^{-44}$ g is not compatible with the present measurement or with geomagnetic measurements. Schrödinger set an upper limit by examining the modified Ampere law:

$$\Delta \times \mathbf{H} = \frac{4\pi}{c}\rho - \mu^2\mathbf{A} + \frac{1}{c}\frac{\partial\mathbf{E}}{\partial t}$$

Here, the additional term acts like "vector current" and by making a fit of geologic data, Schrödinger arrived at a finite rest mass. However, he pointed out that this effect might be produced by "positive or negative particles revolving around the earth at same distances in the equatorial plane."

The discovery of the Van Allen radiation belt required a recalculation of this effect. Goldhaber and Nieto [49] used a recent survey and set the most accurate upper limit. A similar technique is to look at the magnetic field as a function of the distance from the earth using satellite data. An experimental term enters for a finite mass that would be distinguishable from the normal $\frac{1}{r^3}$ dependence of a magnetic dipole. Gintsburg set an upper limit using this method.

Dayhoff [50] suggested that one might measure a rest mass of photon by designing a low-frequency oscillator from an inductor–capacitor (LC) network. The expected frequency can be calculated from Maxwell's equations, and this may be used to give an effective wavelength for photons of that frequency. He claimed that one would have a measure of the dispersion relationship at low frequencies. Williams [51] calculated the effective capacitance of a spherical capacitor using Proca equations. This calculation can then be generalized to any capacitor with the result that a capacitor has an additional term that is quadratic in the area of the plates of the capacitor. However, this term is not exactly the one that Dayhoff referred to. But it seems to be a very close description of it. One can add two identical capacitors C in parallel and obtain the result

$$C_2 = 2C\left(1 - \frac{\mu^2 A}{4\pi}\right)$$

where A is the area of one of the capacitors C and $\mu = \frac{m_\gamma c}{\hbar}$. It should be noted that the capacities of the two capacitors have identical dimensions but do not add linearly. The results of Williams' [51] measurement expressed in the form of the photon rest mass squared is

$$\mu^2 \simeq (1.7 + 2.0)(1 + 0.33) \times 10^{-19}\,\text{cm}^{-2} \quad \text{or}$$

$$\simeq (2.1 \pm 2.5) \times 10^{-94}\,\text{g}^2 \quad \text{with} \quad \mu = \frac{m_\gamma c}{\hbar}$$

This result is statistically consistent with the assumption that the photon mass is identically zero. But this can also be expressed as a deviation from the $\frac{1}{r^2}$ law of either of the form

$$\frac{1}{r^{2+q}} \quad \text{or} \quad \left(\frac{1}{r^2}\right)\left(1 - \frac{\beta}{r}\right)$$

with $q = (4.7 \pm 5.5) \times 10^{-16}$; $\beta = (1.4 \pm 1.7) \times 10^{-14}$ cm.

In 1988, Riis et al. [52] observed a direction-dependent anisotropy of light in the direction of the apex of the 2.7 K microwave background radiation in the universe. These data are consistent with nonzero photon mass. The upper bound on photon mass was estimated as $m_\gamma \sim 10^{-65}$ g. A compilation of laboratory data [53] established that the photon mass should not exceed 10^{-24} eV or even 10^{-26} eV.

But now, if there is a ample evidence of nonzero photon mass, the question of absorption or emission amplitudes for longitudinal photon has to be answered in a consistent manner. Goldhaber and Nieto [49] showed that these are suppressed in comparison with their transverse counterparts by a factor $\frac{m_\gamma c^2}{\hbar v}$. The corresponding rates and cross sections are suppressed by the square of this factor. The quantum mechanical matrix element for ordinary transverse photon is given by $T_f(x, y) = \langle f | \hat{J}_{x,y} | i \rangle$ for a photon-induced transition to an arbitrary state f, where i is the initial target state. The corresponding matrix for a longitudinal photon is

$$T_f^{(z)} = \left(\frac{m_\gamma c^2}{\hbar v}\right) \langle f | \hat{J}_z | i \rangle$$

where v is its frequency.

For $m_\gamma \sim 10^{-52}$ kg (as considered by Goldhaber and Nieto)

$$c \simeq 3 \times 10^8 \, \text{m/s}$$
$$h \sim 10^{-34} J/s$$
$$v \sim 10^{14} \text{s}^{-1}$$

and for comparable transverse and longitudinal matrix elements it is seen that

$$\frac{T_f^{(z)}}{T_f^{(x,y)}} \sim 10^{-15}$$

This result shows why spectral absorption and emission of longitudinal photons of spin zero and frequency v are never observed in the usual infrared, visible, and ultraviolet regions of the electromagnetic spectrum.

However, in the limit $v \to 0$ (i.e., as the frequency of the longitudinal photon goes to zero)

$$\frac{m_\gamma c^2}{\hbar^2} \to \frac{m_\gamma c^2}{h v_0} = 1$$

where $h v_0 = m_\gamma c^2$ for nonzero photon rest mass m_γ. This means that at frequencies comparable with

$$v_0 = \frac{m_\gamma c^2}{h} \sim 10^{-7} \, \text{Hz}$$

and for $m_\gamma \sim 10^{-52} \, \text{kg}$, $T_f^{(z)} = T_f^{(xy)}$ which thus implies that the absorption and emission of longitudinal photon could become observable.

I. Longitudinal Solutions and Nonzero Photon Mass

Several authors have found additional solutions to the relativistic wave equations. These solutions can be summarized as follows:

1. Lehnert and Roy [10] found axial magnetic field component in the direction of propagation of photon considering the axisymmetric wave modes in Maxwell's equations with space charge in vacuo.
2. The $\mathbf{B}^{(3)}$ (Evans–Vigier field) [54], which was obtained as a cross-product of the transverse modes of electromagnetism: $\mathbf{B}^{(1)} \times \mathbf{B}^{(2)} = iB^{(0)}\mathbf{B}^{(3)*}$ and cyclics
3. The $E = 0$ solution of the Maxwell's $j = 1$ equations [55]
4. The generalized solution of Maxwell equations in terms of potentials [56,57]
5. Non-plane-wave solutions of the Klein–Gordon equation using unconventional basic functions and "coupling ansatz"

Here, we shall mainly discuss those solutions that are related to nonzero photon mass only. In this framework, we got three kinds of solutions:

1. *EM wave*: conventional transverse electromagnetic wave
2. *S wave*: a longitudinal purely electric space-charge wave
3. *EMS wave*: a hybrid nontransverse electromagnetic space-charge wave

By choosing the axisymmetric wave modes, we can derive a mass formula for photon [10]. The main implication of the Evans–Vigier field is that the individual photon has three degrees of polarization, the longitudinal one accompanied by the "ghost field" $\mathbf{B}^{(3)}$, which has no energy or linear momentum, and is generated from the angular momentum of the photon.

Within Maxwellian theory, the "ghost field" $\mathbf{B}^{(3)}$ is related to the usual wave fields $\mathbf{B}^{(1)}, \mathbf{B}^{(2)}$ by a cyclical Lie algebra:

$$\mathbf{B}^{(1)} \times \mathbf{B}^{(2)} = iB^{(0)}\mathbf{B}^{(3)*} = iB^{(0)}\mathbf{B}^{(3)}$$

$$\mathbf{B}^{(2)} \times \mathbf{B}^{(3)} = iB^{(0)}\mathbf{B}^{(1)*} = iB^{(0)}\mathbf{B}^{(2)}$$

$$\mathbf{B}^{(3)} \times \mathbf{B}^{(1)} = iB^{(0)}\mathbf{B}^{(2)*} = iB^{(0)}\mathbf{B}^{(1)}$$

Here, $\mathbf{B}^{(1)}$ and $\mathbf{B}^{(2)}$ are complex conjugate wave fields (the usual magnetic components in circular polarization of the electromagnetic plane wave). In the standard Maxwellian theory of electrodynamics these are the only two polarizations considered: left and right circularly polarized plane waves. But these cyclical relations indicate that if the field $\mathbf{B}^{(3)}$ were zero, $\mathbf{B}^{(1)}$ and $\mathbf{B}^{(2)}$ would vanish. The Evans–Vigier field demands that even in classical electrodynamics, there are three polarizations and $\mathbf{B}^{(3)}$ is a real physical field that should be observed in magneto-optical phenomena. The $\mathbf{B}^{(3)}$ is related to the longitudinal degrees of freedom.

Evans showed that the existence of $\mathbf{B}^{(3)}$ is consistent with finite photon mass, m_γ in Proca equation. In nonrelativistic approximation Proca equation can be written as $\nabla^2 \mathbf{A} = \xi^2 \mathbf{A}$, where ξ is related to photon mass m_γ. Taking $\mathbf{B} = \nabla \times \mathbf{A}$, we see that $\nabla^2 \mathbf{B} = \xi^2 \mathbf{B}$ is the same as the Proca equation, because $\nabla^2(\nabla \times \mathbf{A}) = \xi^2 \nabla \times \mathbf{A}$, that is, $\nabla \times \delta^2 \mathbf{A} = \nabla \times \xi^2 \mathbf{A}$; the solution is found to be

$$\mathbf{B}^{(3)} = B^{(0)} e^{-\xi^2} \mathbf{k}$$

Evans and Vigier claimed that since $\xi \sim 10^{-26} \, \mathrm{m}^{-1}$ for a photon mass of 10^{-68} kg, this is identical for all practical purposes with the result

$$\mathbf{B}^{(3)} = B_0 \mathbf{k} = \frac{\mathbf{E}^{(1)} \times \mathbf{E}^{(2)}}{(iE_0 c)}$$

from the classical Maxwell's equations. Many interesting results are found in this framework which are described in detail in two books [54,59]. The $\mathbf{B}^{(3)}$ field theory also predicts the existence of space charge in vacuo as considered by Lehnert and Roy independently. It requires a systematic effort to find a close connection between Evans–Vigier theory and the extended electromagnetic theory as proposed and developed by Lehnert.

Roscoe [57] studied Poincaré-invariant magnetic vector potential formalism and showed that there are two distinct kinds of magnetic vector transverse wave that give rise to the general free solutions. The first kind is a propagating transverse wave; the second kind is a stationary longitudinal wave. In this framework, the longitudinal waves does not give rise to macroscopic electromagnetic effects at all (i.e., $\mathbf{E} = \mathbf{B} = 0$).

This analysis leads to the deduction of the existence of a massive vector boson field that is formally expressed as the dual of the Maxwell field and has been shown to be connected to the Maxwell field. Here, the massive vector boson is to be interpreted as nonzero mass of photon.

Following Roscoe, one can get a system of equations as

$$(n \cdot n)\mathbf{A}_0 = (\mathbf{n} \cdot \mathbf{A}_0)\mathbf{n}$$

with

$$n = (n_1, n_2, n_3, n_4)$$
$$\mathbf{A} \equiv (A_1, A_2, A_3)$$

where $\mathbf{A}_0$ is a constant 3-vector. Two possible cases may be found as follows:

1. *The transverse wave*: $\mathbf{n} \cdot \mathbf{A}_0 = 0$. In this case, one has a nontrivial solution if $n \cdot n = 0$. Consequently, the solution is given by

$$\mathbf{A}_1 = \mathbf{A}_0 \exp(i\, n \cdot X); \qquad n \cdot n = 0, \qquad \mathbf{n} \cdot \mathbf{A}_0 = 0$$

   where $X = (x_1, x_2, x_3, x_4)$. This corresponds to a transverse wave propagating with the speed c. It coincides with solutions that arise from the conventional formalism when the Coulomb gauge is chosen.

2. *The longitudinal wave*: $\mathbf{n} \cdot \mathbf{A}_0 \neq 0$. In this case, $n \cdot n = \mathbf{n} \cdot \mathbf{n}$, which can happen if $n_4 = 0$. We now get the equation as

$$(\mathbf{n} \cdot \mathbf{n})\mathbf{A}_0 = (\mathbf{n} \cdot \mathbf{A})\mathbf{n}$$

   This has a solution of the form $\mathbf{A}_0 = \alpha \mathbf{n}$ for arbitrary α. The solution is

$$\mathbf{A}_L = \alpha \mathbf{n} e\,(i\, n \cdot \mathbf{X}) \quad \text{with} \quad \mathbf{X} \equiv (x_1, x_2, x_3)$$

This corresponds to a longitudinal stationary wave. It gives rise to $\mathbf{E} = \mathbf{B} = 0$. It is clear from Roscoe's analysis that the magnetic vector potential supports two kinds of waves in free space: a propagating transverse wave that corresponds exactly to the Coulomb gauge solutions of the conventional formalism and a stationary longitudinal wave that has no counterpart in the conventional formalism. A general solution can be constructed as

$$\mathbf{A}_{\text{wave}} = \mathbf{A}_T(\mathbf{X}, ct) + \mathbf{A}_L(\mathbf{X})$$

where $\mathbf{A}_T$ and $\mathbf{A}_L$ are the transverse wave propagating with speed c and stationary longitudinal wave, respectively. It should be noted that in this framework, the component $\mathbf{A}_L$ gives rise to a zero electromagnetic field ($\mathbf{E} = \mathbf{B} = 0$) and so no electromagnetic effect is propagated. Since the stationary wave cannot pass through an arbitrarily placed charged particle, we can get an effect only if the charged particle passes through it. A new magnetic effect has been found due to this longitudinal wave. There exists a magnetic field component in the direction of current flow when a line current flows. Taking the Fourier component of the line current $\mathbf{J}_r = \mathbf{J}_0 \exp(in \cdot X)$, we have the Fourier component of the magnetic vector potential $\mathbf{A}_\mu = (\alpha \mathbf{J}_0 + \beta \mathbf{n}) \exp(in \cdot X)$, where

$$\alpha = \frac{1}{n \cdot n}, \qquad \beta = \frac{\mathbf{n} \cdot \mathbf{J}_0}{n \cdot n(n \cdot n - \mathbf{n} \cdot \mathbf{n})}$$

This β term does not occur in the conventional formalism, so that $\mathbf{A}_\mu$ is conventionally aligned with the direction of current, $\mathbf{J}_0$. Similarly, in the case of a steady current, conservation of charge requires $\mathbf{n} \cdot \mathbf{J}_0 = 0$, so that $\beta = 0$ automatically, which again leads to the alignment of $\mathbf{A}_\mu$ with the current direction. Again as $\mathbf{B} = \nabla \times \mathbf{A}$, so $\mathbf{B}_n \perp (\alpha \mathbf{J} - 0 + \beta \mathbf{n})$.

In the conventional formalism, as the β term does not appear, $\mathbf{B}_n \perp \mathbf{J}_0$. Now, in Roscoe's framework, as $\beta \neq 0$, $\mathbf{B}_n$ is not perpendicular to the current flow, and therefore has a component in the direction of the current flow. It has been shown that the magnetization effects similar to the inverse Faraday effect (IFE) can be expected for appropriate polarization states of the transmitted radiation. Moreover, a massive vector boson can be constructed from the electromagnetic field so that it can be interpreted only as a nonzero mass photon. Here, the model suggested for photon can be interpreted as a bound system with discrete mass and frequency states. This may have important role in explaining redshift phenomena.

Múnera and Guzman [56] obtained new explicit noncyclic solutions for the three-dimensional time-dependent wave equation in spherical coordinates. Their solutions constitute a new solution for the classical Maxwell equations. It is shown that the class of Lorenz-invariant inductive phenomena may have longitudinal fields as solution. But here, these solutions correspond to massless particles. Hence, in this framework a photon with zero rest mass may be compatible with a longitudinal field in contrast to that Lehnert, Evans, and Roscoe frameworks. But the extra degrees of freedom associated with this kind of longitudinal solution without nonzero photon mass is not clear, at least at the present state of development of the theory. More efforts are needed to clarify this situation.

IV. STOCHASTIC BACKGROUND

Amelino-Camelia [60] suggested an interesting possibility of considering spacetime foam as a fundamental stochastic gravity wave background. Lots of new literature appear day by day emphasizing the foamy structure of spacetime on very small scale, such as around Planck scale. It is proposed that the quantum fluctuations affecting distances in the conventional picture of spacetime foam manifest themselves in the operation of modern gravity wave detectors in a way that mimics a stochastic gravity wave background. Amelino-Camelia observed that the power spectrum of the star in noise [61] that would be induced in gravity wave detectors is the most convenient way to characterize models of foam-induced distance fluctuations.

Besides such developments in the context of foamy spacetime and quantum gravity, there is another development regarding the structure of spacetime near Planck scale from a different perspective. Menger [62] introduced the concept of statistical geometry studying the new developmnets of quantum theory of measurements. His main idea was that as we approach more and more to smaller lengths, the error in measurement will be greater with that respect to the measurements in larger scales. So it would be more realistic if one introduced a distribution function of the intervals instead of a fixed interval. Genarally, if we take the average with respect to this distribution function, we get the usual distance defined in geometry as Euclidean or non-Euclidean geometrics.

In this geometry the points are considered not as primary entities but rather as lumps of primordial elements that are not further resolvable. Here, the concept of probability is introduced so that the same two objects are sometimes treated as identical and sometimes as distinguishable. In this way Menger solved the Poincaré dilemma of distinguishing between transitive mathematical and intransitive physical relations of equality. These lumps may be the seat of elementary particles or the size of the strings. In this geometry we have two basic notions: (1) the concept of hazy or fuzzy lumps and (2) the statistical notion.

Frechet [63] made an abstract formulation of the notion of distance in 1906. Hausdorff [64] proposed the term "metric space," where he introduced the function d that assigns a nonnegative real number $d(p, q)$ (the distance between p and q) to every pair (p, q) of elements (points) of a nonempty set S. A metric space is a pair (S, d) if the function d satisfies several conditions, such as triangle inequality. In 1942, Menger [65] proposed that if we replace $d(p, q)$ by a real function F_{pq} whose value is $F_{pq}(x)$ for any real number x, this can be interpreted as the probability that the distance between p and q is less than x. Since probabilities can be neither negative nor greater than 1, we have

$$0 \geq F_{pq} \leq 1 \tag{71}$$

for any real x. Menger defined statistical metric space as a set S with an associated set of probability distribution functions F_{pq} that satisfy the following conditions:

$$F_{pq}(0) = 0$$

If $p = q$, then

$$F_{pq}(x) = 1 \quad \text{for all} \quad x > 0$$

If $p \neq q$, then

$$F_{pq}(x) < 1, \quad \text{for some} \quad x > 0$$

and

$$F_{pq}(x + y) \geq T(F_{pr}(x), F_{qr}(y))$$

for all p,q,r in S and all real numbers x, y. Here T is a function from the closed unit square $[0, 1] \times [0, 1]$ into the closed unit interval $[0, 1]$.

One can think of this statistical or probabilistic metric space in terms of measure-theoretic model of probability theory. Here one begins with random variables on a given probability space. Then, of course, we get a different formulation of the probabilistic metric spaces. Spacek [66] was the first to look at the subject from this point of view. He proposed the term *random metric space* rather than *probabilistic metric space* and discussed the relationship between both terms. Stevens [67] in his doctoral dissertation, tried to modify Spacek's approach. The main idea behind Stevens' approach lies in the fact that one has a set S and a collection P of measuring rods and chooses a measuring rod d from P at random and uses it to measure the distance between two given points p and q of S. Using this idea, Stevens defined the distribution function F_{pq} and showed that the metrically generated space so obtained is a Menger space.

Menger started with a probability distribution function rather than with random variables. This is related to the fact that the outcome of any series of measurements of the values of a nondeterministic quantity is a distribution function and the probability space is in principle unobservable. This point of view indicates a nonclassical behvior.

Sherwood [68] approached the problem from a different point of view. Following the distribution-generated space as introduced by Schweizer and Sklar [69], Sherwood proposed the concept of E-space. In E-space, the points are functions from a probability space (Ω, A, P) into a metric space (M, d). For each pair (p, q) of functions in the space, the function $d(p, q)$ is defined as $(d(p,q))(\omega) = d(p(\omega), q(\omega))$ for all ω in ohms. This is assumed to be a random variable on (ω, A, P). The function F_{pq} is considered to be distribution function of this random variable, so that for any real x, $F_{pq}(x) = P(\omega, \Omega \,|\, (d(p, q)) (\omega) < x)$. In this way, F_{pq} can be regarded as the probability that the distance between p and q is less than x. Sherwood showed that every E-space is a Menger space.

Distribution-generated spaces play significant role in the small scale structure. The main idea is as follows. Let S be a set. With each point p of S, associate an n-dimensional distribution function G_F whose margins are in P. Then associate a $2n$-distribution function H_{pq} with each pair (p, q) of distinct points such that

$$H_{pq}(\vec{u}, (\infty, \ldots, \infty)) = G_p(\vec{u}) \tag{72}$$

$$H_{pq}((\infty, \ldots, \infty), \vec{v}) = G_q(\vec{v}) \tag{73}$$

for any $\vec{u} = (u_1, \ldots, u_n)$ and $\vec{v} = (v_1, \ldots, v_n)$ in R^n.
Let $Z(x)$ be a cylinder in R^n, where

$$Z(x) = \{\mathbf{u}, \mathbf{v} \quad \text{in} \quad R^{2n} \, | \, |\mathbf{u} - \mathbf{v}| < x\}$$

for any $x > 0$. Now define F_{pq} in P^+ via

$$F_{pq}(x) = \int_Z (x) \, dH_{pq} = P_{H_{pq}}(Z(x)) \tag{74}$$

Here, if the elements of S are thought of as "particles," then for any Borel set A in R^n, the integral $\int_A dG_p$ is naturally interpreted that the particle p is in the set A and $F_{pq}(x)$ as the probability that the distance between the particles q is less than x. Then we can construct probabilistic metric space. In this approach, the interesting concept is the concept of "clouds" or "cloud spaces (C-spaces)". A function g from R^n into R^+ is an n-dimensional density if the function G defined on R^n by

$$G(\vec{u}) = \int_{((-\infty, \ldots, -\infty), \vec{u}))} g(\vec{v}) \, d\vec{v} \tag{75}$$

is an n-dimensional distribution function. If G is an n-dimensional distribution function and if there is an n-dimensional density g that satisfies relation (75), then G is absolutely continuous and g is a density of G. If p is a point in a distributed-generated space over R^n such that G_p is absolutely continuous, then any density g_p of $G - p$ may be visualized as a "cloud" in R^n—a cloud whose density at any point of R^n measures the relative likelihood of finding the particle p in the vicinity of that point.

A C-space over R^n is homogeneous if either of the followings holds: (1) all points are singular or (2) all points are nonsingular and there is a spherically symmetric unimodal density g such that $g_{pq} = g$ for all pairs (p, q) of distinct points of S.

A C-space over R^n is semihomogeneous if there exists an n-dimensional density g such that it is spherically symmetric and unimodal with center and mode $\vec{0}$. If $\sigma_{pq} > 0$, then, for all $\vec{u}$ in R^n, we can write

$$g_{pq}(\mathbf{u}) = \frac{1}{\sigma_{pq}^n} g\left(\frac{\mathbf{u}}{\sigma_{pq}}\right) \tag{76}$$

In a semihomogeneous C-space

- If σ_{pq} is the same for every pair (p, q) of distinct points, then it is homogeneous space.
- If a semihomogeneous C-space contains at most one singular point, then σ is a metric.

There are some interesting properties of C-space that are very useful for Planckian regime. Before elaborating these properties, let us discuss the moments and metrics in C-space.

A. Moments and Metrics

Let (S, F) be a C-space. The function $d(\beta)$ can be defined on $S \times S$ for β in $(0, \infty)$ as

$$d_\beta(p, q) = \begin{cases} m(\beta) F_{pq} & \beta \ in(0, 1] \\ \left(m^{(\beta)} F_{pq}\right)^{1/\beta} & \beta \ in[1, \infty) \end{cases} \tag{77}$$

Here, we define the moment $m^{(\beta)}(F)$ of order β of F as

$$m^{(\beta)} F = \int_R^+ x^\beta \, dF(x) \tag{78}$$

Now let us state the following important theorem.

Theorem 1. Let $\beta > 1$, and suppose that $\int_0^\infty h(u) \, du < \infty$.

Proof:

1. For a fixed in $[0, \infty)$, M_n^β is a nondecreasing function of σ and

$$in f_\sigma M_n^\beta(\sigma, s) = M_n^\beta(0, s) - s^\beta \tag{79}$$

2. For a fixed σ in $[0, \infty)$, M_n^β is a nondecreasing convex function of s and

$$in f_s M_n^\beta(\sigma, s) = M_n^\beta(\sigma, 0) = \sigma^\beta \int_0^\infty u^\beta h(u) \, du \tag{80}$$

3. For any σ, s in $[0, \infty)$

$$s \leq \left(M_n^\beta(\sigma, s)\right)^{1/\beta} \leq s + \sigma\left(\int_0^\infty u^\beta h(u)\, du\right)^{1/\beta} \tag{81}$$

4. Hence for fixed σ

$$\lim_{s\to\infty} \frac{1}{s}\left(M_n^\beta(\sigma, s)\right)^{1/\beta} = 1 \tag{82}$$

5. if $\int_0^\infty u^2 h(u)\, du < \infty$, then

$$\left(M_n^1(\sigma, s)\right)^2 \leq M_n^2(\sigma, s) = s^2 + \sigma^2 \int_0^\infty u^2 h(u)\, du \tag{83}$$

6. whence for $s > 0$

$$M_n^1(\sigma, s) \leq s + \frac{\sigma^2}{2s} \int_0^\infty u^2 h(u)\, du \tag{84}$$

B. Discussion

It is evident from the equations presented above that the *Frechet–Minkowski metrics* $d_{(\beta)}$ associated with semihomogeneous C-spaces over R^n have a remarkable structure. At small distances this metric is non-Euclidean in nature and the distance between any two distinct points p, q is not smaller than a fixed positive multiple of σ_{pq}. However, it becomes Euclidean in the asymptotic region. It appears from the preceding picture that if we consider C-space as a space of clouds (that may move around), we observe on one hand that as the Euclidean distance $d(c_p, c_q)$ between two centers of the clouds of p and q approaches 0, any *Frechet–Minkowski* distance between p and q remains greater than a positive number, and on the other hand, that when it is large, $d(c_p, c_q)$ is a good estimate of the distance between the clouds themselves. Again, the ratio of the standard deviation of F_{pq} to the mean of F_{pq}, a quantity that measures the relative uncertainty in a probabilistic determination of the distance between the clouds (or particles) p and q decreases to 0. This implies that the "haziness" of the distance between p and q, is predominant when their clouds are close together and becomes virtually insignificant when their clouds are sufficiently far apart. In this sense, the probabilistic metric of a semihomogeneous C-space, just each of the associated *Frechet–Minkowski* metrics, is asymptotically Euclidean.

It should be noted that spacetime as a set of hazy lumps or clouds is considered as a network of relations of factual items or things. In this sense we emphasize the relational view of geometry.

It is clear from the preceding picture of statistical geometry that the geometry around the Planck scale is closely related to the geometry of hazy lumps. The "foamy" spacetime of Wheeler et al. should also be related to this kind of geometry.

In fact, there exists a certain suspicion among the scientific community that nature may be discrete or rather that it "behaves discretely" on the Planck scale. But even if one is willing to agree with this "working philosophy," it is far from evident what this vague metaphor actually means or how it should be implemented into a concrete and systematic inquiry concerning physics and mathematics in the Planck regime.

There are basically two overall attitudes as to "discreteness on the Planck scale;" one starts (to a greater or lesser degree) from continuum concepts (or more specifically, concepts that are more or less openly inspired by them) and then tries to detect or create modes of "discrete behavior" on very fine scales, typically by imposing quantum theory in full or in part on the model system or framework under discussion. We call this the "top–down approach."

There are prominent and very promising candidates in this class, including string theory and loop quantum gravity. The *spin network* is a more recent version (or rather, aspect) of the latter approach. As these approaches are now widely known, we refrain from citing from the vast corresponding literature. We recommend instead more recent reviews as on the latter approach, containing some cursory remarks about the former together with a host of references, including those cited by Smolin [70] and Rovelli [71] and, a beautiful introduction to the conceptual problems of quantum gravity in general by Isham [72].

In the following we undertake to describe how macroscopic spacetime (or rather, its underlying *mesoscopic* or *microscopic* substratum) is supposed to emerge as a *superstructure* of a *web of lumps* in a dynamical cellular network. We call this the "bottom–up approach." In doing this, two strands of research are joined, which, originally, started from different directions. The one is the *cellular network* and *random graph* approach, developed by Requardt [73], the other, the *statistical geometry of lumps*, a notion originally coined by Menger and co-workers and further developed by various groups. (see e.g. *Menger, Roy, Schweizer*). It is worth mentioning that Einstein himself was not against such a grainy substratum underlying our space-time continuum (see the essay by Stachel [74]).

The point where these different strands meet is as follows. In one dynamical network approach, of Manfred [73] macroscopic spacetime is considered to be a coarse-grained emergent phenomenon (called an *order parameter manifold*). It is assumed to be the result of some kind of *geometric phase transition* (very much in the spirit of the physics of self-organization). This framework was developed in quite some detail [75]. We argued in Ref. 76 that what we consider to be the elementary building blocks of continuous spacetime—the so-called *physical points*—are, on a finer scale, actually densely entangled subclusters of *nodes* and *bonds* of the underlying network or graph. We also called them *cliques* (which denote in graph theory the maximal complete subgraphs or maximal subsimplices of a given graph).

We further argued there that the substructure of our spacetime manifold consists in fact of two stories: (1) the primordial network, dubbed by us QX; and (2) overlying it, the web of lumps or cliques, denoted by ST, which can also be viewed as a coarser mesoscopic network with the cliques or lumps as *supernodes* and with *superbonds* that connect lumps having a nonvoid overlap. This corrrespondence suggests the possibility of relating the lumps or cliques with the lumps occurring in the approach of Menger et al.

One should, however, note that the two philosophies are not entirely the same. In Ref. 76, and in related works we argued that the lumps emerge from a more primordial discrete dynamical substratum, where the lumps have a specific internal structure. In the approach of Menger et al. (at least as far as we can see) they figure as the not-further-resolvable building blocks of spacetime if one approaches the so-called Planck regime discussed above, namely, from the continuum side. In other words, the former approach is more bottom–up-oriented while the latter one is more top–down. Such not-further-resovable scales of spacetime (where the ordinary continuum picture ends) was, of course, also speculated by quite a few other people, most notably Wheeler.

Our personal working philosophy is that spacetime at the very bottom (i.e., near or below the notorious Planck scale) resembles or can be modeled as an evolving information processing cellular network, consisting of elementary modules (with, typically, simple internal discrete state spaces) interacting with each other via dynamical bonds that transfer the elementary pieces of information among the nodes. Thus, the approach shares the combinatorial point of view in fundamental spacetime physics that has been initiated by Penrose [77]. It is a crucial and perhaps characteristic extra ingredient of our framework that the bonds (i.e., the elementary interactions) are not simply dynamical degrees of freedom (as with the nodes, their internal state spaces are assumed to be simple) but can a fortiori, depending on the state of the local network environment, be switched on or off, that can be rendered temporarily active or inactive! This special ingredient of dynamics hopefully allows the network to perform *geometric phase transitions* into a new ordered phase displaying a certain *two-story structure*, to be explained below. This conjectured emergent geometric order can be viewed as kind of a discrete *protospacetime* or *pregeometry* carrying metrical, causal, and dimensional structures.

It should be noted that we have to envisage two types of distances at two levels (of the two-story structure): one on the level of the primordial network and the other on the level of the web of hazy lumps. Normally, if we consider the Euclidean distance measure, this distance tends to zero as the centers of the lumps approach each other. On the other hand, if we take, for instance, the

Frechet–Minkowski distance measure at the small scale, this distance appro-aches a constant value as the two lumps approach each other. This infinitesimal regime signals the devitation from the Euclidean geometry on small scales. In the asymptotic distance regime the Frechet–Minkowski distance becomes Euclidean distance. This phenomenon may play a significant role in the Planck regime.

This may perhaps shed some light on the better understanding of quantum mechanics as well as on the concurrent top–down approaches mentioned above. We have expounded the relational view of geometry; that is, we regard *geometry* not as an a priori given structure or *receptacle* but rather consider it as an emergent network of certain elementary constituents. We note in passing that this point of view has a venerable history of its own, beginning with Leibniz.

V. CONCLUSIONS

The astrophysical evidences posses new challenges to test one of the basic tenets of modern physics: constancy of speed of light. To estimate energy-dependent speed of light unambigously, we need to reanalyze the data with more rigirous statistical techniques, and more refined experiments are also needed to make any conclusive progress.

The energy-dependent speed of light is associated with the effect of the medium on the propagation of photon. The fluctuating refractive index of the medium induces this kind of the energy dependence. This kind of the medium has been considered, such as quantum gravity and the Maxwell vacuum with nonzero conductivity. So, it to make distinction between the contributions from standard model predictions as well as from other theories is needed.

The fluctuation of the spacetime metric plays a significant role on the microscopic scale, especially in Planck-scale physics. Here also, various models have been proposed regarding the origin and effect of this kind of fluctuation. More recent attempts to use gravity-wave interferometers deserve serious attention, as this will not only verify the theoretical models of spacetime fluctua-tions but also help us bulid up the geometric structure at Planck regime. This kind of geometric structure and the pregeometric notion as introduced in the context of fuzzy lumps and underlying random graphs may play a significant role in twenty-first-century physics.

References

1. G. Amelino-Camelia, J. Ellis, N. E. Mavromatos, D. V. Nanopoulos, and S. Sarkar, *Nature* **393**, 323 (1998); M. Rees, astro-ph/9701162; J. Ellis, N. E. Mavromatos, and D. V. Nanopoulos, gr-qc/9904068.
2. J. A. Wheeler, in B. S. de Witt and C. M. de Witt (Eds.), *Relativity, Groups and Topology*, (Gordon & Breach, New York, 1963); S. Hawking, D. N. Page, and C. N. Pope, *Nucl. Phys.* **B 170**, 283 (1980); L. Garay, *Phys. Rev.* **D58**, 124015 (1988).

3. S. Roy, *Statistical Geometry and Application to Microphysics and Cosmology*, Kluwer, Dordrecht, 1998.

4. J. Ellis, K. Farakos, N. E. Mavromatos, V. A. Mitsou, and D. V. Nanopoulos, Astro-ph/9907340.

5. G. Amelino-Camelia, J. Ellis, N. E. Mavromatos, and D. V. Nanopoulos, *Int. J. Mod. Phys. A* **12**, 607 (1997).

6. H. Yu and L. H. Ford, gr-qc/0004063.

7. A. Cambell-Smith, J. Ellis, N. E. Mavromatos, and D. V. Nanopoulos, hep-th/9907141.

8. R. Gambini and J. Pullin, *Phys. Rev.* **D59**, 124021 (1999); L. H. Ford, *Phys. Rev.* **D51**, 1692 (1995).

9. S. Covino et al., astro-ph/9906319.

10. B. Lehnert and S. Roy, in *Extended Electromagnetic Theory: Space-Charge in Vacuo and the Rest Mass of the Photon*, World Scientific Series in Contemporary Chemical Physics, Vol. 16, World Scientific, Singapore, 1998.

11. Y. Jack Ng and H. van Dam, gr-qc/9911054.

12. A. Abramovici et al., *Phys. Lett. A* **218**, 157 (1996).

13. K. Menger, *Proc. Natl. Acad. Sci.* (USA) **37**, 227 (1951).

14. S. Roy, *Acta Applicandae Math.* **26**, 209 (1992).

15. L. Diosy, *Phys. Lett A* **120**, 377 (1997).

16. B. Schaefer, astro-ph/9810479; S. D. Biller et al., gr-qc/9810044; P. Kaaret, astro-ph/9903464.

17. W. S. Paciesas et al., astro-ph/9903205.

18. OSSE Collaboration, http://www. astro. nwu. edu/astro/osse/bursts/.

19. J. R. Norris et al., *Astrophys. J.* **459**, 393 (1996).

20. A. V. Olinto, astro-ph/0003013; G. Sigl, astro-ph/0008364.

21. J. Ellis, N. E. Mavromatos, and D. V. Nanopoulos, *Gen. Rel. Grav.* **32**, 127 (2000); *Phys. Rev.* **D61**, 027503 (2000).

22. T. Kiffune, astro-ph/9904164; R. Aloisio, P. Blasi, P. L. Ghai, and A. F. Grillo, *Phys. Rev.* **D62**, 053010 (2000); R. J. Protheroe and H. Meyer, astro-ph/0005349.

23. D. Fergion and B. Mele, astro-ph/9906451.

24. R. J. Protheroe and H. Meyer, astro-ph/0005349.

25. N. E. Mavromatos, gr-qc/0009045.

26. K. Greisen, *Phys. Rev. Lett.* **66**, 748 (1966); G. T. Zatsepin and V. A. Kuzmin, *JETP* **4**, 78 (1966).

27. J. Ellis, N. E. Mavromatos, and D. V. Nanopoulos, *Int. J. Mod. Phys.* **A13**, 1059 (1998).

28. L. D. Landau and E. M. Lifshitz, *Classical Theory of Fields*, Pergamon, New York, 1975, Vol. 2.

29. T. Piran, *Phys. Rep.* **314**, 575 (1999).

30. J. Ellis, N. E. Mavromatos, and D. V. Nanopoulos, gr-qc/9904068.

31. Th. Kaluza, *Sitzungsber. Preuss. Akad. Wiss., Phys. Math. Kl.* 996 (1921); O. Klein, *Z. Phys.* **37**, 895 (1926).

32. H. Yu and L. H. Ford, *Phys. Rev.* **D60**, 084023 (1999); gr-qc/9904082.

33. T. Applequist and A. Chodos, *Phys. Rev. Lett.* **50**, 141(1983); *Phys. Rev.* **D28**, 772 (1983).

34. N. Arkani-Hamed, S. Dimopoulos, and G. Dvali, *Phys. Lett.* **B429**, 263 (1998); *Phys. Rev.* **D59**, 086004 (1999).

35. A. Campbell-Smith, J. Ellis, N. E. Mavromatos, and D. V. Nanopoulos, *Phys. Lett.* **B46**, 11 (1999); hep-th/9907141.

36. S. Majid, *J. Math. Phys.* **34**, 2045 (1993).

37. J. Schwenk and J. Wess, *Phys. Lett.* **B291**, 273 (1992); M. Fichtmüller, A. Lorek , and J. Wess, hep-th/9511106.

38. P. Podlés and S. L. Woronowicz, *Commun. Math. Phys.* **178**, 61 (1996).

39. P. Podlés, *Commun. Math. Phys.* **181**, 569 (1996).

40. S. Esposito, *Classical $v_g \neq c$ Solutions of Maxwell's Equations and the Tunneling Photon Effect*, preprint, 1997.

41. L. Fuli, *Nuovo Cimento Lett.* **31**, 289 (1981).

42. R. A. Monti, *Phys. Essays* **9**, 234 (1996).

43. D. F. Bartlett and T. R. Corle, *Phys. Rev. Lett.* **55**, 99 (1985).

44. G. Nimtz et al., *J. Phys.* (France) **4**, 565 (1994).

45. L. Brillouin, *Wave Propagation and Group Velocity*, Academic, New York, 1960.

46. J. P. Vigier, *Present Experimental status of the Einstein-de Broglie Theory of Light, Proc. ISQM Workshop on Quantum Mechanics* (Tokyo, 1992).

47. J. C. Maxwell, *A Treatise on Electricity and Magnetism*, Vol. 2, Dover, New York, 1954.

48. G. Feinberg, *Science* **166**, 879 (1969).

49. A. S. Goldhaber and M. M. Nieto, *Phys. Rev. Lett.* **21**, 567 (1968).

50. E. S. Dayhoff, "On the possibility of electromagnetic dispersion in vacuum," unpublished, 1957.

51. E. R. Williams, *Phys. Rev. Lett.* **26**, 721,1651 (1971).

52. E. Riis et al., *Phys. Rev. Lett.* **60**, 81 (1988).

53. Particle Data Group, review of particle properties, *Phys. Rev.* **D50** (Part 1) (1994).

54. M. Evans and J. P. Vigier, *The Enigmatic Photon*, Vols. 1,2, Kluwer, Dordrecht, 1994.

55. D. V. Ahluwalia and D. J. Earnst, *Mod. Phys. Lett. A* **7**, 967 (1992).

56. H. A. Múnera and O. Guzman, *Apeiron* **3–4**, 63 (1996).

57. D. F. Roscoe, *Maxwell's Equations as a Consequence of the Orthogonality between Irreducible Two Index Representation of the Poincaré*, preprint, 1997.

58. H. Múnera et al., *Rev. Colomb. Phys.* **27**, 215 (1995).

59. M. W. Evans, J. P. Vigier, Sisir Roy, and S. Jeffers, *The Enigmatic Photon*, Vols. 3,4, Kluwer, Dordrecht, 1996.

60. G. Amelino-Camelia, gr-qc/0001100.

61. P. R. Saulson, *Fundamentals of Interferometric Gravitational Wave Detectors*, World Scientific, Singapore, 1994.

62. K. Menger, in P. A. Schilpp (Ed.), *Albert Einstein Philosopher Scientist*, Cambridge Univ. Press, London, 1970.

63. M. Frechet, *Rend. Circ. Mat. Palermo* **22**, 1 (1906).

64. F. Hausdorff, *Grundzuge der Mengenlehre*, Veit und Comp, Leipzig, 1914.

65. K. Menger, *Proc. Natl. Acad. Sci.* (USA) **28**, 535 (1942).

66. S. Spacek, *Czech. Math. J.* **6**, 72 (1956).

67. R. R. Stevens, *Fund. Math.* **61**, 259 (1968).

68. H. Sherwood, *J. Lond. Math. Soc.* **44**, 441 (1969).

69. B. Schweizer and A. Sklar, *Pacific J. Math.* **10**, 313 (1960).

70. L. Smolin, "The future of spin networks," gr-qc/9702030.

71. C. Rovelli, "Loop quantum gravity," gr-qc/9710008.

72. C. J. Isham, "Structural issues in quantum gravity," gr-qc/9510063 (lecture given at the GR14 Conference, Florence, 1995).

73. M. Requardt, "Spectral analysis and operator theory on (infinite) graphs. . .," JPA (in press); math-ph/0001026; M. Requardt: "(Quantum) space-time as a statistical geometry of lumps in random networks," *Class. Quant. Grav.* **17**, 2029 (2000); gr-qc/9912059.

74. J. Stachel, "Einstein and quantum mechanics," in A. Asthekar and J. Stachel (Eds.), *Conceptual Problems of Quantum Gravity*, Einstein Studies, Vol. 2, Birkhäuser, Boston, 1999.

75. T. Nowotny and M. Requardt, "Pregeometric concepts on graphs and cellular networks," invited paper, *J. Chaos, Solitons and Fractals* **10**, 469 (1999); hep-th/9801199.

76. M. Requardt and Sisir Roy, (Quantum)Space-Time as a Statistical Geometry of Fuzzy Lumps and the Connection with Random Metric Spaces – to be published in Classical & Quantum Gravity (2001).

77. R. Penrose, in T. Bastin (Ed.), *Quantum Theory and Beyond*, Cambridge Univ. Press, Cambridge, UK, 1971.

ENERGY FOR THE FUTURE: HIGH-DENSITY CHARGE CLUSTERS

HAROLD L. FOX

Editor, Journal of New Energy and President, Emerging Energy Marketing Firm, Inc., Salt Lake City, Utah

CONTENTS

Modern Nonlinear Optics, Part 3, Second Edition, Advances in Chemical Physics, Volume 119,
Edited by Myron W. Evans. Series Editors I. Prigogine and Stuart A. Rice.
ISBN 0-471-38932-3   © 2001 John Wiley & Sons, Inc.

623

I. BACKGROUND

A. Valve Metals

During the early days of the development of digital computers, the need for display devices was evident. Prior to the invention and commercialization of the light-emitting diodes, there was considerable research and development activities using valve metals. The term *valve metal* is believed to have been used to designate metals capable of being used for filaments or to heat cathodes in early radio tubes (known in England as *valves*). A valve metal is defined, herein, as a metal whose oxide is luminescent and in which the luminescence increases with increased applied voltage. Aluminum, zirconium, and several other metals were investigated [1]. A literature search revealed about 60 papers on this area of research. One of the most interesting aspects of this R&D effort was the discovery that when the voltage was high enough (in the range of two hundred volts) the metal–metal oxide layer began to emit sparks. As the goal of this activity was increased illumination, no researchers expressed interest in the sparking phenomenon. As sometimes happens in scientific investigation, some of the most fruitful discoveries are shunned because the observed phenomenon is not expected by the researcher or research team. One experimenter, however, did make some chemical analysis of the solution after the observed sparking and found an unexpected element. The element found was not expected because this element was not a part of the initial experimental solution or contained in the electrodes. "Obviously contamination," was the explanation. Thus it was that one of the most important discoveries (charge cluster transmutation) in the history of energy science was left to others to discover many years later.

B. Shoulders' Discovery

Kenneth R. Shoulders, one of the most astute experimental scientists known to this author, was the person who has discovered and reported extensively on high-density charge clusters [2]. His book was written in 1987, prior to the time that the first patent applications were being filed with the U.S. Office of Patents and Trademarks. An intelligent patent attorney suggested that this technology might be classified by the government examiners in the patent office even though there was no government funding of this type of research. Therefore, as soon as the patent was filed, copies of the book [2] were mailed to a reasonable number of fellow scientists in many foreign countries. The information sent, if it had been published prior to the filing of the patent application, could have prevented the inventions from being patented.

The government agents who examine new patent applications did choose to classify this important new discovery and so notified the patent attorney. When informed that this was privately funded, unclassified research with an extensive report that had been sent to many countries, the attorney was asked for a complete list of all addressees. Unfortunately, the mailing list was unobtainable due to a problem in the computer. Three days and an estimated $10,000 later (in legal fees), the secrecy classification was canceled. There is a message here for all researchers who are developing important new technology with private funds. The United States is one of the few countries that has the legal right to essentially confiscate new inventions that are produced with private (as contrasted with government) funds.

The first patent on high-density charge cluster applications was dated May 21, 1991 [3]. This was the first high-density charge cluster (also called EVs) patent to issue and states: "An EV passing along a traveling wave device, for example, may be both absorbing and emitting electrons. In this way, the EV may be considered as being continually formed as it propagates. In any event, energy is provided to the traveling wave output conductor, and the ultimate source of this energy appears to be the zero-point radiation of the vacuum continuum." It is the judgment of this author that this is the first patented device that claims to tap space energy (zero-point energy, vacuum energy, etc.) to obtain more energy output than energy input into the device. The patent [3] provides data showing about 30 times as much output electrical energy as compared to input electrical energy. Shoulders has stated that as much as 100 times the input energy has been obtained in laboratory experiments. The same type of technology can also produce excess thermal energy. In fact, one cannot get the excess electrical energy without producing addition thermal energy.

C. Gleeson's Discovery

Stan Gleeson was a high school graduate and, until his death, worked in Cincinnati, Ohio. Working on a crude wooden workbench in a corner of a welding shop, Stan Gleeson discovered and developed a different method of creating and applying high-density charge clusters. (It has not been scientifically proved that these charge clusters produced in a water-based electrolyte are the same as charge cluster produced in low-pressure gases under near vacuum conditions. It is the author's judgment that the sparking of the valve metals produces short-lived charge clusters.) Gleeson found that by proper use of this underwater sparking of the electrode that some excess thermal energy could be created. However, that effect did not appear to be as promising as using this technology to transmute radioactive elements.

Supported by funds from the Holloman Brothers (a private company that specializes in piping installations in new and rebuilt factories), Gleeson and friends demonstrated that naturally radioactive thorium could be stabilized in a special type of electrochemical cell (see Fig. 1). Using moderate pressure and voltages in the range of 200–300 V, it was demonstrated that more than half of the radioactive thorium could be transmuted into other (usually more stable) elements. The transmutation measurements were made by hiring university or commercial equipment capable of measuring and reporting on small amounts of elements in the before/after processing of the water-based thorium electrolyte. Over 100 such experiments were conducted and proper samples submitted for elemental analysis. Gleeson and associates reported on their experimental results at the second conference on Low-Energy Nuclear Reactions held in Austin, Texas in 1996 [4]. This discovery has also been independently replicated by Liversage [5].

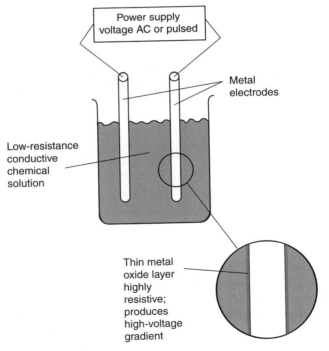

Figure 1. High-voltage gradient across metal oxide layer.

Several months of independent laboratory studies were made at the laboratory of the Fusion Information Center, Inc. at the University of Utah Research Park under the direction of chief scientist, Dr. Shang-Xian Jin. (Fusion Information Center has recently merged with Trenergy, Inc. and moved to a laboratory building in Salt Lake County, Utah) During these investigations, as much as 90% of the radioactive thorium was removed from the water-based thorium electrolyte. This work was done with permission and encouragement of Stan Gleeson and the Holloman brothers. A report on this work was presented at a meeting of the American Nuclear Society [6].

For those who wonder if some science stems from some type of unmeasured source of inspiration, this important discovery by a scientifically uneducated high school graduate is remarkable. This author prefers to report about scientific facts. One of the best definitions of a scientific facts is *the close agreement of a series of observations of the same phenomena.* Some scientists properly insist that there must also be independent observations of the same phenomena. With these definitions, Gleeson's work is a scientific fact of unusual importance. Independent results were obtained at Trenergy's laboratory and also at one of the universities in Cincinnati, Ohio. There are other kinds of facts that are not scientific. One of these facts may be the claim by Gleeson that he received spiritual instructions for his series of remarkable discoveries. Regardless of the source of inspiration, Gleeson's discoveries are expected to be of considerable commercial importance.

D. Ilyanok's and Mesyats' Discoveries

While visiting in the Republic of Belarus, Dr. Alexander M. Ilyanok was introduced to this author. Ilyanok displayed a mockup of a new type of display screen. With the help of a translator it became apparent that Ilyanok was proposing one of the embodiments of the charge cluster applications that was specifically included in Shoulders' first patent [3]. Ilyanok was asked if he had ever heard of the work of Kenneth Shoulders and replied in the negative. In subsequent discussions, it was determined that Ilyanok had independently discovered that charge clusters are quantized in size [7]. This quantization is also reported by Shoulders [2]. The individual clusters are normally 1 μm diameter but range in size from 0.5 to 3 /μm in diameter. However, many (about 20–80) of these charge clusters arrange into a necklace shape that is about 20 μm in diameter (see Fig. 2). At higher energy levels, the clusters arrange into sizes that are about 50 μm in diameter. Some of Shoulders' work did produce 50-μm clusters [2]. However, the abrupt demise of such a large charge cluster can produce an electromagnetic pulse (EMP) that can damage sensitive equipment not even connected to the experiment. Shoulders has chosen to work more with 20-μm charge clusters. Ilyanok wanted to determine the next quantized size larger and, therefore, has done considerable experimental work with 50-μm

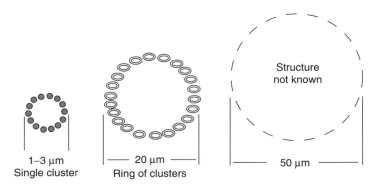

Figure 2.   Charge cluster quantization.

charge clusters. The next quantized level has not, as yet, been found. However, some preliminary analytical work suggests that the charge clusters have stability criteria that should allow for electron clusters up to the size of ball lightning [8].

More recently, Dr. G. A. Mesyats, a Russian scientist working with high-energy vacuum discharges, discovered high-density charge clusters and named them *ectons* [9]. Mesyats presented his paper, "Ecton processes at the cathode in a vacuum discharge," at the 1996 XVIIth International Symposium on Discharges and Electrical Insulation in Vacuum. Kenneth Shoulders was also in attendance at that meeting and informed Mesyats privately that the discovery and a variety of embodiments of high-density charge clusters was patented in the United States in 1991. However, the remarkable historic fact is that four independent inventors and scientists, working in different laboratories in four locations in three countries had, over a period of several years, discovered important aspects of the same new-energy technology.

II.   CHARGE CLUSTER THEORY

A.   General

It is the normal course of academic science to protect (usually by the filing of patent applications) and then publish information about new discoveries. In research activities funded by private funds and where there is no "publish or perish" doctrine, the academic urge to publish is often missing. Neither Kenneth Shoulders nor his former (nor current) private funder has had any strong desire to publish, except to discourage some unknown government agent from classifying the discoveries (and essentially destroying commercial value of the inventions). The result is that there have been few publications about this phenomena until this author began writing about this new technology.

B. Publications

The late Dr. Petr Beckman made an attempt in 1991 to describe a theory about the formation of electron clusters [10]. In the same year, Richard W. Ziolowski and Michael K. Tippett wrote an article in which some attempt was made to analytically describe charge clusters in a suspected spherical shape in the article, "Collective effect in an electron plasma system catalyzed by a localized electromagnetic wave" [11]. However, to date, the best analytical work on charge clusters has been accomplished primarily by Dr. Shang-Xian Jin [8]. Dr. Jin's first attempt was to describe the charge cluster in terms of a spherical collection of electrons. However, the observed stability was not predicted by a spherical model. With a suggestion from this author that a toroid model might demonstrate stability, Dr. Jin was able to show that a toroid structure with electrons circulating around the periphery could create a magnetic field that would support the circulation of the electrons. In addition, Jin worked out stability criteria for a range of sizes of charge clusters. As was suspected, it is analytically suggested that sizes up to ball lightning can be stable. However, there remains the problem of showing analytically the quantization of charge clusters. This observed phenomena is not predicted by the current analytical work. Therefore, this analytical work must be considered as preliminary and incomplete.

C. Charge Cluster Quantization

It is suggested that the work developed by Dr. Peter Graneau (and later in conjunction with his son, Dr. Neal Graneau, especially the work on exploding bridge wires) may have some importance in improving the current theoretical model for high-density charge clusters. In exploding bridge wires, Graneau has shown that a strong high-voltage, high-amperage pulse applied to a conducting wire could explode the wire into reasonably uniform fragments. Although Graneau's work has been published in scientific journals, this experimental challenge to our current electromagnetic equations has not been seriously pursued. Some of the various publications about this topic are listed in several references [12–15]. In this latest Graneau presentation, one of the accompanying photographs showed a bright spot, which this author believes to be strong evidence of a change cluster formation. It is suggested that the same (or similar) electromagnetic forces that can chop a wire into relatively uniform pieces is possibly the phenomena that creates high-density charge clusters that are relatively uniform in their smallest sizes (0.5–3 μm).

D. A Simple Model

The following is a relatively unsophisticated verbal description of the formation of charge clusters as an approach to a model. For a practical visual picture, think of an ocean wave moving toward shore on a sloping ocean floor rising to the beach. The unencumbered top of the wave can move faster than the bottom of the wave that is influenced by the rising ocean floor. The wave, if strong enough, can curl over as it moves toward the shore and provide a hollow center within which a skilled surfboarder can surf. Similarly, think of a pulse of electromagnetic energy on the surface of a dielectric that can provide a similar roll wave of electrons. The curling of the electrons, having suspected velocities of fractional light speeds, creates a strong magnetic field within the curl. Some, as yet unexplained phenomena, chops this roll of electrons into short segments and the resulting inner magnetic fields pull these short curls into toroids as the produced north and south magnetic fields quickly find each other. This admittedly inadequate model at least serves as a possible visual picture of how the charge cluster toroids might be formed. Reality may be considerably different. In the judgment of the author, the person who completes an adequate model of the high-density charge cluster and explains the various quantized sizes deserves to be nominated for a Nobel prize.

E.   The Charge Cluster Combined with Positive Ions

The high-density electron charge cluster (HDCC) is characterized by very high local electric and magnetic fields. The local strength of these electric and magnetic fields is believed to be stronger than the fields made by larger apparatus in the laboratory. As a result, if HDCC are produced in a low-pressure gas atmosphere (such as hydrogen) the gas in the vicinity of the HDCC is ionized to produce positive ions (such as protons).

In the case of hydrogen, the protons produced near a charge cluster are, of course, attracted to the highly negative charge cluster. Because of the huge size of the protons as compared to an electron, the end result is that only about one proton is attracted to the charge cluster for each hundred thousand to a million electrons. Therefore, the combined charge cluster is still highly negative and is attracted toward a positive anode in essentially the same way as a single electron would be attracted. Therefore, the combined charge cluster will be accelerated by the voltage on the anode. An important application of the combined charge clusters will be discussed in following Section III.

F.   Important Discoveries Expected

As with any new discovery, it is expected that there will be many doctoral theses and many new and important discoveries concerning high-density charge clusters. A thorough understanding of the nature and behavior of these HDCCs is far from complete. As more is learned of the nature of such clusters of electrons, there will be additional understanding of the nature of space energy and probably even of elements. It is an interesting historic fact that four different scientific groups from three different countries would independently discover this new electron phenomena in the same decade.

Kenneth Shoulders, the first to discover high-density charge clusters, later wrote a lengthy paper discussing some of his findings and some of his interpretations [16]. Shoulders is most likely the world's leading authority on HDCC. In his article he refers to "soliton behavior." [A strict definition of a soliton is a localized nonlinear wave that regains asymptotically (as $t \rightarrow$ +infinity) its original ($t \rightarrow$ negative infinity) shape and velocity after interacting with any other localized disturbance; the only long-term effect on the soliton from the interaction is a phase shift from *Encyclopedia of Physics, 2nd ed.*] In Shoulders' discussion of permittivity, it is appropriate to review the definition of permittivity as the ability of a dielectric to store electrical potential energy under the influence of an electric field. Permittivity is measured by the ratio of the capacitance of a capacitor with the material as dielectric to its capacitance with vacuum as dielectric. The following is the summary of Shoulders' [16] paper.

Highly organized, micrometer-sized clusters of electrons or EVs (Shoulders' designation for high-density charge clusters), having soliton behavior, with electron populations on the order of Avogadro's number[1] are represented as the necessary function for modifying the permittivity of space in a downward direction. The state of existence for this entity reduces its expressed charge by many orders of magnitude below that calculated for the same number and volume of uncontained electrons. The EV is shown to exist in at least two distinct modes of charge masking, with one of them, the *black EV*, being virtually undetectable using sensitive methods. A form of inertial propulsion will be discussed that arises from the inertial rectification affects available by modulating the state of the EV, thereby the permittivity of space and concomitant inertia or effective mass of material moving through space. It will be shown that the same type of permittivity change, through EV modulation, can achieve a unidirectional current flow and that this gives rise to methods for generating monopole effects and vector potentials useful for communication outside the usual current loop generating them. A form of pseudoparticle entanglement arising from the tight and pseudoquantized coupling between the EV structures will be considered. Complex organisms are discussed that are composed entirely of EV structures that are self-formed at electronic rates without using either mechanical or chemical methods. Some speculations will be made on the benefits of operating such complex entities in regions of greatly reduced permittivity. A condensed-matter dissolution technique will be shown that is capable of cold dissociation of refractory materials into a low-viscosity fluid. The root process for energy conversion methods resembling "cold fusion" are reviewed and shown to likely spring from the same EV technology capable of producing a modified space permittivity. Experimental methods for testing effects on time at greatly reduced levels of permittivity will be considered.

---

[1]The number of molecules in a gram-molecule or about 6.1×10^{23} (similar to the density of electrons in a charge cluster).

The reader will appreciate the perceived complexity of the high-density charge clusters as suggested in the above summary. There is certainly a considerable amount of new phenomena to be explored in the further development and control of high-density charge clusters. Just as certain is the claim that no present theory or current concepts from classical physics fully explains this new discovery and it manifold experimental observations. More important discoveries are expected.

G.   Difficulties in Producing Excess Electrical Energy

The first patent to issue on high-density charge clusters [3] includes data to show that about 30 times as much energy output was achieved, as compared with the energy input to create a charge cluster. More than ten years have passed since that important discovery. It is appropriate to ask why such an important invention has not reached the market and been the source of electrical power for many of our electrical appliances. The reasons for this technology being slow in reaching the market can be described.

The creation of a high-density charge cluster (HDCC) requires either a suitably large electric pulse that is of the order of a nanosecond in pulsewidth or, at least, a nanosecond rise time on a longer pulse. Nanosecond pulses for microelectronics are achievable with voltages of a few volts. Creating nanosecond pulses with the much higher voltages required to create HDCC is much more difficult. Therefore, one of the problems in commercializing this technology is to resolve the problem of creating nanosecond rise times or nanosecond pulses.

To obtain an electric output from HDCC, it is appropriate to have a coil or printed-circuit equivalent in which an electric pulse can be induced by a traveling charge cluster. The time of transit available for the charge clusters to induce electrical output is strongly related to the cathode-to-anode voltage and the distance over which the charge cluster must travel. The velocity of an electron (or a charge cluster) is proportional to the voltage and not to the distance traveled. However, the voltage induced in a coil is proportional to the length of coil and the time it takes for an electron cluster to traverse that space. The longer the travel time, the wider the induced pulse that will be produced.

The amount of excess electrical energy produced is the ratio of the output pulse to the input pulse. For a given configuration of a charge cluster emitter, coil, and anode (Fig. 3), the excess electrical energy has to be computed essentially as the area under the induced output pulse compared to the corresponding area under the delivered input pulse. Therefore, for a given configuration of an HDCC device, the only way to increase the output/input ratio is to shorten the input pulse (see Fig. 4).

Of course, a commercial device is not reasonable unless the rate at which the HDCC can be formed and controlled and the induced electrical charges (onto the parts of the HDCC device) are drained off to prepare for the next HDCC.

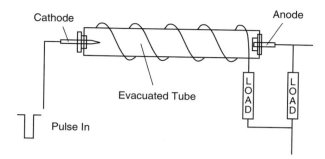

Figure 3. Energy conservation using high charge density.

Previously, the available repetition rate was relatively low. Currently, the repetition rate achieved is about 10,000 Hz [17]. Considering the size of the 20-μm charge cluster, its ability to induce current in an output circuit, and the repetition rate of the clusters, it is relatively easy to determine the expected excess power that can be produced (see Fig. 3). For example, a well-designed charge cluster device is expected to produce about one watt of excess output electrical power. As depicted in the patent [3], it is expected to require an array of HDCC devices to achieve meaningful levels of electrical power (see Fig. 5).

Another type of development problem is that in the design of an HDCC device, a small change in dimensions of any of the three major parts of the device (emitter, launcher, anode assembly) requires adjustments to other parts of the device (see Fig. 3). With the current very small amount of money being spent on these several problems, progress is relatively slow. At some stage it is expected that this technology will attract considerable attention from some of the world's large corporations. Then, there is also expected to be a much more rapid development of the HDCC technology with the higher levels of funding.

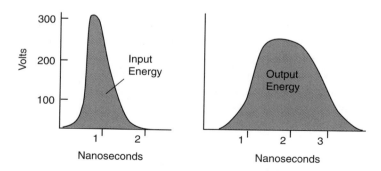

Figure 4. Energy increase using charge clusters.

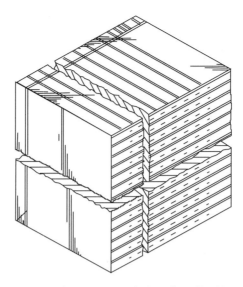

Figure 5. Arrary of charge cluster devices (from Shoulders' patent).

III.   THE PARTICLE ACCELERATOR

One of the remarkable uses for high-density charge clusters (HDCCs) is to ionize selected gases, combine the positive ions with the highly negative HDCC, and accelerate the combined cluster to fractional light velocities. The following description is adapted from a paper by Jin and Fox [18].

A.   Summary

Conventional ion accelerators use high voltages to accelerate ions (especially protons) to high velocities (impact energies) and use these ions to strike target materials. Some scientists and government officials have proposed increasing the funding for such particle accelerators to provide means by which high-level radioactive materials can be stabilized. The following paragraphs discuss the role of high-density charge clusters (HDCC) and their ability to carry positive ions as the basis for a greatly-improved particle accelerator for the on-site stabilization of high-level radioactive wastes, such as spent-fuel pellets from nuclear power plants.

B.   Introduction

Since the early 1930s, the development of conventional particle accelerators has ranged from Van de Graaff generators to the Synchrotron and other models. The problem with these accelerators, in terms of size and cost, is that the electric potential used is limited to about one million volts per meter. Beyond this limit the electric field would produce an electrical breakdown within the supporting structure of the accelerator (especially at the accelerating gap). In addition, such accelerators require large, multiple magnetic systems for ion-beam focusing.

In *collective-ion accelerators*, the above restrictions on accelerating potentials are removed. This change is accomplished by the use of the *collective field effects* of a large number of negative electrons that are used to accelerate a small number of positive ions to high energies. This approach makes it possible to produce compact, high-gradient, strong-field, high-energy, ion accelerators. As shown in the following chart, the Wakefield accelerator, the electron ring accelerator, and the intense relativistic electron beam (IREB) accelerators (such as the IFA, ARA, and CGA accelerators) are based on the use of a large number of - electrons to accelerate a small number of protons. Conceptually, the limitations of 1×10^6 V/m of conventional accelerators are not limiting. Potentials of 1×10^9 V/m can be used.

Particle Accelerators

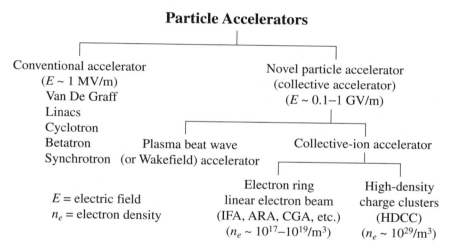

The class of collective-ion accelerators can be divided into two types: (1) the linear electron beam accelerators or electron ring, and the newest type, high-density charge cluster accelerators. It has been shown analytically that the HDCC accelerators can provide an electron density about 10^{10} times higher than the conventional collective-ion accelerator. The result is that very high collective acceleration effects are predicted to be achieved. These accelerators have broad potential applications, such as the following:

Nuclear physics research

Material science studies

Medical radiography

Ion-beam inertial fusion

Most important; stabilization of radioactive nuclear wastes (by transmutation)

C. The High-Density Charge Clusters

The high-density charge clusters (HDCC) consist of micrometer-sized clusters of electrons having soliton-like behavior together with an electron number density approximately equal to Avogadro's number (the number of atoms in a solid). These intense clusters of electrons are typically created by the application of a short pulse of a few hundred to tens of thousands volts to a cathode positioned adjacent to a dielectric (see Fig. 3). The charge cluster is produced at the pointed cathode and travels to the anode at velocities dependent on the anode voltage. The charge clusters were first discovered and developed into new inventions by Kenneth Shoulders, and several patents have been issued [3,19,20]. Later, Alexander M. Ilyanok [7] (a Belorussian scientist) independently discovered the same type of clusters. More recently, a similar discovery was made and reported by G. A. Mesyats [9] who called his discovery "Ectons."

A typical individual charge cluster ranges in size from 0.5 to ∼3.0 µm, and most of the HDCC is about 1 µm in diameter. Additional energy creates a larger number of these small clusters, which then arrange themselves into a ring of clusters with a ring diameter close to 20 µm. With even higher input energy another quantized level of 50 µm is formed [2]. The individual charge clusters are believed to be toroidal in form [8]. An analytical investigation to determine the self-equilibrium of the moving charge clusters and the conditions under which a dynamic equilibrium state could exist has been accomplished by Jin and Fox [8]. This analysis of a plasma fluid description provided, for the first time, some analytical criteria in which a charge cluster could exist. This analysis showed that the HDCC is a toroidal electron vortex and that such an electron cluster could exist at various combinations of electron densities, electron velocities, and cluster sizes. Not as yet shown by the analysis is the experimental evidence that the charge cluster rings are quantized.

As shown in Fig. 6, the charge cluster exhibits certain characteristics that are unusual in the study of electromagnetics. The typical size of a formed cluster ring is 20 µm. The electron density is computed to be about 10^{29} electrons per cubic meter. The estimated local electric field at the surface of a typical charge cluster is calculated to be 10^{13}–10^{15} V/m and the associated magnetic field strength, ∼10^6 T.

The charge cluster, when produced in a low-pressure gas environment, will immediately ionize the local gas molecules. These newly produced positive ions are then trapped in the highly negative potential well of the charge cluster where they cling to and travel with the charge cluster. This combined cluster is then

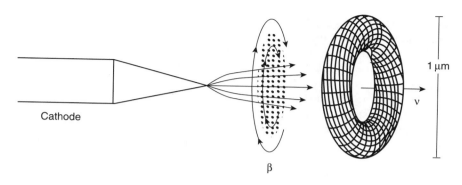

Typical size: 1– 20 μm
Electron density: ~10^{29}/m
Ion density: ~10^{23} to 10^{24}/m
Estimated electric field (surface) ~2×10^{13} to 10^{15} V/m
Estimated magnetic field (surface) ~10^6 Tesla
Electrostatic potential energy (for 1mm size) ~33 J

Figure 6.   Characteristics of high-density charge clusters.

accelerated by the voltage gradient to the anode. The local ion density can be as high as 10^{23}–10^{24} ions per cubic meter. It is important to note that this combined charge cluster can then be accelerated to high energies similar to the acceleration of an individual electron because the combined charge cluster has about one million electrons for each clinging positive ion.

The electron density in terms of the number of electrons per cubic centimeter in the linear electron-beam accelerators ranges from 10^{11} to 10^{13} electrons cm^{-3} (10^{17} to 10^{19} electrons per cubic meter). Under proper design conditions, high-density charge cluster accelerators can collectively accelerate positive ions. It is important to note that the HDCC accelerator can provide an electron density of about 10^{23} electrons cm^{-3} (about 10^{29} electrons per cubic meter). The important discovery is that the HDCC technology provides the means by which particle accelerators can achieve an electron density that is about at least one million times (6 orders of magnitude) larger than any previous technology.

Another important feature of the charge clusters are their inherent ability to ionize dielectric materials (gases, liquids, and solids) and the ability to attract and transport positive ions. A properly designed system (patent pending) can provide for acceleration of such transported positive ions to any reasonable energy levels. It is important to understand that the highly negative electron cluster provides for a high density of the attracted positive ions. The combined charge cluster, therefore, contributes to a resulting high density of transported positive ions. For example, in a low-pressure hydrogen gas atmosphere protons are created and attach themselves to the charge clusters. The proton density can achieve very high local densities because of the enormously high local electron density. Because of the unusual nature of the HDCC accelerator, the usual magnetic field to focus the positive-ion beam is not required.

When an early decision was made by the Department of Energy to package, transport, and store high-level radioactive wastes in Yucca Mountain, the DOE officials had no knowledge of this development. The proper consideration of transmutation by standard particle accelerators was deemed to be prohibitively expensive. Geologic storage was considered to be the most cost-effective means of handling the spent-fuel pellets from nuclear power plants. See the report from the DOE-funded study with the National Research Council [21]. When one considers the cost of and/or the development time for a new type of particle accelerator to be used for the stabilization of high-level radioactive wastes, the HDCC technology has huge advantages. The accelerator potentials used to add sufficient kinetic energy to positive ions (carried by HDCC) is relatively low. For example, the combined charge cluster (HDCC plus carried positive ions) can be accelerated to about one-tenth the speed of light using potentials near 5000 V. At potentials of 50,000 V protons can be accelerated to velocities that would require over 90 million V in standard accelerators. The cost advantages of the HDCC positive-ion accelerator is obvious.

D. Collective-Ion Acceleration

The HDCC can be used to accelerate positive ions. For an ion-loaded HDCC of sufficiently high holding power, the rate of energy gain of the ion in an axial (z) electric field E is

$$\frac{dW_{i(HDCC)}}{dz} = eEM_i/Y_c m_e \left(\frac{1 - ZN_i/N_e}{1 + N_i M_i/Y_c N_e m_e} \right) \tag{1}$$

where M_i and m_e are the ion and electron rest mass, N_i and N_e are the ion and electron number in the HDCC, Z is the charge state of the ion, $Y_c = (1 - (v_e/c)^2)^{-1/2}$ is the relativistic factor, v_e is speed of the HDCC, and c is speed of light. In the case of very small ion loading with $ZN_i/N_e \ll Y_c m_e/M_i$, Eq. (1) reduces to

$$\frac{dW_{i(HDCC)}}{dz} = eEM_i/Y_c M_e \tag{2}$$

or after integration we have

$$W_{i(HDCC)} = \frac{eVM_i}{Y_c m_e} = \left(\frac{M_i}{Y_c m_e} \right) W_e \tag{3}$$

TABLE I
The Proton (Deuteron) Energy Accelerated by HDCC

Applied Voltage (kV)	The Kinetic Energy of Proton (Deuteron (MeV)
1.000	1.836 (3.672)
2.500	4.590 (9.180)
3.000	5.508 (11.016)
5.000	9.180 (18.630)
10.000	18.360 (36.720)

where V is the applied potential difference and W_e is electron kinetic energy. In the same potential difference the energy gain of an ion is

$$W_i = ZeV \qquad (4)$$

Comparing the Eqs. (3) and (4) we have

$$\frac{W_{i(\text{HDCC})}}{W_i} = \frac{M_i}{ZY_c m_e} = \frac{1836A}{ZY_c} \qquad (5)$$

where A is the atomic weight of the ion. This result means that the ion acceleration by HDCC is about $1836A/Z$ times more effective than a conventional accelerator. As an example, Table I shows the applied voltage and the kinetic energy of proton (deuteron) collectively accelerated by HDCC.

Figure 7 is a graphic description of the kinetic energy required by a deuteron to produce D-D, D-T, and D-helium-3 nuclear reactions. The bottom of the chart depicts the required deuteron kinetic energy level in thousands of electron volts. The x-axis coordinate is labeled from 10^0 to 10^3 kilo-electronvolts. The y axis is labeled in terms of the nuclear reaction cross section. Three types of nuclear reaction curves are depicted. Note that each curve rises to a maximum and then decreases in value. The D-D curve is shown with its maximum value at about 1000 keV. Considering the use of a typical ion accelerator, electric potentials ranging from about 10 to 10^6 keV are used.

When using an HDCC accelerator, the required electric potential for having the same nuclear events is dramatically reduced. Note that for the maximum reaction rates for the D-D nuclear events, the maximum occurs at about 272 Volts (lower x-axis label). Some of the desired nuclear reactions to stabilize (transmute) specific highly radioactive species into stable elements will, of course, require that the combined HDCC be accelerated using potentials up to 70 keV.

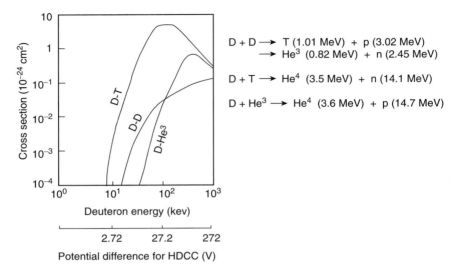

Figure 7. Cross section of D-D, D-T, and He^3 reactions as functions of deuterom energy.

To demonstrate the enormous difference between particle accelerators, Table II has been prepared. Several different parameters have been compared for the intense relativistic electron-beam (IREB) Accelerator and the HDCC Accelerator. The comparative parameters of electron energy, current density, current rise time, pulselength, beam radius, and electron density are provided. Note that in all cases, the HDCC accelerator would provide an improved performance. However, because of the small size of the combined charge clusters, the overall current (as contrasted to current density) may be smaller for a single-charge-cluster emitter. One method to improve the overall current is to use an array of charge cluster emitters as shown in Fig. 5. Also, with the use of special materials, specially designed cathodes, and a pulsed power supply, the average current could be greatly increased. It is certain that further development of this

TABLE II
Comparison of Some Characteristic Parameters between IREB and HDCC

	Electron Energy, W_e (keV)	Current Density, J (A/cm^2)	Current Rise Time, t_r (ns)	Pluse Lenght, t_b (ns)	Beam Radius, r_b (cm)	Electron Density, n_e (cm^{-3})
IREB	$(1-8) \times 10^3$	$10^3 - 10^4$	10–50	50–200	1–6	$10^{11} - 10^{13}$
HDCC	1–10	$10^8 - 10^9$	≤ 1	10–50	$(1-10) \times 10^{-4}$	6×10^{23}

type of HDCC accelerator will provide equivalent positive-ion currents at a much lower cost than the current particle accelerator technology.

Table III provides some of the experimentally determined measurements obtained by using IREB collective ion accelerators. All of the listed experiments, except the last, used accelerated protons. The last entry is for helium ions. A similar series of experiments is needed using HDCC accelerators. In addition, it may be of considerable interest to use hydrogen, deuterium, helium, and nitrogen gases for the positive-ion sources. The ability of the HDCC to ionize and transport positive ions at high local densities and at relatively low costs promises to make this new technology an effective research tool.

E.   Experimental Evidence of HDCC-Produced Nuclear Transmutation

Kenneth and Steve Shoulders report an experiment in which a previously deuteron-loaded palladium cathode was subjected to the impact of a charge cluster [22]. Where the charge cluster impacted the deuteron-loaded palladium a visually-evident, explosive-like reaction occurs (Fig. 8). The palladium cathode was then subjected to an X-ray analysis of the impact crater (see chart in Fig. 8). Typically, the X-ray analysis shows a considerable number of elements not seen when scanning the nearby palladium surface. Such elements as oxygen, calcium, silicon, and magnesium are detected in the exploded region where a charge cluster impacted the palladium.

As explained more fully by Shoulders [22], there is a level of energy required in the impacting charge cluster before such nuclear reactions occur. It is expected that this energy level will be different for a variety of nuclear target materials. One of the required studies is the effect of the combined charge cluster (electrons plus carried positive ions) effect on specific targets as depicted in Fig. 9. It is expected that considerable differences will be found in the ability to cause nuclear reactions in various high-level nuclear wastes using different positive ions. Proton, deuteron, and alpha particle accelerator studies will need to be performed at various impact energy levels.

One unknown in particle bombardment is the effect of the accompanying very high density of electrons. It is expected that the highly negative charge cluster (as it approaches a target material) will cause a local repelling away of all electrons, leaving only the local nuclei plasma to be impacted by the carried positive ions. To what extent this action will facilitate (or impede) nuclear reactions has yet to be experimentally determined.

It is of considerable interest to note that charge clusters can be formed in aqueous solutions and used to target dissolved radioactive materials. In experiments using low-level, naturally radioactive thorium, a considerable reduction of thorium from the solution has been achieved [6]. Charge clusters can be produced in air under various pressures [23]. However, not all arcs and sparks

TABLE III
Experimental Evidence of Naturally Occurring Collective-Ion Acceleration by Intense Electron-Beam Injection into Neutral Gas

	Electron Beam (Beam Density: $\sim 10^{10}$–10^{13}/cm^{13})					Ions Accelerated			Ion Acceleration Efficiency $\eta = (NW_i)(W_eI_et_b/e)^{-1}$	Ref.
W_e(MeV)(cm)	I_e(kA)	t_r(ns)	t_b(ns)	r_c	Type	W_i(MeV)	N			
1	0.2	200	—	80	3.81	—	0.8	10^{13}–10^{14}	$(4 \times 10^{-4}$–$4 \times 10^{-3})$	8,9
2	0.25	200	—	—	3.8?	—	0.1–2.1	10^{12}–10^{14}	—	9,10
3	0.5	160	15	50	3.8?	—	~ 1	10^{13}	(4×10^{-4})	10,11
4	0.5	>40	15	55	6	—	2.5–3.5	—	—	12,13
5	0.65	15–20	15	50	—	—	1–3.8	10^{12}	2×10^{-4}	14–19
6	0.65	145	10	—	3.8?	—	~ 1.8	10^{12}–10^{14}	—	10
7	0.75	100	10	90	2.54,3.81	—	3–7	10^{12}–10^{13}	$(7 \times 10^{-5}$–$2 \times 10^{-3})$	20
8	1.0	110	—	50	3.8	—	1.8–2.2	10^{13}	(10^{-5})	10,11
9	1	115	10	90	2.54	—	2–12	3×10^{10}	(6×10^{-6})	20,21
10	1–1.4	160	—	80	3.81	—	~ 1.5	10^{13}–10^{14}	$(2 \times 10^{-4}$–$2 \times 10^{-3})$	15,30
11	0.45–1.35	32–150	8	55	2.54,3.81	—	2–14	—	—	23–26
12	1.3	35	10	40	1.25	—	4.8	2×10^{12}	(8×10^{-4})	27,28
13	1.3	50	35	50	2.5	—	<4.5	$\sim 10^{13}$	(5×10^{-4})	29
14	1.7	30	10	50	1.25	—	4.5–6.5	2.8×10^{12}	(1×10^{-3})	30–32
15	1.8	75	60	90	0.5	—	1–5	10^{11}–10^{12}	—	33–35
16	2–2.3	15–20	6	45	0.64,1.25	—	2–5	—	(3×10^{-4})	36–40
17	2	80	20	125	5.1	—	4–14	—	—	41,42
18	3	55,80	20–25	125	5.1,2.55	—	3–8	—	—	42,43
19	5	38–40	20	125	0.65,2.55	—	4–16.5	2×10^8	—	42,43
*20	8	230	70	200	15	—	18–40	10^{10}	(8×10^{-8})	44

Key: W_e—peak electron energy; I_e—peak electron current; t_r—current rise time; t_v—voltage rise time; t_b—pulselength; r_c—cathode radius; W_i—peak ion energy; N—ion number; η—ion acceleration efficiency.

Source: C. L. Olson and U. Schumacher, *Springer Tracts in Modern Physics*, Vol. 84, *Collective Ion Acceleration*, Springer-Verlag, New York, 1979.

*Using helium ions (alpha particles).

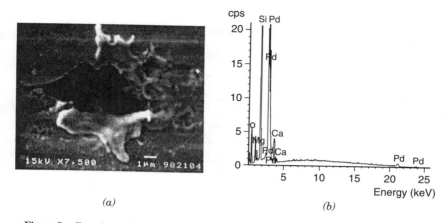

(a)   *(b)*

Figure 8.   Experimental evidence of nuclear reactions produced by high-density charge clusters (HDCCs): (a) HDCC strike on a deuterium-loaded palladium foil; (b) X-ray analysis of the crack illustrated above, showing new materials produced.

are carriers of charge clusters. It is well understood that the best production, control, and use of HDCC will be at relatively low pressures.

One of the best ways to transmute radionuclides is with intense high-energy protons collectively accelerated by HDCC being focused on a target such as

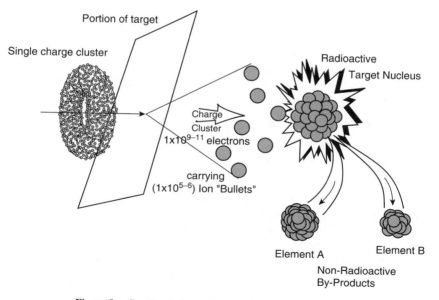

Figure 9.   Combined charge cluster and ions impacting target.

lithium or tungsten that would generate high-energy neutrons. These neutrons would be thermalized and focused on a target to transmute radionuclides.

F.   Conclusions

HDCC accelerators have the potential to provide positive-ion acceleration for creating nuclear reactions at electric potentials heretofore deemed to be much too low to produce nuclear reactions. Therefore, the conclusions of the report published in *Nuclear Wastes: Technologies for Separations and Transmutation*, are no longer valid [21]. *There now is a technology that is more cost-effective than geologic storage for the handling of high-level radioactive wastes.* It appears that on-site stabilization of high-level, radioactive wastes is feasible within the same time-frame required for the preparation of geologic storage facilities. The estimated potential cost savings are enormous.

IV.   OTHER PATENTED NEW ENERGY DEVICES

A.   Introduction

The author has spent over 12 years searching the world for energy devices that have commercial value. Over a hundred various devices have been investigated including a variety of rotating machines. Almost none of these devices have been funded with government funds. The reason for the general lack of government funding is the firm scientific knowledge that it is not possible to produce more energy output than energy input *unless there is a source of that energy.* However, there are a variety of private individuals whose education has not been sufficient to convince them that energy over-unity devices, which lack a new-energy source, were not possible. Many of these privately funded inventors would not agree with this author's limitation on new-energy sources as being one of the following:

1. A new type of chemical energy
2. Energy from low-energy nuclear reactions
3. Energy from the abundant energy in space

This third potential source of new-energy devices is, in general, not accepted by the academic scientific community. Where there is a knowledge of the various papers discussing space energy (vacuum energy, zero-point energy, etc.), it is generally assumed that there is no practical way of tapping this enormous energy source. As a result, there is no known government funding and very little known corporate funding of research and development to tap the energy of space. However, during the fall of 2000, a draft proposal for a proposed Breakthrough Energy Physics Research Program Plan was distributed as "Draft for agency comment only." This plan is intended to provide the groundwork within the U.S.

Department of Energy for the funding of a variety of new-energy and new-transportation physical concepts.

Heretofore, few members of the academic community have accepted any of the new-energy discoveries. That academic acceptance is believed to be subject to rapid change. There is a tongue-in-cheek saying that "scientists will believe what they are paid to believe." It is to be expected that where there are government funds allocated for the study of breakthrough energy physics, there will be an increased academic interest in new-energy discoveries.

A major development in the search for tapping the energy of space has been the various articles authored by Puthoff and friends [24–26]. For these articles about zero-point energy, it was assumed that there is an enormous energy in space everywhere. In fact, this huge source of energy is predicted by the development of quantum mechanics. This dense-energy concept is discussed by many authors, for example by William Tiller [27]. By beginning with the assumption of enormous space energy, Puthoff and co-authors have been able to develop mathematical descriptions for gravity [24], inertia [26] and also to show that the apparent perpetual motion of the electron orbiting the hydrogen atom is explained by the interchange of energy between the electron in its ground state and the energy of space [25]. The clear explanation for these previously little understood phenomena has been a major contribution to science. It should be reported that Dr. Harold Aspden also provided a mathematical description of gravity some years earlier but the paper received less attention than it deserved [28]. In addition, Aspden also discovered and reported on an unexpected inertia effect [29].

In a later paper, Puthoff et al. have shown that the energy of space can be tapped [30]. However, this paper does not specify how a practical tapping of space energy can be achieved. Prior to this paper by Puthoff, Shoulders had already discovered a method of using HDCC to tap the energy of space. Puthoff, himself, also added to the HDCC technology [31].

The search for new-energy devices recognizes that in addition to new chemical discoveries and low-energy nuclear reactions, the tapping the energy of space is an acceptable energy source for one or more new energy devices.

B. Energy from Rotating Machines

A large number of rotating machines have been examined. With very few exceptions, none of the rotating machines were found to be over unity despite strong claims by the inventors. In nearly all cases, the rotating machines used either an input or provided an output that was not sinusoidal. There were often pulse-like or distorted wave forms used or produced by these machines. However, the main error made by nearly all such inventors was to use standard AC meters to measure the input and output energy. Such instruments, designed for accurate measurements of sinusoidal alternating current, are not accurate when used with nonsinusoidal wave forms.

However, there have been a couple of rotating machines that have produced more energy out than energy input. One is the large motor generator designed by Tewari in which the 50-cycle 200 V AC input provides a larger power of high amperage but low voltage DC output [32]. A second machine that produces a relatively small amount of over-unity performance is described by Harold Aspden [33]. If this author is correct, then such rotating machines must use either (or both) high rotational speeds or very strong magnetic fields. In any case, it is not expected that any type of large rotating machine will prove to have the 3 : 1 output/input ratio that this author deems required for commercial success. However, other inventions have shown the ability to meet this 3 : 1 output/input energy requirement.

C.   Energy from Low-Energy Nuclear Reactions

The saga of cold fusion, more properly called "low-energy nuclear reactions," has a fascinating history in the United States. A few years after the initial announcement of the discovery of cold fusion, this author had collected over 3000 papers from over 200 laboratories in exactly 30 countries. Over 600 of these papers reported successful replications or improvements on the original Fleischmann–Pons discovery [34]. However, due to a well-conceived, well-funded, and well-conducted effort by adherents to the hot-fusion technology, cold fusion in the U.S. has been discredited [35].

The important concept about cold fusion is described in the following highly probable explanation for most of the excess thermal energy produced.

D.   Explaining Cold Fusion

At least some of the effects observed in cold fusion electrochemical cells can be explained by the production of high-density charge clusters (HDCC) as shown by Kenneth Shoulders. The deuterium loaded into the palladium cathode causes the cathode to become stressed and brittle (hydrogen embrittlement). The result is that the palladium cracks at random times and places. The cracking of a crystal lattice with its severing of trillions of ionic bonds creates a high voltage across the crack, which quickly shorts out (short-circuits) through the conducting palladium. This type of short pulse of high voltage is precisely the condition required for the production of an HDCC. The cluster emits from the cathode, picks up and carries deuterons, is accelerated to the anode side of the crack, and crashes into the anode side with sufficient force to produce nuclear reactions. A detailed analysis of the impact craters produced by the 20-μm HDCC impact onto a palladium-loaded electrode has been demonstrated by Shoulders to explain the observed impact craters and the new elements produced [22].

Obviously, if one knows how to make HDCC directly, it is of little value to go through the electrochemical process to produce charge clusters at random times and random places. In the judgement of this author, this cold-fusion type of low-energy nuclear reaction is not sufficiently robust to be a major commercial success.

E.   The Koldamasov Low-Energy Nuclear Reactor

Koldamasov works in a Russian facility that develops and produces equipment for use by the nuclear power industry. An observation of excess thermal energy being emitted led to the invention and patenting of a new type of nuclear reactor [36]. This nuclear reactor uses input electrical energy to power a frequency-tunable ultrasonic crystal which is placed at one end of a cylinder of a mixture of light and heavy water. In the cylinder is a short, solid cylinder made of a special dielectric material. There is at least one small-diameter hole (about one millimeter) in this dielectric material. When the proper frequency of input ultrasonic energy is applied, the surging of the water through the small hole produces a very high (reported as 200,000–300,000 V) electric potential developed locally in the dielectric material. The end result is the production of a considerable amount of thermal energy plus a relatively small amount of emitted neutrons.

The inventor has patented this novel nuclear reactor and cites the production of as much as 40 kW of thermal power from the use of about 2 kW of input electrical energy. If the reported results are independently replicated, this discovery and invention is deemed to be of considerable commercial importance.

F.   The Burns Device

David Burns, an Irish-born inventor living in Scotland, has invented another type of new-energy device. At the time of writing this report, only early (1994) experimental data were available. However, the device uses a property of smooth versus rough material housed in a vacuum tube and subjected to electron bombardment. Under proper experimental conditions, as explained in the patent application, there can be a considerable temperature difference between the smooth and rough plates. With the use of sufficiently efficient thermoelectric devices, it is reported possible to provide more electrical energy output as than electrical energy input. The latest report for this type of device is promising. However, considerable development is expected to be necessary before a commercial device is achieved.

G.   The Mills Device

Dr. Randell Mills has invented, tested, improved, and submitted several patent applications on a new type of new-energy device. At least one of the patent applications has proceeded to the point where the *Patent Gazette* indicates that a patent is being issued. Dr. Randell Mills claims that the excess thermal energy produced is obtained from the collapse of the hydrogen atom. There is, as yet, no widespread acceptance of his explanation of the source of the excess thermal energy. However, there have been a variety of independent replications of the effect on which his discovery is based [37].

Beginning with a liquid electrochemical system, Dr. Mills has progressed to the use of gaseous media with a reported considerable improvement in overall performance. Apparently, this invention is capable of providing a considerable increase in thermal energy output as compared to electrical energy input. An energy output/input ratio of well over 100 has been reported.

The reduction of the *orbit* of the hydrogen atom below its normal ground state produces a new type of hydrogen that Mills has termed a *hydrino*. These *hydrinos* can be chemically combined with other elements to produce new molecules that reportedly have material properties different from those of their hydrogen counterparts. For example, a lubricant with improved lubricity has been reported. In the judgment of this author, the most important discovery made by Dr. Mills is believed to be methods of creating new hydrogen-based compounds. It is expected that the current BlackLight Power (registered trademark) company that has been established and funded will become a commercial success.

H. The patented AquaFuel Device by Richardson

An inventor, William H. Richardson, Jr., has invented a new type of energy-producing device (U.S. Patent 5,435,274) [38], which is used to produce a product called *AquaFuel* (registered trademark). The AquaFuel is reported to use water and a mixture of organic substances to produce a burnable gas. The reported performance provides more energy output from the burning of the AquaFuel than supplied by the electrical energy used to produce this combustible gas by electrolysis. Commercial applications of this invention have been announced. The source of the excess energy apparently stems from the energy provided in some of the organic materials that are used in the water-solution that is electrolyzed to produce the combustible gas.

There is a somewhat similar but different method developed by Dr. Ruggero Santilli that also produces a combustible gas. Details of Santilli's method are provided in a special issue of the *Journal of New Energy* [39].

I. The Motionless Electromagnetic Generator (MEG)

A Web-posted letter has announced:

> Magnetic Energy Ltd. announces its successful development of an electromagnetic generator with no moving parts, normal EM input and normal EM output, and having a COP (coefficient of performance) = 5.0 for the first unit and COP = 10.0 for the second unit. The MEG is in patent-pending status, and is presently available for licensing [40].

The five co-inventors of the MEG are T. E. Bearden, Ph.D.; James C. Hayes, Ph.D.; James L. Kenny, Ph.D.; Kenneth D. Moore, B.S.; and Stephen L. Patrick, B.S. The intellectual property rights to this new-energy device (MEG) are assigned to Magnetic Energy Ltd. (925 Tascosa Dr., Huntsville, AL 35802; email: <jlkenny@lbtrltd.com>). Dr. James Lee Kenny is the managing partner of Magnetic Energy Ltd.

Reportedly, the MEG can be fabricated using only conventional tools and techniques already used by many major manufacturers. All the parts and materials to construct an MEG are available commercially. On October 18, 2000, its development was also announced to selected staff personnel of two Senate Committees. A 69-page MEG technical report, presently on a private DOE Website for DOE evaluation, explains the principles and concepts used by this new-energy generator. Arrangements are also in process for formal independent certification tests by a major university under U.S. government auspices, followed by full formal independent replication under proprietary release agreement.

J.   Other New-Energy Devices

There are other new-energy devices that are being developed. There are many claims made for new-energy devices. In some cases, the details are being kept highly confidential for various reasons. For example, one of the new-energy inventors claims to have made a disclosure to a major automobile company and believes that a new development by that company was based on his disclosure. Other independent inventors have deemed their discoveries so important that they are fearful that they might be the subject of persecution by the big oil companies. This author, during a period of over 12 years, of working in this new-energy field has found no evidence of any threats to persons by any energy-related group. The only active negative efforts have been the ill-chosen attack on cold fusion and Dr. Mills new-energy discoveries by a self-proclaimed spokesperson for the American Physical Society [35]. It is strongly believed, by this author that such attacks on any new-energy devices will rapidly fade. The reason for the change of attitude is the final acceptance by the Department of Energy that a part of their role is to be involved in "breakthrough energy physics research" [43].

V.   SUMMARY

Seldom is a new discovery made in isolation. The history of invention is replete with dramatic new discoveries that suddenly appear to have been anticipated by more than one inventor. It appears that the developments in a variety of new technologies are culminating in several new-energy devices that can provide multiple sources for abundant, clean, and inexpensive means to provide both thermal and electrical energy. Most of the several new inventions presented here meet the desired qualifications by providing more than three times the energy output as compared with energy input. In addition, each of these devices has (or will have) a reasonable scientific explanation as to the source of the output energy. Some of these new-energy sources are new or unexpected and include new chemical energy (Mills' device); low-energy nuclear reactions (e.g., Koldamasov's device); and devices that tap the energy of space (Shoulders' discovery).

Not all scientists are ready to accept the explanation that the enormous energy of space (the so-called zero-point energy) can be transferred into useful energy. This dramatic new discovery will take time to be explained in the college textbooks or taught in high school classrooms. However, as the expected commercialization of some of these devices occurs in 2001, there will be a rapid development of serious academic explanations for and the acceptance of such new technology. Perhaps of more importance to the acceptance of new-energy phenomena is the establishment of the Department of Energy's Breakthrough Energy Physics Research Program [43].

Many of these inventions will be licensed, replicated, improved, developed, manufactured and placed into commercial applications by a variety of commercial entities. The potential of tapping into a $4.5 trillion annual fossil-fuel energy market [41] is expected to create a rapidly expanding number of new-energy companies. The realization that humankind need no longer be tied to the production, distribution, and marketing of coal, oil, and natural gas products (the fossil fuels) will have the greatest economic impact of any previous technological innovation. While oil-rich countries may suffer, the countries lacking in fossil-fuel energy sources are expected to greatly benefit. As the burning of fossil fuels decreases, the world's atmosphere is expected to benefit. Fossil fuels will then be perceived as the valuable assets they are as sources of chemical feedstocks.

The ultimate result will be the rapid development and proliferation of devices that tap the energy of space. The knowledge that space everywhere can be tapped for its energy will hasten the development of human endeavors to reach out to the nearby planets. When coupled with new methods of overcoming gravity, new methods of pushing against space without throwing away mass (the old rocket engines), then mankind will be able to move around in space and establish interplanetary commerce. However, for the near future, we here on earth can soon get off the grid.

The near-term impact of the commercialization of new-energy devices will be relatively rapid. However, it is still expected to take a decade to penetrate the current $4.5 trillion annual fossil-fuel market to 10% [41]. Hundreds of billions of dollars will be spent in the manufacturing and marketing of new-energy devices and systems. It is estimated that for the first decade, the energy provided from these new-energy sources will be sufficient only to meet the world's increased demands for energy. Thereafter, the world can expect improvements in the atmosphere, reduction in energy costs, some rearrangements in international money flow, and, in general, an improved new-energy world.

For a book-length discussion of new-energy devices and their predicted impact on a variety of the world's markets, this author has written a book that is soon to be published [42].

Acknowledgments

The author is indebted to hundreds of scientists, engineers, and inventors who have shared their discoveries and inventions by writing and submitting articles to be published in the *Journal of New Energy*. In addition, the author is indebted to the shareholders of Trenergy, Inc. and Emerging Energy Marketing Firm, Inc. for the author's privilege of editing the *Journal of New Energy*, for traveling to various international energy conferences, and for the support to locate, evaluate, and perform tests on many new-energy devices.

References

1. A. Bhadkamkar & H. Fox, *J. New Energy* **1**(4), 62–67 (1997).

2. K. R. Shoulders, *EV, A Tale of Discovery*, 1987 (published and available from the author, P.O. Box 243, Bodega, CA 94922–0243.

3. K. R. Shoulders, U.S. Patent 5,018,180 (May 21, 1991).

4. R. Bass, R. Neal, S. Gleeson, and H. Fox, *J. New Energy* **1**(3), 81–87 (1996). 6 refs, 1 fig, 1 table.

5. R. R. Liversage, "Third-party verification of Cincinnati group's thorium transmutation process," *Infinite Energy*, **13–14** (Special Double Issue) (March–June 1997).

6. H. Fox and S.-X. Jin, "Low-energy nuclear reactions and high-density charge Clusters," paper presented at the annual meeting of the American Nuclear Society, Nashville, TN, June 9, 1998, *J. New Energy*, **3**(2–3), 56–67 (1998).

7. Personal discussion with Dr. Alexander M. Ilyanok.

8. S.-X. Jin and H. Fox, *J. New Energy*, **1**(4), 5–20 (1996).

9. G. A. Mesyats, "Ecton processes at the cathode in a vacuum discharge," *Proc. XVIIth Int. Symp. Discharges and Electrical Insulation in Vacuum*, Berkeley, CA, July 1996, p. 720–731.

10. P. Beckman, *Galilean Electrodyn.* **1**(5), 55–58 (Sept.–Oct. 1990).

11. R. W. Ziolowski and M. K. Tippett, *Phys. Rev. A* **432**(6), 3066–3072 (1991).

12. P. Graneau, *Phys. Lett.* **97A**(6), 253–255 (1983).

13. P. Graneau and N. Graneau, *Newton versus Einstein, How Matter Interacts with Matter*, Carlton Press, 1993.

14. P. Graneau and N. Graneau, *Newtonian Electrodynamics*, World Scientific Publishing, River Edge, NJ, 1995.

15. P. Graneau, "Solar energy liberation from water by electric arcs," paper presented at 1st Int. Conf. Future Energy, Bethesda, MD, April 29, 1999.

16. K. R. Shoulders, "Permittivity transitions," *J. New Energy*, **5**(2), pp. 121–137, 9 figs., 6 ref.

17. Private communication with Kenneth Shoulders.

18. S.-X. Jin and H. Fox, *J. New Energy*, **4**(2), 96–104 (1999).

19. K. R. Shoulders, U.S. Patent 5,054,046 (Oct. 1, 1991).

20. K. R. Shoulders, U.S. Patent 5,054,047 (Oct. 1, 1991.)

21. *Nuclear Wastes: Technologies for Separations and Transmutation*, Committee on Separations Technology and Transmutation Systems, Board on Radioactive Waste Management, Commission on Geosciences, Environment, and Resources, National Research Council, published by National Academy Press, Washington, DC, 1996 by the National Academy of Sciences.

22. K. Shoulders and S. Shoulders, *J. New Energy*, **1**(3), 111–121 (1996).

23. K. Shoulders and S. Shoulders, "Charge clusters in action," paper presented at 1st Int. Conf. Future Energy, Bethesda, MD, April 29, 1999 (printed in the conference proceedings, 1999, by authors; also available are the photographs on a CD/ROM, price $20).

24. H. E. Puthoff, *Phys. Rev. A*, **39**(5), 2333–2342 (1989).

25. H. E. Puthoff, *Phys. Rev. D* **35**(10), 3266–3269 (1987).

26. B. Haisch, A. Rueda, and H. E. Puthoff, *Phys. Rev. A* **49**(2), 678–604 (Feb. 1994).

27. William A. Tiller, *Science and Human Transformation: Subtle Energies, Intentionality and Consciousness*, Pavior Publishing (Walnut Creek, CA), 1997 (see pp. 46–47 and *Energy density* in index).

28. H. Aspden, *Phys. Essays* **4**, 13–19 (1991).

29. H. Aspden, *New Energy News* **2**(10), 1–2 (Feb. 1995); see also Hal Fox, "The Aspden effect," pp. 2–3.

30. D. C. Cole and H. E. Puthoff, *Phys. Rev. E* **48**, 1562–1565 (1993).

31. H. E. Puthoff et. al., U.S. Patent 5,208,844 (May 4, 1993).

32. P. Tewari, in M. Albertson (ed.), *Proc. Int. Symp. New Energy* (April 16–18, 1993), pp. 291–303; *J. New Energy* **5**(1), 105–122 (2000).

33. H. Aspden, "Power from magnetism," *Energy Science Report 1*, 1994; "Power from magnetism: Over-unity motor design," *Energy Science Report 9*, 1996.

34. H. Fox and M. Swartz, "Progress in cold nuclear fusion—metanalysis using an augmented database," paper presented at ICCF-5, 1995 (not published in ICCF proceedings, available from authors).

35. R. Park, comments in many of his email letters circulated to Department of Energy officials and others. These emails originate from the Washington, D.C. office of the American Physical Society. Lately such emailed opinions have carried the admission that the information presented does not reflect the official opinions of the APS but should.

36. A. I. Koldamasov, "Energy release with the Wolgodonsk reactor," paper presented at New Enegy Technolgies on the Millennium Transition Symp., Zurich, Sept. 16, 2000.

37. R. L. Mills and S. P. Kneizys, *Fusion Technol.* **20**, 65–81 (Aug. 1991).

38. W. H. Richardson Jr., U.S. Patent 5,435,274 (July 25, 1995) (the AquaFuel device).

39. R. M. Santilli, (Special Issue, edited by Hal Fox), *J. New Energy* **5**(1), 283–287.

40. T. Bearden et al.; see the following Website for further detailed information: http://jnaudin.free.fr/html/megv1.htm.

41. H. Fox and S.-X. Jin, *J. New Energy* **4**(4) (2000).

42. H. Fox, *Space Energy Impact in the 21st Century* (in press) (see especially Chap. 9; those interested may contact the author for further information on publication at <halfox@slkc.us-west.com>).

43. Staff, *Breakthrough Energy Physics Research (BEPR) Program Plan*, Oct. 2000, *Draft for Agency Comment Only.*

THE SUPERLUMINAL THEORY AND EFFECTS

PETAR K. ANASTASOVSKI

*Department of Physics, Faculty of Technology and Metallurgy,
Saints Cyril and Methodius University, Skopje,
Republic of Macedonia*

DAVID B. HAMILTON

U.S. Department of Energy, Washington, DC

CONTENTS

I. INTRODUCTION

To explore and explain superluminal effects, a definition for superluminal dynamics must be determined. After that, it is necessary to establish the main concepts of the theory of superluminal relativity (SLRT).

Modern Nonlinear Optics, Part 3, Second Edition, Advances in Chemical Physics, Volume 119,
Edited by Myron W. Evans. Series Editors I. Prigogine and Stuart A. Rice.
ISBN 0-471-38932-3   © 2001 John Wiley & Sons, Inc.

Superluminal dynamics is the dynamics of particles in motion with $v > c$ where c is the speed of light in vacuum. To establish the basic concepts of SLRT, it is necessary to introduce superluminal transformation. Superluminal transformation gives the relation between the magnitudes of the frame of reference where $v < c$ only is possible, with the same magnitudes in the frame of reference where $v > c$ is possible. Superluminal transformation correlates with Galilean and Lorentz transformations.

In this work, Einstein's theory of special relativity (SRT) is fully accepted, with the supposition that it is valid in the region of the spacetime where $v < c$ is possible only. There are many experimental proofs that support the concepts of SRT, which justify the main postulate $c = \text{constant}$. However, none of the experimental proofs for the validity of the special relativity concepts have led to the fundamental postulate $c = \text{constant}$ being accepted as a physical law. It still remains a postulate, that is, an assumption. It is a justified assumption for the theory of special relativity, but still an assumption only [1–3].

It is worth mentioning here, that the factor

$$\gamma = \frac{1}{\sqrt{1 - (v^2/c^2)}} \tag{1}$$

which is essential in the theory of special relativity, is taken from the Lorentz transformation, and has emerged as a result of the quest for vacuum properties or more specifically, from the search for a connection between electromagnetism and vacuum properties. Also, it has to be pointed out, that the main postulate for the ultimate velocity of the traveling particles is connected with speed of light in the vacuum. These two facts show clearly, that the whole SRT is based on a supposition for existing of certain properties of the vacuum. All the performed and observed experiments, which verify SRT, also verify the existence of the supposed vacuum properties.

Vacuum properties that are directly connected with the propagation of light in the vacuum are vacuum permeability, $\mu_0 = 1.2566 \times 10^{-6}\,\mathrm{m\,kg\,C^{-2}}$, and vacuum permittivity, $\varepsilon_0 = 8.8544 \times 10^{-12}\,\mathrm{N^{-1}\,m^{-2}\,C^2}$.

These magnitudes are defined by the observation of a charge traveling in the vacuum. The photon has no charge; however, light has electromagnetic properties, and therefore these magnitudes determine the conditions for light propagation in the vacuum.

According to the Maxwell equations, the velocity of light in the vacuum is

$$c = \frac{1}{\sqrt{\mu_0\,\varepsilon_0}} \tag{2}$$

This explicitly shows the connection between electromagnetic vacuum properties and principles of special relativity.

It is interesting that vacuum properties are determined and connected only by the properties of charged particles, and consequently only the properties of charged particles determine the electromagnetic properties of the vacuum. Particles that travel through the vacuum have another important property: mass. However, this property and its magnitude are completely neglected. Our position is that the vacuum should have properties, that are connected with the mass of the particles, as well. A treatise on quantum mass theory (QMT) [4] elaborates on such properties of the vacuum.

The main supposition of the superluminal relativity theory (SLRT) presented here (see also Ref. 5) is that besides the vacuum properties covered by the special relativity theory (SRT) and corresponding observed phenomena, there exist some other vacuum properties as well, which are additional to the first ones, but that allow the possibility for $v > c$.

For the region of the space, where vacuum properties allow $v > c$, SRT concepts will be extended by modifying the main factor γ in to γ' [5]. According to this, the assumption for $v < c$, and all the consequences of that in SRT are valid only in the range of the spacetime where certain vacuum properties prevail, while, in the region of the spacetime where some other vacuum properties are dominant, $v > c$ should be possible. This will have important consequences for the main physical laws, that is, conservation laws. The starting assumption in this work is that conservation laws are preserved in this new frame of reference.

One of the main tasks in Ref. 5 is to formulate one of the most important laws in physics, the energy conservation law, in a spacetime where $v > c$ is possible, and to find the connection between two regions of the space, where different vacuum properties prevail. The other main task is to find the magnitudes, which will determine the vacuum properties of this new region of the spacetime.

The results of the analysis justify the validity of these newly offered hypotheses and suggest performing experiments, which will support the theory and analysis presented. In the last section of this work, the analysis based on the SLRT is presented, which explains experimentally observed superluminal effects [6,7]. Among the other proofs for the validity of SLRT, this could be considered as a direct proof for the main concepts of this theory.

The new proposed deuteron model is founded on the principles of SLRT and QMT. In Ref. 4, where QMT is presented, it is shown that, *if the electron in the hydrogen atom is excited to the state of the potential quantum number, n = 794 then, the electron turns into a positron.* The consequence is very unusual; the hydrogen atom turns into a system of one proton and one positron, which is undoubtedly a very *odd example of CP violation.* This has been obtained as a result of theoretical analysis based on the QMT principles. If this is experimentally proved, then *atoms with very unusual physical characteristics* will certainly be obtained, and a rather exotic regime of matter could be expected.

It is worthwhile mentioning here the reports for existence of hydrogen-like atoms, besides hydrogen itself [8,9]: positronium (Ps)—composition of $e^+ e^-$ muonium (Mu)—composition of $\mu^+ e^-$, μ-mesohydrogen—composition of $p\mu^-$, and π-mesohydrogen—composition of $p\pi^-$. It is obvious that all these rather new atoms are made up of one positively charged particle and one negatively charged particle, while the hypothetical proton–positron system is supposed to be made up of two positively charged particles. That is the reason why this system would be the example of very odd CP violation, according to the current theories. In the SLRT presented here, a new proposed concept for charge conjugation–parity (CP) conservation, offers an alternative approach to the cases of CP violation.

According to the hereby used theory of superluminal relativity [5], nuclear forces are explained by Newton's gravitational law and Einstein's general theory of relativity [10], with the gravitational constant defined and determined by the quantum mass theory [4], for masses and distances characteristic for nuclear structures.

The results of the analysis based on SLRT give a new insight into nuclear forces and structures and offer explanation of superluminal effects.

II. THE THEORY OF SUPERLUMINAL RELATIVITY

A. Principles of Superluminal Relativity

The theory of superluminal relativity is based on the following principles [5]:

Principle 1. The laws of physics must be of such a nature that they apply to reference systems in any kind of motion relative to the mass distribution of the Universe; that is, the laws of nature are the same for all free-moving observers independent of their velocity.

Principle 2. The presence of bodies in general has influence in spacetime continuity.

Principle 3. There is equivalency between particles' masses and energies.

Principle 4. The speed of light in a vacuum relative to an observer is not constant; thus $c \neq$ constant. The speed of light in a certain region of spacetime depends on the vacuum properties, which prevail in that region. If the electromagnetic properties are dominant, the speed of light is $c =$ constant. If the mass properties of the particles and bodies in general, are dominant in the vacuum of a certain region of spacetime, then the speed of the light is $c' \neq c$.

Principle 5. The speed of the particles can be larger than the speed of light in the vacuum described by the principle 4, that is, $v > c$.

Principle 6. The real magnitudes in the frame of reference S' with relative velocity, $v > c$ to the frame of reference S, are virtual magnitudes for the observer O in the frame of reference S. There is a constant M_c that connects the magnitudes of these two frames of reference, S' and S, and that determines the region of the spacetime where the mass properties of the vacuum prevail. The virtual magnitudes in the system S, which the constant M_c turns into the real ones, satisfies principle 1; that is, the energy conservation law is preserved.

B. Superluminal Transformation

In Einstein's special theory of relativity [1,2], the Galilean transformation had to be replaced by the Lorentz transformation, so that the speed of light would be invariant or independent of the relative motion of the observers—in particular, because the assumption $t' = t$ is no longer correct. In the Lorentz transformation the time is $t' \neq t$.

Since we have made another, also fundamental assumption $v > c$, it is necessary to determine the corresponding transformation [5].

Figure 1 shows two coordinates systems, S and S', where the axes YZ and $Y'Z'$ are parallel and axes X and X' coincide and point in the direction of their relative motion. We shall assume that both observers, that is, in O and O', set their clocks so that it is $t = t' = 0$ when they start to observe the mutual motion of their systems [5].

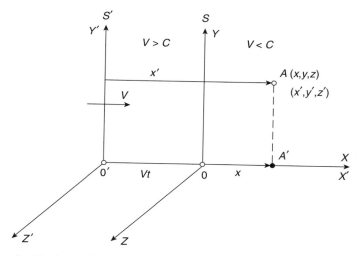

Figure 1. The frames of reference S and S' in relative translational motion. In the frame of reference S, Lorentz transformation and special relativity principles are valid. In the frame of reference S', superluminal transformation and SLRT principles are valid.

As is shown in Fig. 1, the system S' is approaching the system S with relative speed, $v > c$. Let us assume that, at time $t = t' = 0$, a flash of light with speed c is emitted from the origin O in the system S, toward the point A', and simultaneously a flash of light with speed c' is emitted from the origin O' in the same direction of propagation with the light from the origin O [5]. The observer at point A' will detect simultaneously both light signals only if $c' > c$.

Superluminal transformation equations, which connect these two frames of references S' and S, are [5]

$$x' = \sqrt{1 - \frac{c^2}{v^2}}(x + vt) \tag{3}$$

$$y' = y \tag{4}$$

$$z' = z \tag{5}$$

$$t' = \frac{1}{\sqrt{1 - c^2/v^2}}\left[t + \sqrt{\frac{c^2(v^2 - c^2)}{v^4}}x\right] \tag{6}$$

The factor of superluminal transformation is

$$\gamma' = \sqrt{1 - \frac{c^2}{v^2}} \tag{7}$$

The superluminal transformation in its essence is Lorentz transformation modified for assumption of possibility $v > c$.

C. The Basic Superluminal Effects

Two curves are shown in Fig. 2, one for the function $\gamma' = f(c/v)$ from SLRT, and the other one, for the function $\gamma = f(v/c)$ from SRT [5]. In fact, these two curves

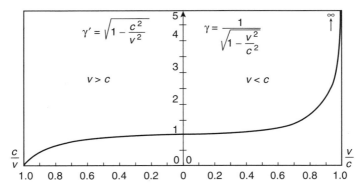

Figure 2. Functions $\gamma' = f(c/v)$ and $\gamma = f(v/c)$ with common ordinate and common O points of the axes.

form one curve that starts from $\gamma' = 0$ and reaches $\gamma = \infty$. The resultant curve formed by γ and γ' shows that SRT covers the phenomena, which are taking place in the frame of reference where $v < c$ is possible only, and SLRT covers the phenomena taking place in the frame of reference where $v > c$ is possible. *Hence, these two theories—SRT and SLRT—do not exclude each other (i.e., are not mutually exclusive); on the contrary, they together form a consistent system.* This is accomplished by including the concepts of the general relativity theory (GRT) in the SLRT.

In Table I, for comparison, we present consequences of Lorentz transformation in SRT, and superluminal transformation, used in SLRT. According to the data in this table, when a particle travels with $v > c$, the superluminal effects are opposite the effects expressed by SRT.

In SRT there is length contraction, while in SLRT there is length dilation of the particle in motion with corresponding velocities. In SRT there is time dilation, while in SLRT there is time contraction of the processes, which are taking place in corresponding frame of reference. In SRT mass of the particle increases with its acceleration, while in SLRT mass of the particle decreases with its acceleration, when it is in motion with $v > c$, and it is given by the expression [5]

$$m = m_0 \sqrt{1 - \frac{c^2}{v^2}} \tag{8}$$

where m_0 is the rest mass of the particle.

Total energy of the particle in SRT is the sum of the rest energy and the kinetic energy of the particle, while in SLRT, the total energy of the particle in motion with $v > c$, is [5]

$$E_t = E_0 - E_k \tag{9}$$

where E_0 is the rest energy of the particle, and [5]

$$E_k = m_0 c^2 \left(1 - \sqrt{1 - \frac{c^2}{v^2}} \right) \tag{10}$$

TABLE I
The Consequences of Lorentz and Superluminal Transformation

Lorentz Transformation ($v < c$)	Superluminal Transformation ($v > c$)
Length contraction: $L_{motion} < L_{rest}$	Length dilation: $L_{motion} > L_{rest}$
Time dilation: $T_{motion} > T_{rest}$	Time contraction: $T_{motion} < T_{rest}$
$m_{motion} > m_{rest}$	$m_{motion} < m_{rest}$
$E_{total} = E_{rest} + E_{kin}$	$E_{total} = E_{rest} - E_{kin}$

The magnitude E_k, which is labeled as kinetic energy is in fact the released energy from the particle in motion with $v > c$, and is equivalent to the loss of the particle's mass.

Here we have shown the basic superluminal effects, but in the following sections we shall elaborate on more complex superluminal effects.

III. THE SUPERLUMINAL EFFECTS IN NUCLEAR STRUCTURES

In Ref. 5 the author proposed a new deuteron structure model (Fig. 3), where the deuteron is comprehended as the $p - n$ system with two mesons in the nuclear structure, which take part in the binding energy and in the formation of nuclear forces. The results of the analysis show that all the participating objects in the deuteron, that is, two nucleons and two mesons, are rotating in a circle with the same radius.

The proton and neutron are rotating with same peripheral velocity $v_2 = 4.53 \times 10^8$ m/s, and the peripheral velocity of muons is $v_1 = v_\mu = 3.96 \times 10^8$ m/s.

Muons are rotating in the same circle with protons and neutrons, but with opposite direction. Each muon is emitted from one of the nucleons and is absorbed by the other one.

These are superluminal effects in nuclear structure, which are verified by very good accordance between computed and observed values of two important magnitudes of this nucleus: the binding energy and the magnetic moment.

The deuteron binding energy is [11,12]

$$E_d = 2.22 \, \text{MeV} \qquad (11)$$

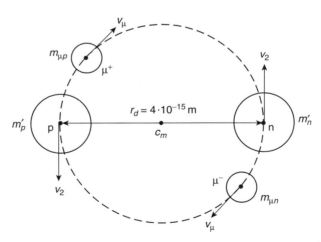

Figure 3. The new model of deuteron structure with proton, neutron, μ^- and μ^+.

and the computed value obtained by SLRT analysis is

$$E_d = 2.23 \, \text{MeV} \tag{12}$$

with accordance of 0.5%.

The observed value of deuteron magnetic moment is

$$\mu_d = 0.8734 \, \text{nm} \tag{13}$$

and the computed value obtained by SLRT analysis is[1]

$$\mu_d = 0.87339 \, \text{nm} \tag{14}$$

with accordance of 0.001%.

Because both versions, which have been elaborated on in Ref. 5—one with muons and the other with pions—give almost identically good results of the analysis, the question arises as to which version prevails, and what arguments may help us reach such a conclusion. The important thing is that the new deuteron model proposed here implies the participation of mesons in the deuteron structure. Whether they are μ mesons, that is muons, or π mesons, that is, pions, does not have any influence at all on any of the main computed magnitudes, which characterize the new deuteron model. What is really important is the fact that the newly proposed deuteron model offers a completely new insight into nuclear forces and nuclear structures in general.

We do not find it productive to involve our analysis into further speculations that will include the decay processes of muons and pions, and their mean lives in order to find out which version is more acceptable, that with muons or with pions. It is well known that the muon's mean life changes as a function of its velocity; however, there are no data regarding what happens with decay processes with particles that reach velocities $v > c$ [11].

However, the results obtained for de Broglie waves of the particles in the deuteron nucleus are in favor of presence of muons rather than pions, in the deuteron. The reason for this conclusion is that there is resonance between a nucleon's and a muon's de Broglie waves [5], which would not be the case if pions were taken as participating objects in the deuteron.[2]

We hope that this question, among the others that arise from all the analysis presented, will be an inspiration for further investigation based on the newly presented ideas for nuclear forces and structures and that it will lead to new types of experiments.

[1]Computation of deuteron magnitudes by hadronic mechanics is presented in Ref. 13.
[2]The possibility for resonance between photons with de Broglie waves of the particles is elaborated on in Refs. 4 and 14.

IV. SPACETIME CURVATURE AROUND NUCLEONS

A. The General Theory of Relativity

In 1916, Einstein published his work "The fundamentals of general relativity" [10], 11 years after he published his theory of special relativity [1,2]. Later, in 1954, he published a work to explain the differences and connections between special and general relativity [15]. In this work he gives the exact formulation of general relativity, with the following two postulates:

1. The laws of physics must be of such a nature that they apply to reference systems, in any kind of motion, relative to the mass distribution of the universe.

2. The principle of equivalence, where all bodies at the same place in a gravitational field experience the same acceleration.[3]

Both postulates are adopted in the theory of superluminal relativity [5].

According to Einstein, instead of a referent body, the Gaussian coordinate system should be used. Einstein states that "To the fundamental idea of the principle of general relativity corresponds the next statement: *All Gaussian coordinate systems are equally valid for formulations of the general laws of nature*" [15].

Furthermore, Einstein states:

> SRT is valid for Galilean ranges, which means for ones, where a gravitational field is absent. The Galilean reference body is used as a reference body, that is the same rigid body with such a chosen state of motion relative to it, so that the Galilean postulate for uniform, straight-line motion of *an individual* material, is valid. However, in gravitational fields there are no rigid bodies with Euclidean properties; the notion of rigid referent bodies has no application in the GRT. The gravitational fields influence the work of the clocks in such a way that physical definition of time strictly by the clock is no longer so evident as in the SRT [15].

This is the main reason why the Gaussian four-dimensional system is more convenient for the GRT, and consequently, *the laws of nature should not be dependent on the chosen frame of reference*. This is exactly what is proven by the theory of superluminal relativity [5].

The first postulate of general relativity, and Einstein's explanation of it, is very important for the superluminal relativity.

In contrast to the SRT concept that the presence of any kind of bodies does not influence the properties of space and time, GRT demands that bodies influence spacetime and one another. Reference 5 elaborates on spacetime curvature

---

[3]More recent evidence indicates that there is a nonlinear relationship; however, we are too close and the relative mass that we can test is too small for us to observe.

(STC) in GRT and predicts the possibility for existence of STC around nucleons and atoms by using SLRT. Reference 16 elaborates on the STC around nucleons. The consequences of STC in electrodynamics are elaborated on in Refs. 17–23.

In particular, the prediction by a validation of GRT that a beam of light will bend in the presence of a gravitation field is of interest here. The speed of light in the gravitational field is [10]

$$c' \approx c \left(1 - \frac{2Gm}{rc^2} \right) \tag{15}$$

where

$$G = 6.67 \times 10^{-11} \, \text{N} \, \text{m}^2 \, \text{kg}^{-2} \tag{16}$$

is the gravitational constant, m is mass of the sun, r is the shortest distance between the light's path and the center of the sun, and c is the speed of light in vacuum. The deflection of the light from its direction propagation is given by Ref. 10 as

$$\alpha_s = \frac{4Gm}{rc^2} \tag{17}$$

In this case, Einstein computed that light from a certain star passing close to the sun's surface would be deflected by the sun's local spacetime curvature by a factor of 1.7 seconds of arc, that is, $\alpha_s = 1.7''$. The effect of STC near massive bodies was verified during the eclipse of the sun in 1919 when the following values were measured: $\alpha_s = 0.8''$ and $\alpha_s = 1.8''$ [10,24,25].

According to Einstein's explanation, half of this deflection is caused by the mass interaction between the sun and the passing photons of the light, and half of it is by the spacetime curvature [10,24,25].

B. Quantum Gravity

Quantization of the gravitation field is applied in the following analysis requiring a brief description of quantum gravity. There is incompatibility between GRT and SRT. Misner et al. [26] note that the question of SRT constantly assumes the absence of gravitational fields, which makes SRT contrary to our reality. Gravity is ignored in the SRT because of the difficulties that the gravitational fields presented on the foundations of SRT at the time of its development. After meeting these difficulties, one can appreciate the STC methods that Einstein introduced to overcome them [10].

According to Hawking and Rocek [27], GRT has its own shortcomings. They state that the Newtonian theory of gravity is very successful in predicting planetary and stellar orbits, but because it implied that gravitational effects propagate with instant velocity, it was incompatible with the local validity of SRT. This difficulty was overcome in 1916 with the formulation of GRT in which the gravitational field was represented using STC. Since that time, the predictions of GRT have been found in excellent agreement with observations. However, GRT is incomplete in at least two ways: (1) it doesn't relate gravity to interactions and matter fields that occur in physical theories, and (2) it is a purely classical theory, whereas all other fields seem to be quantized [27–29].[4]

The necessity to quantize the gravitational field has become more urgent, as the inevitable result of GRT's classical treatment is spacetime singularities [27]. Many theories have been developed to show how to quantize the gravitational field, but to briefly show "the necessity to quantize the gravitational field," we will consider how Unrich [28] tackles the problem. The fundamental equation of GRT is

$$G_{\mu\nu} = 8\pi\langle T_{\mu\nu}\rangle \tag{18}$$

Choosing units so that G equals h equals c equals 1. The left-hand side of this equation represents the geometry of spacetime, while the right-hand side is dependent on the rest of the matter of the universe. The left-hand side is "classical," and an ordinary function of the spacetime points, while the right-hand side is a quantity, which depends on quantum operators. Thus the two sides are different and cannot be set equal to one another [28]. Isham [29] prefers to write Eq. (18) in the following form:

$$G_{\mu\nu} = T_{\mu\nu}(\text{matter}, g) \tag{19}$$

However, Isham found this treatment also to be inadequate and presented the following modification:

$$G_{\mu\nu} = \langle T_{\mu\nu}(\text{matter}, g)\rangle \tag{20}$$

Here the $\langle\ \rangle$, denotes the expectation value of the quantized system in some suitable state [28,29].

So far both Unrich and Isham have the same approach to the quantization of the gravitational field, which leads to the conclusion that the "gravitational field should be introduced as a dynamical variable rather than as a fixed background" [4,28,29].

---

[4]Note that in Part 1 (11th chapter) of this compilation (Vol. 119) Dr. Sachs has submitted a reformulation of GRT that does relate field theories of gravity and is not prone to singularities. However, although this work takes a quantization approach, we do not see this work as a violation of GRT, but an agreement.

Instead of the symbol $\langle \ \rangle$, which denotes only "the expectation value of the quantized system in some suitable state" can be obtained [29], we introduce, using Ref. 5, that quantization of the gravity of proton-neutron system is achieved, by obtaining numerical values for gravitational magnitude G'_n. By this modification of G'_n, "gravitational field is now introduced as a dynamic variable rather than as a fixed background," in agreement with Unrich [28].

The concept that quantization of gravity can be achieved if the gravitational field is introduced as a dynamic variable [28,29], is applied in the QMT [4] and in the SLRT [5]. Using the references to QMT and SLRT, spacetime can be divided into three generalized ranges with different vacuum structures:

1. The range determined by $r < \lambda_{ce}$, where the close proximity of the nucleons and their spacetime curvature affects the local gravitation attractive force, making $G'_n = f(n)$ valid.

2. The range determined by $Qr_0 > r > \lambda_{ce}$, where antigravitational, or repulsive, force develops as a result of the influence of the nucleus with electrons, making G' valid.

3. The range determined by $r > Qr_0$, returns us to the familiar form of the gravitational attractive force that exists between atoms and molecules. This range is beyond the effect of local spacetime curvature on G caused by the nucleons.

Here, r is distance from the center of the observed mass, λ_{ce} is the Compton wavelength of the electron, r_0 is the Bohr radius for the hydrogen atom, and Q is an integer larger than n, where $n = 5$ is the quantum number of the last electron shell.

The primary assumption in QMT [4] is that an antigravitational force exists that is equal to Coulomb's attractive force

$$|F_m| = |F_e| \tag{21}$$

where F_m is antigravitational force and F_e is Coulomb's attractive force, between electron and proton, when they are on the ground energy level, for $n = 1$ in hydrogen atom.

Equation (21) yields [4][5]

$$G' = 1.49 \times 10^{29} \, \text{N} \, \text{m}^2 \, \text{kg}^{-2} \tag{22}$$

In the analyses presented, the STC method is used to support the concept for new comprehension of nuclear forces and structures offered in Ref. 5.

---

[5] An erroneous value of G' is given in Ref. 4 [in Eq. (55), on p. 23], which may confuse the readers.

C. Energy Levels in the Superluminal Frame
of Reference, Where $v > c$

According to the second assumption of the SLRT [5], the vacuum properties of the spacetime determined by the boundaries λ_{ce} and λ_{cp}, influence the properties of the nucleus, and consequently influence on the nuclear reactions as well, where λ_{ce} is electron Compton wavelength and λ_{cp} is proton Compton wavelength [5].

The results of the analysis presented in Table II show that this spacetime can be divided into six energy levels, which correspond to the six values of the gravitational constant G'_n, determined by quantum number n, and to six distances r. These energy levels are determined for a proton–neutron system [5].

In Table II, energy levels for two distances $r_5 = 6.36 \times 10^{-16}$ m and $r_6 = 1.417 \times 10^{-16}$ m which are under the limit λ_{cp} are also presented. This table covers six energy levels altogether. In Table II and positions of λ_{ce} and λ_{cp}, the proton radius r_p, and the proton–neutron distance r_d in the deuteron, are also presented.

The values for G'_5 and G'_6 are computed for distances less than the radius of the proton. The reason is that each nucleus and consequently each particle have

TABLE II
Six Spacetime Energy Levels Where $v > c$

Principal Quantum Number n	$G'_n (\mathrm{Nm^2\,kg^{-2}})$	Subquantum Number, l	r(m)	α_k
			$\lambda_{ce} = 2.4262 \times 10^{-12\,a}$	
0	$G' = 1.49 \times 10^{29}$	0	$r' = 1.1688 \times 10^{-12}$	34.11''
		0.5	$r_{0.5} = 5.513 \times 10^{-13}$	16.08''
1	$G'_1 = 3.314 \times 10^{28}$	0	$r_1 = 2.6 \times 10^{-13}$	34.11''
		0.7	$r_{1.7} = 9.08 \times 10^{-14}$	21.73''
2	$G'_2 = 7.373 \times 10^{27}$	0	$r_2 = 5.78 \times 10^{-14}$	34.11''
3	$G'_3 = 1.64 \times 10^{27}$	0	$r_3 = 1.28 \times 10^{-14}$	34.11''
	$G'_d = 4.6 \times 10^{26}$	0.7	$r_{3.7} = 4.494 \times 10^{-15}$	21.73''
			$r_d = 4.0 \times 10^{-15\,b}$	
4	$G'_4 = 3.64 \times 10^{26}$	0	$r_4 = 2.86 \times 10^{-15}$	34.11''
		0.5	$r_{4.5} = 1.35 \times 10^{-15}$	16.08''
			$\lambda_{cp} = 1.32 \times 10^{-15\,c}$	
			$r = 8.13 \times 10^{-16\,d}$	
5	$G'_5 = 8.12 \times 10^{25}$	0	$r_5 = 6.37 \times 10^{-16}$	34.11''
6	$G'_6 = 1.80 \times 10^{25}$	0	$r_6 = 1.41 \times 10^{-16}$	34.11''

[a]Electron Compton wavelength.
[b]Distance between nucleons in deuteron.
[c]Proton Compton wavelength.
[d]Proton radius.

surface thickness. We shall cite deShalit and Feshbach here—because the nuclear density nor the particle density change abruptly from their nominal values to zero outside the nucleus and particle, there is a finite region called the *nuclear surface* or *particle surface.* The width of that region labeled s is defined to be the distance over which the density drops from 0.9 of its value at $r = 0$–0.1 of that value. Empirically s is a constant, that is, $s = 2.4$ fm [11,12].

Hence, we have computed G'_5 and G'_6 for distances, which go slightly beneath the surface thickness of the proton.

The values of the gravitational constants are determined by the expression [5]

$$G'_n = \frac{G'}{(4.495)^n} \tag{23}$$

where n is an integer from 1 to 6, and could be considered as the principal quantum number for this system. The latter equation can be expressed by the fine structure constant α

$$G'_n \approx \frac{G'}{(1/30\alpha)^n} \tag{24}$$

where $G' = 1.49 \times 10^{29}$ N m^2 kg^{-2}. The value of G' is determined in Ref. 4. The fine structure constant is

$$\alpha = \frac{2\pi e}{hc} \tag{25}$$

which describes the coupling of any elementary particle carrying the elementary charge e to the electromagnetic field and h is Planck's constant.

Hence, we may introduce the fine structure constant for nuclear systems, in this case for the proton–neutron system. This constant could be considered as a magnitude that expresses properties of the proton–neutron system and the surrounding space. This magnitude

$$\alpha' = \frac{1}{30\alpha} \tag{26}$$

is the *nuclear fine-structure constant*, which determines the structure of the vacuum as a function of the principal quantum number n. It determines the energy levels in the space determined by λ_{ce} and λ_{cp} between the proton and the neutron.

Thus, Eq. (23) becomes

$$G'_n \approx \frac{G'}{(\alpha')^n} \tag{27}$$

The gravitational constant G'_n and its values given in Table II suggest the possibility of the existence of a structure of the vacuum related to the gravitational properties of the particles.

The distances r_n, which correspond to certain gravitational constants G'_n, are determined by either of the following equations:

$$r_n = \frac{r'}{(4.495)^n} \tag{28}$$

$$r_n = \frac{r'}{(\alpha')^n} \tag{29}$$

D. Modified Einstein Equation for Deflection of Light Near the Sun as Applied for Protons

Einstein's equation for deflection of light near the sun is modified for nucleon, in our case for proton, when the gravitational constant G in Eq. (17) is substituted, according to Eq. (27), by the gravitational constant

$$G'_n \approx \frac{G}{(\alpha')^n}$$

The values of this constant for quantum numbers from $n = 0$ to $n = 6$ are presented in Table II. Then Eq. (17) becomes [5]

$$\alpha_k = \frac{4G'_n m_p}{r_k c^2} \tag{30}$$

where α_k is the angle of deflection for light near the proton, for the closest distance of light's path from the center of the proton, which corresponds to the total quantum number k; m_p is proton mass; c is speed of light in vacuum; and r_k is the shortest distance between the light's path and the center of the proton, determined by the expression

$$r_k = \frac{r'}{(4.495)^k} \tag{31}$$

where $k = n + l$ is the total quantum number, n is principal quantum number, and l is subquantum number with the values $0, 0.1, 0.2, 0.3, \ldots, 0.8, 0.9$.

Before we use Eq. (30) to determine the deflection angles of the light near the proton in the range λ_{ce}–λ_{cp}, it is necessary to find out how this equation can be applied for distances $r > \lambda_{ce}$, where gravitational constant G' is valid.

For the metric space determined by the distances $Qr_0 > r_p > \lambda_{ce}$, the gravitational constant is $G' = 1.49 \times 10^{29}$ N m^2 kg^{-2}; thus, Eq. (30) becomes

$$\alpha_p = \frac{4G'm_p}{r_p c^2} \tag{32}$$

where r_p is the shortest distance of the light's path from the center of the proton in the range determined above.

Figure 4 shows the curve of the light's deflection angles for characteristic distances from the center of the proton: r', λ_{ce}, $0.1r_0$ and r_0, where r_0 is the Bohr radius in the hydrogen atom. The curve shows that for distance r_0 the angle of light deflection is approaching zero. It is worth mentioning that r_0 here has a new meaning. *The Bohr radius in hydrogen atom is actually the distance from the center of the proton where the curved spacetime ends.*

Now, we will continue the analyses by applying Eq. (30) in the range λ_{ce}–λ_{cp}, where vacuum properties are quantized and gravitational constant is turning into a gravitational magnitude with values determined by the principal quantum number n.

For $l = 0, k = n = 0$ the Eq. (30) becomes

$$\alpha_0 = \frac{4G'm_p}{r'c^2} \tag{33}$$

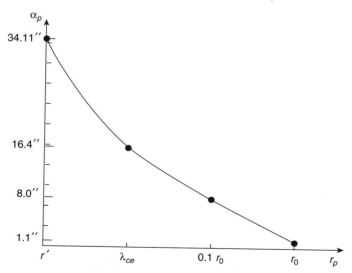

Figure 4. Presentation of the function $\alpha_p = f(r_p)$ for $r_p > r' = 1.1688 \times 10^{-13}$ m.

Table II shows that α_k, for the values of k, from $n = 0$ to $n = 6$, with $l = 0$, has the same value:

$$\alpha_k = \alpha_0 = 34.11''(\text{arcseconds}) \tag{34}$$

The ratio between this angle of light deflection α_k near the surface of the proton and the angle of light deflection near the surface of the sun, α_s is

$$K_\alpha = \frac{\alpha_k}{\alpha_s} = 20.064 \tag{35}$$

This equation (35) shows that light, passing near the proton's surface, has angle of deflection 20 times the angle of deflection of the light passing near the sun.

Determining the angle of light deflection near the proton could be considered as a new method for determining the vacuum properties and presence of mass, which could be experimentally verified and will be discussed later on. By using this method we shall determine the angles of the light's deflection near the proton, in the metric space determined by the total quantum number k. Figure 5a shows the curve $\alpha_k = f(k)$. Figure 5b is a 3D presentation of Fig. 5a. The sphere at the center of the 3D diagram represents the proton out of the scale of proportion.

The curve in the Fig. 5a shows the periodic nature of the STC around a proton. It is important to emphasize that the gravitational magnitude G'_n is valid for distances corresponding to the quantum numbers from $k = n$ to $k = (n-1) + l$, where $l \neq 0$.

By using this method it will be shown that the STC expresses distribution of the intrinsic vacuum energy [5], and in Section V it will be used to explain the phenomena of superluminal effects in an experiment with cesium atomic gas [6,7].

It is possible to choose any particular distance between the light's path and the center of the proton, in the range of λ_{ce}–λ_{cp}, in order to explore the energy distribution in the STC. The light, that is, individual photons, can be used as a kind of probe for detecting the energy distribution in the STC. Besides the photons in the visible range, X and gamma rays can be used.

In Table II values for α_k are presented for distances that correspond to $k = n$ and for distances corresponding to $k = n + l$ when G'_n is valid for distances corresponding to $k = n$ and $k = (n-1) + l$, when $l \neq 0$.

Figure 6a presents the values of α_k for chosen distances and presents a curve, that is formed by extreme values of α_k. Fig. 6b is a 3D presentation of periodical function of $\alpha_k = f(k)$ from Figure 6a. The sphere at the center of the 3D diagram represents the proton with exaggerated scale and proportion.

The curve presented in the Fig. 6a should be considered periodic because it represents the vacuum properties of non-Euclidean spacetime. Here the axes of

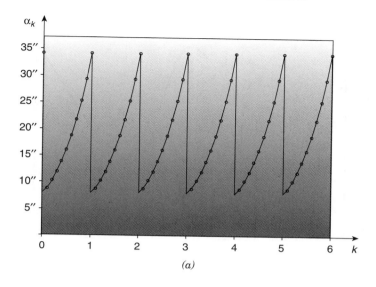

(a)

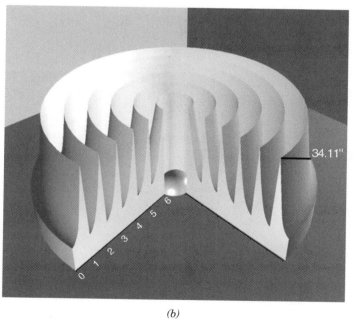

(b)

Figure 5. (a) Presentation of the curve $\alpha_k = f(k)$; (b) 3D presentation of part (a). The sphere at the center of the 3D diagram represents the proton in out-of-scale proportion.

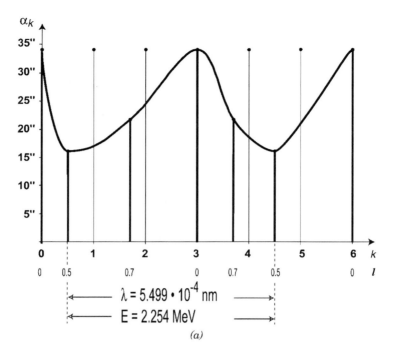

$$\lambda = 5.499 \cdot 10^{-4} \, \text{nm}$$

$$E = 2.254 \, \text{MeV}$$

(a)

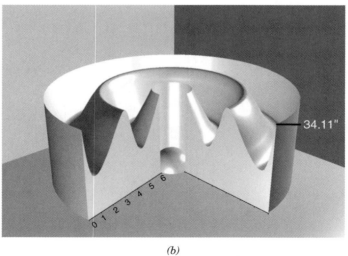

(b)

Figure 6. (a) Presentation of $\alpha_k = f(k)$ for chosen distances determined by l. (b) 3D presentation of the periodic function $\alpha_k = f(k)$ from part (a). The sphere at the center of the 3D diagram represents the proton in out-of-scale proportion.

the coordinate system are straight lines, while in non-Euclidean spacetime there are no straight lines. This figure actually shows that STC around a proton has a complex nature. This is so because it is not possible for non-Euclidean figures to be presented by Euclidean geometry, which is evident in Figs. 5a and 6a. Such graphic presentations have to be taken conditionally. The wavelength of the function presented in Fig. 6a is

$$\lambda_{st} = 5.499 \cdot 10^{-4} \, \text{nm} \tag{36}$$

The quantum of the energy associated with this wavelength is

$$E_{st} = 2.254 \, \text{MeV} \tag{37}$$

The binding energy of deuteron nucleus is

$$E_d = 2.22 \, \text{MeV} \tag{38}$$

The accordance between these two values is 1.5%. This computation shows that spacetime curvature around a proton is characterized by the wavelength λ_{st} and energy E_{st}.

The region of spacetime curvature around proton, which is covered by this computation, is determined by the total quantum numbers from $k = 0$ to $k = 6$. The result shows that this region has potential energy of the order of MeV. The proton-neutron system, which creates deuteron, is in the energy level determined by the total quantum number $k = 3$ while the binding energy of this nucleus is 2.23 MeV.

The analysis presented above shows that the value of the deuteron's binding energy computed by the magnitudes, which characterize the spacetime curvature around proton, has excellent accordance with experimentally observed values. Therefore, it is justified to consider these results as experimentally verified, what proves that the spacetime curvature around proton exists.

The structure of the vacuum, determined by the gravitational field around the sun, depends on the constellation of the planets and the sun itself, which form the solar system. The main magnitude, which expresses this gravitational field, is the gravitational constant G. In the same manner, the vacuum structure determined by the gravitational fields around the proton depends on the proton itself and on the particles around it. In our case the vacuum structure was determined for two-particle system, that is, proton-neutron system. When the distance between these two nucleons is $r_d = 4.0 \times 10^{-15}$ m, they form the deuteron. The space between these two particles in the region $\lambda_{ce}-\lambda_{cp}$ has vacuum structure determined by the gravitational magnitude G'_n.

The ratio between gravitational constants G'_n and G yields very interesting equation,

$$\frac{G'_n}{G} = K_\alpha \frac{m \cdot r_{pn}}{m_p \cdot r} \tag{39}$$

In this equation gravitational magnitudes G'_n and G, which represent corresponding vacuum structures, are connected with masses of the sun (m) and the proton (m_p), and distances, where phenomena of light deflections are taking place. The constant, which establishes this connection, is K_α, the ratio of angles of deflections of the lights passing near the surfaces of the proton and the sun, respectively.

The results of the analysis of the STC around a proton show that there is an energy distribution around a proton as a result of vacuum structure and presence of the mass in that vacuum.

Let us consider the possibility of an electron crossing the threshold determined by r_0 and entering the STC region determined by λ_{ce}–λ_{cp}. In such a case there will be a possibility for the electron to fall into one of the energy holes of the STC. The hydrogen atom with an electron in such a position will exhibit some exotic properties and certainly will be an unstable system. Because it is an unstable system, such an atom will eject the electron. The ejected electron will have an energy that is an order of MeV, corresponding to the energy hole of the STC, from which the electron was ejected. An electron can have such an amount of energy only if it moves with superluminal velocity [5]. *This is superluminal effect caused by STC around atoms.* This electron will produce an electromagnetic field, which corresponds to the energy of the electron.

The process of proton–electron interaction in the STC, described here is in fact interaction between their masses, but the outcome of that interaction is the production of the electromagnetic field. This hypothetical process shows a direct connection between the phenomena covered by the theory of general relativity and electrodynamics phenomena [5,16,20,21].

V. WKD (WANG–KUZMICH–DOGARIU) SUPERLUMINAL EFFECTS

The main assumption in SRT is that speed of any moving object cannot exceed that of the light in vacuum (c) [1–3]. The authors of the experiment with superluminal light propagation, L. J. Wang, A. Kuzmich, and A. Dogariu (WKD), have declared: "The group velocity of a laser pulse in this region exceeds c . . ." [6,7]. The authors explain superluminal light propagation in cesium atomic cell by anomalous dispersion of light, which is defined by group velocity index $n_g = n + \nu(dn/d\nu) < 1$ and by group velocity $V_g = c/n_g > c$.

For frequency $v = 3.5 \times 10^{14}$ Hz and for narrow frequency region of $\Delta v = 1.9$ MHz, the group-velocity index has the value $n_g = -330(\pm 30)$ in the example presented in Refs. 6 and 7. In the cited references the group velocity V_g, is not computed. This is crucial magnitude in anomalous dispersion because if $V_g > c$, then the consequence is superluminal light propagation. However, for this example $V_g = -9.09 \times 10^5$ m/s or $V_g < c$. The conclusion is that there is no superluminal light propagation, which is opposite to the observed phenomenon. Therefore there is no consistency between experimentally observed results and the theory presented in the same references [6,7], used to explain this phenomenon.

The presented analysis is based on the SLRT concepts [5], and the results show that $V_g > c$ for discussed example and for many others in the visible spectrum of light. This provides theoretical verification of the observed superluminal light propagation in WKD experiment [30].

In Refs. 6 and 7 the authors declare:

It has been mistakenly reported that we have observed a light pulse's group velocity exceeding by a factor of 300. This is erroneous. In the experiment, the light pulse emerges on the far side of the atomic cell sooner than it had traveled through the same thickness in vacuum by a time difference that is 310 folds of the vacuum transit time.

In our approach we shall assume that, $c_s = 310c = 9.3 \times 10^{10}$ m/s. If we take that an electron in cesium atomic cell is in motion with velocity c_s, Eq. (10), for this electron yields $E_{ek} = E_{\text{released}} = 2.7$ eV. Using the expression $E_{ek} = hv$, the frequency that corresponds to this quantum of energy is $v = 6.56 \times 11^{14}$ Hz. There are two wavelengths, which correspond to this frequency. One is in the SRT frame of reference, $\lambda_b = c/v = 457.13$ nm, which is in the blue region of the visible spectrum of light. The other one, in the SLRT frame of reference $\lambda_s = c_s/v = 1.417 \cdot 10^5$ nm, is in the range of electromagnetic waves, known as *medium radiofrequencies* [30].

The results obtained, which show that the same electron can emit photons with wavelength λ_b in SRT, and electromagnetic waves with λ_s in SLRT, suggest the existence of anomalous dispersion of light in cesium cell. These results, together with the results of further analysis, show that anomalous dispersion is consequence of superluminal phenomena in cesium cell, not vice versa, as the authors of the Refs. 6 and 7 claim.

By using Eq. (10) from the SLRT, for the released energy of the particle, in this case, electron, in motion with $c_s > c$, are obtained superluminal functions presented in Fig. 7. In the graphic presentation of this figure, c_s is superluminal speed of the electron, and superluminal speed of propagation of electromagnetic waves, λ_s is superluminal wavelength of electromagnetic waves, n_s is superluminal refractive index, and v is the frequency of the photons from the visible spectrum of the light, incident to cesium cell.

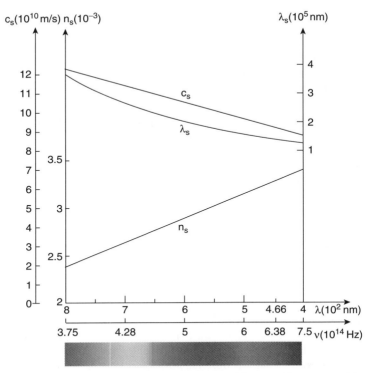

Figure 7. Presentation of superluminal functions $c_s = f(\nu), \lambda_s = f(\nu)$, and $n_s = f(\nu)$, where ν are the frequencies from the visible spectrum of light.

We shall use the superluminal functions from the Fig. 7 to compute V_g for the example given in the Refs. 6 and 7. If the frequency range is determined by the frequency from the example $\nu = 3.5 \times 10^{14}$ Hz and frequency $\nu = 3.75 \times 10^{14}$ Hz, the group velocity index is $n_{gs} = 3.1 \times 10^{-3}$. Then, for group velocity is obtained $V_{gs} = 9.677 \times 10^{10}$ m/s or $V_{gs} = 322.5c$, which is $V_{gs} > c$ *The consequence is superluminal propagation of light.* Hence, by using the concepts of SLRT we have shown that even for the same example from Refs. 6 and 7, group velocity verifies that superluminal propagation of light in the cesium cell is observed.

Using several other examples, which cover the whole visible spectrum of light, for the mean value of group velocity we obtain $\bar{V}_{gs} = 9.65 \cdot 10^{10}$ m/s, or $\bar{V}_{gs} = 321.6c$, which is 3.7% different from "a time difference that is 310 folds of vacuum transit time" [30].

All this shows that the SLRT applied here is consistent with observed results of the WKD superluminal experiment; therefore, this experiment should be considered as a direct proof for the validity of SLRT.

The results of this analysis show that anomalous dispersion of light in a cesium cell is a consequence of superluminal motion of electrons and superluminal propagation of electromagnetic waves. The Feynman diagram, presented in Fig. 8, is used in the analysis, to explain the phenomena that are taking place in cesium atomic cell and that cause *superluminal effects* [30].

Besides the photon-electron interactions, it will be assumed that the main phenomenon, that is taking place in the cesium cell, is an electron–cesium atom interaction. This interaction is assumed to be the result of the existence of STC around nucleons and atoms [5,16]. The momentum $\vec{p}_{STC}$ will express the presence of STC around cesium atoms.

Vertex A. Electron with momentum $\vec{p}_{e1}$ interacts with a photon with momentum $\vec{p}_{\lambda1}$, from the visible spectrum of light. This interaction produces an electron with momentum $\vec{p}_{e\lambda1}$.

Vertex B. Electron with momentum $\vec{p}_{e\lambda1}$ interacts with momentum $\vec{p}_{STC}$. An important assumption is made here. It has already been mentioned that in Ref. 5 the possibility for existence of STC around nucleons and atoms is predicted, and elaborated on in Ref. 16. It is assumed that in the electron–cesium atom interactions, distortion of the STC around the atom is taking place. As a result of this STC distortion, the momentum $\vec{p}_{STC}$ is created, which is added to the electron momentum $\vec{p}_{e\lambda1}$. The resultant momentum of the electron then will be $\vec{p}_{es} = \vec{p}_{e\lambda1} + \vec{p}_{STC}$. As a result of STC distortion, the electron will be in motion with faster than light speed in the STC between cesium atoms. Hence, at the vertex B, the superluminal process is taking place.

Vertex C. It was stated already before, that an electron with speed c_s will emit photon with wavelength $\lambda_s \gg \lambda_1$. That will happen at vertex C.

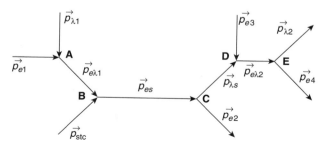

Figure 8. The Feynman diagram presentation of superluminal processes in a cesium cell.

Here, an energy transition of the electron will take place, and residual part of this process will be an electron with momentum $\vec{p}_{e2} = \vec{p}_{e0}$. Hence, the superluminal effects are taking place at vertices B and C. The possibility of an electron at this vertex emitting a photon with $\lambda_s \gg \lambda_1$ shows that anomalous dispersion in cesium cell is a result of superluminal processes, not vice versa.

Vertex D. A photon with momentum $\vec{p}_{\lambda s}$ interacts with an electron with momentum $\vec{p}_{e3} = \vec{p}_{e1}$, and the result is an electron with momentum $\vec{p}_{e\lambda 2}$, which is capable of emitting a photon with wavelength λ_2 from the visible spectrum of light.

Vertex E. At this vertex two possible processes may take place:

E1. The electron with momentum $\vec{p}_{e\lambda 2}$ will emit photon with wavelength $\lambda_2 = \lambda_1 = \lambda_b = 457.13$ nm, which is from the visible spectrum of light. According to Fig. 7, this is in the case when $c_s = 310c = 9.3 \times 10^{10}$ m/s. The residual part of this process will be an electron with momentum $\vec{p}_{e4} = \vec{p}_{e3} = \vec{p}_{e1}$.

E2. If at point E the momentum $\vec{p}_{STC}$ is present again, then the whole process described above will be repeated until the photon with wavelength λ_b enters the material of the cesium cell's wall. Hence, in the cesium cell the chain of the abovementioned interactions will take place. The photon with wavelength λ_b will leave the cell as a photon, which propagates with speed c.

The results of the analysis presented here, based on the concepts of the theory of superluminal relativity [5], show that in the superluminal experiment performed by Wang et al. [6,7] the main phenomenon is distortion of curved spacetime [16] around cesium atoms, which produces superluminal processes, and the final effect is anomalous dispersion of light [30].

The final results of the analysis show that an electron can travel and electromagnetic waves can propagate in curved spacetime between cesium atoms, which is faster than the speed of light in vacuum, which has been observed in the WKD superluminal experiment [6,7,30].

References

1. A. Einstein, "Zur Elektrodynamik bewegter Körper," *Annal. Phys.* **17**, 891 (1905).

2. A. Einstein, "Ist die Trägheit eines Körpers von seinem Energiegehalt abhängen?" *Annal. Phys.* **18**, 639 (1905).

3. H. Lorentz, A. Einstein, H. Minkovski, and H. Weyl, *The Principle of Relativity*, Dover, New York (1958).

4. P. K. Anastasovski and T. M. Benson, *Quantum Mass Theory Compatible with Quantum Field Theory*, Nova Science, New York (1995).

5. P. K. Anastasovski, *Superluminal Relativity Related to Nuclear Forces and Structures*; U.S. Department of Energy Website: http://www.ott.doe.gov/electromagnetic/papersbooks.html.

6. L. J. Wang, A. Kuzmich, and A. Dogariu, *Nature* **406** (6793), 277–279 (2000).

7. L. J. Wang, A. Kuzmich, and A. Dogariu, *Demonstraton Gain-Assisted Superluminal Light Propagation* available at Website http://www.neci.nj.nec.com/homepages/Lwan/demo.h.

8. Ye. P. Velikhov et al., *Phisics of 20th Century*, Mir Publishers, Moscow (1987).

9. J. S. Green, in A. G. Maddock (Ed.), *Radiochemistry*, Butterworth, London, p. 251 (1972).

10. A. Einstein, "Grundlage der Allgemeinen Relativitätstheorie" *Ann. Phys.* **49** (Suppl.), 769–822 (1916).

11. A. deShalit and H. Feshbach, *Theoretical Nuclear Physics*, Vol. I, Wiley, New York (1990).

12. K. N. Muhin, *Eksperimentalnaya yadernaya fizika*, Energoatomizdat, Moscow (1983) (in Russian).

13. R. M. Santilli *J. New Energy* **4**(1) (1999).

14. P. K. Anastasovski, *Theory of Magnetic and Electric Susceptibilities for Optical Frequencies*, Nova Science, New York (1990).

15. A. Einstein, *Über die Spezialle und die Allgemeine Relativitätstheorie*, Vieweg Verlag, Braunschweig (1954).

16. P. K. Anastasovski and D. B. Hamilton, *Space-Time Curvature Around Nucleons*; U.S. Department of Energy Website: http://www.ott.doe.gov/electromagnetic/papersbooks.html.

17. B. Lehnert and S. Roy, *Extended Electromagnetic Theory*, World Scientific, Singapore (1998).

18. M. W. Evans and L. B. Crowell, *Classical and Quantum Electrodynamics and the B Field*, World Scientific, Singapore (2000).

19. L. B. Crowell, *Found. Phys. Lett.* **12**, 585 (1999).

20. M. W. Evans et al., AIAS group paper, "Development of the Sachs theory of electrodynamics, *Optik* (in press).

21. M. W. Evans et al., AIAS group paper, "Electromagnetic energy from curved space-time," *Optik* (in press).

22. M. W. Evans et al., AIAS group paper, "Runaway solutions of the Lehnert equations: The possibility of extracting energy from the vacuum," *Optik* (in press).

23. M. W. Evans et al., AIAS group paper, "longitudinal modes in vacuo of the electromagnetic field in riemannian space-time," *Optik* (in press). acknowledgement of receipt (2000).

24. A. Einstein, "Physics and reality," *J. Franklin Inst.* **221**(3) (1936).

25. G. E. Tauber (Ed.), *Albert Einstein's Theory of General Relativity*, Crown, New York (1979).

26. C. W. Misner, K. P. Thorne, and J. A. Wheeler, *Gravitation*, Freeman, New York (1973).

27. S. W. Hawking, M. Rocek (Eds.), *Superspace and Supergravity, Proc Newfield Workshop* Cambridge Univ. Press, Cambridge, UK (1980).

28. W. G. Unrich, in S. M. Christensen (Ed.), *Quantum Theory of Gravity*, Adam Hilger, Bristol (1984).

29. C. J. Isham, in C. J. Isham, R. Penrose, and D. W. Sciama, (Eds.), *Quantum Gravity Oxford Symp.*, Clarendon Press, Oxford (1975).

30. P. K. Anastasovski, *WKD Superluminal Experiment Explained by the Theory of Superluminal Relativity*; available at U.S. Department of Energy Website: http://www.ott.doe.gov/electromagnetic/papersbooks.html.

SUPERLUMINAL EFFECTS AND TACHYON THEORY

FABIO CARDONE

Dipartimento di Fisica Universitá de L'Aquila, L'Aquila, Italy

ROBERTO MIGNANI

Dipartimento di Fisica "E. Amaldi," Universitá degli Studi "Roma Tre," Rome, Italy

CONTENTS

I. INTRODUCTION: EXTENDED RELATIVITY

The possible existence of faster-than-light objects has a long history, which, since the early 1900s, can be traced back in pre-relativistic times) to J. J. Thomson and A. Sommerfeld (see the excellent review in Ref. 1 by Recami for a full historical account). After the advent of special relativity in 1905, light speed (the speed of light) in vacuum was considered as the maximal causal speed, as an upper limit for any velocity. Such a common belief lasted for about half a century, when the problem of faster-than-light particles was reconsidered

Modern Nonlinear Optics, Part 3, Second Edition, Advances in Chemical Physics, Volume 119, Edited by Myron W. Evans. Series Editors I. Prigogine and Stuart A. Rice.
ISBN 0-471-38932-3   © 2001 John Wiley & Sons, Inc.

in a famous paper by Bilaniuk et al. [2]. G. Feinberg called such objects *tachyons* (from the Greek word ταχύς, meaning fast), whereas E. Recami introduced the word *bradyons* (from βραδύς, slow) to denote usual (slower-than-light) particles [1]. After then, studies on phenomena occurring at Superluminal [1,3] ($v > c$) velocities started again. In particular, basic contributions were given by the Italian school on this subject, headed by E. Recami. Its main fundamental result was the generalization of special relativity (SR) to superluminal inertial frames [1,3], whence the term *extended relativity* (ER) given to such a theoretical framework. Let us briefly review its main features (we refer the reader to Refs. 1 and 3 for a detailed discussion).

Extended relativity (ER) simply follows from the two basic postulates of SR:

1. The principle of relativity
2. The homogeneity of spacetime and the isotropy of space

The existence of a (unique) invariant speed follows on a theoretical basis from (2). Such a speed is identified experimentally with the light speed in vacuum, c, which is a *two-side* limit speed, in the sense that one can approach it either from below or from above. Particles can be thus bradyons, tachyons, or photons, and it's possible to define rest frames for the two first classes of objects. The two principles imply (if one does not restrict oneself, a prior, to subluminal velocities) that the fundamental square interval (i.e. the metric tensor) $ds^2 = g_{\mu\nu}dx^\mu dx^\nu = c^2 dt^2 - dx^2 - dy^2 - dz^2$ is invariant except for the sign. The transformations that preserve also the metric sign are the usual, subluminal Lorentz transformations (LT); those changing the metric sign are Superluminal LTs. In two dimensions, the *generalized Lorentz transformations* (boosts) (valid for $-\infty < u < +\infty$) were first introduced by Recami and one of the present authors (RM), and read ($\beta = u/c$) [1,3]

$$\begin{cases} x_0' &= \pm \dfrac{x_0 - \beta x}{\sqrt{|1 - \beta^2|}} \\[4mm] x' &= \pm \dfrac{x - \beta x_0}{\sqrt{|1 - \beta^2|}} \end{cases} \tag{1}$$

Superluminal LTs are somewhat more complicated in four dimensions–they involve imaginary quantities in the components of four-vectors transverse to the direction of relative motion, and suitable reinterpretation procedures [1]. Let us stress that ER, besides being able of consistently describe tachyons and their properties, has interesting implications even for standard (bradyon) physics [1,3].

The interest in superluminal processes has been revived, due to some new experimental results (mainly based on electromagnetic wave propagation) that provided incontrovertible evidence for motions occurring at faster-than-light speed [4,5]. It is just the aim of this review to present the results of these and of other experiments, apparently pointing in the same direction.

Our review does not pretend to be either exhaustive or self-contained. It is only aimed to briefly discussing the main experimental facts (most of which have been found since 1991) whereby the physics of superluminal phenomena—which was regarded until the mid-1990s or so by most physicists as "a waste of time," or worse—is becoming an exciting and promising reality. Its main purpose is to acquaint the reader with such facts, by framing them in the context of the ER, and to provide him with an essential bibliography.

The order of exposition is not casual, but reflects our own opinion on the increasing relevance of the effects discussed. The organization is as follows. In Section II, we discuss the observation of superluminal motions in astrophysical objects. Section III is concerned with the upper bounds on neutrino mass, whose square is systematically found to be negative, thus suggesting the possibility that neutrinos are tachyons. In Section IV we review the evidence for superluminal propagation of evanescent waves in both the microwave range and the optical domain. Section V is devoted to the same evidence of superluminality found for electromagnetic wave propagation in media with anomalous dispersion. In Section VI we discuss the very important subject of superluminal X-shaped waves, which have been experimentally detected both in acoustics and in optics. The widely debated topics of causality violation possibly implied by the observed superluminal motions is briefly considered in Section VII.

II. ASTROPHYSICAL SUPERLUMINAL EXPANSIONS

The astrophysical results concerning Superluminal phenomena are both the oldest and the most uncertain ones. They date back to 1971, when apparent faster-than-light expansions were first reported for the quasars 3C279 and 3C273 [6]. Since then, superluminal motions have been observed in many quasars (and even a few galaxies) [6]. What is observed is that radio-emitting components in the quasars or in the galactic nuclei (ejected from their central sources, probably supermassive black holes) move away from each other with angular separation rates that seem to correspond to Superluminal speeds. The main problem in the interpretation of these data is related to the fact that they critically depend on the distance of the observed source from earth. So such experimental reports have given rise to a number of theoretical debates, concerning the real distance of quasars and a possible noncosmological nature of their redshift [7]. However, quite recently Superluminal motions have been observed for objects ("microquasars") in our own galaxy [8], and of course the estimated distances are less uncertain for them.

A number of "orthodox" models (some of which quite complex!) have been proposed in order to explain such apparent superluminal expansions (a brief account of the main ones can be found in Ref. 1). A description of the astro-physical superluminal motions can of course, be given in terms of the tachyon theory based on extended relativity [1,9]. One of the intriguing results of such superluminal model is that *only one* faster-than light source is needed the observed facts, since it can be shown that such a radiating source is actually seen (after an initial phase of optical "boom," analogous to the acoustic boom produced by a supersonic aircraft) as splitted in *two* objects, receding from each other with superluminal speed $v > 2c$ [9]. We refer the reader to Refs. 1 and 9 for a detailed discussion.

At present, the majority of astrophysicists are in favor of the standard interpretations. In fact, one is unable to give a definite answer to the question if astrophysical "superluminal expansions" are superluminal, indeed [1].

III. ARE NEUTRINOS TACHYONS?

It is well known that the possibility of a nonzero neutrino mass (according to the old hypothesis by Pontecorvo) has found some ground (although not yet definite) support from solar neutrino experiments. The solar neutrino deficit (with respect to theoretical estimates) is in fact explained on the basis of neutrino oscillations, which can occur only if neutrinos are massive. What is instead less known is that all the experiments that place bounds on the mass of the electron neutrino ν_e and the muon neutrino ν_μ find a negative value for the squared neutrino mass, $m_\nu^2 < 0$ [10]. The orthodox explanation of this fact is ascribing it to some systemic effect, such as the underestimation of energy loss. Such results, find, of course, a straightforward explanation if one assumes that *neutrinos are tachyons*. In "naive" tachyon theory, this corresponds to the fact that a tachyon is assumed to carry an imaginary mass; in the tachyon theory based on extended relativity, this is simply a consequence of the space-like dispersion relation for tachyons [1,3]

$$E^2 - p^2c^2 = -m^2c^4 \tag{2}$$

For a detailed discussion of the theoretical and experimental implications of tachyon neutrinos, we refer the reader to Refs. 11 and 1.

IV. PHOTON TUNNELING AND EVANESCENT WAVES

Tunneling of a particle through a potential barrier is a well-known quantum effect, which finds many applications ranging from the tunnel diode to the scanning tunneling microscope. Nevertheless, many controversies still exist on the seemingly simple-sounding question of the time taken by a particle to tunnel. We don't want to enter into such a debate here (for which we refer the interested reader to Refs. 12–14), and only confine ourselves to quote that most of the theoretical definitions of tunneling time do imply that the tunneling process is superluminal. This is essentially related to the *Hartman–Fletcher* (HF) *effect* [15], where the tunneling time is independent of the barrier width d for sufficiently large d. The validity of the HF effect for all the mean (nonrelativistic) tunneling times has been proved in Ref. 16.

A well-known optical analog of the quantum tunneling is provided by total internal reflection. Consider two right-angle glass prisms, with their faces corresponding to hypothenuse separated by a thin airgap of width d, and a light wave incident on the first prism at an angle $\theta > \theta_c$ [where $\theta_c = \sin^{-1}(1/n)$ is the critical angle of total reflection, if $n > 1$ is the refraction index of the glass]. Then, the wave is transmitted in the second prism, beyond the interface gap, only in the form of an exponentially attenuated (evanescent) wave. This phenomenon is formally analogous to the (one-dimensional) tunneling through a potential barrier of height V_0 and width d by a particle with energy $\mathscr{E} < V_0$ (see Fig. 1). The analogy is rooted in the formal identity between the classical Helmholtz equation describing electromagnetic wave propagation and the quantum Schrödinger equation for a particle. The general form of such an equation can be written as

$$\nabla^2 f + \kappa^2(\mathbf{r}) f = 0 \tag{3}$$

where f is any component of the electromagnetic field for the electromagnetic case, or the wavefunction ψ in the quantum case, and the "wavevector" $\kappa(\mathbf{r})$ depends on the (classical or quantum) case and on the specific problem considered.

For a particle of mass m and energy $\mathscr{E}$ in a potential $V(\mathbf{r})$, the (time-independent) Schrödinger equation reads

$$\nabla^2 \psi + \frac{2m}{\hbar^2} [\mathscr{E} - V(\mathbf{r})] \psi = 0 \tag{4}$$

so

$$\kappa = \frac{1}{\hbar} \sqrt{2m[\mathscr{E} - V(\mathbf{r})]} \tag{5}$$

In particular, for a uniform potential barrier of height V_0, inside the barrier κ becomes imaginary for $\mathscr{E} < V_0$.

For a monochromatic wave of frequency ω in an inhomogeneous and isotropic medium of refraction index $n(\mathbf{r})$, the electromagnetic field satisfies Eq. (1) with κ given by

$$\kappa(\mathbf{r}) = \frac{n(\mathbf{r})\omega}{c} \tag{6}$$

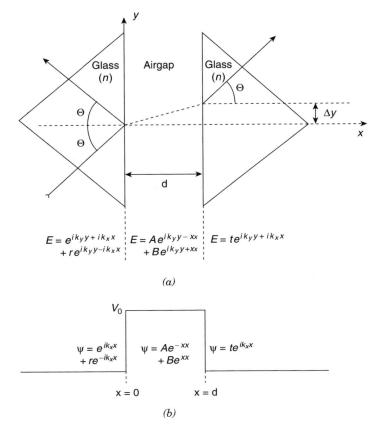

Figure 1. Analogy between total internal reflection and quantum tunneling: (a) two glass prisms separated by an airgap; (b) its quantum barrier analog. (From Nimtz and Heitmann [5].)

whereas in the case of the two-prism system, we have, inside the airgap

$$\kappa = \frac{\omega}{c} \sqrt{1 - n^2 \sin^2 \theta} \tag{7}$$

Finally, for electromagnetic (EM) wave propagation in waveguides, it is

$$\kappa = \frac{n}{c} \sqrt{\omega^2 - \omega_c^2} \tag{8}$$

where ω_c is the cutoff frequency of the waveguide.

The equations above are at the very basis of *the particle–photon tunneling analogy* [17]. The condition $\mathscr{E} < V_0$ in the tunneling of a particle through a potential barrier corresponds to an incident angle $\theta > \theta_c$, for the total internal reflection, and to $\omega < \omega_c$, for an evanescent mode in a waveguide. Therefore, one expects on theoretical grounds that in such cases photon tunneling takes place at superluminal speed. This is exactly what has been observed since 1992 by some experimental groups [18–24]. The observed speed is the *group velocity* v_g of the wavepacket, defined as usual by

$$v_g = \frac{d\omega}{dk} \tag{9}$$

(the phase velocity being given by $v_p = c/n$).

Superluminal photon tunneling has been observed in both the microwave range [18–20] and in the optical domain [21–24] in experiments performed at Cologne [18,19], Florence [20], Berkeley [21], Vienna [22], Orsay [23], and Rennes [24]. Apart from the optical experiment with total internal reflection [23] (in which, as seen before, the barrier is represented by the airgap between the two prisms), two kinds of electromagnetic barriers have been used.

In the case of microwave propagation inside a (rectangular) waveguide, the analogous of an uniform potential barrier for the photon can be implemented by either reducing the cross section of its central part (Fig. 2a) or filling the guide (of uniform cross section) with two dielectric media with refraction indexes n_1, n_2, $n_1 > n_2$ (where the index 2 refers to the central part of the guide) (Fig. 2b). Both kinds of undersized (i.e., under cutoff frequency) waveguides have been used [18]. Another configuration utilized in such a class of experiments was a double-barrier one, i.e a waveguide with two segments of undersized waveguide of different lengths [19]. It was found in this case that the tunneling time does not depend on the length of the two barriers (according to the HF effect), nor on the length of the normal waveguide between them (i.e., in this intermediate region the wave speed becomes infinite).

The second type of barrier is provided by a periodic dielectric multilayer (or "dielectric mirror") structure—a periodic stack of dielectric layers of alternating media with two different refractive indices (Fig. 3a). By virtue of the analogy between Eqs. (3) and (4), such a structure corresponds to the motion of an electron in a periodic potential, described by a Kronig–Penney model. This yields forbidden bandgaps, in which tunneling modes exist (Fig. 3b). This "photonic bandgap" type of barrier has been used, for instance, to study single-photon tunneling [21].

Still another type of evanescent wave has been considered by the Florence group [20]. They worked with microwaves traveling in free space between two horn antennas. The waves were observed to travel at subluminal speed if the antennas faced each other.[1] On the contrary, a superluminal propagation occurs

---

[1]The original aim of the Florence group was just to repeat the experiment by T. K. Ishii and G. C. Giakos, who claimed in 1991 to have observed, by a similar device, superluminal propagation in free space. Such an experiment was basically flawed by the confusion made by the authors between phase and group velocity.

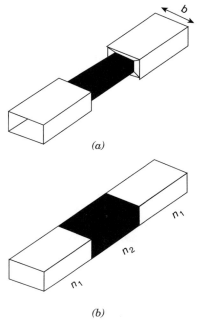

(a)

(b)

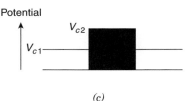

Figure 2. Undersized waveguide used in microwave tunneling experiments: (a) waveguide with central reduced cross section; (b) waveguide filled with two dielectrics with different refractive indexes $n_1 > n_2$; (c) the same quantum potential barrier corresponds to both configurations. (From Nimtz and Heitmann [5].)

(c)

if the receiver is placed perpendicular to the propagation direction. This is because, due to the microwave diffraction out of the square aperture of the launcher, "leaky" evanescent waves propagate in the shadow region (Fig. 4).

In other words, experiments tell us that *evanescent* (i.e., exponentially decaying amplitude) *waves of whatever type do propagate faster than light.*

Needless to say, the prevision of superluminal speeds for evanescent waves was made in the context of ER. It's indeed a straightforward consequence of the fact that an imaginary wavevector.[2] corresponds to a space-like dispersion relation [see Eq. (2)], and therefore to a tachyonic behavior [1,3].

---

[2]This holds (at least formally) also for diffraction leaky waves (like those considered by the Florence group), since diffracted waves can be thought of as evanescent waves with imaginary wavevectors, which represent small departures from the propagation vector of the undiffracted principal wave.

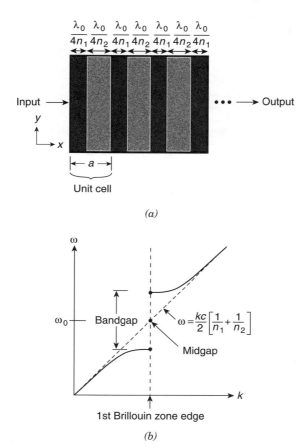

Figure 3. Photonic bandgap: (a) periodic stack of dielectric layers of alternating media with two different refractive indices; (b) the bandgap at the edge of the first Brillouin zone, which provides the analog of the potential barrier in this case. (From Chiao and Steinberg [4].)

V. SUPERLUMINAL OPTICAL PROPAGATION IN MEDIA WITH ANOMALOUS DISPERSION

The propagation of EM waves in a dispersive medium has been investigated for a long time [25]. The group velocity of a wavepacket in such a medium is given by

$$v_g = \frac{c}{n_g} \tag{10}$$

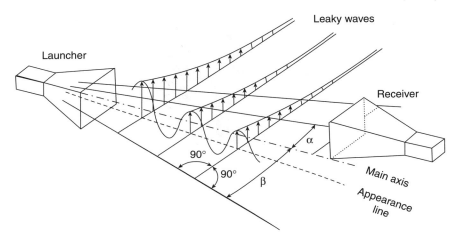

Figure 4. Evanescent "leaky waves" due to diffraction out of the square aperture of the launcher antenna observed as superluminal by the Florence group. (From Nimtz and Heitmann [5].)

where *the group velocity index n_g* is defined as

$$n_g = n(\omega) + \omega \frac{dn(\omega)}{d\omega} \qquad (11)$$

For normal dispersion, it is $(dn/d\omega) > 0$, and $n > 1$, so that $v_g < v_p < c$. However, in regions of anomalous dispersion, $dn/d\omega$ can become large and negative, so that the group velocity can take superluminal, infinite, and even negative values. In order to solve this problem, the accepted and widespread belief (following Sommerfeld and Brillouin [25] is that the very concept of group velocity breaks down in this case. However in 1970 it was shown that [26] (contrary to the common belief on the basis of Ref. 25) the pulse remains essentially undistorted even in regions of anomalous dispersion, and its velocity is just the group velocity, which can indeed become faster than c and negative. In other words, anomalous dispersion allows light in a medium to travel faster than light in vacuum. This was actually observed in two experiments in 1982 [27] and 1985 [28], and in a more recent one [29] (based on gain-assisted linear anomalous dispersion in atomic caesium gas) that received a great deal of attention by the mass media. In the latter two experiments, negative group velocities have been seen.

Let's clarify the meaning of negative group velocity [4,29]. If ℓ is the length of the medium (e.g., of the cell containing the gas), the time taken by the light pulse to propagate is $t = (\ell/v_g) = (\ell n_g/c)$. The propagation time of light in vacuum for the same distance is $t_0 = (\ell/c)$. Therefore, the light pulse leaves the medium at a time delayed by

$$\Delta t = t - t_0 = \frac{\ell}{c}(n_g - 1) \tag{12}$$

For positive group velocities $v_g > 0$, but $n_g < 1$, it is $\Delta t < 0$, and we have not a delay but an advancement; namely, the light pulse leaves the exit face of the cell sooner than if it traversed the same distance in vacuum. For $v_g < 0$, the pulse advancement $-\Delta t = (n_g - 1)(\ell/c)$ is greater than the vacuum transit time t_0. As a consequence, *the pulse appears to leave the gas cell before it enters it.* This is exactly what has been observed in the Princeton experiment [29].

Let us stress that the occurrence of negative times is by no means strange in the context of ER. This is easily seen by the fact that the time sign is not an invariant for space-like intervals under the usual (subluminal) Lorentz transformations, nor for a time-like one under superluminal Lorentz transformations [1,3]. Moreover, whenever an object (either a particle or a wavepacket) overcomes the infinite speed, it appears afterward as its antiobject, traveling in the opposite space direction [1,30], and therefore yielding negative contribution to the tunneling or the traversal time [31].

We want also to stress that (as shown by Chiao and co-workers [32]) there is still another situation in which one expects e.m. wavepacket propagation in a medium at speed higher than c. This occurs for off-resonance pulses through a medium with inverted atomic populations. Experiments aimed at detecting superluminal propagation in such a kind of medium are presently being performed at Berkeley [4].

VI. X WAVES

The last sector of superluminal phenomena has its theoretical grounds in much of mathematical research concerning the solutions of field equations. Starting from the 1915 pioneering work by Bateman [33], who built up explicitly wavelet-type solutions of Maxwell's equations propagating in vacuum with *subluminal* group velocity $v_g < c$, it was shown that all the relativistic, homogeneous wave equations (describing scalar, vector or spinor fields in a medium) do admit solutions endowed with group velocities either less [35–40] or higher [34,41,42] than the ordinary wave speed in the medium considered. Some of the solutions derived in Refs. 33–42 share the relevant property of representing *localized* (or *limited diffraction*) beams [34–39], in the sense that they propagate over very large (theoretically infinite) distances without changing their shape, that is, without appreciable dispersion (whence the term "undistorted progressive waves" [36]). Such beams are also referred to as

"Bessel beams" [39], because their transverse profile is a Bessel function. Because of their peculiar features, Bessel beams have a number of applications in the most disparate sectors, ranging from nondestructive evaluation of materials to medical imaging.

Since 1992, an entirely new class of localized beams has been discovered, first in acoustics [43], and then generalized to the electromagnetic case [44,45]. They differ from the standard Bessel beams because are not monochromatic, but contain multiple frequencies, and are nondispersive in isotropic and homogeneous media (and therefore in free space). They are usually called "X waves," because they are X-shaped in a plane passing through the propagation axis. Moreover (what's of interest to our present aims) X waves do travel at a group velocity higher than the maximal velocity corresponding to the phenomenon considered, i.e faster than sound speed (the speed of sound) in the acoustic case and faster than light speed in the electromagnetic case [43–45].

Both acoustic and electromagnetic X waves have been produced and observed , and checked to be indeed supersonic and superluminal, respectively. The former experiment is due to the very inventors of X waves, Lu and Greenleaf [46]. The evidence for X waves was found at Tartu in 1997 in the optical domain [47] (Fig. 5) and at Florence just in 2000 in the microwave range [48]. Let us stress that (acoustic) X waves have already found application in medical imaging [49].

X waves constitute the most striking example of superluminal effects, because they propagate in free space over large distances. We want to stress that *the existence of superluminal X-shaped waves was predicted by ER*. This is, in our opinion, one of the most impressive among the (verified) ER predictions, and was just put forward as early as 1982 in a fundamental paper by Barut et al. [50]. The point is that, on the basis of ER, an (extended) tachyon is just expected

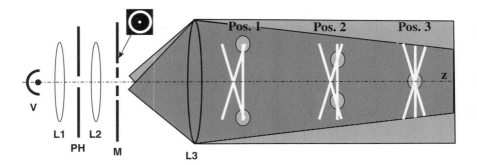

Figure 5. The electromagnetic X waves observed in the Tartu experiment. (From Saari and Reivelt [47].)

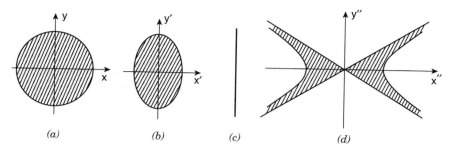

Figure 6. The shape of a spherical particle as its appears in (a) its rest frame, (b) in a frame moving with subluminal (but relativistic) velocity, (c) a frame moving at the light velocity (an unphysical case!), and (d) a superluminal frame. This last figure clearly shows that, according to ER, tachyons are X-shaped objects (waves). A spatial dimension has been dropped for simplicity of representation. (From Barut et al. [50].)

to be seen as a X-shaped object (or better wave). In fact, following Refs. 50 and 1, let us consider a particle, which in general can be thought of as a sphere when at rest (Fig. 6a). If the particle moves at subluminal (but relativistic) speed, it will appear as a rotation ellipsoid (with the minor semiaxis along the direction of motion) due to the Lorentz contraction (Fig. 6b). The world lines of its points form, in Minkowski spacetime, a cylindrical world tube, whose axis is the (straight) world line of the particle center of mass (parallel to the boost direction). For a superluminal motion of the particle, the effect due to the superluminal boost from the rest frame of the particle to the laboratory frame is to change the sign of square intervals, thus transforming the spherical surface into a two-sheeted rotation hyperboloid. Therefore, the (now tachyonic) particle appears as spread on a region bounded by an indefinite double cone and a two-sheeted hyperboloid asymptotic to the cone, both having their common axes along the superluminal boost direction (Fig. 6d). Of course, any section of such a region with a plane passing for this axis (which is the propagation direction of the particle) is X-shaped. Namely, *tachyonic particles* were predicted by ER to appear exactly as X waves.

VII.   DO SUPERLUMINAL PHENOMENA VIOLATE CAUSALITY?

There is presently no serious doubt about the actual evidence for superluminal group velocities, provided by the experiments on photon tunneling and X-shaped waves. On the contrary, it is still a subject of debates and controversies the question whether such a superluminal behavior does imply violation of Einstein causality (and of SR). A thorough discussion of this problem would require a paper by itself, and so we will confine ourselves to its main aspects.

Einstein's causality in its standard formulation implies that no information can be transmitted faster than light. Therefore, the first question to be answered is another there is *actual* transport of information in the observed superluminal processes. A number of different opinions have been expressed about this point [4,5,48,51,52]. For instance, as to the evanescent waves, the most common opinion is that their superluminal propagation does not imply transmission of a signal. This is essentially because, when transmitting an EM wave, there is an initial transient state that is associated with the propagation of precursors, which arrive before the principal signal. As shown by Sommerfeld and Brillouin [25], the speed of such precursors cannot be larger than c. However, such a result is strictly valid only for an infinite frequency spectrum, so the occurrence of precursors could be avoided by making recourse to a finite frequency band [48,51]. But a computer simulation dealing with the transients associated with superluminal evanescent waves seems to contradict such a statement [53]. However, as to the X waves (although a lot of experimental work needs to be done), it's difficult to think they don't carry information, as they are localized beams, which already find numerous practical applications.

An answer to the causality problems seemingly involved in superluminal motions can be found in extended relativity. Indeed, one has to take in due consideration, within the framework of ER, the Stückelberg–Feynman–Sudarshan *reinterpretation* (or *switching*) *principle* (SWP) [1,3] (which can be even regarded as a third postulate, on the same footing of SR postulates 1 and 2 listed in Section I). It states that any object moving backward in time, and carrying negative energy, must be reinterpreted as its antiobject with positive energy moving forward in time. A careful and suitable application of SWP allows one to solve all the causal paradoxes involving superluminal motions (provided one takes into account tachyon mechanics, and the fact that, a priori, the description of a given process *is not required to be Lorentz-invariant*) [54]. In conclusion, irrespective of whether the observed superluminal propagations do carry information or not, they do not jeopardize relativistic causality.

Acknowledgments

It's both a pleasure and a duty to acknowledge that we are enormously indebted to Erasmo Recami (by the way, a very old good friend of RM) for his precious contribution. Writing this review would have been impossible without his valuable support, continuous exchange of ideas, and bibliographical advice. We even asked him to be a co-author, but he is presently hardly working just on the fascinating and far-reaching frontier subjects on superluminal physics, which we tried to review here. We hope he will be able to join us in a near future for a wider project. We are also grateful to Vladislav S. Olkhovsky for stimulating discussions, and, needless to say, we warmly thank Myron Evans for his kind invitation to write this review.

References

1. E. Recami, *Riv. Nuovo Cim.* **9**, 1 (1986), and references cited therein.

2. O. M. Bilaniuk, V. K. Deshpande, and E. C. G. Sudarshan, *Am. J. Phys.* **30**, 718 (1962).

3. E. Recami and R. Mignani, *Riv. Nuovo Cim.* **4**, 209 (1974), and references cited therein.

4. R. Y. Chiao and A. M. Steinberg, *Tunneling Times and Superluminality,* Vol. 37 in E. Wolf (Ed.), *Progress in Optics,* Elsevier Science, New York, 1997, p. 346, and references cited therein.

5. G. Nimtz and W. Heitmann, *Prog. Quantum Electr.* **21**, 81 (1997), and references cited therein.

6. See, e.g., J. A. Zensus and T. J. Pearson (Eds.), *Superluminal Radio Sources,* Cambridge Univ. Press, Cambridge, UK, 1987, and references cited therein.

7. See, e.g., H. Arp, *Quasars, Redshifts and Controversies,* Interstellar Media, Berkeley, CA, 1987, and references cited therein.

8. I. F. Mirabel and L. F. Rodriguez, *Nature* **371**, 46 (1994); S. J. Tingay et al., *Nature* **374**, 141 (1995).

9. E. Recami, A. Castellino, G. D. Maccarrone, and M. Rodonò, *Nuovo Cim.* B **93**, 119 (1985); R. Mignani and E. Recami, *Gen. Rel. Grav.* **5**, 615 (1974).

10. See, e.g., M. Baldo Ceolin, "Review of neutrino physics," in *Proc. XXIII Int. Symp. on Multiparticle Dynamics* (Aspen, CO, Sept. 1993); E. W. Otten, *Nucl. Phys. News* **5**, 11 (1995).

11. E. Giannetto, G. D. Maccarrone, R. Mignani, and E. Recami: *Phys. Lett.* B **178**, 115 (1986).

12. E. H. Hauge and J. A. Støvneng, *Rev. Mod. Phys.* **61**, 917 (1989).

13. V.S. Olkhovsky and E. Recami, *Phys. Rep.* **214**, 339 (1992).

14. R. Landauer and Th. Martin, *Rev. Mod. Phys.* **66**, 217 (1994).

15. T. E. Hartman, *J. Appl. Phys.* **33**, 3427 (1962); J. R. Fletcher, *J. Phys. C* **18**, L55 (1985).

16. J. Jakiel, V. S. Olkhovsky, and E. Recami, *Phys. Lett. A* **248**, 156 (1998).

17. S. Bosanac, *Phys. Rev. A* **28**, 577 (1983); Th. Martin and R. Landauer, *Phys. Rev. A* **45**, 2611 (1992); R. Y. Chiao, P. G. Kwiat, and A. M. Steinberg, *Physica B* **175**, 257 (1991); A. Ranfagni, D. Mugnai, P. Fabeni, and G. P. Pazzi, *Appl. Phys. Lett.* **58**, 774 (1991).

18. A. Enders and G. Nimtz, *J. Phys. I* (France) **2**, 1693 (1992); **3**, 1089 (1993); *Phys. Rev. E* **48**, 632 (1993); G. Nimtz, A. Enders, and H. Spieker, *J. Phys. I* (France) **4**, 1 (1994); W. Heitmann and G. Nimtz, *Phys. Lett. A* **196**, 154 (1994).

19. A. Enders and G. Nimtz, *Phys. Rev. B* **47**, 9605 (1993); G. Nimtz, A. Enders, and H. Spieker, *J. Phys. I* (France) **4**, 565 (1994).

20. A. Ranfagni, P. Fabeni, G. P. Pazzi, and D. Mugnai, *Phys. Rev. E* **48**, 1453 (1993); D. Mugnai, A. Ranfagni, and L. S. Schulman, *Phys. Rev. E* **55**, 3593 (1997); D. Mugnai, A. Ranfagni and L. Ronchi, *Phys. Lett. A* **247**, 281 (1998).

21. A. M. Steinberg, P. G. Kwait, and R. Y. Chiao, *Phys. Rev. Lett.* **71**, 708 (1993).

22. Ch. Spielmann, R. Szipocs, A. Singl, and F. Krausz, *Phys. Rev. Lett.* **73**, 2308 (1994).

23. Ph. Balcou and L. Dutriaux, *Phys. Rev. Lett.* **78**, 851 (1997).

24. V. Laude and P. Tournois, *J. Opt. Soc. Am. B* **16**, 194 (1999).

25. A. Sommerfeld, *Phys. Z.* **8**, 841 (1907); L. Brillouin, *Wave Propagation and Group Velocity,* Academic, New York, 1960.

26. C. G. B. Garrett and D. E. McCumber, *Phys. Rev. A* **1**, 305 (1970).

27. S. Chu and S. Wong, *Phys. Rev. Lett.* **48**, 738 (1982).

28. B. Segard and B. Macke, *Phys. Lett. A* **109**, 213 (1985).

29. L. J. Wang, A. Kuzmich, and A. Dogariu, *Nature* **406**, 277 (2000).

30. R. Mignani and E. Recami, *Int. J. Theor. Phys.* **12**, 299 (1975).

31. V. S. Olkhovsky, E. Recami, F. Raciti, and A. K. Zaichenko, *J. Phys. I* (France) **5**, 1351 (1995).

32. R. Y. Chiao, *Phys. Rev. A* **48**, R34 (1993); R. Y. Chiao, A. E. Kozhekin, and G. Kurizki, *Phys. Rev. Lett.* **77**, 1254 (1996).

33. H. Bateman, *Electrical and Optical Wave Motion*, Cambridge Univ. Press, Cambridge, UK, 1915.

34. J. A. Stratton, *Electromagnetic Theory*, McGraw-Hill, New York, 1941.

35. G. Toraldo di Francia, *Nuovo Cim. Suppl.* **9**, 426 (1952).

36. R. Courant and D. Hilbert, *Methods of Mathematical Physics,* Vol. 2, Wiley, New York, 1966.

37. J. N. Brittingham, *J. Appl. Phys.* **54**, 1179 (1983).

38. R. W. Ziolkowski, *J. Math. Phys.* **26**, 861 (1985); *Phys. Rev. A* **39**, 2005 (1989); ibid. **44**, 3960 (1991); R. W. Ziolkowski, D. K. Lewis, and B. D. Cook, *Phys. Rev. Lett.* **62**, 147 (1989).

39. J. Durnin, *J. Opt. Soc.* **4**, 651 (1987); J. Durnin, J. J. Miceli Jr., and J. H. Eberly, *Phys. Rev. Lett.* **58**, 1499 (1987); *Opt. Lett.* **13**, 79 (1988).

40. A. O. Barut and H. C. Chandola, *Phys. Lett A* **180**, 5 (1993).

41. A. O. Barut and A. Grant, *Found. Phys. Lett* **3**, 303 (1990); A. O. Barut and A. J. Bracken, *Found. Phys.* **22**, 1267 (1992).

42. I. M. Besieris, A. M. Shaarawi, and R. W. Ziolkowski, *J. Math. Phys.* **30**, 1254 (1989); R. Donnelly and R. W. Ziolkowsky, *Proc. Roy. Soc. Lond. A* **440**, 541 (1993); S. Esposito, *Phys. Lett. A* **225**, 203 (1997); W. A. Rodrigues Jr. and J. Vaz Jr., *Adv. Appl. Cliff. Alg.* **S-7**, 457 (1997).

43. J.-Y. Lu and J. F. Greenleaf, *IEEE Trans. Ultrasonics Ferroelectrics Frequency Control* **39**, 19 (1992).

44. R. W. Ziolkowski, I. M. Besieris, and A. M. Shaarawi, *J. Opt. Soc. Am. A* **10**, 75 (1993).

45. E. Recami, *Physica A* **252**, 586 (1998); See also J.-Y. Lu, J. F. Greenleaf and E. Recami, *Limited Diffraction Solutions to Maxwell (and Schrödinger) Equations*, Report INFN/FM-96/01 (INFN, Frascati, Oct. 1996).

46. J.-Y. Lu and J. F. Greenleaf, *IEEE Trans. Ultrasonics Ferroelectrics Frequency Control* **39**, 441 (1992).

47. P. Saari and K. Reivelt, *Phys. Rev. Lett.* **79**, 4135 (1997).

48. D. Mugnai, A. Ranfagni, and R. Ruggeri, *Phys. Rev. Lett.* **84**, 4830 (2000).

49. J.-Y. Lu, H.-H. Zou and J. F. Greenleaf: *Ultrasound in Medicine and Biology* **20**, 403 (1994); *Ultrasonic Imaging* **15**, 134 (1993).

50. A. O. Barut, G. D. Maccarrone, and E. Recami, *Nuovo Cim. A* **71**, 519 (1982).

51. G. Nimtz, "Evanescent modes are not necessarily Einstein casual," *Europ. Phys. J. B* (in press).

52. E. Recami, F. Fontana, R. Garavaglia: *Int. J. Mod. Phys. A* **15**, 2793 (2000).

53. A. P. L. Barbero, H. E. Hernández-Figueroa, and E. Recami, "On the propagation speed of evanescent modes,"*Phys. Rev. E* (in press). See also E. Recami, H. E. Hernández-Figueroa, and A. P. L. Barbero, *Ann. der Phys.* **7**, 764 (1998).

54. See, e.g. E. Recami, *Found. of Phys.* **17**, 239 (1987), and references therein.

TOPOLOGICAL APPROACHES
TO ELECTROMAGNETISM

TERENCE W. BARRETT

BSEI, Vienna, Virginia

CONTENTS

I. INTRODUCTION

Topology addresses those properties, often associated with *invariant qualities*, which are not altered by continuous deformations. Objects are topologically equivalent, or *homeomorphic*, if one object can be changed into another by bending, stretching, twisting, or any other continuous deformation or mapping. Continuous deformations are allowed, but prohibited are foldings which bring formerly distant points into direct contact or overlap, and cutting—unless followed by a regluing, reestablishing the preexisting relationships of continuity.

The continuous deformations of topology are commonly described in differential equation form and the quantities conserved under the transformations commonly described by differential equations exemplifying an algebra describing operations that preserve that algebra. Evariste Galois (1811–1832) first gave the criteria that an algebraic equation must satisfy in order to be solvable by radicals. This branch of mathematics came to be known as Galois or *group theory.*

Beginning with G. W. Leibniz in the seventeenth; L. Euler in the eighteenth; B. Reimann, J. B. Listing, and A. F. Möbius in the nineteenth; and H. Poincaré

Modern Nonlinear Optics, Part 3, Second Edition, Advances in Chemical Physics, Volume 119, Edited by Myron W. Evans. Series Editors I. Prigogine and Stuart A. Rice.
ISBN 0-471-38932-3   © 2001 John Wiley & Sons, Inc.

in the twentieth centuries, *analysis situs* (Riemann) or *topology* (Listing) has been used to provide answers to questions concerning what is most fundamental in physical explanation. That question itself implies the question concerning what mathematical structures one uses with confidence to adequately "paint" or describe physical models built from empirical facts. For example, differential equations of motion cannot be fundamental, because they are dependent on boundary conditions that must be justified—usually by group theoretical considerations. Perhaps, then, group theory[1] is fundamental.

Group theory certainly offers an austere shorthand for fundamental transformation rules. But it appears to the present writer that the final judge of whether a mathematical group structure can, or cannot, be applied to a physical situation is the topology of that physical situation. Topology dictates and justifies the group transformations.

So for the present writer, the answer to the question of what is the most fundamental physical description is that it is a description of the topology of the situation. With the topology known, the group theory description is justified and equations of motion can then also be justified and defined in specific differential equation form. If there is a requirement for an understanding more basic than the topology of the situation, then all that is left is verbal description of visual images. So we commence an examination of electromagnetism under the assumption that topology defines group transformations and the group transformation rules justify the algebra underlying the differential equations of motion.

Differential equations or a set of differential equations describe a *system* and its evolution. Group symmetry principles summarize both invariances and the laws of nature independent of a system's specific dynamics. It is necessary that the symmetry transformations be continuous or specified by a set of parameters which can be varied continuously. The symmetry of continuous transformations leads to conservation laws.

There are a variety of special methods used to solve ordinary differential equations. It was Sophus Lie (1842–1899) in the nineteenth century who showed that all the methods are special cases of integration procedures which are based on the invariance of a differential equation under a continuous group of symmetries. These groups became known as *Lie groups*.[2] A symmetry group

---

[1]Here we mean the kind of groups addressed in Yang–Mills theory, which are *continuous* groups (as opposed to *discrete* groups). Unlike discrete groups, continuous groups contain an infinite number of elements and can be differentiable or analytical [1].

[2]If a topological group is a group and also a topological space in which group operations are continuous, then *Lie groups* are topological groups that are also analytical manifolds on which the group operations are analytic. In the case of *Lie algebras*, the parameters of a product are analytic functions of the parameters of each factor in the product. For example, $L(\gamma) = L(\alpha)L(\beta)$ where $\gamma = f(\alpha, \beta)$. This guarantees that the group is differentiable. The Lie groups used in Yang–Mills theory are *compact* groups, i.e., the parameters range over a closed interval.

of a system of differential equations is a group that transforms solutions of the system to other solutions [2]. In other words, there is an invariance of a differential equation under a transformation of independent and dependent variables. This invariance results in a diffeomorphism on the space of independent and dependent variables, permitting the mapping of solutions to solutions [3].

The relationship was made more explicit by Emmy (Amalie) Noether (1882–1935) in theorems now known as *Noether's theorems* [4], which related symmetry groups of a variational integral to properties of its associated Euler–Lagrange equations. The most important consequences of this relationship are that (1) conservation of energy arises from invariance under a group of time translations, (2) conservation of linear momentum arises from invariance under (spatial) translational groups; (3) conservation of angular momentum arises from invariance under (spatial) rotational groups, and (4) conservation of charge arises from invariance under change of phase of the wave function of charged particles. Conservation and group symmetry laws have been vastly extended to other systems of equations, such as the standard model of modern high-energy physics, and also, of importance to the present interest: soliton equations. For example, the *Korteweg de Vries* "soliton" equation [5] yields a symmetry algebra spanned by the four vector fields of (1) space translation, (2) time translation, (3) Galilean translation, and (4) scaling.

For some time, the present writer has been engaged in showing that the spacetime topology defines electromagnetic field equations [6–11]—whether the fields be of force or of phase. That is to say, the premise of this enterprise is that a set of field equations are valid only with respect to a set defined topological description of the physical situation. In particular, the writer has addressed demonstrating that the A_μ potentials, $\mu = 0, 1, 2, 3$, are not just a mathematical convenience, but—*in certain well-defined situations*—are measurable, that is, physical. Those situations in which the A_μ potentials are measurable possess a topology, the transformation rules of which are describable by the SU(2) group (see paragraphs 1–5 in the following list or higher order groups). The algebras are described as follows:

1. *SU(n) Group Algebra.* Unitary transformations, U(n), leave the modulus squared of a complex wavefunction invariant. The elements of a U(n) group are represented by $n \times n$ unitary matrices with a determinant equal to ± 1. Special unitary matrices are elements of unitary matrices that leave the determinant equal to $+1$. There are $n^2 - 1$ independent parameters. SU(n) is a subgroup of U(n) for which the determinant equals $+1$.

2. *SL(2,C) Group Algebra.* The special linear group of 2×2 matrices of determinant 1 with complex entries is SL(2,C).

3. *SU(2) Group Algebra.* SU(2) is a subgroup of SL(2,C). There are $2^2 - 1 = 3$ independent parameters for the special unitary group SU(2) of 2×2

matrices. *SU(2)* is a Lie algebra such that for the angular momentum generators, J_i, the commutation relations are $[J_i, J_j] = i\varepsilon_{ijk}J_k$, $i, j, k = 1, 2, 3$. The SU(2) group describes rotation in three-dimensional space with 2 parameters (see below). There is a well-known SU(2) matrix relating the Euler angles of O(3) and the complex parameters of SU(2) is:

$$\cos\left[\frac{\beta}{2}\right]\exp\left[\frac{i(\alpha+\gamma)}{2}\right] \quad \sin\left[\frac{\beta}{2}\right]\exp\left[\frac{-(\alpha-\gamma)}{2}\right]$$

$$-\sin\left[\frac{\beta}{2}\right]\exp\left[\frac{i(\alpha-\gamma)}{2}\right] \quad \cos\left[\frac{\beta}{2}\right]\exp\left[\frac{-i(\alpha+\gamma)}{2}\right]$$

where α, β, γ are the Euler angles. It is also well known that a homomorphism exists between O(3) and SU(2); the elements of SU(2) can be associated with rotations in O(3); and SU(2) is the *covering group* of O(3). Therefore, it is easy to show that SU(2) can be obtained from O(3). These SU(2) transformations define the relations between the Euler angles of group O(3) with the parameters of SU(2). For comparison with the above, if the rotation matrix $R(\alpha, \beta, \gamma)$ in O(3) is represented as:

$$\begin{pmatrix} \cos[\alpha]\cos[\beta]\cos[\gamma]-\sin[\alpha]\sin[\gamma] & \sin[\alpha]\cos[\beta]\cos[\gamma]+\cos[\alpha]\sin[\gamma] & -\sin[\beta]\cos[\gamma] \\ -\cos[\alpha]\cos[\beta]\sin[\gamma]-\sin[\alpha]\cos[\gamma] & -\sin[\alpha]\cos[\beta]\sin[\gamma]+\cos[\alpha]\cos[\gamma] & \sin[\beta]\sin[\gamma] \\ \cos[\alpha]\sin[\beta] & \sin[\alpha]\sin[\beta] & \cos[\beta] \end{pmatrix}$$

then the orthogonal rotations about the coordinate axes are

$$R_1(\alpha) = \begin{pmatrix} \cos[\alpha] & \sin[\alpha] & 0 \\ -\sin[\alpha] & \cos[\alpha] & 0 \\ 0 & 0 & 1 \end{pmatrix} \quad R_2(\beta) = \begin{pmatrix} \cos[\beta] & 0 & -\sin[\beta] \\ 0 & 1 & 0 \\ \sin[\beta] & 0 & \cos[\beta] \end{pmatrix}$$

$$R_3(\gamma) = \begin{pmatrix} \cos[\gamma] & \sin[\gamma] & 0 \\ -\sin[\gamma] & \cos[\gamma] & 0 \\ 0 & 0 & 1 \end{pmatrix}$$

An isotropic parameter, ϖ, can be defined:

$$\varpi = \frac{x - iy}{z}$$

where x, y, z are the spatial coordinates. If ϖ is written as the quotient of μ_1 and μ_2, or the homogeneous coordinates of the bilinear transformation, then

$$|\mu_1'\mu_2'\rangle = \begin{bmatrix} \cos\left[\frac{\beta}{2}\right]\exp\left[\frac{i(\alpha+\gamma)}{2}\right] & \sin\left[\frac{\beta}{2}\right]\exp\left[\frac{-(\alpha-\gamma)}{2}\right] \\ -\sin\left[\frac{\beta}{2}\right]\exp\left[\frac{i(\alpha-\gamma)}{2}\right] & \cos\left[\frac{\beta}{2}\right]\exp\left[\frac{-i(\alpha+\gamma)}{2}\right] \end{bmatrix}|\mu_1\mu_2\rangle$$

which is the relation between the Euler angles of O(3) and the complex parameters of SU(2). However, there is no unique one-to-one relation, for two rotations in O(3) correspond to one direction in *SU(2)*. There is thus a many-to-one or homomorphism between O(3) and SU(2). In the case of a complex 2-dimensional vector (u, v):

$$\begin{pmatrix} u' \\ v' \end{pmatrix} = \begin{pmatrix} \cos\left[\frac{\beta}{2}\right]\exp\left[\frac{i(\alpha+\gamma)}{2}\right] & \sin\left[\frac{\beta}{2}\right]\exp\left[\frac{-(\alpha-\gamma)}{2}\right] \\ -\sin\left[\frac{\beta}{2}\right]\exp\left[\frac{i(\alpha-\gamma)}{2}\right] & \cos\left[\frac{\beta}{2}\right]\exp\left[\frac{-i(\alpha+\gamma)}{2}\right] \end{pmatrix}\begin{pmatrix} u \\ v \end{pmatrix}$$

and if we define

$$a = \cos\left[\frac{\beta}{2}\right]\exp\left[\frac{i(\alpha+\gamma)}{2}\right]$$

$$b = \sin\left[\frac{\beta}{2}\right]\exp\left[\frac{-(\alpha-\gamma)}{2}\right]$$

then

$$|\mu_1'\mu_2'\rangle = \begin{bmatrix} a & b \\ -b^* & a^* \end{bmatrix}|\mu_1\mu_2\rangle$$

where

$$\begin{bmatrix} a & b \\ -b^* & a^* \end{bmatrix}$$

are the well-known SU(2) transformation rules. Defining $c = -b^*$ and $d = a^*$, we have the determinant:

$$ad - bc = 1 \quad \text{or} \quad aa^* - b(-b^*) = 1$$

Defining the (x, y, z) coordinates with respect to a complex 2D vector (u, v) as

$$x = \frac{1}{2}(u^2 - v^2), \qquad y = \frac{1}{2i}(u^2 + v^2), \qquad z = uv$$

then SU(2) transformations leave the squared distance $x^2 + y^2 + z^2$ invariant. Every element of SU(2) can be written as

$$\begin{bmatrix} a & b \\ -b^* & a^* \end{bmatrix}, \qquad |a|^2 + |b|^2 = 1$$

Defining

$$a = y_1 - iy_2, \qquad b = y_3 - iy_4$$

the parameters y_1, y_2, y_3, y_4 indicate positions in SU(2) with the constraint

$$y_1^2 + y_2^2 + y_3^2 + y_4^2 = 1$$

which indicates that the group SU(2) is a 3D unit sphere in the 4D y-space. This means that any closed curve on that sphere can be shrunk to a point. In other words, SU(2) is *simply connected*. It is important to note that SU(2) is the quantum mechanical "rotation group."

4. *Homomorphism of O(3) and SU(2).* There is an important relationship between O(3) and SU(2). The elements of SU(2) are associated with rotations in 3D space. To make this relationship explicit, new coordinates are defined:

$$x = \frac{1}{2}(u^2 - v^2); \qquad y = \frac{1}{2i}(u^2 + v^2); \qquad z = uv$$

Explicitly, the SU(2) transformations leave the squared 3-dimensional distance $x^2 + y^2 + z^2$ invariant, an invariance which relates 3D rotations to elements of SU(2). If a,b of the elements of SU(2) are defined

$$a = \cos\frac{\beta}{2}\exp\frac{i(\alpha+\gamma)}{2}, \qquad b = \sin\frac{\beta}{2}\exp\frac{-i(\alpha-\gamma)}{2}$$

then the general rotation matrix $R(\alpha, \beta, \gamma)$, can be associated with the SU(2) matrix

$$\begin{pmatrix} \cos\frac{\beta}{2}\exp\frac{i(\alpha+\gamma)}{2} & \sin\frac{\beta}{2}\exp\frac{-i(\alpha-\gamma)}{2} \\ -\sin\frac{\beta}{2}\exp\frac{i(\alpha-\gamma)}{2} & \cos\frac{\beta}{2}\exp\frac{-i(\alpha+\gamma)}{2} \end{pmatrix}$$

by means of the Euler angles. It is important to note that this matrix does not give a unique one-to-one relationship between the general rotation matrix $R(\alpha, \beta, \gamma)$ and the SU(2) group. This can be seen if (a) we let $\alpha = 0, \beta = 0, \gamma = 0$, which gives the matrix

$$\begin{pmatrix} 1 & 0 \\ 0 & 1 \end{pmatrix}$$

and (b) $\alpha = 0, \beta = 2\pi, \gamma = 0$, which gives the matrix:

$$\begin{pmatrix} -1 & 0 \\ 0 & -1 \end{pmatrix}$$

Both matrices define zero rotation in 3-dimensional space, so we see that this zero rotation in 3D dimensional space corresponds to two different SU(2) elements depending on the value of β. There is thus a *homomorphism*, or many-to-one mapping relationship between O(3) and SU(2)—where "many" is 2 in this case—but not a one-to-one mapping.

5. *SO(2) Group Algebra.* The collection of matrices in Euclidean 2D space (the plane) which are orthogonal and moreover for which the determinant is $+1$ is a subgroup of O(2). SO(2) is the special orthogonal group in two variables. The rotations in the plane is represented by the SO(2) group

$$R(\theta) = \begin{pmatrix} \cos[\theta] & -\sin[\theta] \\ \sin[\theta] & \cos[\theta] \end{pmatrix} \theta \in \mathscr{R}$$

where $R(\theta)R(\gamma) = R(\theta + \gamma)$. S^1, or the unit circle in the complex plane with multiplication as the group operation is an SO(2) group.

6. *U(n) Group Algebra.* Unitary matrices, U, have a determinant equal to ± 1. The elements of $U(n)$ are represented by $n \times n$ unitary matrices.

7. *U(1) Group Algebra.* The one-dimensional unitary group, or U(1), is characterized by one continuous parameter. U(1) is also differentiable and the derivative is also an element of U(1). A well-known example of a U(1) group is that of all the possible phases of a wavefunction, which are angular coordinates in a 2D space. When interpreted in this way—as the internal phase of the U(1) group of electromagnetism—the U(1) group is merely a circle $(0 - 2\pi)$.

Those situations in which the A_μ potentials are not measurable possess a topology, the transformation rules of which are describable by the U(1) group (see paragraphs 6 and 7 in the above list):

Historically, electromagnetic theory was developed for situations described by the U(1) group. The dynamic equations describing the transformations and inter-relationships of the force-field are the well-known Maxwell equations, and the group algebra underlying these equations is U(1). There was a need to

TABLE I
Maxwell Equations in U(1) and SU(2) Symmetry Forms

	U(1) Symmetry Form (Traditional Maxwell Equations)	SU(2) Symmetry Form
Gauss' law	$\nabla \bullet E = J_0$	$\nabla \bullet E = J_0 - iq(A \bullet E - E \bullet A)$
Ampère's law	$\dfrac{\partial E}{\partial t} - \nabla \times B - J = 0$	$\dfrac{\partial E}{\partial t} - \nabla \times B - J + iq[A_0, E] - iq(A \times B - B \times A) = 0$
	$\nabla \bullet B = 0$	$\nabla \bullet B + iq(A \bullet B - B \bullet A) = 0$
Faraday's law	$\nabla \times E + \dfrac{\partial B}{\partial t} = 0$	$\nabla \times E + \dfrac{\partial B}{\partial t} + iq[A_0, B] = iq(A \times E - E \times A) = 0$

extend these equations to describe SU(2) situations and to derive equations whose underlying algebra is SU(2). These two formulations are shown in Table I. Table II shows the electric charge density, ρ_e, the magnetic charge density, ρ_m, the electric current density, g_e, the magnetic current density, g_m, the electric conductivity, σ, and the magnetic conductivity, s.

In the following sections, four topics are addressed: (1) the mathematical entities, or waves, called *solitons*; (2) the mathematical entities, called *instantons*; (3) a beam—an electromagnetic wave—that is *polarization-modulated over a set sampling interval*; and (4) the *Aharonov–Bohm effect*. Our intention is to show that these entities, waves or effects, can be adequately characterized and differentiated, and thus understood, only by using topological

TABLE II
The U(1) and SU(2) Symmetry Forms of the Major Variables

U(1) Symmetry Form (Traditional Maxwell Theory)	SU(2) Symmetry Form
$\rho_e = J_0$	$\rho_e = J_0 - iq(A \bullet E - E \bullet A) = J_0 + qJ_z$
$\rho_m = 0$	$\rho_m = -iq(A \bullet B - B \bullet A) = -iqJ_y$
$g_e = J$	$g_e = iq[A_0, E] - iq(A \times B - B \times A) + J = iq[A_0, E] - iqJ_x + J$
$g_m = 0$	$g_m = iq[A_0, B] - iq(A \times E - E \times A) = iq[A_0, B] - iqJ_z$
$\sigma = J/E$	$\sigma = \dfrac{\{iq[A_0, E] - iq(A \times B - B \times A) + J\}}{E} = \dfrac{\{iq[A_0, E] - iqJ_x + J\}}{E}$
$s = 0$	$s = \dfrac{\{iq[A_0, B] - iq(A \times E - E \times A)\}}{H} = \dfrac{\{iq[A_0, B] - iqJ_z\}}{H}$

characterizations. Once characterized, the way becomes open for control or engineering of these entities, waves and effects.

II. SOLITONS

A soliton is a solitary wave that preserves its shape and speed in a collision with another solitary wave [12,13]. Soliton solutions to differential equations require complete integrability and integrable systems conserve geometric features related to symmetry. Unlike the equations of motion for conventional Maxwell theory, which are solutions of U(1) symmetry systems, solitons are solutions of SU(2) symmetry systems. These notions of group symmetry are more fundamental than differential equation descriptions. Therefore, although a complete exposition is beyond the scope of the present review, we develop some basic concepts in order to place differential equation descriptions within the context of group theory.

Within this context, *ordinary differential equations are viewed as vector fields on manifolds or configuration spaces* [2]. For example, Newton's equations are second-order differential equations describing smooth curves on Riemannian manifolds. Noether's theorem [4] states that a diffeomorphism,[3] ϕ, of a Riemannian manifold, C, indices a diffeomorphism, $D\phi$, of its tangent[4] bundle,[5] TC. If ϕ is a symmetry of Newton's equations, then $D\phi$ preserves the Lagrangian: $\mathcal{L} \circ D\phi = \mathcal{L}$. As opposed to equations of motion in conventional Maxwell theory, *soliton flows are Hamiltonian flows*. Such Hamiltonian functions define *symplectic structures*[6] for which there is an absence of *local* invariants but an infinite-dimensional group of diffeomorphisms which preserve *global* properties. In the case of solitons, the global properties are those permitting the matching of the nonlinear and dispersive characteristics of the medium through which the wave moves.

---

[3]A *diffeomorphism* is an elementary concept of topology and important to the understanding of differential equations. It can be defined in the following way:

If the sets U and V are open sets both defined over the space R^m; that is, $U \subset R^m$ is open and $U \subset R^m$ is open, where "open" means nonoverlapping, then the mapping $\psi: U \to V$ is an infinitely differentiable map with an infinitely differential inverse, and objects defined in U will have equivalent counterparts in V. The mapping ψ is a diffeomorphism and it is a smooth and infinitely differentiable function. The important point is: conservation rules apply to diffeomorphisms, because of their infinite differentiability. Therefore diffeomorphisms constitute fundamental characterizations of differential equations.

[4]A vector field on a manifold, M, gives a *tangent vector* at each point of M.

[5]A *bundle* is a structure consisting of a manifold E, and manifold M, and an onto map: $\pi : E \to M$.

[6]*Symplectic topology* is the study of the global phenomena of symplectic symmetry. Symplectic symmetry structures have no local invariants. This is a subfield of topology; for an example, see McDuff and Salamon [14].

In order to achieve this match, two linear operators, L and A, are postulated associated with a partial differential equation (PDE). The two linear operators are known as the *Lax pair*. The operator L is defined by

$$L = \frac{\partial^2}{\partial x^2} + u(x,t)$$

with a related eigenproblem:

$$L\psi + \lambda\psi = 0 \tag{1}$$

The temporal evolution of ψ is defined as

$$\psi_t = -\mathscr{A}\psi \tag{2}$$

with the operator of the form

$$\mathscr{A} = a_0 \frac{\partial^n}{\partial x^n} + a_1 \frac{\partial^{n-1}}{\partial x^{n-1}} + \cdots + a_n$$

where a_0 is a constant and the n coefficients a_i are functions of x and t. Differentiating (1) gives

$$L_t\psi + L\psi_t = -\lambda_t\psi - \lambda\psi_t$$

Inserting (2):

$$L\psi_t = -L\mathscr{A}\psi$$

or

$$\lambda\psi_t = \mathscr{A}L\psi$$

Using (1) again

$$[L, \mathscr{A}] = L\mathscr{A} - \mathscr{A}L = L_t + \lambda_t \tag{3}$$

and for a time-independent λ

$$[L, \mathscr{A}] = L_t$$

This equation provides a method for finding $\mathscr{A}$.

Translating the preceding equations into a group theory formulation, in order to relate the three major soliton equations to group theory, it is necessary to examine the *Lax equation* [15] (3) above as a *zero-curvature condition* (ZCC). The ZCC expresses the flatness of a connection by the commutation relations of the covariant derivative operators [16] and in terms of the Lax equation is

$$L_t - \mathscr{A}_x - [L, \mathscr{A}] = 0$$

or [17]

$$\left[\frac{\partial}{\partial x} - L\frac{\partial}{\partial t} - \mathscr{A}\right] = 0$$

or

$$\left(\frac{\partial}{\partial x} - L\right)_t = \left[\mathscr{A}\frac{\partial}{\partial x} - L\right]$$

More recently, Palais [17] showed that the generic cases of soliton—the *Korteweg de Vries equation* (KdV), the *nonlinear Schrödinger equation* (NLS), the *sine–Gordon equation* (SGE)—can be given an SU(2) formulation. In each of the three cases considered below, V is a one-dimensional space that is embedded in the space of off-diagonal complex matrices, $\begin{pmatrix} 0 & b \\ c & 0 \end{pmatrix}$ and in each case $L(u) = a\lambda + u$, where u is a potential, λ is a complex parameter, and $\boldsymbol{a}$ is the constant, diagonal, trace zero matrix

$$\boldsymbol{a} = \begin{pmatrix} -i & 0 \\ 0 & i \end{pmatrix}$$

The matrix definition of $\boldsymbol{a}$ links these equations to an SU(2) formulation. (Other matrix definitions of $\boldsymbol{a}$ could, of course, link $\boldsymbol{a}$ to higher group symmetries.)

To carry out this objective, an inverse scattering theory function is defined [15,16] as

$$B(\xi) = \sum_{n=1}^{N} c_n^2 \exp\left[-\kappa_n \xi\right] + \frac{1}{2\pi} \int_{-\infty}^{+\infty} b(k)\exp[ik\xi]\, dk$$

where $-\kappa_1^2, \ldots, -\kappa_N^2$ are discrete eigenvalues of u, $c_1, \ldots, c_N$ are normalizing constants, and $b(k)$ are reflection coefficients.

Therefore, in a *first case* (the KdV), if $u(x) = \begin{pmatrix} 0 & q(x) \\ -1 & 0 \end{pmatrix}$ and

$$B(u) = \boldsymbol{a}\lambda^3 + u\lambda^2 + \begin{pmatrix} \frac{i}{2}q & \frac{i}{2}q_x \\ 0 & -\frac{i}{2}q \end{pmatrix}\lambda + \begin{pmatrix} \frac{q_x}{4} & \frac{-q^2}{2} \\ \frac{q}{2} & \frac{-q_x}{4} \end{pmatrix}$$

then the ZCC (Lax equation) is satisfied if and only if q satisfies the KdV in the form $q_t = -\frac{1}{4}(6qq_x + q_{xxx})$.

In a *second case* (the NLS), if $u(x) = \begin{pmatrix} 0 & a(x) \\ \bar{q}x & 0 \end{pmatrix}$ and

$$B(u) = a\lambda^3 + u\lambda^2 + \begin{pmatrix} \frac{i}{2}|q|^2 & \frac{i}{2}q_x \\ -\frac{i}{2}\bar{q}_x & -\frac{i}{2}|q|^2 \end{pmatrix}$$

then the ZCC (Lax equation) is satisfied if and only if $q(x,t)$ satisfies the NLS in the form $q_t = (i/2)(q_{xx} + 2|q|^2 q)$.

In a *third case* (the SGE), if

$$u(x) = \begin{pmatrix} 0 & -\frac{q_x(x)}{2} \\ \frac{q_x(x)}{2} & 0 \end{pmatrix} \qquad \text{and}$$

$$B(u) = \frac{i}{4\lambda} \begin{pmatrix} \cos[q(x)] & \sin[q(x)] \\ \sin[q(x)] & -\cos[q(x)] \end{pmatrix}$$

then the ZCC (Lax equation) is satisfied if and only if q satisfies the SGE in the form $q_t = \sin[q]$.

With the connection of PDEs, and especially soliton forms, to group symmetries established, one can conclude that *if* the Maxwell equation of motion that includes electric *and* magnetic conductivity is in soliton (SGE) form, the group symmetry of the Maxwell field is SU(2). Furthermore, because solitons define Hamiltonian flows, their energy conservation is due to their *symplectic structure*.

In order to clarify the difference between conventional Maxwell theory which is of U(1) symmetry, and Maxwell theory extended to SU(2) symmetry, we can describe both in terms of mappings of a field $\psi(x)$. In the case of U(1) Maxwell theory, a mapping $\psi \rightarrow \psi'$ is

$$\psi(x) \rightarrow \psi'(x) = \exp[ia(x)]\psi(x)$$

where $a(x)$ is the conventional vector potential. However, in the case of SU(2) extended Maxwell theory, a mapping $\psi \rightarrow \psi'$ is

$$\psi(x) \rightarrow \psi'(x) = \exp[iS(x)]\psi(x)$$

where $S(x)$ is the action and an element of an SU(2) field defined

$$S(x) = \int A\, dx$$

and A is the matrix form of the vector potential. Therefore we see the necessity to adopt a matrix formulation of the vector potential when addressing SU(2) forms of Maxwell theory.

III. INSTANTONS

Instantons [18] correspond to the minima of the Euclidean action and are pseudoparticle solutions [19] of SU(2) Yang–Mills equations in Euclidean 4-space [20]. A complete construction for any Yang–Mills group is also available [21]. In other words [22, p. 80]

It is reasonable... to ask for the determination of the classical field configurations in Euclidean space which minimize the action, subject to appropriate asymptotic conditions in 4-space. These classical solutions are the instantons of the Yang-Mills theory.

In light of the intention of the present writer to introduce topology into electromagnetic theory, I quote further [22, p. 81]

If one were to search ab initio for a non-linear generalization of Maxwell's equation to explain elementary particles, there are various symmetry group properties one would require. These are (i) *external symmetries* under the Lorentz and Poincaré groups and under the conformal group if one is taking the rest-mass to be zero, (ii) *internal symmetries* under groups like *SU(2)* or *SU(3)* to account for the known features of elementary particles, (iii) *covariance* or the ability to be coupled to gravitation by working on curved space-time.

The present writer applied the instanton concept in electromagnetism for the following two reasons: (1) in some sense, the instanton, or pseudo particle, is a compactification of degrees of freedom due to the particle's boundary conditions; and (2) the instanton, or pseudoparticle, then exhibits the behavior (the transformation or symmetry rules) of a high-energy particle, but without the presence of high energy; thus the pseudoparticle shares certain behavioral characteristics in common (shares transformation rules, hence symmetry rules in common) with a particle of much higher energy.

Therefore, the present writer suggested [8] that the Mikhailov effect [23], and the Ehrenhaft effect (Felix Ehrenhaft, 1879–1952), which address demonstrations exhibiting *magnetic charge-like* behavior, are examples of instanton or pseudoparticle behavior. Stated differently: (1) the instanton shows that there are ways, other than possession of high energy, to achieve high symmetry states; and (2) symmetry dictates behavior.

IV. POLARIZATION MODULATION OVER
A SET SAMPLING INTERVAL

This section is based on [Ref. [24]].

It is wellknown that all static polarizations of a beam of radiation, as well as all static rotations of the axis of that beam, can be represented on a Poincaré sphere [25] (Fig. 1a). A vector can be centered in the middle of the sphere and pointed to the underside of the surface of the sphere at a location on the surface that represents the *instantaneous* polarization and rotation angle of a beam. Causing that vector to trace a trajectory over time on the surface of the sphere represents a *polarization modulated* (and *rotation modulated*) beam (Fig. 1b). If, then, the beam is sampled by a device at a rate that is less than the rate of modulation, the sampled output from the device will be a condensation of *two components* of the wave, which are continuously changing with respect to each other, into *one snapshot* of the wave, at *one location* on the surface of the sphere and *one instantaneous polarization and axis rotation*. Thus, from the viewpoint of a device sampling at a rate less than the modulation rate, a two-to-one mapping (over time) has occurred, which is the signature of an SU(2) field.

The modulations which result in trajectories on the sphere are infinite in number. Moreover, those modulations, at a rate of multiples of 2π greater than

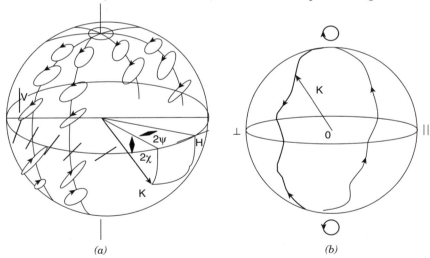

(a)   *(b)*

Figure 1. (a) Poincaré sphere representation of wave polarization and rotation; (b) a Poincaré sphere representation of signal polarization (longitudinal axis) and polarization rotation (latitudinal axis). A representational trajectory of polarization/rotation modulation is shown by changes in the vector centered at the center of the sphere and pointing at the surface. Waves of various polarization modulations $\partial \phi^n / \partial t^n$, can be represented as trajectories on the sphere. The case shown is an arbitrary trajectory repeating 2π. (After Ref. 27.)

1, which result in the return to a single location on the sphere at a frequency of exactly 2π, will all be detected by the device sampling at a rate of 2π as the same. In other words, the device cannot detect what kind of simple or complicated trajectory was performed between departure from, and arrival at, the same location on the sphere. To the relatively slowly sampling device, the fast modulated beam can have "internal energies" quite unsuspected.

We can say that such a static device is *a U(1) unipolar, set rotational axis, sampling device* and the fast polarization (and rotation) modulated beam is a *multipolar, multirotation axis, SU(2) beam*. The reader may ask: how many situations are there in which a sampling device, at set unvarying polarization, samples at a slower rate than the modulation rate of a radiated beam? The answer is that there is an infinite number, because from the point of the view of the writer, nature is set up to be that way [26]. For example, the period of modulation can be faster than the electronic or vibrational or dipole relaxation times of any atom or molecule. In other words, *pulses or wavepackets* (which, in temporal length, constitute the sampling of a continuous wave, continuously polarization and rotation modulated, but sampled only *over a temporal length between arrival and departure time at the instantaneous polarization of the sampler of set polarization and rotation*—in this case an electronic or vibrational state or dipole) have an internal modulation at a rate greater than that of the relaxation or absorption time of the electronic or vibrational state.

The representation of the sampling by a unipolar, single-rotation-axis, U(1) sampler of a SU(2) continuous wave that is polarization/rotation-modulated is shown in Fig. 2, which shows the correspondence between the output space sphere and an Argand plane [28]. The Argand plane, Σ, is drawn in two dimensions, x and y, with $z = 0$, and for a set snapshot in time. A point on the Poincaré sphere is represented as $P(t,x,y,z)$, and as in this representation $t = 1$ (or one step in the future), specifically as $P(1,x,y,z)$. The Poincaré sphere is also identified as a 3-sphere, S^+, which is defined in Euclidean space as follows:

$$x^2 + y^2 + z^2 = 1$$

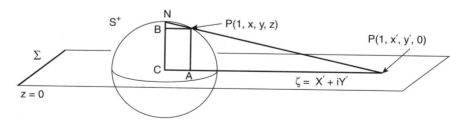

Figure 2. Correspondence between the output space sphere and an Argand plane. (After Penrose and Rindler [28].)

The sampling described above is represented as a mapping of a point $P(1, x, y, z)$ in S^+, and of SU(2) symmetry, to a point $P'(1, x', y', z')$ on Σ, and of U(1) symmetry.

The point P' can then be labeled by a single complex parameter: $\zeta = X' + iY'$. Using the definition

$$z = 1 - \frac{CA}{CP'} = 1 - \frac{NP}{NP'} = 1 - \frac{NB}{NC}$$

then

$$\zeta = \frac{x + iy}{1 - z}.$$

A pair (ξ, η) of complex numbers can be defined

$$\zeta = \frac{\xi}{\eta}$$

and Penrose and Rindler [28] have shown, in another context, that what we have identified as the *pre-sampled* SU(2) polarization and rotation modulated wave can be represented in units of

$$W = \frac{1}{\sqrt{2}} (\xi\xi^* + \eta\eta^*)$$

$$X = \frac{1}{\sqrt{2}} (\xi\eta^* + \eta\xi^*)$$

$$Y = \frac{1}{i\sqrt{2}} (\xi\eta^* - \eta\xi^*)$$

$$Z = \frac{1}{\sqrt{2}} (\xi\xi^* - \eta\iota^*)$$

These definitions make explicit that a complex linear transformation of the U(1) ξ and η results in a real linear transformation of the SU(2) (W, X, Y, Z).

Therefore, a complex linear transformation of ξ and η can be defined:

$$\begin{aligned} \xi \mapsto \xi' = \alpha\xi + \beta\eta \\ \eta \mapsto \eta' = \gamma\xi + \delta\eta \end{aligned} \tag{4a}$$

or

$$\zeta \mapsto \zeta' = \frac{\alpha\zeta + \beta}{\gamma\zeta + \delta} \tag{4b}$$

where α, β, γ and δ are arbitrary nonsingular complex numbers.

Now the transformations, (4a) and (4b), are *spin transformations*, implying that

$$\zeta = \frac{X + iY}{T - Z} = \frac{W + Z}{X - iY},$$

and if a spin matrix, A, is defined

$$A = \begin{pmatrix} \alpha & \beta \\ \gamma & \delta \end{pmatrix}, \quad \det A = 1,$$

then the two transformations, (4a), are

$$A = \begin{pmatrix} \xi' & \xi \\ \eta' & \eta \end{pmatrix}, \tag{5}$$

which means that the spin matrix of a composition is given by the product of the spin matrix of the factors. Any transformation of the (5) form is linear and real and leaves the form $W^2 - X^2 - Y^2 - Z^2$ invariant.

Furthermore, there is a unimodular condition

$$\alpha\delta - \beta\gamma = 1$$

and the matrix A has the inverse

$$A^{-1} = \begin{pmatrix} \delta & -\beta \\ -\gamma & \alpha \end{pmatrix},$$

which means that the spin matrix A and its inverse A^{-1} gives rise to *the same* transformation of ζ even though they define *different* spin transformations. Because of the unimodular condition, the A spin matrix is unitary or $A^{-1} = A^*$, where A^* is the conjugate transpose of A.

The consequence of these relations is that *every proper 2π rotation on S^+—in the present instance the Poincaré sphere—corresponds to precisely two unitary spin rotations*. As every rotation on the Poincaré sphere corresponds to a polarization/rotation *modulation*, then *every proper 2π polarization/rotation modulation corresponds to precisely two unitary spin rotations*. The vector K in Fig. 1b corresponds to two vectorial components; one is the negative of the other. As every unitary spin transformation corresponds to a unique proper rotation of S^+, then any *static* (unipolarized, e.g., linearly, circularly or elliptically polarized, as opposed to polarization-modulated) representation on S^+ (Poincaré sphere) corresponds to a trisphere representation (Fig. 3a). Therefore

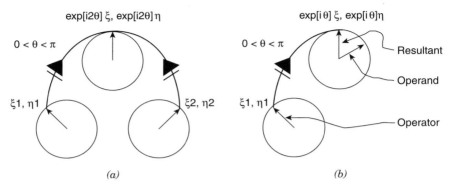

(a)   (b)

Figure 3. (a) Trisphere representation of *static polarization mapping*: $\xi_1, \eta_1; \xi_2 \eta_2 \mapsto e^{i2\theta}\xi$; $e^{i2\theta}\eta$; $0 < \theta < \pi$. Note that a 360° excursion of $\xi_1\eta_1$ and $\xi_2\eta_2$ corresponds to a 360° excursion of $e^{i2\theta}\xi, e^{i2\theta}\eta$; thus, this is a mapping for static polarization. (b) Bisphere representation of *polarization/modulation mapping* (or $\xi_1\eta_1 \mapsto e^{i\theta}\xi, e^{i\theta}\eta$; $0 < \theta < \pi$) exhibiting the property of spinors that, corresponding to two unitary transformations of for instance, 2π, namely, 4π, a null rotation of 2π is obtained. Notice that for a 360° rotation of the resultant (i.e., the final output wave), and with a stationary operand, the operator must be rotated through 720°. Left is after Penrose and Rindler [28].

$A^{-1}A = \pm I$, where I is the identity matrix. *Thus, a spin transformation is defined uniquely up to sign by its effect on a static instantaneous snapshot representation on the S^+ (Poincaré) sphere:*

$$\xi_1, \eta_1; \xi_2, \eta_2 \mapsto e^{i2\theta}\xi, e^{i2\theta}\eta; \qquad 0 < \theta < \pi$$

Turning now to the case of polarization/rotation modulation, or *continuous rotation* of $\xi_1\eta_1; \xi_2\eta_2$: corresponding to a continuous rotation of $\xi_1\eta_1; \xi_2\eta_2$ through 2θ, there is a rotation of the resultant through θ. This correspondence is a consequence of the $A^{-1}A = \pm I$ relation, namely, that if the unitary transformation of A or A^{-1} is applied separately the identity matrix will *not* be obtained. However, if the unitary transformation is applied *twice*, then the identity matrix *is* obtained; and from this follows the remarkable properties of spinors that corresponding to two unitary transformations of, for example, 2π, namely, 4π, one null vector rotation of 2π is obtained. This is a bisphere correspondence and is shown in Fig 3b. This figure also represents the case of polarization/rotation modulation—as opposed to static polarization/rotation.

We now identify the vector, K, in Fig. 1b as a *null vector* defined

$$K = Ww + Xx + Yy + Zz$$

the coordinates of which satisfy

$$W^2 - X^2 - Y^2 - Z^2 = 0$$

where W, X, Y, and Z are functions of time: $W(t), X(t), Y(t)$, and $Z(t)$. The distinguishing feature of this null vector is that phase transformations $\xi \mapsto e^{i\theta}\xi, \eta \mapsto e^{i\theta}\eta$ leave K unchanged, that is, K represents ξ and η only up to phase—which is the hallmark of a U(1) representation.

K thus defines a static polarization/rotation—whether linear, circular or elliptical—on the Poincaré sphere. The ξ, η representation of the vector K *gives no indication of the future position of K*; that is, the representation does not address the indicated hatched trajectory of the vector K around the Poincaré sphere. But it is precisely this trajectory which defines the particular polarization modulation for a specific wave. Stated differently: *a particular position of the vector K on the Poincaré sphere gives no indication of its next position at a later time*, because the vector can depart (be joined) in any direction from that position when only the static ξ, η coordinates are given.

In order to address *polarization/rotation modulation*—not just static polarization/rotation—an algebra is required which can reduce the ambiguity of a static representation. Such an algebra which is associated with ξ, η, and that reduces the ambiguity up to a sign ambiguity, is available in the twistor formalism [28]. In this formalism, polarization/rotation modulation can be accomodated, and a spinor, κ, can be represented not only by a null direction indicated by ξ, η, or ζ, but also a real tangent vector L indicated in Fig. 4.

Using this algebraic formalism, the Poincaré vector—and its direction of change (up to sign ambiguity)—can be represented. A real tangent vector L of S^+ at P is defined:

$$L = \frac{\lambda\partial}{\partial\zeta} + \frac{\lambda^*\partial}{\partial\zeta^*}$$

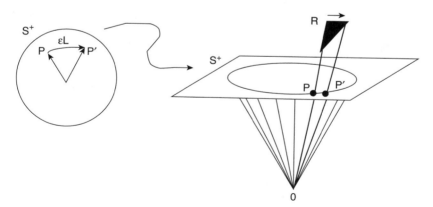

Figure 4. Relation of a trajectory in a specific direction on an output sphere S^+ and a null flag representation on the hyperplane, W, intersection with S^+. (After Penrose and Rindler [28].)

where λ is some expression in ξ, η. With the choice $\gamma = -\left(\frac{1}{\sqrt{2}}\right)\eta^{-2}$

$$L = \left(\frac{1}{\sqrt{2}}\right)\left[\eta^{-2}\left(\frac{\partial}{\partial\zeta}\right) + \eta^{*-2}\left(\frac{\partial}{\partial\xi^*}\right)\right]$$

and thus knowing L at P (as an operator) means that the pair ξ, η is known completely up to sign, or, for any $f(\zeta, \zeta^*)$:

$$\frac{1}{\varepsilon\lim_{\varepsilon\to 0}}(fp' - fp) = Lf$$

Succinctly, the tangent vector L in the abstract space S^+ (Poincaré sphere) corresponds to a tangent vector L in the coordinate-dependent representation S^+ of S^+. L is a unit vector if and only if, K, the null vector corresponding to ξ, η, defines a point actually on S^+. Therefore a plane of K and L can be defined by $aK + bL$, and if $b > 0$, then a half-plane, Π, is defined bounded by K. K and L are both spacelike and orthogonal to each other. In the twistor formalism, Π and K are referred to as a *null flag* or a *flag*. The vector K is called the *flagpole*, its direction is the *flagpole direction* and the half-plane, Π, is the *flag plane*.

Our conclusions are that a polarization/rotation-modulated wave can be represented as a periodic trajectory of polarization/rotation modulation on a Poincaré sphere, or a spinorial object. A defining characteristic of a spinorial object is that it is not returned to its original state when rotated through an angle 2π about some axis, but only when rotated through 4π. Referring to Fig. 5, we see that for the resultant to be rotated through 2π and returned to its original polarization state, the *operator* must be rotated through 4π. Thus a spinorial object (polarization/rotation modulated beams) exists in a different topological space from static polarized/rotated beams due to the additional degree of freedom provided by the polarization bandwidth, which does not exist prior to modulation.

For example, let us consider constituent polarization vectors, $Q^i(\omega, \delta)$, and let C be the space orientations of $Q^i(\omega, \delta)$. A spinorized version of $Q^i(\omega, \delta)$ can be constructed provided the space is such that it possesses a twofold universal covering space C^*, and provided the two different images, $Q_1(\psi, \chi)$ and $Q_2(\psi, \chi)$ existing in C^* of an element existing in C are interchanged after a continuous rotation through 2π is applied to a $Q^i(\omega, \delta)$. In the case we are considering, C has the topology of the SO(3) group. but C^* of the SU(2) group (which is the same as the space of unit quaternions). Thus there is a $2 \to 1$ relation between the SO(3) object and the SU(2) object (Fig. 5).

We may take the $Q^i(\omega, \delta)$ to be polarization vectors (null flags) and C to be the space of null flags. The spinorized null flags, $Q_1(\psi, \chi)$ and $Q_2(\psi, \chi)$, are elements of C^*, i.e., they are spin-vectors. Referring to Figs. 3b and 5, we see

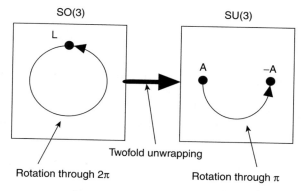

Figure 5. The left side [SO(3)] describes the symmetry of the trajectory K on the Poincaré sphere; the right side describes the symmetry of the associated $Q_1(\psi, \chi)$ and $Q_2(\psi, \chi)$ which are functions of the ψ, χ angles on the Poincaré sphere. (Adapted from Penrose and Rindler [28].)

that each null flag, $Q^i(\omega, \delta)$, defines two associated spin vectors, κ and $-\kappa$. A continuous rotation through 2π will carry κ into $-\kappa$ by acting on (ξ, η). On repeating the process, $-\kappa$ is carried back into κ:

$$-(-\kappa) = \kappa$$

Furthermore, any spin vector, κ_1, can be represented as a linear combination of two spin vectors κ_2 and κ_3

$$\{\kappa_2, \kappa_3\}\kappa_1 + \{\kappa_3, \kappa_1\}\kappa_2 + \{\kappa_1, \kappa_2\}\kappa_3 = 0$$

where { } indicates the antisymmetrical inner product. Thus any arbitrary polarization can be represented as a linear combination of spin vectors.

A generalized representation of spin vectors (and thus of polarization/ rotation modulation) is in terms of components is obtained using a normalized pair, a,b, as a spin frame:

$$\{a, b\} = -\{b, a\} = 1$$

Therefore

$$\kappa = \kappa^0 a + \kappa^1 b$$

with

$$\kappa^0 = \{\kappa, b\}, \qquad \kappa^1 = -\{\kappa, a\}$$

The flagpole of a is $(t + z)/\sqrt{2}$ and of b is $(t - z)/\sqrt{2}$ and can be represented over time in Minkowski tetrad (t,x,y,z) form (t_1 representation) and for multiple timeframes or sampling intervals providing overall $(t_1, \dots, t_n)$ a Cartan–Weyl form representation (Fig. 6) by using sampling intervals that "reset the clock" after every sampling of instantaneous polarization. Thus polarization modulation is represented by the continuous changes in a, b over time or the collection of samplings of a, b over time as depicted in Fig. 6.

The relation to the electromagnetic field is as follows. The (antisymmetrical) inner product of two spin vectors can represented as

$$\{\kappa_1, \kappa_2\} = \varepsilon_{AB}\kappa^A\kappa^B = -\{\kappa_2, \kappa_1\}$$

where the ε (or the fundamental numerical metric spinors of second rank) are antisymmetric:

$$\varepsilon_{AB}\varepsilon^{CB} = -\varepsilon_{AB}\varepsilon^{BC} = \varepsilon_{AB}\varepsilon^{BC} = -\varepsilon_{BA}\varepsilon^{CB} = \varepsilon_A^C = -\varepsilon_A^C$$

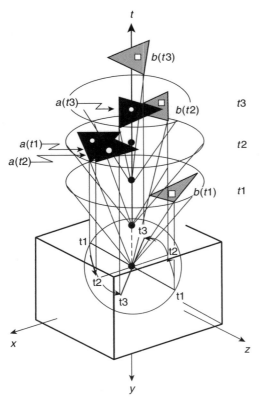

Figure 6. Spin frame representation of a spin-vector by flagpole normalized pair representation $\{a, b\}$ over the Poincaré sphere in Minkowski tetrad (t,x,y,z) form (t_1 representation) and for three timeframes or sampling intervals providing overall $(t_1, \dots, t_n)$ a Cartan–Weyl form representation. The sampling intervals "reset the clock" after every sampling of instantaneous polarization. Thus polarization modulation is represented by the collection of samplings over time. Minkowski form after Penrose and Rindler [28]. This is an SU(2) $Q_i(\psi, \chi)$ in C^* over π representation, *not* an SO(3) $Q^i(\omega, \delta)$ in C representation over 2π. This can be seen by noting that $a \mapsto b$ or $b \mapsto a$ over π, not 2π, while the polarization modulation in SO(3) repeats at a period of 2π.

with a canonical mapping (or isomorphism) between, for instance, κ^B and κ_B:

$$\kappa^B \mapsto \kappa_B = \kappa^A \varepsilon_{AB}$$

A potential can be defined

$$\Phi_A = i(\varepsilon\alpha)^{-1} \nabla_A \alpha$$

where α is a gauge

$$\alpha\alpha^* = 1$$

and ∇_A is a covariant derivative, $\partial/\partial x^A$, but without the commutation property. The covariant electromagnetic field is then

$$F_{AB} = \nabla_A \Phi_B - \nabla_B \Phi_A + ig[\Phi_B, \Phi_A]$$

where g is generalized charge.

A physical representation of the polarization modulated [SU(2)] beam can be obtained using a Lissajous pattern[7] representation (Figs. 7–9).

The controlling variables for polarization and rotation modulation are given in Table III (see page 724). We can note that the Stokes parameters (s_0, s_1, s_2, s_3) defined over the SU(2) dimensional variables, ψ, χ, of $Q_i(\psi, \chi)$ are sufficient to describe *polarization/rotation modulation*, and relate those variables to the SO(3) dimensional variables, $\omega(\tau, z), \delta$, of $Q^i(\omega, \delta)$, which are sufficient to describe the *static polarization/rotation conditions* of linear, circular, left/right-handed polarization/ rotation.

We can also note the fundamental role that concepts of topology played in distinguishing *static polarization/rotation* from *polarization–rotation modulation*.

---

[7]Lissajous patterns are the locus of the resultant displacement of a point that is a function of two (or more) simple periodic motions. In the usual situation, the two periodic motions are orthogonal (i.e., at right angles) and are of the same frequency. The Lissajous figures then represent the polarization of the resultant wave as a diagonal line, top left to bottom right, in the case of linear perpendicular polarization; bottom left to top right, in the case of linear horizontal polarization; a series of ellipses, or a circle, in the case of circular corotating or contrarotating polarization, all of these corresponding to the possible differences in constant phase between the two simple periodic motions. If the phase is not constant, but is changing or modulated, as in the case of polarization modulation, then the pattern representing the phase is constantly changing over the time the Lissajous figure is generated. Named after Jules Lissajous (1822–1880).

V. AHARONOV–BOHM EFFECT

We consider now the Aharonov–Bohm effect as an example of a phenomenon understandable only from topological considerations. Beginning in 1959 Aharonov and Bohm [30] challenged the view that the classical vector potential produces no observable physical effects by proposing two experiments. The one that is most discussed is shown in Fig. 10. A beam of monoenergetic electrons exists from a source at X and is diffracted into two beams by the slits in a wall at Y1 and Y2. The two beams produce an interference pattern at III that is measured. Behind the wall is a solenoid, the B field of which points out of the paper. The absence of a free local magnetic monopole postulate in conventional

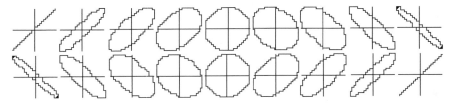

Figure 7. Lissajous patterns representing a polarization modulated electric field over time, viewed in the plane of incidence, resulting from the two orthogonal s and p fields, which are out of phase by the following degrees: 0, 21, 42, 64, 85, 106, 127, 148, 169 (top row); 191, 212, 233, 254, 275, 296, 318, 339, 360 (bottom row). In these Lissajous patterns the plane polarizations are represented at 45° to the axes. In this example, there is a simple constant rate polarization with no rotation modulation. This is an SO(3) $Q^i(\omega, \delta)$ in C representation over 2π, *not* an SU(2) $Q_i(\psi, \chi)$ in $\{C^*\}$ over π.

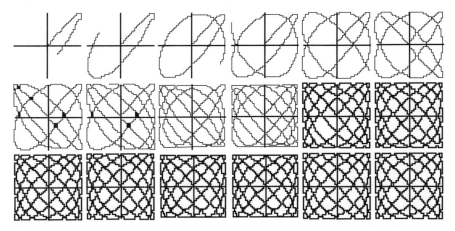

Figure 8. Lissajous patterns representing the polarized electric field over time, viewed in the plane of incidence resulting from the two orthogonal s and p fields. The p field is phase modulated at a rate $d\phi/dt = 0.2t$. In these Lissajous patterns the plane polarizations are represented at 45° to the axes. This is an SO(3) $Q^i(\omega, \delta)$ in C representation over 2π, *not* an SU(2) $Q_i(\psi, \chi)$ in C^* over π.

U(1) electromagnetism ($\nabla \bullet \boldsymbol{B} = 0$) predicts that the magnetic field outside the solenoid is zero. Before the current is turned on in the solenoid, there should be the usual interference patterns observed at III, of course, due to the differences in the two pathlengths.

Aharonov and Bohm made the interesting prediction that if the current is turned on, then, because of the differently directed $\boldsymbol{A}$ fields along paths 1 and 2

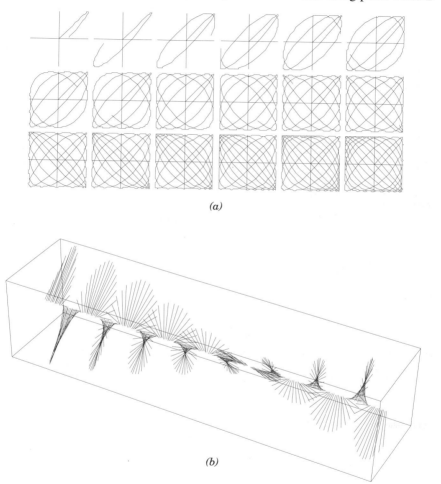

(a)

(b)

Figure 9.   A Lissajous patterns representing the polarized electric field over time, viewed in the plane of incidence, resulting from the two orthogonal s and p fields, which are out of phase by the following degrees: 0, 21, 42, 64, 85, 106, 127, 148, 169 (top row); 191, 212, 233, 254, 275, 296, 319, 339, 360 (bottom row). In these Lissajous patterns, the plane polarizations are represented at 45° to the axes. B: representation of a polarization modulated beam over 2π in the z direction. These are SO(3) $Q^i(\omega, \delta)$ in $\boldsymbol{C}$ representations over 2π, *not* an SU(2) $Q_i(\psi, \chi)$ in $\boldsymbol{C}^*$ over π.

TABLE III
Polarization: Controlling Variables

Field input variables (coordinate axes)	$E_x = a_1 \cos(\tau + \delta_1)$ $E_y = a_2 \cos(\tau + \delta_2)$ $\quad \tau = \omega t - \kappa z$
Field input variables (ellipse axes)	$E_\xi = a\cos(\tau + \delta) = E_x \cos\psi + E_y \sin\psi$ $E_\eta = \pm b\cos(\tau + \delta) = -E_x \sin\psi + E_y \cos\psi$ $\quad \tau = \omega t - \kappa z$
Phase variables	$\delta = \delta_2 - \delta_1; \quad \left(\dfrac{E_x}{a_1}\right)^2 + \left(\dfrac{E_y}{a_2}\right)^2 - 2\dfrac{\cos\delta}{a_1 a_2} = \sin^2\delta$
Auxiliary angle, α	$\dfrac{a_2}{a_1} = \tan(\alpha)$
Control variables	$a_1, a_2, \delta_1, \delta_2$
Resultant transmitted variables and relation of coordinate axes, a_1, a_2, to ellipse axes, a, b	$a^2 + b^2 = a_1^2 + a_2^2$
Rotation	$\tan(2\psi) = (\tan(2\alpha))\cos(\delta) = \dfrac{2a_1 a_2}{a_1^2 - a_2^2}\cos\delta$
Ellipticity	$\sin(2\chi) = (\sin(2\alpha))\sin(\delta); \tan(\chi) = \pm\dfrac{b}{a}$
Rotation	ψ - resultant determined by a_1 and a_2 with δ constant
Ellipticity	χ - resultant determined by δ with a_1 and a_2 constant
Determinant of rotation ψ	a_1, a_2 with δ constant
Determinant of ellipticity χ	δ with a_1, a_2 constant
Stokes parameters	$s_0 = a_1^2 + a_2^2$
	$s_1 = a_1^2 - a_2^2 = s_0 \cos(2\chi)\cos(2\psi)$
	$s_2 = 2a_1 a_2 \cos(\delta) = s_0 \cos(2\chi)\sin(2\psi) = s_1\tan(2\psi)$
	$s_3 = 2a_1 a_2 \sin(\delta) = s_0 \sin(2\chi)$
Linear polarization condition	$\delta = \delta_2 - \delta_1 = m\pi, \quad m = 0, \pm 1, \pm 2, \ldots,$ $\dfrac{E_y}{E_x} = (-1)^m \dfrac{a_2}{a_1}$
Circular polarization condition	$a_1 = a_2 = a; \delta = \delta_2 - \delta_1 = \dfrac{m\pi}{2},$ $m = \pm 1, \pm 3, \pm 5, \ldots, E_x^2 + E_y^2 = a^2$
Right-hand polarization condition	$\sin\delta > 0$ $\delta = \dfrac{\pi}{2} = 2m\pi, \quad m = 0, \pm 1, \pm 2, \ldots,$ $E_x = a\cos(\tau + \delta_1)$ $E_y = a\cos\left(\tau + \delta_1 + \tfrac{\pi}{2}\right) = -a\sin(\tau + \delta_1)$
Left-hand polarization condition	$\sin\delta < 0$ $\delta = -\dfrac{\pi}{2} + 2m\pi, \quad m = 0, \pm 1, \pm 2, \ldots$ $E_x = a\cos(\tau + \delta_1)$ $E_y = a\cos\left(\tau + \delta_1 - \dfrac{\pi}{2}\right) = a\sin(\tau + \delta_1)$

Source: After Born and Wolf [29].

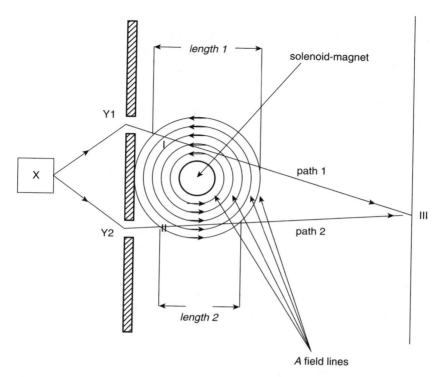

Figure 10. Two-slit diffraction experiment of the Aharonov–Bohm effect. Electrons are produced by a source at X, pass through the slits of a mask at Y1 and Y2, interact with the *A* field at locations I and II over lengths l_1 and l_2, respectively, and their diffraction pattern is detected at III. The solenoid magnet is between the slits and is directed out of the page. The different orientations of the external *A* field at the places of interaction I and II of the two paths 1 and 2 are indicated by arrows following the right-hand rule.

indicated by the arrows in Fig. 10, additional phase shifts should be discernible at III. This prediction was confirmed experimentally [31–38] and the evidence for the effect has been extensively reviewed [39–43].

It is the present writer's opinion that the topology of this situation is fundamental and dictates its explanation. Therefore we must clearly note the topology of the physical layout of the design of the situation that exhibits the effect. The physical situation is that of an *interferometer*. That is, there are two paths around a central location—occupied by the solenoid—and a measurement is taken at a location, III, in the Fig. 10, where there is overlap of the wavefunctions of the test waves that have traversed, separately, the two different paths. (The test waves or test particles are complex wavefunctions with phase.) It is important to note that the overlap area, at III, is the only place where a measurement can take place of the effects of the A field (which occurred earlier and at other locations, I and II). The effects of the A field occur along the two different paths and at locations I and II, but they are *inferred*, and not measurable there. Of crucial importance in this special interferometer, is the fact that the solenoid presents a *topological obstruction*. That is, if one were to consider the two joined paths of the interferometer as a raceway or a loop and one squeezed the loop tighter and tighter, then nevertheless one cannot in this situation—unlike as in most situations—reduce the interferometer's raceway of paths to a single point. (Another way of saying this is that: not all closed curves in a region need have a vanishing line integral, because one exception is a loop with an obstruction.) One cannot reduce the interferometer to a single point because of the existence in its midpoint of the solenoid, which is a positive quantity, and acts as an obstruction.

It is the present writer's opinion that the existence of the obstruction changes the situation entirely. *Without* the existence of the solenoid in the interferometer, the loop of the two paths *can be* reduced to a single point and the region occupied by the interferometer is then *simply connected*. But *with* the existence of the solenoid, the loop of the two paths *cannot be* reduced to a single point and the region occupied by this special interferometer is *multiply connected*. The Aharonov–Bohm effect only exists in the multiply connected scenario. But we should note that the Aharonov–Bohm effect is a *physical* effect and simple and multiple connectedness are *mathematical descriptions* of physical situations.

The topology of the physical interferometric situation addressed by Aharonov and Bohm defines the physics of that situation and also the mathematical description of that physics. If that situation *were not* multiply connected, but simply connected, then there would be no interesting physical effects to describe. The situation would be described by U(1) electromagnetics and the mapping from one region to another is conventionally one-to-one. However, as the Aharonov–Bohm situation is multiply-connected, there is a two-to-one mapping $(SU(2)/Z_2)$ of the two different regions of the two paths to the single region at III where a measurement is made. Essentially, at III a measurement is made of *the differential histories* of the *two* test waves that traversed the *two* different paths and experienced *two* different forces resulting in two different phase effects.

In conventional, that is normal U(1) or simply connected situations, the fact that a vector field, viewed axially, is pointing in one direction, if penetrated from one direction on one side, and is pointing in *the opposite direction*, if penetrated from the same direction, but *on the other side*, is of no consequence at all— because that field is of U(1) symmetry and can be reduced to a single point. Therefore in most cases which are of U(1) symmetry, we do not need to distinguish between the direction of the vectors of a field from one region to another of that field. However, the Aharonov–Bohm situation is not conventional or simply connected, but special. (In other words, the physical situation associated with the Aharonov–Bohm effect has a nontrivial topology). It is a multiply connected situation and of $SU(2)/Z_2$ symmetry. Therefore the direction of the A field on the separate paths is of crucial importance, because a test wave traveling along one path will experience an A vectorial component directed *against* its trajectory and thus be retarded, and another test wave traveling along another path will experience an A vectorial component directed *with* its trajectory and thus its speed is boosted. These "retardations" and "boostings" can be measured as phase changes, *but not at the time nor at the locations of* I *and* II, *where their occurrence is separated along the two different paths*, but *later*, and at the *overlap location* of III. It is important to note that if measurements are attempted at locations I and II in Fig. 10, these effects will not be seen because there is no two-to-one mapping at either I and II and therefore no referents. The locations I and II are both simply connected with the source and therefore only the conventional U(1) electromagnetics applies at these locations (with respect to the source). It is only region III, which is multiply connected with the source and at which the histories of what happened to the test particles at I and II can be measured. In order to distinguish the "boosted" A field (because the test wave is traveling "with" its direction) from the "retarded" A field (because the test wave is traveling "against" its direction), we introduce the notation A_+ and A_-.

Because of the distinction between the A-oriented potential fields at positions I and II—which *are not* measurable and are *vectors or numbers* of U(1) symmetry—and the A potential fields at III—which *are* measurable and are *tensors or matrix-valued functions* of (in the present instance) $SU(2)/Z_2 = SO(3)$ symmetry (or higher symmetry)—for reasons of clarity we might introduce a distinguishing notation. In the case of the potentials of U(1) symmetry at I and II we might use the lowercase, $a_\mu, \mu = 0, 1, 2, 3$ and for the potentials of $SU(2)/Z_2 = SO(3)$ at III we might use the uppercase $A_\mu, \mu = 0, 1, 2, 3$. Similarly, for the electromagnetic field tensor at I and II, we might use the lower case, $f_{\mu\nu}$, and for the electromagnetic field tensor at III, we might use the uppercase, $F_{\mu\nu}$. Then the following definitions for the electromagnetic field tensor are as follows. At locations I and II the Abelian relationship is

$$f_{\mu\nu}(x) = \partial_\nu a_\mu(x) - \partial_\mu a_\nu(x) \tag{6}$$

where, as is well known, $f_{\mu\nu}$ is Abelian and gauge-*invariant*; but at location III the non-Abelian relationship is

$$F_{\mu\nu} = \partial_\nu A_\mu(x) - \partial_\mu A_\nu(x) - ig_m[A_\mu(x), A_\nu(x)] \tag{7}$$

where $F_{\mu\nu}$ is gauge *covariant*, g_m is the magnetic charge density and the brackets are commutation brackets. We remark that in the case of non-Abelian groups, such as SU(2), the potential field *can carry charge*. It is important to note that if the physical situation changes from SU(2) symmetry back to U(1), then $\boldsymbol{F}_{\mu\nu} \rightarrow \boldsymbol{f}_{\mu\nu}$.

Despite the clarification offered by this notation, the notation can also cause confusion, because in the present literature, the electromagnetic field tensor is *always* referred to as $\boldsymbol{F}$, whether $\boldsymbol{F}$ is defined with respect to U(1) or SU(2) or other symmetry situations. Therefore, although we prefer this notation, we shall not proceed with it. However, it is important to note that the $\boldsymbol{A}$ field in the U(1) situation is a *vector or a number*, but in the SU(2) or non-Abelian situation, it is a *tensor or a matrix-valued function*.

We referred to the physical situation of the Aharonov–Bohm effect as an interferometer around an obstruction and it is two-dimensional. It is important to note that the situation is not provided by a toroid, although a toroid is also a physical situation with an obstruction and the fields existing on a toroid are also of SU(2) symmetry. However, the toroid provides a two-to-one mapping of fields in not only the x and y dimensions but also in the z dimension, and *without* the need of an electromagnetic field pointing in two directions $+$ and $-$. The physical situation of the Aharonov–Bohm effect is defined only in the x and y dimensions (there is no z dimension) and in order to be of SU(2)/Z_2 symmetry *requires* a field to be oriented differentially on the separate paths. If the differential field is removed from the Aharonov–Bohm situation, then that situation reverts to a simple interferometric raceway which can be reduced to a single point and with no interesting physics.

How does the topology of the situation affect the explanation of an effect? A typical previous explanation [44] of the Aharonov–Bohm effect commences with the Lorentz force law:

$$\mathscr{F} = e\boldsymbol{E} + e\boldsymbol{v} \times \boldsymbol{B}. \tag{8}$$

The electric field, $\boldsymbol{E}$, and the magnetic flux density, $\boldsymbol{B}$, are essentially confined to the inside of the solenoid and therefore cannot interact with the test electrons. An argument is developed by defining the $\boldsymbol{E}$ and $\boldsymbol{B}$ fields in terms of the $\boldsymbol{A}$ and ϕ potentials:

$$\boldsymbol{E} = -\frac{\partial \boldsymbol{A}}{\partial t} - \nabla\phi, \qquad \boldsymbol{B} = \nabla \times \boldsymbol{A}. \tag{9}$$

Now we can note that these conventional U(1) definitions of $\boldsymbol{E}$ and $\boldsymbol{B}$ can be expanded to SU(2) forms:

$$\boldsymbol{E} = -(\nabla \times \boldsymbol{A}) - \frac{\partial \boldsymbol{A}}{\partial t} - \nabla\phi, \qquad \boldsymbol{B} = (\nabla \times \boldsymbol{A}) - \frac{\partial \boldsymbol{A}}{\partial t} - \nabla\phi. \tag{10}$$

Furthermore, the U(1) Lorentz force law, Eq. (8), can hardly apply in this situation because the solenoid is electrically neutral to the test electrons and therefore $E = 0$ along the two paths. Using the definition of B in Eq. (5), the force law in this SU(2) situation is

$$\mathscr{F} = eE + ev \times B = e\left(-(\nabla \times A) - \frac{\partial A}{\partial t} - \nabla\phi\right) + ev \times \left((\nabla \times A) - \frac{\partial A}{\partial t} - \nabla\phi\right)$$

(11)

but we should note that Eqs. (8) and (9) are *still valid* for the conventional theory of electromagnetism based on the U(1) symmetry Maxwell's equations provided in Table I and associated with the group U(1) algebra. They are *invalid* for the theory based on the modified SU(2) symmetry equations also provided in Table I and associated with the group SU(2) algebra.

The typical explanation of the Aharonov–Bohm effect continues with the observation that a phase difference, δ, between the two test electrons is caused by the presence of the solenoid

$$\Delta\delta = \Delta\alpha_1 - \Delta\alpha_2 = \frac{e}{\hbar}\left(\int_{l_2} A \cdot dl_2 - \int_{l_2} A \cdot dl_1\right)$$

$$= \frac{e}{\hbar}\int_{l_2 - l_1} \nabla \times A \cdot dS = \frac{e}{\hbar}\int B \cdot dS = \frac{e}{\hbar}\varphi_M$$

(12)

where $\Delta\alpha_1$ and $\Delta\alpha_2$ are the changes in the wavefunction for the electrons over paths 1 and 2, S is the surface area, and φ_M is the *magnetic flux* defined as follows:

$$\varphi_M = \int\int A_\mu(x)\, dx^\mu = \int\int F_{\mu\nu}\, d\sigma^{\mu\nu}$$

(13)

Now, we can extend this explanation further, by observing that the local phase change at III of the wavefunction of a test wave or particle is given by

$$\Phi = \exp\left[ig_m \int\int A_\mu(x)\, dx^\mu\right] = \exp[ig_m \varphi_M]$$

(14)

Φ, which is proportional to the magnetic flux, φ_M, is known as the *phase factor* and is *gauge-covariant*. Furthermore, Φ, this phase factor measured at position III, is the *holonomy* of the *connection*, A_μ; and g_m is the *SU(2) magnetic charge density*.

We next observe that φ_M is in units of volt-seconds (V·s) or kg·m^{-2}/
(A s^{-2}) = J/A. From Eq. (12) it can be seen that $\Delta\delta$ and the phase factor, Φ, are
dimensionless. Therefore we can make the prediction that if the magnetic flux,
φ_M, is known and the phase factor, Φ, is measured (as in the Aharonov–Bohm
situation), the magnetic charge density, g_m, can be found by the following
relation:

$$g_m = \frac{\ln(\Phi)}{(i\varphi_M)} \tag{15}$$

Continuing the explanation: as was noted above, $\nabla \times A = 0$ outside the solenoid
and the situation must be redefined in the following way. An electron on path 1
will interact with the A field oriented in the positive direction. Conversely, an
electron on path 2 will interact with the A field oriented in the negative direction.
Furthermore, the B field can be defined with respect to a local stationary
component B_1 that is confined to the solenoid and a component B_2 which is either
a standing wave or propagates:

$$B = B_1 + B_2$$
$$B_1 = \nabla \times A \tag{16}$$
$$B_2 = -\frac{\partial A}{\partial t} - \nabla\phi$$

The magnetic flux density, B_1, is the confined component associated with U(1) $\times$
SU(2) symmetry and B_2 is the propagating or standing-wave component
associated *only* with SU(2) symmetry. In a U(1) symmetry situation, $B_1 =$
components of the field associated with U(1) symmetry, and $B_2 = 0$.

The electrons traveling on paths 1 and 2 require different times to reach III
from X, due to the different distances and the opposing directions of the potential
A along the paths l_1 and l_2. Here we only address the effect of the opposing
directions of the potential A, namely, the distances traveled are identical over
the two paths. The change in the phase difference due to the presence of the A
potential is then

$$\Delta\delta = \Delta\alpha_1 - \Delta\alpha_2 = \frac{e}{\hbar}\left[\int_{l_2}\left(-\frac{\partial A_+}{\partial t} - \nabla\phi_+\right) \cdot dl_2 - \int_{l_1}\left(-\frac{\partial A_-}{\partial t} - \nabla\phi_-\right)dl_1\right] \cdot$$

$$dS = \frac{e}{\hbar}\int B_2 \cdot dS = \frac{e}{\hbar}\varphi_M \tag{17}$$

There is no flux density B_1 in this equation since this equation describes events outside the solenoid, but only the flux density B_2 associated with group SU(2) symmetry; and the " + " and " − " indicate the direction of the A field encountered by the test electrons—as discussed above.

We note that the phase effect is dependent on B_2 and B_1, but not on B_1 alone. Previous treatments found no convincing argument around the fact that whereas the Aharonov–Bohm effect depends on an interaction with the A field outside the solenoid, B, defined in U(1) electromagnetism as $B = \nabla \times A$, is zero at that point of interaction. However, when A is defined in terms associated with an SU(2) situation, that is not the case as we have seen.

We depart from former treatments in other ways. Commencing with a *correct* observation that the Aharonov–Bohm effect depends on the topology of the experimental situation and that the situation is not simply connected, a former treatment then erroneously seeks an explanation of the effect in the connectedness of the U(1) gauge symmetry of conventional electromagnetism, but for which (1) the potentials are ambiguously defined, (the U(1) A field is gauge invariant) and (2) in U(1) symmetry $\nabla \times A = 0$ outside the solenoid.

Furthermore, whereas a former treatment again makes a *correct* observation that the non-Abelian group, SU(2), is simply connected and that the situation is governed by a multiply connected topology, the author fails to observe that the non-Abelian group SU(2) defined over the integers modulo 2, SU(2)/Z_2, is, in fact, multiply connected. Because of the two paths around the solenoid it is this group that describes the topology underlying the Aharonov–Bohm effect [9–11]. SU(2)/$Z_2 \cong$ SO(3) is obtained from the group SU(2) by identifying pairs of elements with opposite signs. For definitions of SO(3) see the following listed paragraphs.

1. *O(n) Group Algebra.* The orthogonal group, O(n), is the group of transformation (including inversion) in an n-dimensional Euclidean space. The elements of O(n) are represented by $n \times n$ real orthogonal matrices with $n(n - 1)/2$ real parameters satisfying $AA^t = 1$.

2. *O(3) Group Algebra.* The orthogonal group, O(3), is the well-known and familiar group of transformations (including inversions) in 3D space with three parameters; those parameters are the rotation or Euler angles (α, β, γ). O(3) leaves the distance squared, $z^2 + y^2 + z^2$, invariant.

3. *SO(3) Group Algebra.* The collection of matrices in Euclidean 3D space which are orthogonal and moreover for which the determinant is $+1$ is a subgroup of O(3). SO(3) is the special orthogonal group in three variables and defines rotations in 3D space. Rotation of the Riemann sphere is a rotation in $\mathscr{R}^3$

or $\xi - \eta - \zeta$ space, for which

$$\xi^2 + \eta^2 + \zeta^2 = 1, \qquad \xi = \frac{2x}{|z|^2 + 1}, \qquad \eta = \frac{2y}{|z|^2 + 1}, \qquad \zeta = \frac{|z|^2 - 1}{|z|^2 + 1}$$

$$z = x + iy = \frac{\xi + i\eta}{1 - \zeta}$$

$$U_\xi(\alpha) = \frac{1}{\sqrt{2}} \begin{pmatrix} 1 & -1 \\ 1 & 1 \end{pmatrix} \begin{pmatrix} e^{i\alpha/2} & 0 \\ 0 & e^{-i\alpha/2} \end{pmatrix} \frac{1}{\sqrt{2}} \begin{pmatrix} 1 & 1 \\ -1 & 1 \end{pmatrix}$$

$$= \begin{pmatrix} \frac{\cos\alpha}{2} & \frac{i\sin\alpha}{2} \\ \frac{i\sin\alpha}{2} & \frac{\cos\alpha}{2} \end{pmatrix} or \pm U_\xi(\alpha) \rightarrow R_1(\alpha)$$

$$U_\eta(\beta) = \frac{1}{\sqrt{2}} \begin{pmatrix} 1 & -i \\ -i & 1 \end{pmatrix} \begin{pmatrix} e^{i\beta/2} & 0 \\ 0 & e^{-i\beta/2} \end{pmatrix} \frac{1}{\sqrt{2}} \begin{pmatrix} 1 & i \\ i & 1 \end{pmatrix}$$

$$= \begin{pmatrix} \frac{\cos\beta}{2} & \frac{-\sin\beta}{2} \\ \frac{\sin\beta}{2} & \frac{\cos\beta}{2} \end{pmatrix} or \pm U_\eta(\beta) \rightarrow R_2(\beta)$$

$$U_\zeta(\gamma) = \frac{1}{\sqrt{2}} \begin{pmatrix} 1 & 0 \\ 0 & 1 \end{pmatrix} \begin{pmatrix} e^{i\gamma/2} & 0 \\ 0 & e^{-i\gamma/2} \end{pmatrix} \frac{1}{\sqrt{2}} \begin{pmatrix} 1 & 0 \\ 0 & 1 \end{pmatrix}$$

$$= \begin{pmatrix} \frac{\cos\gamma}{2} & \frac{-\sin\gamma}{2} \\ \frac{\sin\gamma}{2} & \frac{\cos\gamma}{2} \end{pmatrix} or \pm U_\xi(\gamma) \rightarrow R_3(\gamma)$$

which are mappings from SL(2,C) to SO(3). However, as the SL(2,C) are all unitary with determinant equal to $+1$, they are of the SU(2) group. Therefore SU(2) is the covering group of SO(3). Furthermore, SU(2) is simply connected and SO(3) is multiply connected. A simplification of the above is

$$U_\xi(\alpha) = e^{i(\alpha/2)\sigma_1}, \qquad U_\eta(\beta) = e^{-i(\beta/2)\sigma_2}, \qquad U_\zeta(\gamma) = e^{i(\gamma/2)\sigma_3}$$

where

$$\sigma_1 = \begin{pmatrix} 0 & 1 \\ 1 & 0 \end{pmatrix}, \qquad \sigma_2 = \begin{pmatrix} 0 & -i \\ i & 0 \end{pmatrix}, \qquad \sigma_3 = \begin{pmatrix} 1 & 0 \\ 0 & -1 \end{pmatrix}$$

$\sigma_1, \sigma_2, \sigma_3$ are the Pauli matrices.

The $\Delta\delta$ measured at location III in Fig. 10 is derived from a *single* path in SO(3), because the *two* paths through locations I and II in SU(2) are regarded as a *single* path in SO(3). This path in SU(2)/Z$_2 \cong$ SO(3) cannot be shrunk to a single point by any continuous deformation and therefore adequately describes the multiple-connectedness of the Aharonov–Bohm situation. Because the former treatment failed to note the multiple connectedness of the SU(2)/Z$_2$ description of the Aharonov–Bohm situation, it *incorrectly* fell back on a U(1) symmetry description.

Now back to the main point of this excursion to the Aharonov–Bohm effect—the reader will note that the author appealed to topological arguments to support the main points of his argument. Underpinning the U(1) Maxwell theory is an Abelian algebra; underpinning the SU(2) theory is a non-Abelian algebra. The algebras specify the form of the equations of motion. However, whether one or the other algebra can be (validly) used can be determined only by topological considerations.

VI. SUMMARY

We have attempted to show the fundamental explanatory nature of the topological description of solitons, instantons and the Aharonov–Bohm effect—and hence electromagnetism. In the case of electromagnetism we shown elsewhere that, given a Yang–Mills description, electromagnetism can, and should be, extended in accordance with the topology with which the electromagnetic fields are associated.

This approach has further implications. If the conventional theory of electromagnetism, namely, "Maxwell's theory," which is of U(1) symmetry form, is but the simplest *local* theory of electromagnetism, then those pursuing a unified field theory may wish to consider as a candidate for that unification, not only the simple local theory but also other electromagnetic fields of group symmetry higher than U(1). Other such forms include symplectic gauge fields of higher group symmetry, such as SU(2) and above.

References

1. C. N. Yang and R. L. Mills, *Phys. Rev.* **96**, 191–195 (1954).
2. P. J. Olver, *Applications of Lie Groups to Differential Equations*, Springer-Verlag, 1986.
3. G. Baumann, *Symmetry Analyis of Differential Equations with Mathematica*, Springer-Verlag, 1998.
4. E. Noether, *Nachr. Ges. Wiss. Göttingen, Math.-Phys.* **171**, 235–257 (1918).
5. D. J. Korteweg and G. de Vries, *Philos. Mag.* **39**, 422–443 (1895).
6. T. W. Barrett, *Annal. Fond. de Broglie* **15**, 143–183 (1990).
7. T. W. Barrett, *Annal. Fond. de Broglie* **15**, 253–283 (1990).
8. T. W. Barrett, *Annal. Fond. de Broglie* **19**, 291–301 (1994).
9. T. W. Barrett, in Lakhtakia (Ed.), *Essays on the Formal Aspects of Maxwell's Theory*, World Scientific, Singapore, 1993, pp. 6–86.
10. T. W. Barrett, in T. W. Barrett and D. M. Grimes (Eds.), *Advanced Electromagnetism: Foundations, Theory, Applications*, World Scientific, Singapore, 1995, pp. 278–313.

11. T. W. Barrett, *Specul. Sci. Technol.* **21**(4), 291–320 (1998).

12. T. W. Barrett, in J. D. Taylor (Ed.), *Introduction to Ultra-Wideband Radar Systems*, CRC Press, Boca Raton, FL, 1995, pp. 404–413.

13. E. Infeld and G. Rowlands, *Nonlinear Waves, Solitons and Chaos*, 2nd ed., Cambridge Univ. Press, 2000.

14. D. McDuff and D. Salamon, *Introduction to Symplectic Toplogy*, Clarendon Press, Oxford, UK, 1995.

15. P. D. Lax, *Commun. Pure Appl. Math.* **21**, 467–490 (1968).

16. P. D. Lax, in *Nonlinear Wave Motion, Lectures in Applied Mathematics*, Vol. 15, American Mathematical Society, 1974, pp. 85–96.

17. R. S. Palais, *Bull. Am. Math. Soc.* **34**, 339–403 (1997).

18. R. Jackiw, C. Nohl, and C. Rebbi, *Classical and Semi-classical Solutions to Yang–Mills Theory* (Proc. Banff School), Plenum, 1977.

19. A. Belavin, A. Polyakov, A. Schwartz, and Y. Tyupkin, *Phys. Lett.* **59B**, 85–87 (1975).

20. M. F. Atiyah and R. S. Ward, *Commun. Math. Phys.* **55**, 117–124 (1977).

21. M. F. Atiyah, N. J. Hitchin, V. G. Drinfeld, and Yu. I. Manin, *Phys. Lett.* **65A**, 23–25 (1978).

22. M. F. Atiyah, in *Michael Atiyah: Collected Works*, Vol. 5, *Gauge Theories*, Clarendon Press, Oxford, 1988.

23. T. W. Barrett and D. M. Grimes (Eds.), *Advanced Electromagnetism: Foundations, Theory and Applications*, World Scientific, Singapore, 1995, pp. 291–301.

24. T. W. Barrett, *Annal. Fond. de Broglie* **14**, 37–75 (1989).

25. H. Poincaré, *Théorie Mathématique de la Lumiére*, Vol. 2, Georges Carré, Paris, 1892, Chap. 2.

26. T. W. Barrett, in C. Cormier-Delanoue, G. Lochak, and P. Lochak (Eds.), *Courants, amers, écueils en microphysique*, Fondation Louis de Broglie, 1993, pp. 1–26.

27. T. W. Barrett, U.S. Patent 5,592,177 (Jan. 7, 1997).

28. R. Penrose and W. Rindler, *Spinors and Space-Time*, Vol. 1, *Two-Spinor Calculus and Relativistic Fields*, Cambridge Univ. Press, 1984.

29. M. Born and E. Wolf, *Principles of Optics*, 7th ed., Cambridge Univ. Press, 1999.

30. Y. Aharonov and D. Bohm, *Phys. Rev.* **115**, 485–491 (1959).

31. R. G. Chambers, *Phys. Rev. Lett.* **5**, 3–5 (1960).

32. H. Boersch, H. Hamisch, D. Wohlleben, and K. Grohmann, *Z. Physik* **159**, 397–404 (1960).

33. G. Mollenstedt and W. Bayh, *Naturwissenschaften* **49**, 81–82 (1962).

34. G. Matteucci and G. Pozzi, *Phys. Rev. Lett.* **54**, 2469–2472 (1985).

35. A. Tonomura et al., *Phys. Rev. Lett.* **48**, 1443–1446 (1982).

36. A. Tonomura et al., *Phys. Rev. Lett.* **51**, 331–334 (1983).

37. A. Tonomura et al., *Phys. Rev. Lett.* **56**, 792–795 (1986).

38. A. Tonomura and E. Callen, *ONFRE Sci. Bull.* **12**(3), 93–108 (1987).

39. M. V. Berry, *Eur. J. Phys.* **1**, 240–244 (1980).

40. M. Peshkin, *Phys. Rep.* **80**, 375–386 (1981).

41. S. Olariu and I. I. Popescu, *Rev. Mod. Phys.* **157**, 349–436 (1985).

42. P. A. Horvathy, *Phys. Rev.* **D33**, 407–414 (1986).

43. M. Peshkin and A. Tonomura, *The Aharonov–Bohm Effect*, Springer-Verlag, New York, 1989.

44. L. H. Ryder. *Quantum Field Theory*, 2nd ed., Cambridge Univ. Press, 1996.

AUTHOR INDEX

Numbers in parentheses are reference numbers and indicate that the author's work is referred to although his name is not mentioned in the text. Numbers in *italic* show the pages on which the complete references are listed.

735

Villela, T., 337(11), *381*
Vugumeister, B. E., 473(56), 474-475(56,87), 495(133-134), 497(56,87), *520–521, 523*
Vulpiani, A., 472(12), 476(12), 483(12), *518*

Wadlinger, R. L. P., 340(37-38), 367(37-38), 368(37), 372(38), *382*
Waite, T., 563(114-115), *569*
Wakatani, M., 542(34), *566*
Wallace, B. G., 359(93), *384*
Walther, T., 7(41), 48(41), *178*
Wang, L. J., 5(24), *177*, 657(6-7), 676-677(6-7), 680(6-7), *681*, 692-693(29), *698*
Warburton, F. W., 340(34), 347-348(34), 367(34), *382*
Ward, R. S., 711(20), *734*
Wataghin, V., *193*(244)
Wathaghin, G., *193*(232)
Weinberg, S., 438(9), *467*
Weingard, R., 367(127), *385*
Weiss, C. O., 501(162-163), 513(162-163), *523*
Wells, D. R., 546(45-46), 547(51), 563(113), *567, 569*
Wentzell, A. D., 473-474(57), 487-488(57), 497(57), 500(57), 505(57), *520*
Wess, J., 593(37), *620*
West, B. J., 474(81), *520*
Weyl, S., 656(3), 676(3), *680*
Wheeler, J. A., 104-105(110), *180*, 234(65), *253*, 340(33), 341(44), 347(33), *382*, 556(66), 560(80-81), *567–568*, 572(2), *618*, 665(26), *681*
Wherret, B. S., 481(113), *522*
White, C., 540(25), *566*
Whitehead, J. H. C., 207(43), *252*
Whitney, 347(74), *383*
Whittaker, E. T., 22(64-65), *179*, 198(1), *251*
Widom, A., 359(95), *384*
Wiensenfeld, K., 472(19,33), 476(19,96), 483(19,96), *518–519, 521*
Wiggins, S., 489-490(125), *522*
Wilkens, L., 476(101), *521*
Willemsen, M. B., 474(84), 493(84), 494(84,127-128), *521–522*
Williams, E. R., 605(51), *620*
Willmer, C. N. A., 330(32), *333*
Winterberg, F., 359(96), *384*
Wio, H. S., 474(70), *520*
Wisniewski, S., 474(86), *521*

Witten, E., 200(21-22), *251*, 421(12-13), *467*
Woerdman, J. P., 474(84), 493(84), 494(84,127-128), *521–522*
Wohlleben, D., 725(32), *734*
Wolf, E., 290(29,31), 291(42,52-53), *295*, 345-346(62), 349-350(62), 354(62), 378(62), *383*, 724(29), *734*
Woltjer, L., 210(45), *252*, 538(15-16), *566*
Wong, S., 692(27), *697*
Woronowicz, S. L., 594(38), *620*
Wright, M. H., 511(178), *524*
Wu, K. K. S., 331(37), *333*
Wu, T. T., 89(102), *179*
Wuensche, C. A., 337(11), *381*

Yakhot, V., 533(9), *566*
Yang, C. N., 89(102), *179*, 700(1), *733*
Yariv, A., 472(22), *519*
Yorke, A. J., 500(150), 510(175), *523–524*
Yorke, E., 500(153,155), *523*
Yoshida, Z., 563(104-106), *568*
Yourgrau, W., *189*(177), *190*(187, 196)
Youssaf, M., 88(89), *179*
Yu, A. W., 473(46), 487(46), 500(46), *519*
Yu, H., 573(6), 591-592(32), *619*
Yukawa, H., *194*(249–250), *195*(282)

Zaghloul, H., 544(37,40), 547(49), 552(62), *567*
Zaichenko, K., 693(31), *698*
Zamorani, G., 331(38), *333*
Zaric, A., *181*(22)
Zatsepin, G. T., 584(26), *619*
Zeleny, W. B., 357(85), *383*
Zeni, J. R., 563(115), *569*
Zensus, J. A., 685(6), *697*
Zevic, D., *181*(22)
Zewail, A. H., 474(86), *521*
Zhou, T., 472(36), 476(100), *519, 521*
Zhu, S., 473(46), 487(46), 500(46), *519*
Zhuanag, Y., 472(22), *519*
Zhukov, E. A., 472(29), 474(83), 480(29), 483(115), 486(29), 493(83), *519–522*
Ziolowski, R. W., 629(11), *651*, 693(38,42), 694(44), *698*
Zivanovic, Dj., 145), *188*(140
Zou, H.-H., 694(49), *698*
Zou, X. Y., 5(24), *177*
Zucca, E., 331(38), *333*
Zwanzig, R., 505(170), *524*

SUBJECT INDEX